Understanding Weather and Climate

Third Edition

Edward Aguado
San Diego State University

James E. Burt
University of Wisconsin–Madison

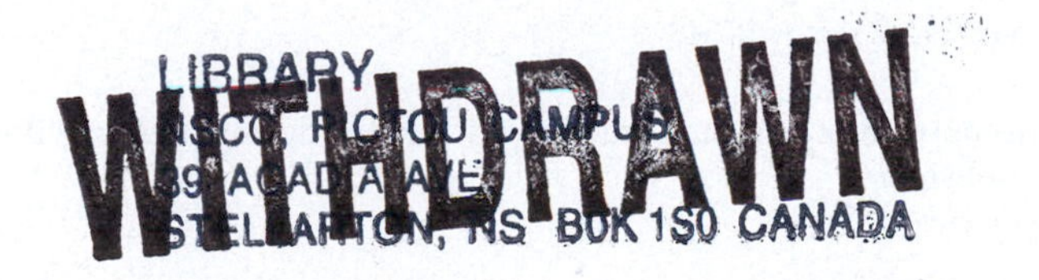

Pearson Education, Inc.
Upper Saddle River, New Jersey 07458

Library of Congress Cataloging-in-Publication Data

Aguado, Edward.
Understanding weather and climate / Edward Aguado, James E. Burt.—3rd ed.
p. cm.
ISBN 0-13-101582-6
1. Atmospheric physics. 2. Weather. 3. Climatology. I. Burt, James E. II. Title.

QC861.2.A27 2004
551.5—dc21

2003052830

Executive Editor: Daniel Kaveney
Vice President of Production and Manufacturing: David W. Riccardi
Executive Managing Editor: Kathleen Schiaparelli
Editor in Chief of Development: Carol Trueheart
Development Editor: Marcia Youngman
Production Editor: Tim Flem/PublishWare
Assistant Managing Editor for Science Media: Nicole Bush
Media Editor: Chris Rapp
Associate Editor: Amanda Griffith
Editorial Assistant: Margaret Ziegler
Marketing Manager: Christine Henry
Director of Creative Services: Paul Belfanti
Director of Design: Carole Anson
Art Director: John Christiana
Managing Editor, AV Production & Management: Patricia Burns
AV Project Manager: Adam Velthaus
Art Studio: Artworks
Manager, Production Technologies: Matt Haas
Production Manager: Ronda Whitson
Illustration Supervisor: Kathryn Anderson
Illustrators: Royce Copenheaver, Stacy Smith
Art Quality Assurance: Pamela Taylor, Timothy Nguyen
Cartography: MapQuest.com
Manufacturing Manager: Trudy Pisciotti
Manufacturing Buyer: Alan Fischer
Photo Research Director: Melinda Reo
Photo Research Administrator: Beth Boyd
Photo Researcher: Truitt and Marshall
Interior/Cover Design: Joseph Sengotta
Cover Photo: Thunderstorm Clouds and Lightning, © Don Hammond/CORBIS
Production Assistant: Nancy Bauer
Composition: PublishWare

Pearson Education, Inc.
Upper Saddle River, New Jersey 07548

Printed in the United States of America

10 9 8 7 6 5 4 3 2

ISBN 0-13-101582-6

Pearson Education Ltd., *London*
Pearson Education Australia Pty., Limited, *Sydney*
Pearson Education Singapore, Pte. Ltd
Pearson Education North Asia Ltd., *Hong Kong*
Pearson Education Canada, Ltd., *Toronto*
Pearson Educación de Mexico, S.A. de C.V.
Pearson Education—Japan, *Tokyo*
Pearson Education Malaysia, Pte. Ltd.

To Lauren, William, and Babsie June

—EA

To my parents, Martha F. Burt
and Robert L. Burt

—JEB

Brief Contents

Contents

6 Cloud Development and Forms 168

7 Precipitation Processes 200

Part Three Distribution and Movement of Air 225

8 Atmospheric Circulation and Pressure Distributions 226

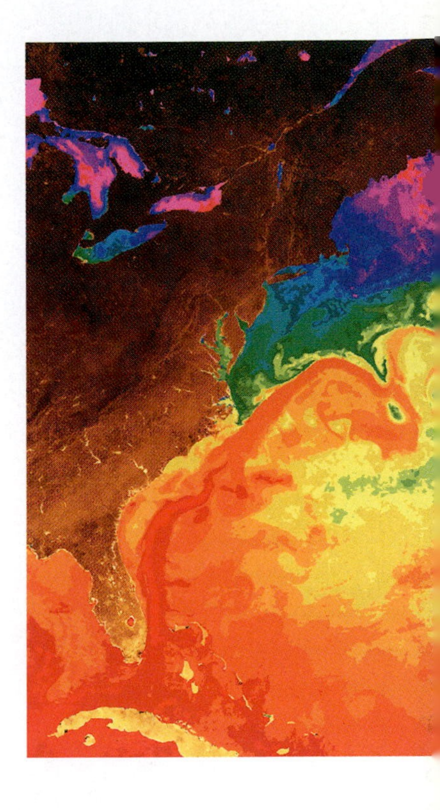

9 Air Masses and Fronts 264

Part Four Disturbances 287

10 Mid-latitude Cyclones 288

11 Lightning, Thunder, and Tornadoes 316

12 Tropical Storms and Hurricanes 356

Part Five Human Activities 383

13 Weather Forecasting and Analysis 384

14 Human Effects: Air Pollution and Heat Islands 424

Part Six Current, Past, and Future Climates 447

15 Earth's Climates 448

16 Climate Changes: Past and Future 474

Part Seven Special Topics and Appendices 511

17 Atmospheric Optics 512

CD-ROM Contents

Tutorials

Solar Geometry

- Simplifying Assumption
- Daylength Variations
- Solar Position Variations
- Effects of Solar Position

Global Energy Balance

- Radiation
- Convective Fluxes
- Energy Balance Concepts
- Achieving Global Balance

Atmospheric Moisture

- Saturation
- Measures of Moisture
- Temperature-Moisture Relations

Adiabatic Processes

- Fundamentals
- Dry Adiabatic Lapse Rate
- Transition to Saturation
- The Saturated Process

Atmospheric Stability

- Lapse Rates and Buoyancy
- Stable and Unstable Air
- Conditional Instability
- Potential Instability
- Controls on Stability

The Coriolis Force

- Rotation Around the Vertical
- Rotation and Apparent Deflection
- Slow vs. Fast Wind
- Arbitrary Motions

Atmospheric Forces and Motion

- Newton's 2nd Law
- Equation of Motion
- Pressure Gradient Force
- Coriolis Force
- Geostrophic Flow
- Gradient Flow
- Friction
- Combination of Forces

Upper-Level Winds and Pressure

- What is Geopotential?
- How Do Height Differences Arise?
- Height and Pressure Relations
- Ridges and Troughs
- Views of the 500 mb Surface
- Waves in the Westerlies
- 500 mb Movie

Cyclones and Anticyclones

- Fundamentals
- Fronts and Cyclones
- Cyclone/Anticyclone Formation
- Cyclones: 3-D Flow

Interactive Exercises

Weather in Motion

Weather Images

Media Library

See CD-ROM for contents

Preface

Meteorology is perhaps the most dynamic of all the earth sciences. In no other sphere do events routinely unfold so quickly, with so great a potential impact on humans. Some of the most striking atmospheric disturbances (such as tornadoes) can take place over time scales on the order of minutes—but nevertheless have permanent consequences. Wind speeds of several hundred kilometers per hour accompany the most violent storms, and large-scale extreme events with attendant widespread destruction are common. Furthermore, even the most mundane of atmospheric phenomena influence our lives on a daily basis (for instance, the beauty of blue skies or red sunsets, rain, the daily cycle of temperature).

Atmospheric processes, despite their immediacy on a personal level and their importance in human affairs on a larger level, are not readily understood by most people. This is probably not surprising, given that the atmosphere consists primarily of invisible gases, along with suspended, frequently microscopic particles, water droplets, and ice crystals. In this book, our overriding goal is to bridge the gap between abstract explanatory processes and the expression of those processes in everyday events. We have written the book so that students with little or no science background will be able to build a nonmathematical understanding of the atmosphere.

That said, we do not propose to abandon the foundations of physical science. We know from our own teaching experience that physical laws and principles can be mastered by students of widely varying backgrounds. In addition, we believe one of meteorology's great advantages is that reasoning from fundamental principles explains so much of the field. Compared to some other disciplines, this is one in which there is an enormous payoff for mastering a relatively small number of basic ideas.

Finally, our experience is that students are always excited to learn the "why" of things, and to do so gives real meaning to "what" and "where." For us, therefore, the idea of forsaking explanation in favor of a purely descriptive approach has no appeal whatsoever. Rather, we propose merely to replace mathematical proof (corroboration by formal argument) with qualitative reasoning and appeal to everyday occurrences. As the title implies, the goal remains understanding atmospheric behavior.

Understanding Weather and Climate is a college-level text intended for both science majors and non-majors taking their first course in atmospheric science. We have attempted to write a text that is informative, timely, engaging to students and easily used by professors.

Distinguishing Features

Scientific Literacy and Currency. We have emphasized scientific literacy throughout the book. This emphasis gives students an opportunity to build a deeper understanding about the building blocks of atmospheric science and serves as tacit instruction regarding the workings of all the sciences. For instance, in Chapter 2 we cover the molecular changes that occur when radiation is absorbed or emitted, items that are often considered a "given" in introductory texts. In Chapter 3 these basic ideas are used to help build student understanding of why individual gases radiate and absorb particular wavelengths of radiation and illustrate how processes operating at a subatomic level can manifest themselves at global scales.

An emphasis on scientific literacy can be effectively implemented only if it is accompanied by careful attention to currency. We believe that two kinds of currency are required in a text: an integration of current *events* as they relate to the topic at hand, and an integration of current *scientific thinking*. For instance, the reader will find discussion of both recent hurricane activity and the most recent theories regarding the mechanisms that generate severe storms. Scientific literacy also calls for attention to language—after all, precision of language is an important distinguishing characteristic of science, one that sets it apart from other intellectual activities. With that in mind, we have tried to avoid some common statements of dubious accuracy, such as "warm air is able to hold more water vapor than cold air."

Media. A fundamental feature of this book is the integration of the classic textbook model with the emerging areas of instructional technology. These nontraditional resources are delivered through the CD provided with the book and via the Internet. The software on the accompanying CD consists of several components. Perhaps most fundamental to our approach, the CD features nine computer tutorials covering basic principles of atmospheric science. The software modules have undergone considerable testing and have been used successfully by thousands of students. They rely heavily on three-dimensional diagrams and animations to present material not easily visualized using conventional media. In choosing topics for the modules, we have emphasized material that is both difficult to master and has the potential to benefit from computer technology. We made no attempt to cover every chapter in the modules.

The software modules follow a tutorial style, with explanations and new vocabulary introduced incrementally, building on what was presented earlier in the modules and what was presented in the text. The tutorials are best used as a supplement to assigned readings. Students and professors will notice that the book and the tutorials are linked. First, the tutorials are described in Media Enrichment sections found at the end of every chapter. In addition, CD icons in the book margins (example shown to the right) indicate that the topic under discussion is covered in a tutorial as well. A numeric subtitle indicates the section of the tutorial covering that topic. For example, the icon to the right indicates tutorial 2, section 2, subsection 1. We advise that you first view a tutorial in its entirety. If additional review is needed, you can use the section number to move directly to the section under discussion. The tutorials are also linked back to the text. The icon shown at right (taken from a tutorial) is used to locate places where the book provides more detailed or background information about the topic at hand.

Section 2.2.1

In addition to the tutorials, the CD for the third edition of the book contains other useful resources, including:

- *Weather in Motion* movies, depicting events and phenomena discussed in the text. Examples include a satellite movie showing clouds and temperature across the globe, three-dimensional simulations of thunderstorm development, and animations depicting variations in Earth's orbit. Like the tutorials, each movie is described in the Media Enrichment section at the end of every chapter.
- *Weather Images*, providing additional illustration of weather phenomena. These include photographs, satellite images, and computer diagrams complementing the text. Each is described in a Media Enrichment section.
- *Media Library* resources, consisting of additional images, movies, and animations. These are intended for self-guided browsing by the student and are therefore not explicitly mentioned in the text. A short description of each is found on the CD.
- *Interactive Exercises*, which are short activities produced by Gregory J. Carbone of the University of South Carolina. These modules cover important topics, such as hurricanes and Earth–Sun relations, and are described in the Media Enrichment sections. Expanded versions of the modules are presented with *The Lab Manual for Atmospheric Science* by Gregory J. Carbone, which is available at a discount when packaged with this book.

The CD-ROM for this third edition of the text has been improved dramatically, with a more logical and easy-to-use design and navigational structure. In addition, the software has been moved outside of a browser environment. This move provides many advantages, foremost among them the fact that now no installation is required (beyond QuickTime) in order to use the CD. This should greatly ease the use of the CD in a lab situation and will relieve users from troublesome issues of browser versioning.

We should emphasize that although the computer resources are tightly integrated with the book, a computer-equipped lab is not required. All of the resources have been designed for stand-alone use, without supervision by an instructor or TA. Extensive knowledge of computers is not assumed for either instructors or students. We must also emphasize that the computer applications described above are intended to supplement rather than replace more traditional teaching tools. In fact, the book is written so that instructors who choose not to use computers at all can assign the text without needing to supply any "missing" information or alternate activities.

The Internet site **http://www.prenhall.com/aguado** includes review exercises, quantitative exercises, and other materials that allow users to query the Internet for timely atmospheric data and the ability to file one's exercise electronically. Instructors may choose to annotate and return exercises by computer or may prefer to simply grade from screen copy. (Of course, print functions are available as well for those requiring paper copy.)

Instructor Flexibility. During the writing process, we have enjoyed interacting with many of our colleagues who teach courses in weather and climate on a regular basis. It was especially interesting to see how little consensus exists regarding topic order (truth be told, the authors of this book don't agree on the optimal sequence). With this in mind, we tried to minimize the degree to which individual chapters depend on material presented earlier. Thus, instructors who prefer a chapter order different than the one we ultimately chose will not be disadvantaged. In this third edition we are using a novel approach to the sequencing of chapters on atmospheric moisture and pressure. The introductory chapter on pressure now resides ahead of the chapters on moisture. This facilitates student learning with regard to vapor pressure, saturation, and the importance of surface convergence into low pressure regions as a mechanism for cloud formation. Chapters on the distribution of pressure and wind patterns follow the discussion of moisture, as they have in previous editions.

Emphasis on Forecasting. In addition to a comprehensive chapter on the topic, this text contains numerous examples of how physical principles are employed in weather forecasting. We have included several discussions of the use of thermodynamic diagrams in weather forecasting and analysis. These charts are extremely valuable but not immediately comprehensible to most students. To alleviate this problem, we introduce thermodynamic diagrams in a sequential fashion. That is, their use for plotting vertical temperature profiles is presented in the chapter on temperature. We expand upon this in the chapter on atmospheric moisture to show how various measures of humidity can also be determined with the aid of the charts. Thus, instructors can teach their students how to use these diagrams without inundating them with excessive detail all at once.

Readability. In contrast to the more formal scientific style used in many science textbooks, we have chosen to adopt more casual prose. Our goal is to present the material in language that is clear, readable, and friendly to the student reader. We employ frequent headings and subheadings to help students follow discussions and identify the most important ideas in each chapter. As a rule, we keep technical language to a minimum.

Focus on Learning. The chapters offer a number of study aids:

- *Key Terms.* Key terms in each chapter are printed in boldface when first introduced. Most are also listed at the end of each chapter, along with the page number on which each first appears. All key terms are defined in the glossary at the end of the book.
- *Focus on the Environment Boxes.* These boxes highlight environmental issues as they relate to the study of the atmosphere.

- *Physical Principles Boxes.* More mathematical in nature than the rest of the text, these boxes accommodate students who have a more quantitative interest in the topic. An understanding of the material in these boxes is not essential to an understanding of the material presented in the body of the text.
- *Special Interest Boxes.* These boxes highlight interesting topics related to the discussion at hand.
- *Forecasting Boxes.* These describe how the principles discussed in the chapter can be used in forecasting and often include simple "rules of thumb" that help students make their own forecasts.
- *Chapter Summary.* Each chapter concludes with a chapter summary that highlights the main points in the chapter.
- *Review Questions.* At the end of each chapter, you will find a list of questions about the subject of that chapter. These review questions test reading comprehension and can be answered from information presented in the chapter. We have greatly expanded this section for the third edition.
- *Quantitative Problems.* The *Understanding Weather and Climate* Web site features quantitative exercises to accompany each chapter. If you choose to work these problems, the computer will grade them and provide you with immediate feedback.
- *Critical Thinking Questions.* For the third edition we have added these questions to the end of each chapter. These questions require students to use the thought process and material presented in the chapter to work out answers relevant to real-world questions.
- *Problems and Exercises.* Another feature new to the third edition, these questions encourage students to work out solutions to numerical questions to gain a better understanding of chapter material.
- *Useful Web Sites.* This is a unique feature we have included to the current edition, describing relevant and interesting Web sites with valuable information or current data at the end of each chapter.
- *Software Tutorial Icons.* Throughout the book, the software icons (see the list above) will appear where the topic under discussion is reviewed in the tutorial modules that accompany the book. We suggest that you use the tutorial if you are having trouble understanding this topic using the text alone.

Supplements

The authors and publisher have been pleased to work with a number of talented people to produce an excellent supplements package for the text. This package includes the traditional materials that students and professors have come to expect from authors and publishers, as well as some new kinds of supplements that involve electronic media.

For the Student

- *Integrated CD Resources.* The CD includes software tutorials that are explicitly linked to the book. They contain interactive exercises, animations, three-dimensional diagrams, and review quizzes. As mentioned, the tutorials cover the most difficult material presented. In addition, the CD contains *Weather in Motion* animations, *Weather Images, Interactive Exercises,* and the *Media Library,* as described previously.
- *Internet Support.* The *Understanding Weather and Climate* Web site gives students the opportunity to further explore the book's topics using the Internet. The site contains numerous review exercises (from which students get immediate feedback), exercises to expand one's understanding of atmospheric science, and resources for further exploration. Please visit the site at **http://www.prenhall.com/aguado**

- *Study Guide* (0-13-101583-4). Written by experienced educators, Robert V. Rohli of Louisiana State University and Thomas W. Schmidlin of Kent State University, the study guide helps students identify the important points from the text and then provides them with review exercises, study questions, self-check exercises, and vocabulary review.
- *Rand McNally Atlas of World Geography* (0-13-959339-X). This atlas includes 126 pages of up-to-date regional maps and 20 pages of illustrated world information tables. It is available FREE when packaged with *Understanding Weather and Climate.* Please contact your local Prentice Hall representative for details.

For the Professor

- *Transparencies* (0-13-101592-3). More than 200 full-color acetates of illustrations from the text are available free of charge to qualified adopters.
- *Instructor's Resource CD-ROM* (0-13-101584-2). The Instructor's Resource CD-ROM provides high-quality electronic versions of photos and illustrations from the book, as well as customizable PowerPoint lecture presentations, images and animations extracted from the student tutorials, and the instructor's manual and test item file presented in Microsoft Word. Images are high-resolution, low-compression, 16-bit jpeg files. To further guarantee classroom projection quality, all images are manually adjusted for color, brightness, and contrast. For easy reference and identification all images are organized by chapter. This powerful presentation tool is available at no cost to qualified adopters of the text.
- *The New York Times Themes of the Times—Geography*. This unique newspaper-format supplement features recent articles about geography from the pages of *The New York Times*. This supplement, available at no extra charge from your local Prentice Hall representative, encourages students to make connections between the classroom and the world around them.
- *Instructor's Manual* (0-13-101598-2). The instructor's manual, authored by experienced educator Anthony J. Vega of Clarion University, is intended as a resource for both new and experienced instructors and includes a variety of lecture outlines, answers to the end-of-chapter exercises, additional source materials, teaching tips, advice about how to integrate visual supplements, and various other ideas for the classroom.
- *Test Item File.* An extensive array of test questions accompanies the book. These questions are available in hard copy (0-13-101595-8) and also on disks formatted for Windows or Macintosh (0-13-101596-6).

Acknowledgments

The authors would like to extend our deepest appreciation to our editor and friend, Dan Kaveney, for his astute management of the entire project and for the careful thought he gave to the numerous issues that arose along the way. His support has been unfailing and his judgments superb, and he demonstrated preternatural resilience in the face of the authors' inclination to extend every deadline to the last possible minute. Numerous people at Prentice Hall were particularly supportive, including Carol Trueheart, Editor-in-Chief of Development; Margaret Ziegler, Editorial Assistant for Geography; and Amanda Griffith, Associate Editor for Geography.

We greatly appreciate the assistance of our development editor, Marcia Youngman, who provided excellent feedback on revised and new material for this edition. Jerry Marshall very ably researched the photos for the text. We thank Chris Rapp for his expert help with the CD and WWW materials. Tim Flem of PublishWare did a terrific job of tracking details and putting the project into its final form.

We also benefited greatly from the advice of many professional educators and meteorologists. Mark Moede of the San Diego office of the National Weather Service provided valuable information on the day-to-day activities of weather forecasters.

We thank researchers and staff at government agencies and institutions throughout the world for creating many of the images and movies on the accompanying CD and for their willingness to make them freely available to use in projects like this. The following deserve special mention in this regard: Space Science and Engineering Center, University of Wisconsin-Madison; Greg Carbone, University of South Carolina; and Paul Morin, University of Minnesota. We also thank Chris Grahl and d'Vinci Interactive for their creative work in addressing a number of difficult technical issues related to the CD.

We must also offer special thanks to the many colleagues who spent valuable time and energy preparing in-depth reviews of our early efforts, many of whom have continued in this role through multiple revisions. In that regard we are particularly grateful to Robert Rohli of Louisiana State University, who read the entire manuscript with exceptional care and made many excellent suggestions. Additionally, we thank the following reviewers:

Rafique Ahmed, *University of Wisconsin–LaCrosse*
Greg Bierly, *Indiana State University*
Mark Binkley, *Mississippi State University*
Gerald Brothen, *El Camino College*
David P. Brown, *University of Arizona*
Adam W. Burnett, *Colgate University*
Gregory Carbone, *University of South Carolina*
R. E. Carlson, *Iowa State University*
Donna J. Charlevoix, *University of Illinois, Urbana–Champaign*
Christopher R. Church, *Miami University of Ohio*
John H. E. Clark, *Penn State University*
Andrew Comrie, *University of Arizona*
Arthur (Tim) Doggett, *Texas Tech University*
Dennis M. Driscoll, *Texas A & M University*
Rex J. Hess, *University of Utah*
Jay S. Hobgood, *Ohio State University*
Edward J. Hopkins, *University of Wisconsin–Madison*
Scott A. Isard, *University of Illinois*
Eric Johnson, *Illinois State University*
Scott Kirsch, *University of Memphis*
Daniel James Leathers, *University of Delaware*
Gong-Yuh Lin, *California State University–Northridge*
Thomas L. Mote, *University of Georgia*
Gerald R. North, *Texas A & M University*
Jim Norwine, *Texas A & M–Kingsville*
John E. Oliver, *Indiana State University*
Robert V. Rohli, *Louisiana State University*
Steven A. Rutledge, *Colorado State University*
Arthur N. Samel, *Bowling Green State University*
Hans Peter Schmid, *Indiana University*
Brent Skeeter, *Salisbury State University*
Stephen Stadler, *Oklahoma State University*
S. Elwynn Taylor, *Iowa State University*
Mingfang Ting, *University of Illinois*
Graham Tobin, *University of South Florida*
Paul E. Todunter, *University of North Dakota*
Barry Warmerdam, *Kings River Community College*
Thompson Webb III, *Brown University*
Thomas B. Williams, *Western Illinois University*
Morton Wurtele, *University of California–Los Angeles*
Douglas Yarger, *Iowa State University*

Integrated Media

A fundamental feature of this book is the integration of the classic textbook model with emerging areas of instructional technology. A CD-ROM accompanies every copy of this book and contains instructional resources that are closely keyed to the textbook. The CD has been carefully designed to be simple to use. No installation is required.

Integrated Text and Media

The software on the CD-ROM is tightly integrated with the textbook using both icons from the text, and by explaining how to use the accompanying media to get the most out of the text in a "Media Enrichment" section featured at the end of every chapter.

Useful Web Sites

http://aa.usno.navy.mil/data/docs/AltAz.html
Some very useful astronomical data, especially with regard to Earth-Sun relationships. Interactive format allows you to request data for particular cities or latitude-longitude coordinates. Includes precise sunrise/sunset data as well as solar angle data.

http://aa.usno.navy.mil/data/docs/EarthSeasons.html
Exact dates and times of perihelion, aphelion, equinoxes, and solstices.

http://www.srrb.noaa.gov/highlights/sunrise/sunrise.html
Fully interactive site that allows you to get time of sunrise/sunset for a choice of world cities or for precise latitude-longitude coordinates. Also gives solar declination and several other items of Earth-Sun position information, complete with glossary to explain their meanings.

Media Enrichment

Tutorial
Solar Geometry
This tutorial covers the geometry of Earth's orbit and its effects on solar radiation. It uses three-dimensional diagrams to show how variations in day length and solar position arise and includes animations depicting the processes of beam spreading and depletion.

Interactive Exercise
Earth–Sun Geometry
Use this program to see how solar position varies seasonally from place to place, and how these variations translate into radiation at the surface. Compare one latitude with another, noting differences in the seasonal range of incoming radiation. Think about the likely effect on temperature at the two locations. Experiment with latitudes in both hemispheres, as well as locations near the equator. Using your latitude as a reference, speculate about differences in thermal conditions.

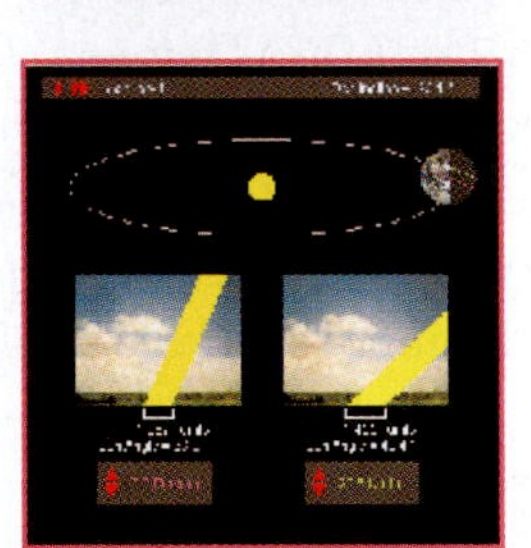

The CD-ROM

The CD-ROM features Tutorials, Student Interactive Exercises, and illustrative movie loops—all keyed to the text. The tutorials contain interactive animations, three-dimensional diagrams, and review exercises, and cover topics such as:

- Solar Geometry
- Global Energy Balance
- Atmospheric Moisture
- Adiabatic Processes
- Atmospheric Stability
- The Coriolis Force
- Atmospheric Forces and Motion
- Upper-Level Winds and Pressure
- Cyclones and Anti-Cyclones.

Tutorials

Perhaps most fundamental to the approach are the nine computer tutorials covering basic principles of atmospheric science. These modules employ three-dimensional diagrams and animations to present material not easily visualized using conventional media. The software modules have undergone considerable testing and have been used successfully by thousands of students.

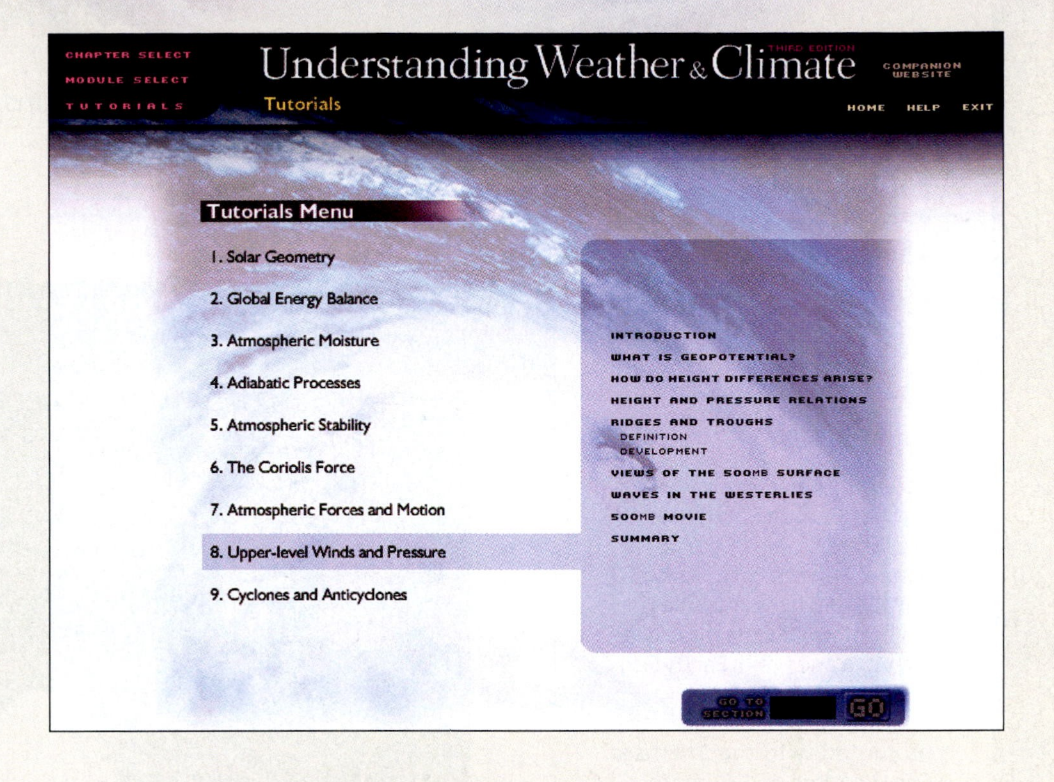

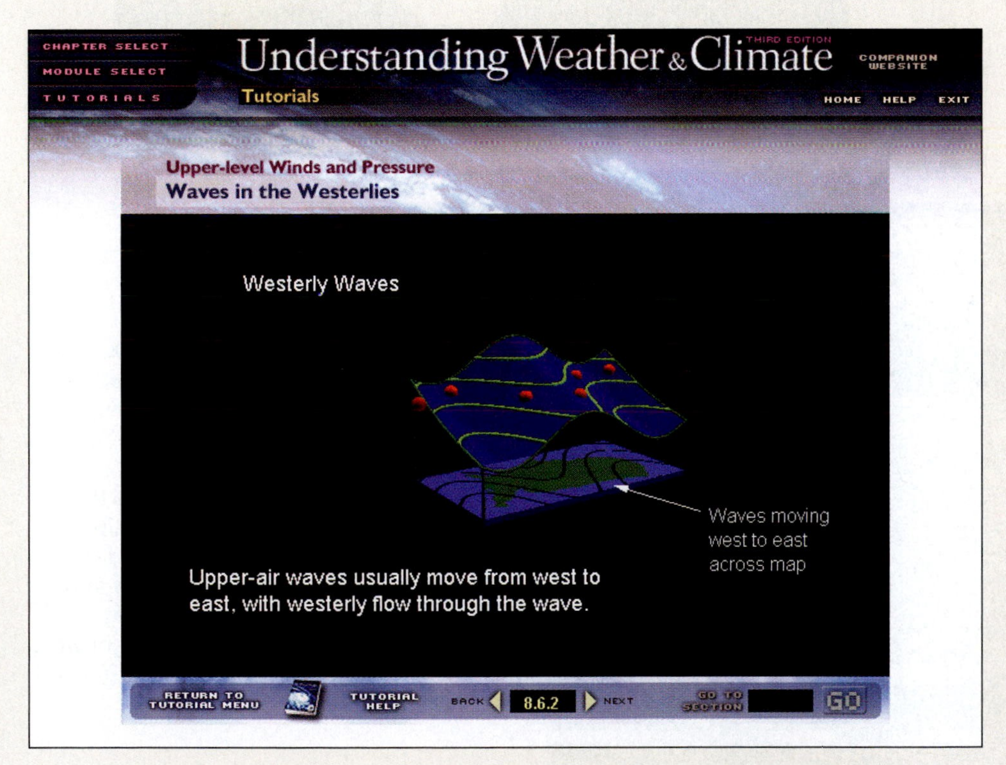

New to the third edition CD:

- The third edition CD features a new software tutorial about atmospheric stability.
- The CD that accompanies the third edition has been thoroughly upgraded so that no installation will be required. The CD runs directly off of the CD-ROM drive, and will not require the installation of any browser plug-ins.

Integrated Media

In addition to the tutorials, the accompanying CD-ROM contains four other types of resources for students and professors:

- Interactive Exercises
- Weather in Motion movies
- Weather Images, and
- A Media Library.

Interactive Exercises

The *Interactive Exercises* are a series of short activities produced by Greg Carbone of the University of South Carolina. These modules cover important topics such as hurricanes and Earth-Sun relations. Expanded versions of the activities are presented with the *Exercises For Weather and Climate, Fifth Edition* lab manual by Greg Carbone, which is available at a discount when packaged with this book.

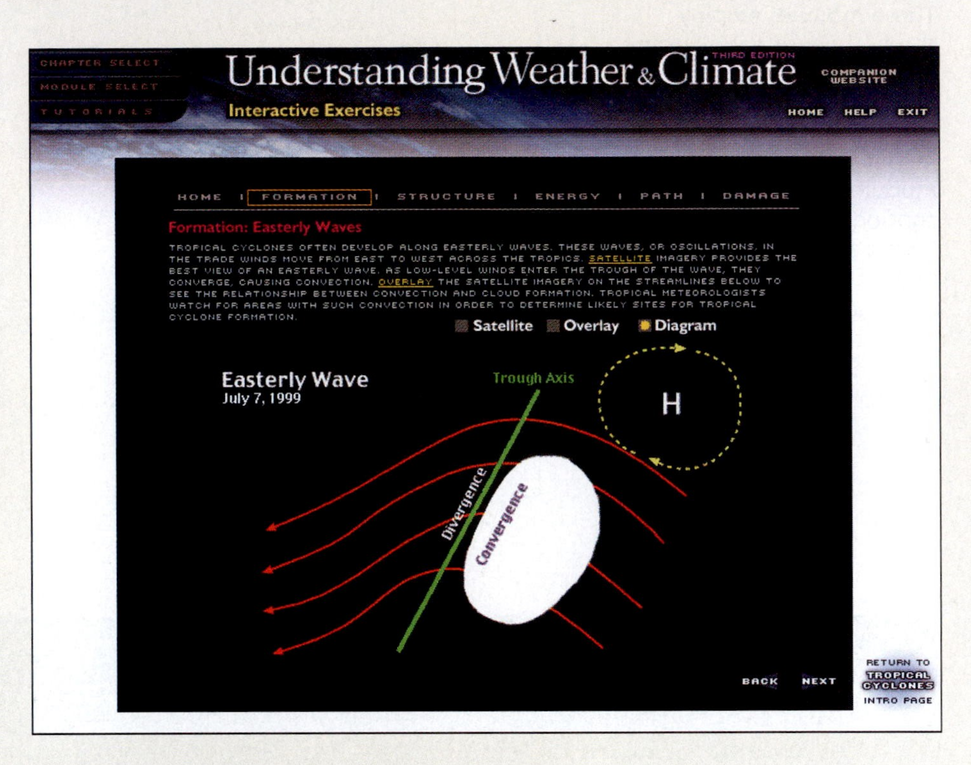

Movies

Weather in Motion movies depict events and phenomena discussed in the text. Examples include a year-long satellite movie showing clouds and temperatures across the globe, three-dimensional simulations of thunder-storm development, and animations depicting variations in Earth's orbit.

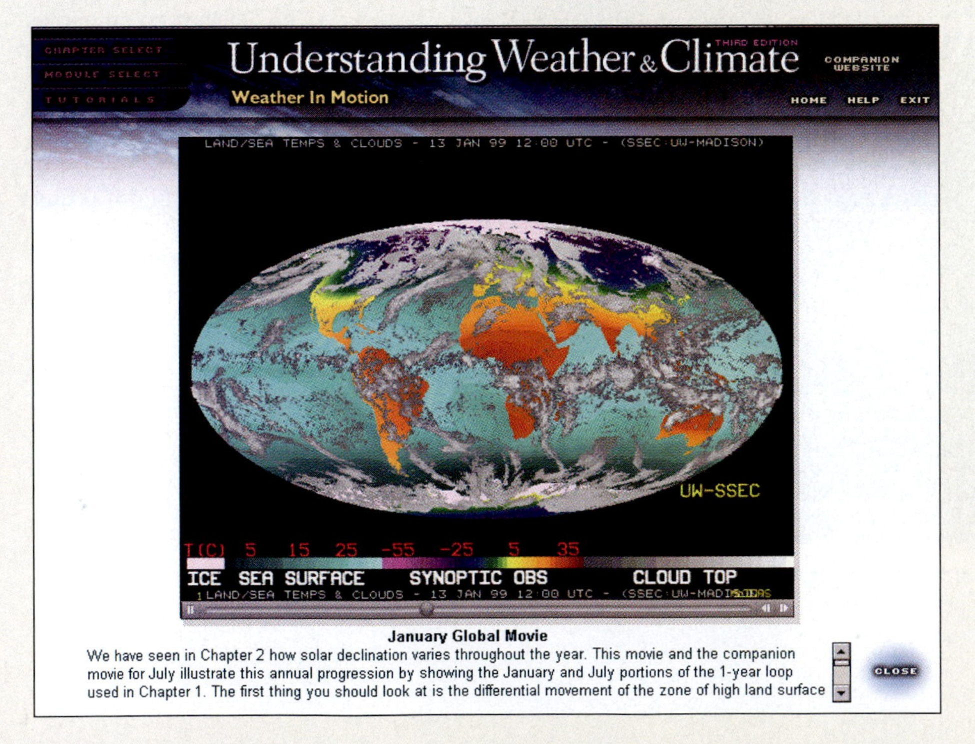

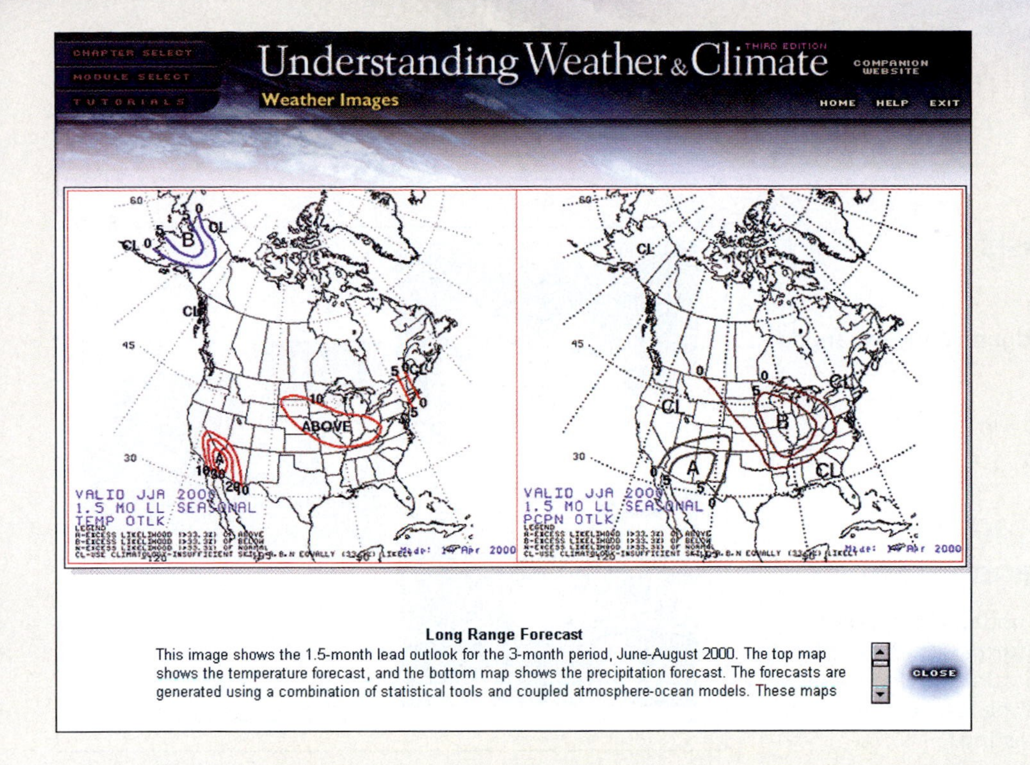

Weather Images

Weather Images provide additional illustrations of weather phenomena using photographs, satellite images, and computer diagrams. These images are keyed to the text.

The *Media Library* provides additional images, movies, and animations. These are intended for browsing by the student and instructor, and are therefore not specifically mentioned in the text. A short description of each is found on the CD.

New to the Third Edition

Expanded Emphasis on Forecasting

The third edition features significantly expanded materials on:
- weather forecasting
- meteorological variables, and
- thermodynamic diagrams.

(a) (b) (c)

▲ Figure 1-21
A typical National Weather Service office (a) and workstation (b). Two AWIPS graphical display monitors (c).

Completely Updated for Currency of Fact and Event

The authors have carefully updated the text to include the most current information. Examples include the recent reclassification of Hurricane Andrew as Category 5, and the latest technology employed in weather forecasting.

Atmospheric Optics

New chapter on Atmospheric Optics.

Reorganized Material

The chapter, "Atmospheric Pressure and Wind" now precedes the chapter, "Atmospheric Moisture." This re-organization better prepares students to understand the basic concepts of vapor pressure and uplift mechanisms.

Review Questions

1. Why is it difficult to define an absolute top of the atmosphere?
2. What are the homosphere and the heterosphere?
3. What is the difference between the permanent and variable gases of the atmosphere? Which gases are the most important in terms of their contribution to the total mass of the atr
6. What is ozone, and why is it both beneficial and harmful to life on Earth?
7. What are aerosols? Are they formed only by human activities or are they also naturally occurring?
8. How do photosynthesis, respiration, and decay affect the carbon dioxide balance of the atmosphere?

Critical Thinking

1. Severe weather can produce major economic damage and loss of life. How do you think the threats vary from one place to another? Rate your home location in terms of risk for each type of severe weather.
2. The Kyoto Treaty that limits carbon dioxide emissions has not been ratified by all countries that originally signed the treaty. What are the pros and cons of doing so?
3. Volcanic eruptions continue to occur and outgas water vapor, carbon dioxide, and other gases. Do you think that this will be a significant factor in increasing the concentration of these gases over the next century? Why or why not?
4. Temperatures usually decrease with height in the troposphere but increase with height in the stratosphere. Why do the two layers have such different profiles?
5. The thermosphere has extremely high temperatures, but a person exposed to the thermosphere would rapidly freeze. Explain the apparent contradiction in terms of what you know about heat and temperature.

Problems and Exercises

1. The National Climatic Data Center (NCDC) has a Web site at http://lwf.ncdc.noaa.gov/oa/climate/severeweather/extremes.html that provides a wealth of information on extreme weather and climate events. Examine the site to
3. Keep a record of daily weather in your area and download current weather maps from one of the available Web sites. Do you notice any patterns on the maps that tend to be associated with particular weather conditions?

End-of-Chapter Materials

End-of-chapter materials have been significantly expanded to include more questions, and to include critical thinking problems.

Laboratory Manual

Exercises for Weather and Climate

Fifth Edition

Greg Carbone

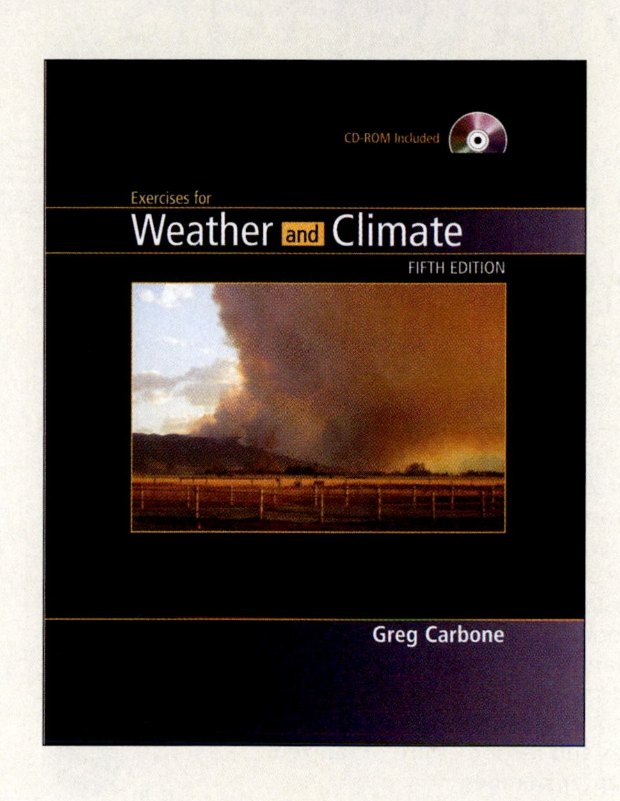

Appropriate for use with any introductory meteorology or weather and climate text, this laboratory manual consists of 18 exercises that present introductory principles in weather and climate to the student. Collectively, the exercises combine data analysis, problem solving, and experimentation; the questions are designed to encourage critical thinking about the fundamental topics of atmospheric processes.

New features of the manual include:

- **A new version of GeoClock** in Lab 3 to show geographical and seasonal variations in sunlight
- **New questions presented in Lab 6** prompt students to make the connections between the moisture measures
- **An updated section in Lab 8** on the forces affecting cloud velocity and terminal velocity
- **A complete revision of Lab 9:** Atmospheric Motion, employs a different approach to presenting Coriolis force, upper-air, and surface winds, and integrating real-world data to illustrate these concepts
- **The integration of Hurricane Floyd as a case study in Lab 13 on hurricanes,** and also a look at the role of hurricane forecasts

Accompanying Solutions Manual contains all solutions to the labs (0-13-101512-5)

Take 20% off Lab Manual price when valuepacked with *Understanding Weather and Climate, Third Edition* (0-13-104253-X)

Supplemental Materials

Lecture Presentation and Testing Aids

Instructor's Resource CD (IRCD) (0-13-101584-2)
The Instructor's Resource CD-ROM provides high-quality electronic versions of photos and illustrations from the book—as well as customizable Power Point lecture presentations, the Instructor's Manual and Test Item File in MS Word format. New to this edition are Authorware Animations, as presented on the Student CD, for use in your lecture presentation. To further guarantee classroom projection quality, all images are manually adjusted for color, brightness, and contrast. For easy reference and identification, all images and resources are organized by chapter number.

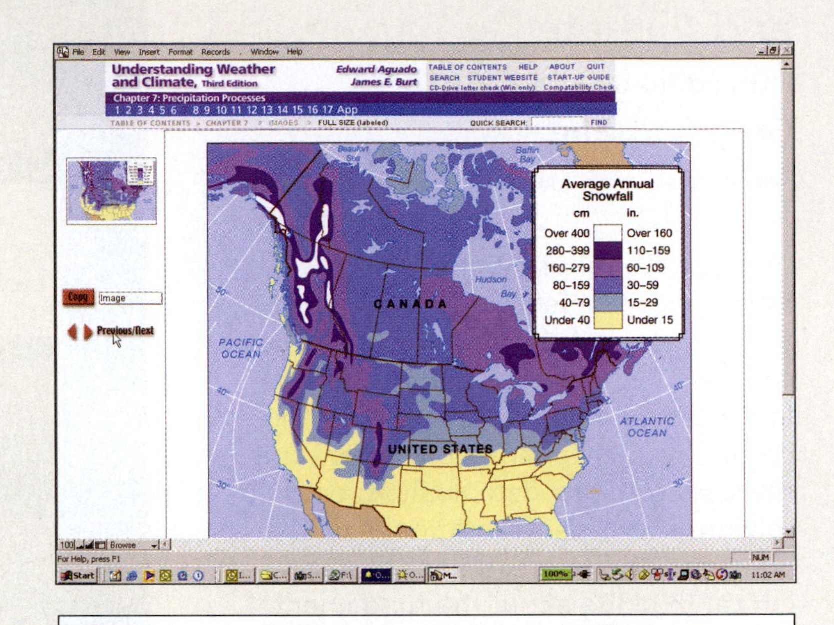

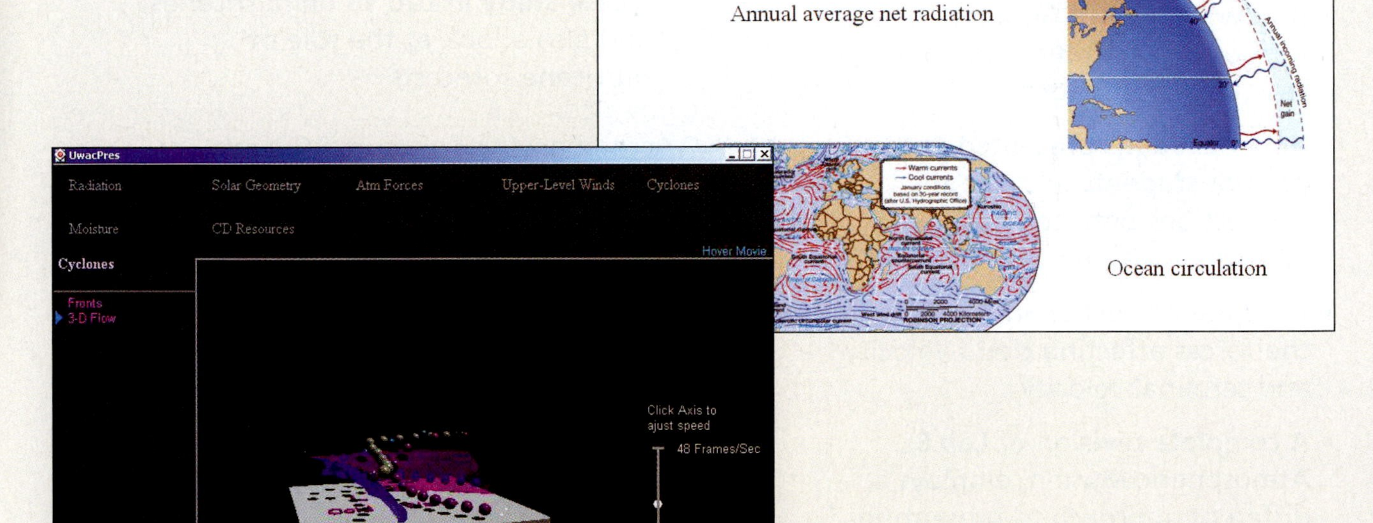

Instructor Resources

Transparencies (0-13-101592-3)
Includes 200 photographs and illustrations from the text, enlarged for excellent classroom visibility.

Instructor's Manual (0-13-101598-2)
Includes a brief chapter overview discussing important concepts with references to text figures, Transparencies, Student CD, and Companion Website, and answers to the End of Chapter Review questions and Critical Thinking problems.

Test Item File (0-13-101595-8)
Includes over 80 questions—multiple choice, true/false and discussion questions per chapter.

Computerized Testing (0-13-101596-6)
Computerized test generator lets you view and edit test bank questions, transfer questions to tests, and print the tests in a variety of customized formats.

Student Learning Aids/Packages

Study Guide (0-13-101583-4)
Includes Chapter Objectives and Outline, Vocabulary Review, Self-Tests, Questions for Thought, Quantitative Problems and Answers, and Research Paper Ideas for each chapter. New to this edition are questions and a How-to-Guide on reading and interpreting graphs. *Take 50% off the Study Guide price when valuepacked with text (0-13-104254-8)*

Rand McNally Atlas of World Geography (0-13-959339-X)
Get the Atlas free when valuepacked with the text (0-13-104255-6)

Research Navigator
Available at no cost with the purchase of any Pearson Education college textbook, the Research Navigator™ gives students online access to thousands of articles from both academic journals and the popular media. Articles are organized by discipline and fully searchable.

Online Study Guide
www.prenhall.com/aguado
In partnership with the book, *Understanding Weather & Climate,* this text-specific Web site provides a forum where professors and students can expand their investigation of meteorology and atmospheric science.

PART ONE

Energy and Mass

Looking north at Atigun Pass, Alaska, on July 4, 1977, just before midnight. Because Atigun Pass is located north of the Arctic Circle, the midnight sun can be seen much of the summer.

Part of the damage from the May 3–4, 1999, tornado outbreak in Oklahoma and Kansas.

chapter 1

Composition and Structure of the Atmosphere

SEVERE WEATHER STATEMENT
NATIONAL WEATHER SERVICE NORMAN OK
7:31 PM CDT MON MAY 3 1999

LARGE DAMAGING TORNADO MOVING THROUGH OKLAHOMA CITY METRO. A LARGE TORNADO HAS CAUSED EXTENSIVE DAMAGE IN SOUTHERN PORTIONS OF THE OKLAHOMA CITY METRO AREA. AT 7:31 PM THE TORNADO WAS ENTERING SOUTHERN OKLAHOMA COUNTY JUST EAST OF INTERSTATE 35 AND SOUTH OF CROSS ROADS MALL. PERSONS IN SOUTHEAST OKLAHOMA CITY AND MIDWEST CITY ARE IN DANGER! IF YOU LIVE NEAR THESE AREAS TAKE IMMEDIATE TORNADO PRECAUTIONS! THE TORNADO WAS MOVING NORTHEAST. THIS IS AN EXTREMELY DANGEROUS AND LIFE-THREATENING SITUATION. IF YOU ARE IN THE PATH OF THIS LARGE AND DESTRUCTIVE TORNADO . . . TAKE COVER IMMEDIATELY.

BULLETIN - EAS ACTIVATION REQUESTED
TORNADO WARNING
NATIONAL WEATHER SERVICE NORMAN OK
10:17 PM CDT MON MAY 3 1999

THE NATIONAL WEATHER SERVICE IN NORMAN HAS ISSUED A

*TORNADO WARNING FOR NOBLE COUNTY IN NORTHEAST OKLAHOMA UNTIL 10:45 PM CDT.

*AT 10:17 PM CDT A ONE-MILE WIDE TORNADO WAS REPORTED NEAR MULHALL, MOVING NORTHEAST AT 30 MPH.

THIS TORNADO WILL CROSS INTERSTATE 35 NEAR OR SOUTH OF PERRY. RESIDENTS IN NOBLE COUNTY NEED TO TAKE COVER NOW. PERRY IS IN THE PATH OF THIS STORM.

THIS IS AN EXTREMELY DANGEROUS AND POTENTIALLY LIFE-THREATENING SITUATION. IF YOU ARE IN THE PATH OF THIS LARGE AND DESTRUCTIVE TORNADO, TAKE COVER IMMEDIATELY.

TORNADOES ARE ESPECIALLY DANGEROUS AT NIGHT BECAUSE THEY ARE HARD TO SEE. TAKE COVER NOW. IF A BASEMENT IS NOT AVAILABLE, MOVE TO AN INTERIOR ROOM OR HALLWAY ON THE LOWEST FLOOR. LEAVE MOBILE HOMES AND VEHICLES FOR REINFORCED SHELTER. STAY AWAY FROM WINDOWS.

During the afternoon and evening of May 3, 1999, dozens of emergency advisories like these two were issued to residents of the south central United States as one of the worst tornado outbreaks in decades punished the area. Worst hit was central Oklahoma, where at least 57 tornadoes killed 44 people and injured 748 others. In Kansas another 5 people were killed and 11,000 homes or businesses were damaged or destroyed. Over the next few days, the same storm that spawned the tornadoes slowly worked its way to neighboring states, killing 5 more people in Texas and Tennessee.

The storm system contained 11 major storm elements, called *supercells*, that followed a southwest-to-northeast path. Each supercell produced at least 1 tornado, with the most destructive supercell yielding 22 in a 3-hour period while covering a distance of 61 kilometers (km), or 38 miles (mi). The largest of its tornadoes was categorized as an F5, the most destructive possible, measuring as much as 1 km (0.6 mi) wide and staying on the ground for about 4 hours. So powerful is an F5 tornado, another was not to be experienced in North America for at least another 3 years.*

*A violent tornado that hit southern Maryland on April 28, 2002, originally thought to be an F5, was later officially determined to have had peak winds of "only" 419 km per hour (260 mph), just below the 420 km/hr threshold of an F5 tornado. The Maryland tornado killed 3 people and injured 93 others.

CHAPTER OUTLINE

Weather in Motion

Satellite Movies of the May 3–4, 1999 Tornado Cluster

As terrible as the outbreak was, things would have been very much worse had it not been for the extensive forecasting, tracking, and warning capabilities of the National Weather Service. One study based on long-term observations suggests that as many as 700 people might have perished had it not been for the recent modernization of the Weather Service. People such as Air Force Captain John Millhouse would readily agree with that notion. Having heard one of the tornado warnings, he and his family moved to the basement of their home. When they emerged unscathed, only two walls of the house were left standing; the rest was totally destroyed.

Not all weather phenomena are as dramatic as tornadoes, but it is indisputable that weather exerts a tremendous impact on our day-to-day lives: we revel in the beauty of a sunrise, marvel at the power of hurricanes, rely on the rain to nourish our gardens, and complain about the heat and cold. And weather can throw a wrench into our travel plans—in the year 2000, the whims of the atmosphere caused more than 300,000 United States aviation delays! Table 1–1 summarizes recent estimates of annual financial and human costs of various types of extreme weather events within the United States during the 1990s. Though hurricanes and tornadoes can exact a terrible human cost, extreme heat and cold are the biggest killers in the United States, with floods and hurricanes producing the greatest financial costs. Table 1–1 actually understates the toll exerted by weather. Though severe weather kills about 900 people annually, that figure is compounded by an estimated 6000 weather-related traffic fatalities that accrue mostly from nonsevere events.

Despite its immediacy, most of us know relatively little about how and why the atmosphere behaves as it does. In the pages that follow, we hope to provide an account of both the how and the why, in ways that will lead you to an understanding of the underlying physical processes. As you will see, the **atmosphere** is a mixture of gas molecules, microscopically small suspended particles of solid and liquid, and falling precipitation. **Meteorology** is the study of the atmosphere and the processes (such as cloud formation, lightning, and wind movement) that cause what we refer to as the "weather." This chapter introduces the most basic elements of meteorology, laying the foundation for much of what follows in the rest of the book.

Table 1–1 • Averages of Annual Fatalities and Financial Costs of Weather Events in the United States During the 1990s

Event	Annual Mean Number of Fatalities	Annual Mean Loss (adjusted to 1999 dollars)
Floods	98	$5,300,000,000
Hurricanes	21	$5,400,000,000
Winter Storms	57	$329,000,000
Tornadoes	56	$777,000,000
Extreme Heat	282	$85,000,000
Extreme Cold	292	$368,000,000
Lightning	69	$38,000,000
Hail	1	$938,000,000
Total	**876**	**$13,000,000,000**

Source: Data from Pielke and Carbone, Bulletin of American Meteorological Society, March 2002.

◀ Figure 1–1
Despite its appearance from the surface, the atmosphere is extremely thin relative to the rest of Earth.

The Thickness of the Atmosphere

Every child has wondered, "How high is the sky?" There is no definitive answer to that question, however, because Earth's atmosphere becomes thinner at higher altitudes. A person in a rising hot-air balloon would be surrounded by an atmosphere that gradually becomes less dense. At some height, the air becomes so thin that the balloonist would pass out from a shortage of oxygen—but there would still be an atmosphere. At an altitude of 16 km, or 10 mi, the density of the air is only about 10 percent of that at sea level, and at 50 km (30 mi) it is only about 1 percent of what it is at sea level. Even at heights of several hundred kilometers above sea level, there is some air and, hence, an atmosphere. However, because no universally accepted definition exists of how much air in a given volume constitutes the presence of an atmosphere (is there an atmosphere if there is only, say, one molecule of air per cubic kilometer?), we have no way to establish its upper boundary.

Viewed from Earth's surface, the atmosphere appears to be extremely deep. In reality, however, most of the atmosphere is contained within a relatively shallow envelope surrounding the oceans and continents. Let's assume for the sake of discussion that the upper limit of the atmosphere occurs at 100 km (60 mi) above sea level (in fact, 99.99997 percent of the atmosphere is contained below this height). By comparing this 100-km thickness with the 6500-km (4000-mi) radius of Earth, we see that the depth of the atmosphere is less than 2 percent of Earth's thickness from the center. This is evident in Figure 1–1, an image of Earth and its atmosphere taken from space. The top of the thunderstorm cloud probably has an altitude of about 12 km (7.5 mi), but when viewed from space, it appears to hug the ground. Though impressive when we look up at them, those clouds are no thicker than the skin of an apple, comparatively speaking.

Given the shallowness of the atmosphere, its motion over large areas must be primarily horizontal. Indeed, with some very notable exceptions, horizontal wind speeds are typically a thousand times greater than vertical wind speeds. However, we cannot overlook the wind's vertical motions. As we shall see, even small vertical displacements of air have an important impact on the state of the atmosphere. Paradoxically, the least impressive motions—vertical—despite being hardest to detect and forecast, turn out to be the most important in determining much of atmospheric behavior.

Despite its relative shallowness, the atmosphere contains considerable mass. The total mass of the atmosphere, 5.14×10^{15} kilograms (kg),* is equivalent to 5.65 billion million tons—the amount of water that would fill a lake the size of California to a depth of 13 km (7.7 mi).

Composition of the Atmosphere

The atmosphere is composed of a mixture of invisible gases and a large number of suspended microscopic solid particles and water droplets. Molecules of the gases can be exchanged between the atmosphere and Earth's surface by physical processes,

*5,140,000,000,000,000 kg

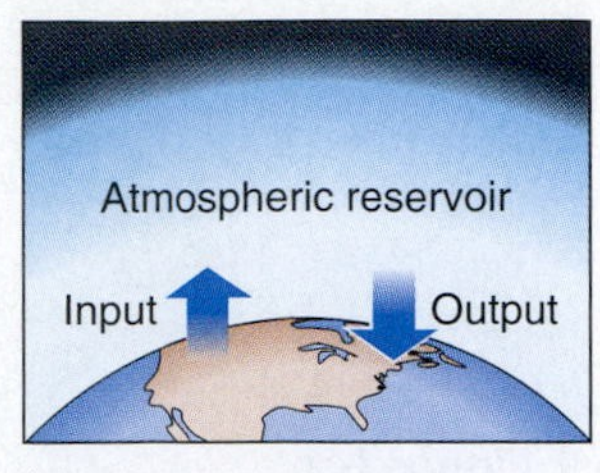

(a)

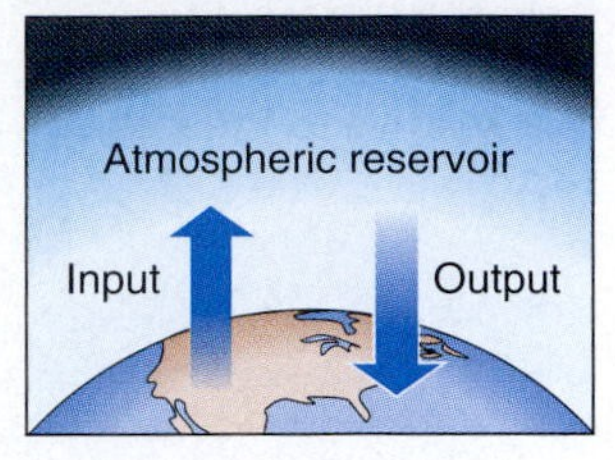

(b)

▲ **Figure 1–2**
The atmosphere can be thought of as a reservoir experiencing constant input and output by surface exchange and/or internal processes. If the inputs and outputs occur at the same rate, there is no net change in the content of the gas. In (a) the arrows depict a slow rate of exchange of a hypothetical gas. As a result, any molecule of that gas can be expected to remain for a long time before being cycled out of the atmosphere. In (b) the reservoir size is the same, but the exchange rate is much more rapid, so the gas has a shorter residence time.

such as volcanic eruptions, or by biological processes, such as plant and animal respiration. Molecules can also be produced and destroyed by purely internal processes, such as chemical reactions between the gases.

Consider a gas that is constantly being cycled between the atmosphere and Earth's surface (Figure 1–2). If we think of the atmosphere as a reservoir for this gas, the gas concentration in the reservoir will remain constant so long as the input rate (the rate at which the gas moves from ground to atmosphere) is equal to the output rate (the rate at which the gas moves from atmosphere to ground). Under such conditions, we say that the concentration of the gas exists in a *steady state*.

Although the atmospheric concentration of a gas remains constant under steady-state conditions, individual molecules stay in the atmosphere for only a finite period of time before they are removed by whatever output processes are active. The average length of time that individual molecules of a given substance remain in the atmosphere is called the *residence time*. The residence time is found by dividing the mass of the substance in the atmosphere (in kilograms) by the rate at which the substance enters and exits the atmosphere (in kilograms per year). Thus, gases that are rapidly exchanged between Earth's surface and the atmosphere have brief residence times, as do gases that have relatively low atmospheric concentrations.

Figure 1–2 illustrates the concept of a steady state and residence time. Parts (a) and (b) show the same mass constituting the "atmospheric reservoir," with the length of the input and output arrows indicating the rates at which gases are put into and removed from the reservoir. In (a), the input and output rates are equal, indicating a steady state. Both rates are small, however, meaning that any particular molecule of the gas has a long residence time. In (b), there is again a steady state, but the greater rate of input and output relative to reservoir size leads to a shorter residence time.

Atmospheric gases are often categorized as being permanent or variable, depending on whether or not their concentration is stable. **Permanent gases** are those that form a constant proportion of the atmospheric mass, whereas **variable gases** are those whose distribution in the atmosphere varies in both time and space.

Permanent gases account for the greater part of the atmospheric mass—99.999 percent—and occur in a constant proportion throughout the atmosphere's lowest 80 km (50 mi). Because of its chemical homogeneity, this region within 80 km of Earth's surface is called the **homosphere**. For most purposes, we consider the homosphere virtually the entire atmosphere.

Above the homosphere is the **heterosphere**, where lighter gases (such as hydrogen and helium) become increasingly dominant with increasing altitude. Because its composition varies with altitude, the heterosphere contains no truly permanent gases. Given the small mass at these altitudes, we won't pursue these and other details of the heterosphere but instead will deal almost exclusively with the homosphere.

The Permanent Gases

The homosphere is composed mostly of nitrogen and oxygen, with small amounts of the inert gases argon and neon and even smaller amounts of several other gases (Table 1–2). Atmospheric **nitrogen** occurs primarily as paired nitrogen atoms bonded together to form single molecules denoted N_2. Nitrogen gas has a molecular weight of 28.02, meaning that on average the mass of one N_2 molecule is slightly in excess of that of a combined total of 28 protons and neutrons. Almost all N atoms contain 7 protons and 7 neutrons, but some are heavier, having 8 neutrons. As a result, the average molecular weight of N_2 is slightly greater than 28.

Nitrogen is a stable gas that accounts for 78 percent of the volume of all the permanent gases, or 75.5 percent of their mass (the difference arises because the gases have different molecular weights but occupy the same volume under similar conditions). The processes that add and remove nitrogen from the atmosphere occur very slowly, so it has a very long residence time—42 million years.

Despite the fact that it makes up more than 75 percent of the atmosphere, N_2 is relatively unimportant in terms of most meteorological and climatological

Table 1–2 • Permanent Gases of the Atmosphere

Constituent	Formula	Percent by Volume	Molecular Weight
Nitrogen	N_2	78.08	28.01
Oxygen	O_2	20.95	32.00
Argon	Ar	0.93	39.95
Neon	Ne	0.002	20.18
Helium	He	0.0005	4.00
Krypton	Kr	0.0001	83.8
Xenon	Xe	0.00009	131.3
Hydrogen	H_2	0.00005	2.02

processes. (However, as we shall discuss later, nitrogen-bearing gases, such as nitrous oxide, are quite important to Earth's climate.)

The second most dominant gas, **oxygen** (O_2), constitutes 21 percent of the volume of the atmosphere and 23 percent of its mass. Oxygen is crucial to the existence of virtually all forms of life. Like nitrogen, the oxygen molecules of the atmosphere consist mostly of paired atoms, called *diatomic oxygen*. Their residence time is about 5000 years. Together, nitrogen and oxygen account for 99 percent of all the permanent gases, with **argon** making up most of the remainder. Removal processes are so slow for argon that its residence time is extremely long.

Variable Gases

The variable gases account for only a small percentage of the total mass of the atmosphere (Table 1–3). Despite their relative scarcity, some of these gases impact the behavior of the atmosphere—and even your own physical comfort.

Water Vapor The most abundant of the variable gases, **water vapor**, occupies about one-quarter of 1 percent of the total mass of the atmosphere. Because the source of water vapor in the atmosphere is evaporation from Earth's surface, its concentration normally decreases rapidly with altitude, and most atmospheric water vapor is found in the lowest 5 km (3 mi) of the atmosphere.

Water is constantly being cycled between the planet and the atmosphere in what is called the **hydrologic cycle**. Earth is sometimes referred to as the "water planet" because three-quarters of its surface is covered by oceans, ice sheets, lakes, or rivers. In addition, much water exists beneath Earth's surface both in saturated (groundwater) formations and in partially saturated upper soil layers. Water continuously evaporates from both open water and plant leaves into the atmosphere, where it eventually condenses to form liquid droplets and ice crystals. These liquid and solid particles are removed from the atmosphere by precipitation as rain, snow, sleet, or hail. Because of the rapidity of global evaporation, condensation, and precipitation, water vapor has a very short residence time of only 10 days.

We all know how damp and muggy the air feels when the water vapor content is high and how parched our skin can get when the air is dry (not to mention

Table 1–3 • Variable Gases of the Atmosphere

Constituent	Formula	Percent by Volume	Molecular Weight
Water Vapor	H_2O	0.25	18.01
Carbon Dioxide	CO_2	0.037	44.01
Ozone	O_3	0.01	48.00

the occurrence of "bad hair days" when the moisture content is extremely high or low). Despite the wide range of physical comfort levels we experience in response to variations in water vapor content, the actual range of water vapor content is really quite limited. Near Earth's surface, the water vapor content ranges from just a fraction of 1 percent of the total atmosphere over deserts and polar regions to about 4 percent in the Tropics. This means that at most we would find that 4 out of every 100 air molecules are water vapor. (Outside of the Tropics, water vapor content does not usually exceed 2 percent.) At higher altitudes, water vapor is even rarer, so that its overall contribution to atmospheric mass is quite small.

Despite being a relatively small portion of the atmosphere, water vapor is extremely important. Not only is it the source of the moisture needed to form clouds, but it is also a very effective absorber of energy emitted by Earth's surface. (We describe radiant energy in the next chapter.) Its ability to absorb Earth's thermal energy makes water vapor one of the "greenhouse gases" we discuss in Chapter 3.

It is extremely important to keep in mind that water vapor is not the same as small droplets of liquid water. Water vapor exists as individual gas molecules. Unlike the molecules of liquids and solids, water vapor molecules are not bonded together. In that regard, water vapor is similar to N_2, O_2, and other atmospheric gases. Unlike other gases, however, it readily changes phase into liquid and solid forms both at Earth's surface and in the atmosphere.

Although water vapor is an invisible gas, there are satellite systems that can detect the amount of water vapor in the air and display its varying content on images (Figure 1–3). These images typically show large changes over very short distances, even more so than other variable gases. Displays of water vapor content can be quite valuable to forecasters, who use the images to determine broad-scale wind patterns in the middle and upper atmosphere. The images also help meteorologists identify the boundaries between adjoining bodies of air.

Carbon Dioxide Another important variable gas is **carbon dioxide**, CO_2. We do not physically sense variations in the amount of carbon dioxide present in the atmosphere, as we do with water vapor; and yet, as you will see shortly, those changes can't be ignored. Increases in the carbon dioxide content of the atmosphere may have some important climatic consequences that could greatly affect human societies.

Carbon dioxide currently accounts for 0.037 percent of the atmosphere. When a gas occupies such a small proportion of the atmosphere, we often express its content as parts per million (ppm) rather than percent (parts per hundred). Thus, the current atmospheric concentration of CO_2 is about 370 ppm. It is supplied to the

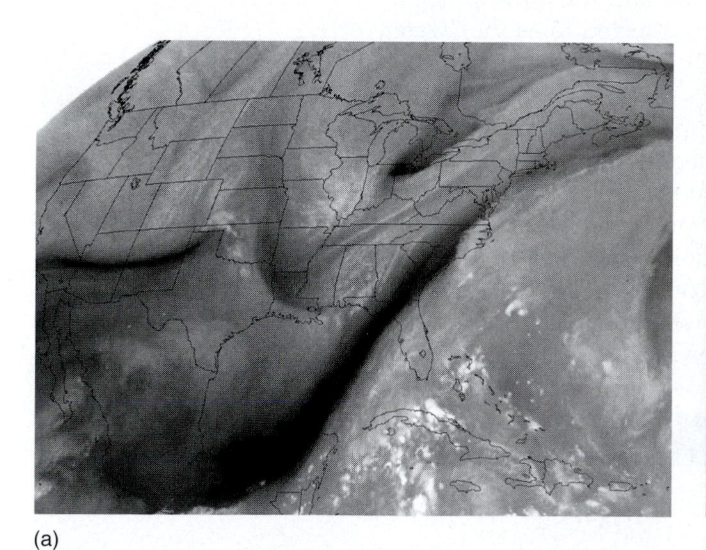
(a)

(b)

▲ **Figure 1–3**
Paired satellite images based on a system that detects the presence of water vapor (a) and one that gives a visible rendition of cloud tops (b). Notice that though the two patterns are somewhat similar, the water vapor image shows a broader distribution of moisture than does the image showing actual clouds.

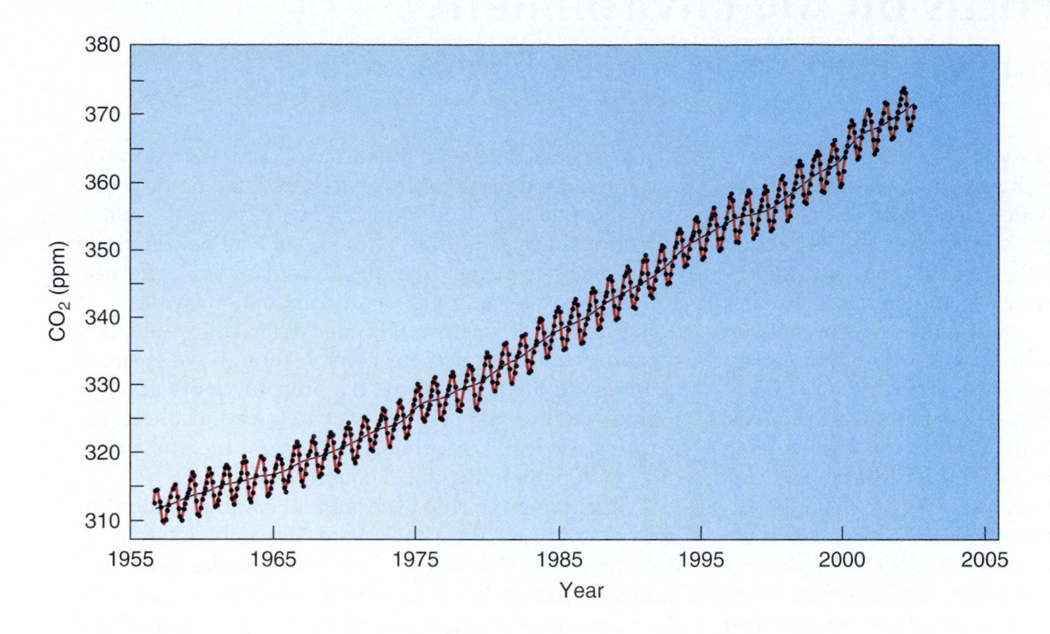

◀ **Figure 1–4**
The carbon dioxide content of the atmosphere has been steadily increasing over the last half century due to human activities. The data are obtained from the Mauna Loa Observatory, with the dots depicting average monthly values and revealing the seasonal cycle in the growth and decay of plants. The smooth line corresponds to annual average values, illustrating the long-term trend.

atmosphere by plant and animal respiration, the decay of organic material, volcanic eruptions, and natural and *anthropogenic* (human-produced) combustion. Carbon dioxide is removed from the atmosphere by **photosynthesis**, the process by which green plants convert light energy to chemical energy. Photosynthesis uses energy from sunlight to convert water absorbed by the roots of plants and carbon dioxide taken in from the air into the chemical compounds known as carbohydrates, which both support plant processes and ultimately nourish the animal kingdom. In this process, oxygen molecules are released into the atmosphere as a by-product (see *Box 1–1, Focus on the Environment: Photosynthesis, Respiration, and Carbon Dioxide*).

In recent decades, the rate of carbon dioxide input to the atmosphere has exceeded the rate of removal, leading to a global increase in concentration. Figure 1–4 plots the record of carbon dioxide content obtained from the Mauna Loa Observatory. (Taken from a 3400-m, or 11,150-ft, elevation on one of the Hawaiian Islands, Mauna Loa's observations are considered representative for the Northern Hemisphere.) Since the 1950s, the concentration of CO_2 has increased at a rate of about 1.8 ppm per year. The increase has occurred mainly because of an increase in anthropogenic combustion and, to a lesser extent, deforestation of large tracts of woodland (Figure 1–5). The increase has received considerable scientific and media

◀ **Figure 1–5**
Burning rain forest in Guatemala. Deforestation by humans has contributed to the increase of carbon dioxide in the atmosphere because it reduces the amount of plant material that can perform photosynthesis.

1–1 Focus on the Environment

Photosynthesis, Respiration, and Carbon Dioxide

Without the process of photosynthesis, Earth would have an entirely different atmosphere—and would likely be without life as we know it. Through photosynthesis, plants utilize light energy from the Sun to make food. Because it requires sunlight, photosynthesis occurs only during the day. It also requires chlorophyll, an organic substance found in green plants and some single-celled organisms. Photosynthesis converts solar energy, water, and carbon dioxide into simple carbohydrates. These can then be converted to complex carbohydrates, starches, and proteins, all of which supply plants with the material for their own growth, while plants in turn provide the basic nutrients for grazing and browsing animals.

In addition to photosynthesis, another important exchange of gases takes place through the leaves of plants—**respiration**. Respiration provides the mechanism by which plants obtain the oxygen they need in order to perform their metabolic processes. (For animals, respiration is synonymous with breathing. Though plants don't have lungs, they take in oxygen through their leaves.) Unlike photosynthesis, respiration occurs during both day and night. A plant must "fix" carbon into its cells in order to grow; to obtain this carbon, the photosynthesis rate in the plant must exceed the respiration rate. Thus, during periods of plant growth more carbon dioxide is being removed from the air than is being added through respiration. After a plant dies, however, it no longer takes in CO_2 from the air. Instead, its stored carbon is oxidized and released back to the atmosphere as CO_2 as the plant decomposes. If no major changes occur in the amount and distribution of vegetation, CO_2 intake (for photosynthesis) balances CO_2 output (from respiration and decay) over the course of a year, and the total store of atmospheric CO_2 is unaffected by plant growth.

There is, however, another factor governing the CO_2 balance. Under some circumstances, dead plant material is quickly buried beneath the surface, does not decompose, and therefore does not release its stored carbon back into the atmosphere. Instead, over millions of years the material is transformed into fossil fuels, such as petroleum or coal, which we extract from the ground and burn for heat and energy, yielding CO_2 as a combustion product. In doing so, we release the carbon that otherwise would have remained underground for eons. Although we speak of "adding" to the atmosphere, it's probably more accurate to think of "moving" carbon back to the atmosphere. Regardless, the result is certainly an increase in atmospheric CO_2, with potentially global consequences.

Recognizing the role of fossil fuel combustion on the carbon balance of the atmosphere, 160 of the world's nations met in 1997 at Kyoto, Japan, to formally agree on a multiyear plan to lower CO_2 emissions. On December 11, they agreed to set target reductions for the industrialized countries. Under this plan, averaged over the 5-year period from 2008 to 2012, most of the developed world would reduce CO_2 emissions by 6 to 8 percent of 1990 values. To be legally binding, the plan would not only have to be signed by the representatives of the participating countries, but also formally ratified by their governments. In March 2002 the European Union formally ratified the treaty, and many other nations were expected to follow suit by the end of the year. Ratification by the United States and Canada, however, is far from assured as of April 2002. In the United States, President Bush has decided to withhold submission of the treaty to the Senate for ratification until the world's less-developed countries agree to commit to CO_2 reductions. Likewise, the Canadian government is trying to obtain concessions that would better enable them to meet their target, and Prime Minister Chretien said that he would like to have the proposal ratified "some day."

attention because CO_2 (like water vapor) effectively absorbs radiation emitted by Earth's surface. The CO_2 increases projected to the year 2100 would double existing levels of carbon dioxide and could lead to a warming of the lower atmosphere.

Figure 1–4 shows not only the overall increase in CO_2 levels in the Northern Hemisphere but also shows seasonal oscillations. Specifically, the amount of carbon dioxide in the atmosphere is greatest in the early spring and lowest in late summer. The springtime maximum occurs because during the winter plant growth is slow, so plants take less CO_2 from the air. In addition, leaf litter has been decomposing all winter, which means carbon in the litter has been oxidized to CO_2 and has entered the air. During the period of summer regrowth, carbon is removed from the atmosphere, and so carbon dioxide levels fall.

For every CO_2 molecule removed from the atmosphere by photosynthesis, one O_2 molecule is produced. Atmospheric O_2 does not show any seasonal variation, however, because the photosynthetic contribution is so small relative to the huge atmospheric O_2 reservoir. Carbon dioxide has a residence time of about 150 years.

Ozone The form of oxygen in which three O atoms are joined to form a single molecule is called **ozone**, O_3. This substance is something of a paradox. The small amount of it that exists in the upper atmosphere is absolutely essential to life on Earth; but near Earth's surface it is a major component of air pollution, causing irritation to lungs

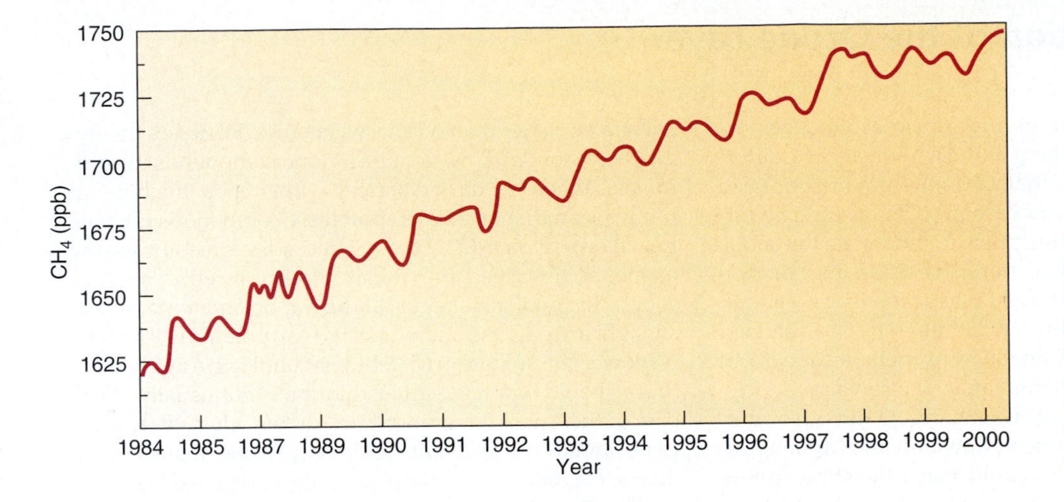

Figure 1–6 Methane is a gas that has been increasing in the atmosphere over recent decades. The smooth line depicts the long-term trend, while the more variable one plots actual values.

and eyes and damage to vegetation. Fortunately, ozone occurs only in minute amounts in the lower atmosphere, so that even in highly polluted urban air its concentration may be only about 0.15 ppm (that is, 15 out of every 100 million molecules are ozone). Aloft, at altitudes of 25 km, its concentration might be 50 to 100 times higher (up to about 15 ppm).

Ozone in the part of the upper atmosphere called the *stratosphere* is vital to life on Earth because it absorbs lethal ultraviolet radiation from the sun. Why is ozone mostly found in the stratosphere rather than at Earth's surface? The answer is complicated, as literally hundreds of chemical reactions govern its abundance. To simplify, it can be said that ozone forms when atomic oxygen (O) collides with molecular oxygen (O_2). Atomic oxygen is produced in the very upper reaches of the atmosphere, but little ozone forms there because of low air density at those altitudes. Nearer Earth's surface (but still high up in the atmosphere), the chances of O atoms colliding with O_2 molecules are greater, so the highest ozone values are found there. When it absorbs ultraviolet radiation, ozone splits into its constituent parts ($O + O_2$), which can then recombine to form another ozone molecule. Through these reactions, ozone is continually being broken down and re-formed to yield a relatively constant concentration in the ozone layer (see *Box 1–2, Focus on the Environment: Depletion of the Ozone Layer*).

The Ozone Hole

Methane Another important variable gas is **methane**, CH_4. The atmospheric concentration of this gas has increased by about 0.01 ppm per year during the last few decades, and its current level is 1.7 ppm (Figure 1–6). Although the cause of this increase is not fully understood, we do know that domesticated cattle release major amounts of CH_4 through their digestive processes. Methane is also emitted from coal mines, oil wells, and gas pipelines and is a byproduct of rice cultivation. The residence time for methane is about 10 years.

Despite its low concentration in the atmosphere, methane is an extremely effective absorber of thermal radiation emitted by Earth's surface. Thus, increases in atmospheric methane levels could play a role in the warming of the atmosphere.

Aerosols

Small solid particles and liquid droplets in the air (excluding cloud droplets and precipitation) are collectively known as **aerosols**.* Although we associate them with polluted urban atmospheres, aerosols are formed by both human and natural processes. Thus, they are ever present, even in areas far from human activity. They normally occur at concentrations of about 10,000 particles per cubic centimeter (cm^3)

*The term **particulate** is often used interchangeably with aerosol. This terminology may be confusing for some people, who would take particulate to mean solid particles only. However, at these microscopically small sizes, it is difficult to make a distinction between the liquid and solid states; thus, there is little reason to draw a fine distinction.

1–2 Focus on the Environment

Depletion of the Ozone Layer

In 1972 Sherwood Rowland and Mario Molina, atmospheric chemists then working at the University of California at Irvine, proposed that certain human-produced chemicals called chlorofluorocarbons (CFCs) could be carried naturally into the stratosphere and damage the ozone layer. CFCs are widely used in refrigeration and air conditioning, in the manufacture of plastic foams, and as solvents in the electronics industry. Although recently banned in the United States and Canada as propellants in aerosol cans, they are still used in this capacity elsewhere.

Knowing that CFCs do not easily react with other molecules in the lower atmosphere, Rowland and Molina proposed that these molecules could reach the stratosphere intact, where they would break down and release free atoms of chlorine (Cl). Under certain circumstances, chlorine atoms can effectively destroy ozone molecules. In the first step of this process, a chlorine atom reacts with an ozone molecule to produce O_2 and chlorine monoxide (ClO). Next, an oxygen atom (O) reacts with ClO, creating another O_2 molecule while freeing the chlorine atom (Cl). Note that the chlorine atom that first reacted with the ozone molecule is still present and capable of reacting again with another ozone molecule. The fact that the chlorine atoms are not consumed in these processes makes them able to repeatedly break down ozone molecules. In fact, as many as 100,000 ozone molecules can be removed from the atmosphere for every chlorine molecule present. Rowland and Molina's theory is now accepted as fact, and the two scientists were rewarded for their work with a Nobel prize in 1995.

The most severe ozone depletion occurs every October (spring in the Southern Hemisphere) and persists for several months. Why is the ozone hole found over the Antarctic during the spring? The answer is fairly complex, but a variety of factors can be cited here. First, air currents surrounding Antarctica isolate the region from the rest of the hemisphere, so there is less mixing with ozone-rich air from the north. Another factor is the unusual chemistry of clouds found in the Antarctic stratosphere. At the very low temperatures found there, clouds are largely made up of nitric acid and water, rather than ordinary water ice. Processes involving these clouds allow certain chlorine compounds to accumulate. When the dark Antarctic winter ends, the burst of ultraviolet radiation breaks apart these compounds to create free chlorine atoms (Cl). These Cl free radicals readily destroy ozone, as described previously. A very recent discovery shows that depletion begins at the rim of Antarctica, where sunlight arrives first, and works its way poleward. Thus depletion begins in June at 65° S but not until late August at 75° S. We should emphasize that chlorine is not usually abundant in the Antarctic stratosphere. Rather, what chlorine exists is found in a form able to destroy ozone, thanks to the unique conditions that take place in the spring (see Figure 1).

Significant stratospheric ozone depletion has also been detected over much of Europe and North America, including a less pronounced hole over the Arctic. Nobody knows how long the ozone will continue to decrease, or how depleted it may become, but in November 1999 satellite and land-based data revealed abnormally low ozone levels over northwestern Europe. Stratospheric zone values were 30 percent below normal for that time of year, resulting in levels nearly as low as those normally observed over the Antarctic.

Government and private industry have taken action to help reduce the amount of CFCs making their way into the atmosphere. In accordance with the Montreal Protocol of 1987 and subsequent conferences, the world's developed countries have ceased production of CFCs, and the developing nations are scheduled to do likewise by 2006. CFCs have lifetimes in the atmosphere of about 100 years, so an immediate reduction in the ozone hole will not occur. But important progress has been made; since 1997 there has been a decline in the amount of chlorine in the stratosphere, and the size of the ozone hole appears to have stabilized. A recent study predicts that with full compliance with the Montreal Protocol, actual decreases in the size of the ozone hole might become detectable sometime between 2015 and 2045. Thus, the curbing of CFC emissions illustrates how solid science, combined with international cooperation, can have major impacts on the protection of the environment.

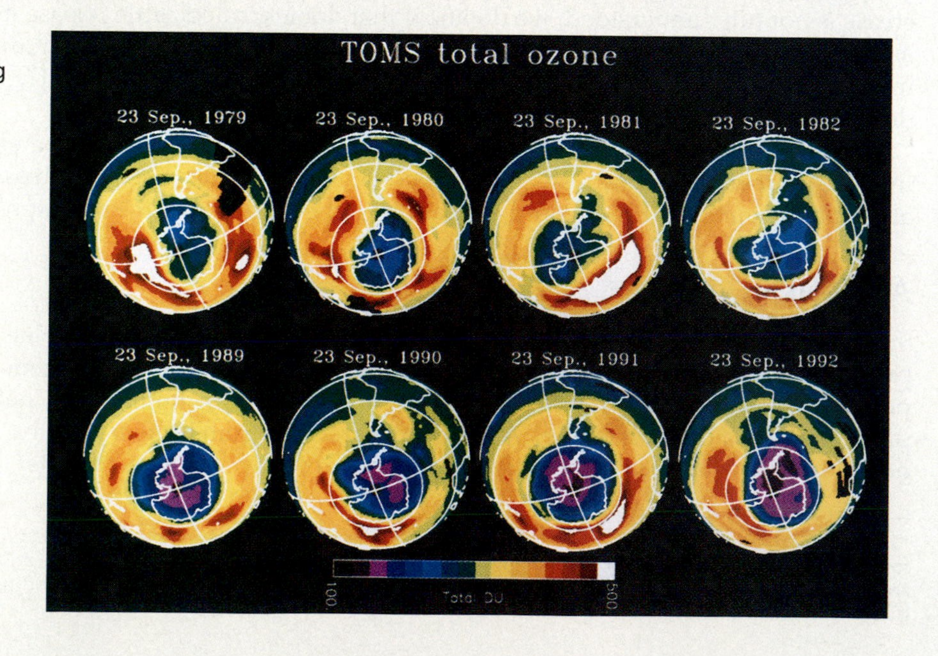

▶ **Figure 1**
A series of satellite images showing a reduction of the stratospheric ozone over Antarctica. The area in red indicates the extent of the "ozone hole."

◀ **Figure 1–7**
A dust storm in Australia.

over land surfaces, which is equivalent to roughly 17,000 particles per cubic inch (in.^{3}). To get an idea of how many aerosols there are in the air, consider that, on average, a person takes in about 1000 cm^3 (1 liter, or 64 in.^{3}) of air with each breath. As a result, each of us draws about 1 trillion aerosols into our lungs several times each minute, or about two tablespoons of solids each day.

The smallest of these particles have radii on the order of 0.1 micrometer (μm; one-millionth of a meter) and are believed to form primarily from the chemical conversion of sulfate gases to solids or liquids. Larger particles are introduced into the air directly as wind-generated dust, volcanic ejections, sea spray, and combustion byproducts. Because these particles are extremely small, most fall so slowly that they are easily kept in suspension in the atmosphere by even the smallest of vertical motions. (The most effective mechanism for the removal of aerosols is their capture by falling precipitation.) Aerosols typically have life spans of a few days to several weeks.

Aerosols have some very important and noticeable effects on the atmosphere. Urban smog, which includes aerosols, severely reduces visibility, and dust storms can reduce visibility to near zero when large volumes of soil are dislodged from the surface (Figure 1–7). Aerosols also play a major role in the formation of cloud droplets because virtually all cloud droplets that form in nature do so on suspended aerosols called **condensation nuclei**. Condensation nuclei are discussed in more detail in Chapter 5 (see *Box 1–3, Focus on the Environment: Aerosols and Climate*).

Vertical Structure of the Atmosphere

As we have seen, the atmosphere has no distinct upper boundary; the air simply becomes less and less dense with increasing altitude. We also know that its composition remains nearly constant up to a height of about 80 km (50 mi)—that is, within the homosphere. Yet despite this gradual change in density and nearly constant chemical composition with height, meteorologists still find it convenient to divide the atmosphere vertically into several distinct layers. Some layers are distinguished by electrical characteristics, some by chemical composition (this is the homosphere/heterosphere distinction discussed earlier), and some by temperature characteristics. Together with the change in density with height, this layering of the atmosphere gives it its **structure**. In this section, we look first at the changing density of the atmosphere, then at its temperature changes, and finally at the electrical properties of the layer called the *ionosphere*.

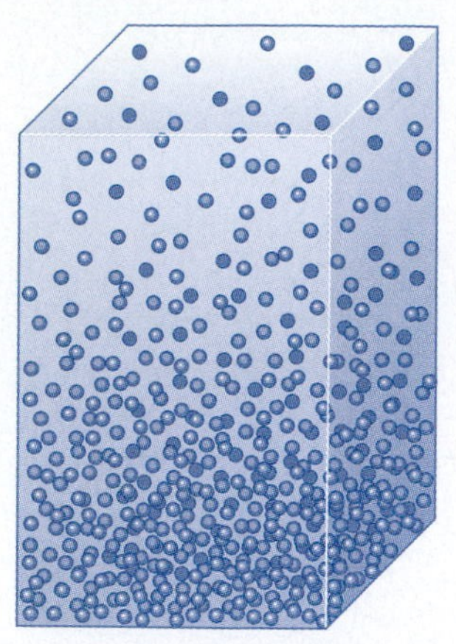

▲ Figure 1–8
Because of compression, the atmosphere is more dense near its base and progressively "thins out" with altitude.

Density

The **density** of any substance is the amount of mass* of the substance (expressed in kilograms) contained in a unit of volume (1 cubic meter, m^3). In gases, individual molecules have no attachment to one another and move about randomly. One characteristic of gases is that no definite limit exists to the amount of mass that can exist in a given volume—molecules can always be added to or removed from the volume, resulting in a density change. Alternatively, the volume containing a fixed mass of gas can decrease (as happens in a bicycle pump as the piston moves inward and compresses the air), resulting in a density increase as the constant mass is squeezed into a smaller and smaller volume.

Like any other assemblage of gases, therefore, the atmosphere is compressible. When we feel the weight of something, we are being subjected to the downward gravitational force exerted by the overlying mass. At lower altitudes, there is more overlying atmospheric mass than is the case higher up. Because air is compressible and subjected to greater compression at lower elevations, the density of the air at lower levels is greater than that aloft. Figure 1–8 illustrates this principle. Despite its simplicity, this concept is a key to understanding concepts presented later.

Air density decreases gradually with increasing altitude. At sea level, the air density is normally about 1.2 kg/m^3. By comparison, at Denver, Colorado, the "Mile High City," the air density is only about 85 percent of that. As a result, punted footballs and batted baseballs experience a corresponding decrease in air resistance and travel farther than they would at sea level.

One way to think about density is in terms of the average distance a molecule travels before colliding with another, called the **mean free path**. Near Earth's surface, the mean free path is a mere 0.0001 millimeter (mm). In contrast, at 150 km (93 mi) above sea level, a molecule travels about 10 m (3.3 ft) before colliding with another; at 250 km (155 mi), a molecule is likely to travel a full kilometer (0.62 mi) before meeting another.

Layering Based on Temperature Profiles

Vertical motions in the atmosphere are usually slow and limited in range because of the shallowness of the atmosphere. Nonetheless, such motions greatly influence the likelihood of cloud development, precipitation, and thunderstorm activity. Air whose temperature decreases rapidly with height is easily moved vertically, while air whose temperature either decreases slowly or increases with height resists such motion. Scientists therefore divide the atmosphere into four layers based not on chemical composition (which is relatively constant throughout most of the atmosphere) but rather on how mean temperature varies with altitude. The average temperature profile shown in Figure 1–9, called the **standard atmosphere**, shows the four layers: troposphere, stratosphere, mesosphere, and thermosphere.

A Global Montage of Clouds

The Troposphere Unless you get a chance to fly in a military jet or supersonic commercial aircraft, you will spend your entire life in the **troposphere**, the lowest of the four temperature layers. The name is derived from the Greek word *tropos* ("turn") and implies an "overturning" of air resulting from the vertical mixing and turbulence characteristic of this layer. The troposphere is where the vast majority of weather events occur and is marked by a general pattern in which temperature decreases with height. Despite being the shallowest of the atmosphere's four layers, the troposphere contains 80 percent of its mass. This is possible, of course, because air is compressible.

*Although a kilogram corresponds to about 2.2 pounds in the English system of measurement used in the United States, there is a difference between the two. A kilogram is a unit of mass. A pound, however, is a unit of weight equal to mass times the acceleration of gravity. On Earth 1 kilogram of a substance has a weight of 2.2 pounds. If the same kilogram is taken to the Moon, its mass will not change but its weight will be only one-sixth of what it was on Earth. This is true because the acceleration of gravity on the moon is one-sixth as strong as the acceleration of gravity on Earth.

1–3 Focus on the Environment

Aerosols and Climate

Suspended particles might be responsible for changes in climate (the long-term state of the atmosphere). They could lead to lower temperatures by reducing the amount of sunlight penetrating the atmosphere, or to higher temperatures by absorbing heat emitted by Earth's surface. Some climatologists have studied increases in the dust content of the atmosphere that follow volcanic activity. Lowering of surface air temperatures has followed several major eruptions, including the June 1991 eruption of Mount Pinatubo in the Philippines (Figure 1). Continuous observations made by NASA's Earth Radiation Budget Experiment (ERBE) satellite have shown that the average surface temperature of the Northern Hemisphere decreased by as much as 1 °C (1.8 °F) a year after the Pinatubo eruption. This decrease in temperature lasted into the spring of 1993.

On an entirely different time scale, a major increase in atmospheric particulates may have led to the extinction of the dinosaurs some 65 million years ago. According to this theory, a large asteroid up to 15 km (9 mi) in diameter may have collided with Earth with an explosive force equivalent to the detonation of millions of hydrogen bombs. Upon the impact, enormous quantities of dust and water would have been ejected into the atmosphere, leading to greatly reduced sunlight and photosynthesis. Eventually, all plant life would have died out and deprived a large number of animal species (including the dinosaurs) of their food supply.

Although the asteroid theory is widely held as the cause of dinosaur extinctions, a conflicting explanation also exists, sometimes called the *Pele hypothesis* (named for the Polynesian volcano goddess). Under this scenario, heavy volcanic activity sometime earlier than 65 million years ago increased carbon dioxide emissions into the atmosphere, and surface vegetation proliferated. When the volcanic activity ended, much of the vegetation died and organic material at the bottom of sea beds decomposed. The massive decomposition reduced oxygen levels below that required by the dinosaurs' respiratory systems and led to their extinction.

Although neither hypothesis can be put forward as undisputed fact, it is interesting that two of the most widely accepted explanations for dinosaur extinctions involve changes in the atmosphere.

▲ **Figure 1**
The eruption of Mount Pinatubo in June 1991 introduced so much material into the stratosphere that temperatures over the Northern Hemisphere were lowered for about 2 years.

The depth of the troposphere varies considerably, ranging from 8 to 16 km (3.6 to 10 mi), with a mean of about 11 km (6.6 mi). The altitude at which the troposphere ends depends largely on its average temperature, being greatest where the temperature is high and lowest in cold regions. The troposphere is therefore thicker over the Tropics than over the polar regions and thicker during the summer than during the winter.

Temperatures vary greatly from bottom to top in the troposphere. The average global temperature is about 15 °C (59 °F) near the ground but only about –57 °C (–71 °F) at the top of the troposphere, an average decrease of about 6.5 °C/km (3.6 °F/1000 ft). Thus, you might feel perfectly comfortable inside an airplane at an altitude of 10 km (33,000 ft), but the temperature outside would likely be in the neighborhood of –50 °C (–58 °F). At the top of the troposphere, a transition zone called the **tropopause** marks the level at which temperature ceases to decrease with height.

An apparent paradox is associated with the trend toward decreasing temperature with height in the troposphere. We know that Earth is heated almost entirely by the Sun, which might make you think temperature should *increase* with

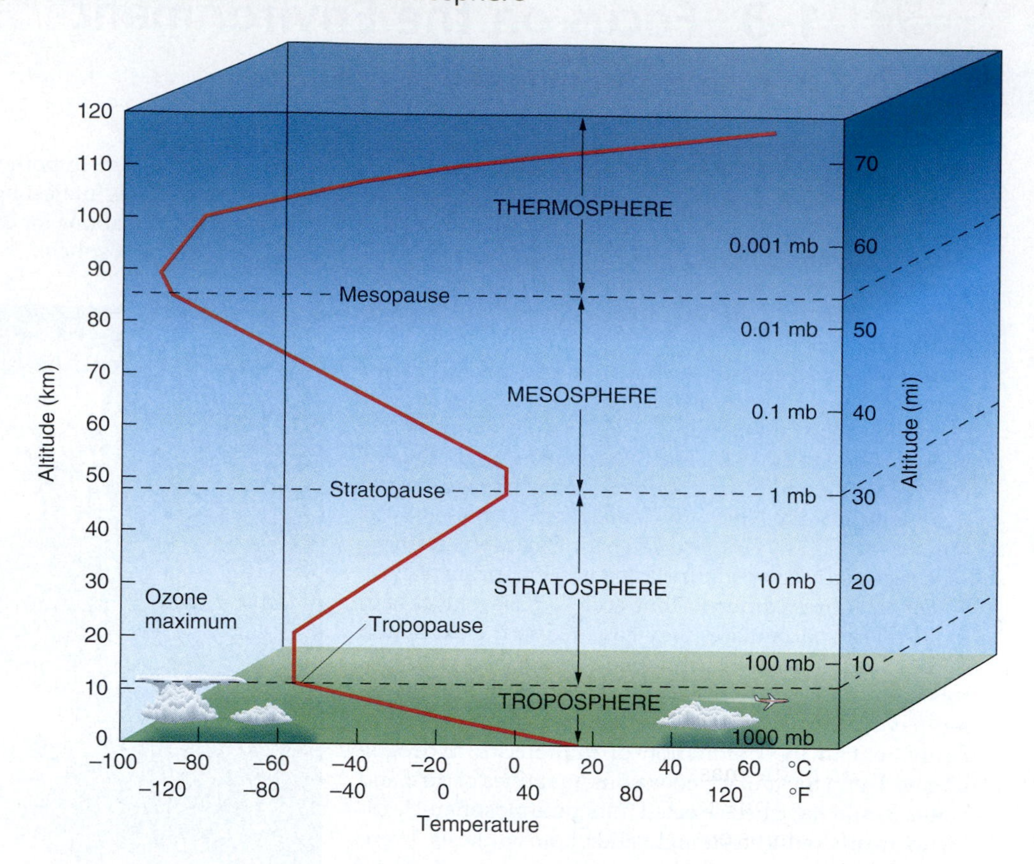

Figure 1–9
The temperature profile of the atmosphere results in four layers based on thermal characteristics.

height during the day (because as we move upward from the surface, we get closer to the Sun). The explanation for this paradox is that the atmosphere is relatively transparent to most types of radiant energy emitted by the Sun. In other words, a large portion of the sunlight passing through the atmosphere is not absorbed and therefore does not contribute greatly to its warming. As we shall see in Chapter 3, the most important direct source of energy for the atmosphere is not downward-moving solar radiation but rather energy emanating upward from Earth's surface.

Despite the strong tendency for temperature to decrease with altitude in the troposphere, it is not uncommon for the reverse situation to occur. Such situations, where temperature increases with height, are known as **inversions**. The presence of an inversion is significant because it inhibits upward motion and thereby allows high concentrations of pollutants to be confined to the lowest parts of the atmosphere.

The Stratosphere Above the tropopause is the **stratosphere**, a name derived from the Latin word for *layer*. Except for the penetration of some strong thunderstorms into the lower stratosphere, little weather occurs in this region (Figure 1–10). In the lowest part of the stratosphere, the temperature remains relatively constant at about –59 °C (–71 °F) up to a height of about 20 km (12 mi). From there to the top of the stratosphere (called the **stratopause**), about 50 km (30 mi) above sea level, the temperature increases with altitude until it reaches a mean value of –2 °C (2 °F).

In the upper stratosphere, heating is almost exclusively the result of ultraviolet radiation being absorbed by ozone. Therefore, as solar energy penetrates downward through the stratosphere, there is less and less ultraviolet radiation available and a resultant decrease in temperature. In the part of the stratosphere where temperature does not vary with height, heating is the result of both absorption of solar ultraviolet radiation and absorption of thermal radiation from below. Thus, as we move up or down in this region, the reduction in solar heat is offset by the increase in heat given off by Earth. The net result is the straight vertical line of no temperature change shown in Figure 1–9.

The stratosphere contains about 19.9 percent of the total mass of the atmosphere. Thus, the troposphere and stratosphere together account for 99.9 percent of

◀ **Figure 1–10**
Most clouds exist in the troposphere, but the tops of strong thunderstorm clouds can extend kilometers into the stratosphere when violent updrafts occur. The flattened area at the top of this cumulonimbus cloud is in the stratosphere.

the total mass of the atmosphere. The fact that gases are so compressible allows the major portion of the mass of the atmosphere to be contained in the two lowest layers. Note that despite being more than triple the depth of the troposphere, the stratosphere contains only one-quarter the mass: troposphere = 80 percent of total atmospheric mass; stratosphere = 20 percent.

Within the stratosphere is the **ozone layer**, a zone of increased ozone concentration existing at altitudes between 20 and 30 km (12 and 18 mi). Despite its name, the ozone layer is not composed primarily of ozone. In fact, at 25 km (15 mi) above sea level, where the percentage of ozone in this region is greatest, its concentration might be only about 10 ppm. Despite its scarcity, ozone is an extremely important constituent of the stratosphere. It is largely responsible for absorbing the solar energy that warms the stratosphere, and it also protects life on Earth from the lethal effects of ultraviolet radiation.

Being well removed from Earth's surface, where evaporation supplies water vapor to the atmosphere, the stratosphere has a very low moisture content. Moreover, the temperature characteristics of the stratosphere inhibit vertical motions that favor the formation of clouds (as discussed in Chapter 6). These conditions inhibit precipitation, and consequently particulates from volcanic eruptions can remain in the stratosphere for many months. Furthermore, the strong winds of the stratosphere can cause its aerosol content to be distributed across the globe, creating a veil of material that can affect the penetration of sunlight to the surface. For a couple of years after the 1991 eruption of Mount Pinatubo in the Philippines, for example, the Northern Hemisphere experienced redder-than-normal sunrises and sunsets as a result of aerosols in the stratosphere.

The Mesosphere and Thermosphere Of the 0.1 percent of the atmosphere not contained in the troposphere and stratosphere, 99.9 percent exists in the **mesosphere**, which extends to a height of about 80 km (50 mi) above sea level. As in the troposphere, temperature in the mesosphere decreases with altitude. Scientists believe that the absorption of solar radiation near the base of the mesosphere provides most of the heat for the layer, which is dispersed upward by vertical air motions.

Above the mesosphere is the **thermosphere**, where temperature increases with altitude to values in excess of 1500 °C. These high temperatures can be misleading, however, if we overlook the distinction between high temperature and high heat content. The temperature of the air is an expression of its kinetic energy, which is related to the speed at which its molecules move. The amount of heat contained in the air reflects not only its temperature but also its mass and *specific heat* (the amount of energy needed to change its temperature by a certain amount). Because there are so few gas molecules in this layer, the air cannot have a high heat content no matter what its temperature is. In fact, the atmosphere is so sparse in the upper reaches of the thermosphere that a gas molecule will normally move as much as several kilometers before colliding with another. Thus, an ordinary thermometer in this part of the atmosphere would have little contact with the surrounding air. Under these circumstances, the concept of temperature loses meaning and cannot be associated with everyday terms such as *hot* and *cold*.

A Layer Based on Electrical Properties: The Ionosphere

The four layers of the atmosphere described previously are delineated on the basis of temperature profiles. An additional layer, called the **ionosphere**, can be defined based on its electrical properties. This layer, which extends from the upper mesosphere into the thermosphere, contains large numbers of electrically charged particles called *ions*. **Ions** are formed when electrically neutral atoms or molecules lose one or more electrons and become positively charged ions or gain one or more electrons and become negatively charged ions. In the ionosphere, atoms and molecules lose electrons as they are bombarded by solar energy, thus creating positively charged ions and free electrons.

The ionosphere is important, among other reasons, for reflecting AM radio waves back toward Earth, thereby increasing the distance at which broadcasts can be received. As shown in Figure 1–11, the ionosphere is divided into several sublayers, including the D-, E-, and F-layers. The lowest of the three, the D-layer, exists only during the daylight hours and absorbs AM radio waves. When night comes, the D-layer begins to disappear and the E-layer weakens as their free electrons recombine with positively charged ions. The radio waves are then able to reach the F-layer, which reflects radio waves rather than absorbing them and thus redirects the transmission back toward the surface. Earth's surface is also reflective of these waves, so they rebound back toward the ionosphere. The repeated reflection of the radio waves between the F-layer and the surface allows them to overcome the effect of Earth's curvature and greatly extends the distance at which they can be "heard." Perhaps you have listened to an AM radio at night and happened

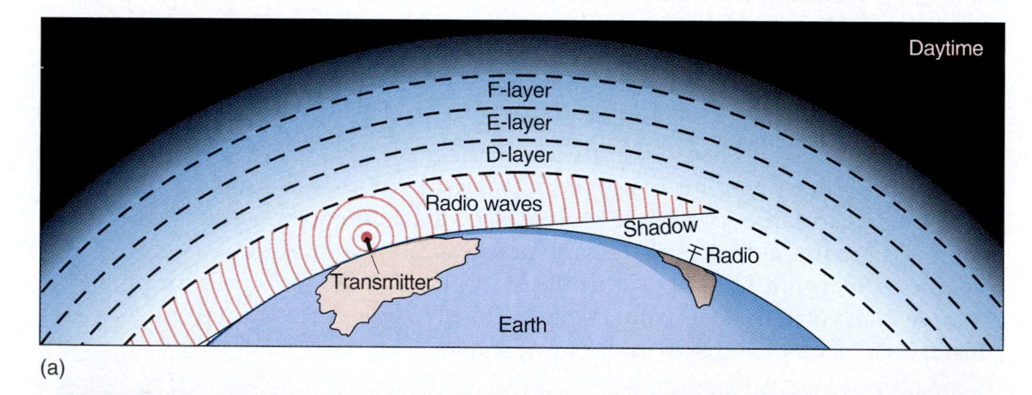

(a)

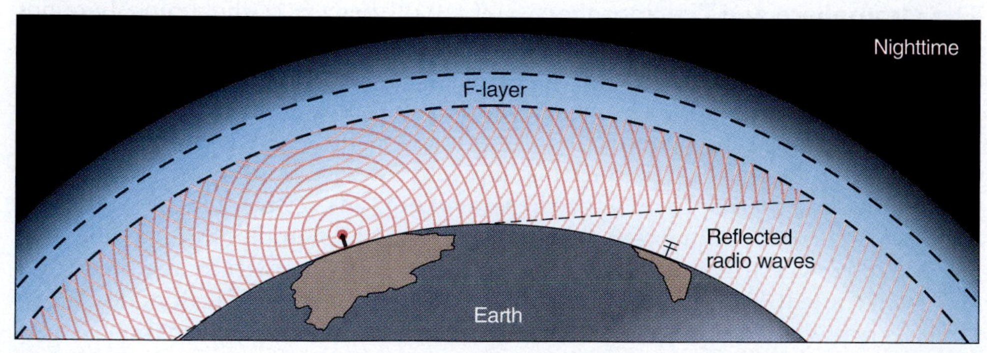

(b)

▶ **Figure 1–11** The ionosphere influences the distance to which AM radio waves can be transmitted. During the day (a), the D-layer absorbs the radio waves. The D- and E-layers break up at night (b), allowing the waves to reach the F-layer. The F-layer reflects the radio waves back toward Earth's surface. Repeated reflections between the E-layer and Earth's surface enable the waves to be picked up by receivers at greater distances at night.

to pick up a distant radio station, only to be surprised by its disappearance the following day. This results from the breakup of the D- and E-layers.

Because of these changes in the range of transmission, radio stations reduce the power of their transmitters at night so that stations on the same frequency do not interfere with one another. Moreover, the Federal Communications Commission (FCC) has designated certain radio stations as "clear channels." Such stations are given unique frequencies (each is the only one in the United States allowed at a given point on the radio dial) and allowed to broadcast from very powerful transmitters. The purpose of such stations is to guarantee that all parts of the country, no matter how far from a city, have access to at least one radio station. Clear channel stations are indeed heard over great distances, and a truck driver going through eastern Colorado may have no difficulty picking up stations from Chicago, Illinois, or Ft. Worth, Texas.

The ionosphere is also responsible for the **aurora borealis**, or northern lights, and **aurora australis**, the southern lights (Figure 1–12). In the ionosphere, subatomic particles from the Sun are captured by Earth's magnetic field (the same field that makes compass needles point to the north). The captured energy causes the gases of the atmosphere to become *excited* (meaning the electrons of the atoms jump to greater orbital distances from their nuclei). When electrons fall back to lower orbits, radiation is emitted. Thus, the aurora does not reflect radiant energy as do clouds but rather emits light that is much like a neon lamp.

▲ **Figure 1–12**
An aurora borealis. Subatomic particles from the Sun are captured by Earth's magnetic field, causing an agitation of molecules and the emission of light with different colors.

Evolution of the Atmosphere

It is generally believed that Earth was formed perhaps 4.5 billion years ago. If an atmosphere formed with Earth, it must have consisted of the gases most abundant in the early solar system—including large amounts of hydrogen and helium, the two lightest elements. Today's atmosphere is much different, composed mostly of nitrogen and oxygen. So where did the original gases go and how were they replaced?

The gases of the earliest atmosphere were simply lost to space during the first half-billion years of the planet's existence. We are not sure exactly how, but two possible answers are widely put forward. For many years, the conventional viewpoint was that Earth's gravitational field would not have been strong enough to keep hold of such an atmosphere, and the light elements therefore escaped into space. To understand why, however, we need a bit of background. Gases surrounding a planet would easily disperse into outer space (much the same way air is dispersed from a bicycle tire when the valve is opened) were it not for the fact that gravity exerts an opposing effect and prevents their escape. However, if molecules move with sufficient speed, known as their **escape velocity**, they can overcome gravity and leave the atmosphere. Light gases are more likely to achieve escape velocity; thus, the hydrogen and helium of the primordial atmosphere were most readily lost.

Recently, an alternate explanation has gained acceptance. According to this view, gases drawn to Earth during its formation would have been removed by collisions between the growing Earth and other large bodies. Some of these other objects, so-called failed planets, would have been as large as Mars. Tremendous energy would have been released by collisions with failed planets, ejecting large amounts of planetary material along with any atmosphere that might be present. (In fact, the moon is believed to have formed by condensation of matter dislodged by just such a collision.) If this hypothesis is correct, removal by slow leakage would have been insignificant.

Over time a new, secondary atmosphere formed, made up of gases released from Earth's interior by volcanic eruptions—a process called **outgassing**. The gases spewed out during volcanic events are predominantly water vapor and carbon dioxide, with lesser amounts of sulfur dioxide, nitrogen, and other gases. Volcanic outgassing was possibly augmented, maybe even dwarfed, by material brought to Earth in small comets on the order of 15 m (45 ft) in diameter. Recent satellite observations suggest that there is a steady rain of these water-bearing "cosmic snowballs," with from 5 to 30 striking Earth at any one time. If they've been falling at this rate since the formation of Earth, they might well be the source of most of the world's water vapor. However it arose, the secondary atmosphere has been greatly transformed, because water vapor and carbon dioxide now constitute only a small portion of the atmosphere.

The removal of the water vapor is easily explained: upon cooling in the prehistoric atmosphere, it readily condensed and precipitated to the surface to form the oceans. Precipitation also contributed to the removal of carbon dioxide from the atmosphere. As raindrops fell through CO_2-rich atmosphere, some CO_2 was dissolved in every drop. (Soft drinks provide an everyday example of how CO_2 is dissolved in water.) As they fell, the CO_2-enriched drops transferred carbon from the atmosphere to the oceans, where it combined with material eroding from the continents and was buried in seafloor sediments.

The transformation to an atmosphere high in oxygen depended on the advent of primitive, anaerobic bacteria (those that survive in the absence of oxygen) about 3.5 billion years ago. These primitive life-forms were the first in a long line of organisms that removed carbon dioxide from the air and replaced it with oxygen (these gases are exchanged freely between the oceans and the atmosphere). Ultimately, plant and later animal material sank to the ocean floor (as it continues to do today), where the organic carbon was locked away in sediments. As a matter of fact, the vast majority of carbon released by volcanoes exists in neither the atmosphere nor the ocean; it is held in carbonate rock formations.

All of these processes gradually led to an increase in atmospheric oxygen at the expense of carbon dioxide. But one other transformation had to take place to create an atmosphere that could support life at the surface. Recall that without an ozone layer in the upper atmosphere, sunlight reaching the surface would contain lethal levels of ultraviolet radiation. Thus, a protective ozone layer had to develop before life could exist outside the oceans. Fortunately, this occurs naturally when the ultraviolet radiation breaks down diatomic oxygen molecules into individual oxygen atoms. The oxygen atoms could then recombine with O_2 in the upper atmosphere to form the ozone layer. Once this happened, the amount of ultraviolet radiation reaching the surface was reduced sufficiently to allow plants to occur on land. This increase in plant cover in turn accelerated the rate at which photosynthesis replaced atmospheric carbon dioxide with oxygen.

The last thing we need to account for is the high concentration of nitrogen in the atmosphere. Although it constitutes a small portion of the material released by outgassing, nitrogen is removed from the atmosphere very slowly. As a result, its concentration has gradually increased to the point that it is now the main constituent of the atmosphere.

Some Weather Basics

All of us are familiar with daily weather forecasts, from which we routinely receive information on the present and predicted state of the atmosphere. In recent years, the amount of weather information available to the public has exploded, most notably via the Internet. Detailed maps and weather reports that were once available only to professional meteorologists can now be accessed by anybody with a computer and a modem. Such access makes learning the fundamentals of weather and climate far more enjoyable, because we can now look at map, satellite, radar, and other resources and see how the principles of meteorology play out on a daily basis. We now present an overview of the fundamentals of weather, along with an introduction to weather maps. We will then follow with a section describing the process by which a meteorologist makes a daily forecast.

Atmospheric Pressure and Wind

Atmospheric **pressure** is one of the most fundamental of weather characteristics, but at the same time it is one you don't directly feel. Unlike temperature and humidity, it is impossible to sense whether the air pressure at the surface is high or low. Yet atmospheric pressure impacts all other aspects of weather. Two generalizations about surface pressure are particularly important. First, air tends to blow away from regions of high pressure toward areas of lower pressure. In other words, it is the horizontal variation in air pressure that generates **winds**. If the pressure were uniform from place to place, the air would be continually calm. The second generalization is that air tends to rise in areas of low surface pressure and sink in zones of high pressure. This is important because rising motions favor the formation of clouds, while sinking motions promote clear skies.

Atmospheric pressure is routinely plotted on maps by the use of lines, called **isobars**. Each isobar connects points having equal air pressure, with the pressure being expressed in units of **millibars** (abbreviated as mb) in the United States and **kilopascals** (kPa) in Canada. Figure 1–13 illustrates how isobars depict the distribution of pressure. Notice that an isobar labeled 1024 encircles most of Idaho. Any point on that line has a pressure that is equal to 1024 mb (which corresponds to 102.4 kPa). Another isobar, indicating a pressure of 1020 mb surrounds the 1024 mb, isobar. Not only do the two isobars tell us what the pressure is at any point along those lines, but they also allow us to infer what the pressure is anywhere between them. Thus, over nearly all of Washington State and Wyoming, which are situated between the two isobars, the air pressure is between 1020 and 1024 mb. The fact that higher pressure exists over Idaho implies that the air flows outward from that region. Just west of the Great Lakes in Figure 1–13 is an area of low pressure, and we can expect that air is flowing into that region from surrounding areas.

Information regarding wind speed and direction can be obtained on weather maps by looking at the **station models**, which contain symbols and numbers giving detailed weather information for particular locations. Notice the station model for Tucson, Arizona, in the southwestern United States. The station model in Tucson contains an open circle with a line pointing toward the southeast, which indicates the direction that the wind is blowing *from*. At the end of the southeastward-pointing line

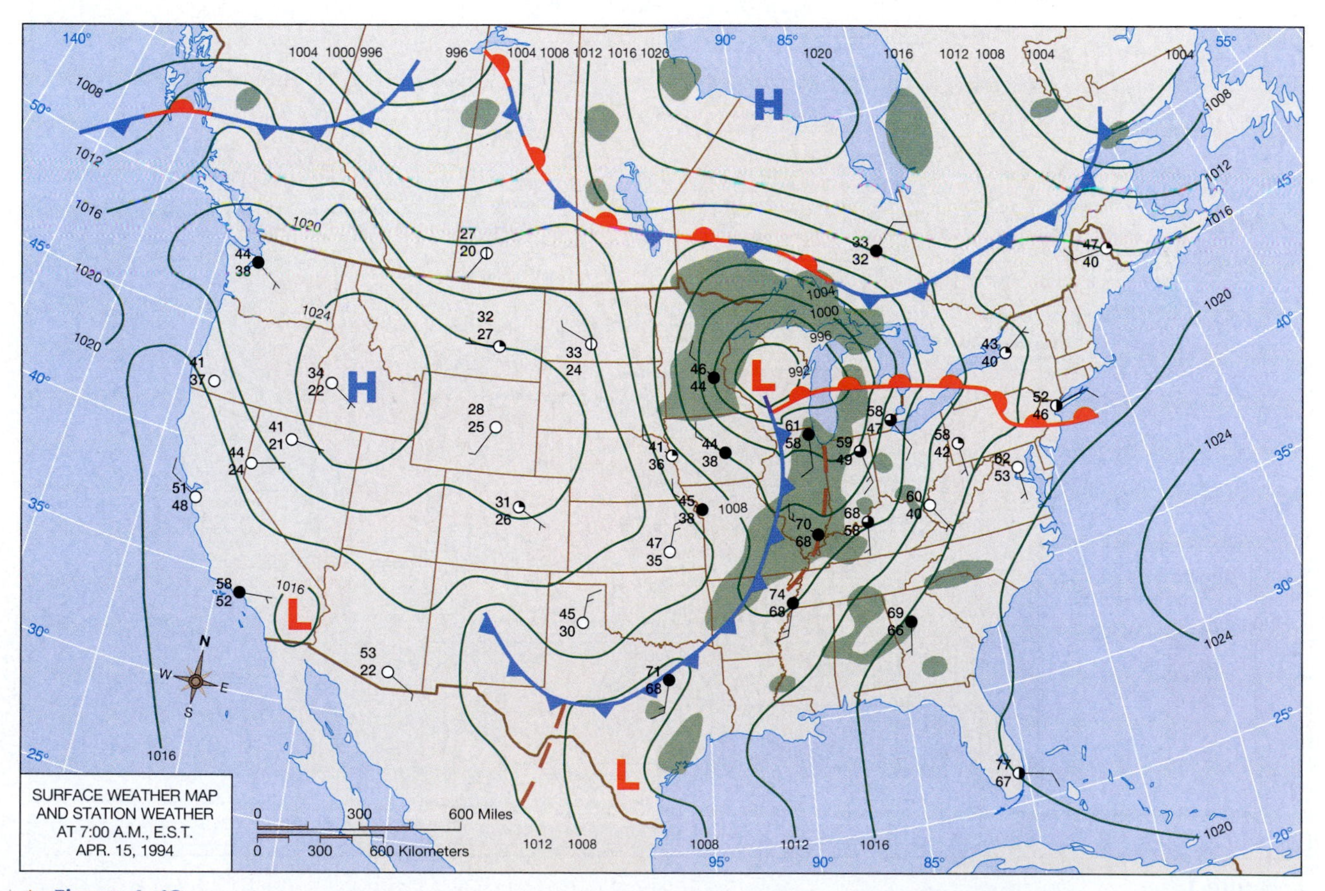

▲ **Figure 1–13**
A typical surface weather map.

1–4 Special Interest

Recent Severe Weather

The atmosphere reminds us of its presence in many ways. While some weather might be merely uncomfortable or inconvenient, there are instances where it can have irreversible impacts on millions of people. Figures 1 and 2 highlight some of the more significant weather events of 2000 and 2001. It did not take long before the new millennium brought with it a major weather event to North America, when on January 22–23, 2000, a large ice storm brought power outages to more than a half million residents of Georgia and South Carolina. Another noteworthy event was a record-breaking rainfall on the eastern part of the big island of Hawaii, when a torrential thunderstorm brought more than 54 cm (24 in.) of precipitation in a 24-hour period. But without doubt the most significant U.S. weather of the year was the extreme heat and drought across the southern and southeastern United States, and the combination of dry, windy conditions that fanned major wildfires across the western part of the country during the spring and summer. The heat wave in the South brought an estimated $4 billion in economic costs, primarily to agriculture and related activities. The summer heat, which was most severe in July, also is believed to have contributed to the deaths of about 140 people. The western wildfires burned more than 7 million acres. Hardest hit was the area of Los Alamos, New Mexico, where fires destroyed more than 400 homes in May of 2000.

Tropical Storm Allison was the biggest North American weather event of 2001. Making landfall at south Texas on June 5, the storm hit the Houston area particularly hard. Between June 5 and 11, the Port of Houston received 96 cm of rain (just under 40 in.)—enough to cause $4.8 billion in damages (Figure 3). But the damage from Allison was hardly limited to Texas. Louisiana experienced some of its worst flooding from a tropical storm in its history. Before it was all over, the storm had drifted up the Atlantic Coast and produced flooding as far northward as Massachusetts

Tropical Storm Allison

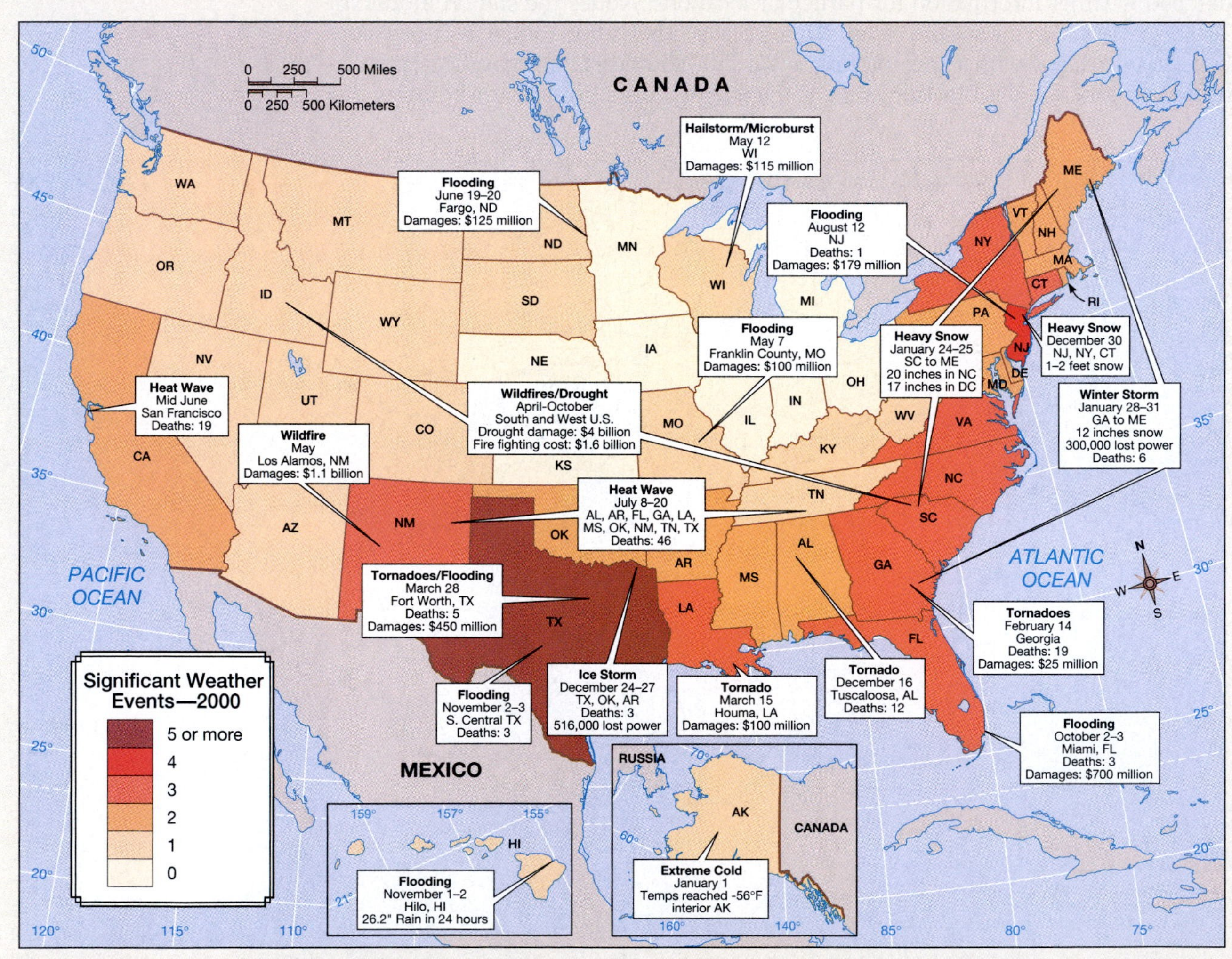

▲ **Figure 1**
Some significant weather events for 2000.

and Rhode Island, with Texas, Mississippi, Florida, and Pennsylvania all having been declared disaster areas.

According to the National Climate Data Center, during the period from 1980 to 2001 the United States experienced 52 weather events that incurred a billion dollars or more in damages. By far the worst of these were the heat/drought events of 1980 and 1988, which resulted in $44 billion and $56 billion dollars in damage, respectively (adjusted to 1998 dollars). More importantly, up to 10,000 people may have died in each of the events, many from heat-related stress. During this period, another six weather events killed at least 100 people nationally: the heat/drought events of 1998, 1999, and 2000; the 1998 severe storm and tornado outbreak in the Southeast; and the blizzard and storm events of 1993 and 1996 along the East Coast. Notice that five of the eight deadliest events of the last 22 years have occurred during the 6-year period of 1996–2001. Atmospheric scientists have expressed concern recently about the potential for more frequent severe weather to occur in the near future due to climate warming. Time will tell if the high incidence of severe weather events is in fact associated with such change or merely an isolated run.

▲ **Figure 3**
Severe flooding due to Tropical Storm Allison.

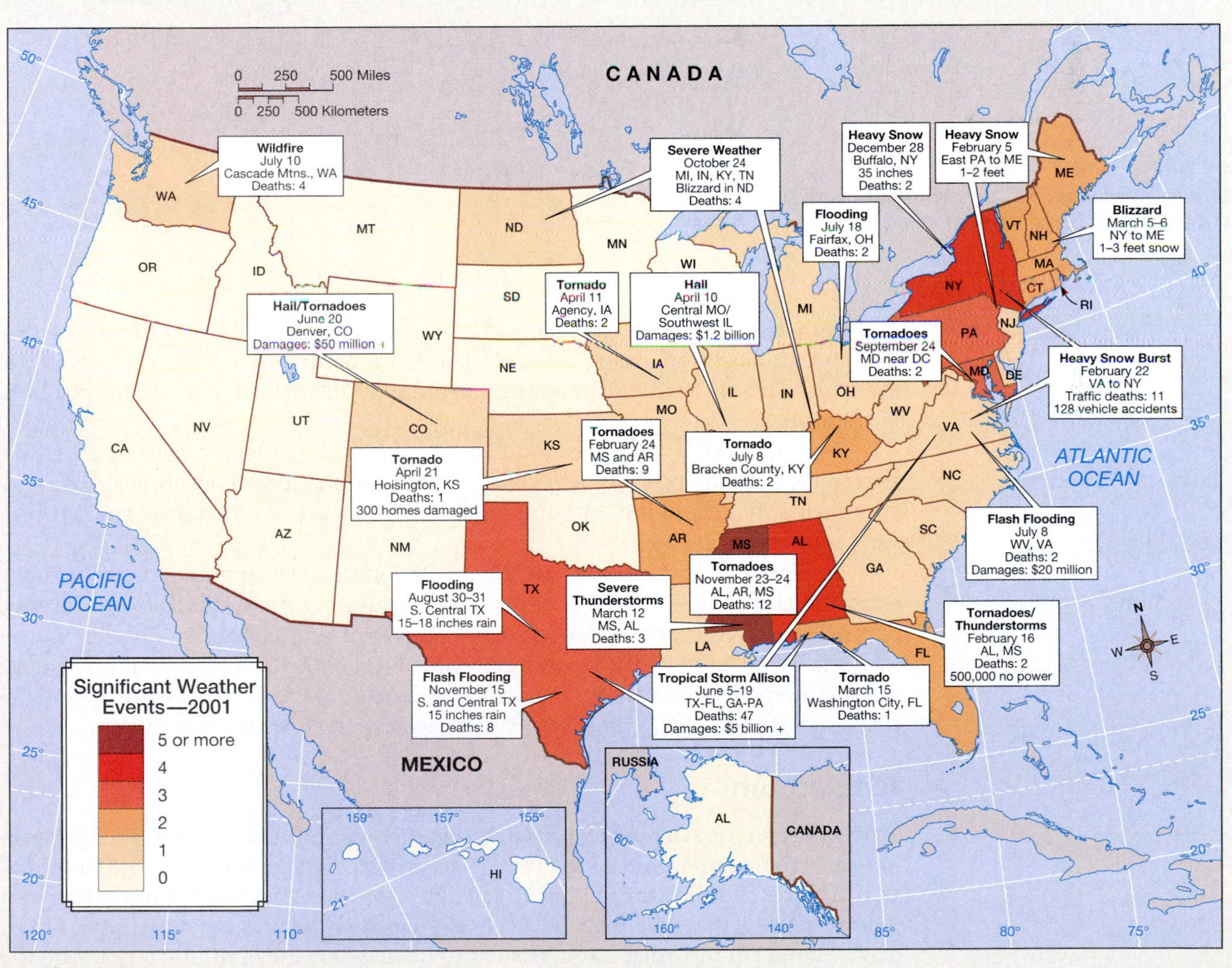

▲ **Figure 2**
Some significant weather events for 2001.

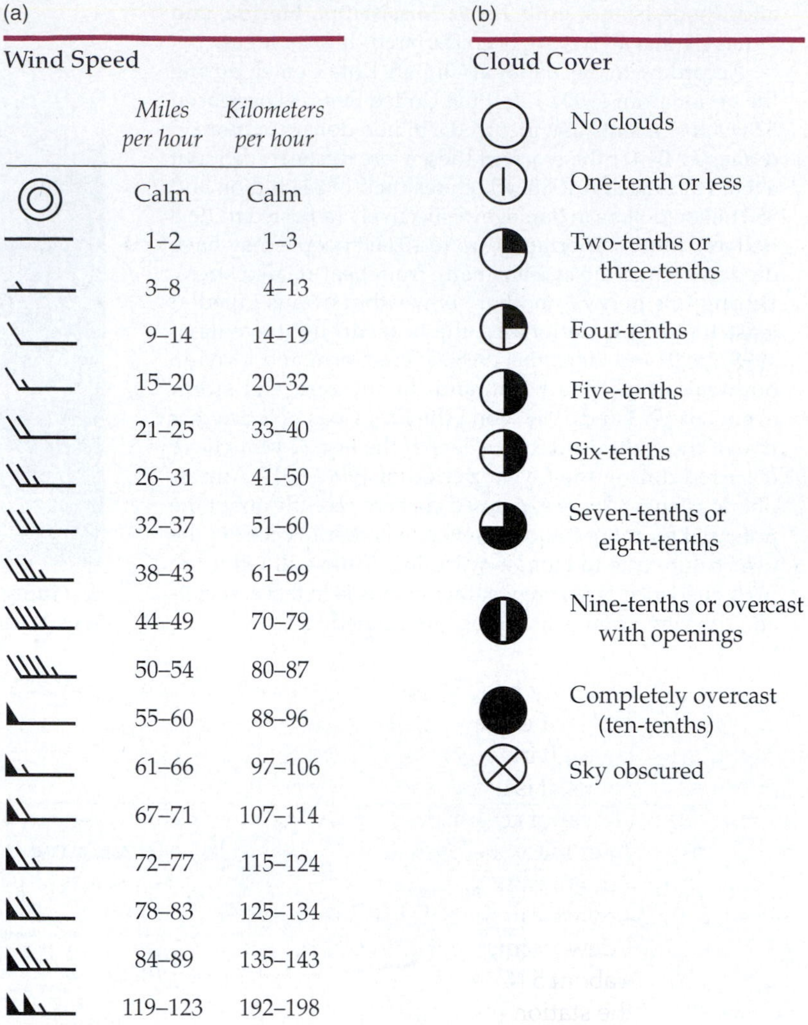

(a)

Wind Speed

Miles per hour	*Kilometers per hour*
Calm	Calm
1–2	1–3
3–8	4–13
9–14	14–19
15–20	20–32
21–25	33–40
26–31	41–50
32–37	51–60
38–43	61–69
44–49	70–79
50–54	80–87
55–60	88–96
61–66	97–106
67–71	107–114
72–77	115–124
78–83	125–134
84–89	135–143
119–123	192–198

(b)

Cloud Cover

- No clouds
- One-tenth or less
- Two-tenths or three-tenths
- Four-tenths
- Five-tenths
- Six-tenths
- Seven-tenths or eight-tenths
- Nine-tenths or overcast with openings
- Completely overcast (ten-tenths)
- Sky obscured

▶ **Figure 1–14**
Station model symbols describing (a) wind speed and (b) percentage of cloud cover.

is a single tick mark. The number and type of tick marks at the end of the line give an approximate wind speed, using the conventions shown in Figure 1–14a. In this example, Tucson has winds coming from the southeast at 14 to 19 km per hour (9–14 mph).

Station models also give cloud cover information. Open circles, such as that shown for Tucson, indicate clear skies, while fully shaded circles indicate overcast conditions. Intermediate amounts of cloud cover are indicated by the patterns shown in Figure 1–14b.

In Figure 1–13, the isobar pattern and station model information verify the generalizations about pressure distributions, wind movement, and cloud cover. The wind barbs show that the air does indeed move outward from the region of high pressure, while air is flowing into the low pressure area west of the Great Lakes. Furthermore, the station models show that much of the area around the high pressure system has clear skies, in contrast to the general overcast near the low pressure system.

Temperature

Temperature is one of the most obvious weather components. Everyday experience indicates that temperatures vary from place to place systematically. In other words, as you take a 50-mile car trip you are not likely to experience temperatures that vary wildly. Instead, you will likely observe only gradual changes in temperature as you drive along. On the other hand, there are times when substantial temperature differences appear over short distances, or when major shifts in temperature occur over short time periods at a particular location. These major changes in temperature often occur due to the presence of **fronts**, fairly narrow boundary zones separating relatively warm and cold air. As you drive across the frontal zones, you will experience notable temperature shifts. Likewise, if a front passes over the place you are at, the temperature will change substantially.

Four types of fronts exist, which are discussed later in this text. For now, let's concern ourselves with two particular types: *cold fronts* and *warm fronts*. Figure 1–13 maps the presence of a cold front (shown as a blue line with triangles) extending southward from the low pressure region over Wisconsin, and a warm front (red line with semicircles) extending eastward from the same location. To the east of the cold front, temperatures are higher than the area to the west of the front. The exact temperatures are shown in degrees Fahrenheit, on the upper left portion of the station models. In this illustration, Chicago, Illinois, has a temperature of 61 °F, while Des Moines, Iowa, to the west, has a temperature of 44 °F.

Figure 1–13 also shows the distribution of precipitation, indicated by the areas in green. Notice that precipitation is occurring to the northwest of the low pressure system over Wisconsin, and also in along the cold front. Precipitation is very common along fronts, so when a front approaches it is not unusual to experience rain or snow in addition to a change in temperature.

Humidity

You have undoubtedly heard the term **relative humidity**. Relative humidity is just one of several ways of expressing the amount of water vapor in the air. (Remember, water vapor is a gas!) It indicates the amount of water vapor present relative to the maximum possible; thus, it is usually reported as a percentage. Though a commonly used indicator of water vapor content, relative humidity has some serious shortcomings. For this reason, another index called the **dew point temperature** (or simply, the dew point) is often preferred. For now let's just say that the higher the dew point, the greater the amount of water vapor in the air. Dew points above about 15 °C (59 °F) or so indicate humid air, and dew points above about 20 °C (68 °F) are very uncomfortable. Dew points less than about 5 °C (41 °F) are relatively dry. Dew point values are given at the bottom left of the station models, just below the temperature readings. In the example given in Figure 1–13, much of the southeastern United States has humid air, while the northeastern United States and eastern Canada are somewhat drier. The driest air in this example occurs over the West.

Planetary Atmospheres

We can now briefly compare the atmosphere of Earth with that of the other planets of our solar system (see Figure 1–15). Several recent space missions have given us increased knowledge of our planetary neighbors, knowledge that not only tells us much about the other planets but might also yield important information on the evolution of Earth's atmosphere. The four planets closest to the Sun—Mercury, Venus, Earth, and Mars—are called the *inner planets*. They are also referred to as the *terrestrial planets* because they have well-defined, solid surfaces.

Mercury, the planet nearest the Sun, has very high temperatures and, because of its small mass, a weak gravitational acceleration. The combination of high temperature and low gravity would allow any atmosphere that might have once existed to escape into space. Thus, it is very easy to describe that planet's atmosphere—it has almost none!

The situation on Venus (shown in Figure 1–16) is quite different. Its atmosphere is some 50 times denser at the surface than Earth's, and its gravitational field is able to hold on to its atmosphere. The Venusian atmosphere consists primarily of carbon dioxide (96.5 percent) and nitrogen (3.5 percent), though clouds of sulfuric acid are found at upper elevations (Figure 1–17). The temperature profile of the lower Venusian atmosphere is similar to that of Earth in that temperature decreases with elevation. Because of its high carbon dioxide content, the atmosphere of Venus is

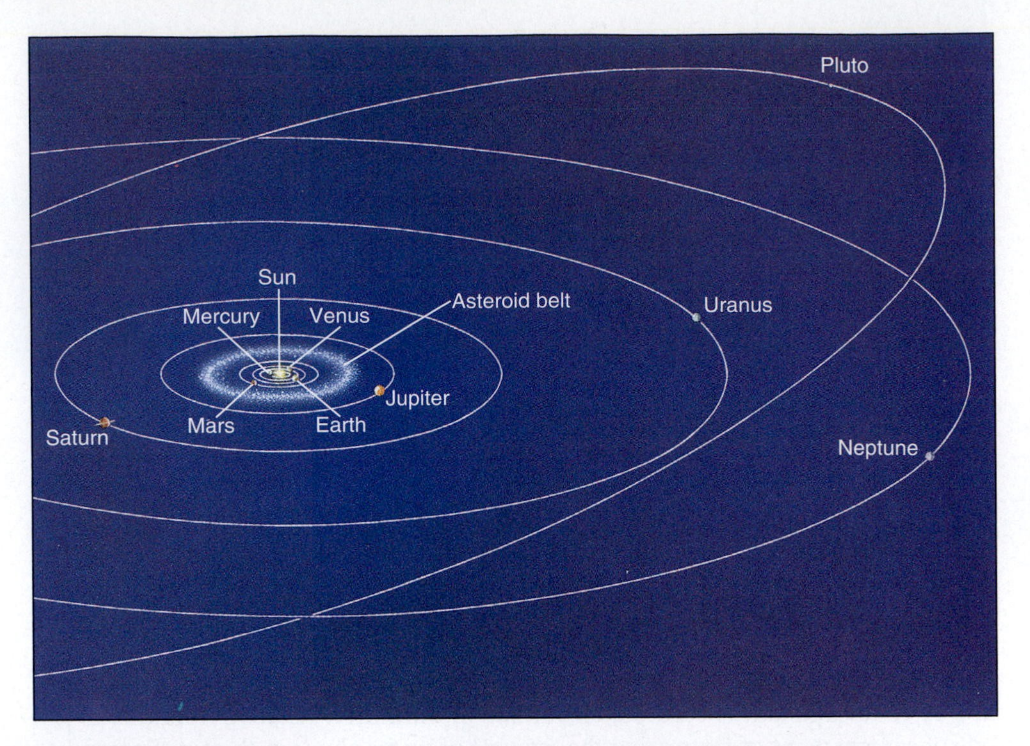

▶ **Figure 1–15**
The planets of the solar system.

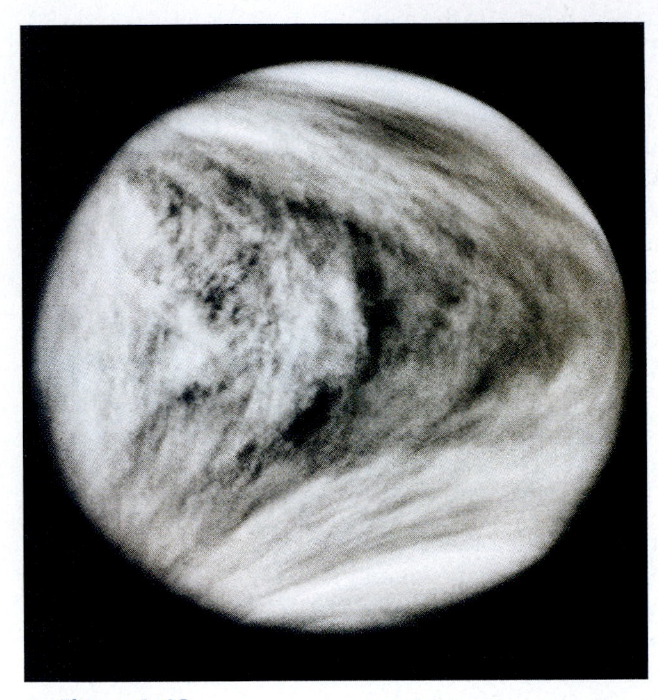

▲ **Figure 1–16**
Venus.

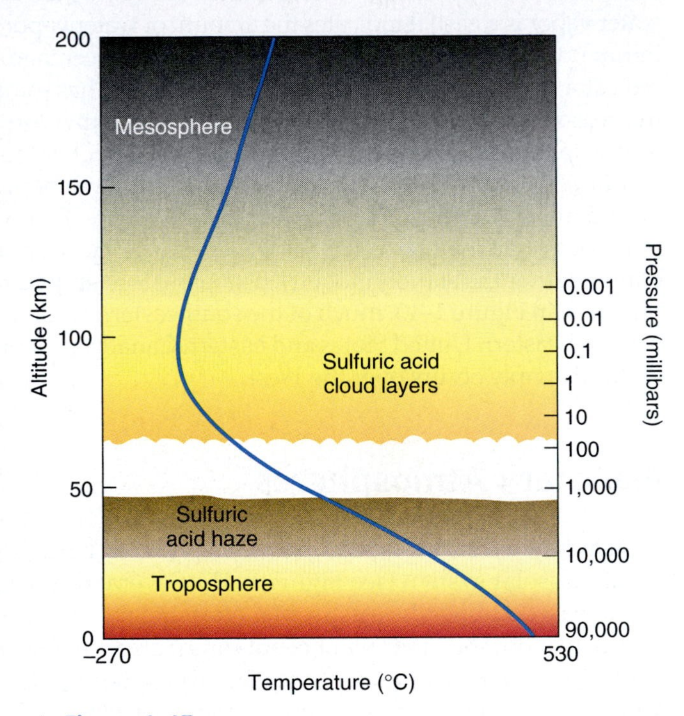

▲ **Figure 1–17**
The mean temperature profile for Venus.

extremely effective at absorbing thermal radiation emitted by the surface. Thus, very little of the energy radiated by the surface escapes into space, and the atmosphere maintains very high temperatures (around 475 °C, or 890 °F), even at night. This represents an extreme example of the "greenhouse effect," found to a lesser extent on Earth and described in greater detail in Chapter 3.

The chemical composition of the atmosphere of Mars is quite similar to that of Venus, with more than 95 percent of its mass consisting of carbon dioxide. The Martian atmosphere also contains a very small amount of water vapor (0.03 percent), which can condense to form the clouds shown in Figure 1–18. Because the density of the Martian atmosphere is only 1150th of Earth's, relatively little of the thermal

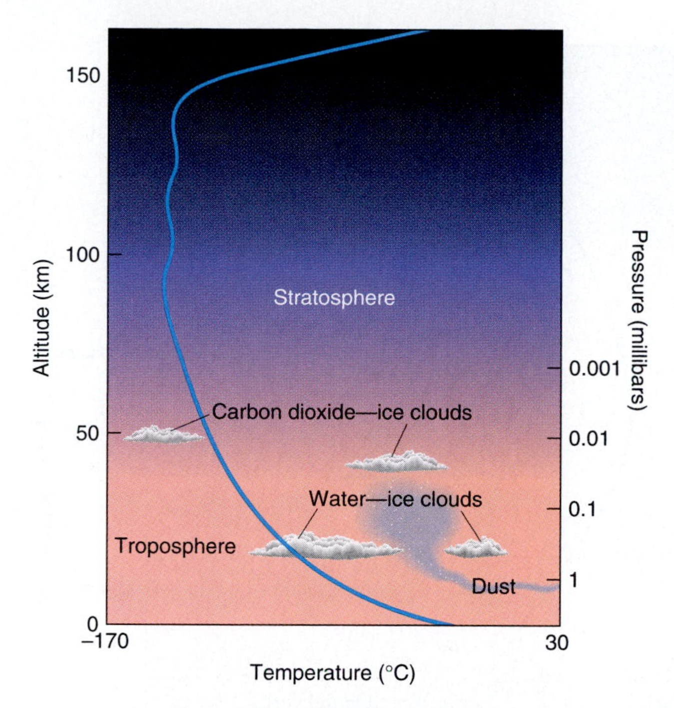

Figure 1–18
The mean temperature profile for Mars.

radiation emitted by the surface of the planet is absorbed and is instead lost to space. This results in rapid cooling at night, with temperatures dropping on the order of 100 °C (180 °F) from those at midday.

In contrast to the inner planets, the *outer* (or *Jovian*) *planets*—Jupiter, Saturn, Uranus, Neptune, and Pluto—do not have solid, well-defined surfaces. Instead, they are composed of hydrogen and helium and have solid or liquid interiors that gradually merge into gaseous outer regions. Thus, it is difficult to clearly define where their atmospheres begin. The atmosphere of each of the Jovian planets formed along with the planet by the gravitational attraction of low-density ices and gases. Figure 1–19 is a *Voyager* image of Jupiter, whose atmosphere is depicted in Figure 1–20.

Figure 1–19
A *Voyager* photo of Jupiter.

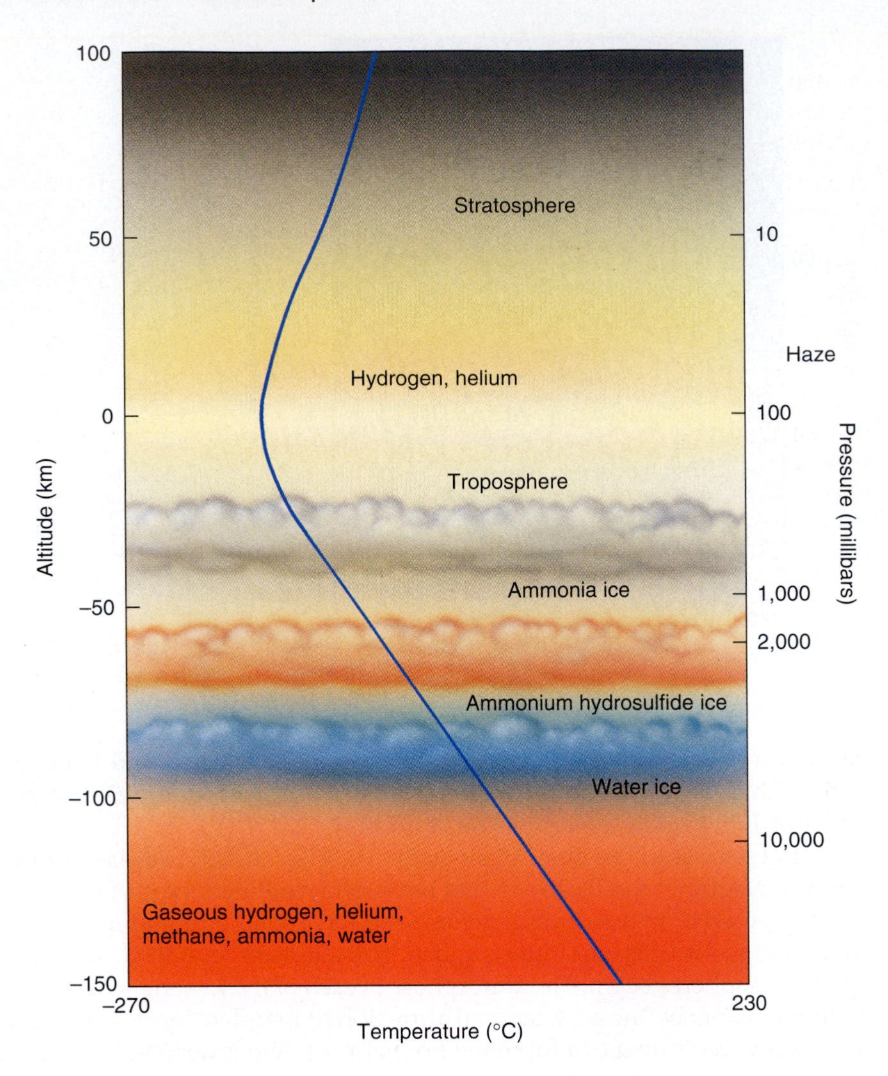

▶ **Figure 1–20**
The mean temperature profile for Jupiter.

Weather Forecasting—Both Art and Science

Weather forecasting is a highly sophisticated and data-intensive endeavor. Forecasters routinely employ state-of-the-art computer hardware and software systems that perform millions of calculations, based on a huge amount of input data and displayed at sophisticated work stations. The meteorologists who work with the information come armed with rigorous university training that includes intensive coursework in mathematics, physics, chemistry, and, of course, meteorology. But the scientists who work at the public and private forecasting agencies also rely heavily on their ability to subjectively analyze the current weather situation, thus making their task a blending of scientific principle and art form. This book includes a complete chapter on weather forecasting, but at this point it is useful to provide an overview of what a forecaster does in the course of his or her shift to produce weather information to the general public and specialized users.

Forecasters for the **National Weather Service (NWS)** in the United States and the **Meteorological Service of Canada (MSC)** work 8-hour shifts in offices that are open 24 hours a day, 7 days a week. All forecasters apply the same scientific underpinnings to their work and make use of the same types of resources, but there is no precisely defined routine by which a forecast is made. Every meteorologist has his or her own preferred set of procedures for analyzing the weather situation—and indeed those procedures may vary considerably under differing conditions. For example, a forecaster concerned with the possibility of near-term tornado development might spend much of his or her time looking at radar images, while one concerned with the possibility of an advancing snowstorm might spend more time assessing the changing pattern shown by satellite images. Furthermore, meteorologists employ numerous rules of thumb that might help them distinguish the likelihood of a given storm producing rain instead of snow, and these rules of thumb often vary by geographic region.

Upon starting his or her shift, a forecaster will likely receive a briefing about the weather of the last 24 hours from the meteorologist whose shift is about to end. The incoming meteorologist will not only be given a verbal description of the current situation but will also examine a large amount of textual and graphic information. Many professional meteorologists can still remember a time when most of the maps and information came in via a facsimile machine that would produce black-and-white output. Each weather map or satellite image would take several minutes to print on a damp sheet of paper, and then it would be hung up on a wall atop the previously printed maps or images. If the meteorologist wanted to examine the changes that occurred in the weather over a certain time period, he or she would have to manually flip through a sequence of maps. It would take a large section of wall to display the various types of maps and images that the forecaster might want to reference. This situation has changed radically over the last several decades.

Figure 1–21 shows a typical National Weather Service office (a) and workstation (b) and (c), at which the forecaster will perform most of the job. Each workstation employs the **Advanced Weather Interactive Processing System** (AWIPS), which allows forecasters to display maps of current weather conditions, output of computer forecast models displayed in map form, satellite and radar images, forecast advisories and discussions from other weather facilities, and a great deal more. Each AWIPS station has two graphical display monitors and a separate alphanumeric display. The graphical display monitor is typically set up to present one large

(a)

(c)

(b)

▲ **Figure 1–21**
A typical National Weather Service office (a) and workstation (b). Two AWIPS graphical display monitors (c).

panel and four smaller panels off to the left side. Forecasters can zoom in on or out of any image, and even superimpose several different types of information simultaneously. For example, a forecaster might set the large panel to show a satellite image of North America, with superimposed arrows depicting the upper-level winds. At the same time, the four smaller panels on the left side of the screen might display the current temperature, humidity, surface winds, and air pressure distributions. With just a few clicks of a mouse, the forecaster can call up just about any information that might help decipher the current weather patterns and help produce a successful forecast. AWIPS can also display all maps and images as movie loops to show the recent movement of significant weather features.

The forecaster goes through a sequence of procedures using AWIPS, depending on what the current conditions are and what type of weather may be developing. The first step is often to determine what type of cloud or precipitation conditions might be in the offing. This is of primary concern not only because of their intrinsic importance but also because such knowledge is necessary before any kind of temperature forecast can be made. (Obviously, a heavy overcast will reduce the amount of sunlight able to reach the surface and can thereby retard daytime warming.) If cloudy conditions currently exist, the question at hand is whether the clouds will move on and be replaced by clear skies. Visual inspection of satellite loops can be particularly useful in these situations. Morning cloud cover can also "burn off" later in the day as the sun gets higher above the horizon. Meteorologists would want to look at the temperature and moisture profiles in these instances to determine just how thick the cloud layers are; if clouds are relatively thin, a greater likelihood exists that they will give way to clear skies. Clear days also include the possibility that clouds will move in from other regions or develop later on and create local rain showers. Vertical profiles showing the trend in temperatures and humidity from the surface to the upper stratosphere can be particularly useful in these situations.

Changes in the weather can occur even if there is little movement of major weather systems. Assume, for example, that a forecaster starting the morning shift at Charleston, South Carolina, determines that no major systems are moving toward or away from the southeastern United States. The meteorologist still has to look for a change in temperatures over the next few days. A good start in such instances would be to compare the current temperature to that of the same time the day before. If the morning temperature is 5 degrees warmer than it was 24 hours earlier, it is likely afternoon temperatures will undergo a similar increase. But additional information can contribute to an even more reliable forecast. Therefore, the meteorologist might compare other conditions, such as the distribution of pressure across the Southeast. A change in such a pattern could, for example, indicate a weakening of the wind blowing from the coast to the inland region, which might promote even further warming.

Whether skies are cloudy or clear, forecasters now rely heavily on the output of computer models. These highly complex programs apply the fundamental laws of atmospheric physics to a huge amount of data and produce maps showing the expected distribution of air pressure, temperature, and other variables at the surface and upper atmosphere. The meteorologist interprets the overall patterns depicted on the maps, infers the overall type of weather that can be expected at a particular location, and issues a forecast that is usable for the general public. For example, if the computer maps indicate a substantial decline in air pressure at the surface coupled with particular upper-level air flows, the forecaster may advise the public of increasing cloudiness and a possibility of precipitation.

Forecast maps produced by the computer models often depict the distribution of weather variables directly as an application of the physical laws. Such a map is shown in Figure 1–22a, showing the predicted distribution of atmospheric pressure across the United States and Canada 24 hours in advance.

Other models rely on statistical methods to assist the forecaster. These products rely on output from the physical models. Thus, the physical models predict the state of the entire atmosphere over a wide area (e.g., North America), and the

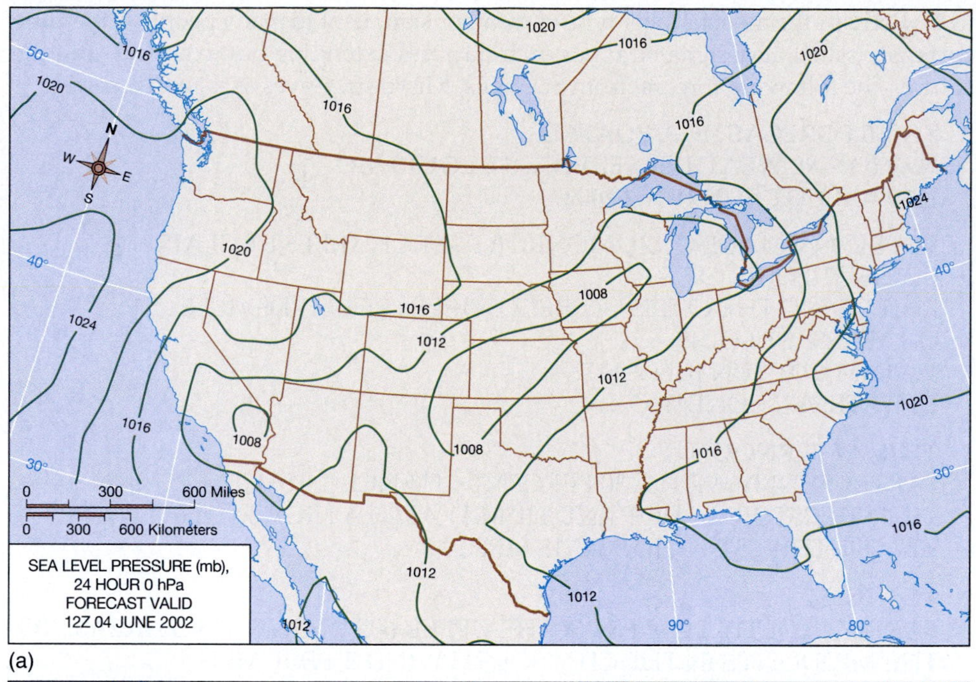

(a)

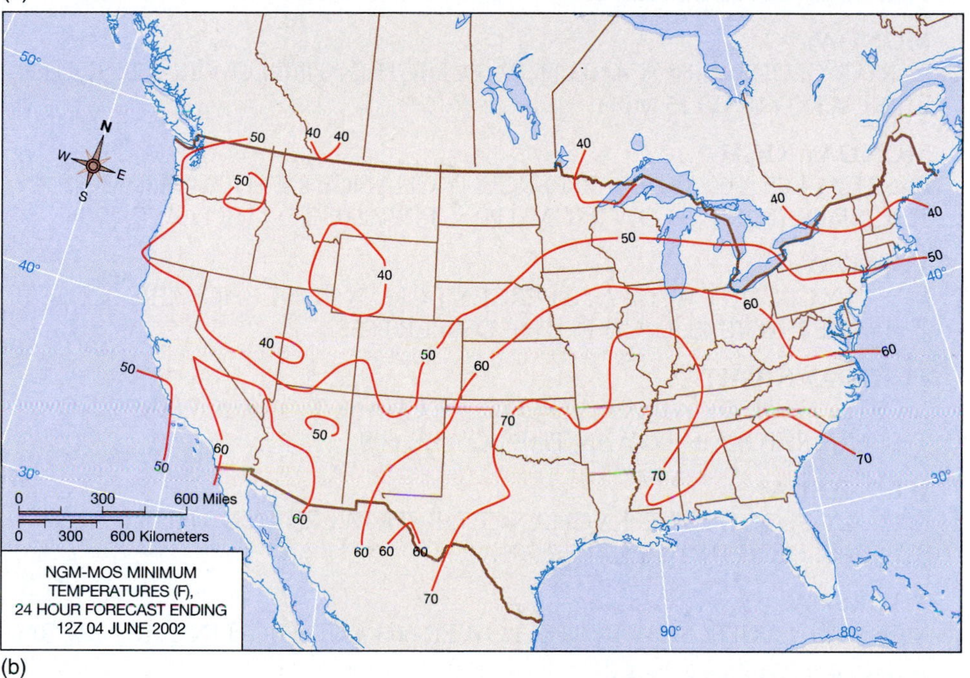

(b)

Figure 1–22
Output from a computer model run showing the predicted distribution of surface air pressure (a), based primarily on the application of physical laws. Weather Service computers can also apply statistical information based on past events to produce maps for further guidance. The map in (b) shows the predicted minimum temperature based on the application of statistical relationships to the weather patterns shown in (a).

statistical models use that information to predict values such as maximum and minimum temperature at particular locations (Figure 1–22b). Though the temperatures predicted by the computer can provide the forecaster with useful guidance, the professional meteorologist knows when to overrule such output in lieu of his or her professional judgment.

Forecasters make more than a single forecast and weather statements over the course of a shift, each intended to serve a particular person or constituency. Numerous types of forecasts are routinely issued at particular times during each shift, and a forecaster's activities are timed to assure these reports are disseminated on time. The scheduled reports include short-term forecasts, more detailed forecast discussions intended for people with a more advanced understanding of meteorology, local aviation advisories, hydrologic information (such as information on potential flooding), and numerous other statements.

Perhaps the most familiar type of weather statement to many people is the **zone forecast**, issued at designated times each day and extending out to a week into the future. The following forecast from St. Louis, Missouri, serves as a good example:

ZONE FORECASTS...UPDATED
NATIONAL WEATHER SERVICE ST LOUIS MO
1130 AM CDT SUN JUN 2 2002

JEFFERSON-MADISON IL-MONROE IL-ST CHARLES-ST CLAIR IL-ST LOUIS-ST LOUIS CITY-
INCLUDING THE CITIES OF...BELLEVILLE...EDWARDSVILLE...ST CHARLES...ST LOUIS
1130 AM CDT SUN JUN 2 2002
UPDATE SKY CONDITION

THIS AFTERNOON
PARTLY SUNNY WITH A 30 PERCENT CHANCE OF SHOWERS AND THUNDERSTORMS. HOT AND HUMID WITH A HIGH IN THE LOWER 90S. SOUTHWEST WIND 5 TO 15 MPH.

TONIGHT
PARTLY CLOUDY. LOW FROM THE UPPER 60S IN OUTLYING AREAS...TO THE MIDDLE 70S IN THE CITY. SOUTH WIND 5 TO 10 MPH.

MONDAY
PARTLY CLOUDY...HOT AND HUMID. HIGH IN THE LOWER 90S. SOUTHWEST WIND 10 TO 15 MPH.

MONDAY NIGHT
PARTLY CLOUDY WITH A 30 PERCENT CHANCE OF SHOWERS AND THUNDERSTORMS. LOW FROM THE UPPER 60S TO THE LOWER 70S.

TUESDAY
MOSTLY CLOUDY WITH A 50 PERCENT CHANCE OF SHOWERS AND THUNDERSTORMS. HIGH IN THE LOWER 80S.

TUESDAY NIGHT
MOSTLY CLOUDY WITH A 50 PERCENT CHANCE OF SHOWERS AND THUNDERSTORMS. LOW IN THE LOWER 60S.

WEDNESDAY
MOSTLY CLOUDY WITH A CHANCE OF SHOWERS AND THUNDERSTORMS. HIGH IN THE UPPER 70S.

THURSDAY
MOSTLY CLOUDY. LOW IN THE LOWER 60S AND HIGH IN THE MID 70S.

FRIDAY AND SATURDAY
PARTLY CLOUDY. LOWS IN THE LOWER 60S AND HIGHS FROM THE UPPER 70S TO THE LOWER 80S.

The above forecast was issued at 11:30 A.M., in accordance with the routine schedule for that particular shift. But on that day conditions were such that the forecaster decided to issue a supplementary, unscheduled report about the potential for severe weather. The following hazardous weather outlook was issued about a half hour after the scheduled zone forecast:

HAZARDOUS WEATHER OUTLOOK NATIONAL WEATHER SERVICE ST LOUIS MO 1205 PM CDT SUN JUN 2 2002

HAZARDS THROUGH TONIGHT... THERE IS A SLIGHT RISK OF SEVERE THUNDERSTORMS THIS AFTERNOON OVER PARTS OF NORTHEAST MISSOURI AS WELL AS WEST CENTRAL AND SOUTH CENTRAL ILLINOIS. THIS SLIGHT RISK AREA IS ALONG AND EAST OF A KIRKSVILLE MISSOURI TO QUINCY TO CENTRALIA ILLINOIS LINE.

DISCUSSION...
A STATIONARY FRONT EXTENDS FROM NORTHEAST MISSOURI TO SOUTH CENTRAL ILLINOIS THIS AFTERNOON. THE ATMOSPHERE IS VERY UNSTABLE SOUTH OF THE FRONT BECAUSE OF THE VERY WARM AND HUMID WEATHER. SCATTERED THUNDERSTORMS MAY DEVELOP ALONG THE FRONT LATER THIS AFTERNOON AND EVENING. A FEW OF THESE STORMS MAY BECOME SEVERE...CAPABLE OF PRODUCING LARGE HAIL...DAMAGING WINDS AND LOCALLY HEAVY DOWNPOURS.

IN ADDITION...A COMPLEX OF THUNDERSTORMS IS EXPECTED TO DEVELOP LATER TODAY OVER EASTERN NEBRASKA AND WESTERN IOWA. THESE STORMS MAY REACH NORTHEAST MISSOURI AND WEST CENTRAL ILLINOIS LATE TONIGHT.

OUTLOOK FOR MONDAY... THERE WILL BE A SLIGHT RISK OF SEVERE THUNDERSTORMS OVER NORTHEAST MISSOURI AND WEST CENTRAL ILLINOIS...GENERALLY ALONG AND NORTH OF A PARIS TO BOWLING GREEN TO PITTSFIELD LINE.

THIS PRODUCT WILL BE ISSUED BETWEEN 6 AND 8 AM DAILY...AND WILL BE UPDATED AS NEEDED. FOR DETAILED INFORMATION ON A PARTICULAR HAZARD...MAKE SURE TO REFER TO THE ZONE FORECASTS AND THE ACTUAL WATCH...WARNING OR ADVISORY PRODUCT.

Some forecast offices require the issuance of routine weather reports particularly important to the local environment. Thus, unlike a forecaster in the St. Louis office of the National Weather Service, one working in Honolulu routinely has to write a marine forecast such as the following:

COASTAL AND OFFSHORE FORECAST FOR HAWAII NATIONAL WEATHER SERVICE HONOLULU HI 6 AM HST SUN JUN 2 2002

SYNOPSIS
HIGH SURF ADVISORY REMAINS IN EFFECT FOR THE SOUTH FACING SHORES OF ALL HAWAIIAN ISLANDS

TODAY
SOUTHEAST WINDS 10 TO 15 KT WEST OF MOLOKAI AND EAST WINDS 10 TO 15 KT ELSEWHERE. SEAS 3 TO 5 FT. SOUTH SWELL 4 FT. ISOLATED WINDWARD SHOWERS.

TONIGHT AND MONDAY
SOUTHEAST WINDS 10 TO 15 KT WEST OF MOLOKAI AND EAST WINDS 10 TO 15 KT ELSEWHERE. SEAS 2 TO 4 FT. SOUTH SWELL 4 FT. ISOLATED WINDWARD SHOWERS.

From these examples, you can see that not all meteorologists have the same set of procedures to follow on a given day. Variations from one forecast office to another affect a meteorologist's schedule, and differing weather conditions may alter the usual routine.

This text is written to provide you with an understanding of how the atmosphere works. It will also introduce you to the very same principles used by forecasters in their day-to-day work. As you read this book, we strongly recommend that you regularly follow daily weather via the news media or through the Internet (we describe several useful Web sites at the end of each chapter and on this book's own Web page at **www.prenhall.com/aguado**). Basic principles and rules of thumb related to forecasting will be introduced in the chapters immediately following this one, as they relate to the chapter topics. The goal here is not only to present the science involved in meteorology but also to demonstrate how weather and climate play an important role in our lives.

Summary

Earth's atmosphere is composed of a mixture of gases and contains an enormous number of suspended solids and liquids called *aerosols*. Three of the gases—nitrogen, oxygen, and argon—occur in nearly constant proportions and constitute the vast majority of the atmospheric mass. Other gases occur in slight amounts and vary considerably in their concentration. These variable gases, especially water vapor and carbon dioxide, can be extremely important to life on Earth.

Four layers of the atmosphere—the troposphere, stratosphere, mesosphere, and thermosphere—can be distinguished on the basis of their temperature profiles, and one—the ionosphere—is designated for its electrical characteristics. Of the first four, the lowest two (the troposphere and stratosphere) contain the vast majority of atmospheric mass. The other two, the mesosphere and thermosphere, account for less than 0.1 percent of the atmosphere's mass and are relatively unimportant for most of the processes described in this book. The ionosphere spans the upper mesosphere and thermosphere. It has important effects on the transmission of AM radio waves and provides the locale for the aurora borealis and aurora australis.

The present atmosphere did not arrive with the formation of the planet some 4.5 billion years ago but rather evolved after the primordial atmosphere was lost to space. The process of outgassing released water vapor and carbon dioxide (along with other gases) from Earth's interior. Ultimately, photosynthesis reduced carbon dioxide levels and contributed oxygen to the atmosphere.

This chapter introduced some of the most significant weather elements and demonstrated how they are depicted on surface weather maps. It also described the atmospheres of other planets in our solar system. Some of the rudiments of weather forecasting were introduced so readers of this book can make use of current weather data to see how the principles of meteorology are expressed on a day to day basis.

Key Terms

atmosphere p. 4
meteorology p. 4
permanent gases p. 6
variable gases p. 6
homosphere p. 6
heterosphere p. 6
nitrogen p. 6
oxygen p. 7
argon p. 7
water vapor p. 7
hydrologic cycle p. 7
carbon dioxide p. 8
photosynthesis p. 9
respiration p. 10
ozone p. 10
methane p. 11
aerosols p. 11
particulate p. 11
condensation nuclei p. 13
structure p. 13
density p. 14
mean free path p. 14
standard atmosphere p. 14
troposphere p. 14
tropopause p. 15
inversion p. 16
stratosphere p. 16
stratopause p. 16
ozone layer p. 17
mesosphere p. 17
thermosphere p. 17
ionosphere p. 18
ions p. 18
aurora borealis p. 19
aurora australis p. 19
escape velocity p. 19
outgassing p. 19
pressure p. 20
wind p. 21
isobar p. 21
millibar p. 21
kilopascal p. 21
station model p. 21
front p. 25
relative humidity p. 25
dew point temperature p. 25
National Weather Service p. 28
Meteorological Service of Canada p. 28
Advanced Weather Interactive Processing System p. 29
zone forecast p. 32

Review Questions

1. Why is it difficult to define an absolute top of the atmosphere?
2. What are the homosphere and the heterosphere?
3. What is the difference between the permanent and variable gases of the atmosphere? Which gases are the most important in terms of their contribution to the total mass of the atmosphere?
4. Given that variable gases are so rare, why are they considered at all?
5. Why has the concentration of carbon dioxide in the atmosphere been increasing over the last century?
6. What is ozone, and why is it both beneficial and harmful to life on Earth?
7. What are aerosols? Are they formed only by human activities or are they also naturally occurring?
8. How do photosynthesis, respiration, and decay affect the carbon dioxide balance of the atmosphere?
9. In what way does the density of the atmosphere vary with altitude?
10. What are the distinguishing characteristics of the troposphere, stratosphere, mesosphere, and thermosphere?
11. What is the tropopause?

12. In which thermal layer of the atmosphere is the ozone layer found? Why is the term "ozone layer" somewhat misleading?
13. What percentage of the total mass of the atmosphere is contained in the troposphere and the stratosphere?
14. Why does the troposphere contain much more mass than the stratosphere, despite the fact that the troposphere is a thinner layer than the stratosphere?
15. How is the ionosphere distinct from the layers of the atmosphere defined by their temperature profiles?
16. What is outgassing and why was it important?
17. Why were anaerobic bacteria important to the evolution of the atmosphere?
18. Briefly describe the effect that variations in pressure exert on other weather elements.
19. What are isobars?
20. What are station models and what useful information do they depict?

Critical Thinking

1. Severe weather can produce major economic damage and loss of life. How do you think the threats vary from one place to another? Rate your home location in terms of risk for each type of severe weather.
2. The Kyoto Treaty that limits carbon dioxide emissions has not been ratified by all countries that originally signed the treaty. What are the pros and cons of doing so?
3. Volcanic eruptions continue to occur and outgas water vapor, carbon dioxide, and other gases. Do you think that this will be a significant factor in increasing the concentration of these gases over the next century? Why or why not?
4. Temperatures usually decrease with height in the troposphere but increase with height in the stratosphere. Why do the two layers have such different profiles?
5. The thermosphere has extremely high temperatures, but a person exposed to the thermosphere would rapidly freeze. Explain the apparent contradiction in terms of what you know about heat and temperature.

Problems and Exercises

1. The National Climatic Data Center (NCDC) has a Web site at **http://lwf.ncdc.noaa.gov/oa/climate/severeweather/extremes.html** that provides a wealth of information on extreme weather and climate events. Examine the site to see how much information is available to you. Do any of the topics relate to weather events in your hometown?
2. Examine the map of National Weather Service Offices at **http://www.wrh.noaa.gov/wrhq/nwspage.html**. Is there a Weather Service office near you? If so, consider visiting the office and seeing firsthand what goes into producing a forecast.
3. Keep a record of daily weather in your area and download current weather maps from one of the available Web sites. Do you notice any patterns on the maps that tend to be associated with particular weather conditions?
4. Look at today's weather map and observe the contrasting weather conditions across the United States and Canada. Do any areas exhibit significant changes in weather from adjacent regions? How well defined are the boundary zones? Repeat this exercise for several days and see if these transition regions show movement.

Quantitative Problems

Students' comprehension of a great many concepts can be enhanced by working out numerical solutions to questions. Go to this book's Web site at **http://www.prenhall.com/aguado/** and click on the Begin button. This will take you to the main page for Chapter 1, which offers a variety of valuable resources. Take a look at the options available. Along with the self-quizzes, links, and other resources on the page, the quantitative exercises should prove particularly valuable.

Useful Web Sites

http://www.weather.com/

Home page for the Weather Channel. Contains current weather information along with reports related to travel, health, severe weather, and many other topics. A good first stop.

http://www.wrh.noaa.gov/wrhq/nwspage.html

Map of all local National Weather Service offices, with links connecting viewer directly to sites. Also lists Uniform Resource Locators (URLs), Internet addresses, for regional and national support offices such as the National Hurricane Center. An excellent site for local information anywhere in the United States and for advisories on severe weather.

http://weatheroffice.ec.gc.ca/canada_e.html

Home page for the Meteorological Service of Canada. Provides forecasts and links to satellite, radar, and many other useful resources.

http://www.wunderground.com/ and http://www.wunderground.com/global/CN_ST_Index.html.

Multipurpose sites that provide detailed information for the United States and Canada, respectively. Offer standard and unusual weather maps and local weather data. Also good outlets for weather information all over the world.

Media Enrichment

Weather in Motion

Satellite Movies of the May 3–4, 1999, Tornado Cluster

The large weather system that produced the severe weather outbreak was tracked by weather satellites. Two movies on this book's CD-ROM provide different perspectives on this storm. The black-and-white movie is based on visible light, and the other is based on infrared radiation (discussed in Chapter 2). The most severe portions of the storm system show up dramatically in both movies. Note that the movie based on visible energy ends at sunset, while the infrared movie continues past sunset.

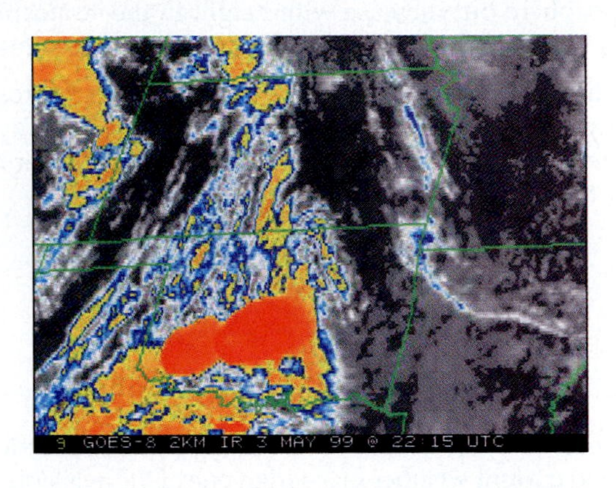

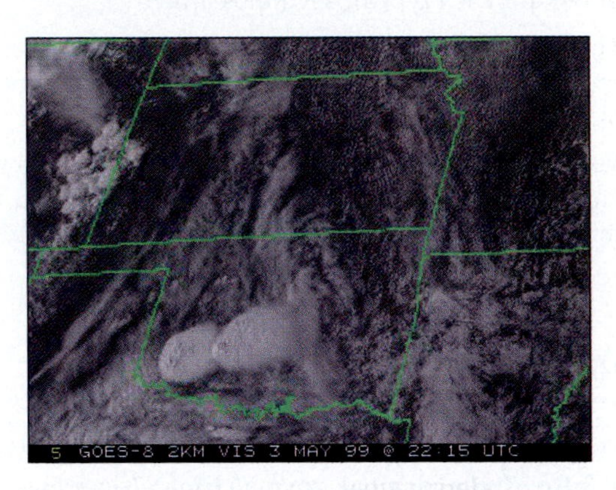

Weather in Motion

A Global Montage of Clouds and Sea-surface Temperature

Although this chapter hasn't emphasized atmospheric motion, it's not too soon to stress the ceaseless movement of Earth's fluid envelope. This movie depicts weather for an entire year (1999), obtained by merging information acquired every 6 hours from three satellite systems. It contains a wealth of information and will be used repeatedly in the text. For now, we draw your attention to just a couple of features. First, notice the distinctly different movement of major cloud systems at various latitudes. In particular, you can see tropical systems drifting mostly from east to west, whereas those at higher latitudes show west-to-east movement. There are also strikingly different shapes to the cloud systems, with tropical versions more or less circular compared to the elongated storm clouds at higher latitudes.

There is a pronounced north–south migration of the band of clouds over the Tropics during the year, especially over Africa. During January the clouds are centered to the south of the equator, but as the movie progresses into July and August, the clouds move into the Northern Hemisphere. The latitudinal movement of the cloud bands is directly related to the seasonal changes caused by Earth–Sun geometry, discussed in Chapter 2. You see this also reflected in sea-surface temperature, as the zone of high temperatures (colored red) moves north and south. Finally, pay close attention to the Atlantic coast of North America during the August and September portions of the movie. Some notable hurricanes that hit the eastern United States show up very well. We will discuss hurricanes and other major weather events throughout this book.

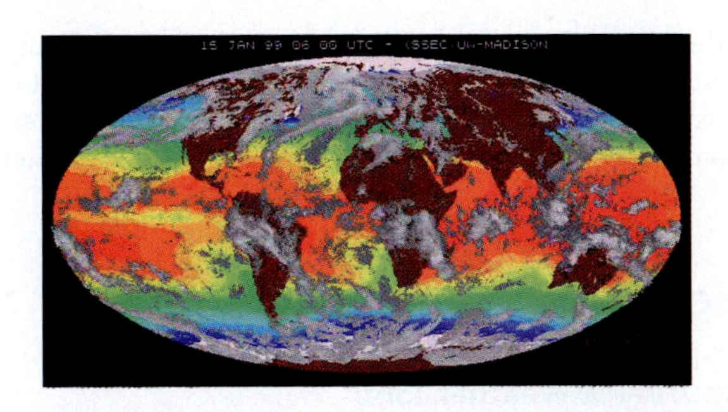

Weather in Motion

The Ozone Hole

This movie shows the progressive development of the Antarctic ozone hole during the Southern Hemisphere spring of 1991. The movie's perspective moves to the Antarctic in late September, when the hole normally begins to develop. By mid-October most of the Antarctic continent is covered by a zone of greatly reduced stratospheric ozone. Notice that the ozone hole is not a stationary feature; instead, its size, shape, and intensity vary throughout the next few months.

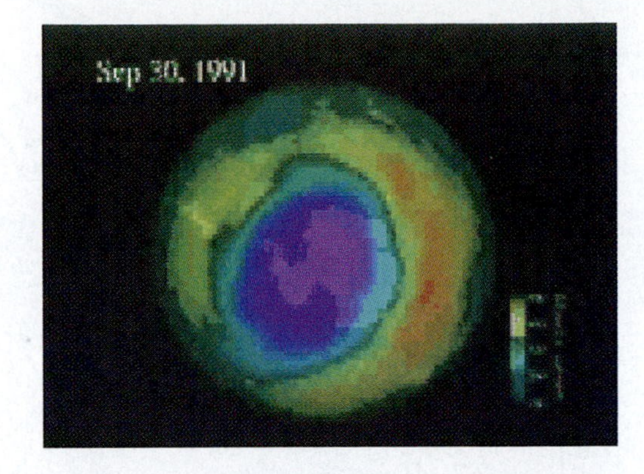

Weather in Motion

Tropical Storm Allison

Tropical Storm Allison brought huge amounts of rain and extreme flooding to Texas and Louisiana, as it hovered over the region for several days. The system then moved to the east coast where it brought floods as far north as New England. This movie tracks the storm from June 4–14, 2001.

Weather Image

First Look at the New Millennium

If you were wondering what North America looked like at the very beginning of the new millennium, here is the first weather satellite image of the year 2000. The image was collected at 15 minutes past midnight, Greenwich mean time (UTC). We see a band of clouds that has recently passed over the eastern United States, along with generally cloudy conditions over eastern Canada. Another extensive region of cloud cover exists over much of the western United States and Canada.

The January 1998 ice storm.

chapter 2

Solar Radiation and the Seasons

Over the period of January 5–9, 1998, residents of a large part of the eastern United States and Canada encountered a rainfall event that greatly affected their lives. Unlike most rain events, the falling drops froze immediately upon hitting the surface to form a continuous layer of smooth, white ice. The ice became so thick that it caused tree branches to break, power lines to snap, and roads to become impassable. The freezing rain was so widespread and the damage so pervasive that millions of residents in New England, New York State, and the Canadian province of Quebec lost electrical power—for more than 2 weeks in some places. In Quebec alone, 3 million people—40 percent of the population—were without power, and damage estimates were in the neighborhood of $1 billion. One of those who endured the outage, Mr. Andre Champagne of Montreal, described it thusly: "It's so dark, it's like I'm on Mars."

The storm had numerous other consequences—all of them bad. In the southeastern and Mid-Atlantic states, the rain caused extensive flooding, in part because of direct runoff into rivers. The percolation of rainwater into the existing snowpack caused rapid melt, which further exacerbated the flooding problem. Four states called in the National Guard. And there was no shortage of unhappy cattle, as the loss of electricity made it impossible for farmers to milk their cows, causing more than a million dollars in agricultural losses. More than 30 people in the United States and Canada died as a result of the storm.

The cold winter storm of January 1998 offers a vivid contrast to the severe weather outbreak that hit the northern United States and southern Canada on May 30 and 31, when tornadoes, hail, and lightning killed at least 17 people. Such storms, though possible any time during the year, are most prevalent in spring. And then there was the summer heat wave of July 1999, which spread its share of misery from eastern Canada down to the southern plains and southeastern portion of the United States. More than 250 people died across the United States over the 2-week episode.

As these examples show, many sorts of severe weather can create widespread havoc, but each is more likely to occur at some particular time of the year. Even in the absence of severe weather, our lives are profoundly influenced by the seasonal cycle that occurs each year. But as important as the changes of the seasons are to everyday activities, many of us do not really understand what causes them. In many cases we incorrectly think we know what causes them. For example, many people believe that variations in the distance between Earth and the Sun are responsible for changes in the seasons, with summer occurring when Earth and the Sun are closest together. But Earth and the Sun are closest to each other on or about January 3—in the midst of the Northern Hemisphere winter! Likewise, the Sun is farthest from Earth on July 3, a time Northern Hemisphere residents associate not with winter, but with long, hot days, barbecues, and fireworks. Obviously, then, there must be another explanation. In this chapter, we describe exactly how Earth's orbit about the Sun produces our seasons. We begin with some of the basics.

CHAPTER OUTLINE

January 1998 Ice Storm

Energy

Energy is traditionally defined as "the ability to do work." This definition isn't entirely accurate, and raises it own questions (What's work?), but it's impossible to do better in just a few words. Rather than travel far afield in search of a precise definition, we'll assume everyone has at least a vague idea of energy as an agent capable of setting an object in motion, warming a teapot, or otherwise manifesting itself in everyday events. The standard unit of energy in the International System (SI) used in scientific applications is the **joule** (abbreviated as J). Although students may be more familiar with the calorie as the unit for energy, the joule is preferred in this context (1 joule = 0.239 calories). A related term, **power**, is the rate at which energy is released, transferred, or received. The unit of power is the **watt** (W), which corresponds to 1 joule per second.

Even the simplest activity requires a transfer of energy. In fact, while you read these words, an energy transfer is occurring as the chemical energy from food you have eaten is converted into the kinetic energy (energy of motion) needed to move your eyes across this line of type. But your body, like any other machine, is not perfectly efficient; it loses some thermal energy (heat) as chemical energy is converted into kinetic energy. Thus, your eye muscles give off heat as they contract and expand.

The same concept applies to our atmosphere. About one two-billionth of the energy emitted by the Sun is transferred to Earth as **electromagnetic radiation**, some of which is directly absorbed by the atmosphere and surface. This radiation provides the energy for the movement of the atmosphere, the growth of plants, the evaporation of water, and an infinite variety of other activities.

Kinds of Energy

Energy can occur in a variety of forms. We often speak of radiant, electrical, nuclear, and chemical energy; but, strictly speaking, all forms of energy fall into the general categories of **kinetic energy** and **potential energy**. These are illustrated in Figure 2–1.

Kinetic energy can be viewed as energy in use and is often described as the energy of motion. Motion can occur on a large scale, as in the movement of an object from one place to another. Examples that occur in nature include falling raindrops (Figure 2–2a), water flowing through a river channel, and grains of dust

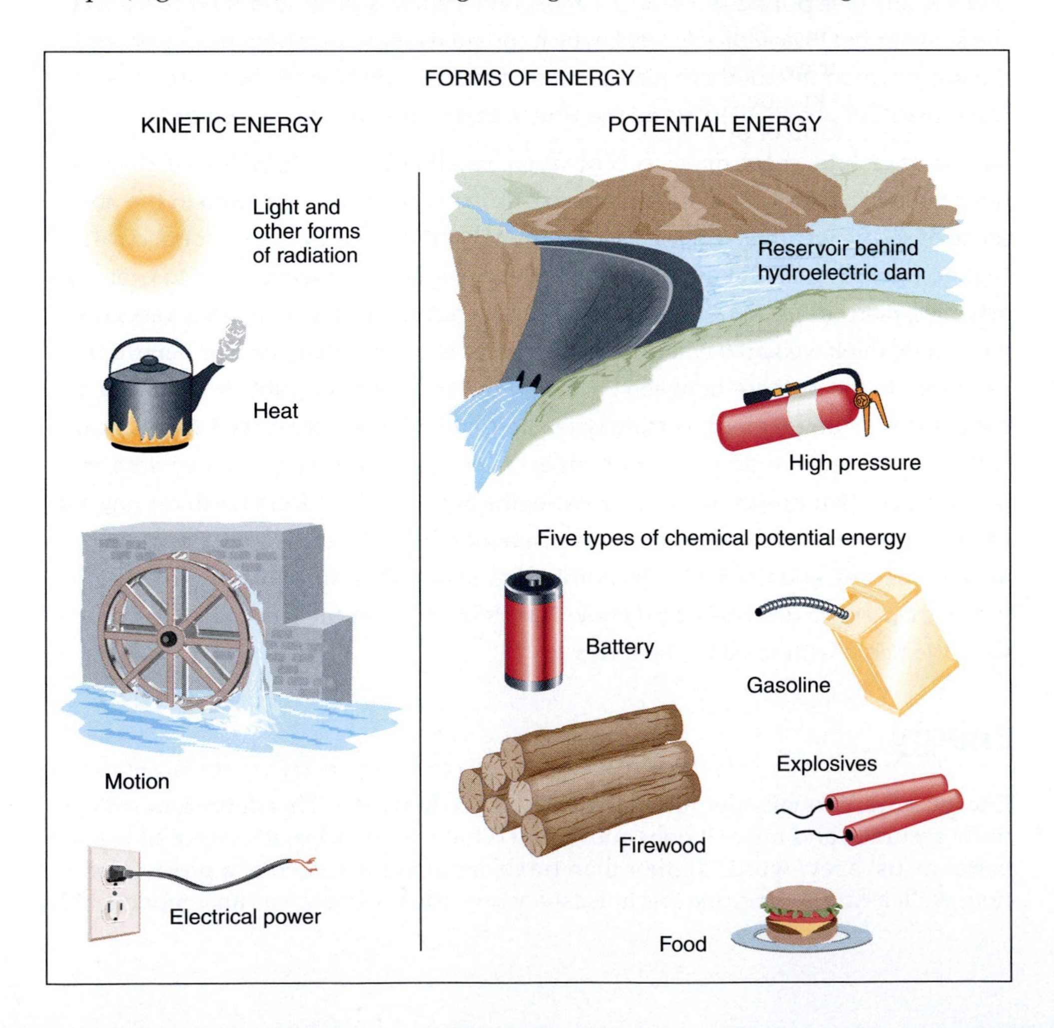

▶ **Figure 2–1** Energy assumes several different forms, but each of these is a form of either kinetic energy (the energy of motion) or potential energy.

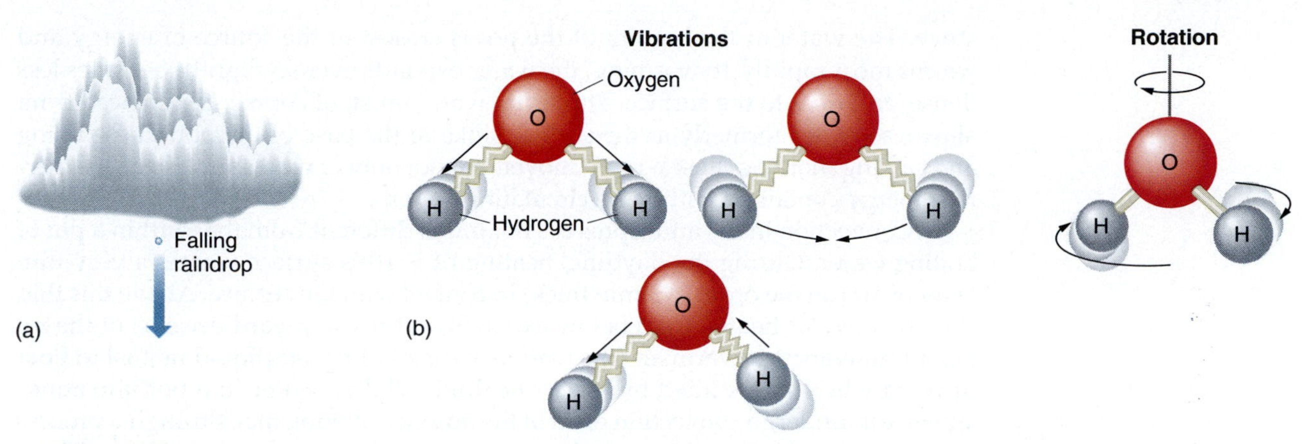

▲ **Figure 2–2**
Kinetic energy can occur as the motion associated with moving objects, such as the falling raindrop in (a), or as molecular vibration or rotation, as depicted for water molecules in (b). The greater the rate of vibration or rotation, the higher the temperature of the substance.

transported by the wind. The motion of kinetic energy can also occur at a microscale, as in the case of molecular vibration or rotation (illustrated for water in Figure 2–2b). A solid object may seem to be standing still, but its molecules are undergoing a certain amount of vibration.

Gas and liquid molecules, in contrast, are not fixed in space but move about randomly (Figure 2–3). In solids, liquids, or gases, the rate of vibration or random movement determines the temperature of the object.

If kinetic energy is energy in use, potential energy is energy that hasn't yet been used. Potential energy can assume many forms. For example, a plant's carbohydrates have potential energy that can be consumed by animals (or by the plant itself) and then metabolized to yield the energy needed for all of its biological activity. When our own bodies metabolize food, we are using this potential energy, converting it to kinetic energy, and releasing heat as a byproduct.

Another form of potential energy results from an object's position. Consider, for example, a cloud droplet that occupies some position above Earth's surface. Like all other objects, the droplet is subject to the effect of gravity. As it falls toward Earth's surface, the object's potential energy is converted to kinetic energy. Obviously, the higher the droplet's elevation, the greater the distance it is capable of falling and the greater its potential energy. It is important to recognize that the droplet did not attain its height by magic, because energy was used to elevate its mass in the first place.

Energy Transfer Mechanisms

Conduction Energy can be transferred from one place to another by three processes: conduction, convection, and radiation. **Conduction** is the movement of heat through a substance without the movement of molecules in the direction of heat transfer. A simple example is a metal rod, one end of which is placed over a campfire. The part of the rod above the flame is warmed, and molecules there gain energy. Some of this is passed to neighboring molecules, which in turn heat adjacent molecules. (The exact mechanism of molecular "passing" depends on the substance—in metals, it is mainly accomplished by electrons.) This process occurs throughout the length of the rod so that after a few moments the entire piece of metal becomes too hot to handle. The transfer of heat from the warmer to the colder part of the rod is conduction. Note that although heat travels through the rod, the molecules that make up the rod do not move. Conduction is most effective in solid materials, but as we will see in Chapter 3, it also is an important process in a very thin layer of air near Earth's surface.

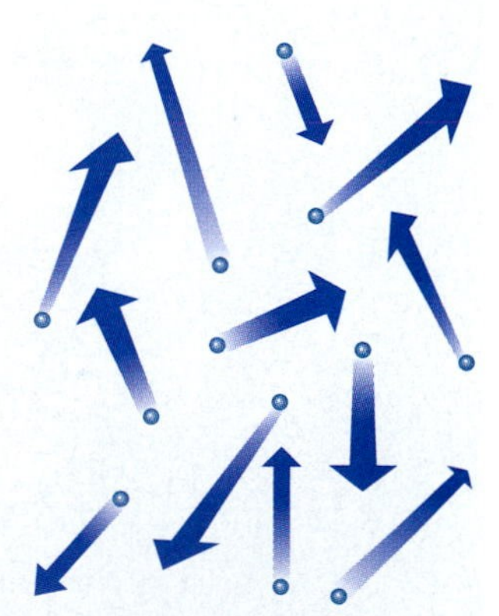

▲ **Figure 2–3**
Gas molecules have no bonds to other molecules (as opposed to solids and liquids) and move about in random motion.

Convection The transfer of heat by the mixing of a fluid is called **convection**. Unlike conduction, convection is accomplished by displacement (movement) of the medium. You can observe this process by watching a pot of water boil on a kitchen stove. The water at the bottom of the pot is closest to the source of energy and warms most rapidly. In warming, the water expands ever so slightly, becomes less dense, and rises to the surface. The rising water must, of course, be replaced from above, so water formerly at the surface sinks to the base of the pot. These rising and sinking motions cause a rapid movement not only of mass, but also of the thermal energy contained within the circulating water.

Convection in the atmosphere is not much different from that within a pot of boiling water. During the daytime, heating of Earth's surface warms a very thin layer of air (on the order of 1 mm thick) in contact with the surface. Above this thin *laminar layer*, air heated from below expands and rises upward because of the inherent **buoyancy** of warm air (the tendency for a light fluid [liquid or gas] to float upward when surrounded by a heavier fluid). Unlike water in a pot, the atmosphere can undergo convection even in the absence of buoyancy through a process called *forced convection*, the vertical mixing that happens as the wind blows. These processes are discussed in more detail in Chapter 3.

Radiation Of the three energy transfer mechanisms, **radiation** is the only one that can be propagated without a transfer medium. In other words, unlike conduction or convection, the transfer of energy by radiation can occur through empty space. Virtually all the energy available on Earth originates from the nearby (in astronomical terms) star we call the Sun, a member of the Milky Way galaxy (Figure 2–4). The atmosphere also has other sources of energy: minute amounts of radiation are received from the billions of other stars in the universe, and some energy reaches the surface from Earth's interior. However, the contribution of these sources is minuscule compared to the energy obtained from the Sun.

We will now examine the characteristics of radiation and the way Earth's orientation affects the radiation received. The spatial and seasonal variations in the receipt of solar energy are not mere abstractions; they are, in fact, the driving force for virtually all the processes discussed in the rest of this book.

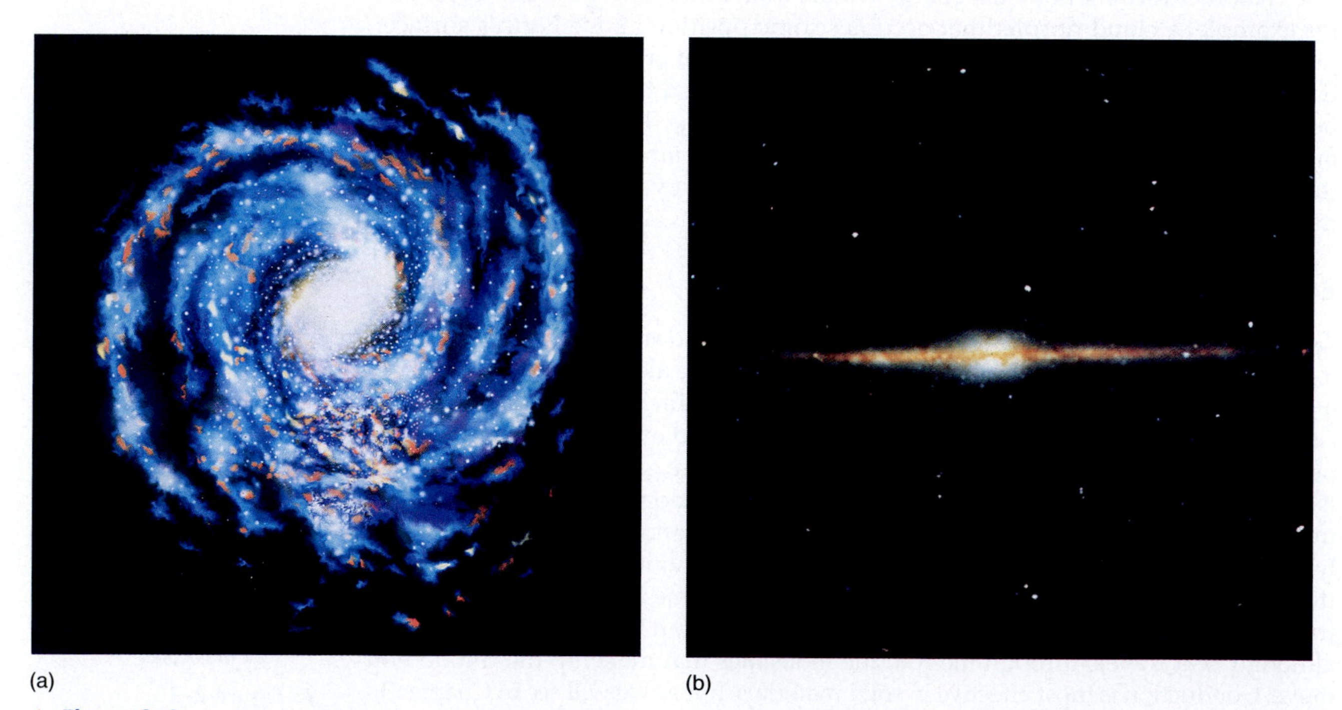

(a) (b)

▲ **Figure 2–4**
Our solar system is part of the Milky Way galaxy, a rather typical galaxy that contains more than 100 billion stars. (a) An artist's rendition. (b) A wide-angle infrared image of the plane and bulge of our Milky Way.

Radiation

Radiation is emitted by all matter. Thus, *everything*—including the stars, Earth, ourselves, and this book—is constantly emitting electromagnetic energy. We are all familiar with electromagnetic energy in many of its forms. We see the environment around us because a certain type of radiation we call *visible light* impinges on our eyes, which then send signals to our brains to produce visual images. A different type of electromagnetic energy is used when we warm a meal in a microwave oven; the radiation agitates the molecules of the food and thereby increases its temperature. Other types of radiation may be less beneficial or even harmful, such as ultraviolet radiation, which can lead to sunburns, malignancies, or even death. Although different types of radiation have different effects, they are all very similar in that they are transmitted as a sequence of waves.

Think of a wave created by a rock tossed into a pond. The wave is revealed by an oscillation in the water surface with alternating crests (high points in the ripple) and troughs (low points). When you observe the regular rise and fall of the surface as the wave passes, you know energy is being transferred. In the case of radiation, the waves are electrical and magnetic oscillations. That is, radiation consists of both an electrical and a magnetic wave. With the proper instruments, we would detect these electrical and magnetic variations—hence the term *electromagnetic radiation*. To put it differently, when an object emits radiation, both an electrical and a magnetic field radiate outward. At a fixed point in space, the strength of both fields rises and falls rhythmically, thereby forming electric and magnetic waves, each with its own crest-to-trough pattern. The electric and magnetic waves are perpendicular to one another, as shown in Figure 2–5. More importantly, the electric and magnetic components are closely coupled—the two rise and fall in unison.

Radiation Quantity and Quality

Section 2.2.1

To describe electromagnetic radiation completely, we need to provide information about the amount of energy transferred (quantity), and the type, or quality, of the energy. This is similar to describing someone's weight, where we might state quantity in pounds and indicate quality using words such as "mostly flab." In the case of radiation, quantity is associated with the height of the wave, or its *amplitude*. Everything else being equal, the amount of energy carried is directly proportional to wave amplitude.

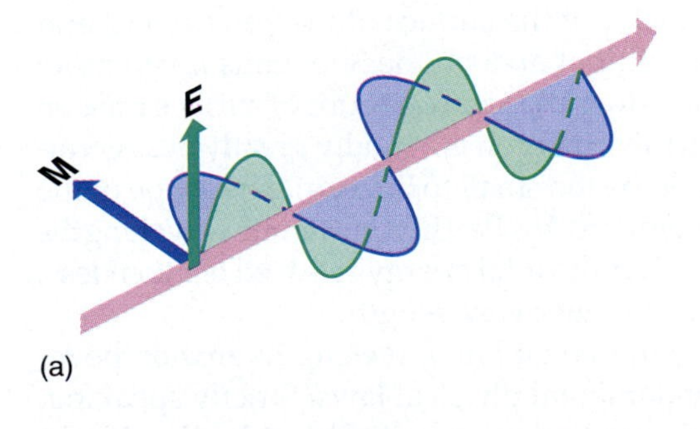

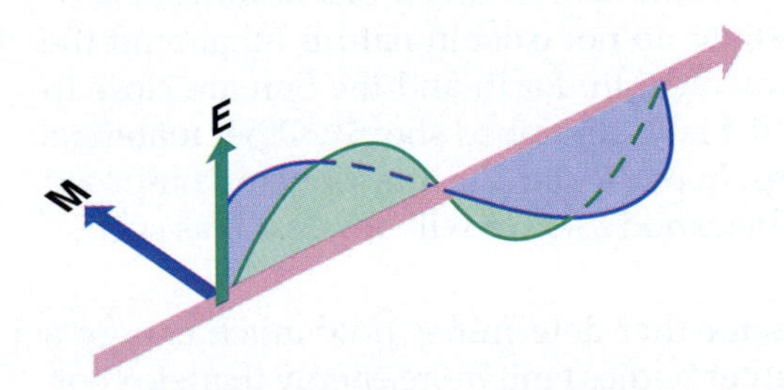

◀ **Figure 2–5**
Electromagnetic radiation consists of an electric wave (E) and a magnetic wave (M). As radiation travels, the waves migrate in the direction shown by the pink arrow. The waves in (a) and (b) have the same amplitude, so the radiation intensity is the same. However, (a) has a shorter wavelength, so it is qualitatively different than (b). Depending on the exact wavelengths involved, the radiation in (a) might pass through the atmosphere, whereas that in (b) might be absorbed.

The quality, or "type," of radiation is related to another property of the wave, the distance between wave crests. Figure 2–5 shows waves of electromagnetic radiation moving in the same direction. All have the same amplitude, but the distance between the individual wave crests is smaller for the waves depicted at the top. The upper waves therefore have a shorter **wavelength**, which is the distance between any two corresponding points along the wave (crest-to-crest, trough-to-trough, etc.). Because of their shorter wavelengths, the waves in Figure 2–5a are qualitatively different, and might produce different effects, than the waves in Figure 2–5b. For example, X-ray radiation has an extremely short wavelength and is able to penetrate soft tissues. On the other hand, ordinary light, having a somewhat longer wavelength, is absorbed by the skin. Compared to everyday objects, the radiation of interest here has very small wavelengths. It is therefore convenient to specify wavelengths using small units called **micrometers** (or **microns**). 1 micrometer—signified by μm—equals one-millionth of a meter, or one-thousandth of a millimeter (0.00004 in.).

All forms of electromagnetic radiation, regardless of wavelength, travel through space at the speed of light, which is about 300,000 km (186,000 mi) per second. At that speed, it takes 8 minutes for energy from the Sun to reach Earth. The energy received from the other, more distant stars takes even longer to arrive at Earth. For instance, radiation from the next nearest star, Proxima Centauri, must travel through space for 4.3 years before reaching us. Though this may seem like a long time, it is minuscule compared to the *billions* of years needed for light from a distant star to arrive at Earth.

Electromagnetic energy comes in an infinite number of wavelengths, but we can simplify things by categorizing wavelengths into just a few individual "bands," as indicated in Figure 2–6 and Table 2–1. The band with the shortest wavelengths consists of gamma rays, with a maximum wavelength of 0.0001 μm. Successively longer wavelength bands include X rays, ultraviolet (UV), visible, near-infrared (NIR), thermal infrared (IR), microwave, and radio waves. Note that there is nothing unique or special about the visible portion of this electromagnetic spectrum other than the fact that our eyes and nervous systems have evolved to be able to sense this type of energy. Except for their wavelengths, visible rays are just like any other form of electromagnetic energy.

Intensity and Wavelengths of Emitted Radiation

All objects radiate energy, not merely at one single wavelength but over a wide range of different wavelengths. Figure 2–7a graphs the intensity of radiation emitted at all wavelengths every second by a square meter of the surface of the Sun (in red) and Earth (in blue). We can readily see that a unit of area on the Sun emits much more radiation (about 160,000 times more) than does the same amount of surface area on Earth (notice that the curve showing Earth's emission is actually greatly exaggerated—if it were drawn to true scale, it would be too small to be seen). The shape of the curve showing the intensity of energy emitted by Earth at different wavelengths (Figure 2–7b) is similar to that of the Sun, but the total energy released is much less, and the peak of the curve corresponds to a longer wavelength.

Of course, the amount of radiation emitted and its wavelengths are not the result of mere chance; they obey some fundamental physical laws. Strictly speaking, these laws apply only to perfect emitters of radiation, so-called **blackbodies**. Blackbodies are purely hypothetical bodies—they do not exist in nature—that emit the maximum possible radiation at every wavelength. Earth and the Sun are close to blackbodies and therefore nearly follow the laws described shortly. Other materials may or may not approximate blackbodies. In particular, the atmosphere, composed mainly of gases, is especially far from a blackbody, so we will not treat it as such.

Stefan-Boltzmann Law The single factor that determines how much energy a blackbody radiates is its temperature. Hotter bodies emit more energy than do cooler ones; thus, not surprisingly, a glowing piece of hot iron radiates more energy than an ice cube. Interestingly, though, the amount of radiation emitted by an object is

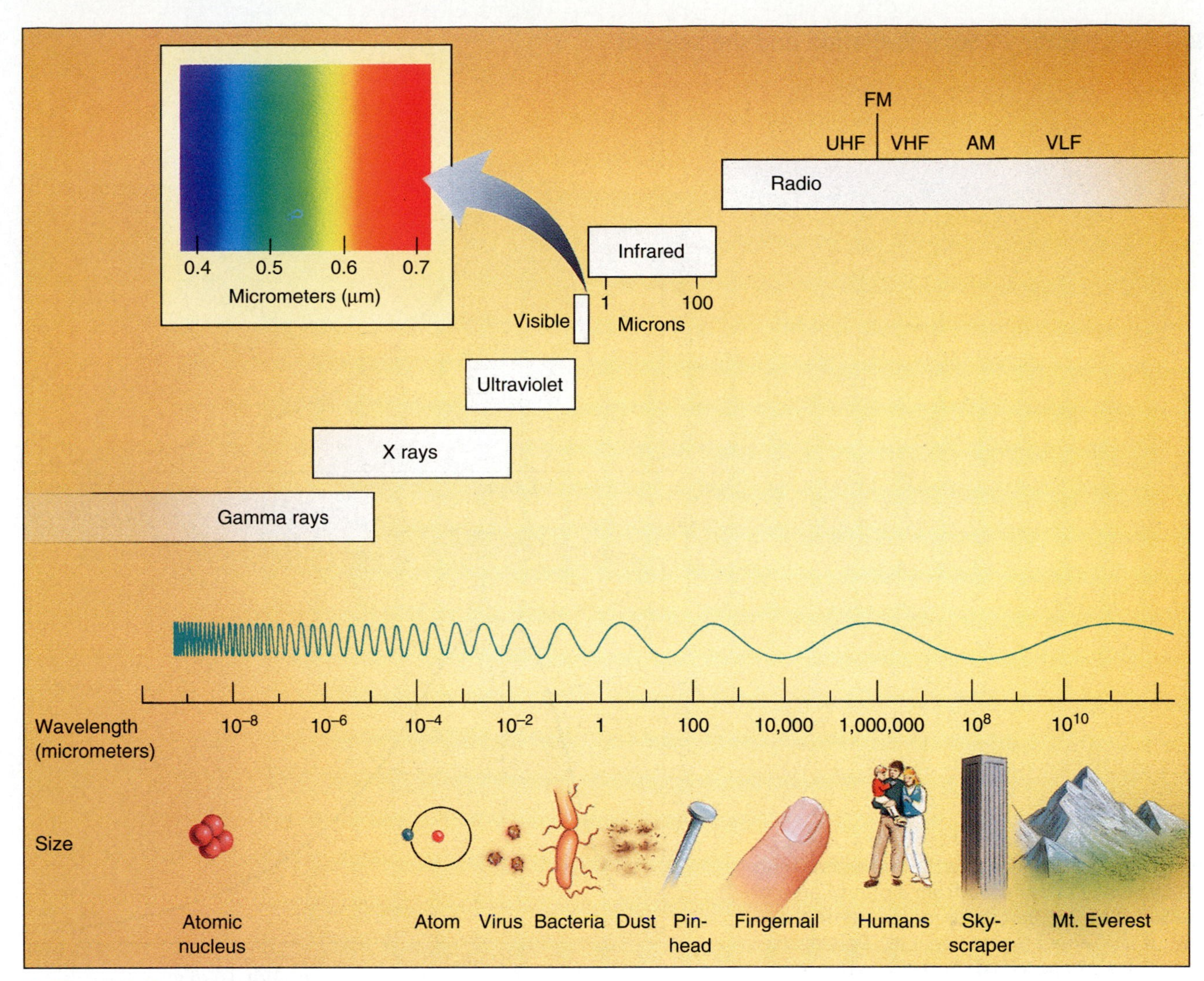

▲ **Figure 2–6**
Electromagnetic energy can be classified according to its wavelength.

Table 2–1 • Wavelength Categorizations

Type of Energy	Wavelength (micrometers)
Gamma	<0.0001
X ray	0.0001 to 0.01
Ultraviolet	0.01 to 0.4
Visible	0.4 to 0.7
Near Infrared (NIR)	0.7 to 4.0
Thermal Infrared	4 to 100
Microwave	100 to 1,000,000 (1 meter)
Radio	>1,000,000 (1 meter)

more than proportional to its temperature. In other words, a doubling of temperature produces *more* than a doubling of the amount of radiation emitted. Specifically, the intensity of energy radiated by a blackbody increases according to the fourth power of its absolute temperature. This relationship, the blackbody version of the **Stefan-Boltzmann law**, is expressed as

$$I = \sigma T^4$$

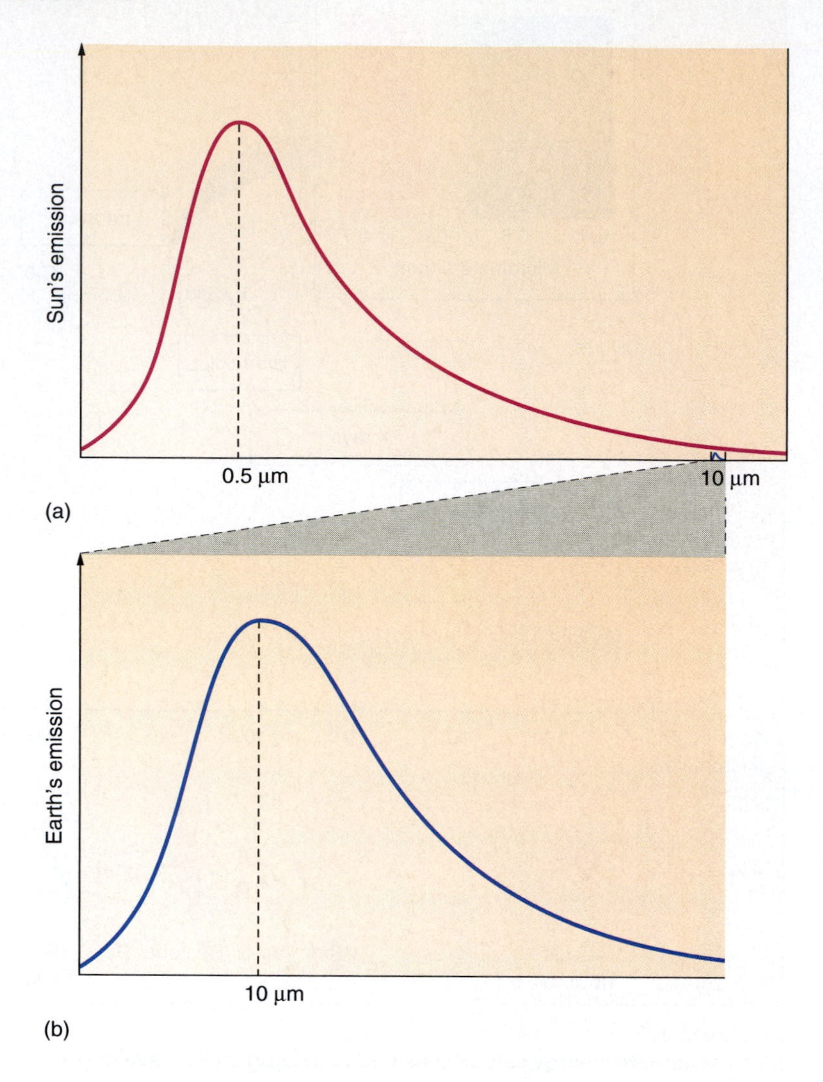

▶ **Figure 2–7**
Energy radiated by substances occurs over a wide range of wavelengths. Because of its higher temperature, emission from a unit of area of the Sun (a) is 160,000 times more intense than that of the same area on Earth (b). Solar radiation is also composed of shorter wavelengths than that emitted by Earth.

where I denotes the intensity of radiation in watts per square meter, σ (Greek lowercase sigma) is the Stefan-Boltzmann constant (5.67×10^{-8} watts per square meter per K^4), and T is the temperature of the body in kelvins (see *Box 2–1, Physical Principles: The Three Temperature Scales*).

Because the intensity of radiation depends on the temperature raised to the fourth power, a doubling of temperature leads to a sixteenfold increase in emission. Solving the Stefan-Boltzmann equation using the mean temperature of Earth's surface (about 290 K, 15 °C, or 59 °F) reveals that a square meter emits about 401 watts of power. In contrast, the surface of the Sun, with its temperature of about 6000 K (5700 °C, or 10,300 °F), emits about 73 million watts per square meter.

Although true blackbodies do not exist in nature, they provide a useful model for understanding the maximum amount of radiation that can be emitted. Most liquids and solids can be treated as **graybodies**, meaning that they emit some percentage of the maximum amount of radiation possible at a given temperature. Whereas some substances (for example, water) are highly efficient at radiating energy, others (for example, aluminum) are less efficient. The percentage of energy radiated by a substance relative to that of a blackbody is referred to as its **emissivity**. Emissivities range from just above zero to just below 100 percent and are denoted by the Greek letter epsilon (ε). By incorporating the emissivity of any body, we derive the complete version of the Stefan-Boltzmann law:

$$I = \varepsilon \sigma T^4$$

The inclusion of the emissivity factor means that the electromagnetic energy emitted by any graybody will be some fraction of what would be emitted by a blackbody. Note that even though the graybody form of the Stefan-Boltzmann law

2–1 Physical Principles

The Three Temperature Scales

We are all familiar with the Fahrenheit and Celsius (or centigrade) scales. Though useful in everyday applications, both of these scales have a serious flaw—they allow for negative values. This is not the case for other units of measurement. For example, buildings don't have negative heights and weights, cars don't travel at negative speeds, and children don't have negative ages. But the existence of negative temperatures gives the impression that substances can have negative heat contents—a situation that is physically impossible. To overcome this problem, scientists use a different scale for the measurement of temperature, called the **Kelvin scale**. In this system, the temperature 0 K is the lowest possible temperature that can exist in the universe (even in Wisconsin!). (Notice that we omit the degree notation with this scale and just refer to the number of kelvins,* K). A temperature of 0 K implies that no molecular vibration is occurring, and it is therefore impossible for subzero temperatures to exist with this scale.

Fahrenheit and Celsius

At one time, the temperature scale used all over the world was the Fahrenheit scale. Invented in the early 1700s by Gabriel Fahrenheit, it assigns values of 32° and 212° to the freezing and boiling points of water.† There are thus 180 Fahrenheit degrees between freezing and boiling. Although the Fahrenheit scale has been replaced in Canada and nearly every other country around the world, in the United States it is still the scale used by the general public.

The other familiar scale for the measurement of temperature is the Celsius scale, named for Anders Celsius, who formulated it in 1742. The Celsius scale assigns values of 0° and 100° to the freezing and boiling points of water, so there are only 100 Celsius degrees between the two points. This means that a Celsius degree is larger than a Fahrenheit degree. Thus, for example, a 2 °C change is larger than a 2 °F change. However, this is *not* to say a temperature expressed in °C is always higher than the same temperature expressed in °F; it can be higher or lower. To convert from Celsius to Fahrenheit, we use the following formula:

$$°F = 9/5\,°C + 32$$

To convert from Fahrenheit to Celsius, we use

$$°C = 5/9(°F - 32)$$

You can use these formulas to verify that −40 °F = −40 °C.

Kelvin

The Kelvin scale is really a modified form of the Celsius scale insofar as the increments of the two are equal. Thus, if the temperature increases 1 degree Celsius, it also increases 1 kelvin. The only difference between the two is the starting point; 0 K corresponds to −273.16 °C. Therefore, conversion from Celsius to Kelvin is simply

$$K = °C + 273.16$$

To convert from Kelvin to Celsius, we use

$$°C = K - 273.16$$

Figure 1 shows the Kelvin, Celsius, and Fahrenheit scales.

It is vitally important to use Kelvin units in the radiation formulas presented in this chapter. The use of Fahrenheit or Celsius temperatures would result in erroneous or, at worst, nonsensical results.

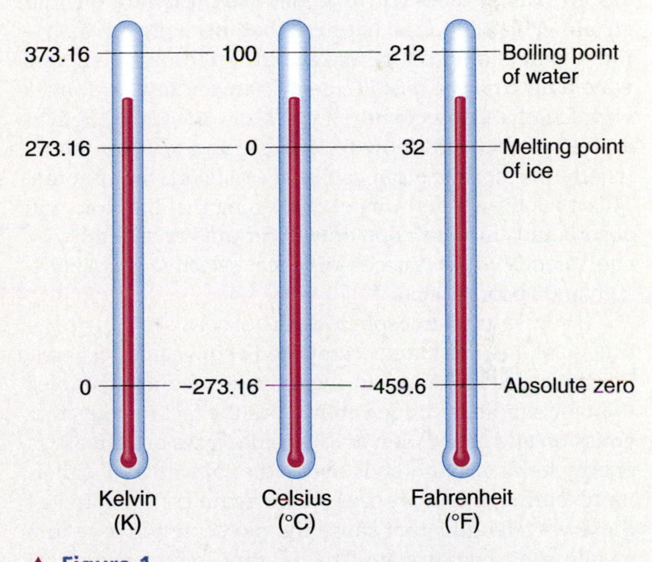

▲ **Figure 1**
The Kelvin, Celsius, and Fahrenheit scales.

*The Kelvin scale is spelled with a capital K; the unit of measurement is spelled with a lowercase k.

†To be precise, these values apply to pure water at sea level. In addition, water doesn't freeze spontaneously at 32 °F, so a better term is "melting point" rather than "freezing point."

shows radiation intensity to be a function of both emissivity and temperature, most natural surfaces have emissivities above 0.9. In most cases, therefore, differences in emission are governed by temperature differences. The atmosphere is an exception to this rule, because emission depends on a number of factors, such as the amount of water vapor in the air. Moreover, for a gas there is tremendous variability in emission with wavelength (see *Box 2–2, Physical Principles: The Nature of Radiation, Absorption, and Emission*); therefore, it's not accurate to think of atmospheric emission as just scaled-down blackbody emission. In other words, the concept of emissivity loses meaning when applied to the atmosphere. The bottom

2–2 Physical Principles

The Nature of Radiation, Absorption, and Emission

We commonly describe electromagnetic energy traveling through space as a sequence of waves, but in some contexts it behaves as a stream of particles. The particle nature of radiation applies at the smallest scale of observation, as when visible light is emitted by a single atom or molecule. When light is emitted, there is a change in the orbital characteristics of the electrons in the emitter. As the orbit changes, a small bundle of energy, called a **photon**, is released.

Using the simplest example, imagine a hydrogen atom (with one proton and a single orbiting electron). The electron is not free to assume just any orbital distance from the nucleus. Instead, it is confined to fixed orbital distances, called *shells*. Each shell is associated with a given energy level; the greater the distance from the nucleus, the greater the energy level. As shown in Figure 1, when sufficient energy is absorbed by the hydrogen atom, it can become "excited," and its electron jumps from its "ground state" to a higher shell. Similarly, if the electron jumps back to its previous energy level, it gives off energy in the form of a photon. Because only a few discrete shells exist, only certain energy changes are possible. This means that a photon emitted by the atom can contain only certain discrete amounts of energy, corresponding to the atom's decrease in energy.

Similarly, an atom is restricted in what photons it can absorb, namely those with energies that push the atom into an allowable state. It so happens that the energy of a photon depends only on its wavelength: photons at shorter wavelengths have more energy than photons at longer wavelengths. Consequently, if you know the wavelength of a photon, you can know its energy, and you can know whether or not the atom can emit or absorb that photon. This is a long-winded way of explaining that the atom will absorb and emit radiation only at certain wavelengths, or, equivalently, that it is necessarily very selective about emission and absorption.

Of course, our atmosphere does not consist of single hydrogen atoms; it is mainly composed of molecules of gases. For these gases, changes in energy level are more complex than for simple hydrogen atoms, but the fact remains that emission and absorption involve a decrease or increase in energy level as a photon is released or absorbed. Furthermore, emission and absorption are again confined to just those wavelengths that cause the molecule to move into an allowable energy state. That is, atmospheric gases, just like hydrogen, are selective absorbers and emitters. This is not true for liquids and solids, which tend to emit and absorb a wide range of wavelengths.

One very important consequence of all of this is that the atmosphere and surface respond differently to radiation. In addition, the atmosphere responds quite differently to radiation of various wavelengths. As will be seen later, these basic principles go a long way toward explaining Earth's climate and climate change.

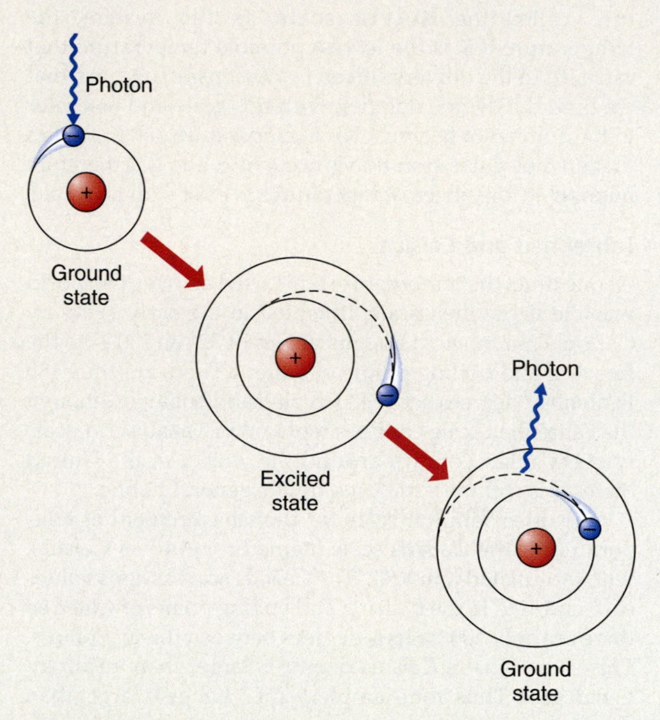

▲ **Figure 1**
Electrons orbit the nucleus of atoms in prescribed zones called *shells*. This figure depicts a single electron orbiting the nucleus of a hydrogen atom. Upon receiving energy, the electron is in an excited state and jumps to its next shell. When the electron returns to its ground state, it releases energy in the form of a photon. Note that the energy emitted by such atoms must occur in discrete packets; at the atomic scale, units of energy are divided into individual parcels.

line, however, is that the atmosphere is certainly not a perfect emitter of radiation, and therefore emits less radiation at any particular temperature than would a blackbody.

Wien's Law As we saw in 2–7, the radiation emitted by the Sun and Earth (or any other body) is not of a single wavelength, nor are all wavelengths emitted in equal amounts. In the case of the Sun, the wavelength emitted more than any other is 0.5 μm, while energy radiated by Earth peaks at about 10 μm. For any radiating body, the wavelength of peak emission (in micrometers) is given by Wien's (pronounced "weens") law:

$$\lambda_{max} = \text{constant}/T$$

where λ_{max} refers to the wavelength of energy radiated with greatest intensity. More specifically, the constant in the preceding equation rounds off to the value 2900 for T in Kelvins and λ_{max} in micrometers. Thus, we can determine the peak wavelength of radiation emitted as:

$$\lambda_{max} = 2900/T$$

Wien's law tells us that hotter objects radiate energy at shorter wavelengths than do cooler bodies. This is not surprising given that shorter wavelengths correspond to higher energies. Hot objects, possessing more thermal energy, necessarily radiate a higher proportion of energy at those shorter, more energetic wavelengths. Thus, for example, solar radiation is most intense in the visible portion of the spectrum, though it emits over a wide range of wavelengths. Most of the radiation has wavelengths less than 4 μm, which we generically refer to as **shortwave radiation**. Of the radiation emitted by the Sun, about 46.5 percent is near and thermal infrared, 46.8 percent is visible light, and 6.7 percent is ultraviolet.

Radiation emanating from Earth's surface and atmosphere consists mainly of that having wavelengths longer than 4 μm. This type of electromagnetic energy is called **longwave radiation**.

As shown in Figure 2–7, hotter bodies radiate more energy than do cooler bodies *at all wavelengths*. For example, the Sun radiates energy with wavelengths of about 0.5 μm most effectively and puts out far less energy at $\lambda = 10$ μm. Nevertheless, the Sun emits more radiation at those wavelengths than does Earth, despite the fact that Earth emits virtually nothing but longwave radiation (see *Box 2–3, Physical Principles: The Sun*).

The Stephan-Boltzmann law and Wien's law have some very useful and interesting applications. You have undoubtedly seen color-enhanced satellite images showing the distribution of clouds across North America, such as the one in Figure 2–8. This type of image depicts the height of cloud tops, which can be used as an indicator of the intensity of precipitation occurring below. The images are obtained by measuring the intensity of infrared radiation emitted by the cloud tops. Colder surfaces radiate less intense energy than do warm bodies. Weather satellites measure radiation intensity to determine the cloud top temperatures across a target region. Because higher clouds tend to be colder than lower level clouds (remember that in the troposphere temperature tends to decrease with altitude), temperatures

Figure 2–8
A color-enhanced satellite image of North America. Cloud-top heights are inferred by the application of radiation laws.

2–3 Physical Principles

The Sun

The Sun may seem special to us, but compared to the 100 billion or more other stars in our galaxy it is not particularly unique. Although stars vary considerably in size, temperature, brightness, and density, the Sun is about average in terms of these characteristics. Obviously, we have no first-hand observations of the solar interior, but based on physical principles we can infer the processes that go on within it. Using this information, we can divide the Sun into three sections (Figure 1).

The Core and Interior

In the innermost portion of the Sun, the **core**, extremely high temperatures (about 15 million °C, or 27 million °F) and high densities lead to the energy-generating process of **nuclear fusion**. In this reaction, hydrogen atoms combine under tremendous heat and pressure to form a smaller number of heavier helium atoms. A certain amount of mass is lost in the process, but radiant energy is released—the same energy that works its way to the solar surface, travels through space, and ultimately warms Earth. The amount of this energy is staggering. Try to imagine the explosion of 100 billion one-megaton hydrogen bombs—that is equivalent to the amount of energy released in the core *every second!*

The Photosphere

The energy produced in the core radiates through the interior of the Sun to the base of the **convection zone**, where upwelling of the solar gases transfers the energy to the relatively thin solar surface, called the **photosphere**. The photosphere is the layer of the Sun that radiates most of its emitted energy and is the portion we actually see as the **solar disk**. Although radiation travels from the photosphere to Earth in only about 8 minutes, the transfer of radiant energy within the Sun is incredibly slow. In fact, it takes about a million years for the energy unleashed in the core to travel to the base of the photosphere; thus, the energy reaching Earth is ancient.

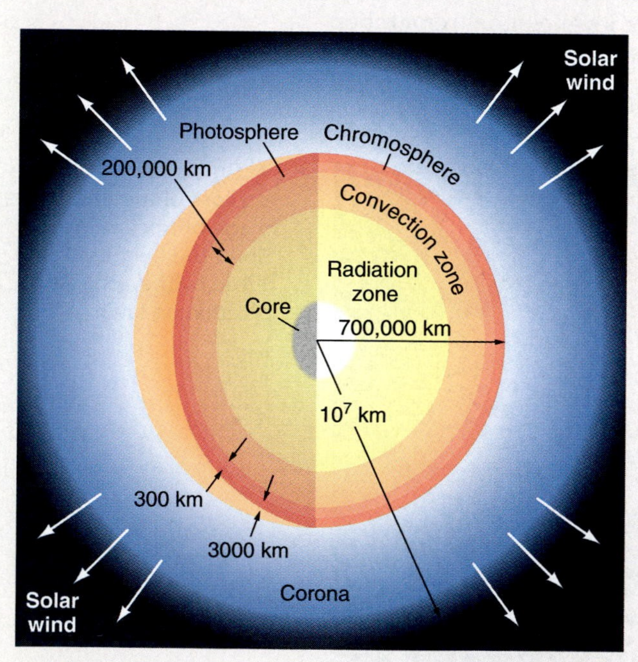

▲ **Figure 1**
Energy is produced in the core of the Sun by nuclear fusion. Within the Sun, the energy is radiated to the base of the convection zone, where mixing transfers the energy upward to the base of the photosphere (the layer of the Sun visible from Earth).

Both the temperature and the density of the photosphere increase with depth. Because of this, the edges of the Sun appear less bright than the center, a phenomenon referred to as **limb darkening** (Figure 2). The lower photosphere is denser and hotter than the upper portion, so it radiates considerably more energy.

Inspection of the photosphere with heavily filtered telescopes reveals that its outer layer is not a uniform, smooth surface. It is, instead, marked by a number of features of varying sizes and lifetimes.

Granules are the ever-present tops of convection cells that transport energy from the base of the photosphere to its surface. These features, analogous to bubbles in a pot of boiling water, are about 1000 km (600 mi) in diameter with life spans on the order of 5 to 10 minutes. At any given time, there are literally millions of these on the surface of the photosphere.

Sunspots (each lasting a few weeks or months) are dark regions on the photosphere with diameters of about 10,000 km (6,000 mi) and temperatures about 1500 °C cooler than the surrounding surface. They form in response to locally strong magnetic fields, a thousand times more intense than those of the surrounding photosphere, which block the upwelling of heat from below.

Records of sunspot activity have been maintained since the days of *Galileo Galilei* (1564–1642), who observed their apparent movement across the solar surface. We now know that sunspots remain fixed in place and appear to move because of the rotation of the Sun (which takes about 24 days and 16 hours to complete one turn of its axis). The number of sunspots tends to peak every 11 years (Figure 3). Although the cycle is usually well defined, long episodes of minimal or unusually high sunspot activity have appeared during historic times. Figure 3, for example, shows a long period of reduced sunspot activity during the seventeenth century. (This gap does not appear in all records.) For many years, scientists have speculated about the possible role of sunspot activity in climate changes on Earth, but these associations have generally been weak and tentative. Because there is no known cause-and-effect mechanism between sunspots and climate, we should be very skeptical about attempts to link them.

Perhaps the most spectacular of solar disturbances are **flares**, intensely hot flashes (perhaps 100 million °C) across the photosphere surface due to magnetic instabilities. Temperatures within flares can achieve a staggering 100,000,000 K, and these features are so explosive that they have been likened to incredibly large bombs exploding on the solar surface. Although they exist for only a matter of minutes, they release a huge amount of energy, particularly in the form of X-ray and ultraviolet radiation.

Chromosphere and Corona

Above the photosphere are the **chromosphere** and the **corona**, the lower and upper parts, respectively, of the

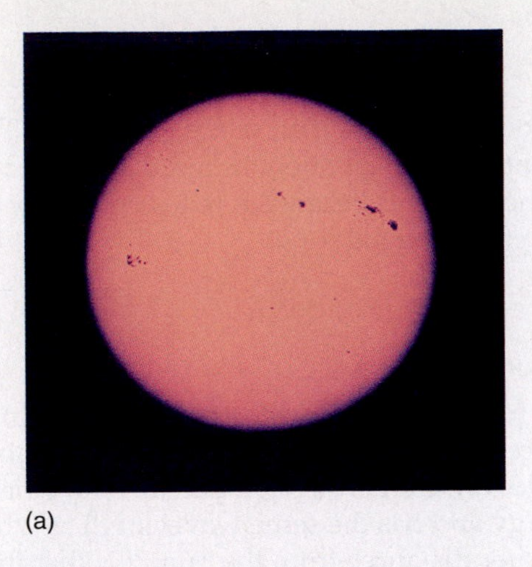

(a)

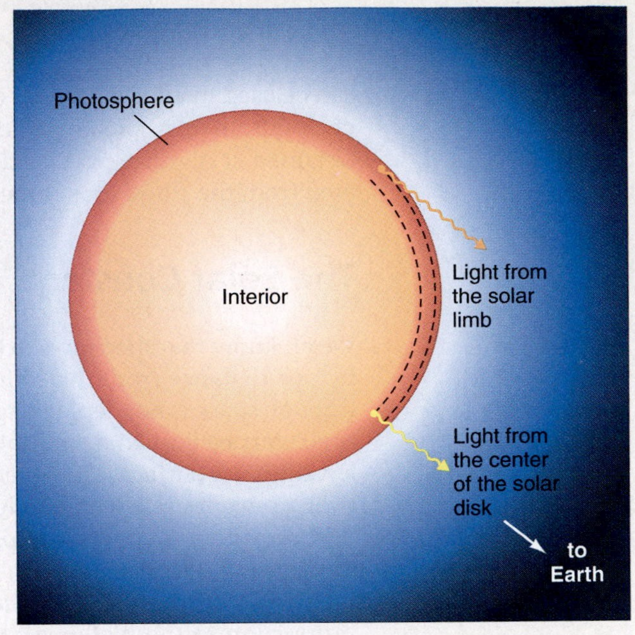

(b)

▲ **Figure 2**
(a) Limb darkening occurs along the margin of the Sun because the density and temperature of the photosphere both increase with depth. (b) Because of limb darkening, the outermost portion of the solar disk radiates less energy and appears darker than the inner portion.

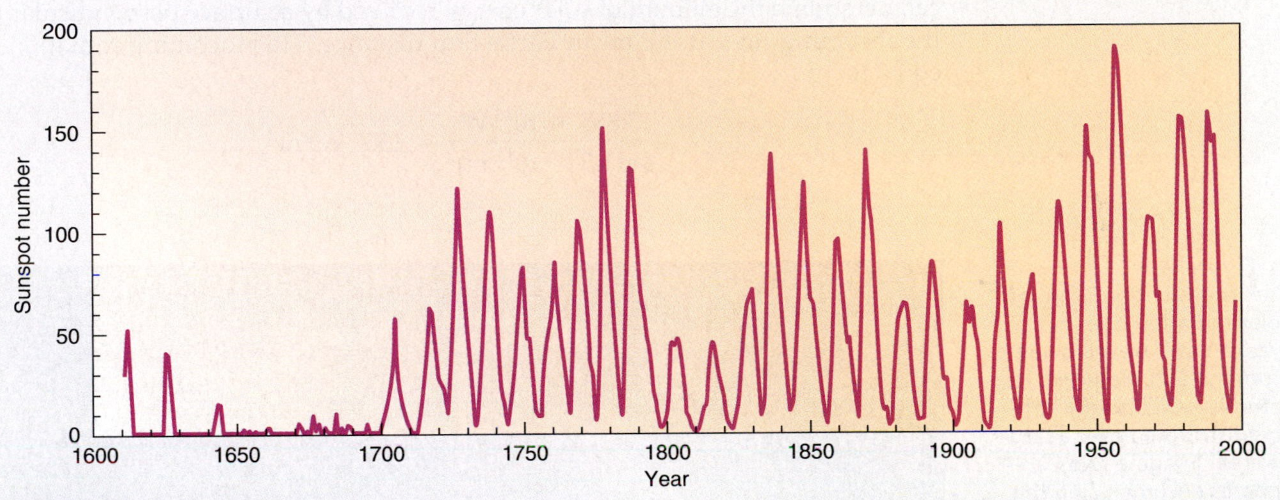

▲ **Figure 3**
Sunspots appear in a somewhat regular manner, with peak occurrences normally observed about every 11 years. Note in this long record, however, that the number of sunspots during the episodes of peak activity varies considerably. Much of the seventeenth century, for example, was a period marked by minimal sunspot activity.

Sun's atmosphere. Against the intensely bright background of the photosphere, the chromosphere is normally not visible. During solar eclipses, however, the moon blocks out the brighter photosphere, so the chromosphere is easily seen. Furthermore, if the eclipse is such that the moon is able to obscure both the photosphere and the chromosphere, the corona will appear, as in Figure 4.

In addition to radiating electromagnetic energy, the corona emits protons, electrons, and other subatomic particles, collectively referred to as the **solar wind**. These particles can be captured by Earth's magnetic field, where they can interact with gases in the ionosphere to produce the aurora (see Figure 1–12). Unusually heavy episodes of solar wind activity can interfere with normal radio and television transmissions.

▲ **Figure 4**
The solar atmosphere, consisting of the chromosphere and corona, is visible only during a total eclipse. In this image, the moon obscures the photosphere and the chromosphere so that only the corona is visible.

can be used to infer the height of the clouds, and thus their relative thickness. Thicker clouds, in turn, usually yield more intense precipitation. Infrared imagery can be obtained at night, as well as during the daytime, because it relies on energy radiated from the cloud tops rather than reflected light.

The Solar Constant

We all know that the Sun is extremely hot and we are protected from its great heat by our distance from the solar surface. But the electromagnetic energy moving through space is not depleted as it moves toward Earth. Radiation traveling through space carries the same amount of energy and has the same wavelength as when it left the solar surface. However, at greater distances from the Sun, it is distributed over a greater area, which reduces its intensity.

Consider a sphere completely surrounding the Sun, whose radius is equal to the mean distance between Earth and the Sun, or 1.5×10^{11} m (Figure 2–9). As the distance from the Sun increases, the intensity of the radiation diminishes in proportion to the distance squared. This relationship is known as the **inverse square law**. By dividing total solar emission (3.865×10^{26} W) by the area of our imaginary sphere surrounding the Sun (the area of any sphere is given as $4\pi r^2$), we can determine the amount of solar energy received by a surface perpendicular to the incoming rays at the mean Earth–Sun distance. This incoming radiation is equal to

$$\frac{3.865 \times 10^{26}\ \mathrm{W}}{4\pi(1.5 \times 10^{11}\ \mathrm{m})^2} = 1367\ \mathrm{W/m^2}$$

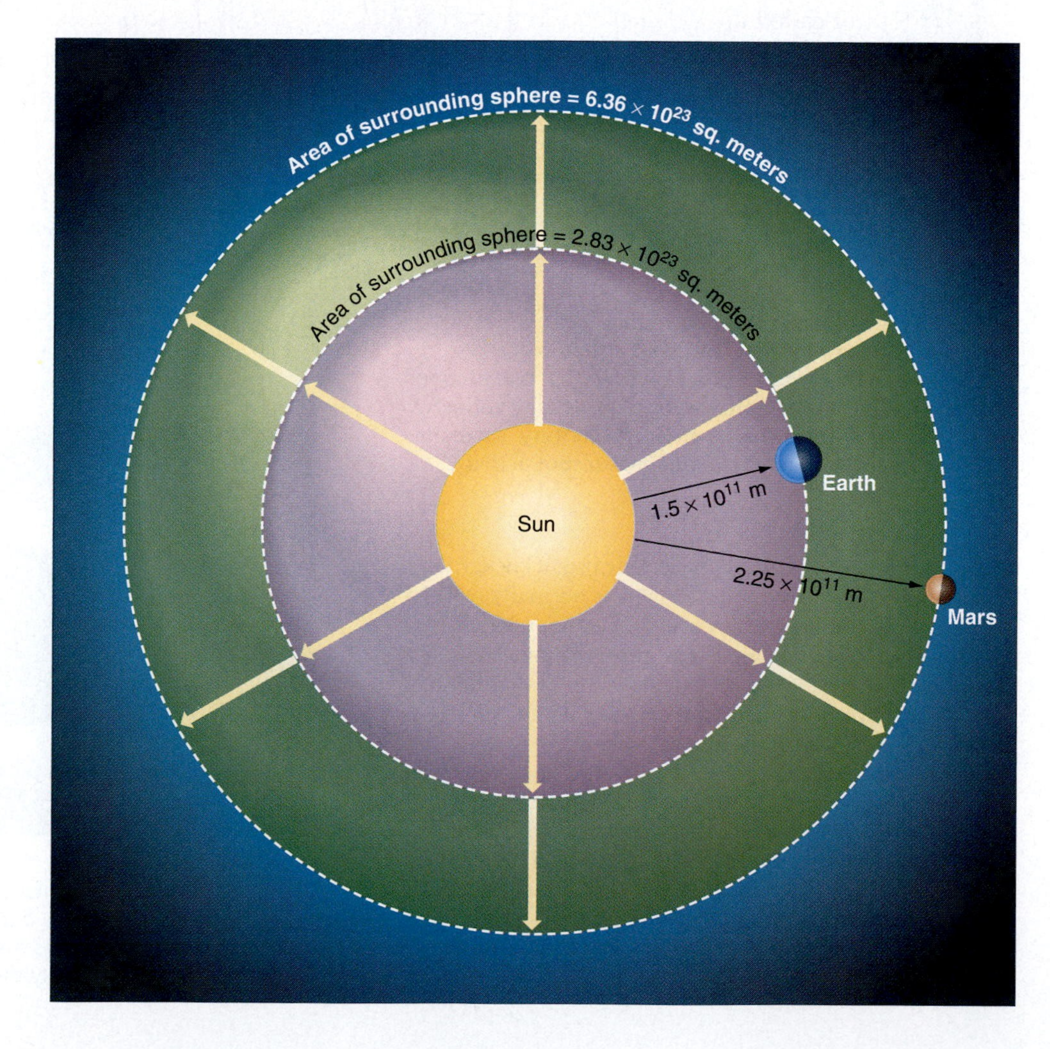

▶ **Figure 2–9**
The intensity of a beam of solar radiation does not weaken as it travels away from the Sun. However, its intensity is reduced when it is distributed over a larger area. Imagine two spheres encompassing the Sun (such as the one with a radius equal to the mean Earth–Sun distance). All the radiation emitted from the Sun would be captured by this surrounding sphere. Now imagine that the surrounding sphere has a radius equal to the mean distance between the Sun and Mars. This sphere is larger than the previous one, so the energy emitted must be distributed over a greater area.

We refer to the value 1367 W/m^2 as the **solar constant** (although minor variations in solar output and other factors allow for some minor departures from this "constant"). For the sake of comparison, using the same procedure we can determine that the solar constant for Mars (2.25×10^{11} m from the Sun) is 445 watts per square meter.

The Causes of Earth's Seasons

Solar Geometry
Section 1.2.1

Although the Sun emits a nearly constant amount of radiation, on Earth we experience significant changes in the amount of radiation received during the course of a year. These variations in energy manifest themselves as the seasons. We also know that the low latitudes (for example, the Tropics and subtropics) receive more solar radiation per year at the top of the atmosphere than do regions at higher latitudes (for example, the Arctic and Antarctic). In this section, we will discuss how Earth's orbit around the Sun and its orientation with respect to incoming radiation influence the seasonal and latitudinal receipt of incoming solar radiation (often called **insolation**, for short).

Earth's Revolution and Rotation

As we know, and as Figure 2–10 shows, Earth orbits the Sun once every 365¼ days as if it were riding along a flat plane. We refer to this imaginary surface as the **ecliptic plane** and to Earth's annual trip about the plane as its **revolution**.

A Solar Eclipse

The orbit is not quite circular but instead sweeps out an elliptical path, so that the distance between Earth and the Sun varies over the course of the year. Earth is nearest the Sun—at a point called **perihelion**—on or about January 3, when Earth–Sun distance is about 147,000,000 km (91,000,000 mi). Earth is farthest from the Sun—at a point called **aphelion**—on or about July 3, when Earth–Sun distance is about 152,000,000 km (94,000,000 mi). Thus, on perihelion Earth is 3 percent closer to the Sun than on aphelion. But because the intensity of incoming radiation varies inversely with the square of Earth–Sun distance (recall the inverse square law), the radiation is almost 7 percent more intense. (As mentioned at the outset of this chapter, however, this variation in intensity is not what causes the change of seasons.)

In addition to its revolution, Earth also undergoes a spinning motion called **rotation**. Rotation occurs every 24 hours around an imaginary line called Earth's axis, connecting the North and South Poles. The axis is not perpendicular to the plane of the orbit of Earth around the Sun but is tilted 23.5° from it. Moreover, no matter what time of year it is, the axis is always tilted in the same direction and always points to a distant star called **Polaris** (the North Star). The constant direction of the tilt means that for half the year the Northern Hemisphere is oriented somewhat toward the Sun, and for half the year it is directed away from the Sun. The changing orientation of the hemispheres with regard to the Sun is the true cause of the seasons—not the varying distance between Earth and the Sun.

It is easier to visualize how the tilt of the axis influences the seasons if we consider a hypothetical situation in which the axis is tilted not 23.5°, but rather a full 90°, as depicted in Figure 2–11. Actually, this is the case for Uranus, so what we describe is not entirely hypothetical. Examine the situation when Earth is in position #1. The

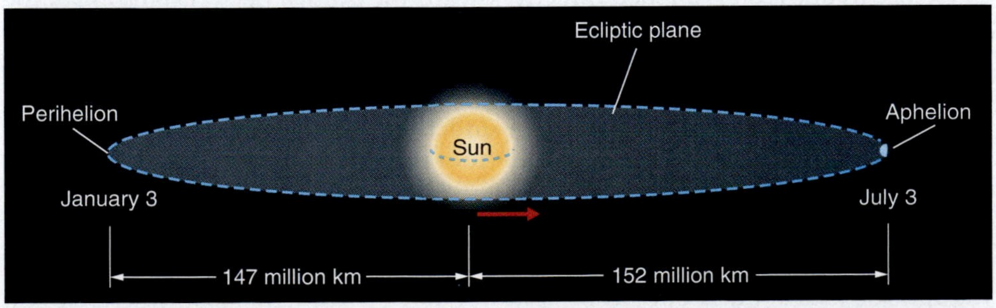

◀ **Figure 2–10** Earth's orbit around the Sun is not perfectly circular but is an ellipse. Earth is nearest to the Sun (perihelion) on about January 3 and farthest away (aphelion) on July 3.

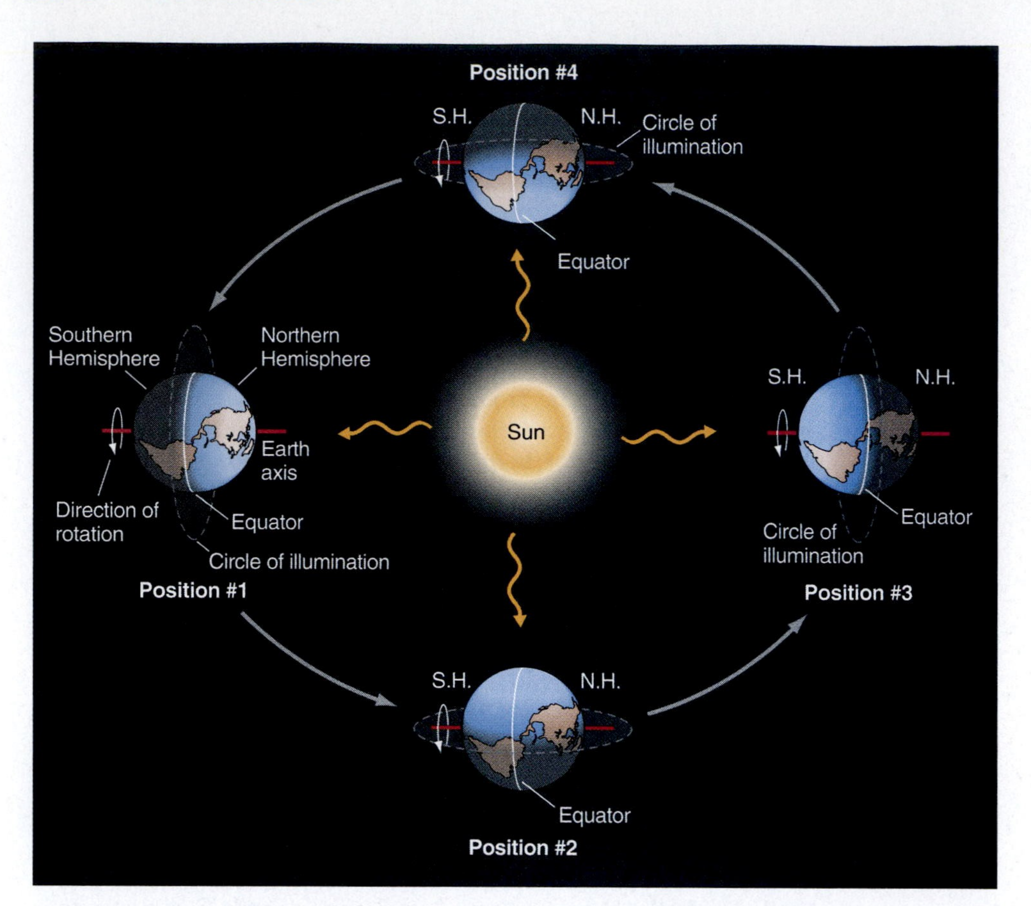

▶ **Figure 2–11**
A hypothetical situation wherein Earth's axis is aligned along the ecliptic plane. In position #1, the Northern Hemisphere receives much energy from the Sun while the Southern Hemisphere is in constant darkness. The situation reverses six months later (position #3). In positions #2 and #4, both hemispheres receive equal amounts of solar energy.

Northern Hemisphere is oriented directly toward the Sun so that it is fully illuminated over the entire 24-hour period of rotation. Meanwhile, the Southern Hemisphere undergoes 24 hours of continual darkness. This situation favors greater warmth in the Northern than in the Southern Hemisphere. Furthermore, a person standing at the North Pole would observe the Sun as being directly overhead during the entire day. Moving away from the North Pole toward the equator, there is a gradual reduction in the angle of the Sun above the horizon, until at the equator the Sun appears to be right on the horizon. South of that line, the Sun is below the horizon and nighttime covers the Southern Hemisphere.

Now refer to position #3, which occurs 6 months after position #1. In this situation, the Southern Hemisphere is in continual sunlight while the Northern Hemisphere is subjected to 24 hours of darkness. Furthermore, someone standing at the South Pole would see the Sun directly overhead, and the apparent position of the Sun would shift toward the horizon for viewers located closer to the equator.

Finally, observe the intermediate positions, #2 and #4. In these two situations, the 90° tilt of the axis is neither toward nor away from the Sun, and the tilt becomes irrelevant to the receipt of insolation. Moreover, in positions #2 and #4 every place on Earth receives 12 hours of daylight and 12 hours of darkness because every latitude is half sunlit and half dark. Finally, note that at noon (when the longitude of any place in question is aligned directly toward the Sun), a person standing at the equator would observe the Sun to be directly overhead. Thus, Earth's revolution causes seasonal changes in the amount of heating of the surface. When either the Northern or Southern Hemisphere is oriented toward the Sun, that hemisphere receives a greater amount of insolation and therefore warms more effectively. Whichever hemisphere is oriented away from the Sun receives less radiation.

Solar Geometry
Section 1.3.1

Solstices and Equinoxes Figure 2–12 shows the true seasonal change in orientation of Earth with respect to the Sun, based on the actual 23.5° tilt of the axis. Although the axis is tilted only 23.5°, and not 90°, the principle just described still

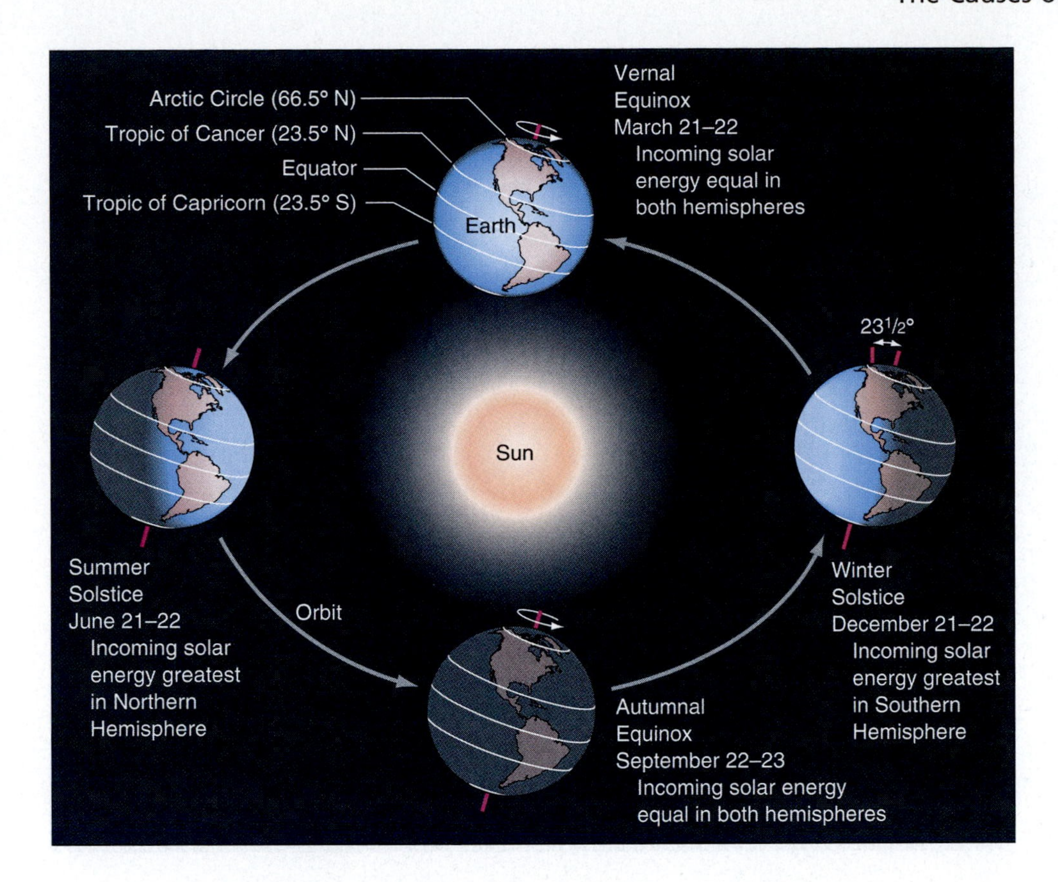

◀ **Figure 2–12** Earth's revolution around the Sun.

applies. During 6 months of the year, the Northern Hemisphere obtains more sunlight than does the Southern Hemisphere; during the other 6 months, the Southern Hemisphere receives a greater amount of insolation. The four positions shown in the diagram represent 4 days that have particular significance.

In the farthest left position in Figure 2–12, the Northern Hemisphere has its maximum tilt toward the Sun. This occurs on or about June 21, which we refer to as the **June solstice** (this is also called the *summer solstice* according to the corresponding Northern Hemisphere season). Although we designate this as the first day of summer, it actually represents the day on which the Northern Hemisphere has its greatest availability of insolation. Six months later (on or about December 21), the Northern Hemisphere has its minimum availability of solar radiation on the **December solstice** (*winter solstice* in the Northern Hemisphere), which is called the first day of winter in the Northern Hemisphere and the first day of summer in the Southern Hemisphere. Intermediate between the two solstices are the **March equinox** (often called the *vernal* or *spring equinox* for the Northern Hemisphere) on or about March 21, and the **September equinox** (called the *autumnal equinox* in the Northern Hemisphere) on or about September 21. On the equinoxes, every place on Earth has 12 hours of day and night (the word *equinox* refers to "equal night"), and both hemispheres receive equal amounts of energy.

Of course, the transitions between the four positions shown in Figure 2–12 do not occur in sudden leaps; instead, a steady progression occurs from one position to the next. As shown in Figure 2–13, the 23.5° tilt of the Northern Hemisphere toward the Sun on the June solstice causes the *subsolar point* (the point on Earth where the Sun's rays meet the surface at a right angle—and where the Sun appears directly overhead) to be located at 23.5° N. This is the most northward latitude at which the subsolar point is located. The fact that the Sun never appears directly overhead poleward of 23.5° N gives that latitude special significance, and we call it the **Tropic of Cancer**.

Likewise, on the December solstice, the sun is directly overhead at 23.5° S, the **Tropic of Capricorn**. On the two equinoxes, the subsolar point is on the equator. Thus, the subsolar point migrates 47° (i.e., between 23.5° N and 23.5° S) over a

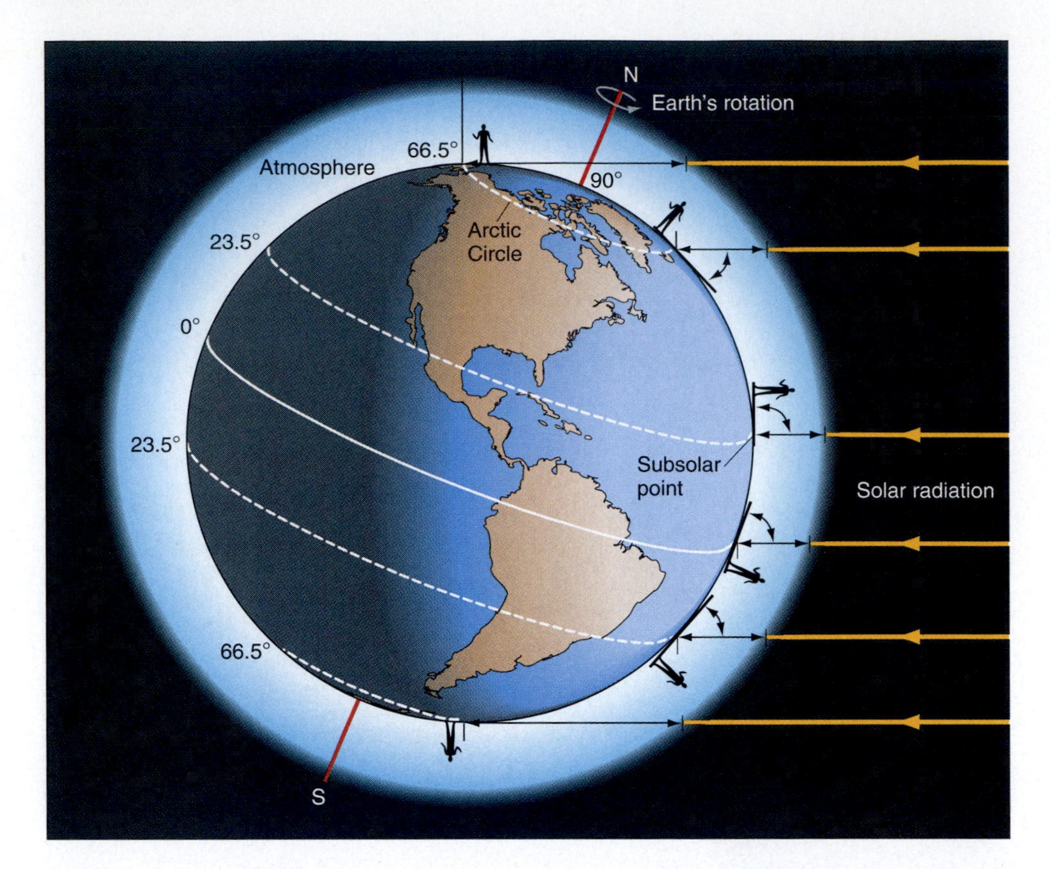

▶ **Figure 2–13**
Because Earth's axis is tilted 23.5°, the subsolar point is at 23.5° N during the summer solstice.

6-month period, and on any particular day it is located somewhere between the Tropics of Cancer and Capricorn. This seasonal movement of the subsolar point is similar to what would happen if instead of Earth orbiting the Sun, its axis were to slowly rock back and forth, toward and away from the Sun.

The latitudinal position of the subsolar point is the **solar declination**, which can be visualized as the latitude at which the noontime Sun appears directly overhead. Figure 2–14 plots the solar declination for several days of the year, while the arrows between the dates indicate the direction toward which the declination is moving at that time of year.

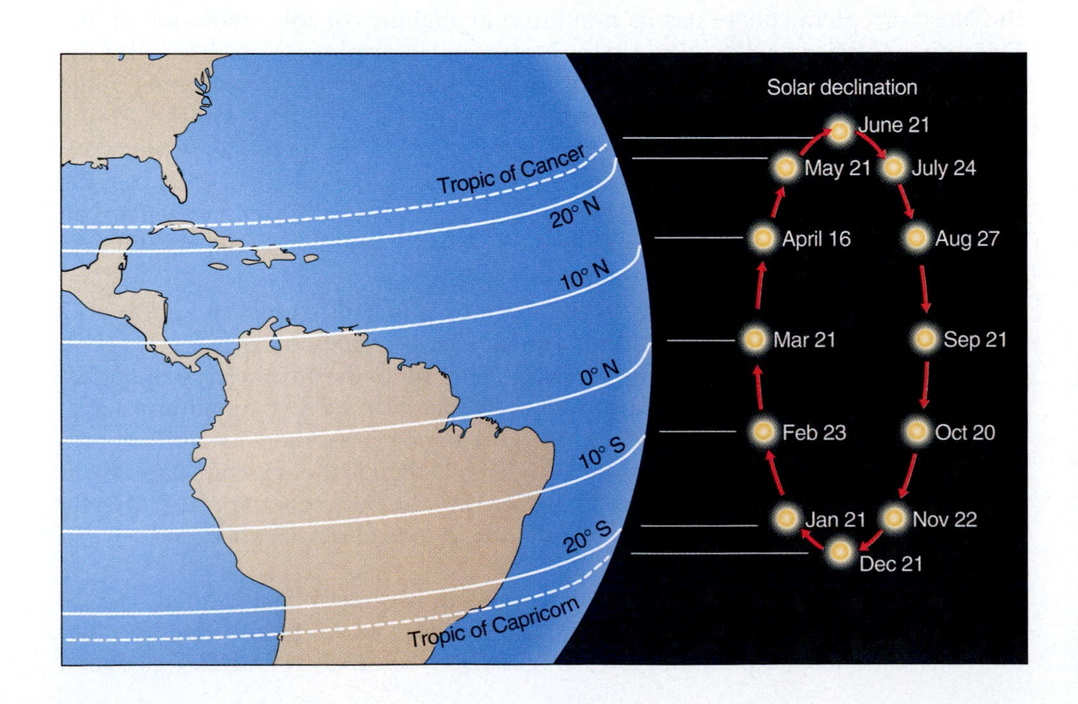

▶ **Figure 2–14**
The solar declination gradually migrates north and south over the course of the year. On the summer solstice (June 21), the subsolar point marks its most northward extent, 23.5° N. Solar declination is 23.5° S on the winter solstice (December 21).

Now we are ready to see how the changing orientation of Earth with respect to the Sun directly affects the receipt of insolation through three mechanisms: (1) the length of the period of daylight during each 24-hour period, (2) the angle at which sunlight hits the surface, and (3) the amount of atmosphere that insolation must penetrate before it can reach Earth's surface.

Period of Daylight One way the tilt of the axis influences energy receipt on Earth is by its effect on the lengths of day and night. We have already seen that if the axis were tilted 90° from the plane of Earth's orbit, there would be 1 day when the entire Northern Hemisphere underwent 24 hours of daylight and an equivalent period 6 months later of continuous darkness. But the axis is only tilted 23.5° from the plane of the orbit, so only the latitudes poleward of 66.5° (that is, 90° minus 23.5°) experience a 24-hour period of continuous daylight or night. These lines of latitude are the **Arctic Circle** (in the Northern Hemisphere) and the **Antarctic Circle** (in the Southern Hemisphere). This is illustrated in Figure 2–15. On the June solstice, any place north of the Arctic Circle has 24 hours of daylight. Just a short distance south of the Arctic Circle, there is almost (but not quite) 24 hours of daylight. Moving toward the equator, the period of daylight decreases until reaching the equator, where the day and night are both 12 hours long. Moving into the Southern Hemisphere, day length shrinks until 66.5° S, where night is 24 hours long. The opposite pattern, of course, holds for the December solstice.

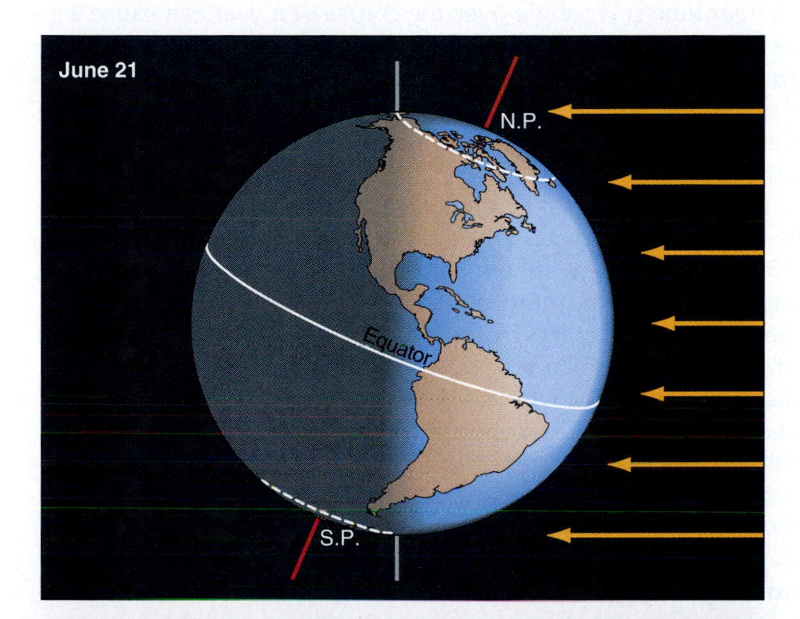

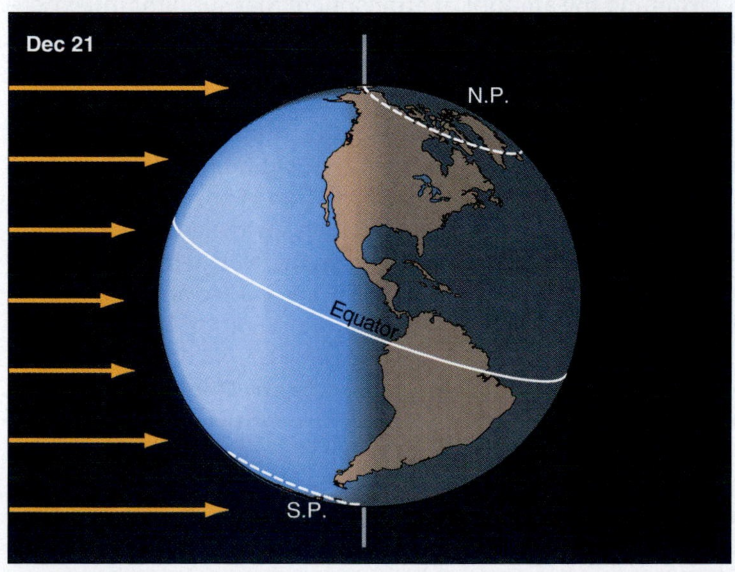

Figure 2–15
On the June solstice (a), every point north of 66.5° N has 24 hours of daylight and every point south of 66.5° S has continual night. During the December solstice (b), the situation is reversed. These latitudes are called the Arctic and Antarctic Circles, respectively.

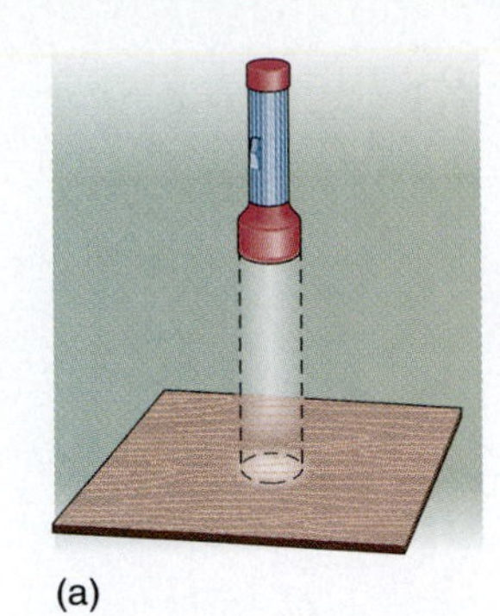
(a)

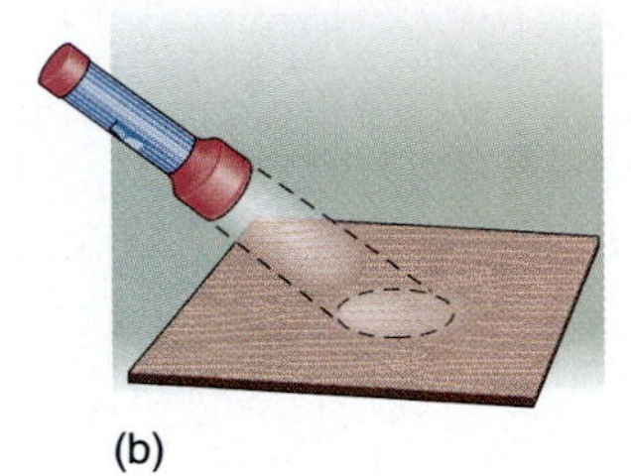
(b)

▲ Figure 2–16
The intensity of radiant energy hitting a surface is affected by its angle of incidence. More direct illumination causes a more intense amount of heating.

Solar Angle Think back to the last time you spent a few hours lying out in the Sun. If you went outside early in the day when the Sun was low above the horizon, you probably did not feel a great deal of warming from its rays. But as the Sun got higher in the sky, it became more effective at warming your body. This change was largely due to a decrease in beam spreading. **Beam spreading** is the increase in the surface area over which radiation is distributed in response to a decrease of solar angle, as illustrated in Figure 2–16. The greater the spreading, the less intense the radiation is. In part (a) of the figure, the incoming light is received at a 90° angle, which concentrates it to a small area and increases its ability to heat the surface. In (b), the rays hit the surface more obliquely and the energy is distributed over a greater area, leading to a less intense illumination (less energy per unit area). Thus, a beam of light is more effective at illuminating or warming a surface if it has a high angle of incidence.

Noontime sun angles for any given latitude can be easily determined if the solar declination is known. To do this, you simply subtract the latitude of a given location from 90° and then add the solar declination. Thus, on either of the equinoxes (solar declination = 0°) Toronto, Ontario (latitude 44° N), has a noontime solar angle of 90° – 44° = 46°. During the June solstice the noontime solar angle is 90° – 44° + 23.5° = 69.5°. Six months later the solar declination = –23.5°, with the negative sign indicating the sun is overhead in the Southern Hemisphere. At this time, the noontime solar angle = 90° – 44° – 23.5° = 22.5°.

The change in noontime solar angle over the course of a year can cause significant differences in the intensity of sunlight hitting the surface due to beam spreading. For example, beam spreading of the noontime sun at Toronto on the December solstice is almost two and a half times greater than on the June solstice. Thus, in the absence of other effects, the midday sun will be less than half as intense in the winter as in the summer.

Differences in the amount of beam spreading have a major role in causing the seasons. During the 6-month period between the March and September equinoxes, any latitude in the Northern Hemisphere has a more direct angle of incidence than does its Southern Hemisphere counterpart. Thus, insolation available to the Northern Hemisphere is subjected to less beam spreading, which promotes greater warming of the surface. During the following 6 months, the situation is reversed and the Southern Hemisphere has, on the whole, higher Sun angles.

Section 1.5.1

Figure 2–17 shows the combined effects of noontime solar angle and length of daylight period for the solstices and equinoxes. On the June solstice, both factors work together to enhance warming in the Northern Hemisphere; on the December solstice, they combine for less effective heating.

Atmospheric Beam Depletion The third way in which the tilt of the axis influences heating is in determining the amount of atmosphere that sunlight must penetrate before reaching the surface. As you can see in Figure 2–18a, insolation approaching the surface at a 90° angle passes through the atmosphere as directly as possible. Compare this to Figure 2–18b, in which sunlight approaches the surface at a low angle (as it does around sunrise or sunset). In this situation, a beam of sunlight must pass through a greater amount of atmosphere. Although the atmosphere is mostly transparent to incoming sunlight, some radiation is absorbed and even more is reflected back to space. The greater the thickness traveled, the more the beam is weakened. Because the solar altitude is lowest in the Northern Hemisphere near the December solstice, at that time more energy is lost due to atmospheric effects than at any other time of the year.

Overall Effects of Period of Daylight, Solar Angle, and Beam Depletion We can now summarize some of the important points regarding controls on energy reaching the surface. We must emphasize that differences in atmospheric transparency or cloudiness have not been considered at all; thus, the discussion implicitly assumes uniform optical properties. In other words, the patterns described below arise solely from geometrical considerations:

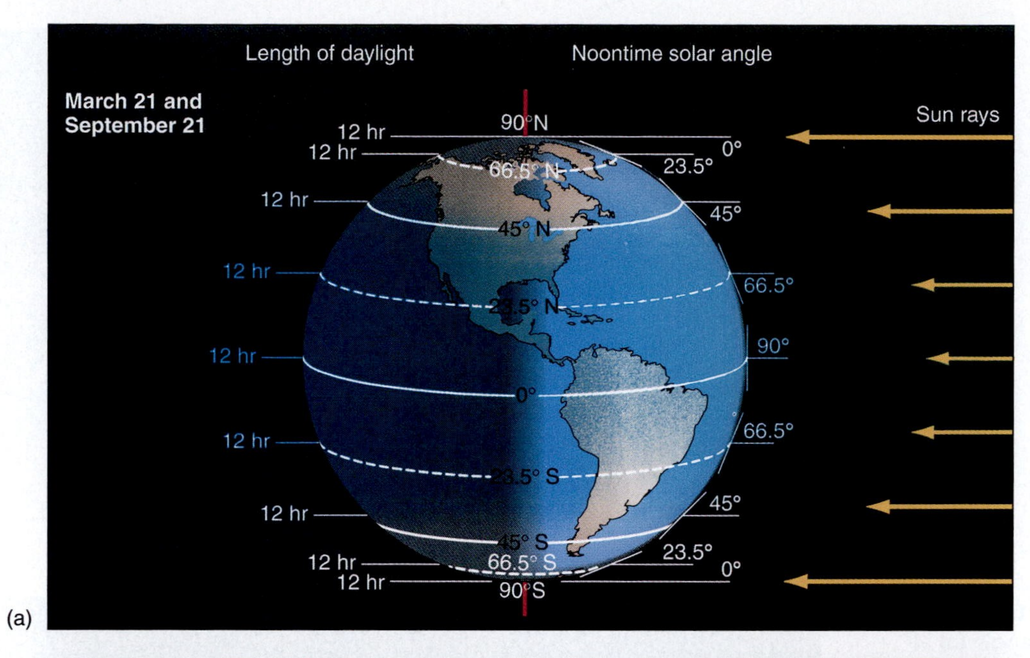

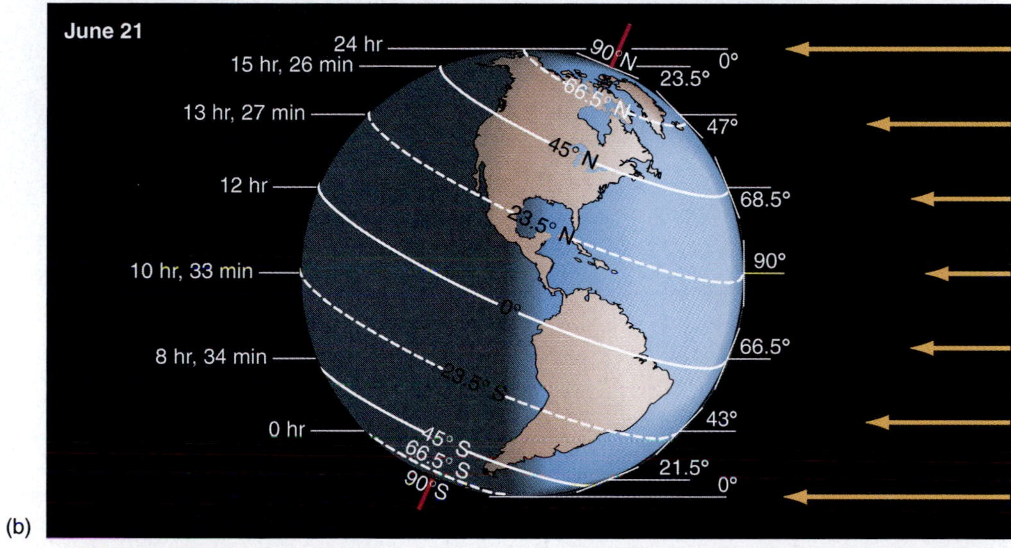

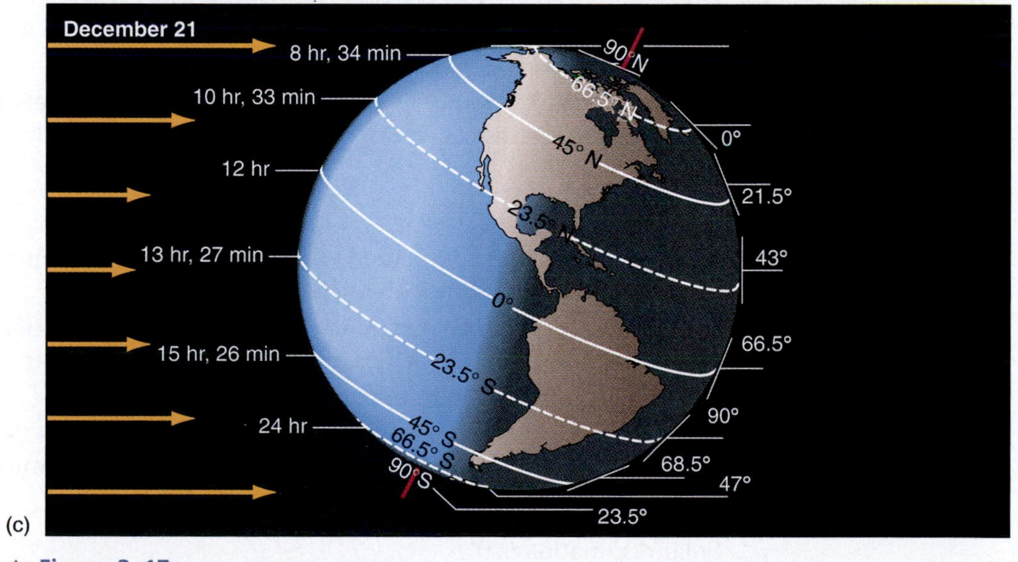

▲ **Figure 2–17**
The length of day (left) and the noon solar angle (right) are shown for the equinoxes (a), June solstice (b), and December solstice (c).

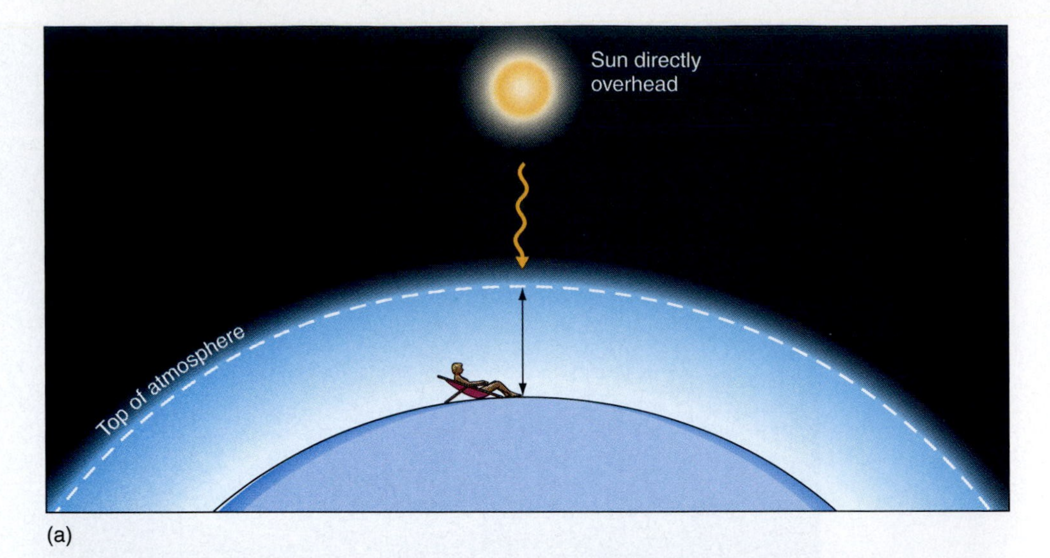

▶ **Figure 2–18**
A high solar angle (a) allows sunlight to pass through the atmosphere with a relatively short path. Lower sun angles such as those near sunrise and sunset (b) require that the energy pass through more of the atmosphere. This increase in atmospheric mass results in a greater depletion of energy than in (a).

1. June solstice:
  a. Solar declination = 23.5° N.
  b. Every latitude in the Northern Hemisphere receives more energy than the corresponding latitude in the Southern Hemisphere.
  c. Every place north of the Arctic Circle receives 24 hours of daylight, and every place south of the Antarctic Circle has 24 hours of night.
  d. The equator receives 12 hours of daylight.
  e. For the Northern Hemisphere as a whole, sunlight travels through less atmosphere than it does in the Southern Hemisphere.

2. December solstice:
  a. Solar declination = 23.5° S.
  b. Every latitude in the Southern Hemisphere receives more energy than does the corresponding latitude in the Northern Hemisphere.
  c. Every place south of the Antarctic Circle receives 24 hours of daylight, and every place north of the Arctic Circle has 24 hours of night.
  d. The equator receives 12 hours of daylight.
  e. For the Southern Hemisphere as a whole, sunlight travels through a lesser amount of atmosphere than it does in the Northern Hemisphere.

3. Equinoxes:
  a. Solar declination is 0°.
  b. Every place on Earth has 12 hours of daylight.
  c. Both hemispheres receive equal amounts of insolation.

Table 2–2 • Variations in Solar Angle and Day Length

	Solar Angle at Noon	Length of Day	Total Radiation for Day (Megajoules/m^2)
December 21			
Winnipeg	16.5°	7 hr, 50 min	7.44
Austin	36.5°	10 hr, 04 min	12.18
June 21			
Winnipeg	63.5°	16 hr, 10 min	37.15
Austin	83.5°	13 hr, 56 min	35.97

January and July Global Movies

Changes in Energy Receipt with Latitude We have seen that seasonal changes in the orientation of Earth with respect to the Sun determine the availability of solar radiation across the globe. Obviously, a large supply of energy is favorable for warm temperatures and a lack of energy leads to cold conditions. But the impact of radiation availability does not end with temperature. As we will see in later chapters, energy receipts also affect the distribution of pressure, which in turn affects winds, cloudiness, precipitation, and other aspects of weather and climate.

Having examined the effects of latitude and time of year separately, we can now look at how their combined effects set up differences in available insolation from one latitude to another. Table 2–2 presents the values of noontime solar angle and length of day for Winnipeg, Manitoba, Canada (50° N), and Austin, Texas (30° N), for the December and June solstices. (Remember that the values shown in the table represent the amount of energy available at the top of the atmosphere. Atmospheric conditions—especially the amount and type of cloud coverage—will reduce the amount of radiation that actually reaches the surface. The processes that affect the receipt of the energy at the surface are described in Chapter 3.) During the winter, Winnipeg has only 62 percent as much available solar radiation as does Austin because of its lower solar angle and the shorter period of daylight. During the June solstice, however, Winnipeg has a slightly greater amount of radiation because its greater period of daylight more than offsets its lower midday solar angle. Thus, during the winter both factors lead to decreasing values of available insolation with latitude. In contrast, during the summer the increasing day length at the high latitudes offsets the effect of lower sun angles. The result is a weak latitudinal gradient in summer, and strong north–south differences in winter. As will be seen in coming chapters, this has profound implications for seasonal temperature distributions and for seasonal changes in the vigor of large-scale atmospheric motion.

Summary

Virtually all the energy available for physical and biological processes originates within the Sun and travels through space as electromagnetic radiation. Such energy is propagated as a series of waves having characteristic wavelengths.

All matter emits electromagnetic energy in a continuum of different wavelengths, but not all substances or objects radiate the same amount of energy, nor do they emit it at the same dominant wavelengths. The Stefan-Boltzmann law tells us that hot objects emit radiation more intensely than do cooler objects, and Wien's law dictates that hotter objects emit radiation at shorter wavelengths. Together, these laws mandate that the Sun puts out hundreds of thousands of times more energy than does Earth, and that the energy from the Sun is emitted at shorter wavelengths, with effectively no overlap.

Despite the fact that the distance between Earth and the Sun varies slightly through the year, these differences exert only a small impact on the seasonal heating of the planet. More important to the distribution of available radiation is the relative orientation of Earth to the Sun, specifically through the effects of solar angle, length of day, and the length of path a beam of radiation must take through the atmosphere before reaching the surface. The patterns described in this chapter deal with the availability of insolation before the atmosphere acts on the incoming radiation. In the next chapter, we describe the atmospheric processes that reduce the intensity of incoming radiation and the transfer mechanisms by which energy is exchanged between the atmosphere and the surface. These processes directly influence global temperature and pressure variations—variations responsible for all other weather phenomena.

Key Terms

energy p. 39
joule p. 40
power p. 40
watt p. 40
electromagnetic radiation p. 40
kinetic energy p. 40
potential energy p. 40
conduction p. 41
convection p. 41
buoyancy p. 42
radiation p. 42
wavelength p. 44
micrometers/microns p. 44
blackbody p. 44
Stefan-Boltzmann law p. 45
graybody p. 46
emissivity p. 46
Kelvin scale p. 47
photon p. 48
shortwave radiation p. 49
longwave radiation p. 49
core p. 50
nuclear fusion p. 50
convection zone p. 50
photosphere p. 50
solar disk p. 50
limb darkening p. 50
granules p. 50
sunspots p. 50
flares p. 50
chromosphere p. 50
corona p. 50
solar wind p. 51
inverse square law p. 52
solar constant p. 53
insolation p. 53
ecliptic plane p. 53
revolution p. 53
perihelion p. 53
aphelion p. 53
rotation p. 53
Polaris p. 53
solstice p. 55
equinox p. 55
Tropics of Cancer and Capricorn p. 55
solar declination p. 56
Arctic and Antarctic Circles p. 57
beam spreading p. 58

Review Questions

1. Describe the different ways kinetic and potential energy may exist on Earth.
2. Conduction and convection are alike in that both transfer heat within a substance. What is the critical difference between them?
3. We have discussed sunlight, X rays, etc., as electromagnetic radiation. Describe radiation as a wave phenomenon, and explain what is meant by "electromagnetic."
4. Why is wavelength important in radiation transfer? That is, when discussing radiation, why isn't it enough to specify the amount or rate of energy transfer?
5. Place the following wavelength bands in correct order of wavelength: visible, X rays, ultraviolet, microwave, infrared.
6. Is there a temperature that has the same value on both the Fahrenheit and Celsius scales? If so, find that temperature. (*Hint:* Draw a graph of °C versus °F.)
7. Convert the following Fahrenheit temperatures to Celsius: –22 °F, 50 °F, 113 °F.
8. Convert the following Celsius temperatures to Fahrenheit: –20 °C, 10 °C, 40 °C.
9. Why is the Kelvin scale superior to the Fahrenheit and Celsius scales in many scientific applications?
10. Describe how the wavelengths and total energy emitted change as the temperature of an object increases.
11. The solar constant is about 1367 W/m^2. If the distance between Earth and Sun were to double, what would be the new value?
12. What is the most important factor responsible for seasons on Earth?
13. Describe the annual march of solar declination.
14. What is the significance of the Arctic and Antarctic Circles?
15. If the solar declination were 10°, where would the Arctic and Antarctic Circles be found? Would this cause a change in the dates of the solstices, equinoxes, and perihelion and aphelion?
16. Pick a day in the Northern Hemisphere winter. Describe the changes in day length and solar position you would encounter if you were to travel from the North Pole to South Pole. Do the same for a day in the Northern Hemisphere summer.
17. Explain why the equator always has 12 hours of sunlight.
18. Explain how changes in solar position influence the intensity of radiation on a horizontal surface.
19. If you were to travel from the equator to the North Pole, on what day would variations in solar radiation be smallest? Why? Explain how day length and solar angle change as you move poleward.
20. Burlington, Vermont, is located at 44.5° N. What is the angle of the noontime Sun on either of the equinoxes and on the solstices?

Critical Thinking

1. Goose down is composed of a large number of very small filaments that separate the air within a parka or sleeping bag into many small packets that do not readily circulate. How does this feature make down such a good insulating material?
2. Some old buildings are warmed by "radiators." Is this a truly accurate descriptor of how the rooms in those buildings are actually warmed?
3. Why is it not truly accurate to describe the energy coming from the Sun as visible radiation?
4. If Earth's speed of rotation were different from what it actually is, would there be a corresponding change in the amount of energy received by the planet as a whole?
5. At noon the solar angle will always be greater at Tucson, Arizona, than at Laramie, Wyoming. Is the same also true for 6 P.M.?
6. Locations near the equator typically have less seasonality than do locations farther away from the Tropics. Explain why this is so.
7. Why is it that the solar angle cannot be considered a single influence on the amount of radiation reaching Earth's surface? Is the situation different for the moon?
8. How might the temperature change in the course of a day differ on east-facing vs. west-facing slopes?
9. At noon at 45° N latitude, the solar angle is 45° above the southern horizon. What would the angle of incidence be on a north-facing slope of 45°? Would the slope of the surface affect both beam spreading and atmospheric path length?
10. Describe the apparent path of the Sun to a person standing at the North Pole on June 22.
11. On the equinoxes a person at the equator would observe the sun to rise exactly to the east, pass directly overhead at noon, and set exactly in the west—all over a 12-hour period. How will this change on the solstices?

Problems and Exercises

1. An instrument measures the radiation emitted from an ocean surface as 365 watts per square meter. What law would you apply to determine the ocean surface temperature? (More advanced question: what would the ocean surface temperature actually be? *Hint*: you will need to rearrange one of the equations given in this chapter.)
2. Assume that a body has an emissivity of 0.9 and a temperature of 300 K. Which would have a greater impact on the intensity of radiation emitted: a 50 percent reduction in the emissivity or a 5 percent reduction in the absolute temperature?
3. Saturn is about 1.42×10^{12} m from the Sun, or about 9.5 times as far from the Sun as Earth is. Calculate the solar constant for Saturn. Do you suppose the distribution of wavelengths of the sunlight received at that distance is different from that which reaches Earth?
4. One of the factors that influences the amount of insolation available is the varying Earth–Sun distance. Using the distances for perihelion and aphelion of 1.47×10^{11} m and 1.52×10^{11} m, respectively, determine the intensity of solar radiation at the top of Earth's atmosphere on those two days.
5. What is the difference in the noontime solar angle between the two solstices at a latitude of 10° N? How does this compare to the range of noontime solar angles at 30° N? Can you think of any significant outcomes that would occur because of this difference?
6. Go to this Web site: **http://aa.usno.navy.mil/data/**. Check the first of the two hypertext options under the category "Positions of the Sun and Moon." This site will allow you to determine the solar angle throughout the course of any date you select. Plot the solar angle of the Sun throughout the daylight period for the solstices and the equinoxes and notice how the pattern changes through the year. Next, for the solstices and equinoxes, plot and compare the differing solar elevations over the course of the day for East Lansing, Michigan; Knoxville, Tennessee; and Gainesville, Florida. What generalizations can you make?

Quantitative Problems

This chapter has introduced some important laws and concepts describing the type and amount of radiation received by Earth. You can enhance your understanding of these laws by solving some quantitative problems given in this book's companion Web site, **http://www.prenhall.com/aguado/**. After entering the site, go to the bottom of the page and select Chapter 2. Then highlight "Quantitative Examples" on the left-hand panel. These brief problems should bolster your comprehension of the Stefan-Boltzmann and Wien's laws and the temperature scales discussed in this chapter.

Useful Web Sites

http://aa.usno.navy.mil/data/docs/AltAz.html

Some very useful astronomical data, especially with regard to Earth–Sun relationships. Interactive format allows you to request data for particular cities or latitude-longitude coordinates. Includes precise sunrise/sunset data as well as solar angle data.

http://aa.usno.navy.mil/data/docs/EarthSeasons.html

Exact dates and times of perihelion, aphelion, equinoxes, and solstices.

http://www.srrb.noaa.gov/highlights/sunrise/sunrise.html

Fully interactive site that allows you to get time of sunrise/sunset for a choice of world cities or for precise latitude-longitude coordinates. Also gives solar declination and several other items of Earth–Sun position information, complete with glossary to explain their meanings.

Media Enrichment

Tutorial

Solar Geometry

This tutorial covers the geometry of Earth's orbit and its effects on solar radiation. It uses three-dimensional diagrams to show how variations in day length and solar position arise and includes animations depicting the processes of beam spreading and depletion.

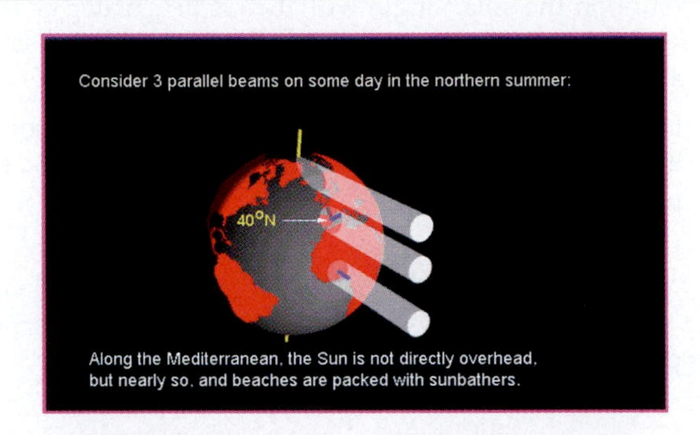

Interactive Exercise

Earth–Sun Geometry

Use this program to see how solar position varies seasonally from place to place, and how these variations translate into radiation at the surface. Compare one latitude with another, noting differences in the seasonal range of incoming radiation. Think about the likely effect on temperature at the two locations. Experiment with latitudes in both hemispheres, as well as locations near the equator. Using your latitude as a reference, speculate about differences in thermal conditions.

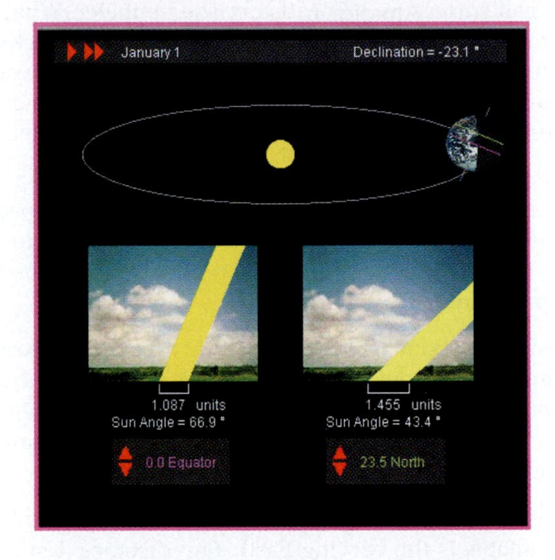

Weather in Motion

January 1998 Ice Storm

This chapter opened with an account of the ice storm that hit the eastern United States and Canada in January 1998. This brief movie shows the movement of the storm as observed by satellite.

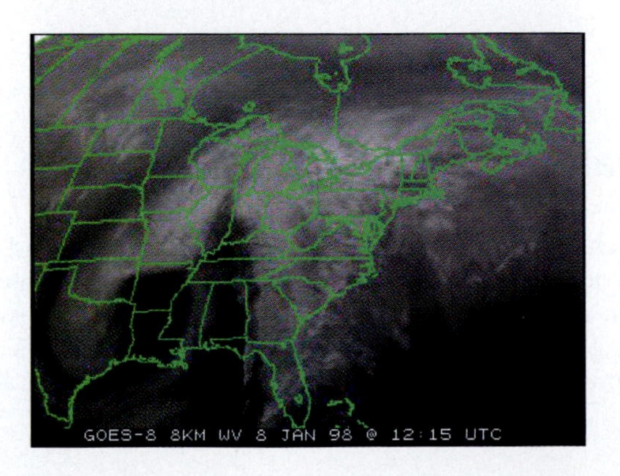

Weather in Motion
A Solar Eclipse

As Earth orbits the Sun, the Moon simultaneously revolves around Earth. Occasionally, the three bodies briefly line up in a straight line. If the Moon is between the Sun and Earth, the Moon's shadow passes across the Earth, forming a solar eclipse. In some instances the Earth–Sun distance is small enough that the center of the Moon's shadow (called the *penumbra*) crosses the Earth in a *total eclipse*. This movie depicts the path of the Moon's shadow across North America on February 26, 1998. Notice that only a small part of the planet experiences a total eclipse, and that the shadow moves quickly. Because of this fast shadow movement, eclipses are short, lasting just a few minutes (712 minutes at most).

Weather in Motion
January and July Global Movies

We have seen in this chapter how the solar declination varies through the course of a year. The movies illustrate this annual progression by showing the January and July portions of the one-year loop used in Chapter 1. The first thing you should look at is the differential movement of the zone of high land surface temperatures associated with the east-to-west migration of sunlight over the course of a day. In January the hottest regions are found in the daytime in the Southern Hemisphere desert regions of Australia, southern Africa, and South America. In July the maximum temperature zone shifts to southern Asia, North Africa, and the southern United States and northern Mexico. Notice also the very dramatic changes in temperature that take place in the higher latitudes, especially over Canada and northern Asia.

A couple of other interesting patterns can be observed in the distribution of clouds and the speed with which they migrate during January and July. During both months, a conspicuous zone of cloud cover extends east–west across the tropics. This zone, called the Intertropical Convergence Zone, or ITCZ (discussed in Chapter 8), is largely the result of surface heating near the equator. Because the zone of most intense heating migrates northward from January to July, this cloud band makes a similar, though much smaller, shift in location.

Another important feature exhibited by the two movies is the differential speed of movement of storm systems outside of the Tropics. Notice that the cloud bands sweeping across North America during January move about twice as rapidly as those in July. The difference occurs because both day length and solar angle give rise to a large gradient in heating across the winter hemisphere. As will be seen later in the book, this leads to vigorous movement of atmospheric systems. In summer, day length and solar position combine to create small differences in heating, resulting in generally lower wind speeds.

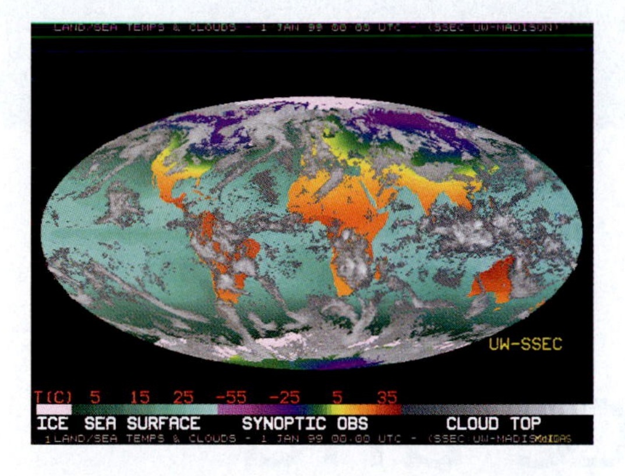

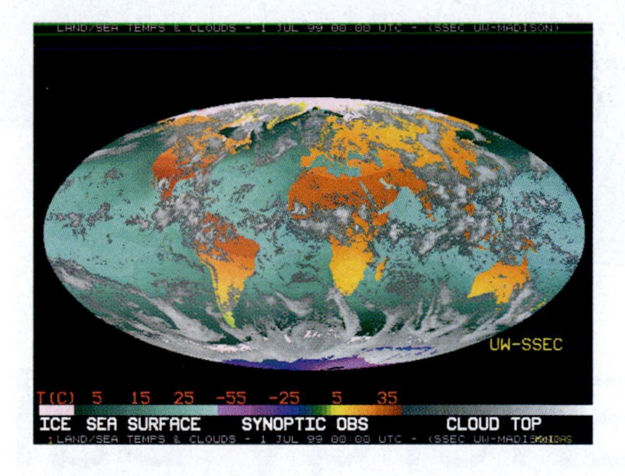

chapter 3

Madison, Wisconsin, at –37 °F.

Energy Balance and Temperature

It was early morning on October 24, 1998, and Tropical Storm Mitch had just intensified into a true hurricane in the western Atlantic. While forecasters were certain the storm would strengthen further and move westward, nobody knew that within a few days Mitch would become the worst hurricane to hit Central America since 1780. Its ultimate devastation was hard to comprehend. The exact death toll from Hurricane Mitch will never be known but likely stands at somewhere between 9000 and 18,000 people. Wind gusts of more than 320 km/hr (200 mph) and rainfall amounts greater than 50 cm (20 in.) ravaged many villages. Worst hit was Honduras, where more than 20 percent of the population was suddenly homeless. The mountainous terrain of Nicaragua was particularly hard hit by devastating mudslides. A farmer, José Morales, recounted his experience with one of the worst of them, the Casitas mudslide in northwestern Nicaragua: "It was a ball of earth and trees, and suddenly I couldn't see houses anymore" (see Figure 3–1).

The misery caused by Hurricane Mitch still continues. The battered nations of the region have had to contend with the threat of cholera, malaria, and dengue fever—not to mention the widespread hunger from the ruined crops. In Honduras about 70 percent of the crops were destroyed and a similar percentage of the transportation infrastructure demolished. Some relief experts have estimated that it will take 15 to 20 years before the effects of the storm will be fully undone.

Like other major hurricanes, Mitch covered an area greater than a quarter of a million square kilometers, lasted for about a week, dumped millions of tons of rainwater, and brought winds capable of wiping out whole villages. Such activity requires an enormous amount of energy and, as we saw in Chapter 2, solar radiation provides

CHAPTER OUTLINE

Figure 3–1
A satellite image of Hurricane Mitch, the deadliest hurricane to hit Central America in the last 200 years.

virtually all of that energy. But there is more to the story, because most of the energy contained in the atmosphere does not accrue by the *direct* absorption of solar radiation. Instead, the majority of the energy comes *indirectly* from the Sun after first having been absorbed by Earth's surface. From there, several processes combine to transfer this absorbed energy to the atmosphere. In this chapter, we examine this energy transfer, which provides the fuel for everyday weather and for catastrophic events such as Hurricane Mitch.

Atmospheric Influences on Insolation

Solar radiation reaching the top of the atmosphere does not pass unimpeded through the atmosphere, but rather is attenuated by a variety of processes. The atmosphere absorbs some radiation directly and thereby gains heat. Another portion disperses as weaker rays going out in many different directions through a process we call *scattering*. Some of the scattered radiation is directed back to space; the remainder is scattered forward as the light we see from the portion of the sky away from the solar disk. In either case, the energy that is scattered is not absorbed by the atmosphere and therefore does not contribute to its heating.

The remaining insolation is neither absorbed nor scattered and passes through the atmosphere without modification, reaching the surface as direct radiation. But not all the energy reaching the surface is absorbed. Instead, a fraction is scattered back to space and, like the radiation scattered by the atmosphere, it does not contribute to the heating of the planet.

These processes—the absorption, scattering, and transmission of solar radiation—directly affect the distribution of temperature throughout the atmosphere. They also explain a number of atmospheric phenomena of everyday interest, such as the blue sky on a clear day or the redness of a sunset. In this section, we explore the processes affecting incoming radiation.

Absorption

Atmospheric gases, particulates, and droplets all reduce the intensity of insolation by **absorption**. It is important to note that absorption represents an energy transfer to the absorber. This transfer has two effects: the absorber gains energy and warms, while the amount of energy delivered to the surface is reduced.

The gases of the atmosphere are not equally effective at absorbing sunlight, and different wavelengths of radiation are not equally subject to absorption. Ultraviolet radiation, for example, is almost totally absorbed by ozone in the stratosphere. Visible radiation, in contrast, passes through the atmosphere with only a minimal amount of absorption. This is of no minor consequence, because if the atmosphere *were* able to absorb all the incoming solar energy, the sky would appear completely dark. Artificial lights would be useless, because their radiation would likewise be absorbed. The very fact that we can see great distances suggests that the atmosphere is not particularly good at absorbing visible radiation, an impression that turns out to be correct.

Near-infrared radiation, which represents nearly half the radiation emitted by the Sun, is absorbed mainly by two gases in the atmosphere—water vapor and (to a lesser extent) carbon dioxide. This is why direct sunlight in the desert feels so hot and shade is so welcome, whereas the apparent temperature difference between standing in direct sunlight and standing in shade is relatively small in humid regions. When the humidity is high, water vapor absorbs a significant portion of near-infrared radiation, thereby reducing the amount of energy available to warm your skin. On dry days, the lack of water vapor allows a greater amount of near-infrared radiation to penetrate the atmosphere and raise your skin temperature.

Reflection and Scattering

The **reflection** of energy is a process whereby radiation making contact with some material is simply redirected away from the surface without being absorbed. The reason we are able to see is that the human eye has the ability to detect the receipt of visible radiation. Visible energy travels in all directions as it is reflected off objects in our field of view. Some of the reflected light comes into contact with our eyes, which in turn send signals to be processed by optical centers in our brains. All substances reflect visible light, but with vastly differing effectiveness. For instance, a fresh patch of snow very effectively reflects visible light, while a piece of coal reflects only a small portion of the visible radiation hitting its surface. The percentage of visible light reflected by an object or substance is called its **albedo**. Furthermore, objects do not reflect all wavelengths equally. A shirt, for example, will appear green if it most effectively reflects wavelengths in the green portion of the spectrum.

Light can be reflected off a surface in a couple of different ways. When light strikes a mirror, it is reflected back as a beam of equal intensity, in a manner known as **specular reflection**. In contrast, when a beam is reflected from an object as a larger number of weaker rays traveling in many different directions, it is called **diffuse reflection**, or **scattering**. When scattering occurs, you cannot see an image of yourself on the reflecting surface as you can in a mirror. Consequently, although a surface of fresh snow might reflect back most of the visible light incident on it, you would not be able to check out your appearance by looking at it. The vast majority of natural surfaces are diffuse rather than specular reflectors.

In addition to large solid surfaces, gas molecules, particulates, and small droplets scatter radiation. Furthermore, although much is scattered back to space, much is also redirected forward to the surface. The scattered energy reaching Earth's surface is thus **diffuse radiation**, which is in contrast to unscattered **direct radiation**. Figure 3–2 illustrates the process of scattering and the transformation of direct radiation to diffuse radiation. You can think of it this way: the blocking of direct radiation is what creates shadows, but a surface in the shadow of the direct radiation is not completely dark because it is illuminated by diffuse radiation. Notice that whether accomplished by a gas molecule, particulate, or droplet, this result is still a scattering process in which the radiation is redirected but not absorbed.

The characteristics of radiation scattering by the atmosphere depend on the size of the scattering agents (the air molecules or suspended particles) relative to the wavelength of the incident electromagnetic energy. Three very general categories of scattering exist: Rayleigh scattering, Mie scattering, and nonselective scattering.

Rayleigh Scattering Scattering agents smaller than about one-tenth the wavelength of incoming radiation disperse radiation in a manner known as **Rayleigh scattering**. Rayleigh scattering is performed by individual gas molecules in the atmosphere. It does not affect all wavelengths of solar radiation equally; rather, it is biased toward shorter wavelengths. Rayleigh scattering is particularly effective for visible light, especially those colors with the shortest wavelengths, so blue light is more effectively scattered by air molecules than is longer-wavelength red light. Furthermore, Rayleigh scattering disperses radiation both forward and backward. Combined with its greater effectiveness in scattering shorter wavelengths, this characteristic leads to three interesting phenomena: the blue sky on a clear day, the blue tint of the atmosphere when viewed from space, and the redness of sunsets and sunrises.

Figure 3–3 illustrates how Rayleigh scattering produces a blue sky. As parallel beams of radiation enter the atmosphere, a portion of the light is redirected away from its original direction. A person looking upward, away from the direction of the Sun, can see some of the scattered light that has been redirected toward the viewer. Because blue light is among the shortest (and therefore most readily scattered) of the visible wavelengths, the scattered radiation contains a higher proportion of blue light than yellow, green, or other longer-wavelength light. Rayleigh scattering occurs at every point in a clear atmosphere and diverts energy toward a viewer from all directions, so no matter where you look on a cloudless day, the sky is blue.

▲ **Figure 3–2**
Scattering is a process whereby a beam of radiation is broken down into many weaker rays redirected in other directions.

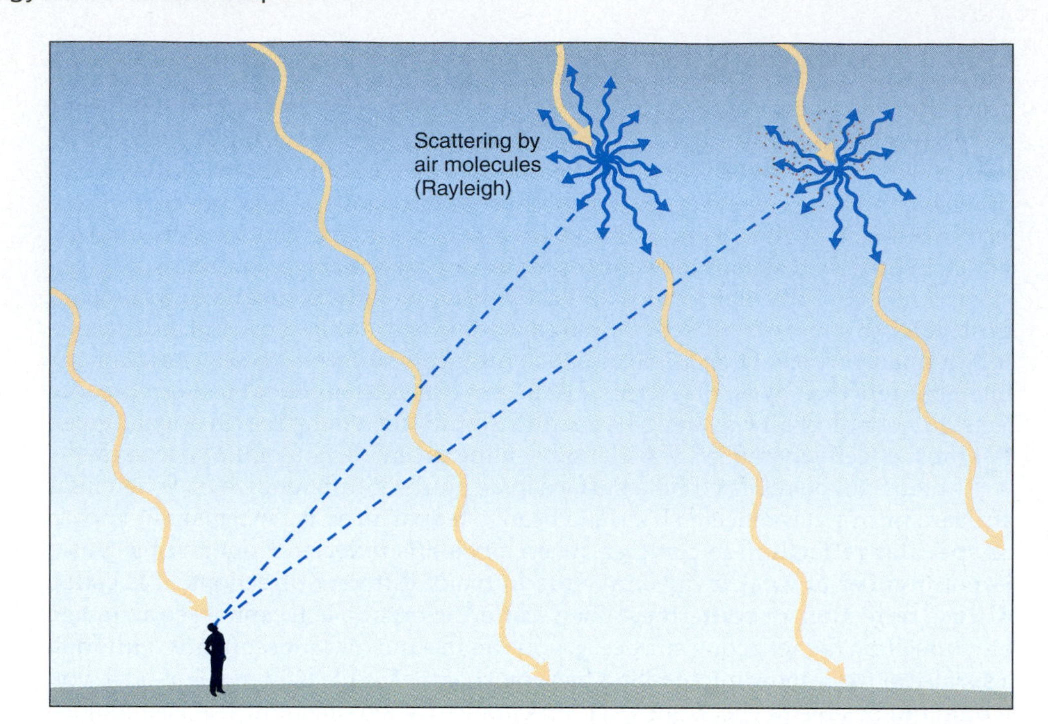

▶ **Figure 3–3**
The sky appears blue because the gases and particles in the atmosphere scatter some of the incoming solar radiation in all directions. Air molecules scatter shorter wavelengths most effectively. Someone at the surface looking skyward perceives blue light, the shortest wavelength of the visible portion of the spectrum.

Of course, not all the incoming radiation is scattered on a clear day. In fact, the amount of diffuse radiation received at the surface under cloudless skies is normally about one-tenth that of the direct radiation.

On the moon, which has no atmosphere, the "sky" appears black (Figure 3–4). As a viewer looks toward the horizon on the Moon, there is no downward scattered light because of the absence of an atmosphere, and the sky appears little different from the way it does at night. All that can be seen is the energy reflected off the lunar surface and Earth.

The same process that leads to the blue sky as seen from the surface also produces the bluish tint of the atmosphere as viewed from space. Like forward scattering, backscattering is biased toward blue wavelengths, so diffuse radiation directed back to space appears blue.

▶ **Figure 3–4**
An "Earthrise" from the moon, as seen by the *Apollo 11* astronauts. Although this photo was taken during the day, the moon has no blue sky. This is due to the absence of an atmosphere to scatter incoming solar radiation. Notice the blue tint of Earth, the result of Rayleigh scatter.

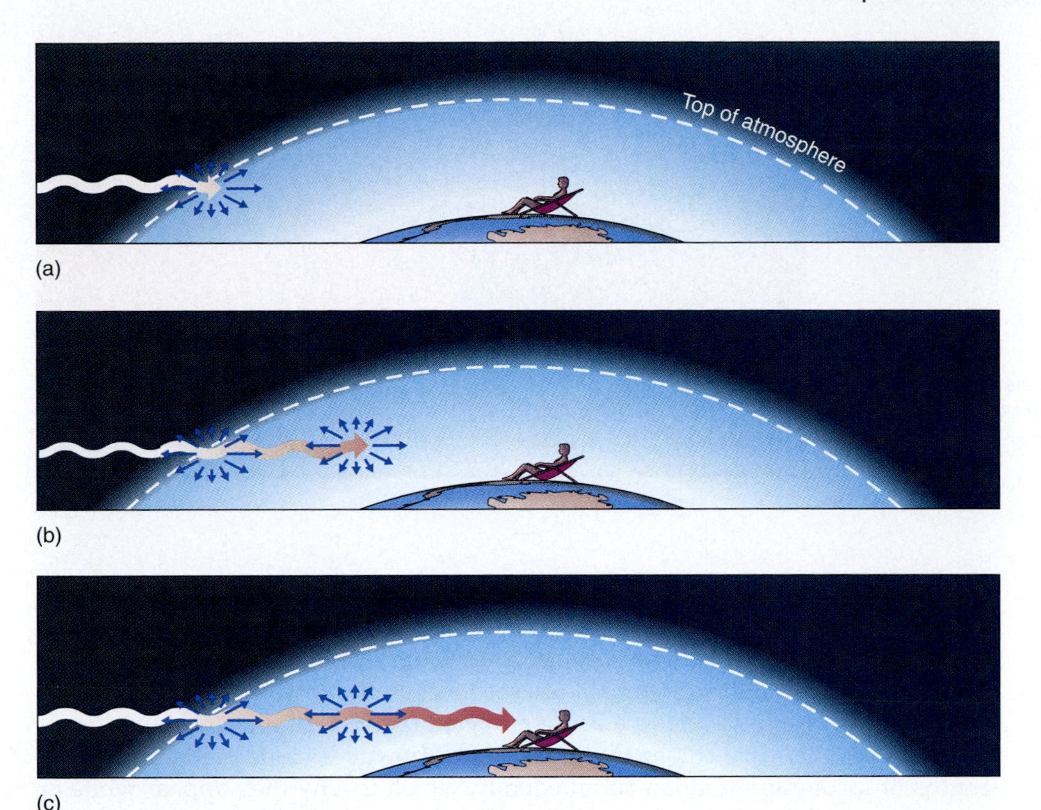

◀ **Figure 3–5**
Sunrises and sunsets appear red because sunlight travels a longer path through the atmosphere. This causes a high amount of scattering to remove shorter wavelengths from the incoming beam radiation. The result is sunlight consisting almost entirely of longer (e.g., red) wavelengths.

Rayleigh scattering is also largely responsible for the redness of sunrises and sunsets, as can be seen in Figure 3–4. Figure 3–5 shows how this happens. When the sun is barely over the horizon, sunlight must travel a greater distance through the atmosphere than it does during the middle of the day, and the longer path increases the amount of Rayleigh scattering. As the direct beam travels its long path, the shortest wavelengths of radiation are depleted, so the longer wavelengths constitute an increasing percentage of the direct sunlight. The sky in the general vicinity of the Sun thereby takes on a reddish tint due to the depletion of the green and blue (shorter wavelength) light.

Mie Scattering Vertical motions in the atmosphere are sufficiently strong that the atmosphere always contains suspended aerosols. This is true not only in cities, which tend to have higher air pollution concentrations, but also in rural areas far removed from urban activities. The microscopic aerosol particles are considerably larger than air molecules and scatter sunlight by a process known as **Mie** (pronounced "mee") **scattering**. Unlike Rayleigh scattering, Mie scattering is predominantly forward, diverting relatively little energy backward to space. Furthermore, Mie scattering does not have nearly the tendency to scatter shorter wavelength radiation that Rayleigh scattering does. Thus, on hazy or polluted days (when there are high concentrations of aerosols) the sky appears gray, as the whole range of the visible part of the spectrum is effectively scattered toward the surface.

Mie scattering causes sunrises and sunsets to be redder than they would due to Rayleigh scattering alone, so episodes of heavy air pollution often result in spectacular sunsets (Figure 3–6). Fires can also trigger enhanced Mie scattering. Residents of the western United States observed this phenomenon firsthand, when major fires burned across the region during the summer of 2002. If a fire is large enough, Mie scattering can be increased great distances downwind. In 1988, for example, fires in Yosemite National Park reddened the sky as far away as Minneapolis, Minnesota. Volcanic eruptions, such as the major eruption of Mount Pinatubo in 1991, can even enhance the color of sunrises and sunsets across an entire hemisphere, as stratospheric winds transport aerosols far from their source.

▶ **Figure 3–6**
The scattering of shorter wavelengths enhances the redness of sunrises and sunsets during episodes of heavy particulate concentrations.

Nonselective Scattering The water droplets in clouds are considerably larger than suspended particulates; therefore, they scatter sunlight in yet another way, behaving more or less like lenses. An isolated water droplet affects various wavelengths of solar radiation differently. You see this whenever you witness a rainbow, which involves each wavelength being refracted (bent) a different amount, hence the bands of individual colors. In the aggregate, however, clouds reflect all wavelengths of incoming radiation about equally, which is why they appear white or gray. Because of the absence of preference for any particular wavelength, scattering by clouds is sometimes called **nonselective scattering**.

Clouds are by far the most important agent in nonselective scattering and exert a tremendous impact on the global receipt of solar radiation by reflecting large amounts of energy back to space.

Transmission

When solar radiation travels through the vacuum of outer space, there is no modification of its intensity, direction, or wavelength. However, when it enters the atmosphere, only some of the radiation can pass unobstructed to the surface. The amount varies greatly, depending on atmospheric conditions. A clear, dry atmosphere might transmit as much as 80 percent of the incoming solar radiation as direct beam radiation without scattering or absorption. This is what you experience on a sunny, unpolluted day with sharp, distinct shadows. In contrast, when it is cloudy or hazy, only a small fraction of solar radiation will reach the surface as direct radiation. Under these conditions, there is both a reduction in the amount of radiation reaching the surface and a shift from direct radiation to diffuse, or scattered, radiation.

The Fate of Solar Radiation

Because Earth's orbit around the Sun is not perfectly circular, there are slight seasonal variations in the availability of insolation, with almost 7 percent more solar energy available on perihelion than on aphelion. Despite this variation, it is useful to think of a constant supply of radiation at the top of the atmosphere and to examine what happens, on average, to this energy. In other words, we need to account for the relative amount of radiation that is transmitted through the atmosphere, absorbed by the atmosphere and surface, and scattered back to space.

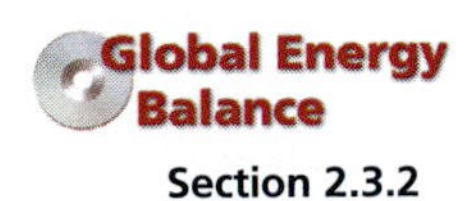

Section 2.3.2

Such an exercise is more than a mere bookkeeping activity, because the amount of radiation absorbed by the atmosphere and surface will greatly influence their temperatures. For the sake of simplicity, we will assume that 100 units of insolation are available at the top of the atmosphere and then compare the amount of energy scattered back to space and absorbed by the atmosphere and surface to

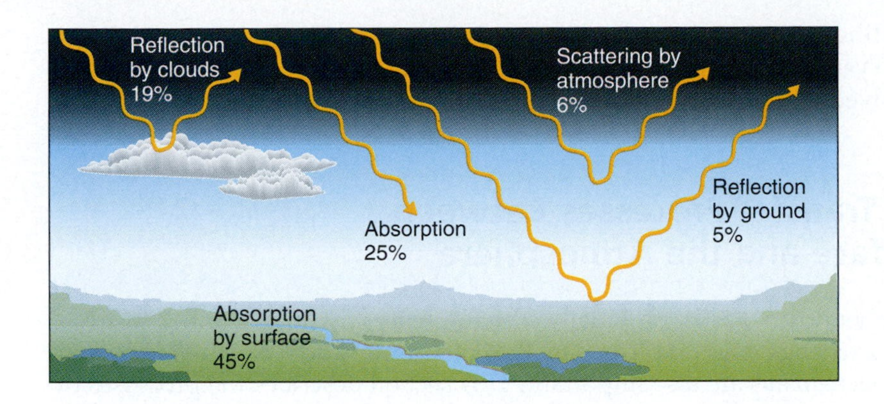

Figure 3–7
Incoming solar radiation available is subject to a number of processes as it passes through the atmosphere. The clouds and gases of the atmosphere reflect 19 and 6 units, respectively, of insolation back to space. The atmosphere absorbs another 25 units. Only half of the insolation available at the top of the atmosphere actually reaches the surface, of which another 5 units are reflected back to space. The net solar radiation absorbed by the surface is 45 units.

these 100 units. Bear in mind that the values presented in this discussion are annual and global averages; they do not necessarily apply to any particular place or time (see Figure 3–7).

As a global average, the atmosphere absorbs 25 of the 100 units available at the top of the atmosphere. Seven of the 25 units are ultraviolet radiation absorbed in the stratosphere by ozone, with most of the remainder being near-infrared radiation absorbed in the troposphere by gases (mostly water vapor). Thus, most of the radiation absorbed by the atmosphere is not visible radiation—a situation that suits us well, because if visible radiation were strongly absorbed by the atmosphere, it would be harder for us to see. Also note that relatively little of the available shortwave radiation is absorbed by clouds; instead, clouds affect the incoming radiation primarily by scattering and reflection.

Though comparatively ineffective at absorbing shortwave radiation, cloud droplets scatter backward a large percentage of the incident energy. Their high albedo not only gives them their bright appearance, which makes them readily visible even from space, but also greatly reduces the amount of energy available for warming the atmosphere and surface. On average, the global cloud cover reflects 19 units of the incoming solar radiation back to space. But clouds are not the only agent of backscatter. Averaged across the globe, the backscattering of radiation by atmospheric gases and aerosols accounts for 6 of the 100 units of insolation at the top of the atmosphere, with Rayleigh scattering more important than Mie scattering (largely because most Mie scattering is directed downward rather than back to space). Together, scattering by clouds and gases reflects back to space 25 units (that is, they yield an atmospheric albedo of 25 percent).

After atmospheric absorption and backscattering, 50 units of insolation are able to reach the surface. But not all the radiation reaching the surface is absorbed, because Earth's surface is not completely black. Of the 50 units incident at the surface, 5 are backscattered to space. Overall, a total of 30 units of solar radiation (25 from the atmosphere and 5 from the surface) are scattered back to space, resulting in a **planetary albedo** of 30 percent. Note that the amount of insolation reflected from the surface is slightly smaller than that backscattered by atmospheric gases. In other words, when viewed from space, the planet shines more from atmospheric reflection than from surface reflection.

The end result of these processes is that the atmosphere absorbs 25 units of energy, while the surface takes in 45 units. If this were the end of the story, we would all be in big trouble, because the constant supply of heat would cause continued warming of the planet. In fact, if this energy were stored in the upper few centimeters of Earth's surface, the surface would be heated at a rate of several hundred degrees Celsius each day!

Obviously, we do not observe the oceans boiling away, nor the land surface melting; thus, the surface must be continually losing energy. The same is true for the atmosphere, and for the Earth–atmosphere system as a whole. In other words, in the absence of climate change, the surface, the atmosphere, and the planetary system must give up as much energy as they obtain. To achieve this energy balance, huge quantities of energy must be transferred from the Earth system, and also within the system between surface and atmosphere. We now discuss the mechanisms involved in the maintenance of Earth's energy balance.

Energy Transfer Processes Between the Surface and the Atmosphere

The atmosphere and surface continually exchange energy with each other. Much of this energy exchange is accomplished by the emission and absorption of radiation, but other mechanisms are also important. This section describes the processes by which the energy transfer occurs.

Surface–Atmosphere Radiation Exchange

Like all other objects at terrestrial temperatures, Earth's surface and atmosphere radiate energy almost completely in the longwave (primarily thermal infrared) portion of the spectrum. Any discussion of longwave energy transfer is somewhat more complex than that of solar radiation, because longwave energy has no obvious beginning or end point.

Longwave radiation emitted by Earth's surface is largely absorbed by the atmosphere. This increases the temperature of the atmosphere, which causes it to radiate more energy outward. The energy radiated by the atmosphere is transferred in all directions, including downward, and so the surface receives a considerable portion of this radiation. This causes further surface heating, which leads to an increase in longwave radiation emission from the surface, which again warms the atmosphere, and so on and so on. In other words, there is an infinite cycle of exchange, with energy constantly transferring back and forth.*

Figure 3–8 describes the globally averaged amount of longwave radiation exchanged between the atmosphere and the surface. Beginning with the surface, we see that 104 units of longwave radiation are emitted up toward the atmosphere, with the greatest part (100 units) being absorbed by the atmosphere. The clear atmosphere absorbs thermal radiation far better than solar radiation, mainly due to the presence of water vapor and carbon dioxide. As shown in Figure 3–9, both of these gases are good absorbers of longwave radiation, with strong absorption bands in the longwave part of the spectrum (see *Box 3–1, Physical Principle: Selective Absorption by Water Vapor and Carbon Dioxide*).

*Although tempting to do so, we should not think of longwave radiation being cycled back and forth, or "bouncing" between the surface and atmosphere. First, natural surfaces do not reflect much longwave radiation. Second, when radiation is absorbed, the energy no longer exists as radiation; having ceased to exist, it can hardly bounce anywhere.

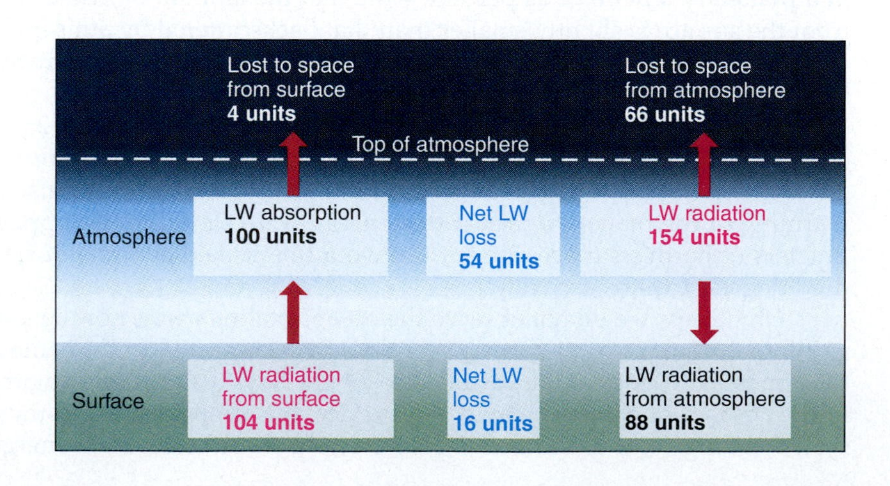

▶ Figure 3–8
The disposition of longwave radiation between the surface and atmosphere. The surface radiates 104 units upward and receives 88 from the atmosphere, for a net longwave radiation loss of 16 units. The atmosphere radiates 154 units and receives 100 from the surface, for a net deficit of 54 units.

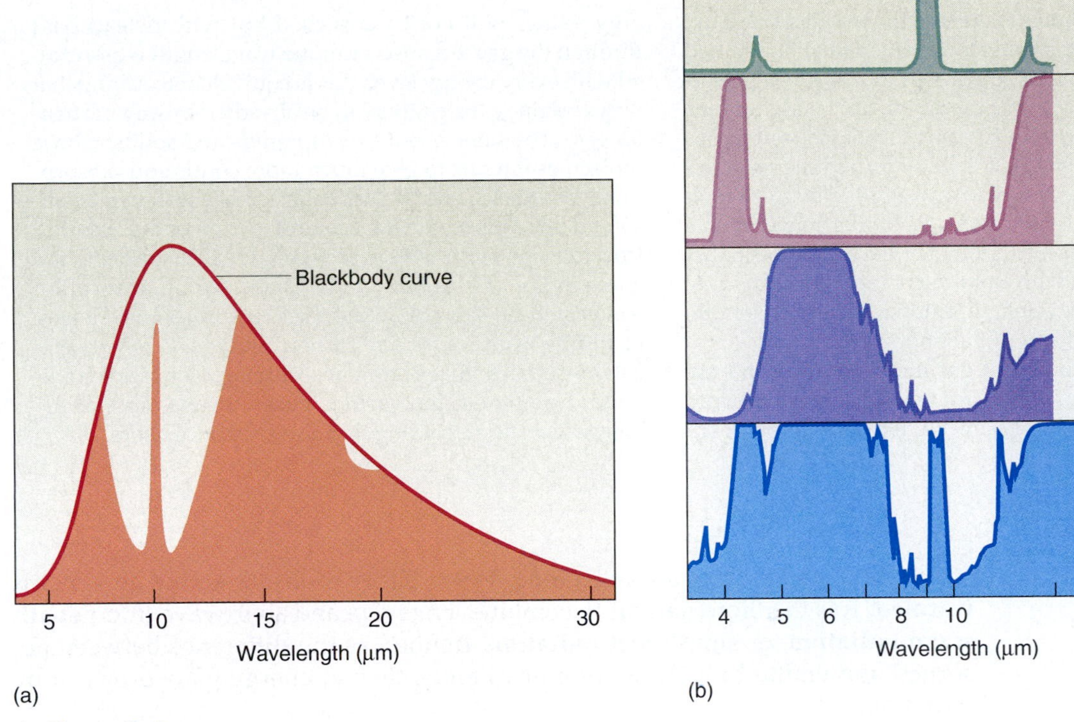

▲ **Figure 3–9**
Earth's surface acts nearly as a blackbody in its emission of radiation (a), but the gases of the atmosphere absorb most of the energy with wavelengths outside of the range of 8 to 11 μm. The shaded area in (a) indicates the energy absorbed by the atmosphere. Figure (b) shows the effectiveness of individual gases in absorbing the energy. The percentage of area shaded indicates the percentage of longwave energy absorbed.

Although water vapor, carbon dioxide, and other greenhouse gases are good at absorbing most wavelengths of longwave radiation, a portion of the longwave spectrum can pass through the atmosphere relatively unimpeded. Interestingly, wavelengths in this band, 8 to 12 μm, happen to match those radiated with greatest intensity by Earth's surface. This range of wavelengths not readily absorbed by atmospheric gases is called the **atmospheric window**. The atmospheric window must not be thought of as a place in the atmosphere or the absence of some gas; it represents just a certain range of wavelengths of special importance to the radiation balance.

Although the gases of the atmosphere are not effective at absorbing the wavelengths in the atmospheric window, clouds (even those of fairly modest thickness) readily absorb virtually all longwave radiation. This explains why cloudy nights do not cool off nearly as rapidly as do clear nights. When the evening sky is overcast, the cloud cover absorbs a large portion of the energy that otherwise would escape to space. Warmed by this energy, the clouds emit longwave radiation downward to the surface and lesser amounts upward to space. The clouds thus act something like a blanket, helping to retain heat.

Refer again to Figure 3–8. The energy emitted by the atmosphere amounts to 154 units, of which 88 are directed down toward the surface and 66 are radiated to space. Notice that the longwave energy lost from the atmosphere exceeds the amount it absorbs from the surface. The difference between absorbed and emitted longwave radiation is referred to as the **net longwave radiation**. For the atmosphere, the net longwave radiation is a negative 54 (100–154) units. Similarly, the surface receives 88 units of longwave radiation, but this amount is exceeded by the 104 units that are radiated, for a net longwave radiation deficit of 16 units.

Although shortwave and longwave radiation undergo different amounts of absorption and reflection, they are not separate entities as far as the heating of the

3–1 Physical Principles

Selective Absorption by Water Vapor and Carbon Dioxide

Water vapor and carbon dioxide are the two gases that are most important for absorbing longwave radiation emitted by the surface. Why are these gases so selective, being nearly transparent for shortwave radiation, but nearly opaque for longwave? Recall from Chapter 2 that isolated atoms have discrete energy states, with only certain energy states possible. As energy is absorbed and emitted by a gas molecule, its energy state rises and falls by discrete amounts from one allowable state to another. We have also seen that the energy associated with a photon of radiation is discrete and depends on its wavelength. Knowing the wavelength, we know the energy level of the photon.

It must be, therefore, that gas molecules absorb only certain photons, namely those that push the molecule into allowable energy states. Photons with higher or lower energy values will not be absorbed but will instead pass through the gas. Because a unique wavelength is associated with every energy level, this is equivalent to saying that only certain wavelengths can be absorbed by any particular gas. (The same is not true of liquids and solids, whose molecules interact to give much more continuous absorption.) Whether or not a particular wavelength can be absorbed depends on the molecular structure of the absorber (the configuration of electrons, etc.). As it happens, the gases in the atmosphere do not have strong absorption bands in the visible part of the spectrum. But some of them, including water vapor and carbon dioxide, do have molecular structures that permit absorption of longwave radiation. Combined, the various gases absorb most of the longwave energy passing through the atmosphere.

atmosphere and surface are concerned. When either is absorbed, the absorber is warmed. It is therefore natural to combine longwave and shortwave into **net all-wave radiation**, or simply **net radiation**, defined as the difference between absorbed and emitted radiation, or equivalently, the net energy gained or lost by radiation.

Figure 3–10 summarizes the net radiation balance for Earth. The atmosphere absorbs 25 units of solar radiation but undergoes a net loss of 54 units of thermal radiation, for a net deficit of 29 units. The surface absorbs 45 units of solar radiation but has a longwave deficit of 16, resulting in a net radiative surplus of 29 units. In other words, the atmosphere has a net deficit of radiative energy exactly equal to the surplus attained by the surface.

If radiation were the only means of exchanging energy, the surplus of radiative energy obtained by the surface would result in a perpetual warming, while the

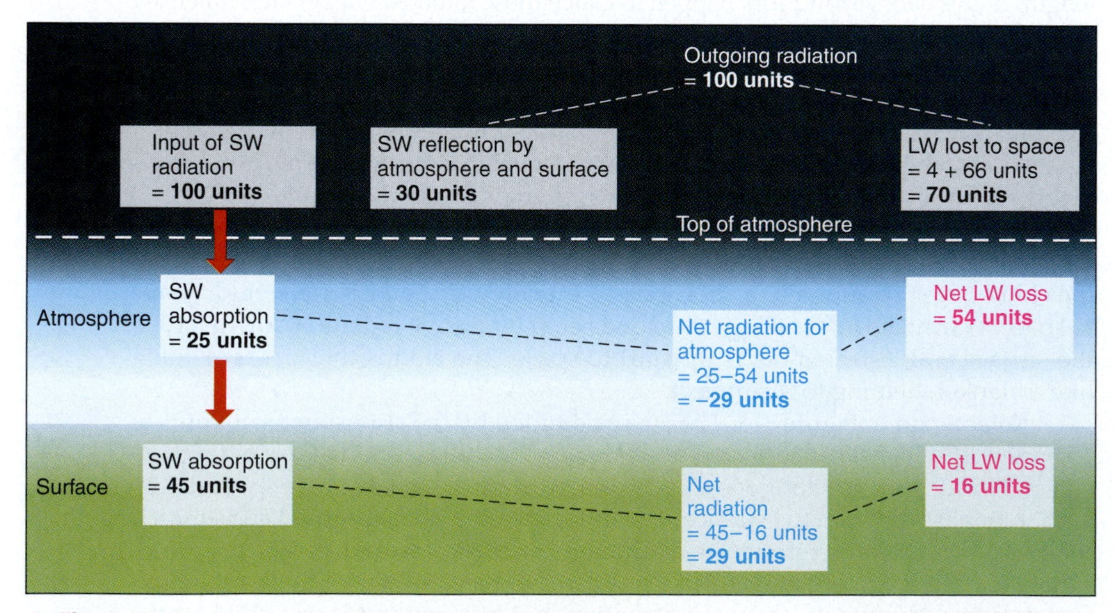

▲ **Figure 3–10**
Net radiation is the end result of the absorption of insolation and the absorption and radiation of longwave radiation. The surface has a net radiation surplus of 29 units, while the atmosphere has a deficit of 29 units.

deficit of the atmosphere would lead to a continual cooling. Eventually our feet would be scorched by a terrifically hot ground while the rest of our bodies would freeze, surrounded by a bitterly cold atmosphere. This, of course, is not about to happen, because energy is transferred from the surface to the atmosphere and within the atmosphere by two other forms of heat transfer: conduction and convection. The net transfer of energy by these two processes allows the radiation surplus at the surface to be eliminated while at the same time offsetting the radiation deficit of the atmosphere.

Conduction

Conduction, described in general terms in Chapter 2, plays an important role in transferring energy near the surface. As radiant energy is absorbed by a solid Earth surface during the middle of the day, a temperature gradient (a rate of change of temperature over distance) develops in the upper few centimeters of the ground. In other words, temperatures near the surface become greater than those a few centimeters below. As a result, conduction transfers energy downward. Warming of the ground during the day also sets up a temperature gradient within a very thin, adjacent sliver of air called the **laminar boundary layer**. Although air is usually highly mobile and capable of being easily mixed, very thin layers on the order of a few millimeters in thickness resist mixing. During the middle of the day, very strong temperature gradients can therefore develop in the laminar boundary layer, through which a substantial amount of conduction can occur. Energy conducted through the laminar boundary layer is then distributed through the rest of the atmosphere by a mixing process called *convection*.

Convection

Convection is a process whereby heat is transferred by the bodily movement of a fluid—that is, a liquid or gas. In contrast to conduction, convection involves the actual displacement of molecules. Unlike conduction, which transfers energy from the surface to the atmosphere, convection circulates this heat between the very lowest and the remaining portions of the atmosphere. The direction of heat transfer is upward when the surface temperature exceeds the air temperature (the normal situation in the middle of the day). At night the surface typically cools more rapidly than the air, and energy is transferred downward. Convection can be generated by two processes in fluids: local heating (free convection) and mechanical stirring (forced convection).

Free Convection **Free convection** is the mixing process related to buoyancy, the tendency for a lighter fluid to float upward when surrounded by a denser fluid. Recall your days as a child, when you would annoy your parents by blowing bubbles through a straw in a glass of milk. As the air was injected into the milk, it would immediately rise upward because of its lesser density and cause turbulent mixing. This was free convection at work.

Free convection (shown in Figure 3–11a) often occurs when a localized parcel of air is heated more than the nearby air. Because warm air is less dense than cold, it is relatively buoyant and rises. On a warm summer day, we can see the effect of free convection by observing a circling hawk (Figure 3–11b) that stays airborne without flapping its wings. This flight is possible because the hawk's wings are designed to catch the rising parcels of buoyant air that carry it upward. Convection can have far more important impacts than helping to keep hawks in the air. (For an illustration, refer to *Media Enrichment* at the end of this chapter.)

Heavy Convection over Florida

Forced Convection **Forced convection** (also called **mechanical turbulence**) occurs when a fluid breaks into disorganized swirling motions as it undergoes a large-scale flow. When water flows through a river channel, for example, it does not flow uniformly, as would a very thick syrup. Instead, the flow breaks down into numerous

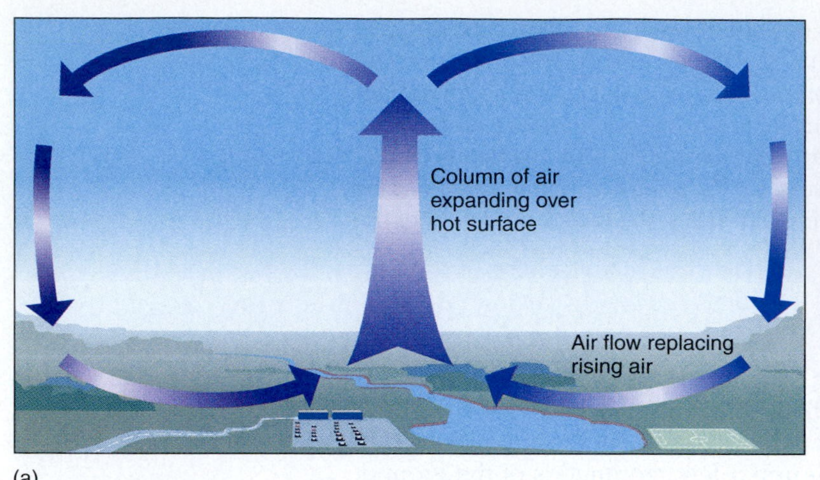

(a)

(b)

▲ **Figure 3–11**
Convection (a) is a heat transfer mechanism involving the mixing of a fluid. In free convection, local heating can cause a parcel of air to rise and be replaced by adjacent air. Free convection can create updrafts able to keep a hawk airborne (b) without it having to flap its wings.

▶ **Figure 3–12**
Forced convection. Air is forced to mix vertically because of its low viscosity (ability to be held together) and the deflection of wind by surface features.

eddies. Forced convection in the atmosphere is shown in Figure 3–12. Horizontally moving air undergoes the same type of turbulence. Instead of moving as a uniform mass, the air breaks up into numerous small parcels, each with its own speed and direction, that are superimposed on the larger-scale flow. Because there is a strong vertical component to the eddy motions, the forced convection helps to transport energy from the top of the laminar boundary layer upward during the day.

Generally speaking, higher wind speeds generate greater forced convection. Mechanical turbulence is also enhanced when air flows across rough surfaces (for example, forests and cities) rather than smooth ones such as glaciers. Both free and forced convection transfer two types of energy: sensible heat and latent heat.

Section 2.3.1

Sensible Heat The transfer of energy as **sensible heat** is easily understood. When energy is added to a substance, an increase in temperature can occur that we physically sense (hence the term *sensible*). This is what you experience when you rest outside on a warm, sunny day; the increase in your skin temperature results from a gain in sensible heat. The magnitude of temperature increase is related to two factors, the first of which is **specific heat**, defined as the amount of energy needed to produce a given temperature change per unit mass of the substance. In SI units, specific heat is expressed in joules per kilogram per kelvin. Everything else being equal, a substance with high specific heat warms slowly, because much energy is required to produce a given temperature change. Likewise, it also takes longer for a substance with a high specific heat to cool off, assuming the same rate of energy loss.

The temperature increase resulting from a surplus of energy receipt also depends on the *mass* of a substance. Not surprisingly, a given input of heat results in a greater rise in temperature if it is applied to only a small amount of mass. For example, compare the amount of energy needed to boil water for a cup of tea to that needed to take a warm bath. In just a few minutes, a single burner on a kitchen stove can have the water ready for the tea, but your hot water heater must supply considerably more energy for the large amount of bath water. These relationships are shown in Figure 3–13.

Sensible heat travels by conduction through the laminar boundary layer and is then dispersed upward by convection. Through these mechanisms, 8 of the 29 units of net radiation surplus for the surface are transferred to the atmosphere, where they help to offset the net radiation deficit. The remaining 21 units are transferred to the atmosphere by the convection of heat in another form.

Latent Heat **Latent heat** is somewhat less intuitive to understand than sensible heat. It is the energy required to change the phase of a substance (that is, its state as a solid, liquid, or gas). In meteorology we are concerned almost exclusively with the heat involved in the phase changes of water.

Recall that all physical processes require energy. The evaporation of water and the melting of ice are no exceptions to this rule—for either process, energy must be supplied. In the case of melting ice, the energy is called the *latent heat of fusion*. For the change of phase from liquid to gas, the energy is called the *latent heat*

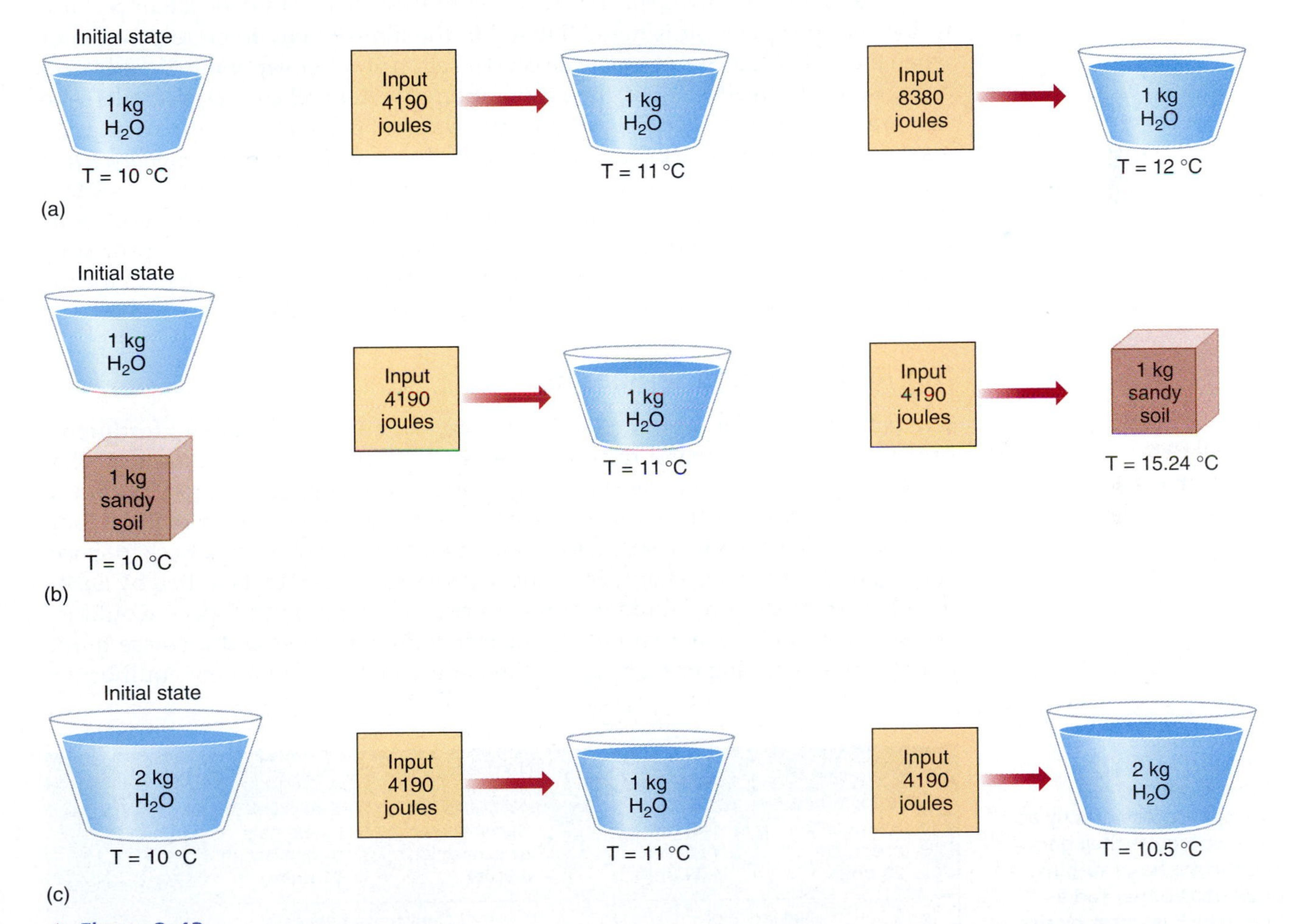

▲ **Figure 3–13**
The heat content of a substance depends on several factors. In (a) the input of 4190 J of energy to a kilogram of water increases its temperature 1 °C, while a doubling of the energy input causes twice as much heating. The specific heat of a substance also influences the amount of temperature change resulting from an input of energy. In (b) the application of 4190 J to 1 kg of sandy soil produces more than five times the increase in temperature than it would for a kilogram of water. The amount of mass also affects the temperature change accruing for a given energy input. Note in (c) that the temperature increase for 2 kg of water is half as much as that for 1 kg.

of evaporation. It takes seven and a half times more energy (2,500,000 joules) to evaporate a kilogram of liquid water than it does to melt the same amount of ice (335,000 joules). Although both forms of latent heat can be important locally, on a global scale the latent heat of evaporation is far more important.

When radiation is received at the surface, it can raise the temperature of the land or the water. If water happens to exist at the surface (or can be brought up from below the surface through the root systems of plants), some of the energy that might have been used to increase the surface temperature is instead used to evaporate some of the water. This results in a smaller temperature increase than would occur for a dry surface. You have probably experienced this while walking barefoot on a hot pavement. If the pavement is watered down, the surface cools as energy is taken from the ground and used for evaporation. The amount of energy consumed can be quite large, as much as 90 percent of absorbed solar radiation for a completely wet surface.

Let's use another common example to illustrate the concept of latent heat. We all know that perspiration is a mechanism that lowers our body temperatures and keeps us from overheating. But how does it work? Clearly, it's not a matter of sweat being cold—it's as warm as the body producing it. The reason sweat cools a person is latent heat. As you exercise, the heat produced as a byproduct causes your body temperature to rise. However, if your skin is covered with water and that water is free to evaporate, some of the energy produced by your body is used to evaporate the moisture rather than increase your body temperature.

The energy needed to evaporate water or melt ice is said to be latent because it does not disappear. It is held, "latent" in the atmosphere, to be released later when the reverse process occurs—the condensation of water vapor into liquid cloud or fog droplets. In effect, then, the evaporation of water makes energy available to the atmosphere that otherwise would have warmed the surface. It thus acts as an energy transfer mechanism, taking heat from the surface to the atmosphere. On a global average basis, the amount of energy transferred to the atmosphere as latent heat amounts to 21 units, which makes latent heat considerably more important as a mode of heat transfer than sensible heat (8 units). Perhaps this isn't surprising, given that the planet is mostly covered by ocean. The mean annual values of the net radiation components, latent and sensible heat, are depicted in Figure 3–14.

Net Radiation and Temperature

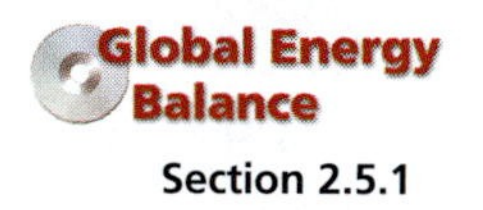

Section 2.5.1

The balance between incoming and outgoing radiation is not merely fortuitous; physical laws dictate that it must be the case. To understand why, consider what would happen if the Sun were suddenly to increase its radiative output and raise the temperature of Earth. Because warm bodies radiate more energy than do cooler ones, the planet would respond by emitting more longwave energy back to space. As long as the incoming energy from the Sun were to exceed that emitted by Earth, the global temperature would continue to rise, and emission to space would increase accordingly. Eventually the planetary temperature would increase to the point where outgoing energy equaled incoming energy, and a new equilibrium

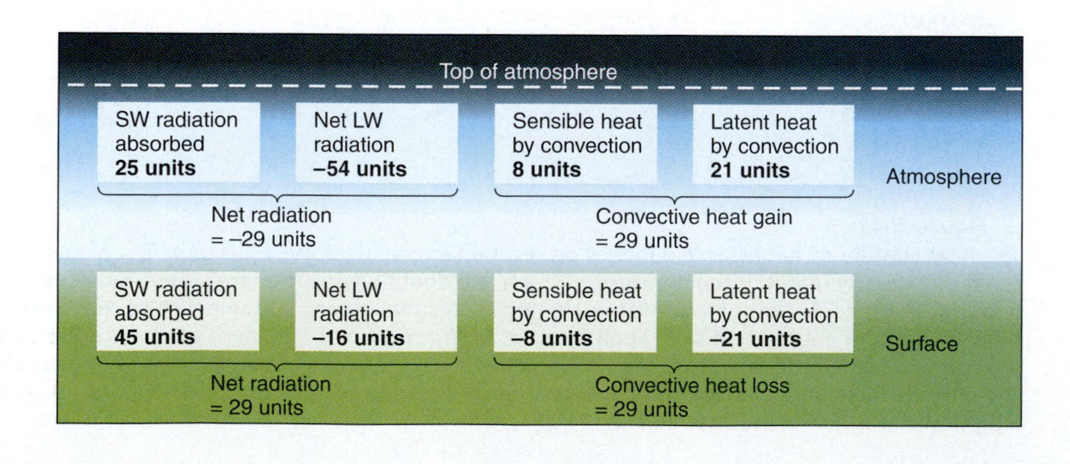

▶ Figure 3–14
Both the surface and atmosphere lose exactly as much energy as they gain. The surface has a surplus of 29 units of net radiation, which is offset by the transfer of sensible and latent heat to the atmosphere. The atmosphere offsets its 29 units of radiation deficit by the receipt of sensible and latent heat from the surface.

temperature would be established. The individual values in Figure 3–14 would all be changed, but the global inputs and outputs of radiation would still sum to zero.

So far we have discussed the average annual input and output of radiation over the entire globe. But the gains and losses of radiative energy are also responsible for the variations we observe in temperature during the course of a day. Consider what happens over a 24-hour period on a cloudless day. At sunrise the Sun appears just above the horizon. Because of the low solar angle, the insolation reaching the surface is weak due to beam spreading and the long path length the rays must travel to penetrate the atmosphere (discussed in Chapter 2). At noon the Sun achieves its greatest angle above the horizon and the surface receives its greatest input of solar radiation. From noon to sunset, the solar angle decreases and the receipt of solar radiation diminishes. But experience tells us that the warmest period of the day does not normally occur at noon; rather, it is sometime in the early afternoon, usually between 2:00 and 4:00 P.M.

To explain the lag between maximum insolation and temperature, we must consider the other energy transfer mechanisms: longwave radiation, convection, and conduction into the ground. Surface temperature increases as long as energy gained by the surface is greater than that which is lost. Although solar radiation begins to decrease after noon, the surface energy budget continues to run a surplus until later in the day, at which point the temperature begins to fall. Exactly when this occurs depends on local conditions, including the amplitude of the solar heating curve, wind speed, surface wetness, and the heat capacity and conductivity of the ground, to name a few.

A similar pattern occurs on an annual basis. The Northern Hemisphere experiences its greatest amount of solar radiation, but not its highest temperatures, on the June solstice. For about 4 to 6 weeks following the solstice, the amount of incoming radiation continues to exceed longwave radiation lost to space. The resultant positive value of net radiation leads to a continual hemispheric warming until sometime in July or August, when maximum temperatures are achieved.

Latitudinal Variations

The overall balance between incoming and outgoing radiation discussed in previous sections applies to the planet as a whole, but not to any particular place. Figure 3–15 shows that the balance between incoming and outgoing radiation varies with latitude. On an annual basis, the Earth surface–atmosphere system gains more radiation

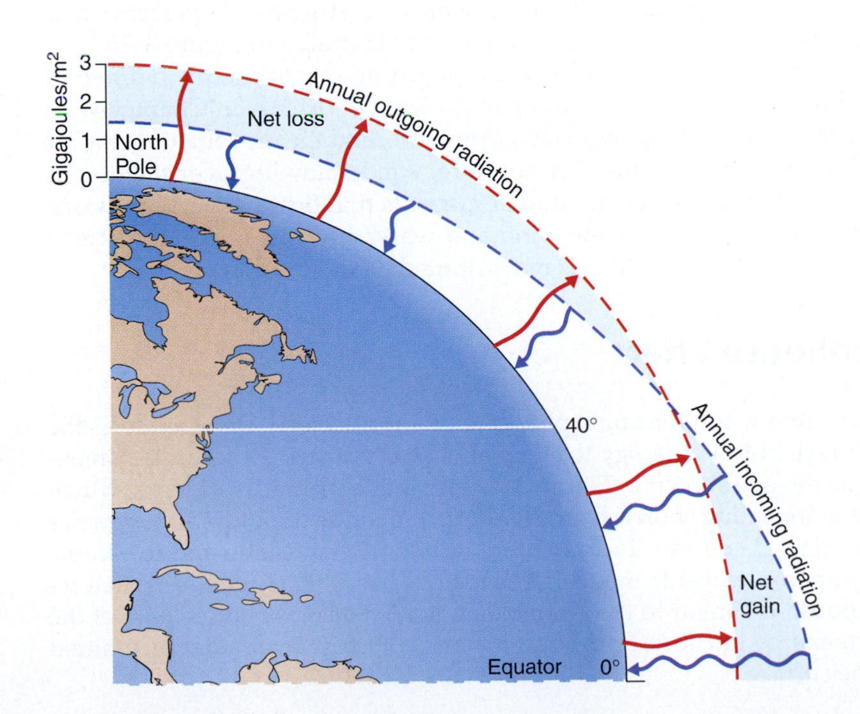

◀ **Figure 3–15**
Annual average net radiation values for the atmosphere and surface combined. At latitudes greater than about 38° north and south, a radiant energy surplus exists. Poleward of these latitudes, the atmosphere and surface lose more radiant energy than is gained.

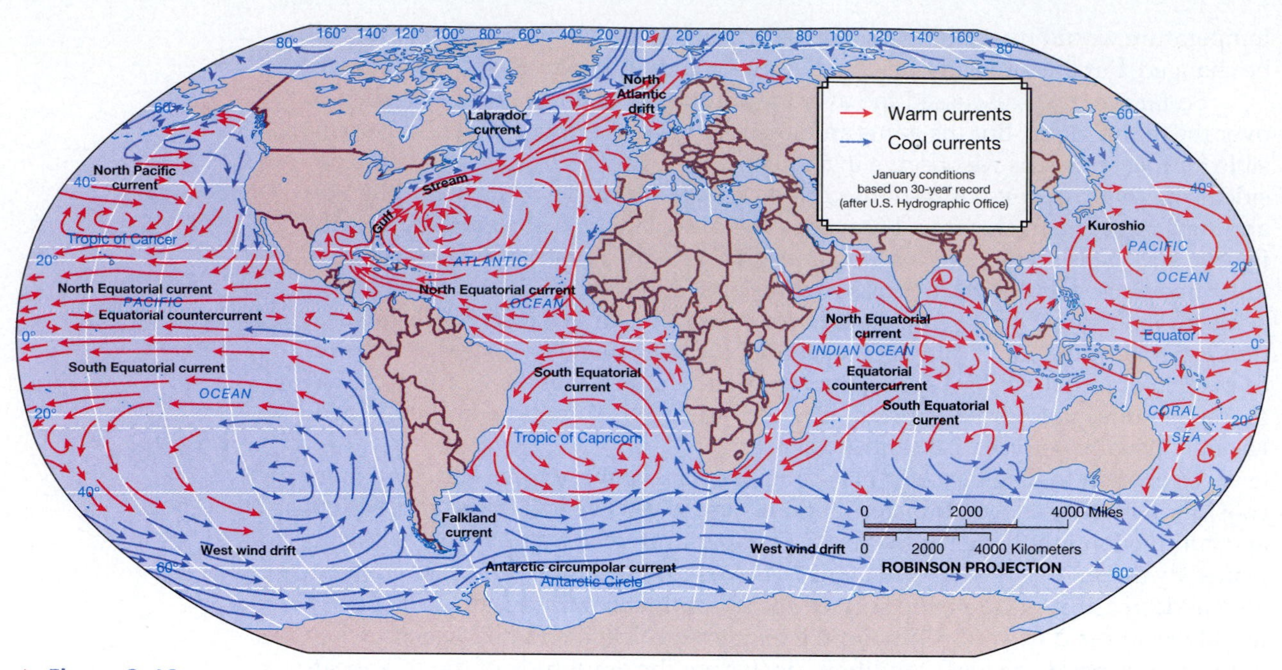

▲ **Figure 3–16**
The circulation of ocean currents. Those moving warm water are depicted by red arrows, those moving cold water by blue arrows.

than it loses between about 38° north and south latitudes, and it loses more than it gains poleward of these two parallels. The boundary between zones of radiant energy surpluses and deficits migrates seasonally. During the Northern Hemisphere summer, most of the area north of about 15° S gains more radiant energy than it loses. During the Northern Hemisphere winter, most areas south of about 15° N take in more radiation than they emit.

If no other processes were involved, the net gain of radiation between 38° north and south latitudes and the deficits outside this zone would cause the Tropics and subtropics to undergo continual heating and the extratropical regions to constantly cool. This does not occur, however, because the energy surplus at low latitudes is offset by the horizontal movement, or **advection**, of heat poleward. This transfer is accomplished primarily by the global wind systems (75 percent), and secondarily by the oceanic currents (25 percent), as illustrated in Figure 3–16.

Wind systems are generated because unequal amounts of energy in different places cause differences in pressure to occur. As we will discuss in Chapter 4, air tends to move from zones of higher to lower pressure, and the movement of the air is manifested as horizontal winds. Furthermore, winds blowing over the oceans exert a drag on the surface water and initiate currents that flow in nearly the same direction as the overlying air. As the wind and water currents move, they carry with them their internal heat, which is redistributed across the globe.

The Greenhouse Effect

The interactions that warm the atmosphere are often collectively referred to as the **greenhouse effect**, but the analogy to a greenhouse is not strictly accurate. Greenhouses, such as the one shown in Figure 3–17, are made primarily of glass, which is transparent to incoming shortwave radiation but opaque to outgoing longwave radiation. The glass therefore allows in more radiation than is allowed to escape, causing the temperature inside the structure to be warmer than outside. In that regard, a greenhouse is similar to the atmosphere, which also transmits most of the incoming solar energy but absorbs the vast majority of longwave radiation emitted upward by the surface.

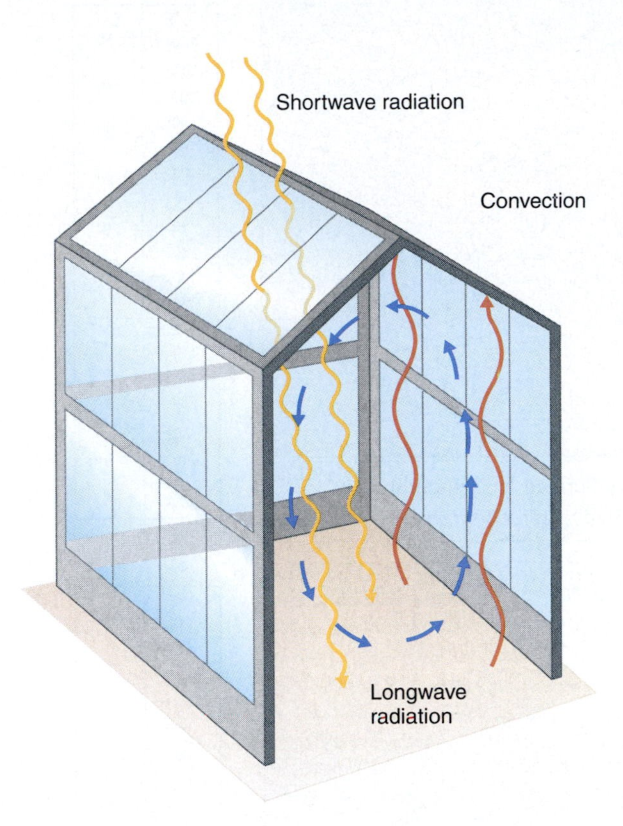

◀ Figure 3–17
The air inside a greenhouse is warmer than that outside because glass allows solar radiation to enter but is opaque to outgoing longwave radiation. It also precludes the movement of heat away from the surface by convection. This latter effect makes the action of a greenhouse different from that of the atmosphere. Therefore, the term "greenhouse effect" is not completely appropriate when applied to the atmosphere.

The analogy breaks down, however, when we incorporate the effect of convection. A greenhouse not only reduces the loss of energy by longwave radiation but also prevents the loss of sensible and latent heat by convection. In contrast, the greenhouse gases of the atmosphere don't impede the transfer of latent and sensible heat. Thus, it would be more accurate if the term "greenhouse effect" were replaced by "atmospheric effect."

If the atmosphere had none of the "greenhouse gases" that absorb outgoing longwave radiation, Earth would be considerably colder, on average, and the temperature would oscillate wildly from day to night. In fact, without greenhouse gases and clouds, Earth's surface would have an average temperature of –18 °C (0 °F), rather than the observed mean temperature of 15 °C (59 °F). The greenhouse effect keeps Earth warmer by absorbing most of the longwave radiation emitted by the surface, thereby warming the lower atmosphere, which in turn emits radiation downward. These days you find considerable discussion in the media about the possibility of climatic warming due to the anthropogenic increase of greenhouse gases such as carbon dioxide and methane. Although a few scientists still doubt the roll of increasing greenhouse gasses as an agent of global warming, the importance of greenhouse gases in the present-day climate is indisputable. We discuss the climate change issue in Chapter 16.

Global Temperature Distributions

One of the most immediate and obvious outcomes of radiation gain or loss is a change in the air temperature. Figure 3–18 shows the mean air temperature distributions for January and July and the differences between the two.

Each line on the maps, called an **isotherm**, connects points of equal temperature. Several patterns of large-scale temperature are apparent in (a) and (b). First, as expected, temperatures tend to decrease poleward in both hemispheres. Second, the latitudinal temperature gradient is greatest in the hemisphere experiencing winter (that is, the Northern Hemisphere in January and the Southern Hemisphere in July). The strong gradients in the winter hemisphere occur because the midday sun

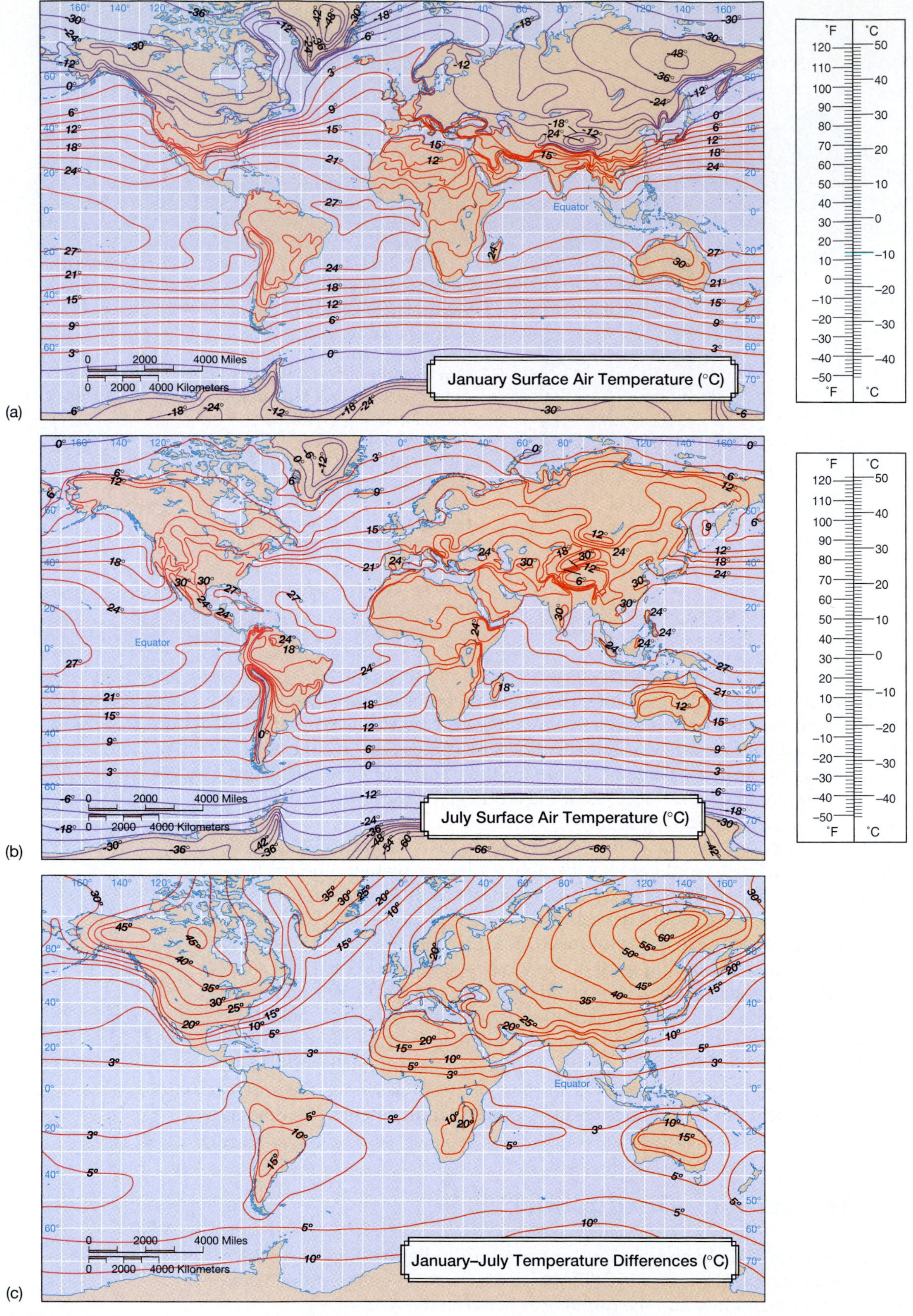

▲ **Figure 3–18**

Distribution of mean January (a) and July (b) surface air temperatures. The difference in temperature between the two months is shown in (c).

angles and the length of day both decrease with latitude. During summer, the lower midday sun angles at the higher latitudes are offset by longer days, so the temperature gradients are relatively weak.

The third feature of the maps is that the isotherms shift poleward over land in the hemisphere experiencing summer and shift equatorward during the winter. In other words, temperatures over land tend to be warmer in the summer than over adjacent water bodies, and colder during winter. Finally, the Northern Hemisphere has a steeper temperature gradient in its winter (a) than the Southern Hemisphere does in its winter (b). This is because the Southern Hemisphere consists of a much greater proportion of ocean than land. As you can see from (c), land masses have much greater annual ranges in temperature than do ocean bodies, for reasons discussed later in this chapter.

Influences on Temperature

Certain geographical factors combine to influence temperature patterns across the globe. These factors include latitude, altitude, atmospheric circulation patterns, local conditions, continentality, and ocean current characteristics along coastal locations.

Latitude

Most people know that outside the Tropics, the annual mean temperature decreases with latitude—Santa Claus, living at the North Pole, must have a wardrobe suitable for extremely cold conditions. Not only does latitude influence the average temperature, it also affects seasonal patterns. As described in Chapter 2, the tilt of Earth's axis influences the amount of solar radiation available at any latitude on any particular day. Within the Tropics, there is relatively little annual variation in the length of day and the midday solar angle, so energy receipts exhibit little change through the course of the year. Outside the Tropics, the noontime solar angles exhibit a range of 47°, with the lowest solar angles coinciding with the periods of shorter days. As a result, the availability of incoming radiation (and therefore the temperature) is more variable as distance from the equator increases.

Altitude

Any point within the atmosphere has some particular *altitude* (that is, its height above mean sea level). Altitude is not synonymous with *elevation*, the distance above sea level for a land surface. For example, some particular city may have an elevation of 1000 m above sea level, while air that is 1000 m above the city would have an altitude of 2000 m. Altitude and elevation both deal with position relative to sea level, but the former relates to points within the atmosphere while the latter refers to the position of a land surface.

As was shown in Figure 1–10, temperatures in the troposphere typically decrease with altitude above sea level. This occurs because the surface is the primary source of direct heating for the troposphere, and increased altitude implies a greater distance from the energy source. Figure 3–19 contrasts day and night temperatures at three locations. Position A is located a couple of meters (about 6 ft) above the surface at sea level. Position B is 3000 m (about 10,000 ft) directly above A, and Position C is 3000 m above sea level but just a couple of meters above the mountain surface.

During the middle of the day, Position *A* responds to the absorption of solar radiation at the surface and warms as the latter transfers energy upward by convection and the emission of longwave energy. Position *B*, located a considerable distance from the surface, undergoes virtually no warming. Position C, although at the same altitude as *B*, is nearer to the primary source of warming, and its daytime temperature rises appreciably.

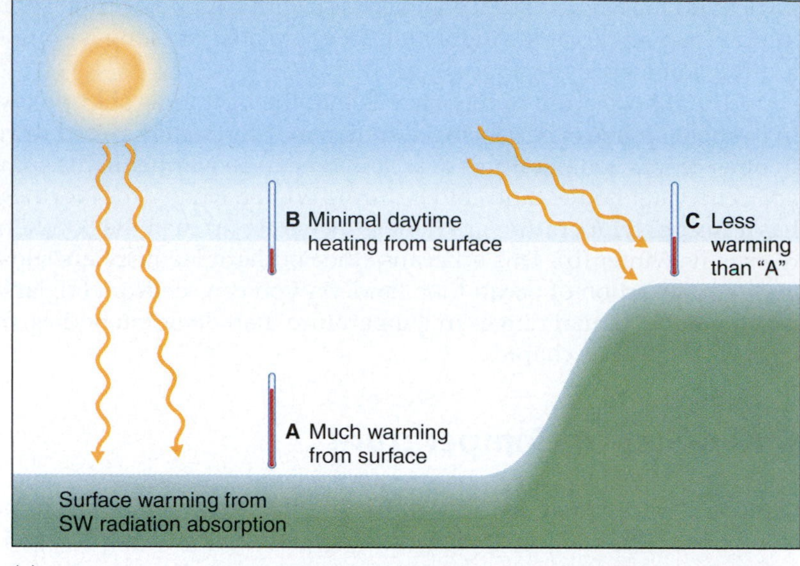

(a)

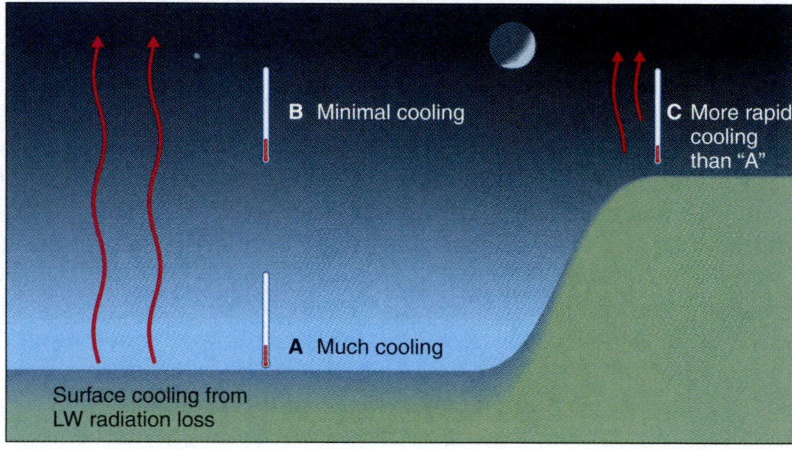

(b)

▶ **Figure 3–19**
Effects of elevation and altitude on daily temperature patterns.

At night, the surface below Position *A* cools by the emission of longwave radiation. The air 6 m above the surface also undergoes a lowering of temperature in response to the cooling of the underlying surface. At Position *B*, the air undergoes little cooling because of its distance from the surface. Position *C* chills rapidly because the sparse atmosphere above does not effectively absorb the outgoing radiation from the surface. Cooling is often enhanced by rapid evaporation into the drier high-altitude air. Perhaps while on a mountain camping trip you've noticed how rapidly a hot meal becomes stone cold? This is a consequence of strong evaporation and weak atmospheric counterradiation, compared to lower elevations. The overall effect of elevation above sea level and altitude above the ground, therefore, is that cooling and warming cycles are minimal high above the surface, and the air just above a high-elevation surface will undergo greater cycles of cooling and warming than will air at the same altitude but farther above the surface. It is not uncommon that the nighttime loss of longwave radiation at the surface can lead to lower temperatures near the surface than aloft. This reversal of the normal pattern in the troposphere (lower temperatures with increasing distance from the surface) is known as an **inversion**. Inversions have important characteristics that are described in Chapter 6.

3–2 Physical Principles

Earth's Equilibrium Temperature

If Earth had no atmosphere, and therefore no greenhouse effect (see page 82), the mean temperature of the planet would be much colder than what we experience. Using the principles discussed thus far, we can easily estimate the magnitude of the greenhouse effect. To do so, we will compute the equilibrium temperature for a planet having no atmosphere. By comparing the computed and observed temperatures, we will determine the importance of the atmosphere's influence on Earth's temperature.

First, assume the planet acts as a blackbody with regard to longwave radiation, that the planetary albedo is 30 percent, and that the solar constant is 1367 watts per square meter. If Earth were a flat disk perpendicular to incoming radiation, each square meter would receive 1367 joules/second. But Earth is not a flat disk; it is a sphere, the surface area of which is four times larger than that of a disk of the same radius. Thus the intensity of radiation averaged over the sphere is one-fourth as large as for the imaginary disk. Because of this, each square meter of Earth receives 1367/4, or 341 watts/m^2. Given the planetary albedo of 30 percent, it must be that 70 percent of this incoming radiation is absorbed. In other words, total absorbed radiation is

$$1367 \text{ watts/m}^2 \times 0.25 \times 0.7 = 239.2 \text{ watts/m}^2$$

The planet must lose exactly as much energy as it gains, and the intensity of radiation for a blackbody is determined by rearranging and applying the Stefan-Boltzmann law. Recall that the Stefan-Boltzmann law for a blackbody states that

$$I = \sigma T^4$$

We know, however, that the intensity of radiation for the planet without an atmosphere must be 239.2 watts per square meter. We therefore rearrange the equation to solve for T, rather than I, to get

$$T^4 = I/\sigma$$

which can be reduced to

$$T = (I/\sigma)^{0.25}$$

Using the values, $\sigma = 5.67 \times 10^{-8}$ watts/(m^2 K^4) and $I = 239.2$ watts/m^2, the equilibrium temperature works out to 254.9 K (–18.3 °C, 0 °F). Thus, the mean temperature of Earth would be far colder without an atmosphere.

Note that our calculation is highly simplified and somewhat questionable. For example, we used 30 percent for the planetary albedo, but that value arises in part from the albedo of the atmosphere, which our imaginary planet lacks. Should we therefore have used present-day surface albedo in the computation? Perhaps, but surface albedo on the real Earth is in part the result of temperature, the very thing we are trying to compute! Ideally, we would treat albedo as a variable, allowing it to respond to changes in temperature.

We see that even a beginning question about the global effect of greenhouse gases raises complications not easily addressed by a simple model. Given this, it is not surprising that realistic computer models of atmospheric behavior are enormously complex, requiring huge computer resources. Nonetheless, computer models have become indispensable for daily weather forecasting (Chapter 13) and tell us much of what we know about potential climatic changes due to human activity (Chapter 16).

Atmospheric Circulation

As we will discuss in Chapter 8, an organized pattern of mean atmospheric pressure and air flow across the globe strongly influences the movement of warm and cold air, with a direct effect on temperature. These large-scale circulation patterns also influence the development of cloud cover, which has an indirect effect on temperature. Subtropical areas (latitudes 20° to 30° in both hemispheres), for example, tend to be regions of minimal cloud cover, and insolation passing through the atmosphere undergoes less attenuation on its way to the surface. In contrast, equatorial regions are often cloudy in the afternoon and experience a greater attenuation of incoming solar radiation. The result is that the highest temperatures on Earth tend to occur not at the equator but in the subtropics. Many other patterns of atmospheric circulation affect regional temperatures.

Contrasts Between Land and Water

Because the atmosphere is heated primarily from below, it should be no surprise that the type of surface present influences air temperature. The most important surface influence arises because of contrasts between land and water. Water bodies are far more conservative than land surfaces with regard to their temperature, taking longer to warm and cool when subjected to comparable energy gains and losses.

San Francisco, California, and St. Louis, Missouri, together provide an interesting example of the effect of **continentality**—the effect of an inland location that favors greater temperature extremes. They are at similar latitudes and elevations and both are subject to a predominantly west-to-east airflow. Yet San Francisco, situated along the Pacific Coast, has more moderate temperatures than does its inland counterpart. During July, for example, its mean temperature is 15 °C (59 °F), while St. Louis averages 26 °C (79 °F). The temperature difference is similar but runs in the opposite direction in January when San Francisco has an average temperature of 10 °C (50 °F) and St. Louis a mean of 0 °C (32 °F).

Four reasons cause water bodies to be more conservative than land masses with regard to temperature:

1. The specific heat of water is about five times as great as that of land.
2. Radiation received at the surface of a water body can penetrate to several tens of meters deep and distribute its energy throughout a very large mass. In contrast, the insolation absorbed by land heats only a very thin, opaque surface layer.
3. The warming of a water surface can be reduced considerably because of the vast supply of water available for evaporation. Because much energy is used in the evaporative process, less warming occurs.
4. Unlike solid land surfaces, water can be easily mixed both vertically and horizontally, allowing energy surpluses from one area to flow to regions of lower temperature.

Warm and Cold Ocean Currents

Global Sea Surface Temperatures—Climatology

Global Sea Surface Temperatures—Actual

Figure 3–16 depicts the location of warm and cold ocean currents. The warm currents typically move poleward in the western portion of the ocean basins near the east coats of continents in the middle latitudes, carrying large amounts of energy with them. Similarly, along the eastern margins of oceans, cold ocean currents dominate in the middle latitudes. Where the water temperatures are high, heat is transferred to the atmosphere and promotes higher air temperatures. Thus, the existence of a warm ocean current offshore can cause a location along the east coast of a continent to have higher temperatures than would a cold current offshore along a west coast.

Compare, for example, Los Angeles, California, and Charleston, South Carolina, two coastal locations at similar latitudes and elevations. Summers are considerably warmer at Charleston than at Los Angeles, largely (but not entirely) because the temperature of the ocean off Charleston is higher than the temperature of the ocean off Los Angeles. The high water temperatures of the Gulf of Mexico and the western margin of the Atlantic Ocean allow the transfer of an enormous amount of heat to the atmosphere, and the average July temperature in Charleston of 31 °C (88 °F) is substantially warmer than that of Los Angeles (23 °C; 73 °F). In winter, temperatures are lower at Charleston than at Los Angeles because the westerly winds blowing toward Los Angeles are subject to the moderating effects of the Pacific, whereas Charleston's winter temperatures are affected by colder prevailing winds passing over the continental interior. Thus, the influence of atmospheric circulation, which we noted previously, can interact with the position along the edge of a continent in influencing temperature patterns.

Local Impacts on Temperature

A number of site-specific factors, such as slope orientation and steepness, can influence the temperature characteristics of an area. In the Northern Hemisphere, slopes that are south-facing receive midday sunlight at a more direct angle than do those oriented in other directions, thereby promoting greater energy receipt and higher surface temperatures. The heating of south-facing slopes often results in a greater amount of drying than on the opposite, north-facing slopes. Vegetation patterns often respond to the change in microclimate, with plants intolerant of dry conditions occupying north-facing slopes. Such a pattern is shown in Figure 3–20.

Densely wooded areas also have different temperature regimes than do those devoid of vegetation cover. In a region like that shown in Figure 3–21, a tall, dense vegetation cover reduces the amount of sunlight hitting the surface during the day; and considerable evaporation of water from leaf surfaces occurs. At night the plant canopy reduces the net longwave radiation losses. These factors lead to lower daytime temperatures but warmer evenings.

The effects of vegetation on local climate can often be put to use in a manner that increases human comfort. For example, it is often a good idea to plant deciduous trees on the equatorward side of houses. During the warm summer months, trees can cast shade on houses, thereby keeping interior temperatures down or reducing air conditioning costs. When winter arrives, the loss of the leaves from the deciduous trees minimizes the reduction in sunlight that can help keep the building warm.

Measurement of Temperature

For everyday purposes, the measurement of temperature is a simple and routine procedure in which the expansion and contraction of the fluid in a thermometer is noted. The most accurate thermometers contain mercury, the only metal that exists as a liquid at normal Earth temperatures. Less expensive models are available using dyed alcohol.

It is often useful to know the daily maximum and minimum temperatures. A **maximum thermometer** is very similar to a regular thermometer, but with two differences. Unlike a regular thermometer, it must contain mercury—it cannot use dyed alcohol. The other difference is that in the tube just beyond the bulb is a very narrow constriction (Figure 3–22) that allows the mercury to expand outward when the temperature increases but prevents it from contracting back into the bulb when the temperature decreases. The temperature shown on the maximum thermometer

◀ **Figure 3–20**
Slope aspect is one of the local factors affecting temperature. In the Northern Hemisphere, north-facing slopes typically receive less intense daytime heating and therefore exhibit lower temperatures. This retards the rate of surface evaporation, making more water available for plant life; thus, different types of vegetation are encountered on north- and south-facing slopes.

(a)

(b)

▶ **Figure 3–21**
A dense vegetation cover lowers daytime temperatures because of its shadowing effect on incoming solar radiation (a). At night (b) the forest canopy retards the loss of longwave radiation to space, resulting in higher nighttime temperatures than in the open.

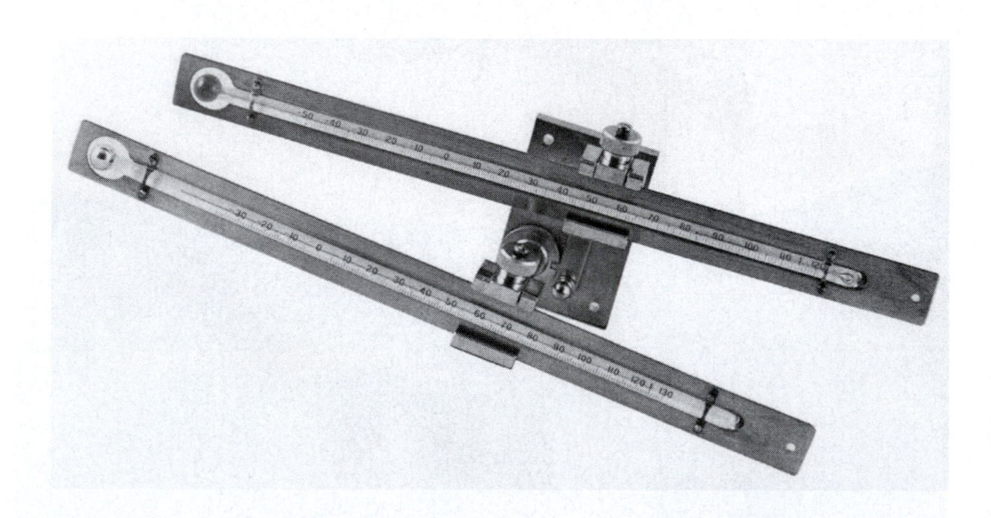

▶ **Figure 3–22**
Maximum thermometers (bottom) have a small constriction that only allows the mercury to move outward from the bulb. Minimum thermometers (top) have a sliding index indicating the lowest temperature since the instrument was last reset.

indicates the highest temperature experienced since the last time it was reset. Resetting the thermometer is easy; the mercury can be forced down simply by shaking the thermometer downward (note that an oral thermometer used at home or in a medical clinic is merely a type of maximum thermometer). If the instrument is mounted on a pivoting base, it can be reset by being spun so that centripetal force propels the mercury back into the bulb.

A **minimum thermometer** (see Figure 3–22) is also similar to a regular thermometer, except that it can only contain dyed alcohol and has within it a small index shaped like a weightlifter's dumbbell. If the index is at the end of the alcohol and the temperature is decreasing, surface tension (the force that holds water molecules together) pulls the index toward the bulb. When the temperature increases, the index remains at its present position as the alcohol expands away from the bulb. Minimum thermometers are mounted horizontally with a latch that can be released to allow the instrument to be inverted. The instrument is reset by turning it upside down, allowing the index to slowly slide down to the end of the alcohol.

Another instrument for the measurement of temperature is the **bimetallic strip**, which consists of two thin strips of different metals bonded together. Because all metals have different rates of expansion and contraction with temperature, one undergoes a greater change in length than the other, causing the strip to bend. A pointer and scale are attached to the bimetallic strip, whose bending is amplified by a lever. When this mechanism is coupled with a rotating drum and a pen, the resulting **thermograph** (Figure 3–23) gives a continuous record of temperature. Thermostats for the heating and air conditioning units of many homes use bimetallic strips to determine the room temperature.

Considerably more sophisticated instruments for the measurement of temperature also exist. One such device is a **thermistor**, an instrument that sends an electrical current through a very thin wire exposed to the air. The temperature of the wire influences its resistance to the electrical current, which allows the thermistor to determine temperature by measuring the reduction in the current flowing through the wire.

◀ Figure 3–23
A thermograph.

3–3 Special Interest

Recent Severe Heat Waves

Summer heat waves are certainly no rarity in the eastern United States and Canada. Sooner or later everybody endures an episode of unpleasantly high temperatures. But heat waves can cause much more than a few days of discomfort—they can kill. One of the most notable heat waves of the last few decades was the relatively brief but severe event in mid-July 1995 in the north-central United States.

Although extremely high temperatures occurred from the Great Plains to the Atlantic Coast, nowhere was the problem more acute than in Chicago, Illinois, where 525 people died from the heat. The heavy mortality resulted from a combination of high temperatures (Midway Airport recorded an all-time high temperature of 41.1 °C, or 106 °F) and unusually high humidities. The heat and humidity combined to make the "apparent temperature" equivalent to 47 °C (117 °F). Though the searing daytime heat created plenty of misery on its own, it is believed that the most important factor leading to the many deaths is the fact that the extreme heat went uninterrupted, with the apparent temperature exceeding 31.5 °C (89 °F) for nearly 48 consecutive hours. (Recent research suggests that such conditions pose a greater danger than do brief periods of more extreme heat.)

Four years later, in 1999, another major July heat wave occurred in the eastern two-thirds of the United States. Once again, Illinois was in the center of the action, with more than half of the 232 fatalities across the Midwest occurring in Chicago. Missouri was the second hardest-hit state, with 61 fatalities. All across the region, power outages occurred from excessive demand, roads buckled, and crops wilted in the field.

As July gave way to August, the heat moved eastward toward the Atlantic states, where it broke numerous weather records. Charleston, South Carolina, had an all-time high temperature of 40.5 °C (105 °F). Augusta, Georgia's high temperature of 39.4 °C (103 °F) came on the sixth consecutive day in which the record for the daily maximum temperature was tied or exceeded. And on August 8, Raleigh-Durham, North Carolina, broke the 100 °F mark (37.8 °C) for the eleventh time that summer.

Though the death tolls were not as high as they were in 1995 and 1999, major heat waves in 2000 and 2001 resulted in numerous fatalities across the southern and southeastern United States. Such events serve as important reminders that about one-third of all deaths directly attributed to weather events in the United States result from extremely high temperatures. The decade of the 1990s was remarkably warm relative to other periods in recorded history. This is particularly noteworthy because the topic of human-induced climatic warming has been a major issue for scientists and policy makers. This matter will be discussed further in later chapters of this book.

Thermistors are fast-response instruments, meaning that they rapidly register changes in the ambient temperature. Compare that to what might happen if you were to put a thermometer (a slow-response device) in a refrigerator for an hour or so and then remove it. Although your kitchen is probably much warmer than the inside of your refrigerator, it would take several minutes before the instrument accurately measured the temperature of the room. When slow response is not a problem, thermometers can be perfectly acceptable instruments. Thermistors or other fast-response instruments must be used in situations where rapid temperature changes may be encountered. Experimental research stations that measure the transfer of sensible and latent heat in the atmosphere near the surface require such instruments. *Radiosondes*, packages of weather instruments carried by balloons, likewise require fast-response instruments as they rapidly ascend through air with highly varying temperature characteristics.

▲ **Figure 3–24** Temperature instruments must be kept in shelters that protect them from the absorption of solar radiation. Note that the box is painted white, has slats to allow the movement of air, and has a door that opens on the north side (in the Northern Hemisphere).

Instrument Shelters

You have no doubt heard statements such as this one: "It was 100 degrees in the shade." This is a common expression but one that might have originated from the Department of Redundancy Department because, to be meaningful, temperature must *always* be measured in the shade. A thermometer exposed to direct sunlight will be warmed by the absorption of insolation and assume a temperature greater than that of the air. But when we refer to "air" temperature, we are really concerned with that of the air—not that of a thermometer. As a result, temperature-measuring instruments should always be kept in an instrument shelter like the one in Figure 3–24.

Because an instrument shelter is designed to reduce the influence of incoming radiation on the instruments, certain design criteria must be met. The shelter should be painted white so that its albedo will be maximized and reduce the absorption of radiation. It should also be paneled with slats rather than solid side walls to permit the free flow of air and the removal of any heat that might otherwise accumulate. The door must be mounted on the north side of the box (in the Northern Hemisphere) so that direct sunlight will not strike the instruments if the door is opened during the middle of the day. Finally, the shelter must conform to a standardized height, so that the thermometers will be mounted at 1.52 m (5 ft) above the ground.

In the United States, temperature is observed hourly at National Weather Service offices and Federal Aviation Agency (FAA) facilities at airports across the country. This network of observers is supplemented by a large number of cooperative agencies (such as U.S. Forest Service stations) and individual volunteers. At the cooperative stations, observations are made once or twice daily, normally in the early morning or midafternoon. Environment Canada is responsible for temperature data acquisition in Canada.

Temperature Means and Ranges

In just about all aspects of daily life, we use descriptive statistics to talk about the things around us. Although the concept of an average, or mean, value of a property is fairly straightforward, applying the concept involves occasional complications. For example, trying to determine a daily mean temperature poses a dilemma—exactly how many times during the day must we measure it to obtain a true mean? We could make observations every hour, every minute, or even every second, with each method giving us a separate value.

The standard procedure is simple—the *daily mean* is defined as the average of the maximum and minimum temperature for a day. The advantage of this method is that the daily mean can be obtained even at weather stations having just the most basic instrumentation; a minimum and maximum thermometer are all we need to compute daily mean temperaturess. The disadvantage of using just the maximum and minimum temperature is that it introduces a bias. Observe the daily temperature pattern for a particular day, shown in Figure 3–25. Notice that the nighttime temperatures remain nearly equal to the minimum throughout most of the night, while afternoon temperatures are near the maximum for only a few hours. If we were to obtain a daily mean by taking 24 hourly spaced observations and dividing by 24, the mean value would be lower than that obtained by using just the maximum and minimum temperatures. In other words, our calculation would be biased. Nonetheless, averaging the maximum and minimum temperatures is the accepted method for obtaining daily mean temperatures.

The *daily temperature range* is obtained simply by subtracting the minimum temperature from the maximum.

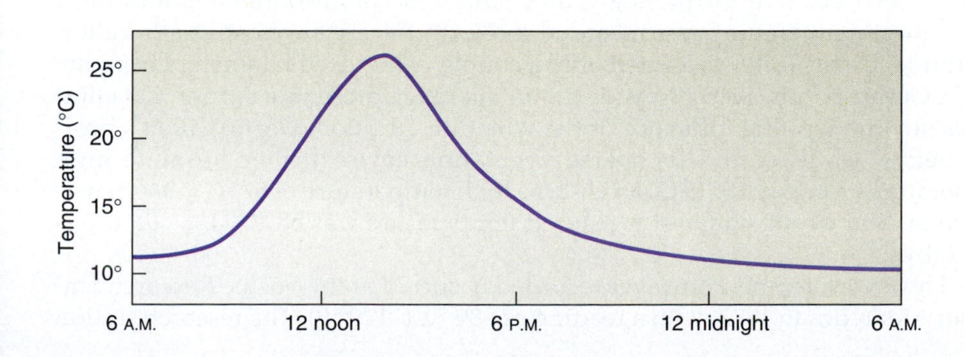

Figure 3–25
A continuous plot of temperature over a 24-hour period with clear skies. Note that the temperature is near that of the maximum for a relatively short time period. In contrast, the air temperature is near the daily minimum throughout most of the night.

3–4 Forecasting

Effects of Cloud Cover and Wind Speed on Temperature

Cloud cover and wind speed are two valuable indicators for predicting maximum and minimum temperatures. Cloud cover has its most obvious effect in reducing the amount of solar radiation received at the ground, causing a reduction in daytime heating. But overcast conditions also retard the nighttime loss of longwave radiation, thereby keeping overnight temperatures higher than they would otherwise be. Thus, persistent, extensive cloud cover tends to suppress daily temperature ranges.

Strong winds also moderate daily temperature ranges. Recall that higher wind speeds promote greater forced turbulence. When turbulence is increased, the enhanced vertical movement causes any small parcel of air immediately in contact with the ground to be quickly displaced upward and replaced by another parcel. The result is that the sensible heat transferred from the ground is distributed to a greater mass of air, which reduces the increase in temperature near the surface. (Not only do strong winds reduce the rate of warming, they also make the air feel colder at any particular temperature, as discussed later in this chapter.) The same process works at night to inhibit cooling. If a lot of forced convection occurs, no parcel of air will remain in contact with the chilled surface long enough to undergo substantial cooling. In the absence of forced convection, a shallow layer of air is subject to cooling for an extended time period, and temperatures can drop rapidly.

The following data from Chicago's Midway Airport on October 25–26, 2001, show just how effectively overcast cloud cover and strong winds can suppress temperature changes.

Time (CST)	Temperature °F (°C)	Wind (mph)
9 A.M.	37.9 (3.3)	WNW 23
8 A.M.	37.9 (3.3)	W 18
7 A.M.	37.9 (3.3)	W 20
6 A.M.	37.9 (3.3)	W 23
5 A.M.	37.9 (3.3)	W 20
4 A.M.	37.9 (3.3)	W 21
3 A.M.	37.9 (3.3)	W 20
2 A.M.	37.9 (3.3)	W 17
1 A.M.	37.9 (3.3)	W 21
12 A.M.	37.9 (3.3)	W 24
11 P.M.	39.0 (3.9)	W 26
10 P.M.	39.0 (3.9)	WNW 26
9 P.M.	41.0 (5.0)	W 16

In this case, the combined effects of cloud cover and vigorous winds allowed only a 1.7 °C (3.1 °F) drop in temperature between 9 P.M. and 9 A.M. This example illustrates why meteorologists always take into account the role of cloud cover and wind in the production of their forecasts.

Having obtained daily mean temperatures for an entire month, we calculate the *monthly mean temperature* simply by summing the daily means and dividing by the number of days in the month. Similarly, an *annual mean temperature* is obtained by summing the monthly means for a year and dividing by 12. The *annual range* is obtained as the difference between the highest and lowest monthly mean temperatures.

Global Extremes

More impressive than mean temperatures are the maximum and minimum values ever recorded. Not surprisingly, they tend to occur at continental locations. The highest temperature ever recorded in North America was at Death Valley, California. Death Valley is located only a couple of hundred kilometers from the Pacific Ocean, but the Sierra Nevada mountain range presents a barrier that eliminates any moderating influence of the water. In addition, Death Valley's position below sea level and its sparse vegetation cover further promote high temperatures. On July 10, 1913, an all-time high temperature of 57 °C (134 °F) was recorded. The world's highest measured temperature was 58 °C (136 °F) at Azizia, Libya.

The lowest temperature ever recorded occurred at the Vostoc Research Station in Antarctica in 1983, with a reading of –89 °C (–129 °F). The research station is located atop thousands of meters of glacial ice, thereby combining the effects of high latitude, a continental locale, and a sparse atmosphere. In North America, the record low temperature of –63 °C (–81 °F) was observed at Shag, Yukon, in February 1947.

Undoubtedly, more extreme temperatures have occurred across the globe and North America at locations without temperature observation stations.

Temperature and Human Comfort

Temperature by itself exerts a major impact on human comfort, but the discomfort caused by high or low temperatures can be compounded by other weather factors. If low temperatures are accompanied by windy conditions, a person's body loses heat much more rapidly than it would under calm conditions, due to an increase in sensible heat loss. Thus, a windy day with a temperature of –2 °C (–1 °F) might feel colder than a calm day at –40 °C (–40 °F). As a result, when temperatures are low, it is common for weather reports to state both the actual temperature and how cold that temperature actually feels, the **wind chill temperature index** (or simply the *wind chill temperature*).

The earliest wind chill temperature index was based on low-temperature research conducted in Antarctica in 1945. Though the original research was not intended to quantify the combined effect of wind and low temperatures on people, the index derived from that work was widely used for more than a half century. But meteorologists in Canada and the United States were aware of certain shortcomings in the index—the most notable of which was that it was inaccurate! People living in very cold environments were familiar enough with low temperatures to know that the wind chill values posted did not accurately describe how cold they really felt. Thus, researchers at the National Weather Service and the Meteorological Service of Canada joined forces and in November 2001 released a new wind chill index based on the results of tests on human volunteers. The six men and six women who provided their services took 90-minute walks on treadmills under controlled temperature and wind conditions. Four external temperature sensors were placed on each volunteer's face to determine surface heat loss. An additional temperature probe was placed inside each participant's cheek to determine the temperature gradient between the external skin and the inside of the mouth, and a small rectal probe was used to monitor body core temperatures. Observations of heat transfer from and within the subjects' bodies enabled the researchers to calculate cooling rates due to varying wind speeds. Tables 3–1 and 3–2 show new wind chill tables based on these experiments.

The new wind chill temperature index has values that differ markedly from those of the original index—usually indicating less severe wind chill than those obtained using the old formula. For example, at a temperature of –15 °C (5 °F) and a wind speed of 65 km/hr (40 mph), the old formula indicated a wind chill temperature of –43 °C (–45.4 °F). Under the same conditions the new index yields a wind chill temperature of –30 °C (–22 °F), a full 13 °C (2 °F) higher than that given by the old formula.

Despite the improvements in the new formula, it still has its shortcomings. For example, the potential warming effect of sunlight on a person's body is not taken into account in the wind chill calculations. Nonetheless, the index provides people with guidance in the way they should dress and the types of activities they should undertake under cold and windy conditions. This is especially important when one considers the fact that extreme cold is the number one cause of fatalities directly attributable to weather.

Just as windy conditions can make low temperatures feel even colder, high humidities can cause warm days to feel oppressively hot. As a result, a heat index has been calculated that incorporates the effect of high atmospheric moisture at high temperatures. This index is discussed in Chapter 5, Atmospheric Moisture.

Table 3–1 • Wind Chill Temperature (°C)

Wind (km/hr)	Temperature (°C) 5	0	−5	−10	−15	−20	−25	−30	−35	−40
5	4	−2	−7	−13	−19	−24	−30	−36	−41	−47
10	3	−3	−9	−15	−21	−27	−33	−39	−45	−51
15	2	−4	−11	−17	−23	−29	−35	−41	−48	−54
20	1	−5	−12	−18	−24	−31	−37	−43	−49	−56
25	1	−6	−12	−19	−25	−32	−38	−45	−51	−57
30	0	−7	−13	−20	−26	−33	−39	−46	−52	−59
35	0	−7	−14	−20	−27	−33	−40	−47	−53	−60
40	−1	−7	−14	−21	−27	−34	−41	−48	−54	−61
45	−1	−8	−15	−21	−28	−35	−42	−48	−55	−62
50	−1	−8	−15	−22	−29	−35	−42	−49	−56	−63
55	−2	−9	−15	−22	−29	−36	−43	−50	−57	−63
60	−2	−9	−16	−23	−30	−37	−43	−50	−57	−64
65	−2	−9	−16	−23	−30	−37	−44	−51	−58	−65
70	−2	−9	−16	−23	−30	−37	−44	−51	−59	−66
75	−3	−10	−17	−24	−31	−38	−45	−52	−59	−66
80	−3	−10	−17	−24	−31	−38	−45	−52	−60	−67

Table 3–2 • Wind Chill Temperature (°F)

Wind (mph)	Temperature (°F) 40	35	30	25	20	15	10	5	0	−5	−10	−15	−20	−25	−30	−35	−40
5	36	31	25	19	13	7	1	−5	−11	−16	−22	−28	−34	−40	−46	−52	−57
10	34	27	21	15	9	3	−4	−10	−16	−22	−28	−35	−41	−47	−53	−59	−66
15	32	25	19	13	6	0	−7	−13	−19	−26	−32	−39	−45	−51	−58	−64	−71
20	30	24	17	11	4	−2	−9	−15	−22	−29	−35	−42	−48	−55	−61	−68	−74
25	29	23	16	9	3	−4	−11	−17	−24	−31	−37	−44	−51	−58	−64	−71	−78
30	28	22	15	8	1	−5	−12	−19	−26	−33	−39	−46	−53	−60	−67	−73	−80
35	28	21	14	7	0	−7	−14	−21	−27	−34	−41	−48	−55	−62	−69	−76	−82
40	27	20	13	6	−1	−8	−15	−22	−29	−36	−43	−50	−57	−64	−71	−78	−84
45	26	19	12	5	−2	−9	−16	−23	−30	−37	−44	−51	−58	−65	−72	−79	−86
50	26	19	12	4	−3	−10	−17	−24	−31	−38	−45	−52	−60	−67	−74	−81	−88
55	25	18	11	4	−3	−11	−18	−25	−32	−39	−46	−54	−61	−68	−75	−82	−89
60	25	17	10	3	−4	−11	−19	−26	−33	−40	−48	−55	−62	−69	−76	−84	−91

Thermodynamic Diagrams and Vertical Temperature Profiles

The distribution of temperatures across Earth's surface is a basic component of daily weather. But temperature variations don't only occur horizontally; temperatures are just as likely to vary in the vertical as well, and those variations can greatly affect atmospheric behavior. Rapid decreases with height, for example, increase the potential for cloud development and even severe thunderstorms. Temperature inversions (increasing temperature with altitude) have their own implications for the atmosphere, including the suppression of atmospheric mixing that can concentrate air pollutants near the surface.

Thermodynamic diagrams, which depict the vertical profiles of temperature and humidity with height above the surface, provide extremely important information to the forecaster. These charts enable forecasters to determine the height and thickness of existing clouds and the ease with which the air can be mixed vertically (an important consideration in the near-term development of clouds and precipitation). The data on the charts are obtained from instrument packages called **radiosondes** that are carried aloft by weather balloons twice a day at weather stations across the globe.

Though extremely valuable, thermodynamic diagrams are not altogether simple to understand, and some of the information one can obtain from them is beyond the scope of this text. Consequently, rather than trying to present a complete guide to the use of these charts all at once, we will introduce various elements of thermodynamic diagrams as they pertain to the individual discussions in this text. As we progress, you will see just how important such charts can be to day-to-day forecasting. For now we will only discuss the plotting of vertical temperature profiles on these charts. This will provide a foundation for further understanding in later chapters.

Thermodynamic diagrams come in several varieties. The easiest to use is the **Stuve** (pronounced STU-vay) diagram (a simplified version is shown in Figure 3–26). The Stuve diagram starts with a rectangular grid, with temperature plotted on the horizontal axis and pressure on the vertical axis. Three things must be noted at the onset, the first of which is that the chart allows us to plot temperature as a function of the *pressure level*—not the height above the surface (the altitude of a particular pressure level—such as the 500 mb level—is not a constant, but varies over

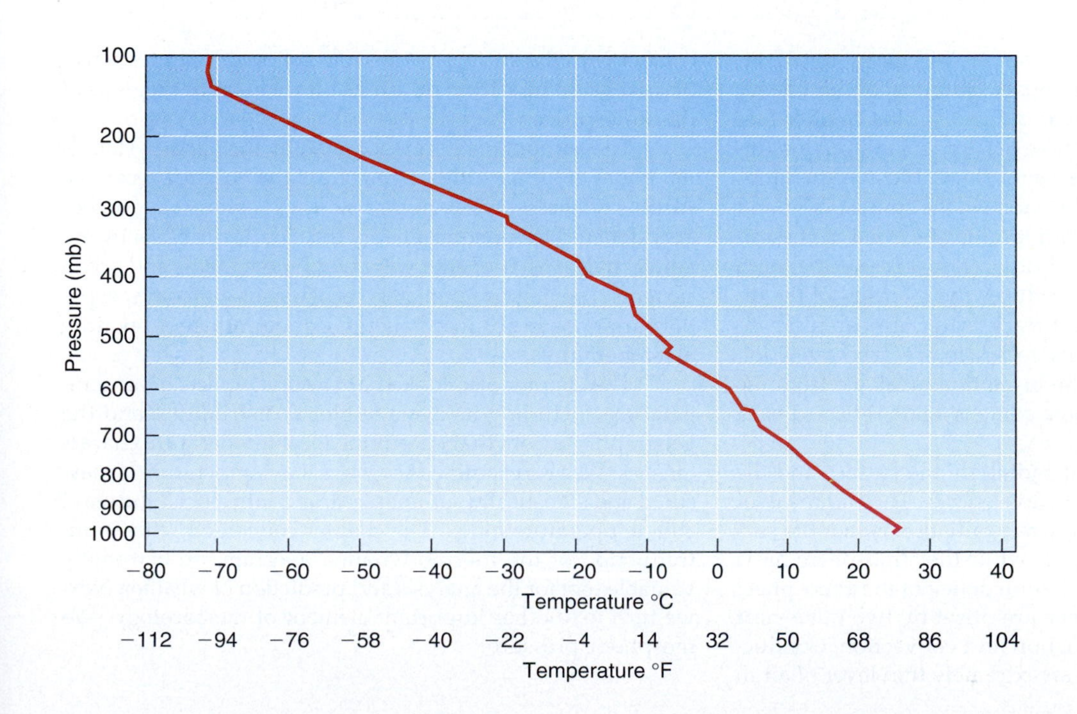

Figure 3–26 A simplified Stuve diagram showing the temperature profile for Slidell, Louisiana, on the morning of June 9, 2002.

time and from one location to another). This convention allows a more direct application of meteorological laws than would a plotting of temperature against altitude.* Second, you should note that the pressure decreases upward along the vertical axis. This is a response to the fact that atmospheric pressure invariably decreases with height from the surface. Finally, you should observe that the pressure is not plotted on a linear scale. Instead, the axis is plotted on a nearly logarithmic scale so that the 1000 mb and 900 mb lines are closer together than are the 300 mb and 200 mb lines. This mimics the way that pressure actually decreases in the real atmosphere.

Let's look at the example of a simplified Stuve in Figure 3–26. The temperature plotting (or *sounding*, as these plots are often called) was based on the ascent of a radiosonde launched at Slidell, Louisiana, June 9, 2002, at 12 noon, Greenwich mean time (7 A.M. CDT). In all, observations were recorded at 52 levels from the surface all the way up to the 100 mb level (well into the stratosphere). The data are available in text form, which is useful for detailed observations, and then plotted automatically onto the Stuve diagram. At the surface, the air pressure was 1015 mb and the temperature was 22.8 °C (73.0 °F). As the radiosonde initially rose from the surface, it recorded increasing temperatures—an inversion—up to a pressure level of 988 mb (which in this case was 238 meters—7781 ft—above the surface). At the top of the inversion, the temperature was 25.0 °C (77.0 °F), or 2.2 °C (4.0 °F) warmer than at the surface. Ground-based inversions such as these are very common at night and during the early morning hours due to the loss of longwave radiation (inversions were briefly discussed earlier in this chapter—we will return to them in greater detail in Chapter 6). Above the inversion layer, the temperature decreases fairly consistently with altitude until about the 140 mb level (located at about 14.7 km—9.1 miles—above sea level), where the air temperature becomes nearly constant with height at about –70 °C (–94 °F). The point at which the air temperature ceases to decrease with height marks the tropopause, which was described in Chapter 1.

*For reference, the height of the average altitude of several pressure levels is often indicated along the left, vertical axis of the diagram.

Summary

We all know that the ultimate source of energy for all atmospheric processes is solar energy. But most of this energy is delivered to the atmosphere indirectly. This chapter first described the processes that take place as solar radiation (insolation) passes through the atmosphere. Most of the ultraviolet radiation is absorbed by ozone in the stratosphere, and much of the near-infrared radiation is absorbed in the troposphere by water vapor and other gases. Scattering, both forward and backward, also reduces the intensity of the direct radiation reaching the surface. After atmospheric absorption and scattering back to space are accounted for, slightly more than half of the insolation available from the Sun reaches the surface, where some is again reflected back and the rest absorbed.

Having absorbed solar radiation, the surface and the atmosphere both radiate longwave energy. The end result of the transfer of solar and longwave radiation is that the surface has a net radiation surplus while the atmosphere has a deficit of equal value. The radiation deficit of the atmosphere and the surplus at the surface are offset by two other heat transfer mechanisms: conduction and convection. Conduction transfers energy across an extremely thin layer of air in contact with the surface, while convection distributes sensible and latent heat from near the surface to higher regions of the atmosphere.

The amount of energy obtained by the Earth system is not equal at all latitudes. Areas near the equator receive a surplus of energy, while more poleward regions have an energy deficit. This unequal energy distribution is offset by the latitudinal transfer of energy by wind movements and ocean currents. The most immediately recognizable outcome of spatial variations in absorbed energy is their influence on global temperatures.

The chapter also looked at various aspects of air temperature, including its average global distribution and the geographic factors that determine local temperature characteristics. We also examined the ways that temperature is measured and statistically summarized, and introduced the wind chill temperature index. The chapter concluded with an introduction of the thermodynamic diagram, an extremely valuable tool for the analysis and prediction of weather. Next we turn to another important element of meteorology—atmospheric pressure.

Key Terms

absorption p. 68
reflection p. 69
albedo p. 69
specular reflection p. 69
diffuse reflection p. 69
scattering p. 69
diffuse radiation p. 69
direct radiation p. 69
Rayleigh scattering p. 69
Mie scattering p. 71
nonselective scattering p. 72
planetary albedo p. 73
atmospheric window p. 75
net longwave radiation p. 75
net radiation p. 76
laminar boundary layer p. 77
free convection p. 77
forced convection p. 77
mechanical turbulence p. 77
sensible heat p. 78
specific heat p. 78
latent heat p. 79
advection p. 82
greenhouse effect p. 82
isotherm p. 83
inversion p. 86
continentality p. 88
maximum thermometer p. 89
minimum thermometer p. 91
bimetallic strip p. 91
thermograph p. 91
thermistor p. 91
wind chill temperature index p. 95
thermodynamic diagram p. 97
radiosonde p. 97
Stuve p. 97

Review Questions

1. Explain how the absorption and scattering of radiation in the atmosphere affect the receipt of solar radiation at the surface.
2. Which two gases are most effective at absorbing longwave radiation?
3. How do specular reflection and diffuse reflection differ?
4. What does the term *albedo* mean?
5. What characteristics of Rayleigh scattering cause it to create a blue sky?
6. What properties of Mie scattering distinguish it from Rayleigh scattering?
7. Why are overcast days typically gray?
8. What is the numerical value of Earth's planetary albedo?
9. Which type of scattering accounts for the majority of Earth's planetary albedo?
10. Describe quantitatively how much solar radiation is absorbed and reflected by Earth's atmosphere and surface.
11. What is the atmospheric window?
12. Why is it incorrect to state that longwave radiation bounces back and forth between clouds and the surface?
13. Explain why the incoming and outgoing radiation for the Earth system (radiation entering and leaving the top of the atmosphere) must be equal to each other.
14. How do conduction and convection work together to transfer heat upward?
15. What is the difference between free convection and forced convection?
16. Describe sensible and latent heat.
17. How do the net input and output of radiation vary with latitude?
18. Which two processes transport energy from zones of radiation surplus to zones of radiation deficit?
19. Why does the term *greenhouse effect* inaccurately describe how the atmosphere is heated?
20. Discuss how geographic factors such as latitude and altitude influence the distribution of temperature across Earth's surface.
21. How do the various instruments used to observe temperature work?
22. Explain how daily, monthly, and annual mean temperatures are computed from observed temperatures. Discuss some of the factors that can bias resulting values.
23. Describe the horizontal and vertical scales on Stuve diagrams.

Critical Thinking

1. Shorter wavelength radiation is more subject to Rayleigh scattering than is longer wave radiation. Explain how this might affect the importance of facing directly toward the Sun in the pursuit of an even suntan.
2. Even on cloudy days excessive exposure can lead to a danger of sunburn. What does this imply about clouds' effect on ultraviolet radiation?
3. Desert areas are often photographed with spectacular sunsets. Can you think of any reasons why they may be more inclined to have particularly flashy skies at dusk?
4. Our eyes are sensitive only to wavelengths between 0.4 and 0.7 micrometers. Would the sky appear any different on clear days if our eyes could also perceive wavelengths

between 0.2 and 0.4 micrometers? How would the ratio of perceived diffuse to direct radiation change?

5. Would you expect both the Northern and Southern Hemispheres to have the same average albedo? What factors might cause the two hemispheres to reflect different percentages of insolation back to space?
6. Melting snow often occurs more rapidly in wooded areas immediately adjacent to trees than in nearby openings. What type of energy transfer processes could lead to this effect?
7. Net radiation values in the summer may be higher in forested areas than desert areas, despite the higher temperatures in the desert. How can this be?
8. The ratio of sensible to latent heat loss from a surface is called the Bowen ratio. How do you suppose Bowen ratios might differ between desert, wooded, and urban landscapes?
9. Clouds can reduce the amount of insolation reaching Earth's surface, but they can also reduce the amount of longwave radiation from the surface that escapes to space. How might this affect maximum and minimum temperatures? Do you think that all types of clouds would produce similar effects?
10. Figure 3–16 shows the mean distribution of ocean currents. It is believed that climate change, through a variety of mechanisms, could cause a shift in the position of some currents. Can you identify any land regions whose climate could be vulnerable to shifts in nearby currents? Are there any localities whose climates could cool even if the average global temperature were to warm?
11. Instrument shelters protect thermometers from the heating effect of absorbed sunlight. Is it also true that shelters protect the thermometers from the chilling effect of wind?
12. An orchard farmer hears a weather forecast for overnight low temperatures to hover just above the freezing point of 0 °C (32 °F), but with wind chill temperatures expected to drop significantly lower. Will the wind chill increase the possibility of frost damage? Why or why not?

Problems and Exercises

1. Go to the Web site **http://rredc.nrel.gov/solar/old_data/nsrdb/redbook/atlas/**. Select mean solar radiation as data type and horizontal flat plate for instrument orientation. Examine the maps of average solar radiation for January, April, July, and October. Describe each of the patterns. To what extent do you think Earth–Sun relationships affect the distribution relative to the effect of cloud cover and other weather elements?
2. Go to **http://earthobservatory.nasa.gov/Observatory/Datasets/netflux.erbe.html** and observe the seasonal shift in net radiation for the surface and atmosphere. What is the most obvious pattern? Are there significant differences between land bodies and adjacent oceans?
3. The latent heat for water is 2,500,000 joules per kg, and the specific heat of water is 4190 joules per kg per degree Celsius of temperature change. Assume that a kg of water of water begins with a temperature of 20 °C (68 °F). Compare the amount of energy needed to bring the water up to the boiling point to the amount of energy needed to evaporate the same amount of water.
4. View the maps of mean minimum temperatures in January and mean maximum temperatures in July at **http://www.climatesource.com/map_gallery.html.** Assess the relative importance of latitude, elevation, continentality, ocean current, and local conditions to these distributions.

Quantitative Problems

This chapter has described the energy balance of the atmosphere and the factors that influence global temperatures. The companion Web site for this book, **http://www.prenhall.com/aguado/**, offers a brief quiz in which you can test your knowledge of the material by answering several quantitative problems. The brief exercise should help you better understand equilibrium temperatures, seasonal effects of cloud cover on energy receipt, and relative expenditures of surface energy on latent and sensible heat transfer. After entering the site, go to the bottom of the page and select Chapter 3. Then highlight the Quantitative Examples line on the left-hand panel.

Useful Web Sites

http://rredc.nrel.gov/solar/old_data/nsrdb/redbook/atlas/

Provides monthly and annual maps of solar radiation across United States. For the material relevant to this chapter, be sure to check the box for *Horizontal Flat Plate* as instrument orientation mode.

http://earthobservatory.nasa.gov/Observatory/Datasets/netflux.erbe.html

Allows you to observe the shift in net radiation over multiyear period. Maps are simultaneously matched with other biophysical distributions (such as vegetation) across the globe to show how they vary with net radiation. Monthly averages of net radiation can also be viewed individually or as an annual loop at **http://itg1.meteor.wisc.edu/wxwise/museum/a2/a2net.html**.

http://lwf.ncdc.noaa.gov/oa/climate/severeweather/temperatures.html

Features maps of temperature records for the United States. Also contains numerous special reports on temperature conditions.

http://www.ems.psu.edu/wx/usstats/uswxstats.html

Contains some very interesting maps, such as the current distribution of temperature across the 48 conterminous United States. Also provides the current temperature and other values averaged over the 48 conterminous United States.

http://www.srh.noaa.gov/FTPROOT/FFC/html/metcalc.shtml

Provides wind chill and temperature index converters.

Media Enrichment

Tutorial

Global Energy Balance

This tutorial covers the planetary heat budget and energy exchange between the surface and atmosphere. It reviews the wave nature of radiation, emission laws, and convective heat flux and shows how energy balance is achieved for the surface, the atmosphere, and the Earth–atmosphere system as a whole.

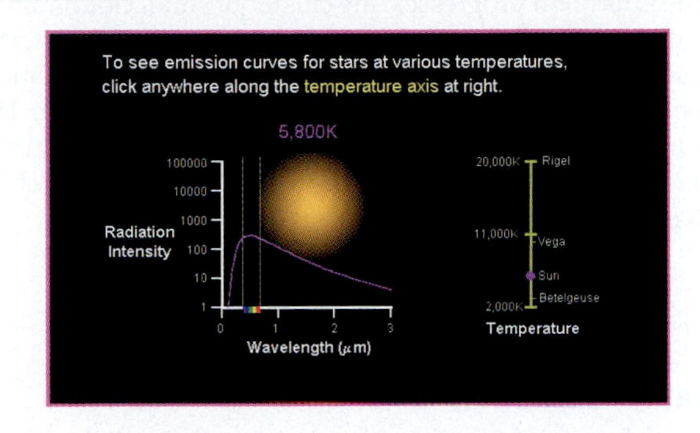

Interactive Exercise

Shortwave Radiation

Surface radiation budgets respond to many atmospheric and geographic factors. Use this program to examine the effects of cloud cover, surface albedo, latitude, season, and time of day on shortwave radiation. Note the relative effects of cloud cover. Are the effects the same regardless of cloud height? As albedo increases, what is the effect on shortwave radiation? According to the exercise, does latitude influence the relative importance of clouds? How about time of day? Do you think the real world exhibits the same behavior? (*Hint:* Imagine sunlight striking a cloud at various solar angles.)

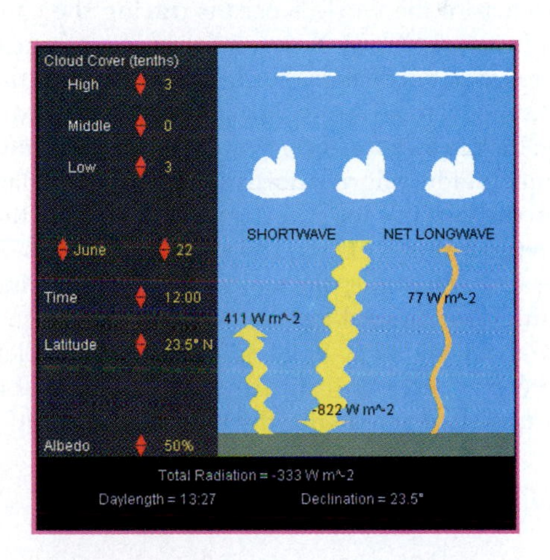

Interactive Exercise

Longwave Radiation

This exercise uses the Stefan-Boltzmann law to compute radiation emitted by the surface at various temperatures. Incoming longwave radiation is computed using statistical relationships between cloud amount and radiation. As cloud amount increases, what happens to incoming longwave radiation? What is the physical explanation for this? Of the three cloud types (low, medium, high), which has the greatest effect on incoming longwave radiation? If you increase low cloud to 100 percent (10 tenths), incoming radiation increases. At what temperature would outgoing longwave radiation exactly balance that amount?

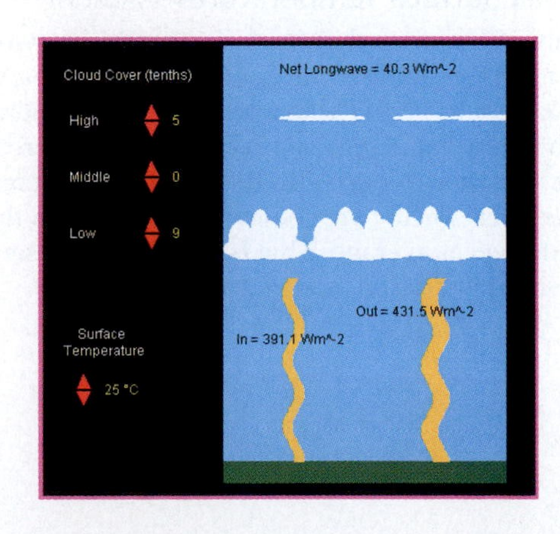

Weather in Motion

Heavy Convection over Florida

Earlier in this chapter, we discussed the role of convection in transferring heat between the surface and the atmosphere. This has far more than just theoretical importance, because convection frequently leads to the formation of deep clouds that can yield heavy precipitation, lightning, hail—and even tornadoes.

In this movie, we see the formation of a major thunderstorm over southern Florida. The scene begins in the early morning with predominantly clear skies across the peninsula. (Subtract 4 hours from the UTC time at the bottom of the movie to convert to EDT.) Shortly after noon, a narrow band of small, isolated clouds emerges along the Atlantic Coast, along with a broad area of scattered clouds across the state. As the afternoon progresses, the cluster of small clouds along the coast gradually grows into a single larger thunderstorm. The situation changes rapidly at about 4 P.M., with an almost explosive growth of the thunderstorm along the coast. The first major cell of intense weather is followed by several larger ones, and by early evening the state has been fully obliterated from view.

This movie shows a large part of the spectrum of thunderstorm activity over just a brief time period. The small clouds that dominated the early part of the afternoon formed in response to localized heating that caused relatively small parcels of air to rise. Such clouds are commonplace, especially in the summer. But there are a number of conditions that must be present for the small, fair-weather clouds to grow into major storms capable of violent weather. These conditions include a humid atmosphere near the surface and certain changes in temperature with height above the surface. These factors are discussed in Chapters 5–7.

Weather in Motion

Global Sea Surface Temperatures—Climatology

This movie shows the average distribution of sea surface temperatures across the world's oceans during the course of a year, beginning in June. Temperatures are warmest in the tropical regions and generally decrease with latitude. The distribution of temperatures does not remain fixed in time, however. Notice that for about 6 months prior to late August the warmer waters migrate northward, which is a lagged response to the varying solar declination. Also note the difference in temperatures off the west and east coasts of North America. At low latitudes the temperatures are warmer along the Atlantic coast than along the Pacific. In contrast, the east coast of Canada and New England tend to have cooler waters than the west coast of North America, because of the cool temperatures associated with the Labrador current.

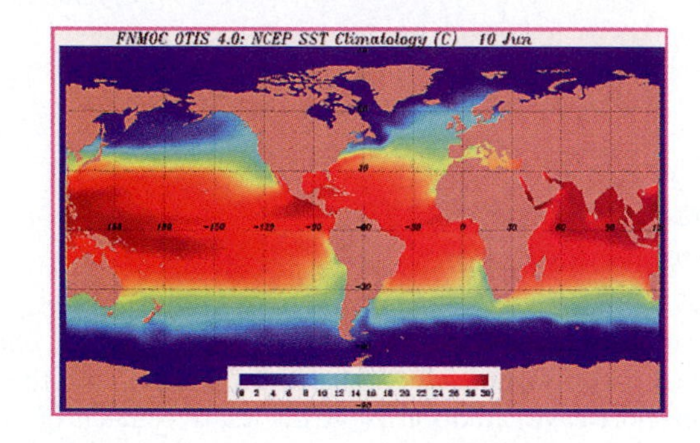

Weather in Motion

Global Sea Surface Temperatures—Actual

A companion movie to the previous one, this movie shows actual sea surface temperatures observed from November 1998 to November 1999. The temperature distribution is very similar to the climatological average, although there is more fine scale variation. Notice also that the temperatures along the east tropical Pacific are somewhat cooler than those for the climatological average. This is due to the presence of a La Niña situation, described in Chapter 8.

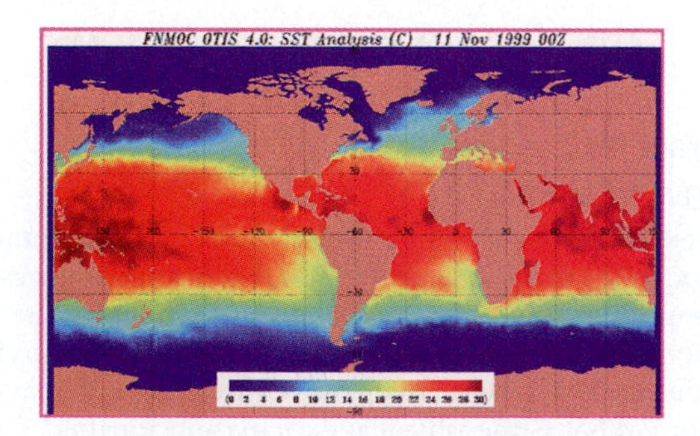

Weather Image

Ocean Currents

This image provides a unique depiction of ocean currents in the Pacific Ocean and a small portion of the western Atlantic, showing sea surface temperatures and the direction of flow. The cold California current is apparent along the extreme eastern Pacific, as is the warm, westward-moving current along a wide portion of the tropical Pacific. The North Pacific current is also evident, transporting warm water toward the Gulf of Alaska. Note that a small portion of the Gulf Stream can be seen off the east coast of North America.

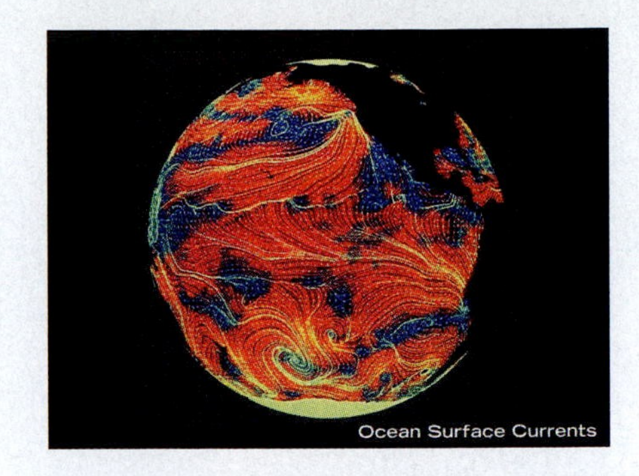

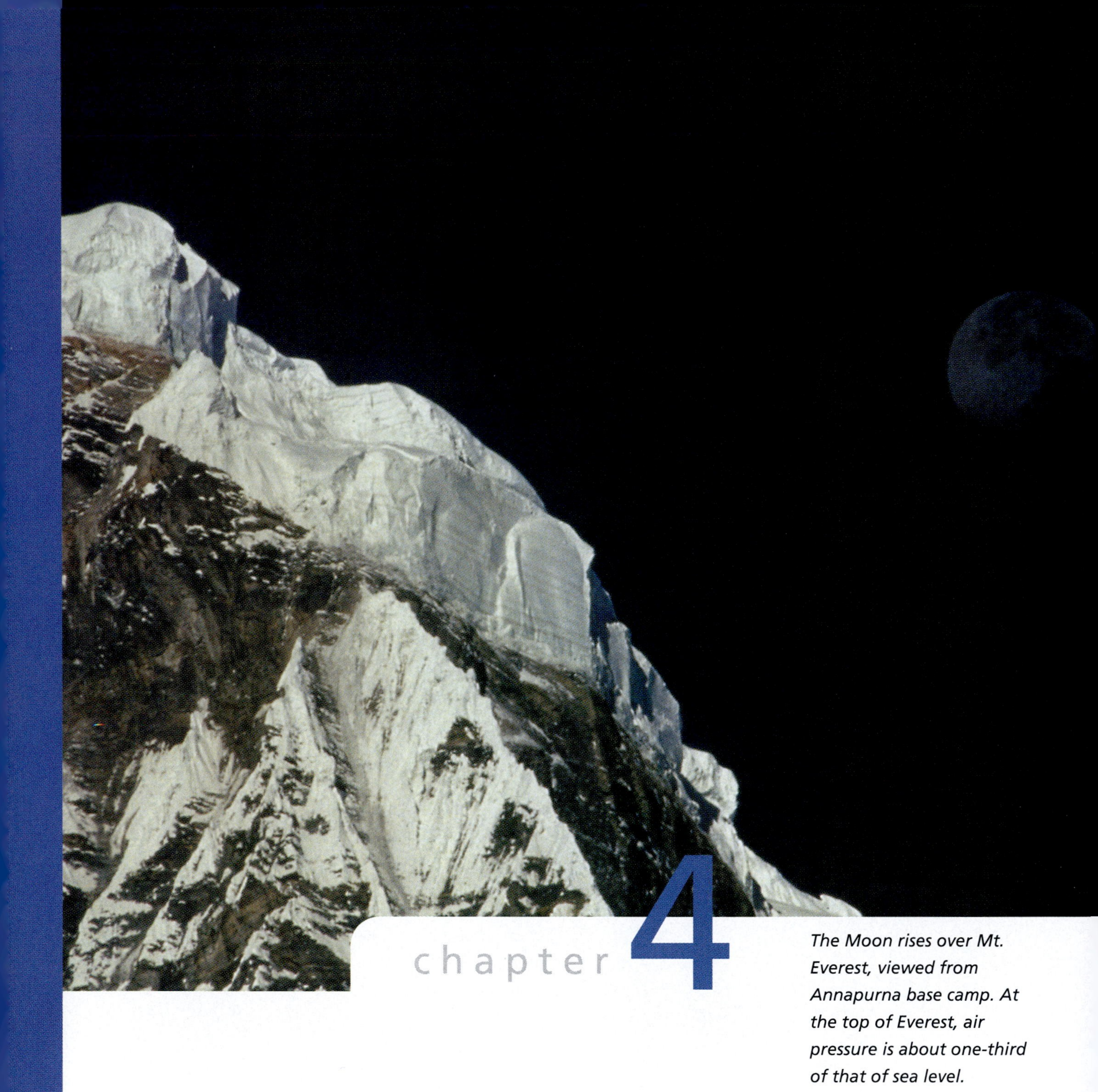

chapter 4

The Moon rises over Mt. Everest, viewed from Annapurna base camp. At the top of Everest, air pressure is about one-third of that of sea level.

Atmospheric Pressure and Wind

Wind can have a direct impact on how we feel even under the mildest of weather conditions. A gentle breeze can make a hot afternoon more comfortable, or it can make a winter night bitterly cold. But occasionally, the movement of air can impact our lives in far more substantial ways. Residents of eastern Washington and Oregon were reminded of this on September 25, 1999, when winds up to 135 km/hr (85 mph) created a blinding dust storm that triggered a spate of accidents. Over the course of the day, 6 people were killed and 23 injured, and miles of highway were completely shut down for hours. The worst of the multiple-vehicle accidents occurred along Interstate Highway 84 in northeastern Oregon. Fifty-eight-year-old Harold Fell described his experience: "An 18-wheeler passed us and the next thing I knew, it was stopped dead in the road. It just loomed up in front of me." Fell and his wife suffered only minor injuries, but they undoubtedly realized how lucky they were when they surveyed the accident scene and realized that 4 people were dead amid the wreckage of the 16 vehicles involved.

As concerned as we sometimes are with wind conditions, few of us pay much attention to a closely associated component of weather—*atmospheric pressure*. After all, how many times have you canceled a picnic because the pressure was too low? Or how many people do you know who have special clothes they wear only on days of high pressure?

Although seldom considered in everyday life, air pressure deeply affects other weather variables that have much more immediate impact. For example, horizontal variations in atmospheric pressure are directly responsible for the motion of the wind. And because air descends in areas of high surface pressure and rises in regions of low surface pressure, differences in air pressure strongly influence the likelihood of cloud formation and precipitation.

This chapter introduces the basic concepts of atmospheric pressure and its vertical and horizontal distributions. We discuss the relationship between pressure and other atmospheric variables and the processes that create horizontal and vertical variations in pressure. With this foundation, we can go on to discuss storm patterns in later chapters.

CHAPTER OUTLINE

The Concept of Pressure

The atmosphere contains a tremendous number of gas molecules being pulled toward Earth by the force of gravity. These molecules exert a force on all surfaces with which they are in contact, and the amount of that force exerted per unit of surface area is **pressure** (see *Box 4–1, Physical Principles: Velocity, Acceleration, Force, and Pressure*). Of course, the concept of pressure is not confined to meteorology but rather is fundamental to all the physical sciences. In most physical science applications, the standard unit of pressure is the **pascal** (Pa), but in the United States meteorologists use the **millibar** (mb), which equals 100 Pa. Canadian meteorologists use yet another unit, the **kilopascal** (kPa), equal to 1000 Pa, or 10 mb. For purposes of comparison, air pressure at sea level is typically roughly 1000 mb (100 kPa)—or more precisely, 1013.2 mb.

To understand the characteristics of pressure, refer to Figure 4–1, which depicts a sealed container of air. The enclosed air molecules move about continually and exert a pressure on the interior walls of the container (a). The pressure of the air is proportional to the rate of collisions between the molecules and walls. We can increase the pressure two ways. The first way is by increasing the density of the air by either pumping more air into the container or by decreasing the volume of the

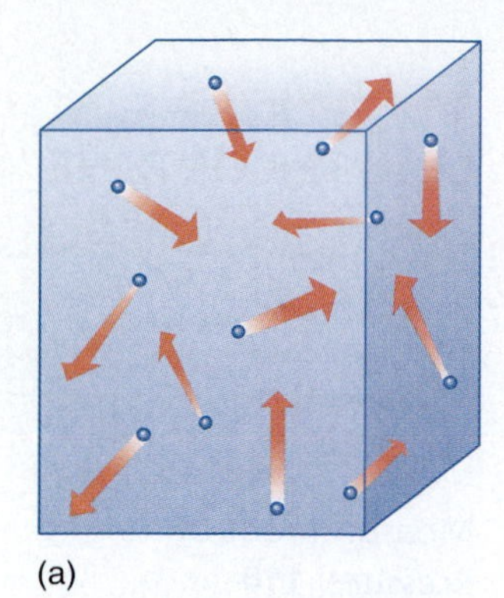

(a)

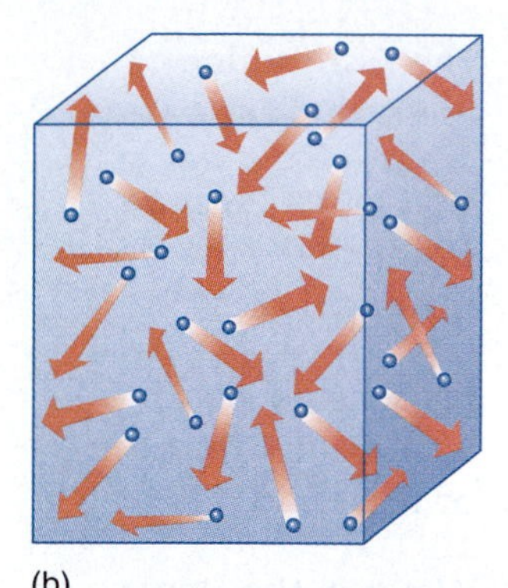

(b)

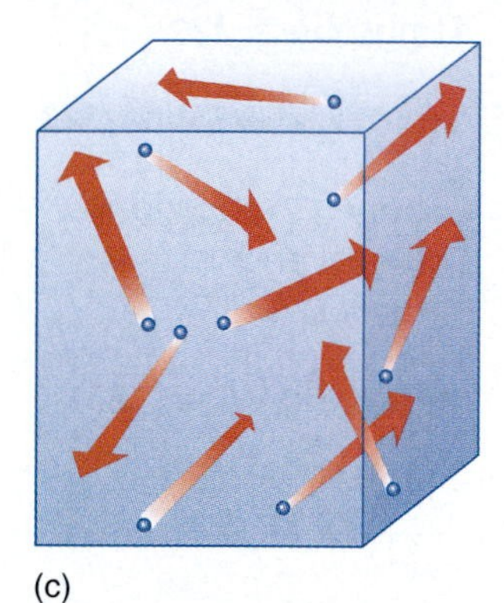

(c)

▲ **Figure 4–1**
The movement of air molecules (indicated by the dots with the red arrows) within a sealed container exerts a pressure on the interior walls (a). The pressure can be increased by increasing the density of the molecules (b) or increasing the temperature (c). The speed of the molecules (and therefore the temperature) is indicated by the degree of redness and length of the arrows.

container (b). The second is by increasing the air temperature, in which case the molecules exert higher pressure because they are moving more rapidly (c). Thus, pressure reflects both the density and temperature of the gas.

If the air in the container is a mixture of gases (as it is in the atmosphere), each gas exerts its own specific amount of pressure, referred to as its *partial pressure*. The total pressure exerted is equal to the sum of the partial pressures. This relationship is known as **Dalton's law**.

On Earth the container is surrounded by the atmosphere, which exerts pressure on the exterior walls. Now consider what would happen if we were to remove the lid of the container or puncture a hole in its side. If the pressure *outside* the container were greater than that within, the outside air would be forced inward until the pressure equalized. (The force of this equalization is what causes the "whoosh" when you open a vacuum-packed coffee container.) On the other hand, if the pressure were greater *inside* the container, air would be forced outward until the internal pressure decreased to match the surrounding air. In either case, within moments the air pressure exerted on the outside of the container would become exactly equal to that on the inside. This example introduces us to another characteristic of air: it constantly moves to establish an equilibrium between areas of high and low pressure.

What we experience as atmospheric pressure is in fact the mass of the air above us being pulled downward by gravity. In other words, the pressure at any point reflects the mass of atmosphere above that point. As we go upward through the atmosphere, the mass of atmosphere above necessarily decreases, so pressure must also decrease. We feel this change, for example, in a rapidly rising elevator, where the decrease in pressure can lead to "popping" of the ears. Note that pressure is unique among atmospheric variables in that it always decreases vertically. Other variables (such as temperature, moisture, and density) do not necessarily behave this way. Despite the fact that the atmosphere is pulled downward by the force of gravity, air pressure is exerted equally in all directions—up, down, and sideways. Revisit the case of the sealed container we just described, with a greater pressure inside than outside. It doesn't matter whether the container is punctured along one of its sides, on its bottom, or on its top; the greater pressure within still forces the air outward.

Here is another way to visualize the fact that air pressure is exerted equally in all directions. Hold your arm directly out from your body. Air pushes on your arm, not only down but also along its side and almost equally on its underside.* If pressure were applied only downward, the weight of the air would be so great that even the strongest person would be unable to extend her arm outward. How great would this weight be? Under normal conditions at sea level, the force would equal about 14.7 pounds on every square inch—quite a load for even a short arm!

Vertical and Horizontal Changes in Pressure

To understand the wind, we need to measure and compare the differences in pressure that arise in different locations, since these differences produce horizontal movements of air. But the job is complicated by the fact that elevation varies from place to place. (Recall that high elevations have lower pressure simply because there is less overlying air.) If we used just surface measurements for comparisons, it would be impossible to separate the effects of elevation from the true pressure differences that lead to wind. To overcome this problem, meteorologists use the concept of sea level pressure.

Surface pressure is the pressure actually observed at a particular location, whereas **sea level pressure** is the pressure that would exist if the observation point were at sea level. Because most land surfaces are above sea level, surface pressure is almost always the lower of the two. Compare, for example, the surface pressures

*The pressure on the bottom of your arm is very slightly greater than at the top. This is because pressure always decreases with altitude, and the top of your arm is a few centimeters higher than the bottom. This difference is extremely small, however, and can be disregarded in this example.

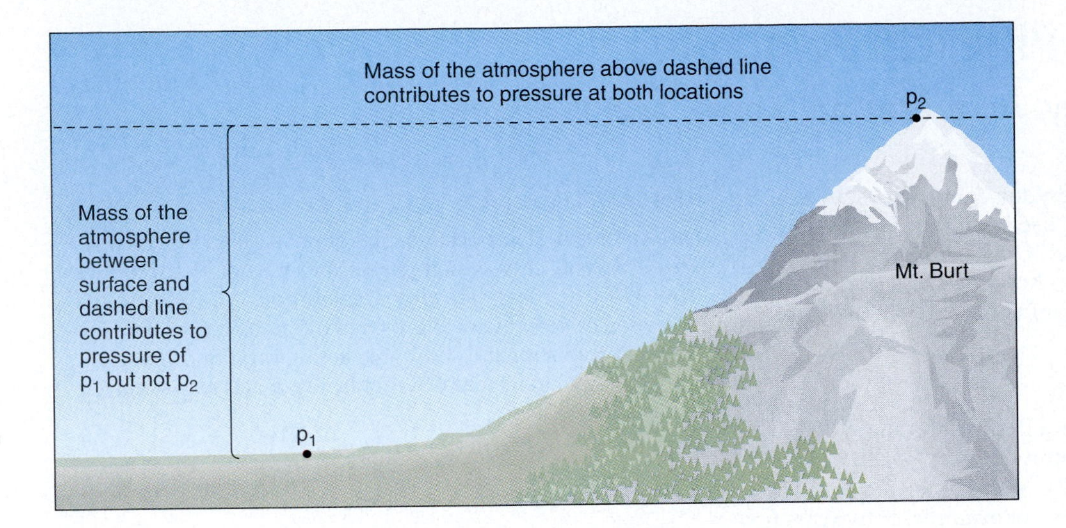

Figure 4–2
Because atmospheric pressure is a response to the weight of the overlying atmosphere, it always decreases with elevation. The pressure at the top of the mountain, p_2, is less than that at the base of the mountain, p_1, because of the greater amount of overlying air. For this reason it is important to distinguish *surface air pressure* from *sea level pressure*.

at the tall mountain peak and the nearby valley in Figure 4–2. Although the atmosphere is uniformly distributed over the area, surface pressure at the mountain location is considerably less than at the valley.

Sea level pressure allows us to compare pressure at different locations, taking into account differences in elevation. For locations not very much above sea level, we can get a good indication of sea level pressure by assuming a uniform change in pressure with elevation. At an elevation of 150 m (500 ft), for example, we add 14 mb to the surface pressure to obtain the sea level pressure (roughly a 1 mb increase for every 10 m). For high-elevation sites, however, this method is unreliable because we must account for compressibility of the atmosphere.

As shown in Figure 4–3, pressure does not decrease with height at a uniform rate. Instead, it decreases most rapidly at low elevations and gradually tapers off at greater heights. For example, from sea level to 1 km (0.6 mi) in elevation, the average pressure decreases about 100 mb, but between 9 and 10 km it drops only half

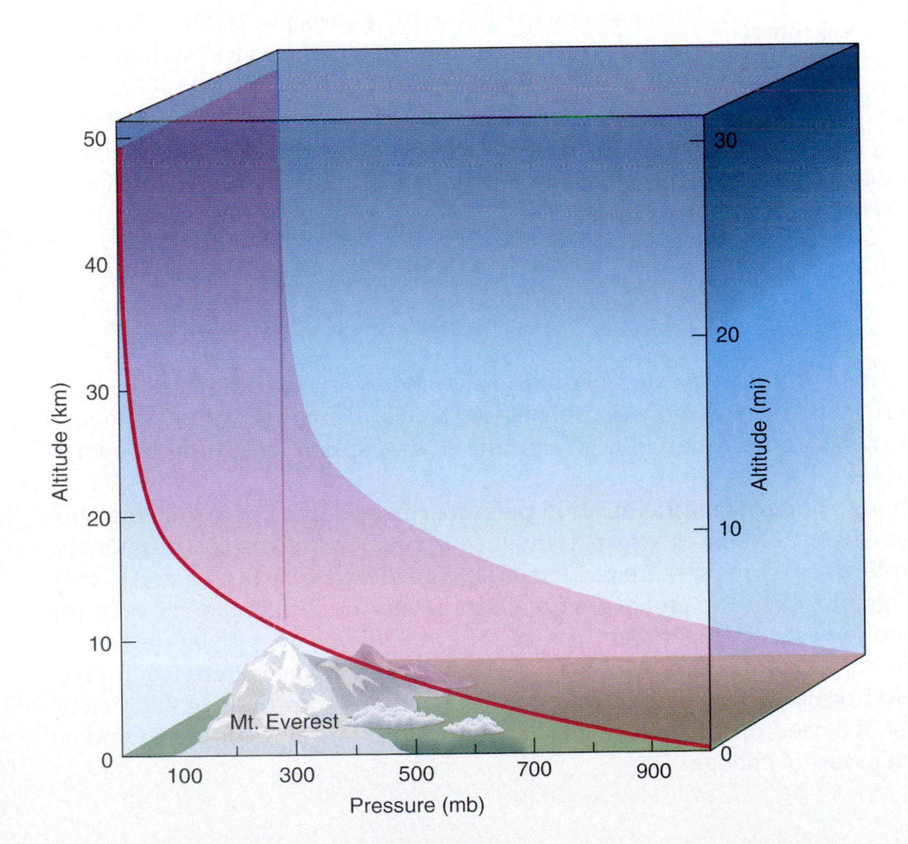

Figure 4–3
Pressure decreases with altitude by about half for each 5.5 km (3.3 mi).

4–1 Physical Principles

Velocity, Acceleration, Force, and Pressure

It is quite common in everyday conversation to hear the terms *force* and *pressure* used interchangeably, just as *velocity* and *speed* are often considered synonymous. In the language of science, however, intermixing these terms can lead to great confusion. Let us look briefly at how they differ.

Velocity and Acceleration

Any object that moves has a particular **speed**, defined as the distance traveled per unit of time. Speed is related to, but not the same as, velocity. **Velocity** incorporates direction as well as speed. Think, for example, of two cars traveling at 20 meters per second (44 mph) but moving in opposite directions. Though they have the same speed, their velocities are not equal because of their different directions. This distinction is crucial for understanding our next quantity, **acceleration**, the change in velocity (not speed) with respect to time.

Because velocity includes both speed and direction, a change in either speed or direction is an acceleration. For example, consider a car that at one moment in time travels at 20 m/sec; one second later, the same car has a speed of 19 m/sec; one second later, the speed is 18 m/sec, and so on. As each second goes by, the car's speed decreases 1 m/sec (note that acceleration can be either positive or—as in this example—negative).

An acceleration can also occur as a change in direction with respect to time, even for an object whose speed does not change. A car traveling at a constant speed but gradually turning undergoes an acceleration, just like a car whose speed is changing.

In meteorology there is one particular acceleration of utmost importance—**gravity** (*g*). This acceleration, 9.8 m/sec/sec (32.1 ft/sec/sec), is nearly constant across the globe. There is a slight decrease in *g* from equator to pole, and also a very small difference in *g* from the surface to the upper atmosphere. For most applications, however, these variations in *g* are so slight they can be ignored.

Force and Pressure

One of the most important tenets of physical science is Newton's Second Law, which relates the concept of **force** (denoted *F*) to mass (*m*) and acceleration (*a*). Specifically, Newton's Second Law tells us that the acceleration of an object is proportional to the force acting on it and inversely proportional to its mass. Symbolically, this is expressed as

$$a = \frac{F}{m}, \text{ or}$$

$$F = ma$$

Imagine that a fully loaded 18-wheeler truck is stopped at a red traffic light next to a bicycle. As the light turns green, both begin to accelerate at the same rate. It's easy to see that if the two remain right next to each other, the much more massive truck requires a larger force (and more powerful "engine"). Likewise, if two bodies with equal mass are subjected to different forces, the one subjected to the greater force will undergo a greater acceleration.

Keep in mind that *F* in the equation above is the net force acting on the object. If various forces are acting simultaneously, they must all be considered together to determine the acceleration; we must account for both the magnitude and direction of each. As we will see with regard to falling raindrops (Chapter 7), forces acting in opposite directions reduce the net force and resulting acceleration, sometimes to zero.

Let's apply Newton's Second Law to our atmosphere. The atmosphere contains 5.14×10^{18} kg of mass. (To get an idea of what 5.14×10^{18} kg weighs, picture a million boxcars each containing a billion elephants.) Multiplying the mass of the atmosphere by the acceleration of gravity, we determine that the force exerted on the atmosphere is about 5.0×10^{19} newtons (a newton [N] is the unit of force it takes to accelerate 1 kg one meter per second every second).

as much. This nonlinearity exists because, as we know, air is compressible. With atmospheric mass packed more densely at low levels, a small elevation change at low levels takes you through a large amount of atmosphere, resulting in a large pressure drop.

Although horizontal differences in pressure can be sufficient to generate violent winds, these differences are very small compared to the vertical differences. For example, a sea level pressure of 1050 mb is considered very high, yet it is only 4 percent greater than the global average. Moreover, the difference between the highest and lowest sea level pressures over North America on a given day might amount to only about 25 mb. And even this small percentage difference in pressure would normally be realized over a distance of many hundreds of kilometers. In contrast, we need only go to the top of a modest hill or tall building to find an equivalent pressure change.

Force divided by the area on which it is exerted equals **pressure.** So dividing this force of 5.0×10^{19} newtons by the surface area of Earth gives us the average force per unit area, or average surface pressure of about 10.132 newtons per square centimeter. This is equivalent to 1013.2 mb, or about 14.7 pounds per square inch.

Having made the distinction between force and pressure, we now should address the question of how this distinction applies to the atmosphere. The answer is that despite the nearly constant total force of the atmosphere, its gases are not uniformly distributed across the planet. Greater pressure exists where the atmospheric column contains a greater number of molecules, and lower pressures are found where less air occupies the column. Just how these differences in pressure arise is considered later in this chapter; for the time being, the important point is that surface pressure reflects the mass of atmosphere within the column, as shown in Figure 1.

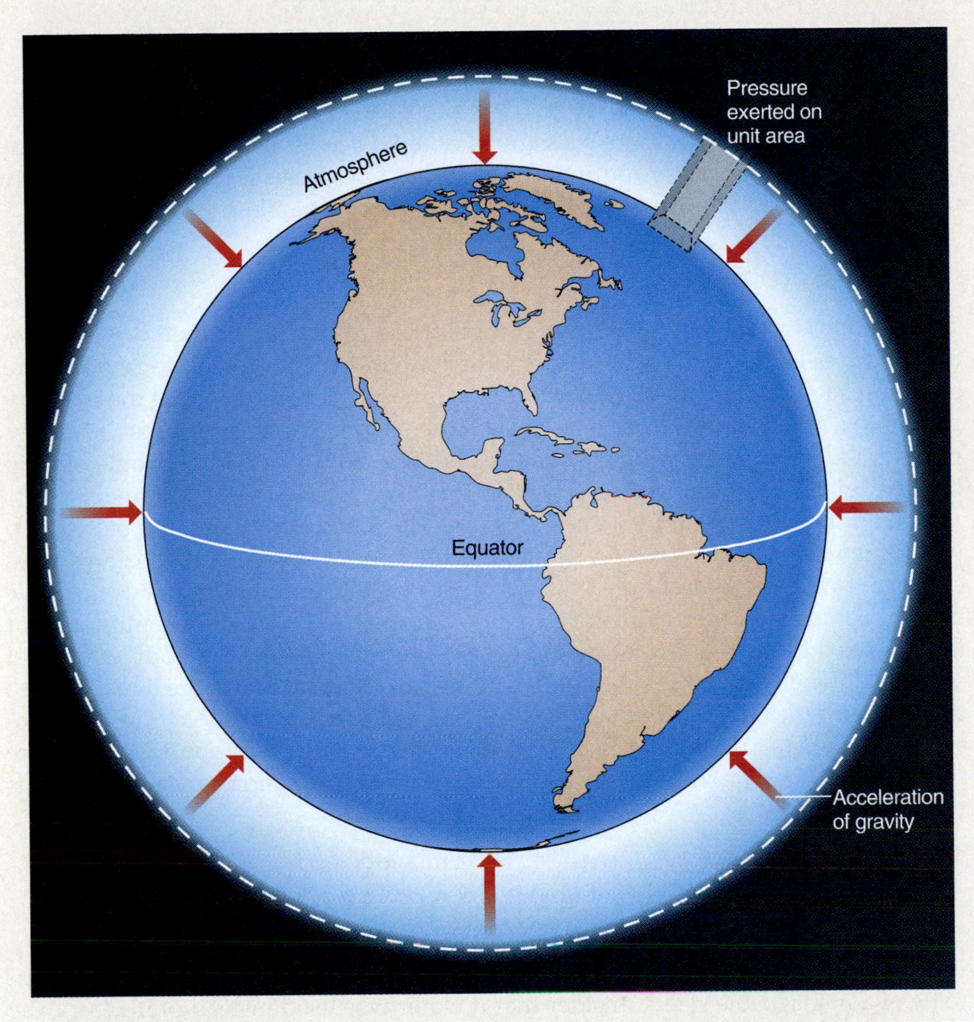

Figure 1
The downward force of the atmosphere is equal to the mass of the entire atmosphere times the acceleration of gravity. Because the amount of mass and the acceleration of gravity are constant through time, the force of the atmosphere does not change. Pressure is defined as the amount of force exerted per unit of area. Thus, the shaded area shown in the figure experiences a certain amount of pressure. Pressure varies because the mass of overlying air varies from place to place and time to time.

The Equation of State

Everyday experience indicates that gases tend to expand when heated and become denser when cooled. This suggests that temperature, density, and pressure are related to one another. As a matter of fact, their relationship is quite simple. It is described by the **equation of state** (also called the **ideal gas law**),

$$p = \rho RT$$

in which p is pressure expressed in pascals, ρ (the Greek letter rho) is density in kilograms per cubic meter, R is a constant equal to 287 joules per kilogram per kelvin, and T is temperature (in kelvins). To put the equation in words, it tells us that if the air density increases while *temperature is held constant*, the pressure will increase. Similarly, at constant density, an increase in temperature leads to an increase in pressure.

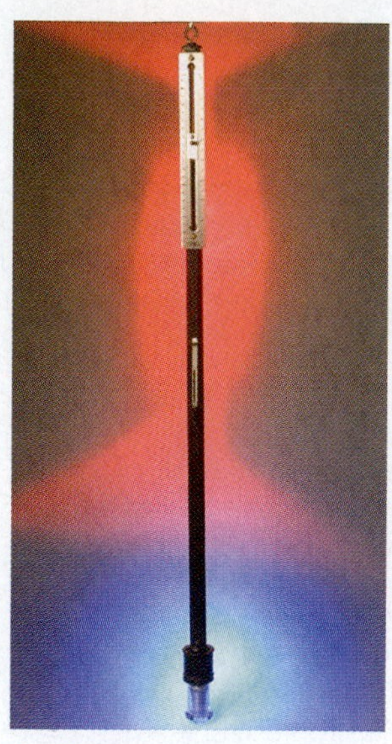

▲ **Figure 4–4**
A mercury barometer.

This point leads us to a common confusion. Among the general public (and some students), there is often a temptation to look for a simple relationship between temperature and pressure. We might hear "the pressure is falling, it's going to get cold," or perhaps "it's hot—must be high pressure." But the equation of state says that blanket statements such as these are not correct. A third variable, density, can overwhelm changes in the other two variables. For example, in the plains of North America, high pressure following a winter storm often brings bitterly cold temperature, not soothing warmth. Although counterintuitive for some, such an occurrence is entirely consistent with the equation of state.

Measurement of Pressure

Any instrument that measures pressure is called a *barometer.* Two types of barometers are most common for routine observations: one consisting of a tube partially filled with mercury, and another that uses collapsible chambers.

Mercury Barometers

The standard instrument for the measurement of pressure is the **mercury barometer** (Figure 4–4), invented by Evangelista Torricelli in 1643. It is a simple device made by filling a long tube with mercury and then inverting the tube so that the mercury spills into a reservoir. Although the tube is turned over completely, it does not empty. Instead, the air pushes downward on the pool of mercury and forces some of it up into the tube. The greater the air pressure, the higher the column of mercury.

Barometric pressure is often expressed as the height of the column of mercury in a barometer, which at sea level averages 76 cm (29.92 in.). This measure is inconsistent with the concept of pressure, however, because pressure does not have units of length. In other words, expressing barometric pressure in centimeters or inches is as incongruous as stating somebody's age as "30 miles per hour" or weight as "$1.99!" The length measurements obtained from a barometer are a response to the atmospheric pressure but are not direct observations of pressure itself. Meteorologists prefer a unit that measures force per unit area, such as pounds per square inch or millibars. The simple conversion formulas for converting barometric heights to millibars are

$$1 \text{ cm} = 13.32 \text{ mb and}$$
$$1 \text{ inch} = 33.865 \text{ mb}$$

Mercury is an excellent fluid for use in a barometer because it is extremely heavy, with a density 13.6 times greater than that of water. This feature allows the instrument to be of a manageable size. Consider that if water was used instead of mercury, the column of water would need to be about 10 m (33 ft) tall to counterbalance the weight of the atmosphere. On the other hand, although a three-story water-filled barometer would be less than portable, it would make for very precise measurements because even small changes in pressure would translate into large height changes.

Corrections to Mercury Barometer Readings

One of the most important tools of a meteorologist is the weather map, which among other things plots the distribution of air pressure across the surface. Before barometer data can be used on the map, however, three corrections must be made to compensate for local factors that affect the readings.

The first correction compensates for the influence of elevation we described earlier in this chapter. If surface pressure values were plotted on weather maps, they would give a false representation of the distribution of the atmosphere. This happens because high elevations have lower surface pressures than do low elevations, even if the sea level pressures are the same. To standardize the observations, we must convert surface pressure readings to sea level values. For a station situated 100 m (328 ft) above sea level, about a centimeter (0.4 in.) is

4–2 Physical Principles

Variations in Density

Perhaps you have wondered how much air weighs. The air around you has a particular density, and any volume of air contains a certain amount of mass. Changes in the density of air affect many everyday phenomena. For example, the density of the atmosphere influences how much lift a plane gets as it accelerates down a runway in preparation for takeoff. Likewise, automobile fuel injectors must account for variations in density to deliver the right mixture of gasoline and air into the car's engine. The density of air can even affect the amount of resistance the air exerts on a batted baseball, thereby influencing its distance traveled.

But are the variations in density really substantial? We can use the equation of state to see exactly how much variations in temperature affect the air's density. To do this, let's first rearrange the equation to the following:

$$\rho = p/RT,$$

Let's now also compare the air density for two situations: a warm day with a temperature of 308 K (35 °C or 95 °F) and a cold one with a temperature of 278 K (5 °C or 41 °F). For consistency we will assume that the pressure is 100,000 Pa (1000 mb; 100 kilopascals) in both instances.

Applying the equation of state for the warmer day, we find that the density of the air is

$$\rho = \frac{100{,}000 \text{ Pa}}{287 \text{ J kg}^{-1} K^{-1} \times 308\, K} = 1.13 \text{ kg/m}^3.^*$$

When we lower the air temperature to 278 K (5 °C or 41 °F), the equation yields an air density of

$$\rho = \frac{100{,}000 \text{ Pa}}{287 \text{ J kg}^{-1} \text{K}^{-1} \times 278 \text{ K}} = 1.25 \text{ kg / m}^3,$$

*To get the units to balance, you must reduce the units of pascals (Pa) and joules (J) to their fundamental dimensions. Thus, $\text{Pa} = \text{kg m}^{-1} \text{sec}^{-2}$ and $\text{J} = \text{kg m}^2 \text{sec}^{-2}$.

This is nearly 11 percent greater than the density the air had on the warmer day—a nontrivial amount.

In addition to temperature and pressure, the humidity of the air also exerts an influence (although only a very minor one) on density. Let's see how. Molecular oxygen (O_2) and nitrogen (N_2) make up most of the mass of the atmosphere and exist in a constant proportion. Other, lesser constituents of the atmosphere are present in different amounts at different places and times, and because each has its own unique molecular weight (an expression of the relative amount of mass for molecules), their relative abundance can slightly affect the density of the atmosphere. Among these gases, water vapor usually accounts for about 1 percent of the atmospheric mass. Intuitively, we might assume that a greater humidity would favor a denser atmosphere. Actually, just the opposite is true.

Compare the amount of mass contained in individual molecules of water vapor and of the most abundant atmospheric gases. The molecular weights of nitrogen and oxygen are 28.01 and 32.00, respectively, and the mean molecular weight of the dry atmosphere is 28.5. Water vapor, on the other hand, has a molecular weight of only 18.01. Thus, as the proportion of the air occupied by water vapor increases, an accompanying reduction in the mean molecular weight of the atmosphere must occur. In other words, all other things being equal, humid air is less dense, "lighter" than dry air.

How important are changes in humidity on the density of the atmosphere? You might expect that because of the small amount of moisture in the atmosphere they are relatively minor, and you would be right. Incorporating the effect of varying moisture content requires a small modification to the equation of state. Calculations using the revised formula show that at 15 °C (59 °F), air density declines by only 0.6 percent for a 1 percent increase in water vapor (from dry air to 1 percent water vapor).

added, corresponding to about 13 mb. At higher elevations, a much greater adjustment might be needed. Thus, at Denver, Colorado (the "Mile High City"), for example, the correction is about 16 cm (6.24 in.), or 213 mb.

The second correction deals with the similarity between a mercury barometer and a thermometer. Just as the mercury in a thermometer expands with increasing temperature, so does the mercury in a barometer. The expansion reduces the density of the fluid and requires that it attain a greater height to offset the weight of the atmosphere. In other words, on a hot day the height of the mercury column is greater than on a cold day, even if the atmospheric pressure is the same. For this reason, mercury barometers always have a thermometer attached to determine the temperature of the instrument, and a correction table tells us what height the mercury column would be if the temperature were at the standard value of 0 °C (32 °F). At normal room temperature, this correction is small, requiring the subtraction of about 2.5 mm (0.01 in.).

The third correction accounts for the slightly varying acceleration of gravity with latitude. To standardize the readings from all latitudes, we convert them to what they would be if the local gravity were equal to that at 45° north or south, or midway between the equator and poles. The latitudinal changes in gravity are small, however, and corrections are usually on the order of 0.25 mm (0.01 in.).

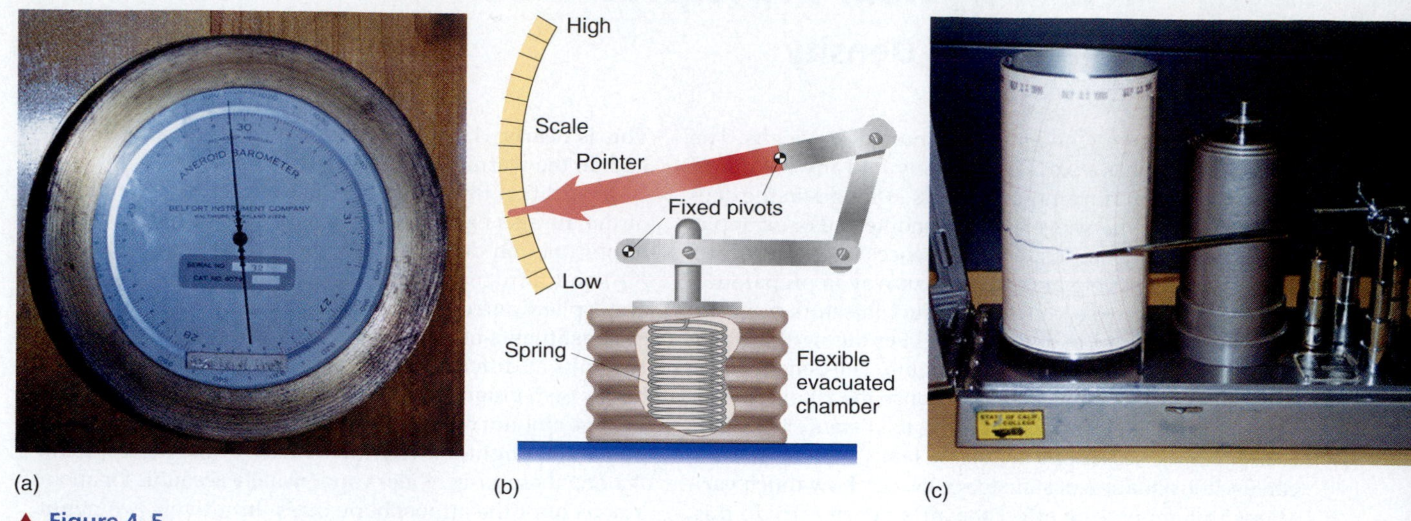

▲ Figure 4–5
An aneroid barometer (a) and its workings (b). A barograph (c).

Aneroid Barometers

Mercury barometers are precise instruments, but they are also expensive and inconvenient to relocate. An alternative instrument for the observation of pressure is the **aneroid** (meaning "without liquid") **barometer** (Figure 4–5a). Aneroid barometers are relatively inexpensive and can be quite accurate. They contain a collapsible chamber from which some of the air has been removed (b). The weight of the atmosphere presses on the chamber and compresses it by an amount proportional to the air pressure. A pointing device connected to a lever mechanism indicates the air pressure.

Aneroid barometers, which are often found in homes, must be calibrated when first installed. The user simply finds out the current sea level pressure and sets the instrument by turning a small screw on the back of the casing. Because there is no expandable fluid in an aneroid barometer, the instrument requires no temperature correction. Furthermore, the effects of altitude and latitude are already accounted for when the instrument is first calibrated. Thus, once calibrated, an aneroid barometer gives the sea level pressure without corrections or adjustments.

Sometimes it is useful to have a continual record of pressure through time. Aneroid devices that plot continuous values of pressure are called **barographs** (Figure 4–5c). A rotating drum (usually set to one rotation per week) turns a chart so that a pen traces a permanent record of the changing pressure.

The Distribution of Pressure

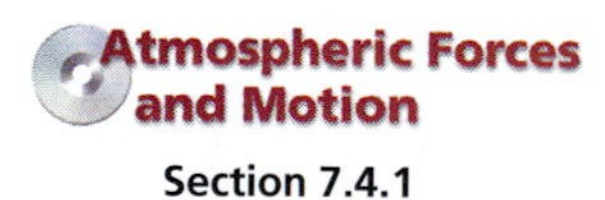

Section 7.4.1

The distribution of sea level pressure across the globe is an extremely important and highly variable characteristic of the atmosphere. In order to visualize this distribution, meteorologists plot lines called isobars on weather maps.

Each **isobar** is drawn so that it connects points having exactly the same sea level pressure, and locations between any two isobars have pressures between those represented by the two lines. Isobars are drawn at intervals of 4 mb on U.S. surface weather maps, so the pressure difference between adjacent isobars is the same everywhere on the map. The advantage of this is that the distance between adjacent isobars provides information about how rapidly pressure changes from one place to another. In other words, the spacing of the isobars indicates the strength of the **pressure gradient**, or rate of change in pressure, in the same way that spacing of isotherms reveals temperature gradients. A dense clustering of isobars indicates a steep pressure gradient (a rapid change in pressure with distance), while widely spaced isobars indicate a weak gradient.

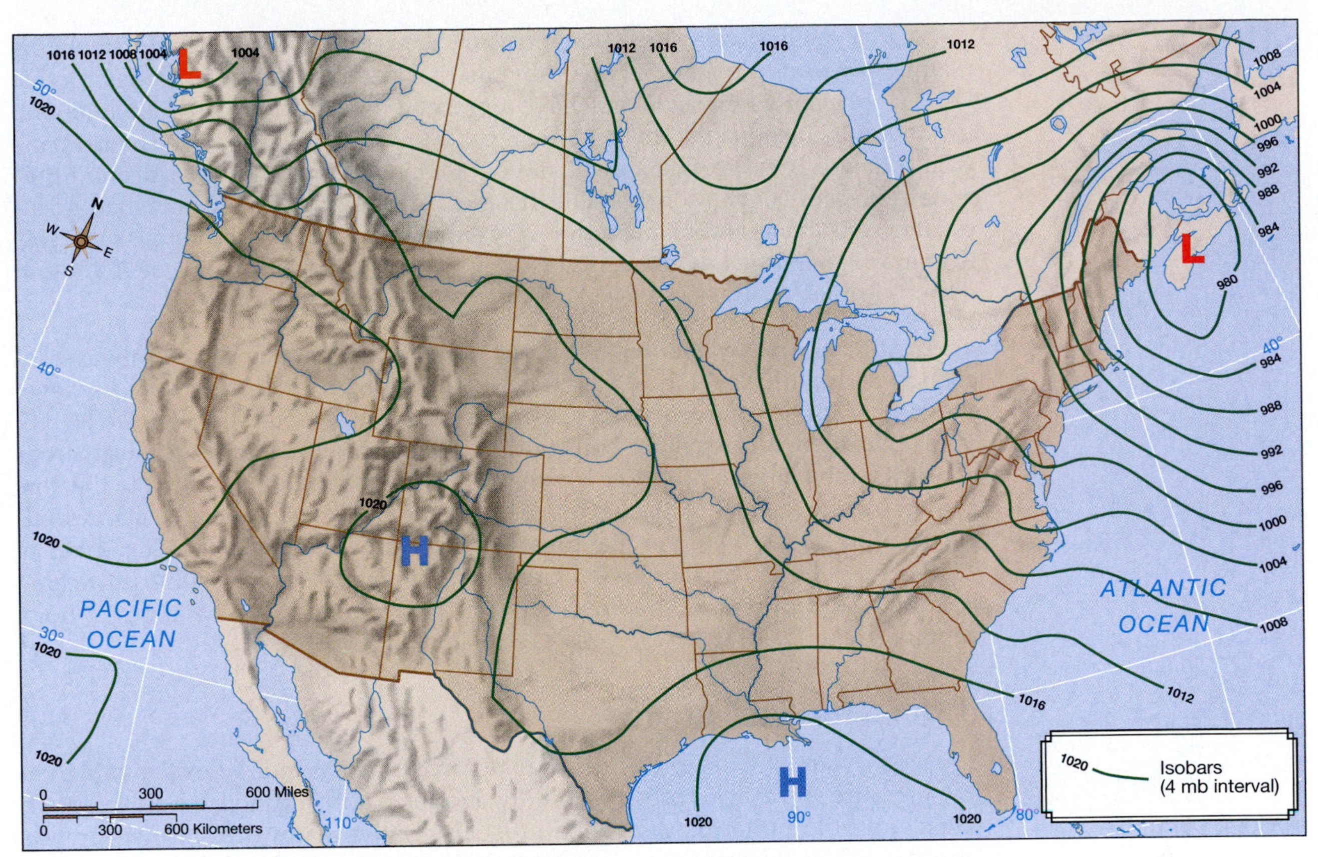

▲ **Figure 4–6**
A weather map showing the distribution of sea level air pressure on March 4, 1994. Note that the pressure is relatively low over the northeastern United States and eastern Canada. Also note that the highest and lowest pressure on the map are only within about 4 percent of each other.

By way of example, Figure 4–6 maps the sea level pressure distribution as it existed on March 4, 1994. The pressure over New England and southeast Canada was lower than over most of the West, and the strongest pressure gradient was over eastern North America.

Pressure Gradients

Pressure gradients provide the impetus for the movement of air we call *wind*. Imagine two people pushing against each other. The person who exerts the greater force pushes the other one back, and the greater the difference in force applied, the faster the pushed person will move. The same concept applies to air. If the air over one region exerts a greater pressure than the air over an adjacent area, the higher-pressure air will spread out toward the zone of lower pressure as wind. The pressure gradient gives rise to a force, called the **pressure gradient force**, which sets the air in motion. For pressure gradients measured at constant altitude, we use the term *horizontal* pressure gradient force, and call the resulting motion *wind*. Everything else being equal, the greater the pressure gradient force, the greater the wind speed.

Horizontal Pressure Gradients The map of sea level pressure shown in Figure 4–6 is fairly typical, having low- and high-pressure areas of average magnitude. Notice that the changes in pressure across the map are small. The lowest pressure observed is about 977 mb, while the highest is about 1021 mb. This 44 mb difference represents a mere 4 percent or so of the average pressure. Note also that the physical distance separating the areas of highest and lowest pressure is about 3000 km (1800 mi). In the most general sense, then, the pressure gradients across the map are on the order of 40 mb per 3000 km, or about 1 mb per 75 km. Clearly, on a continental scale at least, pressure gradients are usually small.

On a smaller scale, horizontal pressure gradients can be much greater. Hurricanes, for example, have steep gradients that produce violent and destructive winds. Yet even a hurricane may have a pressure in its interior only about 50 mb less than that just outside the storm, some 300 km (180 mi) away. Such a hurricane would have a pressure change of 1 mb per 6 km, yielding only a 5 percent difference in pressure over a considerable distance. The point is that even at their greatest, horizontal pressure gradients are very small relative to the mean pressure. This is in marked contrast to vertical pressure gradients, wherein a drop of 50 mb can occur within a vertical distance of only half a kilometer (0.3 mi).

Vertical Pressure Gradients We've seen that atmospheric pressure always decreases with altitude. Notice, for example, in Figure 4–3 that the mean sea level pressure of 1013.2 mb decreases to 500 mb at an altitude of 5640 m (about 18,000 ft). Thus, the average vertical pressure gradient in the lower half of the atmosphere is about 500 mb per 5640 m, or just less than 1 mb per 10 m. Compare that to the horizontal pressure gradient of an average hurricane, which we saw to be about 1 mb per 6000 m. The *average* vertical pressure gradient in this example is 600 times greater than the *extreme* horizontal pressure gradient associated with a hurricane! In sum, vertical pressure gradients are very much greater than changes in horizontal pressure.

Hydrostatic Equilibrium

You already know that a pressure gradient force causes wind to flow from high to low pressure and that air pressure rapidly decreases with altitude. Given these two facts, you might infer that the wind must always blow upward. If this were the case, it would have troublesome implications for humans on the surface, who would suffocate as all the air around us literally exploded out to space in response to the vertical pressure gradient force.

Before you panic, however, consider a second relevant fact: Gravity pulls all mass, including the atmosphere, downward. Then why doesn't the atmosphere collapse all the way down to the point where we would be able to breathe only by getting on our hands and knees and sucking up the air that has fallen to the surface? Because the vertical pressure gradient force and the force of gravity are normally of nearly equal value and operate in opposite directions, a situation called **hydrostatic equilibrium**.

When the gravitational force exactly equals the vertical pressure gradient force in magnitude, no vertical acceleration occurs. When the gravitational force slightly exceeds the vertical pressure gradient force, downward motions result. Such downward motions are always very slow. On the other hand, the upward-directed pressure gradient force sometimes greatly exceeds the gravitational force, and updrafts in excess of 160 km/hr (100 mph) can develop. This is what you see when large clouds rapidly grow as thunderstorms develop.

The Role of Density in Hydrostatic Equilibrium Although the gravitational and vertical pressure gradient forces are normally almost in balance, the exact value of each varies from place to place and time to time. The downward gravitational force on a volume of air is proportional to its mass (remember that force = mass × acceleration), so a dense atmosphere experiences a greater gravitational force than does a sparse atmosphere. Thus, if a dense atmosphere is to remain in hydrostatic equilibrium, it must have a greater vertical pressure gradient force to offset the gravitational force.

Examine the two identical columns of air in Figure 4–7a. If the column on the right is heated, as shown in (b), its upward expansion causes it to have lower density. There is no change in mass, so the pressure is still 1000 mb at the surface. However, the 500 mb level pressure is higher than before and the vertical pressure gradient is smaller. In other words, the heated column has lower density than the unheated column, leading to a smaller vertical pressure gradient. This simple statement has some very important implications for horizontal motions and should be

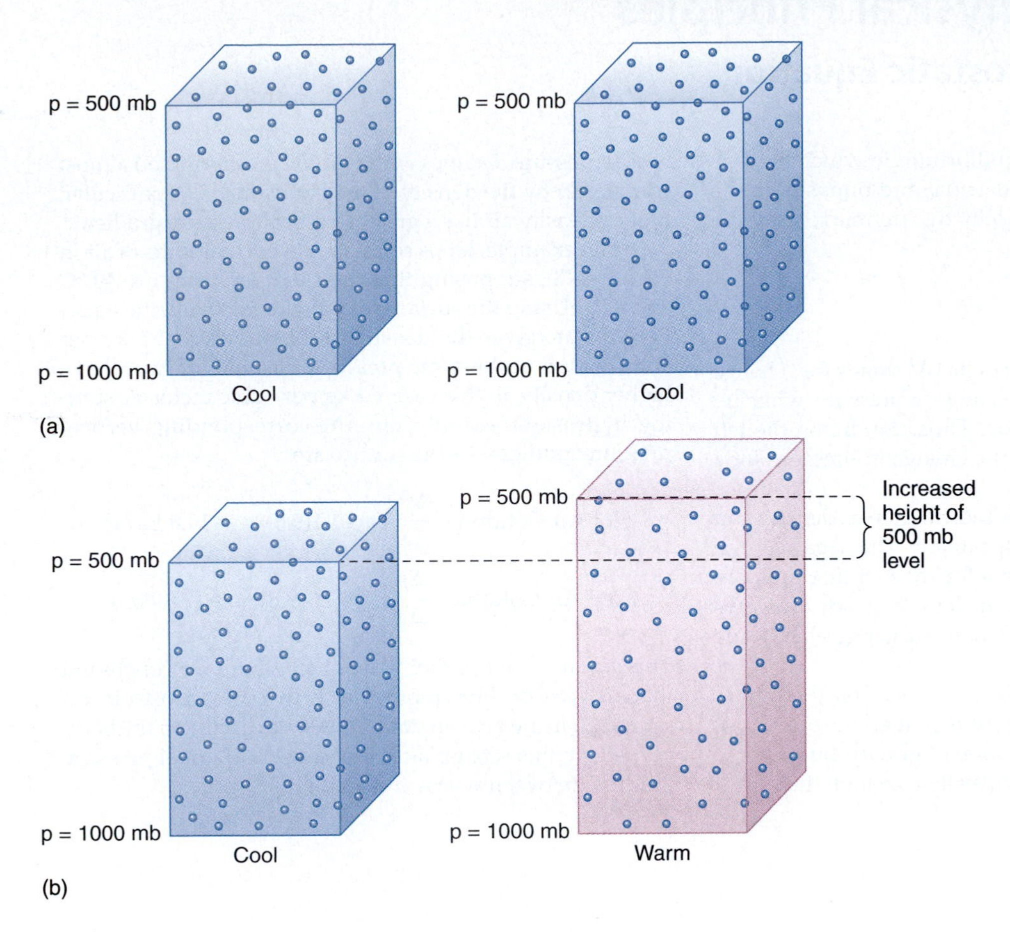

Figure 4–7 Two columns of air with equal temperatures, pressures, and densities (a). Heating the column on the right (b) causes it to expand upward. It still contains the same amount of mass, but it has a lower density to compensate for its greater height. Because the pressure difference between the base and top is still 500 mb, the vertical pressure gradient is smaller.

thoroughly understood before continuing. (For mathematical details on the relation between density and the vertical pressure gradient, see *Box 4–3, Physical Principles: The Hydrostatic Equation.*)

Horizontal Pressure Gradients in the Upper Atmosphere

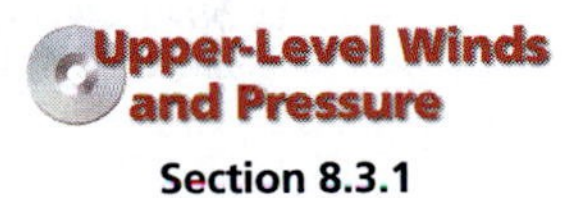

Section 8.3.1

As we have already seen, atmospheric pressure decreases more rapidly with height in a cold, dense air column. Looking at Figure 4–7b, you can see that at the height of the "top" of the cold column, where the pressure is 500 mb, the pressure is greater than 500 mb in the warm column. In other words, there is a horizontal pressure gradient in the middle atmosphere, with lower pressure over the cold column. Equivalently, the height of the 500 mb level is lower in the cold column. Where a horizontal pressure gradient exists, there must also be a slope in the isobars, with heights decreasing toward colder air. It so happens that the horizontal pressure gradient force is proportional to the slope of the isobars. If we know the slope, we know the pressure gradient force.

Figure 4–8 shows the distribution of the 500 mb level in an idealized atmosphere, with a gradually decreasing temperature toward the North Pole (assuming for simplicity's sake that sea level pressure is uniform). Notice that the decrease in temperature toward the pole causes a decrease in the height of the 500 mb level, with the surface sloping downward toward colder air. On the "ground" of the diagram are contour lines showing the height of the 500 mb surface. The contour labels tell how high you must go to find a pressure of 500 mb. For example, if you stand on a line labeled 5500 m, the 500 mb surface is 5.5 km above you. The heights decrease toward the north; thus, the contour values also decrease northward.

Figure 4–9 shows a real 500 mb map for May 3, 1995. The height contours are labeled in decameters (units of 10 meters); thus, heights range from 5880 m in the

4–3 Physical Principles

The Hydrostatic Equation

The concept of hydrostatic equilibrium (in which the vertical pressure gradient force is equal and opposite to the gravitational force) can be succinctly summarized by the **hydrostatic equation**:

$$\frac{\Delta p}{\Delta z} = -\rho g$$

By convention, the Greek letter delta (Δ) stands for "change in." In this case, Δp refers to a change in pressure, while Δz refers to the change in altitude. Thus, $\Delta p / \Delta z$ on the left side of the equation refers to the change in pressure for a unit of increase in altitude.

We have met the symbols ρ and g before as density and the acceleration of gravity, respectively. The negative sign on the right-hand side accounts for the fact that pressure decreases with height; that is, the left-hand side is always negative. For the two sides to balance, the right-hand side must also be negative.

Thus, the hydrostatic equation states that the rate at which pressure decreases with height equals the product of the air density and the acceleration of gravity. But because the acceleration of gravity is virtually constant, the rate at which pressure declines with altitude is determined almost completely by the density of the atmosphere. In particular, higher-density air has a greater vertical pressure gradient.

As an example, let us compare the two columns of air in Figure 4–7b, supposing that their temperatures are 40 °C and 0 °C. Using the surface pressure of 1000 mb, the equation of state gives the density of the hot air as 1.1 kg per cubic meter. At the same pressure, the cold air must have higher density, in this case 1.3 kg per cubic meter. Assuming hydrostatic equilibrium, the corresponding vertical pressure gradients at the surface are

$$\text{Hot Air Column} \quad \frac{\Delta p}{\Delta z} = -1.1(9.8) = -10.8 \text{ Pa/m}$$

$$\text{Cold Air Column} \quad \frac{\Delta p}{\Delta z} = -1.3(9.8) = -12.8 \text{ Pa/m}$$

This confirms our earlier reasoning, where we concluded that pressure declines more rapidly in a cold, dense air column than in a warm air column. As we discuss in the body of the text, this sets up an upper-level horizontal pressure gradient between warm and cold air.

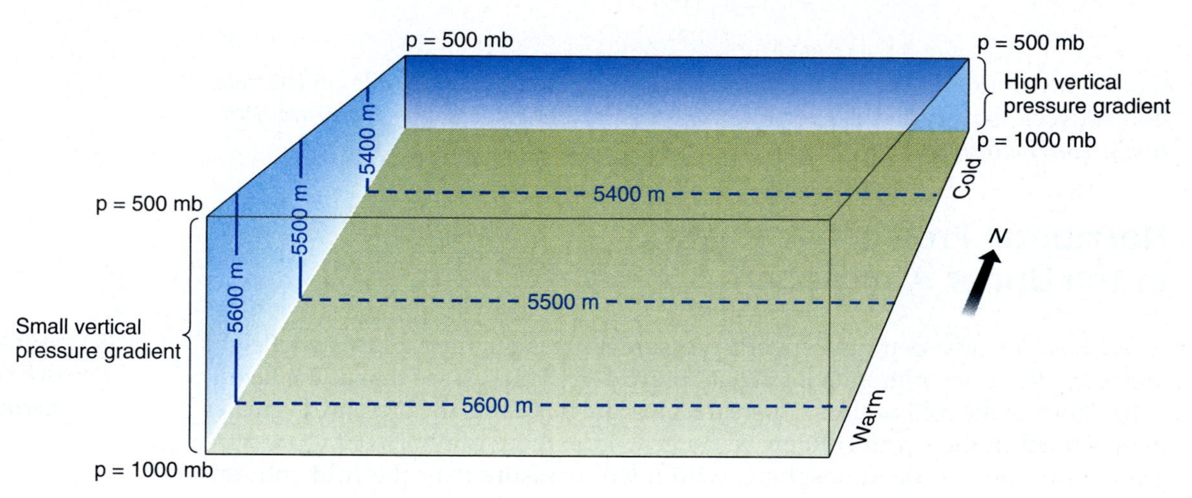

▲ **Figure 4–8**
The gradual poleward decrease in mean temperature results in denser air occurring at high latitudes. As indicated by the hydrostatic equation, pressure drops more rapidly with height at high latitudes and lowers the height of the 500 mb level. The dashed lines depict the height of the 500 mb level as they would be drawn on a 500 mb weather map.

south to 5220 m in the extreme northwest. Contours for 500 mb maps are drawn at 60 m intervals (every 6 decameters). Overall, the pattern is consistent with a decrease in height from south to north, following the temperature gradient from south to north.

Where the height contours are close together, the pressure gradient force is large. Thus, on that day, upper-level winds were strongest over Newfoundland, with speeds above 160 km/hr (100 mph). Over Nebraska, the pressure gradient force was much weaker, and winds were only 35 km/hr (22 mph).

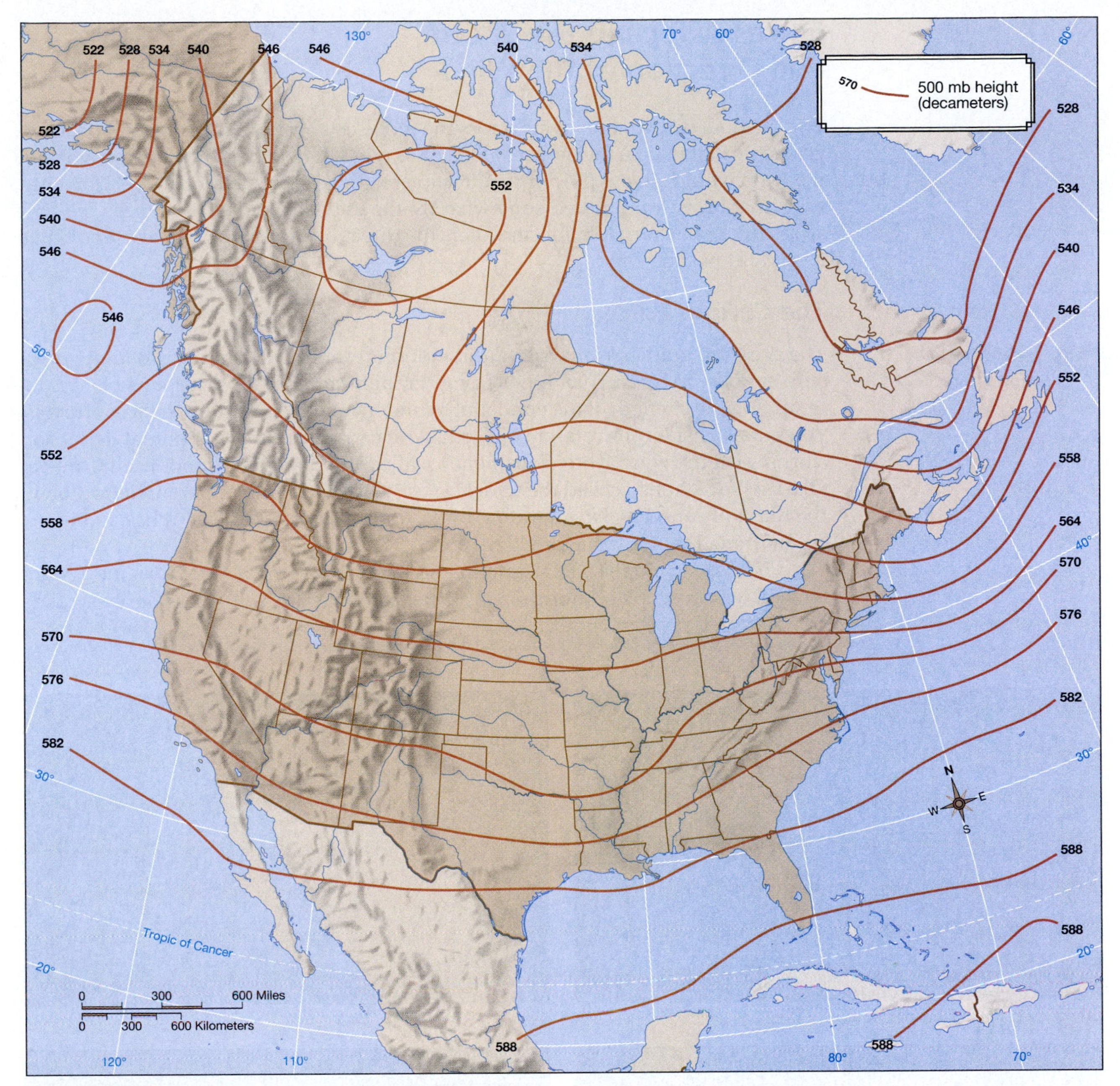

▲ **Figure 4–9**
The distribution of the height of the 500 mb level on May 3, 1995. The height contours are labeled in units of decameters.

Notice that the range in heights is only about 660 m from highest to lowest, a change of about 10 percent across a huge distance. Like horizontal pressure gradients at the surface, these upper atmosphere gradients are small and the pressure surfaces are nearly horizontal. Nevertheless, these weak gradients can produce significant accelerations and high winds, especially in the upper atmosphere where friction is nearly absent and density is low.

Upper-level maps are produced twice each day by the National Weather Service and similar agencies in other countries. They are extremely important for weather forecasting, as we will discuss in Chapter 13. In addition to the maps of the 500 mb level, similar maps are produced for the 850, 700, and 300 mb levels. These correspond to conditions at about 1500, 3000, 10,000, and 13,000 m above sea level (5000, 10,000, 33,000, and 43,000 ft), respectively. Keep in mind that these maps depict the varying heights of these individual pressure levels.

Forces Affecting the Speed and Direction of the Wind

The unequal distribution of air across the globe establishes the horizontal pressure gradients that initiate the movement of air as wind. If no other forces were involved, the wind would always flow in the direction of the pressure gradient force. However, the situation is complicated somewhat by the effect of two other forces. The first arises from planetary rotation and alters the direction of the wind. The second force, friction, slows the wind.

The Coriolis Force

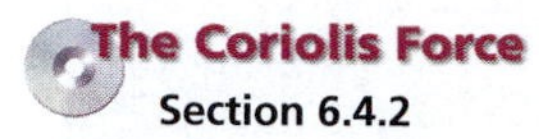
The Coriolis Force
Section 6.4.2

The pressure gradient force sets air in motion from higher pressure to lower pressure, and the magnitude of the pressure gradient force is most responsible for determining the strength of the wind. But to understand the direction the wind travels, you must consider the rotation of Earth, which gives rise to an apparent deflection (turning) of the wind. The phenomenon, called the **Coriolis** (pronounced Core-ee-OH-liss) **force*** (also called the *Coriolis effect*), also causes an apparent deflection in the flight of cannonballs, migrating birds, and jet aircraft. In fact, it has an impact on anything that moves in any direction.

To understand the effect Earth rotation has on moving objects, imagine a counterclockwise rotating platform, such as the one in Figure 4–10. (This figure is taken

*For G. G. de Coriolis (1792–1843), who gave the first mathematical explanation for this phenomenon.

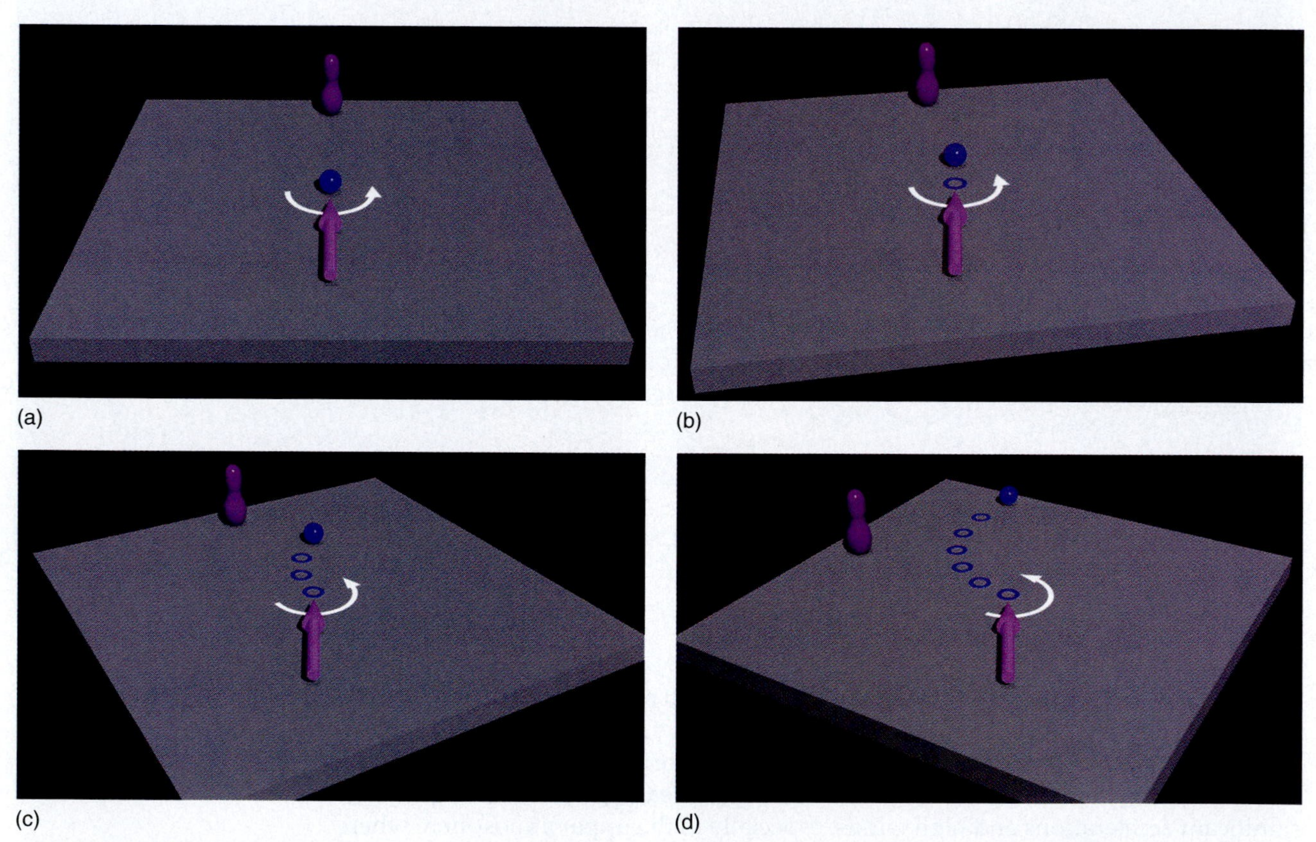

▲ **Figure 4–10**
As an object moves along a rotating surface, its motion appears to curve away from the target. The ball in the center of the counterclockwise-rotating platform in (a) is about to move toward the target at the top of the figure. As the ball moves toward the target, the rotation of the platform causes the target to move away from its original position. This continues as the ball moves away from the center (b) and (c), so that by the time it nears the edge of the platform, the ball appears to have curved to its right (d). Because all points on Earth (except along the equator) undergo some rotation, all moving objects experience this apparent displacement, which is ascribed to a force called the *Coriolis force*. This force not only acts upon projectiles, but it acts on the movement of the atmosphere as well.

4–4 Physical Principles

The Coriolis Effect

In describing wind and other motions, convention says we take the surface as a reference frame. For example, when we say there is a 10-meter-per-second wind, we mean the air is moving past the surface at 10 m/sec. Because the surface rotates, we are describing motions relative to a rotating reference frame. The result is that an object moving in a straight line with respect to the stars appears to follow a curved path on Earth's surface, as can be seen in Figure 4–10. In Figure 4–10a, the ball is at the center of a counterclockwise-rotating platform. As it moves toward the pin (b) and (c), the platform rotates beneath the ball. By the time the ball reaches the edge (d), the platform has turned considerably and so has the pin at which the ball was aimed. Using the platform as a reference, the ball appears to have been deflected to the right, although we know the ball traveled in a straight path.

The magnitude of the Coriolis force is determined by the rate at which the planet rotates (which is constant), the speed of the object (or wind) as it moves across the surface, and latitude. More precisely, it can be expressed as

$$F_C = 2\Omega v \sin \varphi$$

where F_C is the Coriolis force, Ω is Earth's rotation rate (1 revolution per day), v is wind speed, and φ is latitude (note that at the equator, $\sin \varphi = 0$ and there is no Coriolis force). As written here, the equation gives the Coriolis force per unit mass, that is, the force per kilogram of moving air. Combining the information with Newton's Second Law, which tells us that acceleration is force per unit mass, we see that F_C is an acceleration—specifically, the Coriolis acceleration. That said, from here forward we won't draw a distinction between the Coriolis acceleration and the Coriolis force, but instead will use the two interchangeably.

directly from the Coriolis Force Tutorial on the CD that came with this text. The tutorial shows actual motions relative to a rotating platform, and thereby illustrates the phenomenon in a way that cannot be duplicated by static figures. *You are strongly encouraged to make full use of the tutorial.*) As a projectile moves in a straight direction, it sweeps a curved path relative to the turning platform beneath it. This is exactly the same phenomenon that would be encountered if two people on a rotating carousel were to observe a ball passing just overhead. Even if the ball were initially moving directly toward one of the people, that person would be unable to catch the ball because she or he would have been moved out of the ball's path by the carousel's rotation. To the people on the platform, the ball would appear to be turning to its right (assuming that the platform rotates counterclockwise), even though it is in fact moving in a straight line.

The same phenomenon can easily be imagined across Earth's North Pole. Because the Northern Hemisphere rotates in a counterclockwise direction (albeit much more slowly than does a typical carousel), an object moving across the pole would exhibit the same apparent curvature. This also applies to the wind, so air moving in a straight line across the North Pole appears to us on Earth to be turning to the right. Does such a deflection occur at the equator? The answer is no, because the planet's 24-hour rotation imparts no twisting motion at the equator. Instead, any point on the surface of the equator travels a 40,000 km (24,000 mi) sweep around Earth's axis (as is shown in the Coriolis tutorial on this book's CD). In between the equator and the poles, there is a gradual increase in the strength of the Coriolis force with increasing latitude. The same relationship between latitude and the magnitude of the Coriolis force holds true in the Southern Hemisphere, with the only difference being that the apparent deflection is to the left, rather than to the right, because that hemisphere rotates in a clockwise direction.

The magnitude of the Coriolis force increases with wind speed. This may seem counterintuitive at first, but it is nonetheless important to recognize. The Coriolis force involves the deflection that occurs over a given increment of time (say, one second)—not the amount of deflection that occurs during the time an object moves from one particular location to another. Consider this: if an object has the minimum speed possible (zero), it would undergo no deflection. A very slow-moving object might be deflected slightly, but over any brief increment of time the deflection would be minimal. A fast-moving object would cover a considerable distance over the same time interval and therefore be subject to a greater deflection.

4–5 Physical Principles

The Equation of Motion

Whenever a horizontal pressure gradient exists, air tends to move from the high-pressure area to the low. However, it seldom blows directly from high to low, because other forces come into play that affect both the direction and speed of the wind.

Newton's Second Law tells us that the acceleration experienced by a parcel of air depends on the net force acting on it, so we need to add all the forces together and consider the direction and magnitude of each. For horizontal motion, we need to account for the combined effects of the pressure gradient, friction, and the Coriolis effect.

A convenient way of summarizing these three forces is to use the **equation of motion.** If we think in terms of the forces acting on a unit mass of air, say 1 kg, the equation of motion can be written

$$\Delta v / \Delta t = F_p + F_C + F_f$$

where F_p stands for pressure gradient, F_C stands for the Coriolis effect, and F_f stands for friction. Recall that a force per unit mass is an acceleration ($F/m = a$); thus, the three terms on the right of the equation are all accelerations: the acceleration due to the pressure gradient, the Coriolis acceleration, and frictional acceleration. On the left is change in velocity per unit time, which is likewise an acceleration.

Simply put, the equation of motion says that the acceleration experienced by a mass of air is the sum of the accelerations due to the three forces just mentioned.

The equation of motion is an expression of the conservation of momentum. Just as we know temperature changes don't arise out of nothing, we know that accelerations can't mysteriously appear but rather arise from a combination of forces.

Very often the individual terms in the equation of motion nearly cancel one another, so the acceleration of the air is usually much smaller than the individual accelerations. This is not by chance. Rather, the accelerations interact with each other in remarkably subtle ways that greatly limit the resulting net acceleration. To take a simple example, friction increases with increasing wind speed, everything else being equal. If the opposite were true, the wind would be much stronger.

We need to make two final points about the equation of motion. First, note that the equation gives the acceleration of the air, not the velocity. The accelerations might sum to zero, but that does not mean there is no wind. It only means that the wind is not changing in speed or direction. Second, all the accelerations in our equation are horizontal; thus, the equation provides information about horizontal, not vertical, motions. Of course, vertical motions are subject to Newton's Second Law, too, but we have not considered them in our version of the equation of motion.

The equation of motion can be seen as a kind of accounting principle, one of the so-called fundamental, or "primitive," equations of the atmosphere. It lies at the core of the large computer programs used in weather forecasting and in general circulation models used to study climate change on global scales. In an introductory text such as this, it serves a conceptual purpose, encouraging the student to look for the forces behind changes in the wind.

After examining the CD tutorial on this topic, you should have a solid grasp of the following four fundamental characteristics of the Coriolis force:

1. The Coriolis force deflects all moving objects, regardless of their direction of motion. The deflection is to the right in the Northern Hemisphere and to the left in the Southern Hemisphere.
2. The Coriolis force is zero at the equator and increases with increasing latitude, reaching a maximum level at the poles.
3. The Coriolis force acting on any moving object increases with the object's speed.
4. The Coriolis force changes only the direction of a moving object, never its speed.

The Coriolis force is not large relative to other commonly encountered forces. To produce detectable effects, the Coriolis force must act over relatively long periods of time. It is important mainly for the motion of objects traveling long distances, such as the air circulating around a hurricane. For motions that occur over short distances, its effect is negligible. Thus, although a basketball shot undergoes a minor deflection because of Earth's rotation, the deflection is so small that it cannot be used as an excuse for a miss. For the same reason, the Coriolis force does not materially influence the motion of water spiraling into a bathtub or kitchen drain, contrary to what people commonly believe. Whether the water spirals clockwise or counterclockwise is usually determined by an asymmetry in the shape of the basin, or by an initial direction of spin imparted by the water supply. The Coriolis acceleration is present, but it cannot produce significant changes in direction because it has so little time to act before the water reaches the drain.

Friction

The other factor that influences the movement of air is **friction**. Air in contact with the surface experiences frictional drag, which decreases wind speed. Air just above, in contact with the slower-moving surface layer, likewise experiences frictional drag, but from the underlying air rather than from the surface. As this layer slows down, air at higher levels is similarly affected. The effects of friction therefore originate from the surface but are found throughout the lower atmosphere.

Generally speaking, friction is important within the lowest 1.5 km (1 mi) of the atmosphere—often called the **planetary boundary layer** or just the **boundary layer**. Friction lowers the wind speed for a given pressure gradient, which reduces the Coriolis force as well. In contrast, air in the **free atmosphere**, above about 1.5 km, experiences negligible friction. In the absence of friction, winds in the upper atmosphere are fundamentally simpler. We begin our discussion, therefore, with a description of upper-level wind patterns.

Winds in the Upper Atmosphere

Figure 4–11 illustrates the simplest type of pressure pattern that can exist in the free atmosphere. In this case the height contours are straight and parallel to one another, with a pressure gradient force directed northward. Assume that a balloon with a density equal to the surrounding air is tethered to a pole in the free atmosphere. Although the balloon is held in place, it nonetheless is susceptible to the pressure gradient force acting on it.

Figure 4–12 shows what happens when the connecting cord is severed. Initially, the parcel had no movement and therefore no Coriolis acceleration. But after the cord is cut (a), the pressure gradient force propels the balloon slowly toward the low-pressure area. As the pressure gradient force causes the balloon to move faster, there is an accompanying increase in the Coriolis force, and the balloon accelerates farther to its right (in the Northern Hemisphere) (b) and (c). Eventually, the air flows parallel to the height contour lines (d). At this point, the Coriolis force and pressure gradient balance one another, so no net force is acting on the air.

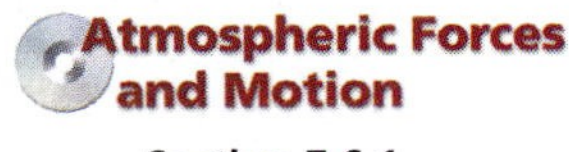

Section 7.6.1

From here on, the balloon moves parallel to the height contours, which in this case is a straight line with a constant speed. In other words, the air flow becomes unaccelerated, with unchanging speed and direction. Such nonaccelerating flow is called the **geostrophic flow** (or **geostrophic wind**), and it occurs when the pressure gradient force equals the Coriolis force. Geostrophic flow occurs only in the upper atmosphere where friction is absent and only the Coriolis and pressure gradient forces apply.

You may wonder why the airflow in Figure 4–12d does not turn all the way back toward the high-pressure area. The simplest answer is that if it were to move in that direction, it would have to flow against the pressure gradient force. This would slow the air down, reduce the Coriolis force, and thereby cause the flow to turn back to its left. Likewise, if it were to turn northward, the air would receive a

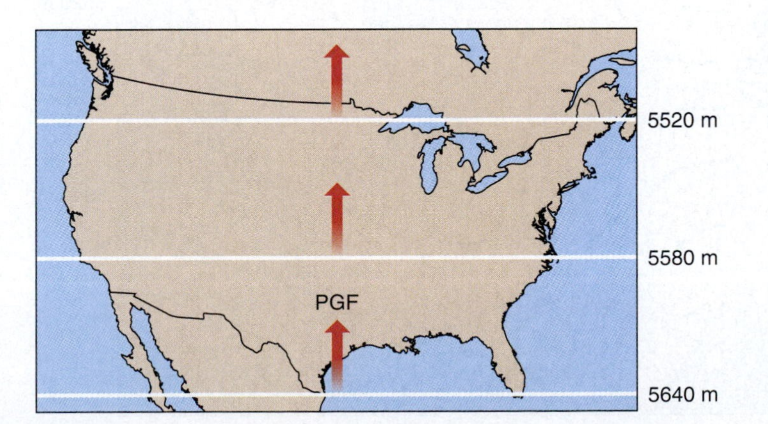

Figure 4–11
The pressure gradient force is directed from south to north in this hypothetical distribution of the height of the 500 mb level. On a nonrotating planet, this would cause air to move from south to north.

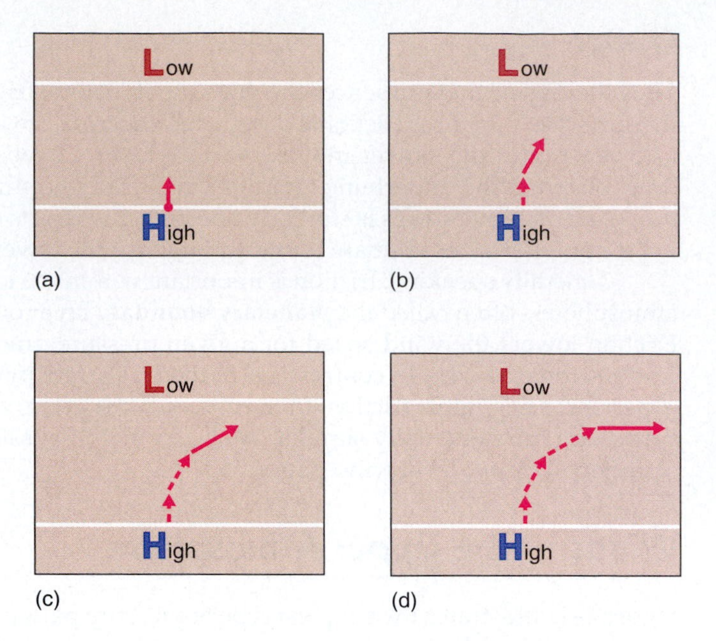

▶ **Figure 4–12**
Geostrophic wind. Assume that there is a stationary parcel of air in the upper atmosphere subjected to a south-to-north pressure gradient force (a). If the parcel is tethered to an imaginary pole, no movement of the parcel can take place. Once the imaginary cord is cut, the horizontal pressure gradient accelerates the parcel northward (b). Initially, when the wind speed is low (as indicated by a short arrow), the Coriolis force is small. As the parcel speeds up (longer arrow), the strength of the Coriolis force increases and causes greater displacement to the right (c). Eventually, the wind speed increases the Coriolis force sufficiently to cause the air to flow perpendicular to the pressure gradient force.

boost from the pressure gradient force. This would lead to a stronger Coriolis acceleration, which would turn the air back to the south. We see that geostrophic flow is stable, meaning that once established, it is not easily disrupted. (For a more detailed look at this process, please refer to the Atmospheric Forces and Motion Tutorial on this book's CD-ROM.)

Supergeostrophic and Subgeostrophic Flow

Changing Wind Patterns at Five Levels of the Atmosphere

The situation shown in Figure 4–11 is a simple one in which the pressure gradient force is uniform, with straight and parallel contours throughout the region. Such situations do occur in nature, but they are the exception rather than the rule. A more common pressure distribution is the type shown in Figure 4–13, in which the height contours curve and assume varying distances from one another. In the absence of friction, the air flows parallel to the contours, for the same reasons as in geostrophic flow. But this type of flow is not truly geostrophic because it is constantly changing direction and therefore undergoing an acceleration. In order for the air to follow the contours, there must be a continual mismatch between the pressure gradient and Coriolis forces. Meteorologists refer to this movement as **gradient flow** (or as **gradient wind**). Like geostrophic flow, gradient flow develops only in the absence of friction, and the wind flows perpendicular to the pressure gradient. In fact,

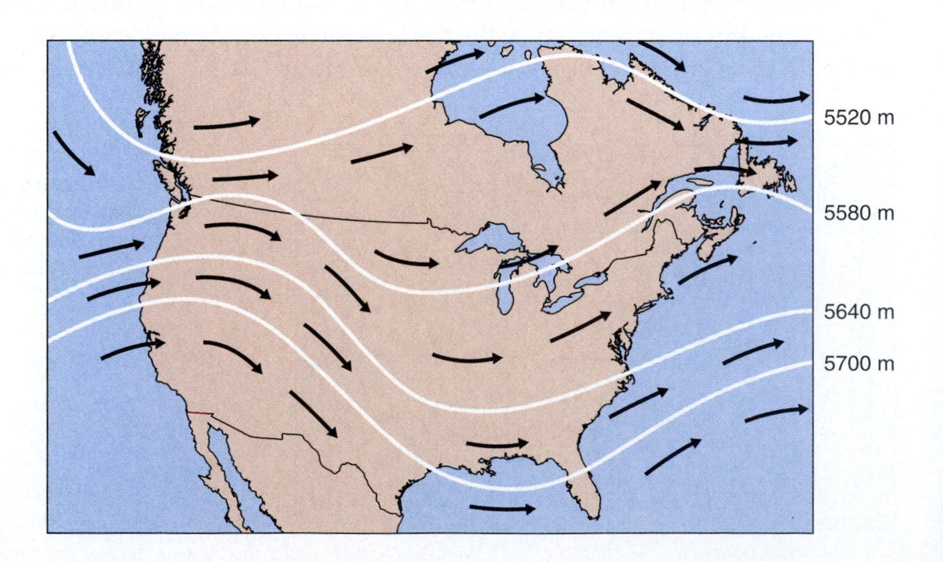

▶ **Figure 4–13**
Gradient wind. Gradient flow occurs in the upper atmosphere where the flow is unaffected by friction from the surface. Note that geostrophic flow is a special case of gradient flow, where there is no acceleration.

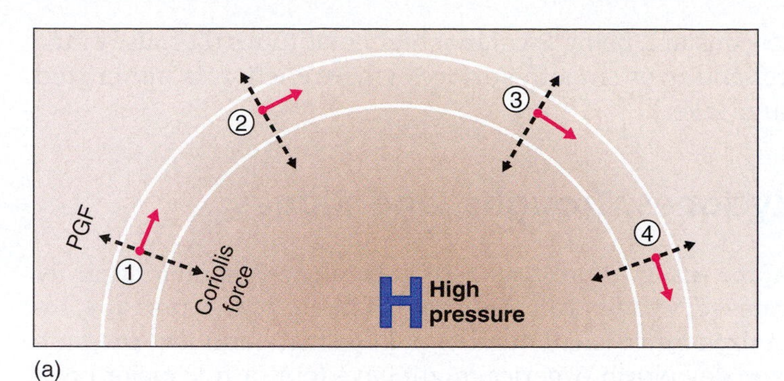

(a)

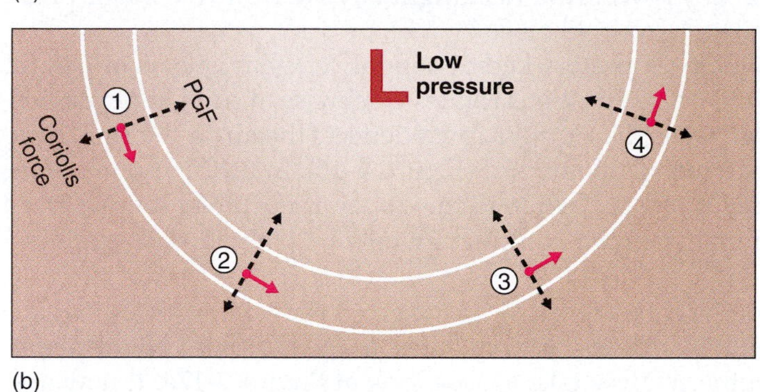

(b)

◀ **Figure 4–14**
Supergeostropic flow (a) occurs in the upper atmosphere around high-pressure systems. As the air flows, it is constantly turning to its right. This turning motion occurs because the Coriolis force has a greater magnitude than the pressure gradient force (as represented by the length of the dashed arrows). Observe the changing direction of the four solid arrows 1 through 4. Subgeostrophic flow (b) occurs in the upper atmosphere around low-pressure systems. The pressure gradient force is greater than the Coriolis force and the air turns to its left in the Northern Hemisphere.

geostrophic flow is simply a special case of gradient flow, arising if the contours happen to be straight and parallel.

Figure 4–14a examines gradient air flow around a circular region of high pressure in the upper atmosphere of the Northern Hemisphere. The pressure gradient force (PGF) is directed away from the center of high pressure while the Coriolis force is directed inward, as shown by the respective arrows. Here the air does not flow in a straight path; rather, it turns to the right and remains parallel to the pressure gradient. In order for it to turn right, the Coriolis force must exceed the pressure gradient force. However, for this to happen, the wind speed must be higher than what would occur under geostrophic conditions. Given the same pressure gradient force, the wind flows faster around a high-pressure region than it does where the height contours are straight, because a larger Coriolis force is required to keep the flow turning to its right. This situation in which the Coriolis force exceeds the pressure gradient force and causes the air to turn is called **supergeostrophic flow**.

The opposite situation is illustrated in Figure 4–14b, where the air rotates counterclockwise around a zone of low pressure. The pressure gradient force directs the flow inward toward the low pressure, and the Coriolis force turns the air to the right. Once again, the air flows parallel to the height contours, but the forces do not balance one another—if that happened, the air would move in a straight line. If the air turns counterclockwise, it must be the case that the Coriolis force is weaker than the pressure gradient force. The existence of a weaker Coriolis force demands that the wind flow more slowly than it would if it were geostrophic. Such flow is said to be **subgeostrophic**.

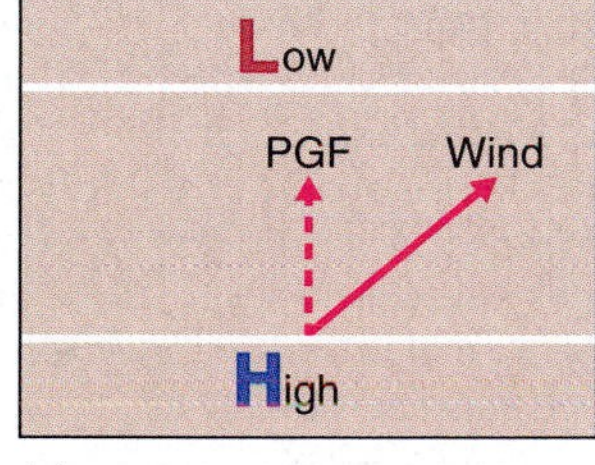

(a)

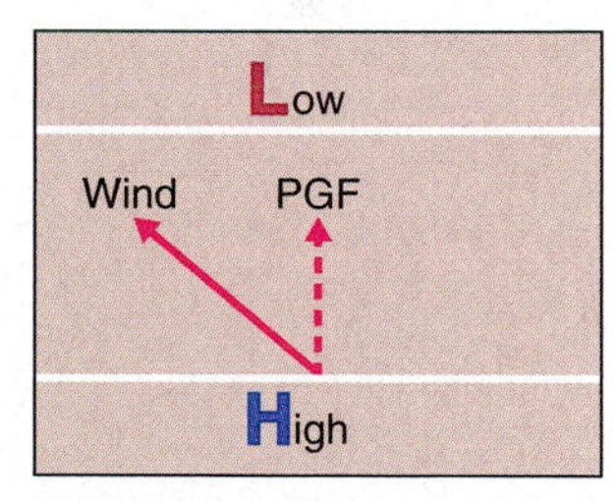

(b)

▲ **Figure 4–15**
Geostrophic flow cannot exist near the surface. Friction slows the wind, so that the Coriolis force is less than the pressure gradient force. Thus, the air flows at an angle to the right of the pressure gradient force in the Northern Hemisphere (a) and to the left in the Southern Hemisphere (b).

Near-Surface Winds

Friction makes winds near the surface slower than those in the middle and upper atmosphere, given equal pressure gradients. The lower wind speeds reduce the Coriolis force and thereby prevent the flow from becoming gradient or geostrophic. Thus, the winds in the boundary layer do not flow parallel to the isobars; rather, they cross the isobars at an angle as they blow from high to low pressure. As always, there is deflection to the right in the Northern Hemisphere and to the left in the Southern Hemisphere (Figure 4–15). The angle of airflow relative to the pressure gradient is not constant, being greater at higher latitudes (because of the stronger Coriolis force) and over smooth surfaces where friction is minimized (such as oceans and large lakes).

Cyclones, Anticyclones, Troughs, and Ridges

Experience tells us that the sea level pressure across the globe is not haphazardly distributed into disorganized, widely scattered zones of high and low pressure. Instead, it is usually organized into a small number of large high- and low-pressure regions. Thus, on a given day North America might have four or five major pressure centers at one time.

Enclosed areas of high pressure marked by roughly circular isobars or height contours are called **anticyclones**. The wind rotates clockwise around anticyclones in the Northern Hemisphere, as the Coriolis force deflects the air to the right and the pressure gradient force directs it outward (Figure 4–16). In the boundary layer the air spirals out of anticyclones (a), while in the upper atmosphere it flows parallel to the height contours (b). In the Southern Hemisphere, the flow is counterclockwise (c) and (d).

Section 7.9.1

Closed low-pressure systems are called **cyclones**. As shown in Figure 4–17, air at the surface rotates counterclockwise into cyclones in the Northern Hemisphere and clockwise in the Southern. Now take a close look at Figure 4–17a. It may appear that the air turns to its *left* as it moves toward the low, but this is not the case. Figure 4–18 explains this apparent contradiction.

The air in position 1 has a pressure gradient force that directs it northward to the center of the low, but the Coriolis force deflects the air to the right so that it ends up in position 2. At position 2, the pressure gradient force still directs the air (now just to the west of due north) toward the center of the cyclone, and again the

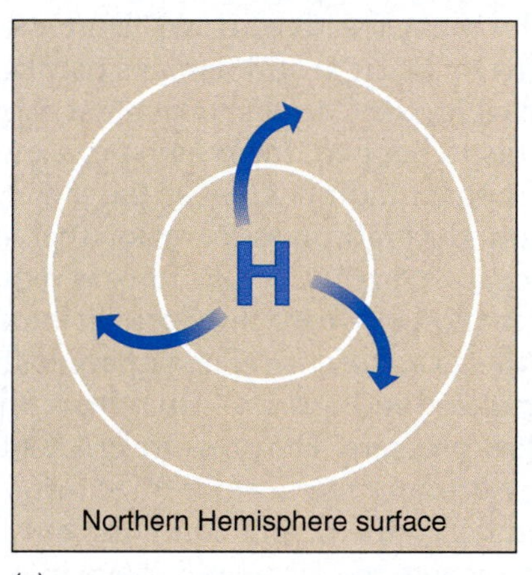

(a)

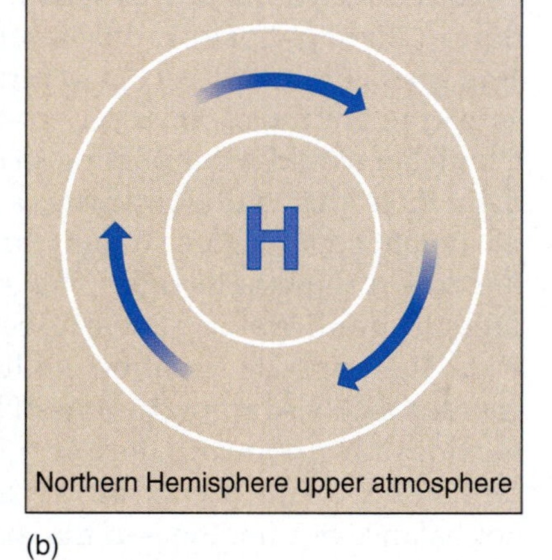

(b)

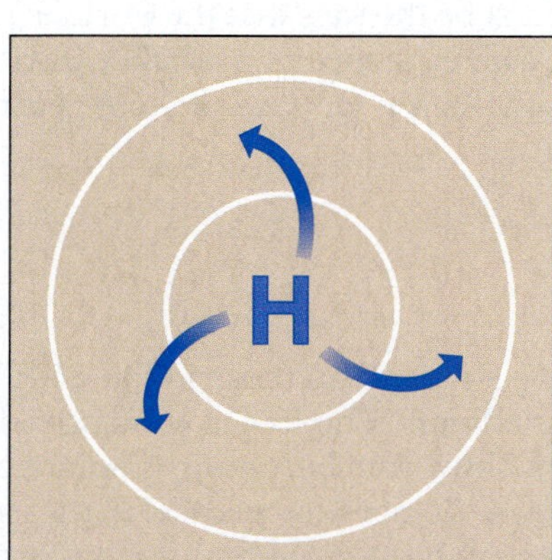

(c)

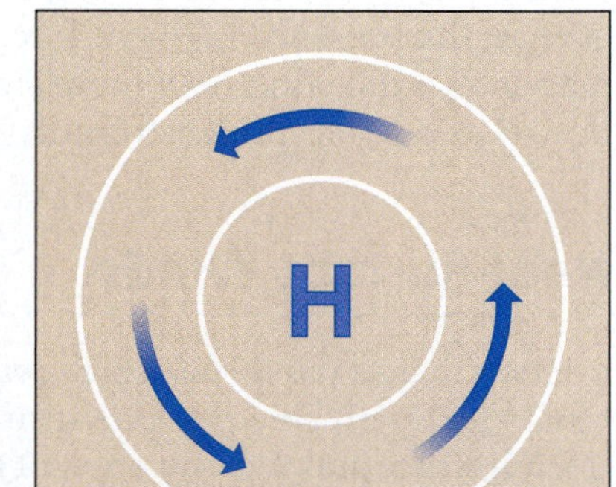

(d)

▶ **Figure 4–16**
Air spirals clockwise out of anticyclones in the Northern Hemisphere (a) and rotates clockwise around the high in the upper atmosphere (b). The flow is reversed in the Southern Hemisphere (c) and (d).

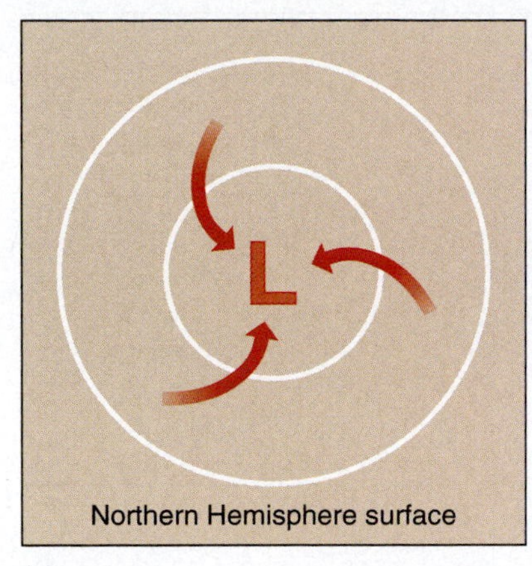

(a)

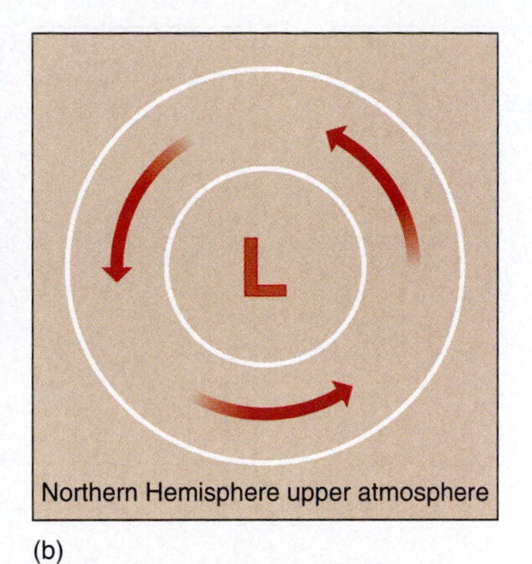

(b)

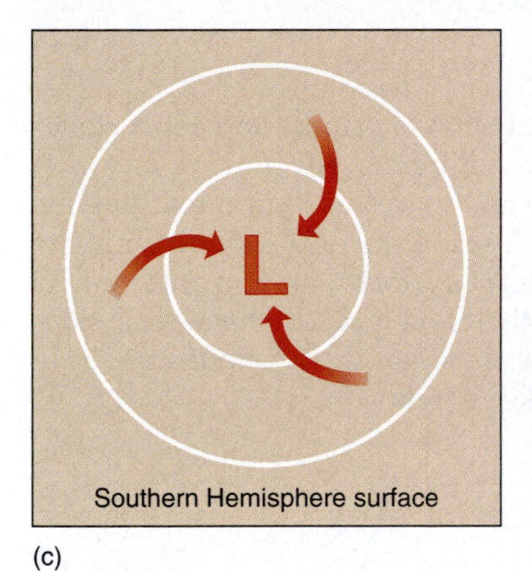

(c)

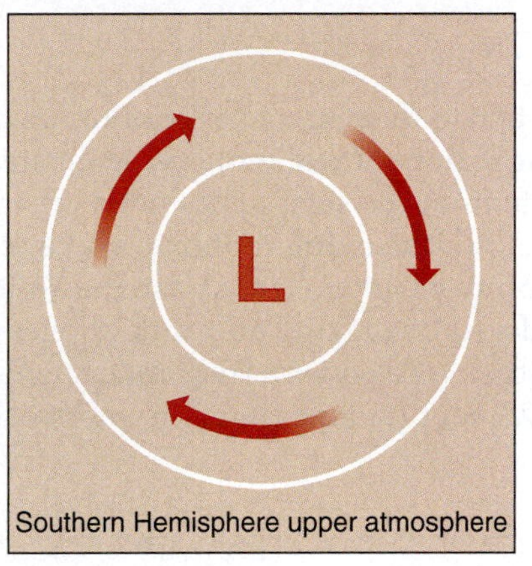

(d)

Figure 4–17
Air spirals counterclockwise into surface cyclones in the Northern Hemisphere (a) and rotates counterclockwise around an upper-level low (b). The flow is reversed in the Southern Hemisphere (c) and (d).

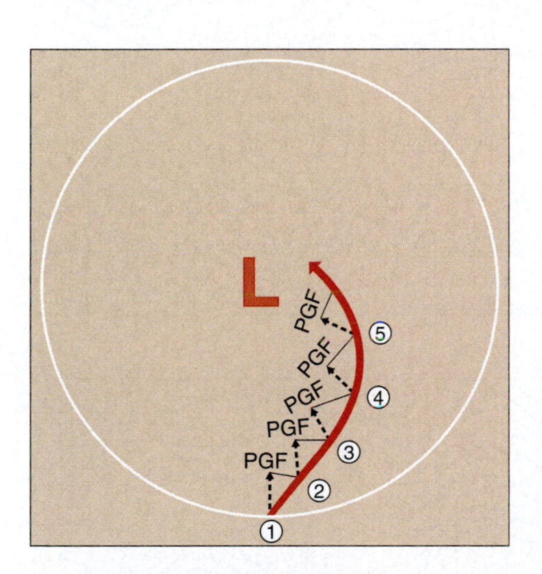

Figure 4–18
The counterclockwise flow of air into surface cyclones in the Northern Hemisphere turns to the left but is nevertheless deflected to the right by the Coriolis force. If there were no Coriolis force, it would cross the isobars at a right angle; instead, it is everywhere deflected to the right.

Coriolis force turns it to the right. This shift in the direction of the pressure gradient gives the trajectory the appearance of being deflected to the left. But at any moment in time, it is accelerated to the right.

The term *cyclone* sometimes causes confusion because people associate it with major tropical storms near southern Asia. In fact, any closed low-pressure system (even one that produces nothing other than gentle breezes and a few clouds) is a cyclone. Although the violent tropical storm goes by the same name, in this discussion we use the term in its most generic sense. We will discuss the more specific type of cyclone in Chapter 12.

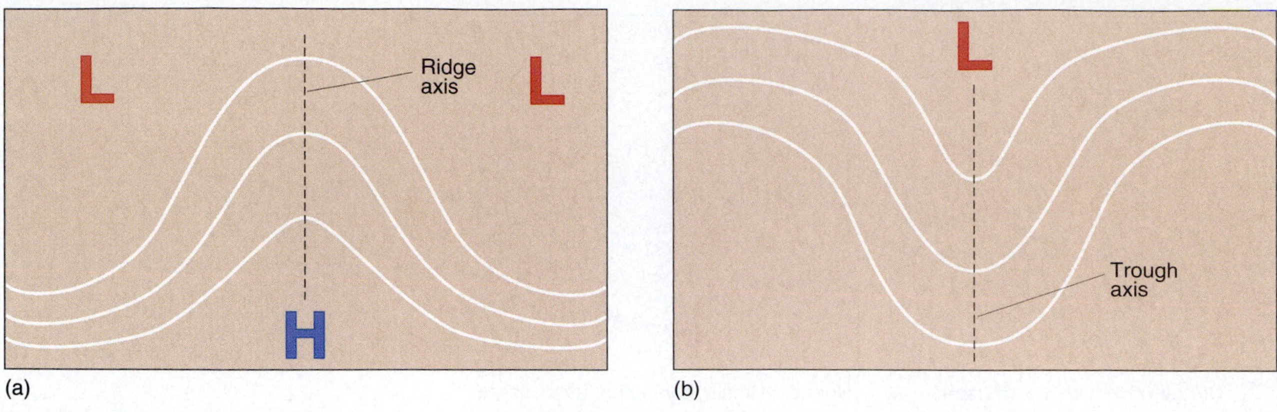

▲ **Figure 4–19**
Elongated zones of high and low pressure are called ridges (a) and troughs (b), respectively.

Upper-Level Winds and Pressure
Section 8.4.1

In addition to their characteristic horizontal winds, cyclones and anticyclones also have particular vertical motions associated with them. Air moving into low-pressure systems at the surface has nowhere to go but upward upon reaching the center of the low. The rising motions, as we will see in the next chapter, promote cloudy or stormy weather. In contrast, the air in anticyclones moving out from the center is replaced by sinking air. Anticyclones therefore typically have clear skies and fair weather. As a general rule, anticyclones are larger than cyclones and have weaker horizontal pressure gradients and weaker winds.

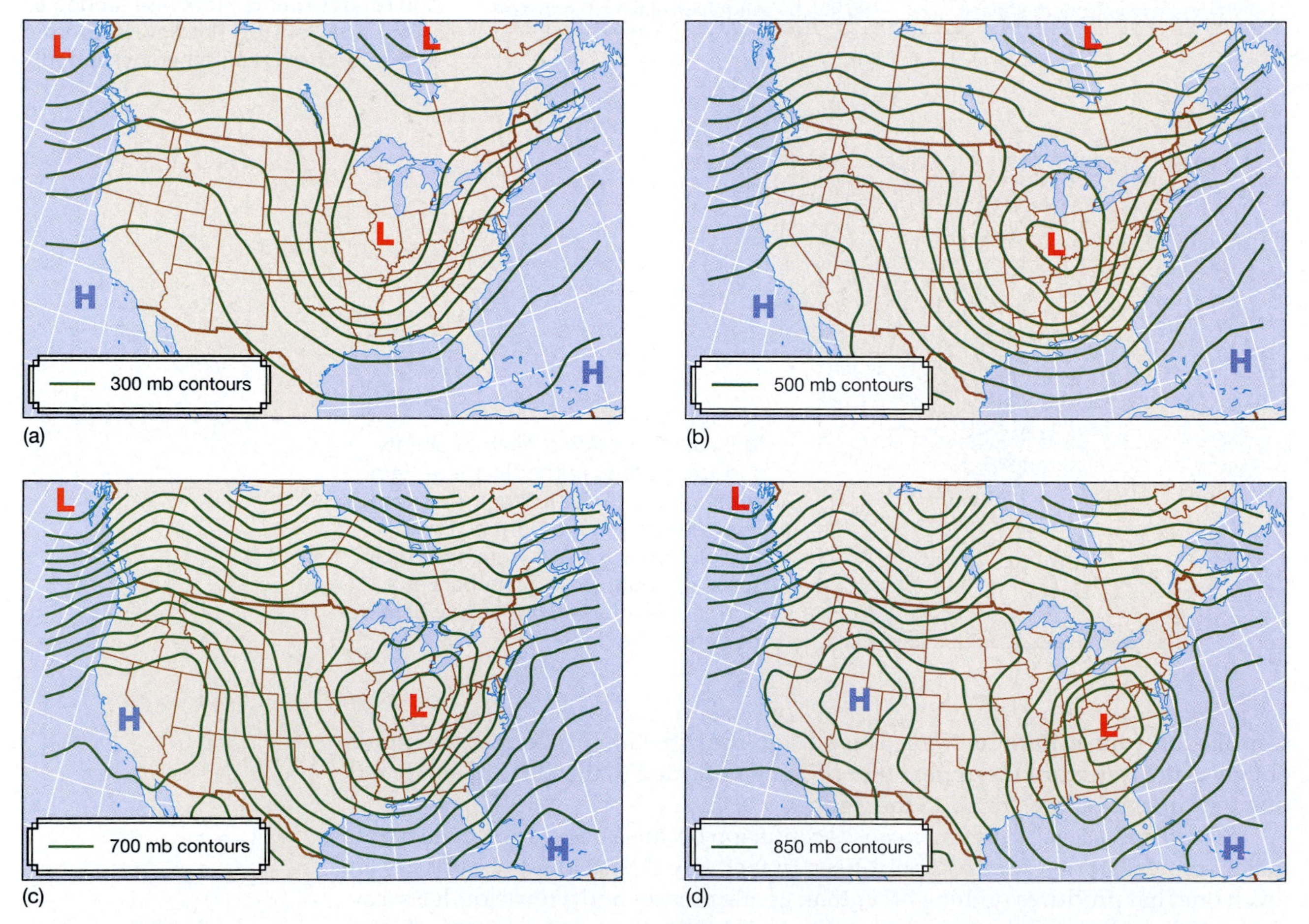

▲ **Figure 4–20**
Maps of the 300, 500, 700, and 850 mb levels for the same date and time. Observe how the pattern of cyclones and anticyclones at the 850 mb level gradually gives way to one of troughs and ridges.

4–6 Special Interest

Strong Winds and the Little Rock, Arkansas, Air Disaster

Aviation in Canada and the United States has a stellar safety record, making air travel within the two countries the safest form of travel. In all of 1998, not a single commercial air traffic fatality occurred within the United States. Not only that, but planes travel millions of miles and undertake many flights under inclement weather. However, that does not mean that bad weather can be ignored, especially during takeoffs and landings. This was clearly the case at 11:50 P.M., CDT, on June 2, 1999, at Little Rock, Arkansas. Despite the fact that a strong thunderstorm was under way and strong winds were being recorded, American Airlines flight 1420 from Dallas/Ft. Worth came in for a landing with 145 people on board.

As the plane touched down, it immediately went into a spin, skidded sideways down the runway for more than a mile, and crashed into a light tower. The plane then broke apart and burst into flames. The death toll, though regrettable, was surprisingly low, with 10 dead. More than 80 people were injured and, miraculously, 51 did not even require hospital treatment.

Investigators immediately focused their attention on the strong winds as the probable cause of the crash. Little Rock was under a severe thunderstorm warning issued a mere 34 minutes prior to the accident, stating that "the main severe weather threat in the warning area is strong and gusty winds." Just 1 minute before the crash, the automated weather sensor adjacent to the runway recorded a wind gust of 140 km/hr (87 mph). Under these conditions, the wisdom of the attempted landing was truly questionable.

Nothing that occurred on June 1 should shake our confidence in our very safe air traffic system. Nonetheless, it would be foolish to completely ignore the reminder that we should never be complacent or overconfident in the presence of severe weather.

Many low- and high-pressure systems occur not as closed cells, but rather as elongated areas called **troughs** (low pressure) and **ridges** (high pressure). Examples are shown in Figure 4–19. There is a tendency for pressure to be distributed as cyclones and anticyclones at the surface and gradually give way to ridges and troughs in the upper atmosphere. This is depicted in Figure 4–20, which shows simultaneously obtained weather maps for the surface and the 300, 500, 700, and 850 mb levels.

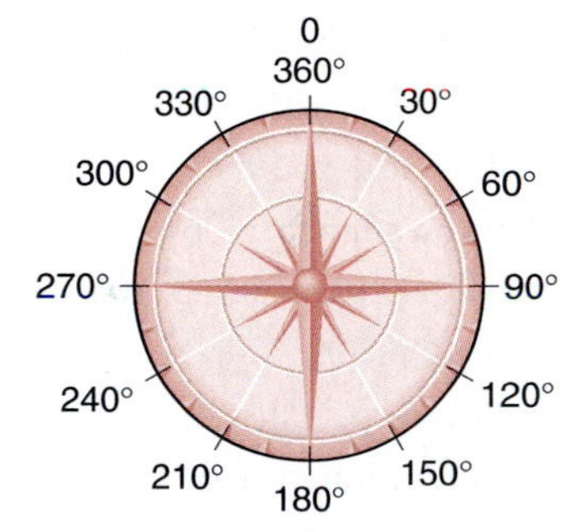

▲ **Figure 4–21**
An azimuth.

Measuring Wind

Wind direction and velocity are measured at all major weather stations. Direction is always given as that from which the wind blows, so that a "westerly" wind is one from the west. It is often expressed by its azimuth. As shown in Figure 4–21, the **azimuth** is the degree of angle from due north, moving clockwise. Thus, due north is represented as 0° (or 360°), east has an azimuth of 90°, south of 180°, and west of 270°. A simple device for observing wind direction is the **wind vane**.

Wind speeds are measured with anemometers. **Anemometers** have rotating cups mounted on a moving shaft. Wind blowing into the cups turns the shaft and generates an electrical current. The strength of the current is proportional to the wind speed, which is then plotted on a strip chart or entered digitally into a computer.

Looking like an airplane without wings, an **aerovane** (Figure 4–22) indicates both wind direction and speed. When the wind changes direction, it pushes against the tail and points the aerovane toward the wind. Like an anemometer, the aerovane has a propeller to generate an electrical current that indicates the wind speed.

Winds in the middle and upper troposphere are every bit as important to weather analysis as surface winds and are especially significant to aviation. Upper-level wind measurements are obtained by tracking the motion of small balloons. Launched twice a day by meteorological offices throughout the world, the balloons carry instrument packages that also provide temperature and moisture information.

▼ **Figure 4–22**
An aerovane.

Summary

The distribution of air pressure across Earth is an extremely important feature of the atmosphere. It sets the winds in motion and determines whether air will rise or sink. Air pressure increases with both the density and temperature of the air, as dictated by the equation of state. Because pressure decreases with altitude, it is important to distinguish between surface air pressure and sea level pressure, the latter being a standard that allows comparisons between different locations. Mercury and aneroid barometers both measure air pressure, which is expressed in units of millibars, pascals, or kilopascals.

The nonuniform distribution of the atmosphere gives rise to horizontal pressure gradients. Although small compared to vertical gradients, horizontal differences in pressure can sometimes produce devastating winds. The vertical pressure gradient force is much larger but is usually offset by a nearly equal but opposite gravitational force, resulting in hydrostatic equilibrium. Under conditions of hydrostatic balance, the vertical pressure gradient is proportional to the density of the atmosphere, as shown by the hydrostatic equation. Spatial variations in the density of the lower atmosphere lead to horizontal pressure gradients in the upper atmosphere that initiate upper-level winds.

Horizontal winds respond to the combined effect of three forces: the pressure gradient force, the Coriolis force, and friction. These combine to form gradient and geostrophic flow in the upper atmosphere and flow across isobars near the surface. The distribution of pressure tends to organize into fairly large-scale patterns of anticyclones, cyclones, ridges, and troughs. Wind vanes, anemometers, and aerovanes each measure different components of the wind velocity.

Key Terms

pressure p. 105
pascal p. 105
millibar p. 105
kilopascal p. 105
Dalton's law p. 106
sea level pressure p. 106
speed p. 108
velocity p. 108
acceleration p. 108
gravity p. 108
force p. 108
equation of state/ideal gas law p. 109
mercury barometer p. 110
barometric pressure p. 110
aneroid barometer p. 112
barograph p. 112
isobar p. 112
pressure gradient p. 112
pressure gradient force p. 113
hydrostatic equilibrium p. 114
hydrostatic equation p. 116
Coriolis force p. 118
equation of motion p. 120
friction p. 121
planetary boundary layer p. 121
free atmosphere p. 121
geostrophic flow p. 121
gradient flow p. 122
supergeostrophic flow p. 123
subgeostrophic flow p. 123
anticyclones p. 124
cyclones p. 124
troughs p. 126
ridges p. 126
azimuth p. 127
wind vane p. 127
anemometer p. 127
aerovane p. 127

Review Questions

1. What is a partial pressure?
2. What is Dalton's law?
3. Why does pressure always decrease with altitude?
4. What is the difference between surface pressure and sea level pressure?
5. How do speed and velocity differ? How do force and pressure differ?
6. What are the equation of state and the hydrostatic equation, and what do they tell us?
7. What two variables determine air pressure?
8. Describe how mercury and aneroid barometers measure air pressure, and explain why corrections need to be used for the observations made from one but not the other.
9. In what way is it misleading to express pressure in inches of mercury?
10. Explain the concept of hydrostatic equilibrium.
11. Explain how air temperature affects the vertical pressure gradient.
12. Explain how the pressure gradient force, the Coriolis force, and friction determine the movement of air in the free atmosphere and in the boundary layer.
13. Describe the roles (if any) that wind speed, latitude, and direction of motion have in determining the magnitude of the Coriolis force.
14. What are geostrophic and gradient flows? Why do they not occur near the surface?
15. What are supergeostrophic and subgeostrophic flows?
16. Define the terms *cyclone, anticyclone, trough,* and *ridge.*
17. Briefly describe the movement of air around cyclones and anticyclones in the Northern and Southern Hemispheres.
18. What do anemometers and aerovanes measure?

Critical Thinking

1. Pressurized cans of shaving cream advise users not to expose the product to excessive heat. What problem might occur if the advice is not followed? Will this potential problem remain throughout the life of the product?
2. On a particular day, the vertical pressure gradient at the surface is –11 pascals per meter. What is the vertical pressure gradient in units of millibars per kilometer? Would you be able to use this gradient to exactly determine the pressure at the top of a building 200 m tall?
3. If a low-pressure region were to instantaneously replace a high-pressure system (assuming normally encountered values of high and low pressure), do you think you would be able to notice the difference by the pressure in your ears?
4. Would a particular pressure gradient produce the same exact wind speed over an Arizona desert that it would over a dense forest of tall trees?
5. The pilot of a small plane wants to fly at a constant height above the surface. Can the pilot fly at a constant pressure level (e.g., 500 mb) to assure the constant height above the ground? Why or why not?
6. A rule of thumb is that the 850 mb level often marks the boundary between the free atmosphere and the boundary layer. Are there parts of North America where this relationship is likely not to be valid?
7. The Coriolis force applies equally to objects moving in any horizontal direction. Do you think the Coriolis effect also affects objects moving directly up or down? If so, how does latitude affect the magnitude of the force?
8. Consider a 90-story skyscraper with high-speed elevators. Would a person ascending from the 46th to the 90th floor undergo the same degree of ear popping as a person going from the 1st floor up to the 45th? Why or why not?

Problems and Exercises

1. Refer to any of the Web sites below that produces surface and 500 mb maps. Examine the current surface map and identify the major cyclones, anticyclones, troughs, and ridges at the surface. Then look at the 500 mb map. Does the same general pattern emerge? Do the troughs and ridges at the 500 mb level occur directly over the corresponding features on the surface map? (Relationships between the surface and 500 mb levels will be discussed further in Chapter 10.)
2. On a daily basis, go to the Weather Channel's Web site at **www.weather.com**. Read the narrative describing the general weather pattern across the United States and identify the most notable weather events occurring across the country. Then look at the surface and 500 mb weather maps. This process will help you to become more familiar with normal pressure distributions and the type of weather often associated with them. As you proceed through this text, the pressure patterns and their association with daily weather will become more meaningful to you.
3. Examine today's 500 mb weather map. You will likely find a trend toward decreasing 500 mb heights with increasing latitude. Are there any exceptions on the map to that general pattern? If so, observe the surface temperatures across North America. Do the temperature patterns have any association with the 500 mb pattern?
4. Observe a surface weather map that plots isobars and station models (**http://www.atmo.arizona.edu/products/wximagery/** is a good source). Do the air flow patterns around cyclones and anticlines shown by the station models completely correspond with the generalizations made in this chapter? If not, why not?

Quantitative Problems

Differences in atmospheric pressure across the globe exert important impacts on all the other important elements of weather. Several quantitative problems are presented in this book's Web site to help you understand the concept of pressure, how it decreases with height, and its sensitivity to changes in moisture. It also provides problems to help illustrate the effect of latitude on the Coriolis force. To get to the problems, enter the Web site and click on Chapter 4. Then click on the "Quantitative Problems" section on the bar at the left.

Useful Web Sites

http://www.atmo.arizona.edu/products/wximagery/

Offers numerous images, maps, and animations. Scroll down to the last two thumbnails to view the current surface and 500 mb weather maps for North America. These maps are plotted with the same conventions that have been used by weather forecasters for many decades. You can also observe an animation of the 500 mb surface for North America (also showing the water vapor distribution) by clicking on the appropriate thumbnail under the heading *GOES Water Vapor Images*.

http://weather.uwyo.edu/upperair/uamap.html

Offers the user the ability to produce upper air maps at different levels for any portion of the globe. Maps are archived for approximately 2 weeks.

http://weather.unisys.com/eta/index.html

Includes maps of the forecasted upper air patterns for a number of upper level pressure surfaces, based on one of the primary forecast models. Output from other models also available from this site. Forecast maps for sea level pressure and expected precipitation also available.

http://www.princeton.edu/%7Eoa/safety/altitude.html

Provides interesting information on how low pressure associated with altitude affects humans.

Media Enrichment

Tutorial

The Coriolis Force

This tutorial will help you visualize Coriolis accelerations—why they appear in Newton's law and how they vary with latitude and wind speed. This is a "must-do" if you doubt any of the four Coriolis properties described earlier in the chapter.

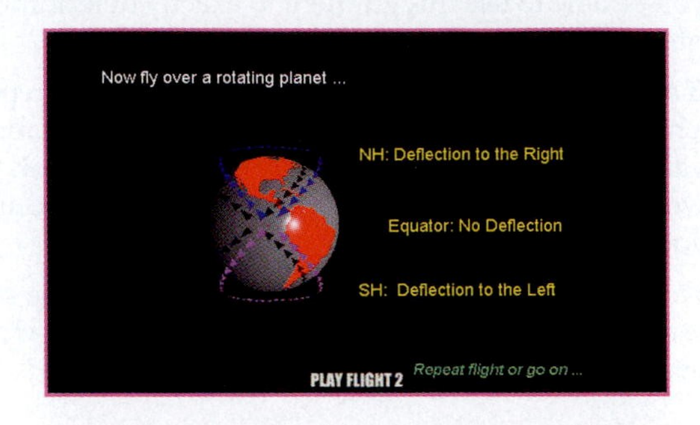

Tutorial

Atmospheric Forces and Motion

This is an extensive review of the equation of motion, showing the joint effects of the various forces involved in horizontal motion. It covers the development of geostrophic flow (and its close relative, gradient flow) and shows wind flow around cyclones and anticyclones. In addition, an interactive exercise lets you experiment with various combinations of friction and pressure gradient to see the resulting effects on wind speed and direction. Animations are used throughout to present many issues difficult to depict with static diagrams.

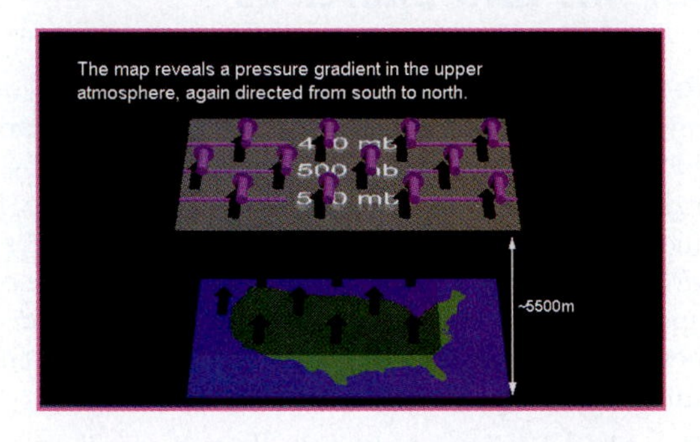

Tutorial

Upper-level Winds and Pressure

This tutorial develops the relationship between upper-level pressure and height fields and shows how ridges and troughs are reflected on contour maps. Rossby waves are presented as common features of the mid-latitudes, with animations used to depict simultaneous movement of the wave and prevailing westerly flow through the wave.

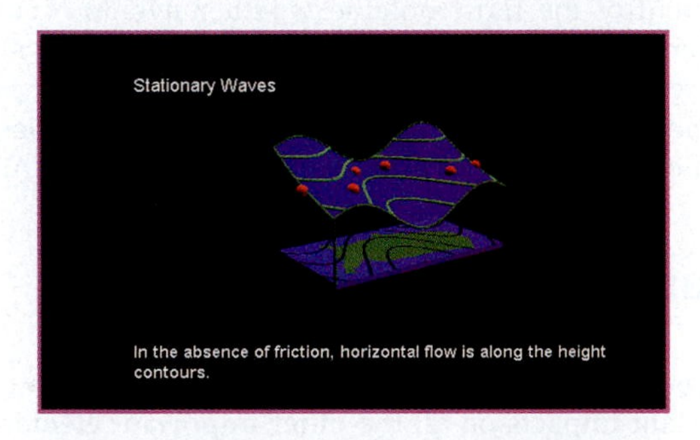

Weather in Motion

Changing Wind Patterns at Five Levels of the Atmosphere

This movie shows wind patterns as they evolve over a 5-day period at five different levels. You can view each of the five "slices" of the atmosphere by clicking the appropriate circle. Notice that the wind field changes its overall shape and migrates in the downwind direction, generally from west to east. These patterns will be discussed further in Chapter 8.

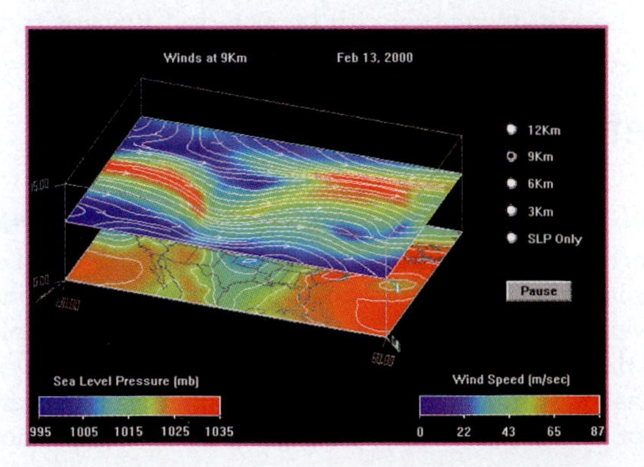

PART TWO

Water in the Atmosphere

Fog descending over East River mountains, Blufield, West Virginia.

chapter 5

Atmospheric Moisture

Liquid water secreted as sweat is converted to water vapor using energy supplied by the player's warm scalp. But the vapor condenses quickly in the cold environment, forming a visible "cloud" around his head. Yet another change in phase occurs in the left side of the photo where the cloud evaporates into drier air downwind.

Despite the snowstorm on the afternoon of January, 9, 1997, air traffic into and out of Detroit Metropolitan Airport continued as normal, just as it had throughout the Midwest. Shortly before 4 P.M., Comair Flight 3272 was approaching for landing. The flight, which had taken off about an hour earlier from Cincinnati/Northern Kentucky International Airport, had proceeded normally until the pilot suddenly lost control of the aircraft. According to witnesses such as Ted Rath, who watched the horrifying scene, the twin-engine plane rolled three times and then crashed into a field 18 miles short of Runway 3. There were no survivors among the 29 passengers and crew.

Air safety experts immediately suspected icing as the cause of the accident. Further evidence, such as the fact that the aircraft had encountered two previous failures of its deicing system, added further support to the suspicion. The National Transportation Safety Board ultimately concluded that a coating of ice on the plane's wings and a low approach speed caused the plane to lose the necessary aerodynamic lift.

Although other crashes have been caused by aircraft icing, these incidents are extremely rare. However, the fact remains that weather conditions can pose significant problems to aviation. Dangerous weather most often takes the form of extreme turbulence or rapidly changing wind conditions. Sometimes even relatively mild conditions can present significant risks, as was the case near Detroit. Even a mere fog bank or layer of overcast clouds can reduce visibility and pose a threat to safe ground and air travel. Despite the fact that clouds and fog are common to our everyday lives, many of us have a weak understanding of how they form.

Water is unique not only because it occurs naturally near Earth's surface in its solid, liquid, and gaseous phases, but also because it can change between these phases in a fairly short period of time. In fact, on average, a molecule of water vapor in the atmosphere will exist in its gaseous phase for only a few days before condensing to liquid droplets or ice crystals to form clouds or fog. Clouds and fog likewise have limited life spans and often evaporate away or precipitate much of their moisture within a few hours of their formation.

This chapter opens Part II, "Water in the Atmosphere," by describing the fundamentals of atmospheric moisture. It lays out the processes by which water can change from one phase to another and describes the common measures by which we express humidity. The chapter also describes the fundamentals involved in fog and cloud formation. Chapter 6 describes the processes of cloud development and the resultant cloud forms. Chapter 7 discusses how cloud droplets grow large enough to fall as precipitation. The topics discussed here are vital to understanding some of the most common weather phenomena, as well as those that sometimes lead to major human impacts.

CHAPTER OUTLINE

Water Vapor and Liquid Water

Although matter in the gaseous phase is highly compressible, the density of a gas cannot be increased to an arbitrarily high level. At some point a limit is reached, forcing a change to liquid or solid state. For one atmospheric gas, water vapor, that limit is routinely achieved at temperatures and pressures found on Earth. (Other gases, such as nitrogen and oxygen, can be liquefied only at very low temperatures.) Air that contains as much water as possible is said to be *saturated,* and the introduction of additional water vapor results in the formation of water droplets or ice crystals. The concept of saturation is fundamental to understanding the processes

January and July Water Vapor Movies

that form clouds and fogs. We begin our discussion with a hypothetical laboratory experiment that describes the general principles of evaporation and condensation. We then apply those principles to the processes that take place in the real atmosphere.

Evaporation and Condensation

Atmospheric Moisture
Section 3.2.1

Figure 5–1 depicts a hypothetical experiment, in which a tightly sealed container is partially filled with pure water (H_2O). Although it may seem obvious at this juncture, let's stipulate that the water in the jar has a perfectly flat surface. Furthermore, assume that at the onset of the experiment the water surface is covered by an impermeable coating, so no water vapor exists in the volume of the container above the water surface. Whether the volume above the water surface contains any air is entirely irrelevant to this experiment. The volume can contain normal air, pure hydrogen, methane, or fumes from French perfume—it can even be a complete vacuum. All that matters with respect to the evaporation/condensation process is that no water vapor be present initially.

Figure 5–1b shows what happens when we remove the covering on the liquid water surface. Without the covering, some of the molecules at the surface can escape into the overlying volume as water vapor. The process whereby molecules break free of the liquid volume is known as **evaporation**. The opposite process is **condensation**, wherein water vapor molecules randomly collide with the water surface and bond with adjacent molecules. At the beginning of our hypothetical experiment, no condensation could occur because no water vapor was present. As evaporation begins, however, water vapor starts to accumulate above the surface of the liquid.

At the early stages of evaporation, the low water vapor content prevents much condensation from occurring, and the rate of evaporation exceeds that of condensation. This leads to an increase in the amount of water vapor present. With increasing water vapor content, however, the condensation rate likewise increases. Eventually, the amount of water vapor above the surface is enough for the rates of condensation and evaporation to become equal, as shown in Figure 5–1c. A constant amount of water vapor now exists in the volume above the water surface due to offsetting gains and losses by evaporation and condensation. The resulting equilibrium state is called **saturation**.

The state of saturation described here can occur whether or not air (or other gases, for that matter) exists in the container. In other words, the water vapor is not "held" by the air (although this erroneous statement is frequently made). Water

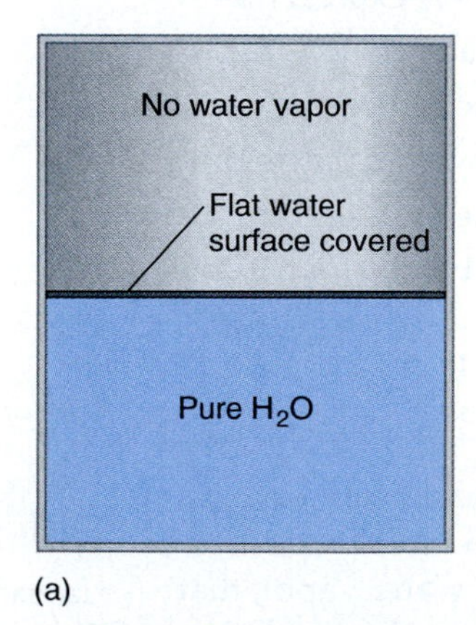

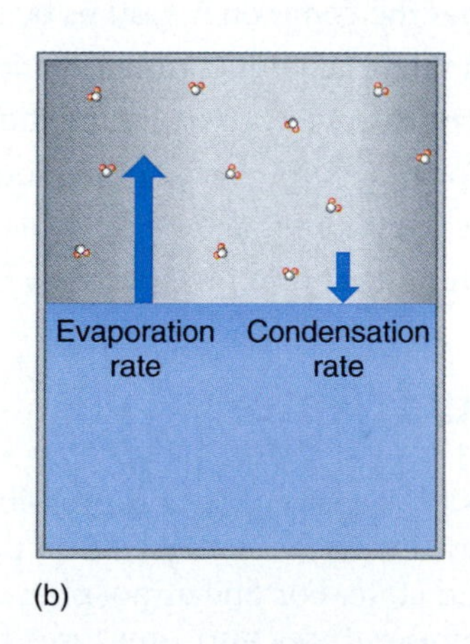

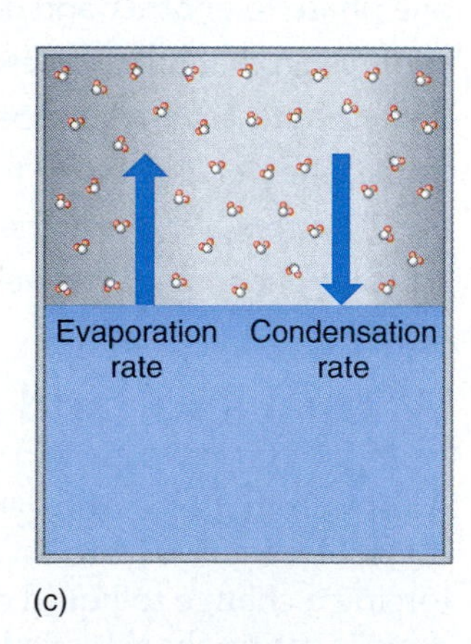

▲ **Figure 5–1**
A hypothetical jar containing pure water with a flat surface, and an overlying volume that initially contains no water vapor (a). When evaporation begins (b), water vapor accumulates in the volume above the liquid water. Initially, no condensation can occur because of the absence of water vapor above the liquid. But as evaporation contributes moisture to the overlying volume, some condensation can occur. Evaporation exceeds condensation for a while and thereby increases the water vapor content. Eventually enough water vapor is above the liquid for condensation to equal evaporation (c). At this point, saturation occurs.

vapor is a gas, just like the other components of the air. Thus, it does not need to be "held" by air any more than the oxygen, nitrogen, argon, and other gases of the atmosphere need to be held by water vapor! When the air is saturated, there is simply an equilibrium between evaporation and condensation; the dry air plays no role in achieving this state. It is also important to realize that the exchange of water vapor and liquid described here also applies to the change of phase between water vapor and ice. The change of phase directly from ice to water vapor, without passing into the liquid phase, is called **sublimation**. The reverse process (from water vapor to ice) is called **deposition**. (Meteorologists sometimes use the word *sublimation* to apply to vapor-to-solid phase changes as well as solid-to-vapor. Because opposite processes should not have the same name, the use of the term *deposition* is preferred for vapor-to-ice changes.)

Indices of Water Vapor Content

Atmospheric Moisture
Section 3.3.1

It is useful to have some way of expressing the moisture content of the air, as it is for any other property of the atmosphere. **Humidity** refers to the amount of water vapor in the air. Humidity can be expressed in a number of ways—in terms of the density of water vapor, the pressure exerted by the water vapor, the percentage of the amount of water vapor that can actually exist, or several other methods. There is no single "correct" measure, but, rather, each has its own advantages and disadvantages, depending on the intended use. All measures of humidity have one thing in common, however—they apply exclusively to water vapor, and not to liquid droplets or ice crystals suspended in or falling through the air. Let's now take a look at these measures.

Vapor Pressure

In Chapters 1 and 4, we saw that the air exerts pressure on all surfaces. Each of the gases that make up the atmosphere contributes to the total air pressure, with the most abundant permanent gases accounting for most of the pressure. Because water vapor seldom accounts for more the 4 percent of the total atmospheric mass, it exerts only a small percentage of the total air pressure. The part of the total atmospheric pressure due to water vapor is referred to as the **vapor pressure**. Like the atmospheric pressure, vapor pressure is commonly expressed in units of millibars (mb) by U.S. meteorologists, and as kilopascals (kPA) by their Canadian counterparts, though in most scientific applications the pascal (Pa) is the preferred unit (100 Pa = 1 mb = 0.1 kPa).

The vapor pressure of a volume of air depends on both the temperature and the density of water vapor molecules (Figure 5–2). If the air temperature is high, water vapor molecules (along with all the other gaseous constituents of the atmosphere) move more rapidly and exert at greater pressure. Similarly, a greater concentration of water vapor molecules means that a greater amount of mass is available to exert pressure. In practice, temperature influences are small compared to density changes, so vapor pressure closely follows changes in the density or abundance of water molecules.

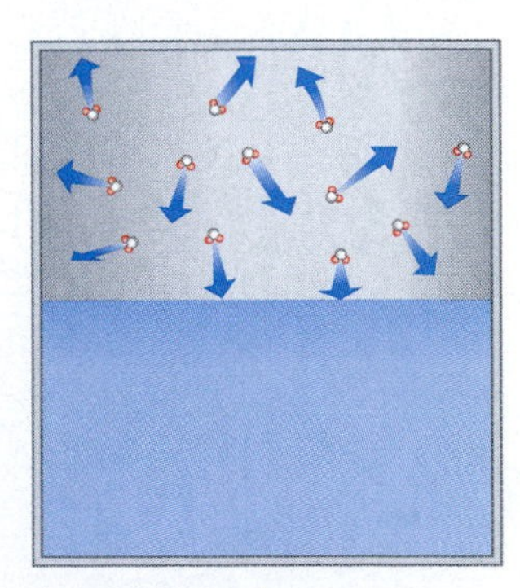

▲ Figure 5–2
The movement of molecules exerts a pressure on surfaces, called *vapor pressure*. The vapor pressure increases with concentration and temperature.

Because there is a maximum amount of water vapor that can exist, there is a corresponding maximum vapor pressure, called the **saturation vapor pressure**. The saturation vapor pressure does not represent the current amount of moisture in the air; rather, it is an expression of the maximum that *can* exist. The saturation vapor pressure depends on only one variable—temperature. Figure 5–3 shows the relationship between saturation vapor pressure and temperature, with higher temperatures having higher saturation vapor pressures. For example, at 40 °C the saturation vapor pressure is 73.8 mb, while at 0 °C it is only 6.1 mb, less than one-tenth as much.

The increase in saturation vapor pressure with temperature is not linear. At low temperatures there is only a modest increase in saturation vapor pressure, but at high temperatures saturation vapor pressure grows rapidly. For example, a 2 °C increase in temperature, from 0 °C to 2 °C, increases the saturation vapor pressure from 6.1 mb to 7.1 mb, only a 1 mb difference. Raising the temperature the same amount from a higher starting point, from 40 °C to 42 °C, raises the saturation vapor pressure by 7.7 mb, from 73.8 to 81.5 mb.

Absolute Humidity

Another measure of water vapor content is the **absolute humidity**, which is simply the density of water vapor, expressed as the number of grams of water vapor contained in a cubic meter of air. Because absolute humidity represents the amount of moisture contained in a volume of air, its value changes whenever air expands or contracts. Thus, for example, if an air parcel expands (as it does when it is heated or lifted upward), its absolute humidity will fall, even though no water vapor is removed from the parcel. Because absolute humidity suffers from this drawback and has no strong advantage over any other index, it is not widely used.

Specific Humidity

Although not normally encountered outside scientific applications, **specific humidity** is a useful index for representing atmospheric moisture. Specific humidity expresses the mass of water vapor existing in a given mass of air. Consider, for example, a volume containing exactly 1 kg of air (at sea level such a volume would be about 0.8 cubic meters, or about 27 cu. ft). Of that kilogram, some number of grams would be water vapor. The proportion of the atmospheric mass accounted for by water vapor is the specific humidity. Most often, specific humidity is expressed as the number of grams of water vapor per kilogram of air.

In fact, we implicitly considered the average specific humidity for the atmosphere when we stated in Chapter 1 that water vapor normally accounts for only about 1 or 2 percent of its mass. Because the water vapor outside the Tropics usually is less than 2 percent of the mass of the air, specific humidities are normally less than 20 grams of water vapor per kilogram of air. Specific humidity, q, is expressed mathematically as

$$q = \frac{m_v}{m} = \frac{m_v}{m_v + m_d}$$

where m_v = the mass of water vapor, m = the mass of atmosphere, and m_d = the mass of dry air (all the atmospheric gases other than water vapor). Unlike vapor pressure, specific humidity is affected to a small degree by atmospheric pressure, because it depends in part on the total mass of the atmosphere, m.

Unlike absolute humidity, specific humidity has the advantage of not changing as air expands or contracts. For example, when a kilogram of air expands, its mass is unchanged (it is still 1 kg), and the proportion that is water vapor is unchanged. As a result, the specific humidity is unaffected. Another favorable characteristic of specific humidity is that it is not temperature dependent. If a kilogram of air contains 1 g of water vapor, it still contains 1 g after heating. For this reason, specific humidity is a good indicator for comparing water vapor in the air at different locations whose air temperatures might be different from each other.

For example, if Toronto, Ontario, has a specific humidity of 10 grams of water vapor per kilogram of air on a given day, and Albuquerque, New Mexico, has 5 g/kg, we can infer that Toronto has twice as much water vapor in the air as does Albuquerque, no matter what their temperatures are. This may not seem very profound, but the direct correspondence between specific humidity and water vapor content does not hold for the more frequently used index of moisture, relative humidity. Thus, specific humidity is a useful measure of water vapor whose only real drawback is the general public's unfamiliarity with the term.

Because there is a maximum amount of water vapor that can exist at a particular temperature, there is likewise a maximum specific humidity. This maximum is called the **saturation specific humidity**. This property is directly analogous to the saturation vapor pressure and increases in the nonlinear manner shown in Figure 5–3.

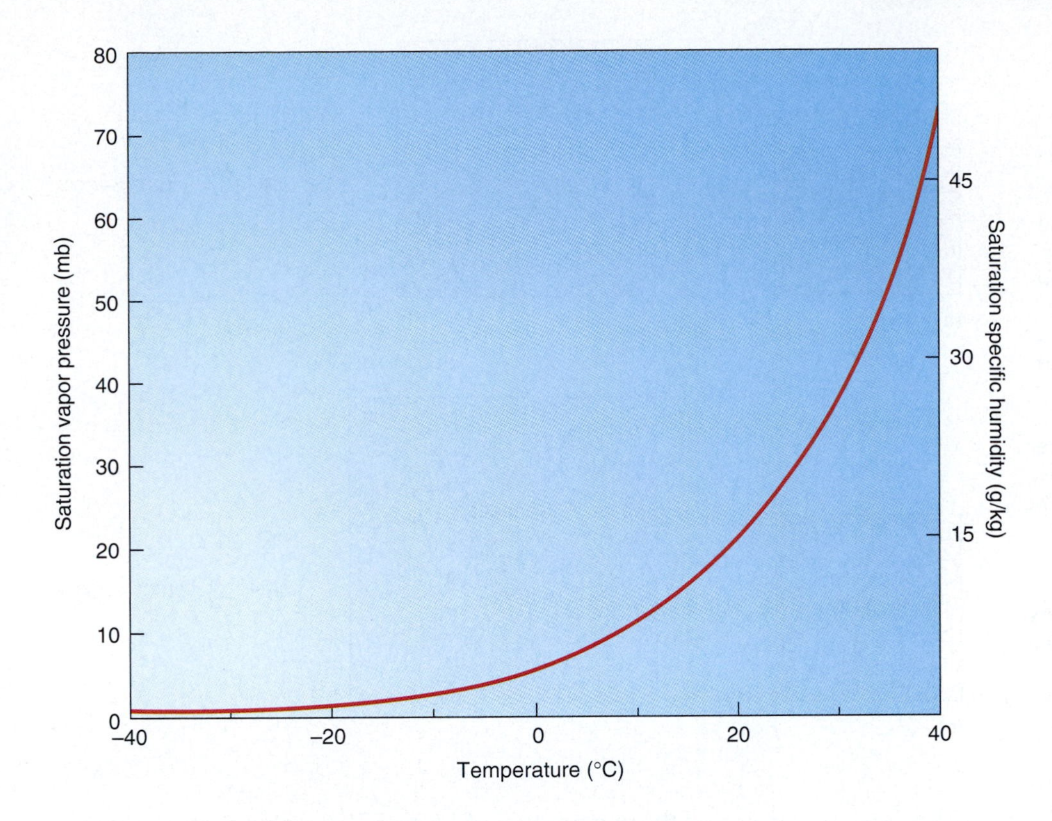

◀ **Figure 5–3**
Saturation vapor pressure and saturation specific humidity as a function of temperature. The curve is steeper at higher temperatures, meaning that saturation vapor pressure is more sensitive to temperature changes when the air is warm.

Mixing Ratio

The **mixing ratio** is very similar to specific humidity. In the case of specific humidity, we express the mass of water vapor in the air as a proportion of *all* the air. In contrast, the mixing ratio, r, is a measure of the mass of water vapor relative to the mass of the other gases of the atmosphere, or

$$r = m_v / m_d$$

(Note that the denominator denotes a mass of *dry* air as opposed to *all* air.) Numerically, the mixing ratio and specific humidity will always have nearly equal values. This is so because the amount of water vapor in the air is always small, so that whether or not it is counted in the denominator hardly changes the ratio.

A simple example should clarify the similarity between specific humidity and mixing ratio. If the specific humidity is 10 grams of water vapor per kilogram of air, the mixing ratio is 10 grams of water vapor per 990 grams of dry air. Note that 10 divided by 990 equals 10.011. In other words, if the specific humidity is 10.0 g/kg, the mixing ratio is only 1.1 percent higher, or 10.011 g/kg.

The use of the mixing ratio as an index of moisture content offers the same advantages as does the specific humidity. Furthermore, just as the maximum specific humidity that can exist is called the *saturation specific humidity*, the maximum possible mixing ratio is called the **saturation mixing ratio**.

Relative Humidity

The most familiar measure of water vapor content is **relative humidity**, RH, which relates the amount of water vapor in the air to the maximum possible at the current temperature. Equivalently, it can be thought of as

$$\text{RH} = (\text{specific humidity}/\text{saturation specific humidity}) \times 100\%$$

In words, relative humidity is the amount of water vapor present relative to the maximum that can exist at a particular temperature (its saturation point). To see how this works, let's refer to the example in Figure 5–4 in which the actual specific

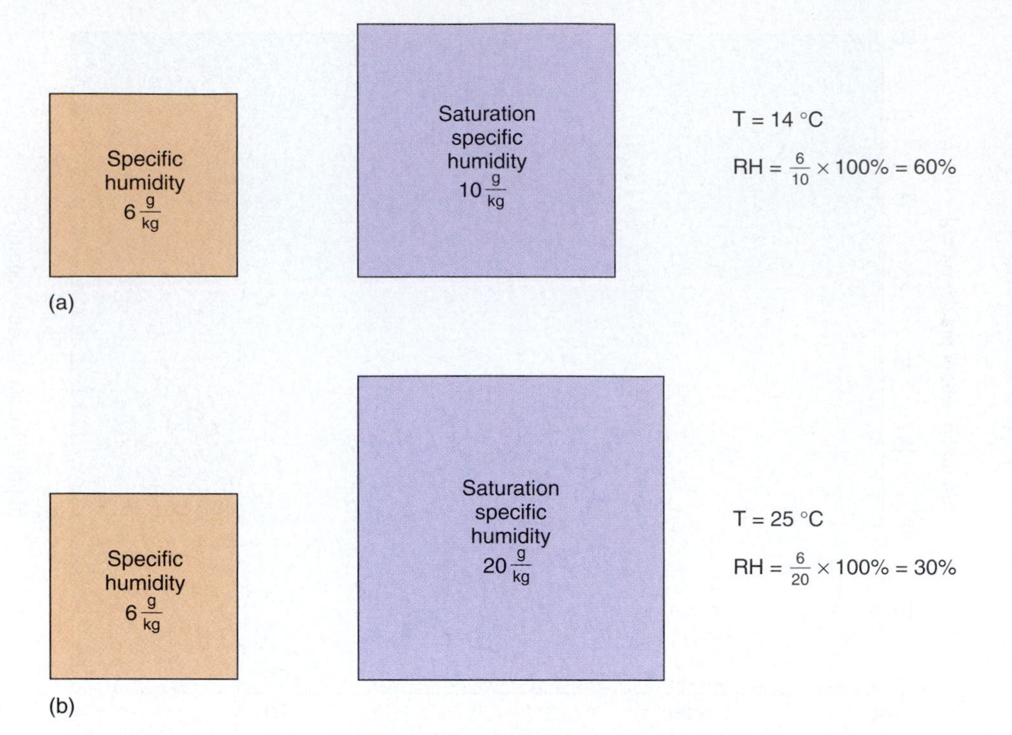

▶ **Figure 5–4**
The relationship of relative humidity to temperature. In (a), the temperature of 14°C has a saturation specific humidity of 10 grams of water vapor per kilogram of air. If the actual specific humidity is 6 grams per kilogram, the relative humidity is 60 percent. In (b), the specific humidity is still 6 grams per kilogram, but the higher temperature results in a greater saturation specific humidity. As a result, the relative humidity is less than in (a), even though the density of water vapor is the same.

humidity is 6 g/kg of air, and the temperature of 14 °C (57 °F) yields a saturation specific humidity of 10 g/kg of air. The relative humidity would thus be

$$RH = \frac{6}{10} \times 100\% = 0.6 \times 100\% = 60\%$$

The relative humidity is not uniquely determined by the amount of water vapor present. Because more water vapor can exist in warm air than in cold air, the relative humidity depends on both the actual moisture content and the air temperature. If the temperature of the air increases, more water vapor can exist and the ratio of the amount of water vapor in the air relative to saturation decreases. Thus, the relative humidity declines even if the moisture content is unchanged. Again referring to the example in Figure 5–4, let's consider what would happen if the amount of water vapor remains constant but the temperature increases from its original 14 °C to 25 °C (77 °F). At the new temperature, the saturation specific humidity increases to 20 grams of water vapor per kilogram of air, and the relative humidity becomes

$$RH = \frac{6}{20} \times 100\% = 0.3 \times 100\% = 30\%$$

The relative humidity decreased even though the amount of water vapor remained constant! This is a significant drawback to any index that is supposed to be a measure of humidity.

Because of its dependence on temperature, the relative humidity will change throughout the course of the day even if the amount of moisture in the air is unchanged. Relative humidity is usually highest in the early morning—not because of abundant water vapor, but simply because the temperature is lower. As the day warms up, the relative humidity typically declines because the saturation specific humidity increases.

The influence of temperature on relative humidity creates another problem—it confounds direct comparisons of moisture contents at different places with unequal temperatures. Consider, for example, a cold morning in Montreal, Quebec, where the temperature is –20 °C (–4 °F) and the specific humidity is 0.7 g/kg. At –20 °C, the saturation specific humidity is 0.78 grams of water vapor per kilogram of air, and the resultant relative humidity is [(0.70 ÷ 0.78) × 100%], or 89.7%. Now compare that to the warmer situation at Atlanta, where the temperature is 10 °C (50 °F) and the specific humidity 6.2 g/kg (nearly nine times greater than at Montreal!). At 10 °C, the saturation specific humidity is 7.7 grams of water vapor per kilogram of air, so the relative humidity is [(6.2 ÷ 7.7) × 100%], or 79.9%. Notice that the relative humidity is lower at Atlanta than at Montreal despite the fact that it contains much more water vapor. This illustrates why relative humidity is a poor choice for comparing the amount of water vapor in the air at one place to that at another.

Some people are quite confused about the true meaning of relative humidity. Some believe the term represents the percentage of the air that is water vapor. This is not correct. To see why, consider an instance in which the relative humidity is 100%. If the air were 100% water vapor, it would include no nitrogen or oxygen, and we would have a difficult time breathing, let alone discussing water vapor content! Another common misperception is about how high the relative humidity can be on a hot, humid day. Many people would estimate that on such a day the relative humidity would be about 99%. But in reality, very hot days never have relative humidities approaching that value. That is because at high temperatures the saturation specific humidity is very much higher than the actual specific humidities likely to be encountered. For example, if the temperature is 35 °C (95 °F), the saturation specific humidity is 36.8 grams of water vapor per kilogram of air. But we have seen that outside the Tropics it is unusual for the specific humidity to exceed 20 g/kg—even when the air is humid. Thus, a 99% relative humidity is not a realistic possibility at that temperature. Indeed, warm days can be extremely uncomfortable even with relative humidities of only about 50%.

Dew Point

A useful moisture index that is free of the temperature relationship just described is the **dew point temperature** (or simply the **dew point**). This quantity may seem confusing at first because it is expressed as a temperature, but it is a simple index to use and easy to interpret. And it is dependent almost exclusively on the amount of water vapor present.

To illustrate the dew point, consider the parcel of unsaturated air in Figure 5–5. Initially the air temperature was 14 °C (58 °F), yielding a saturation specific humidity of 10 grams of water vapor per kilogram of air. The initial specific humidity was 8 g/kg. The relative humidity was therefore 80%. As the air cools, its relative humidity increases, and if the air is cooled sufficiently its relative humidity reaches 100% and it becomes saturated. Any further cooling leads to the removal of water vapor by condensation. The temperature at which saturation occurs is the dew point. In this example, the dew point is 10 °C (50 °F) because that is the temperature at which the saturation specific humidity is 8 g/kg. Notice that even though the relative humidity increased as the temperature decreased, the dew point remained constant at 10 °C.

What would have happened if the specific humidity had remained constant and the temperature had increased from its initial 14 °C? The relative humidity would have decreased, yet the dew point would have remained constant. The dew point would not have changed because eventual cooling of the air to 8 °C still would have led to saturation.

The dew point is a valuable indicator of the moisture content; when the dew point is high, abundant water vapor is in the air. Moreover, when combined with air temperature, it is an indicator of the relative humidity. When the dew point is much lower than the air temperature, the relative humidity is very low. When the dew point is nearly equal to the air temperature, the relative humidity is high. Furthermore, when the air temperature and the dew point are equal, the air is saturated and the relative humidity is 100 percent.

Unlike relative humidity, the dew point does not change simply because air temperature changes. Moreover, if one location has a higher dew point than another, it will also have a greater amount of water vapor in the air, assuming the same air pressure. Once you are familiar with dew point, it is probably the most effective index of water vapor content. Dew points on very humid, hot days are typically in

5–1 Forecasting

Dew Point and Nighttime Minimum Temperatures

Knowledge of the current dew point temperature is a useful tool to the forecaster for the prediction of the following morning's low temperature. If no major wind shifts or other weather changes are anticipated, the minimum temperature will often approximate the dew point. Consider a hypothetical evening with an air temperature of 15 °C (59 °F) and a dew point of 5 °C (41 °F). The spread between the air temperature and the dew point temperature is not very large, and a 10 °C (18 °F) lowering of the air temperature is feasible. If the air temperature does indeed drop to the dew point and there is a light wind, a radiation fog has a good chance of forming. The fog would then inhibit further cooling, because water droplets are extremely effective at absorbing longwave radiation from the surface. Without the loss of radiation, the surface temperature would remain constant, and the overnight low would equal the dew point temperature.

The relationship between dew point and minimum temperature will not hold under certain conditions. The first has to do with the changes in the big weather picture. Imagine, for example, that a mass of warmer air is moving into the forecast region. This large body of air can replace the one present at the time of the forecast and bring with it higher nighttime temperatures. Similarly, the passage of an advancing cold front (briefly described in Chapter 1 and discussed in more detail in Chapter 9) can lead to significant drops in temperature below the current dew point.

Both heavy cloud cover and strong winds inhibit a drop in air temperature, and their presence may keep minimum air temperatures above the dew point. Cloud cover achieves this effect because of its absorption and downward reradiation of longwave energy. Strong winds prevent large temperature decreases at the surface by vertical mixing. A shallow layer of cold air that would otherwise develop is easily disrupted, leading to higher surface temperatures and more uniform temperatures with height.

Minimum temperatures won't go down to the dew point if the difference between the air temperature and the dew point temperature is very large. One can readily see how this might occur if a desert has a high temperature of 45 °C (113 °F) and a dew point of 0 °C (32 °F). Even with calm winds and no cloud cover, a cooling of 45 °C is unlikely over the course of a short, summer night. Though the temperature won't always drop down as low as the dew point, it is always certain that unless a front passes through or the wind direction changes significantly, the minimum temperature will not fall much below the evening dew point.

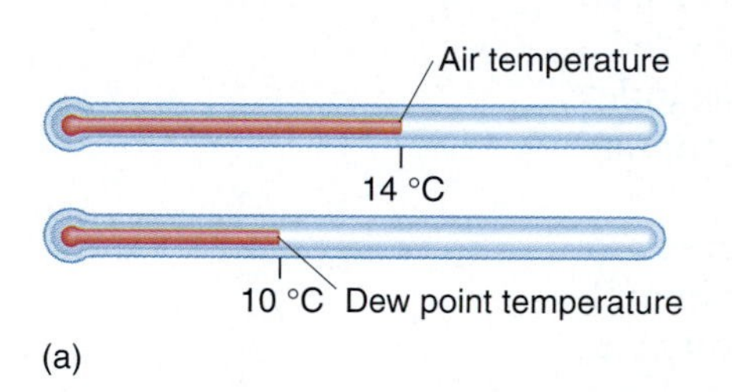

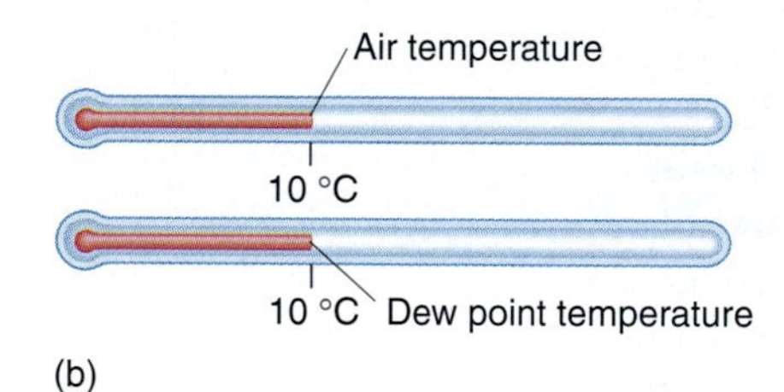

▲ **Figure 5–5**
The dew point is an expression of water vapor content, although it is expressed as a temperature. In (a), the temperature exceeds the dew point and the air is unsaturated. When the air temperature is lowered so that the saturation specific humidity is the same as the actual specific humidity (b), the air temperature and dew point are equal. Further cooling (c) leads to an equal reduction in the air temperature and dew point so that they remain equal to each other.

the low 20s on the Celsius scale (low 70s Fahrenheit). (When you see a dew point of 70 °F or higher, you can plan on a sleepless night unless you have air conditioning.) On comfortable days that are neither humid nor dry, dew points may be in the low teens Celsius (mid-50s Fahrenheit); very dry days can have dew points in the minus 20s or lower on the Celsius scale (0 °F).

The dew point is always equal to or less than the air temperature; under no circumstances does it ever exceed the temperature. So what happens if the air is lowered to the dew point and then cooled further? In that case, the amount of water vapor exceeds the amount that can now exist, and the surplus is removed from the air. This happens by condensation of the air to form a liquid, or by the formation of ice crystals. In either case, the dew point decreases at the same rate as the air temperature, and the two remain equal to each other. This is illustrated in Figure 5–5b and c. When the temperature was lowered to 10 °C in Figure 5–5b, the air became saturated with 8 grams of water vapor per kilogram of air. As the air cooled further to 6 °C (Figure 5–5c), the saturation specific humidity decreased to 6 g/kg. Since the specific humidity, by definition, cannot exceed the saturation specific humidity, 2 grams of water vapor (8 grams minus 6 grams) had to be removed from each kilogram of air by condensation. The removal of water vapor kept the specific humidity equal to that of the saturation specific humidity and also lowered the dew point. Note that when the temperature at which saturation would occur is below 0 °C (32 °F), we use the term **frost point** instead of dew point.

Distribution of Water Vapor

Water vapor gets into the atmosphere either from local evaporation or from the horizontal transport (advection) of moisture from other locations. The importance of advection to the distribution of water vapor is clearly evident in Figure 5–6, which shows the spatial distribution of dew points (in °F and in °C) across the United States in January (a) and July (b). Looking at the eastern two-thirds of the country first, it is clear that for both months the amount of water vapor generally decreases with distance from the Gulf of Mexico. Because of the Gulf's high water temperatures, moisture is readily evaporated into the atmosphere year round, and this moisture can be transported northward. The decline in water vapor content is seen in a north–south direction and also moving westward from about the Mississippi River toward the Rocky Mountains during the summer. During the winter months, the amount of moisture extending into the Great Plains is low and only a minimal amount of east-west variation exists.

The effect of distance from the source of moisture is also evident in the West, with water vapor generally decreasing from the Pacific Coast to the Rocky Mountains. The most rapid drop occurs very near the coast because local mountains block off substantial amounts of moisture from the inland area.

Comparing the two maps, you will note a substantial increase in the amount of water vapor in the air in July over that in January. This should not be surprising because lower temperatures preclude the existence of high water vapor contents. Thus, for example, along the Ohio River Valley the average dew point increases from about –7 °C (20 °F) in January to perhaps 17 °C (63 °F) in July. This is why residents of much of the country, especially those east of the Rockies, are subject to uncomfortable dry skin in the winter—only to find themselves sweating profusely during the summer.

The general patterns described for the United States also apply to most of Canada.

Methods of Achieving Saturation

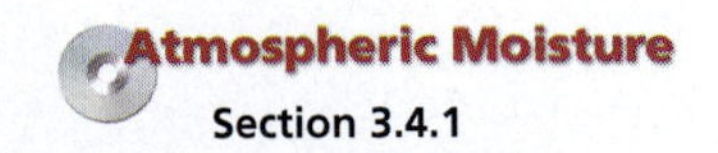

Air can become saturated by any one of three general processes: adding water vapor to the air; mixing cold air with warm, moist air; and lowering the temperature to the dew point. The first of these processes can be seen in your bathroom when you

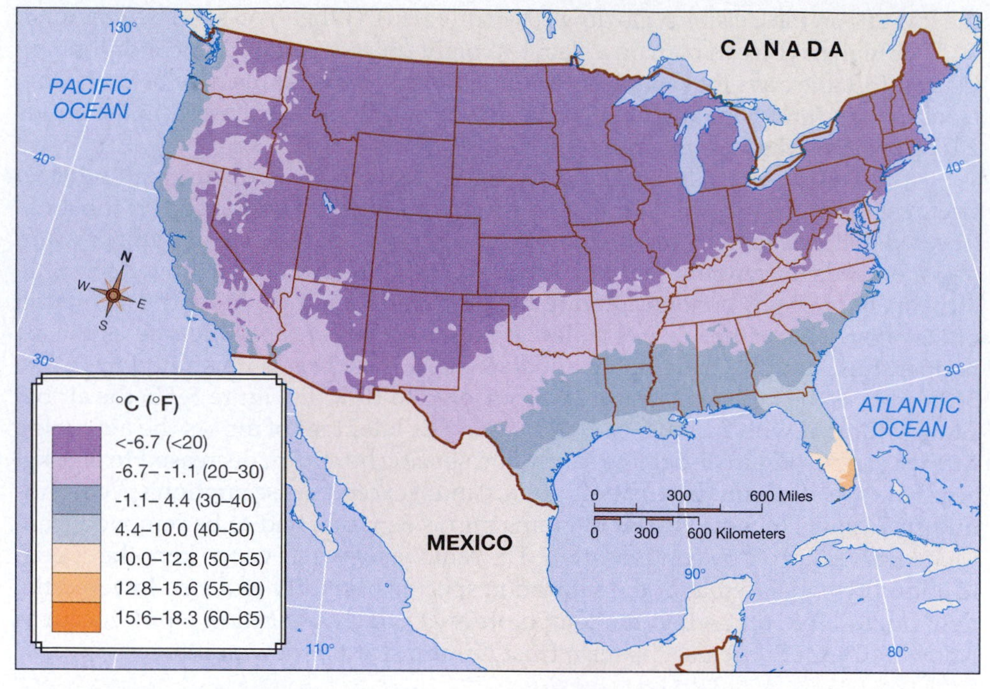

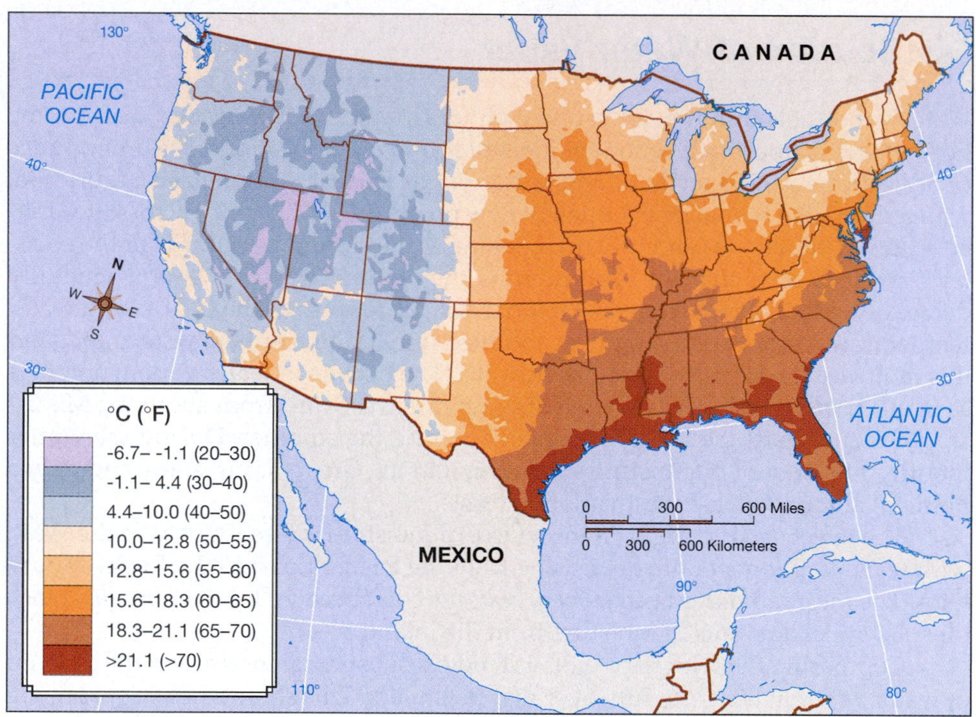

▶ **Figure 5–6**
The average distribution of dew points across the United States in January (a) and July (b).

take a warm shower. The warm water from a shower head evaporates moisture into the air in the room and brings it to the saturation point. Condensation first forms on your mirrors and other surfaces, then a general fog develops. In the natural environment, the evaporation of water from falling raindrops can raise the dew point in the air beneath the cloud from which the rain is occurring. If enough vapor is added to the air to saturate it, a **precipitation fog** forms beneath the cloud.

Condensation from the second process, the mixing of cold with warm moist air, is illustrated in Figure 5–7. Consider the two parcels of unsaturated air highlighted by the boxes. Parcel *A* has a temperature of 0 °C (32 °F) and a specific humidity of 3 grams of water vapor per kilogram of air; Parcel *B* has a temperature of 30 °C (86 °F) and a specific humidity of 20 g/kg. If equal amounts of the two parcels are mixed together, the new parcel has a temperature of 15 °C (59 °F), exactly midway

5–2 Forecasting

Vertical Profiles of Moisture

In Chapter 3, you saw how simplified thermodynamic diagrams can be used to plot the vertical profile of temperature. Because dew point values are likewise expressed as temperatures, they too can be plotted on thermodynamic charts. In fact, by plotting temperatures and dew points simultaneously, you can obtain considerable information on cloud conditions. Refer to the sounding of temperature and dew point in Figure 1, taken from Stapleton Airport, near Denver, Colorado, at midnight, Greenwich mean time, on April 12, 2002.

The example in Figure 1 plots temperature (the curve on the right) and dew point (left). If you contrast this to the profile shown in Chapter 3, taken at Slidell, Louisiana, you will notice that the sounding begins at a much lower pressure—at about the 840 mb level—than the earlier example. The reason for this is very simple. Denver's elevation of 1625 m (5330 ft) causes its surface pressure to be much lower than that at Slidell's nearly sea level location (remember, pressure *always* decreases with elevation).

In Figure 1, the temperature at the surface is 17 °C (63 °F), and the dew point is –4 °C (25 °F). This large difference between the two values indicates that the relative humidity is low (calculated to be 23%). But as distance from the surface increases, temperature decreases more rapidly than dew point. At about the 560 mb level (at a height of about 4850 m, or 15,900 ft, above the surface), the two values become equal to each other and the air is saturated (a slight measurement error accounts for the plotted temperature and dew point values not being exactly equal). The dew point and the air temperature then decrease at the same rate up to about the 460 mb level (about 6500 m, or 21,300 ft, above the surface), above which the temperature once again exceeds the dew point. Because the air is saturated in the layer of air between the 4850 m and 6500 m above the surface, we can infer that a cloud occupies that 1650 m thick layer.

The thermodynamic diagram shown here is slightly more complex than the one in Chapter 3, because it includes an additional set of lines that provides one more type of moisture information. The brown lines that slope gently to the left as they extend upward indicate the specific humidity and the saturation specific humidity at any level. Let's first see how the plot of the air temperature profile can be used to determine the saturation specific humidity at a given pressure level. Notice that at the 700 mb level the air temperature is 2 °C. It so happens that one of the sloping lines (labeled at the bottom of the diagram with the number 7) nearly intersects the temperature profile at the 700 mb level. This indicates that at the air at the 700 mb level has a saturation specific humidity of just under 7 grams of water vapor per kilogram of air.

We can follow a similar procedure to find out what the actual specific humidity is at the 700 mb level. To do this, we follow the dew point profile up to the 700 mb level. The point where the profile crosses the 700 mb line occurs right between the two sloping lines labeled 2 and 4. In other words, the actual specific humidity at the 700 mb level is right between 2 grams of water vapor per kilogram of air and 4 g/kg—with 3 g/kg a very good approximation.

After estimating the actual specific humidity and saturation specific humidity at the 700 mb level, it is easy to use the values to obtain the relative humidity:

$$RH = (\text{specific humidity}/\text{saturation specific humidity}) \times 100\% = (3/7) \times 100\% = 43\%$$

This procedure can be performed at the surface or any other level of the atmosphere.

Later in this chapter, we will see how thermodynamic diagrams can give us important information in forecasting the likelihood of cloud development.

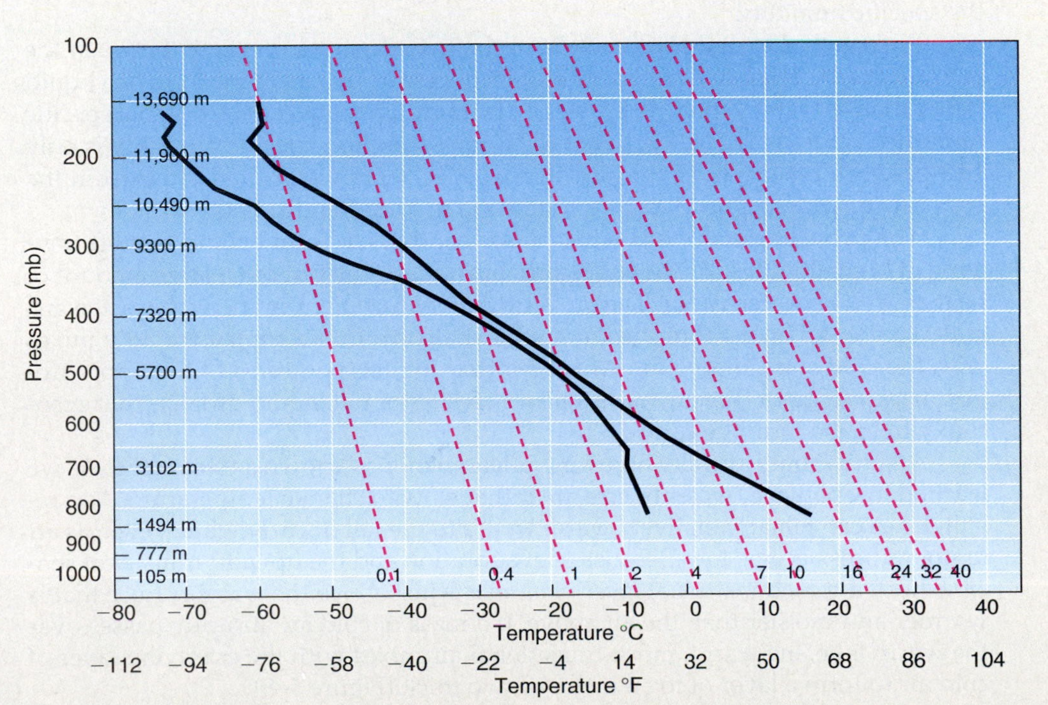

Figure 1 A sounding of temperature and dew point. This chart plots temperature and dew points throughout the troposphere and much of the stratosphere. The slightly sloping brown lines depict values of specific humidity in grams of water vapor per kilogram of air. Meteorologists use these lines along with the plots of temperature and dew point to determine the saturation specific humidity and the actual specific humidity, respectively.

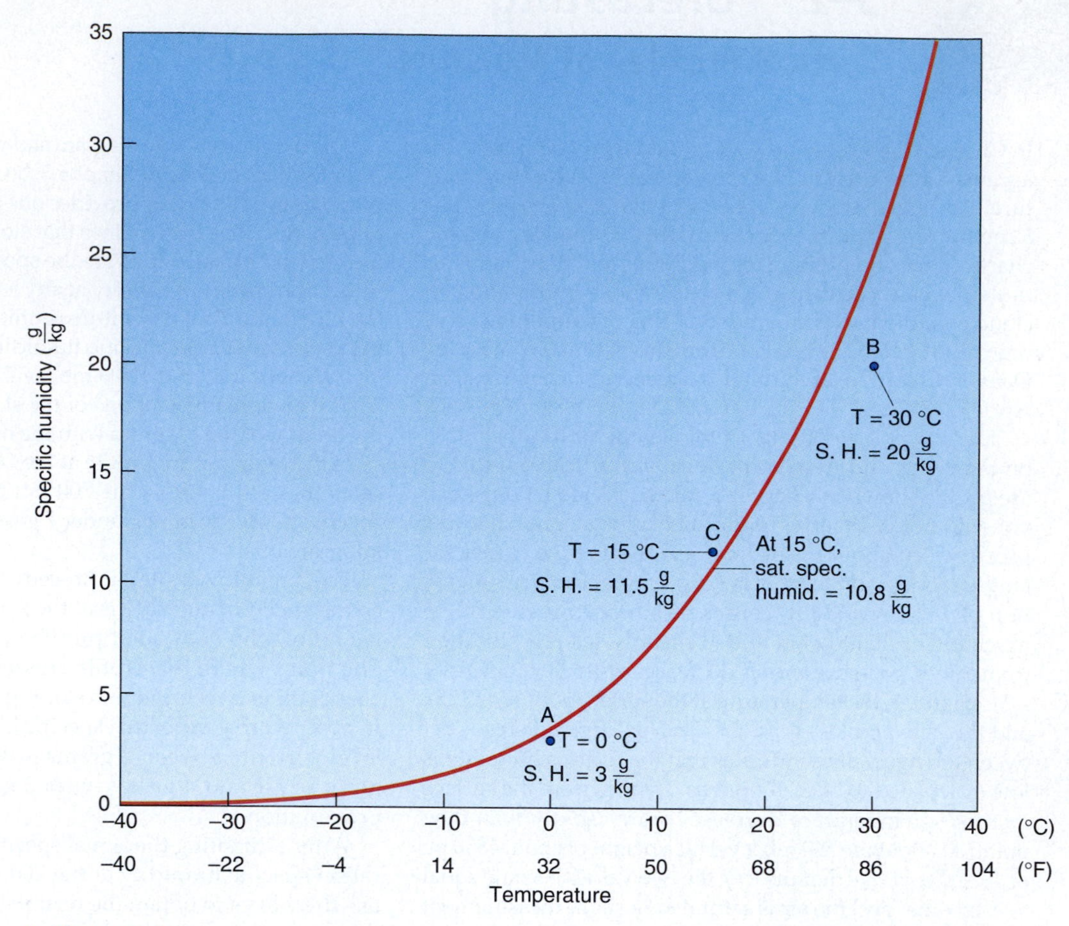

▲ **Figure 5–7**
Saturation by the mixing of warm, moist air with cold air. Parcel *A* has a low temperature and can therefore contain only a small amount of water vapor. The warm parcel (Parcel *B*) has a higher moisture content. Parcel *C* results when Parcels *A* and *B* are mixed together and has a temperature in between those of the original parcels. The amount of moisture in the air is greater than that which can exist at the new temperature, so the excess water vapor condenses.

between the temperatures of the original parcels. However, such is not the case for the specific humidity.

Although the total H_2O content in the mixed parcel is exactly between the amounts of the original parcels, some of the moisture occurs in the form of a liquid rather than as water vapor. Here is why: The midpoint between the original specific humidity values is 11.5 g/kg, but at 15 °C the saturation specific humidity is only 10.8 g/kg. In other words, the air contains 0.7 g more water than can exist in the vapor form. The surplus therefore condenses to form fog droplets.

The preceding process is what causes contrails to form behind aircraft traveling at high altitudes. As the jet engines burn fuel, they put out a large amount of heat as well as water vapor. The air is extremely turbulent in the wake of the aircraft, so the hot, moist (but unsaturated) exhaust from the engines rapidly mixes with the cold surrounding air. At the subfreezing temperatures of the upper troposphere, the vapor directly forms into ice crystals or into liquid droplets that eventually freeze to form the contrail.

A similar but naturally occurring phenomenon is known as **steam fog**. As we learned in Chapter 3, water bodies are rather slow to change temperature. As a result, lakes can remain relatively warm well into the fall or early winter, even as air temperatures become low. Because of evaporation and the upward transfer of sensible heat, a thin, transitional layer of air exists just above the water surface that is warmer and moister than the air above. If a mass of cold air abruptly passes over the warm lake, the warm, moist transitional air mixes with the overlying layer of cold air to form a layer of fog a meter or two thick (Figure 5–8).

◀ Figure 5–8
Steam fog.

Although we can cite several examples of how clouds form by the increase of moisture content or by mixing warm, moist air with cold, dry air, experience shows that most clouds form when the air temperature is lowered to the dew point. There are several ways this cooling can occur, requiring considerable explanation, as we will see later in this chapter. For now we simply note that this third mechanism, atmospheric cooling, is by far the most important process for cloud formation.

The Effects of Curvature and Solution

This chapter opened with a hypothetical experiment in which water in a jar attained an equilibrium between condensation and evaporation. The experiment, assuming pure water with a flat surface, provided a foundation for understanding saturation. But meteorology studies the atmosphere—not what goes on in hypothetical jars. In the real atmosphere, we are concerned with the rates of evaporation and condensation across the surfaces of suspended cloud and fog droplets. Such droplets are neither flat nor made up of pure H_2O. We therefore expand on our discussion of evaporation, condensation, and equilibrium to take into account the effects of curvature and impurity of cloud and fog droplets.

Effect of Curvature

Water droplets exist in nature not as tiny cubes with flat sides, but rather as microscopically small spheres with considerable curvature. Compare the two droplets shown in Figure 5–9. The one on the left is much larger than the one on the right and therefore has less curvature. We could even consider a more extreme example—Earth itself. Most of us are pretty certain by now that the planet is not flat but spherical. However, because of Earth's large size, its curvature is inconspicuous, and a centimeter of distance across its surface essentially forms a straight line. Moreover, a straight-edge ruler can lie flat against its surface. But hold a ruler over a tennis

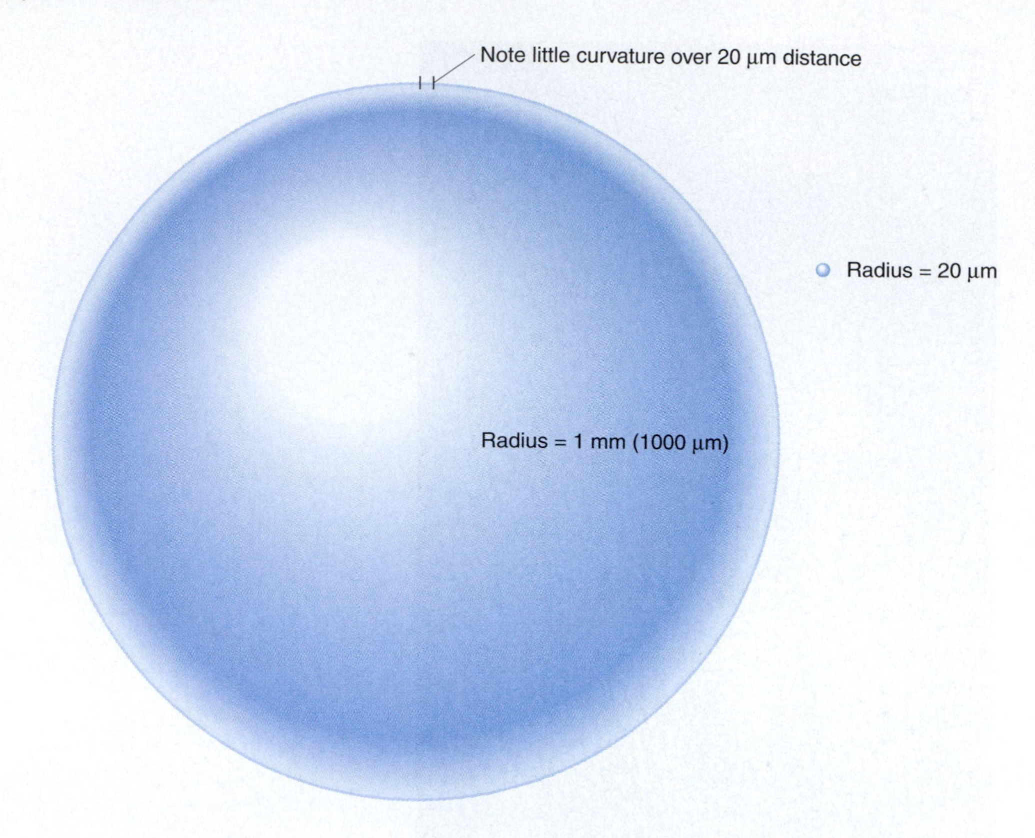

▶ **Figure 5–9**
Large droplets have less curvature than smaller droplets.

ball and only a small part of it is in contact with the ball's surface; it is more strongly curved than Earth.

Cloud droplets are much smaller than tennis balls, of course, and thus exhibit an even greater curvature. But what does curvature have to do with saturation? The answer is that curvature has an effect on evaporation from cloud droplet surfaces and therefore on the vapor pressure necessary for saturation. Put simply, effects arising from surface tension lead to differences in the saturation point, as described shortly.

The graph of saturation vapor pressure versus temperature, shown as Figure 5–3, applies only to flat surfaces of pure H_2O. For curved water surfaces, the evaporation rate is greater. The enhanced rate of evaporation requires that condensation also be increased for the two to remain in balance. Thus, a highly curved droplet of pure water at any given temperature has a higher saturation vapor pressure than indicated in Figure 5–3. Stated another way, highly curved droplets of pure water require relative humidities in excess of 100% to keep them from evaporating away.

Figure 5–10 illustrates the effect of droplet size on the relative humidity needed to maintain an existing droplet of pure H_2O. For very small droplets (those with radii of about 0.1 μm), relative humidities in excess of 300% are needed to achieve an equilibrium between evaporation and condensation. In other words, those droplets require a *supersaturation* of 200%. The degree of supersaturation necessary to maintain a droplet rapidly decreases with increasing droplet size. Droplets with radii larger than 1.0 μm require supersaturations of only about 10%.

If the atmosphere were devoid of any aerosols, condensation would occur only by **homogeneous nucleation**, in which droplets form by the chance collision and bonding of water vapor molecules under supersaturated conditions. Such droplets would necessarily have only a small number of molecules and a high degree of curvature, and therefore only exist at high levels of supersaturation. This process seldom if ever occurs, because certain **hygroscopic** (water-attracting) aerosols in the atmosphere assist the formation of droplets at relative humidities far below those necessary for homogeneous nucleation. The formation of water droplets onto hygroscopic particles is called **heterogeneous nucleation**, and the particles onto which the droplets form are called **condensation nuclei**. When condensation occurs, the condensation nuclei dissolve into the water to form a *solution*.

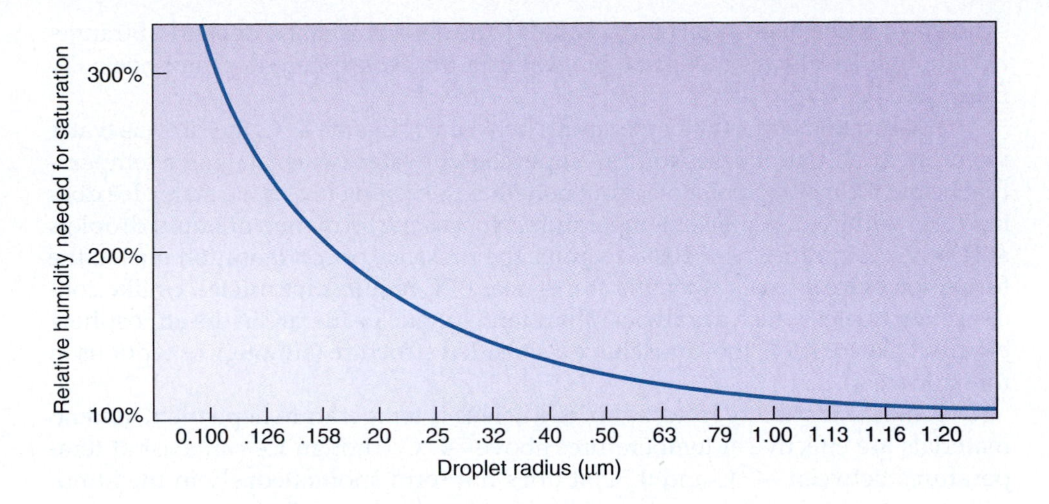

Figure 5–10
Small droplets of pure water require relative humidities above 100% to remain in equilibrium.

Effect of Solution

When materials are dissolved in water, a certain number of the molecules at the surface are those of the *solute* (the material dissolved in the water) rather than H_2O molecules. With fewer H_2O molecules existing at the surface, the rate of evaporation is lower than for pure water. As a result, solutions require less water vapor above the surface to maintain an equilibrium between evaporation and condensation than does pure water.

If condensation in the atmosphere occurred by homogeneous nucleation, newly formed droplets would consist of pure water. But we've seen that in fact droplets actually form by heterogeneous nucleation and the incorporation of the nuclei into the water solution. This has the opposite effect of droplet curvature by lowering the amount of water vapor necessary to keep the droplets in equilibrium. Under most circumstances, this *solute effect* is about equal to that of droplet curvature, and condensation normally occurs at relative humidities near or slightly below 100%.

Although the proportion of aerosols in the air that are hygroscopic is small, the atmosphere contains so many suspended particles that condensation nuclei are always abundant. Some materials are more hygroscopic than others, and large aerosols are generally more effective than smaller ones. Some are even capable of attracting water at relative humidities below 90% and forming extremely small droplets. We observe such droplets as **haze**.

Until very recently it was believed that most condensation nuclei in the atmosphere were natural, consisting mostly of continental dust, sea salt, and aerosols derived from volcanic eruptions, natural fires, and gases given off by marine phytoplankton. Research undertaken during the 1990s now indicates that the role of human activity was previously understated and that anthropogenic sources may account for the majority of condensation nuclei over industrialized areas in the Northern Hemisphere. The effect of human activity on aerosol concentration even extends to oceanic areas in the Northern Hemisphere. In the Southern Hemisphere, human influences are much smaller and are confined mainly to regions of heavy industrial activity and frequent burning. Although salt particles are very hygroscopic, they are less abundant over land than over their oceanic source. Thus, over land, salt particles account mainly for the very largest of cloud condensation nuclei, but they are relatively rare compared to other materials.

Ice Nuclei

So far we have examined the formation of liquid water droplets when air becomes saturated. But saturation can occur at very low temperatures, which suggests that rather than liquid water droplets, ice crystals may form. On the other hand, many of us have walked through fog composed of liquid droplets even though the temperature was below 0 °C. So what really happens when saturation occurs at temperatures below the freezing point? Does this lead to the condensation of liquid droplets or to the deposition of ice crystals? The answer is that it depends. Strangely, although ice always melts at 0 °C, water in the atmosphere does not normally freeze at 0 °C.

If saturation occurs at temperatures between 0 °C and –4 °C, the surplus water vapor invariably condenses to form **supercooled water** (water having a temperature below the melting point of ice but nonetheless existing in a liquid state). Ice does not form within this range of temperatures. Just as the formation of liquid droplets at relative humidities near 100% requires the presence of condensation nuclei, the formation of ice crystals at temperatures near 0 °C requires **ice nuclei**. Unlike condensation nuclei, which are always abundant, ice nuclei are rare in the atmosphere because, like ice itself, they must have a six-sided structure (although exceptions to this rule exist).

A material's ability to act as an ice nucleus is temperature dependent, and no materials are effective at temperatures above –4 °C. Though ice can exist at temperatures between –4 °C and 0 °C, it does not form spontaneously in the atmosphere in this range. (In fact there is little ice in the atmosphere at temperatures above –10 °C.) Thus, between –4 °C and 0 °C, the removal of water vapor occurs only by the condensation of supercooled water.

As temperature decreases, the likelihood of ice formation increases, and at temperatures between about –10 °C (14 °F) and –40 °C (–40 °F) saturation can lead to the nucleation of ice crystals, supercooled droplets, or both. Clouds having temperatures within this range will usually have liquid droplets and solid crystals coexisting, but with a greater proportion of ice at lower temperatures. At temperatures below –30 °C (–22 °F), the cloud will be mostly ice crystals. For temperatures less than –40 °C, saturation leads to the formation of ice crystals only, with or without the presence of ice nuclei. As we will see in Chapter 7, the coexistence of ice crystals and liquid droplets in clouds is extremely important for the development of precipitation outside the tropics.

Among the materials that serve as ice nuclei are components of natural soils called *clays*. Clay materials have a platy structure, microscopic sizes, and strong electrical attractions. These characteristics make them very difficult to dislodge from the surface and incorporate into the atmosphere, which largely explains their scarcity. Clays occur in a variety of compositions, of which the most effective seems to be *kaolinite* (aluminum silicate). Other types of clay serve as ice nuclei but are active only at lower temperatures.

Unfortunately, it is very difficult for cloud physicists to obtain much knowledge about the sources of ice nuclei because they tend to be small in size and rare in number. Observations do seem to show, however, that a few ice crystals contain a foreign ice nucleus. This has led to the idea that ice crystals themselves may be very important ice nuclei. In this model, existing ice crystals break into small fragments, which then act as nuclei for new crystals. Other materials cloud physicists consider possible ice nuclei include fragments of decaying leaves, volcanic material, and even certain bacteria.

Measuring Humidity

Considering the fact that water vapor is an invisible gas mixed together with all the other gases of the atmosphere, you might suspect that its measurement would entail some highly sophisticated instrumentation. Such is not the case. The simplest and most widely used instrument for measuring humidity, the **sling psychrometer** (Figure 5–11a), consists of a pair of thermometers, one of which has a cotton wick around the bulb that is saturated with water. The other thermometer has no such covering and simply measures the air temperature. The two thermometers, called the **wet bulb** and **dry bulb thermometers**, respectively, are mounted to a pivoting device that allows them to be circulated through the surrounding air. If the air is unsaturated, water evaporates from the wet bulb, whose temperature falls as latent heat is consumed. After about a minute or so of circulating, the amount of heat

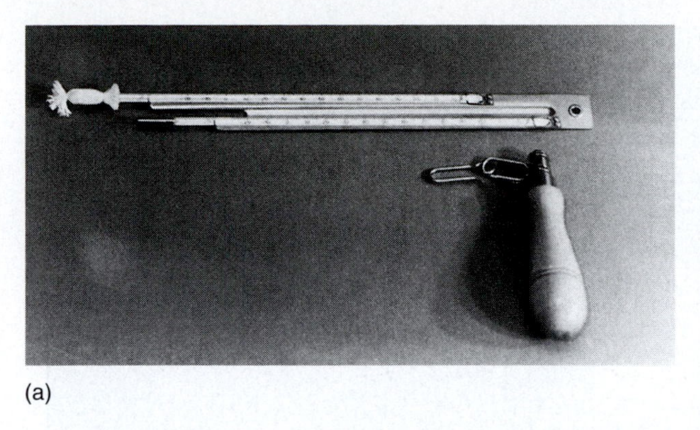

(a)

(b)

▲ Figure 5–11
(a) Sling psychrometer and (b) hygrothermograph.

lost by evaporation is offset by the input of sensible heat from the surrounding, warmer air, and the cooling ceases. Thereafter, the wet bulb maintains a constant temperature no matter how long the instrument is swung around.

The difference between the dry and wet bulb temperatures, called the **wet bulb depression**, depends on the moisture content of the air. If the air is completely saturated, no net evaporation occurs from the wet bulb thermometer, no latent heat is lost, and the wet bulb temperature equals the dry bulb temperature. On the other hand, if the humidity is low, plenty of evaporation will take place from the wet bulb, and its temperature will drop considerably before reaching an equilibrium value. To determine the moisture content, first you note the difference between the dry and wet bulb temperatures. Then, with the use of tables such as Tables 5–1 and 5–2, the dew point, relative humidity, or any other humidity measure is obtained by finding the value corresponding to the row for the air temperature and the column for the wet bulb depression.

Some psychrometers are equipped with fans that circulate air across the bulbs of the two thermometers. These **aspirated psychrometers** save the user the effort needed to sling the thermometers through the air (as well as the aggravation of cleaning up the mess after accidentally striking nearby objects). Another alternative to the sling psychrometer is the **hair hygrometer**, whose basic part is a band of human hair. Hair expands and contracts in response to the relative humidity. By connecting a hair to a lever mechanism, we can easily determine its water vapor content. Often, the hygrometer is coupled with a bimetallic strip and rotating drum to give a continuous record of temperature and humidity. Such a **hygrothermograph** is shown in Figure 5–11b.

High Humidities and Human Discomfort

Temperature is one of the most important weather variables with regard to human comfort, and excessively high and low temperatures account for more North American fatalities than do hurricanes, lightning, floods, and tornadoes combined. But the effects of temperature extremes can be compounded by other factors, such as humidity, the intensity of sunlight, and strength of wind. In Chapter 3 we saw how wind speeds can be combined with low temperatures to create a wind chill index. Similarly, the effect of humidity and high temperatures can be expressed in a **heat index**.

One of the human body's most effective mechanisms in guarding against excessive heat is perspiration. Sweat cools the body because, when released to the surface of the skin, it is free to evaporate into the atmosphere and consume latent

Dry Bulb (°C)	Wet Bulb Depression, °C (Dry Bulb Temperature Minus Wet Bulb Temperature = Wet Bulb Depression)																				
	1	2	3	4	5	6	7	8	9	10	11	12	13	14	15	16	17	18	19	20	21	22
−20	−33																					
−18	−28																					
−16	−24																					
−14	−21	−36																				
−12	−18	−28																				
−10	−14	−22																				
−8	−12	−18	−29																			
−6	−10	−14	−22																			
−4	−7	−22	−17	−29																		
−2	−5	−8	−13	−20																		
0	−3	−6	−9	−15	−24																	
2	−1	−3	−6	−11	−17																	
4	1	−1	−4	−7	−11	−19																
6	4	1	−1	−4	−7	−13	−21															
8	6	3	1	−2	−5	−9	−14															
10	8	6	4	1	−2	−5	−9	−14	−28													
12	10	8	6	4	1	−2	−5	−9	−16													
14	12	11	9	6	4	1	−2	−5	−10	−17												
16	14	13	11	9	7	4	1	−1	−6	−10	−17											
18	16	15	13	11	9	7	4	2	−2	−5	−10	−19										
20	19	17	15	14	12	10	7	4	2	−2	−5	−10	−19									
22	21	19	17	16	14	12	10	8	5	3	−1	−5	−10	−19								
24	23	21	20	18	16	14	12	10	8	6	2	−1	−5	−10	−18							
26	25	23	22	20	18	17	15	13	11	9	6	3	0	−4	−9	−18						
28	27	25	24	22	21	19	17	16	14	11	9	7	4	1	−3	−9	−16					
30	29	27	26	24	23	21	19	18	16	14	12	10	8	5	1	−2	−8	−15				
32	31	29	28	27	25	24	22	21	19	17	15	13	11	8	5	2	−2	−7	−14			
34	33	31	30	29	27	26	24	23	21	20	18	16	14	12	9	6	3	−1	−5	−12	−29	
36	35	33	32	31	29	28	27	25	24	22	20	19	17	15	13	10	7	4	0	−4	−10	
38	37	35	34	33	32	30	29	28	26	25	23	21	19	17	15	13	11	8	5	1	−3	−9
40	39	37	36	35	34	32	31	30	28	27	25	24	22	20	18	16	14	12	9	6	2	−2

Dry Bulb (Air) Temperature

Dew point temperatures

Dry Bulb Temp. (°F)	Wet Bulb Depression (°F)																																	
	1	2	3	4	5	6	7	8	9	10	11	12	13	14	15	16	17	18	19	20	21	22	23	24	25	26	27	28	29	30	31	32	33	34	35
0	−7	−20																																	
5	−1	−9	−24																																
10	5	−2	−10	−27																															
15	11	6	0	−9	−26																														
20	16	12	8	2	−7	−21																													
25	22	19	15	10	5	−3	−15	−51																											
30	27	25	21	18	14	8	2	−7	−25																										
35	33	30	28	25	21	17	13	7	0	−11	−41																								
40	38	35	33	30	28	25	21	18	13	7	−1	−14																							
45	43	41	38	36	34	31	28	25	22	18	13	7	−1	−14																					
50	48	46	44	42	40	37	34	32	29	26	22	18	13	8	0	−13																			
55	53	51	50	48	45	43	41	38	36	33	30	27	24	20	15	9	1	−12	−59																
60	58	57	55	53	51	49	47	45	43	40	38	35	32	29	25	21	17	11	4	−8	−36														
65	63	62	60	59	57	55	53	51	49	47	45	42	40	37	34	31	27	24	19	14	7	−3	−22												
70	69	67	65	64	62	61	59	57	55	53	51	49	47	44	42	39	36	33	30	26	22	17	11	2	−11										
75	74	72	71	69	68	66	64	63	61	59	57	55	54	51	49	47	44	42	39	36	32	29	25	21	15	8	−2	−23							
80	79	77	76	74	73	72	70	68	67	65	63	62	60	58	56	54	52	50	47	44	42	39	36	32	28	24	20	13	6	−7	−53				
85	84	82	81	80	78	77	75	74	72	71	69	68	66	64	62	61	59	57	54	52	50	48	45	42	39	36	32	28	24	19	12	3	−12		
90	89	87	86	85	83	82	81	79	78	76	75	73	72	70	69	67	65	63	61	59	57	55	53	51	48	45	43	39	36	32	28	24	19	11	1
95	94	93	91	90	89	87	86	85	83	82	80	79	78	76	74	73	71	70	68	66	64	62	60	58	56	54	52	49	46	43	40	37	33	29	24
100	99	98	96	95	94	93	91	90	89	87	86	85	83	82	80	79	77	76	74	72	71	69	67	65	63	61	59	57	55	52	50	47	44	41	37
105	104	103	101	100	99	98	96	95	94	93	91	90	89	87	86	84	83	82	80	78	77	75	74	72	70	68	67	65	63	61	58	56	54	51	48
110	109	108	106	105	104	103	102	100	99	98	97	95	94	93	91	90	89	87	86	84	83	81	80	78	77	75	73	72	70	68	66	64	62	60	57
115	114	113	112	110	109	108	107	106	104	103	102	101	99	98	97	96	94	93	92	90	89	87	86	84	83	81	80	78	76	75	73	71	69	67	65
120	119	118	117	115	114	113	112	111	110	108	107	106	105	104	102	101	100	98	97	96	94	93	92	90	89	87	86	84	83	81	80	78	76	75	73
125	124	123	122	121	119	118	117	116	115	114	112	111	110	109	108	106	105	104	103	101	100	99	97	96	95	93	92	90	89	88	86	84	83	81	80
130	129	128	127	126	124	123	122	121	120	119	118	116	115	114	113	112	110	109	108	107	106	104	103	102	100	99	98	96	95	94	92	91	89	88	86

Dew point temperatures

▲ **Table 5–1**
Dew points.

Dry Bulb (°C)	Wet Bulb Depression, °C (Dry Bulb Temperature Minus Wet Bulb Temperature = Wet Bulb Depression)																					
Dry Bulb (Air) Temperature	1	2	3	4	5	6	7	8	9	10	11	12	13	14	15	16	17	18	19	20	21	22
−20	28																					
−18	40																					
−16	48	0																				
−14	55	11																				
−12	61	23																				
−10	66	33	0																			
−8	71	41	13																			
−6	73	48	20	0																		
−4	77	54	32	11																		
−2	79	58	37	20	1																	
0	81	63	45	28	11																	
2	83	67	51	36	20	6																
4	85	70	56	42	27	14																
6	86	72	59	46	35	22	10	0														
8	87	74	62	51	39	28	17	6														
10	88	76	65	54	43	33	24	13	4													
12	88	78	67	57	48	38	28	19	10	2												
14	89	79	69	60	50	41	33	25	16	8	1											
16	90	80	71	62	54	45	37	29	21	14	7	1										
18	91	81	72	64	56	48	40	33	26	19	12	6	0									
20	91	82	74	66	58	51	44	36	30	23	17	11	5									
22	92	83	75	68	60	53	46	40	33	27	21	15	10	4	0							
24	92	84	76	69	62	55	49	42	36	30	25	20	14	9	4	0						
26	92	85	77	70	64	57	51	45	39	34	28	23	18	13	9	5						
28	93	86	78	71	65	59	53	45	42	36	31	26	21	17	12	8	4					
30	93	86	79	72	66	61	55	49	44	39	34	29	25	20	16	12	8	4				
32	93	86	80	73	68	62	56	51	46	41	36	32	27	22	19	14	11	8	4			
34	93	86	81	74	69	63	58	52	48	43	38	34	30	26	22	18	14	11	8	5		
36	94	87	81	75	69	64	59	54	50	44	40	36	32	28	24	21	17	13	10	7	4	
38	94	87	82	76	70	66	60	55	51	46	42	38	34	30	26	23	20	16	13	10	7	5
40	94	89	82	76	71	67	61	57	52	48	44	40	36	33	29	25	22	19	16	13	10	7

Relative humidity values

Dry Bulb Temp. (°F)	Wet Bulb Depression (°F)																																		
	1	2	3	4	5	6	7	8	9	10	11	12	13	14	15	16	17	18	19	20	21	22	23	24	25	26	27	28	29	30	31	32	33	34	35
0	67	33	1																																
5	73	46	20																																
10	78	56	34	13																															
15	82	64	46	29	11																														
20	85	70	55	40	26	12																													
25	87	74	62	49	37	25	13	1																											
30	89	78	67	56	46	36	26	16	6																										
35	91	81	72	63	54	45	36	27	19	10	2																								
40	92	83	75	68	60	52	45	37	29	22	15	7																							
45	93	86	78	71	64	57	51	44	38	31	25	18	12	6																					
50	93	87	80	74	67	61	55	49	43	38	32	27	21	16	10	5																			
55	94	88	82	76	70	65	59	54	49	43	38	33	28	23	19	11	9	5																	
60	94	89	83	78	73	68	63	58	53	48	43	39	34	30	26	21	17	13	9	5	1														
65	95	90	85	80	75	70	66	61	56	52	48	44	39	35	31	27	24	20	16	12	9	5	2												
70	95	90	86	81	77	72	68	64	59	55	51	48	44	40	36	33	29	25	22	19	15	12	9	6	3										
75	96	91	86	82	78	74	70	66	62	58	54	51	47	44	40	37	34	30	27	24	21	18	15	12	9	7	4	1							
80	96	91	87	83	79	75	72	68	64	61	57	54	50	47	44	41	38	35	32	29	26	23	20	18	15	12	10	7	5	3					
85	96	92	88	84	81	77	73	70	66	63	59	57	53	50	47	44	41	38	36	33	30	27	25	22	20	17	15	13	10	8	6	4	2		
90	96	92	89	85	81	78	74	71	68	65	61	58	55	52	49	47	44	41	39	36	34	31	29	26	24	22	19	17	15	13	11	9	7	5	3
95	96	93	89	86	82	79	76	73	69	66	63	61	58	55	52	50	47	44	42	39	37	34	32	30	28	25	23	21	19	17	15	13	11	10	8
100	96	93	89	86	83	80	77	73	70	68	65	62	59	56	54	51	49	46	44	41	39	37	35	33	30	28	26	24	22	21	19	17	15	13	12
105	97	93	90	87	84	81	78	75	72	69	66	64	61	58	56	53	51	49	46	44	42	40	38	36	34	32	30	28	26	24	22	21	19	17	15
110	97	93	90	87	84	81	78	75	73	70	67	65	62	60	57	55	52	50	48	46	44	42	40	38	36	34	32	30	28	26	25	23	21	20	18
115	97	94	91	88	85	82	79	76	74	71	69	66	64	61	59	57	54	52	50	48	46	44	42	40	38	36	34	33	31	29	28	26	25	23	21
120	97	94	91	88	85	82	80	77	74	72	69	67	65	62	60	58	55	53	51	49	47	45	43	41	40	38	36	34	33	31	29	28	26	25	23
125	97	94	91	88	86	83	80	78	75	73	70	68	66	64	61	59	57	55	53	51	49	47	45	44	42	40	38	37	35	33	32	30	29	27	26
130	97	94	91	89	86	83	81	78	76	73	71	69	67	64	62	60	58	56	54	52	50	48	47	45	43	41	40	38	37	35	33	32	30	29	28

Relative humidity values

▲ **Table 5–2**
Relative humidities.

Temp. °F (°C)	Relative Humidity (%)											
	40	45	50	55	60	65	70	75	80	85	90	95
110 (47)	136 (58)											
108 (43)	130 (54)	137 (58)										
106 (41)	124 (51)	130 (54)	137 (58)									
104 (40)	119 (48)	124 (51)	131 (55)	137 (58)								
102 (39)	114 (46)	119 (48)	124 (51)	130 (54)	137 (58)							
100 (38)	109 (43)	114 (46)	118 (48)	124 (51)	129 (54)	136 (58)						
98 (37)	105 (41)	109 (43)	113 (45)	117 (47)	123 (51)	128 (53)	134 (57)					
96 (36)	101 (38)	104 (40)	108 (42)	112 (44)	116 (47)	121 (49)	126 (52)	132 (56)				
94 (34)	97 (36)	100 (38)	103 (39)	106 (41)	110 (43)	114 (46)	119 (48)	124 (51)	129 (54)	135 (57)		
92 (33)	94 (34)	96 (36)	99 (37)	101 (38)	105 (41)	108 (42)	112 (44)	116 (47)	121 (49)	126 (52)	131 (55)	
90 (32)	91 (33)	93 (34)	95 (35)	97 (36)	100 (38)	103 (39)	106 (41)	109 (43)	113 (45)	117 (47)	122 (50)	127 (53)
88 (31)	88 (31)	89 (32)	91 (33)	93 (34)	95 (35)	98 (37)	100 (38)	103 (39)	106 (41)	110 (43)	113 (45)	117 (47)
86 (30)	85 (29)	87 (31)	88 (31)	89 (32)	91 (33)	93 (34)	95 (35)	97 (36)	100 (38)	102 (39)	105 (41)	108 (42)
84 (29)	83 (28)	84 (29)	85 (29)	86 (30)	88 (31)	89 (32)	90 (32)	92 (33)	94 (34)	96 (36)	98 (37)	100 (38)
82 (28)	81 (27)	82 (28)	83 (28)	84 (29)	84 (29)	85 (29)	86 (30)	88 (31)	89 (32)	90 (32)	91 (33)	93 (34)
80 (27)	80 (27)	80 (27)	81 (27)	81 (27)	82 (28)	82 (28)	83 (28)	84 (29)	84 (29)	85 (29)	86 (30)	86 (30)

Category	Heat Index	Possible heat disorders for people in high risk groups
Extreme Danger	130 °F or higher (54 °C or higher)	Heat stroke likely.
Danger	105 – 129 °F (41 – 54 °C)	Muscle cramps and/or heat exhaustion possible with prolonged exposure and/or physical activity.
Extreme Caution	90 – 105 °F (32 – 41 °C)	Muscle cramps and/or heat exhaustion possible with prolonged exposure and/or physical activity.
Caution	80 – 90 °F (27 – 32 °C)	Fatigue possible with prolonged exposure and/or physical activity.

Source: National Weather Service Office, Birmingham, AL.

▲ **Table 5–3**
Heat index.

heat. If the atmosphere has a high moisture content, however, the rate of evaporation is retarded and the loss of latent heat is reduced. In other words, the sweat is unable to effectively do what it is supposed to do. The heat index accounts for this effect (Table 5–3).

The *apparent temperatures* caused by the combination of heat and humidity provide important guidelines for people. At values between 41 °C to 54 °C (105 °F to 129 °F) muscle cramps or heat exhaustion are likely for high-risk people, and even people who are not at high risk face the threat of heat stroke (a potentially fatal increase in the body's internal temperature). Apparent temperatures above 54 °C (129 °F) are considered extremely dangerous, and heat stroke is likely for at-risk people.

The apparent temperatures shown in Table 5–3 should be considered approximate guidelines. Some people react differently to heat, and the index does not account for variables such as exposure to bright sun or wind.

Cooling the Air to the Dew or Frost Point

Although condensation can occur from an increase in the amount of water vapor or from mixing cold air with warm, moist air, the most important mechanism for cloud formation is the lowering of the air temperature to the dew or frost point. We might expect that air temperature will change only in response to gains or losses of energy, but such is not the case. Air temperature changes can occur from two very general classes of processes: those that involve the removal or input of heat, and those that do not. These are referred to as *diabatic* (DIE-a-bat-ic) and *adiabatic* (A-dee-a-bat-ic), respectively.

Diabatic Processes

A **diabatic process** is one in which energy is added to or removed from a system. A pot of water placed over a stove warms diabatically, as does air that is warmed by conduction when in contact with a warm surface. Likewise, air that passes over a cool surface loses energy by conduction into the surface and therefore cools diabatically. Note that the direction of heat transfer is in accordance with the **second law of thermodynamics**, which dictates that energy moves from regions of higher to lower temperatures. Diabatic processes are frequently responsible for the formation of fog but are secondary to adiabatic processes for the development of clouds.

Adiabatic Processes

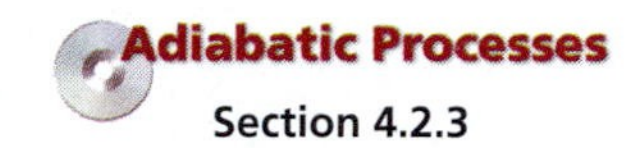

Section 4.2.3

Processes in which temperature changes but no heat is added to or removed from a substance are said to be **adiabatic** (which literally means "not diabatic"). Although changes in temperature without the exchange of heat may at first seem counterintuitive, adiabatic processes are common in the atmosphere and provide the most important mechanism for the formation of clouds.

To adequately understand such processes, we refer to a version of the **first law of thermodynamics**, which states what happens when heat is added to or removed from gases. Specifically, if heat is added, there will be some combination of an expansion of the gas and an increase in its temperature. The law is given in numerical form as

$$\Delta H = p \cdot \Delta\alpha + c_v \cdot \Delta T$$

where ΔH = Heat Added to System, p is the air pressure, $\Delta\alpha$ is the change in volume (positive for expansion and negative for contraction), c_v is the specific heat for air (assuming a constant volume), and ΔT is the change in temperature. (Note that the Greek letter delta, Δ, preceding a symbol represents a change in the value of the quantity.) The first term on the right-hand side of the equation, $p \cdot \Delta\alpha$, is the work performed by the gas as expansion occurs. The second term, $c_v \cdot \Delta T$, refers to the change in internal energy. The important thing for us is that heat added to the air does not simply disappear but rather is apportioned between temperature and volume changes.

The first law of thermodynamics describes the underlying principle of what occurs in the cylinder of an internal combustion engine in an automobile, as shown in Figure 5–12. As the air–fuel mixture burns, it expands and pushes down on the piston (this is the work performed) and ultimately propels the car. In addition, there is an increase in the internal energy of the gas, which we observe as an increase in temperature. (Just ask anyone who has ever burned his or her hands on an exhaust manifold!) The energy unleashed with the combustion of the fuel is therefore manifested as work performed and an increase in temperature. Of course, good automotive design calls for the engine to convert most of the chemical energy to work performed, with little going toward an increase in internal energy. In other words, engine heat represents wasted energy, and a cold exhaust manifold would be the mark of good engineering. The same relationship between heat, temperature, and volume also applies to our atmosphere.

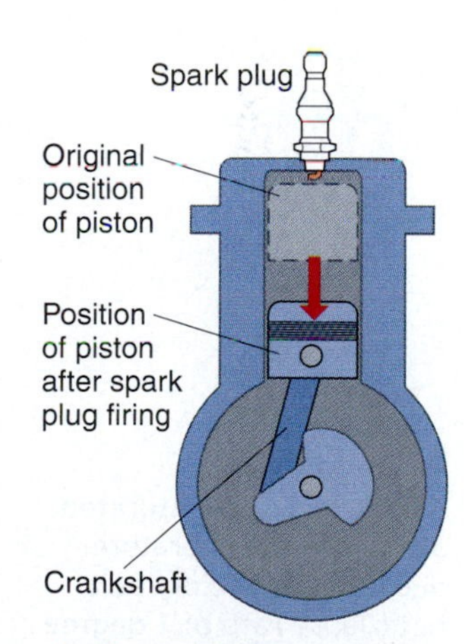

▲ **Figure 5–12**
A four-stroke automobile engine works on the principle invoked by the first law of thermodynamics. The spark plug ignites the fuel. Expansion and an increase in temperature occur from the input of heat released by the burning of the fuel.

An adiabatic process represents a special case of the first law of thermodynamics in which the left-hand side of the equation equals 0 (no heat is added or removed). Substituting 0 for ΔH yields

$$0 = p \cdot \Delta\alpha + c_v \cdot \Delta T$$

which can be rearranged as

$$p \cdot \Delta\alpha = -c_v \cdot \Delta T$$

or

$$-p \cdot \Delta\alpha = c_v \cdot \Delta T$$

Stated in words, the adiabatic form of the first law indicates that if no heat is added or removed from the system, work performed *by* the air (the expansion of the gas) causes a decrease in internal energy (a decrease in temperature), and work performed *on* the gas (compression) leads to warming. Stated even more succinctly, expanding air cools and air undergoing compression is warmed.

This principle applies to what happens when a hand pump is used to inflate a bicycle tire. Compressing the air causes its temperature to increase even though no heat is added, as you can readily observe by touching the base of the pump. Similarly, when air is allowed to escape from the tire valve, its temperature decreases because of expansion, and the jet of outgoing air feels cold. Likewise, blowing on a burnt finger feels good, not because the air in your lungs is cool but rather because the air cools as it expands past your lips.

Figure 5–13 applies this concept to a parcel of unsaturated air that is displaced upward. As the air rises, it encounters lower surrounding pressure, expands, and cools. The rate at which a rising parcel of unsaturated air cools, called the **dry adiabatic lapse rate** (DALR), is very nearly 1.0 °C/100 m (5.5 °F/1000 ft). Thus, a parcel of unsaturated air cools 1 degree Celsius for every 100 meters of ascent, despite the fact that no heat is removed. Likewise, the downward movement of unsaturated air leads to compression and warming at the same rate. It is important to note that the term *lapse* refers to a decrease in temperature with altitude. We therefore leave out the minus sign preceding the 1.0 °C/100 m.

If a parcel of air rises high enough and cools sufficiently, expansion lowers its temperature to the dew or frost point, and condensation or deposition commences. The altitude at which this occurs is known as the **lifting condensation level** (LCL). As the saturated air rises beyond the LCL, expansion continues to lower its temperature, but the cooling is partially offset by the release of latent heat from condensation (or the deposition to ice). Thus, the lifting of saturated air results in a less rapid cooling than that which occurs for unsaturated air. The rate at which saturated air cools is the **saturated adiabatic lapse rate** (SALR), which is about 0.5 °C/100 m (3.3 °F/1000 ft). The term *wet adiabatic lapse rate* is often used interchangeably with saturated adiabatic lapse rate. Unlike the dry adiabatic rate, the SALR is not a constant value but instead varies with temperature, as we explain in *Box 5–3, Physical Principles: The Varying Value of the Saturated Adiabatic Lapse Rate.*

The Environmental Lapse Rate

The adiabatic lapse rates are not to be confused with the **environmental** (or ambient) **lapse rate** (ELR), which applies to the vertical change in temperature through still air. A large mass of air is not likely to have a constant temperature; rather, its temperature usually decreases with altitude in the troposphere. The rate at which

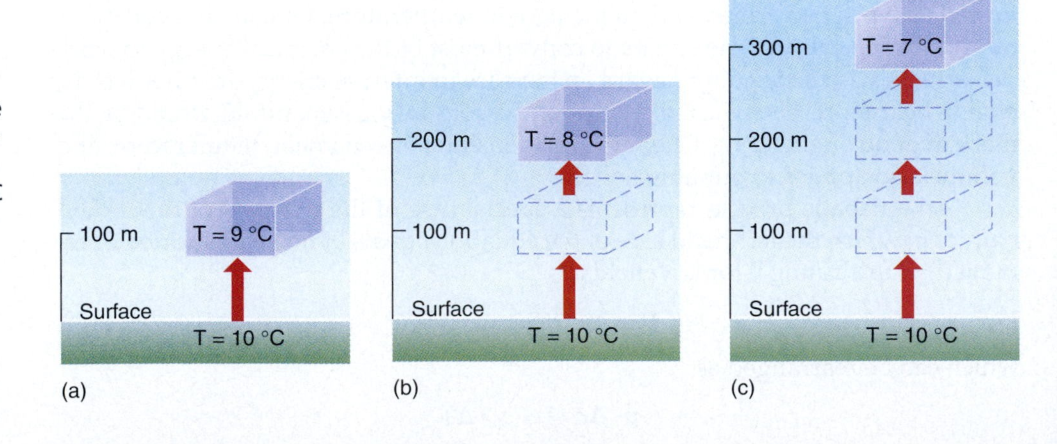

Figure 5–13 As a parcel of unsaturated air rises, its temperature decreases at the dry adiabatic lapse rate of 1 degree Celsius per 100 meters. The parcel shown here has an initial temperature of 10 °C but cools to 7 °C when lifted 300 meters.

5–3 Physical Principles

The Varying Value of the Saturated Adiabatic Lapse Rate

The dry adiabatic lapse rate (DALR) has a constant value of 1 °C/100 m. The saturated adiabatic lapse rate (SALR) is usually about half that value, because the release of latent heat as saturated air rises partially offsets the cooling by expansion. But unlike the value of the dry adiabatic lapse rate, the SALR is not constant. Rather, it depends on the temperature of the saturated air parcel, with higher temperatures causing lower lapse rates. Figure 1 can help us see why this is the case. Recall that if air is saturated, its actual specific humidity is equal to the saturation specific humidity. As a rising parcel of saturated air is cooled adiabatically, the amount of water vapor that can exist decreases and the surplus water vapor is removed by condensation (or deposition).

Now observe what happens when the temperature of warm, saturated air decreases 5 °C (9 °F), from 30 °C to 25 °C. As the air cools, its specific humidity changes from 27.7 grams of vapor per kilogram of air to 20.4—a decrease of 7.3 grams of vapor per kilogram of air. The 7.3 g of water vapor do not simply vanish; rather, they are converted to an equal mass of liquid water. Upon condensation, each gram of water releases 2500 joules of latent heat, for a total of 18,250 J. Compare that to what happens if the temperature undergoes another 5 °C drop in temperature, but this time from 5 °C to 0 °C. At this lower temperature, the 5°of cooling reduces the water vapor content from 5.5 to 3.8 g/kg—only a 1.7-g decrease—and only 4250 J of energy are released to the air. Thus, at low temperatures relatively little latent heat is released to offset the cooling due to expansion, and the SALR is nearly equal to the DALR. When warm, saturated air is lifted, a greater amount of latent heat is available to offset the cooling by expansion, and the SALR assumes a lower value.

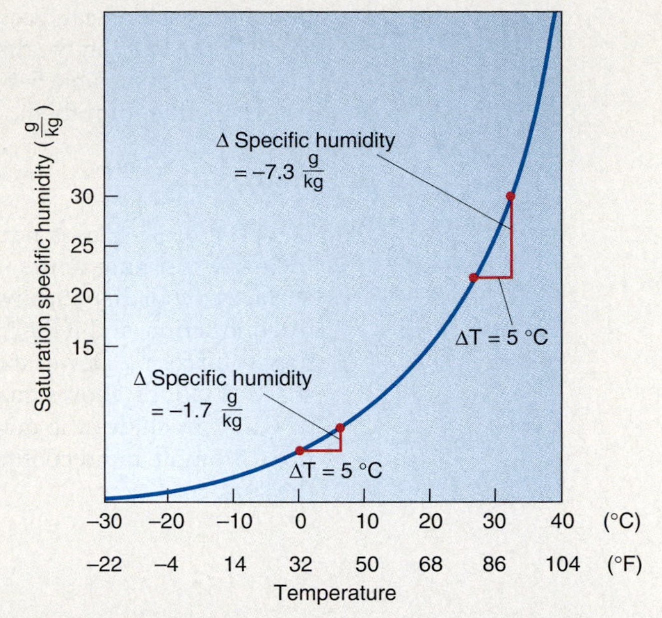

▲ **Figure 1**
Unlike the DALR, the SALR is not a constant value. When warm, saturated air cools, it causes more condensation (and hence more latent heat release) than for cold, saturated air. For example, if saturated air cools from 30 °C to 25 °C (a 5° decrease), the specific humidity decreases from 27.7 grams of water vapor per kilogram of air to 20.4. A 5 °C drop in temperature from 5 °C to 0 °C lowers the specific humidity only 1.7 grams for each kilogram of air. This brings about less warming to offset the cooling by expansion, as well as a greater saturated adiabatic lapse rate.

the ambient temperature decreases with height is called the environmental lapse rate (ELR). The ELR for the troposphere is highly variable. It changes from day to day, from place to place, and even from one altitude to another within the atmospheric column. An atmosphere in which temperature decreases rapidly with elevation is said to have a "steep" environmental lapse rate.

As an analogy, we can contrast the changes in temperature that would occur inside a rising balloon with that of the surrounding air. In Figure 5–14, a balloon rises through the atmosphere, which in this case has an ELR of 0.5 °C/100 m. A thermometer within the balloon would record the temperature change inside the expanding balloon, corresponding to the DALR as long as the air within remains unsaturated. A thermometer attached outside the balloon would record the temperature of the surrounding, nonmoving air, reflecting the environmental lapse rate. As the balloon rises, the temperature within the balloon will be lower than that of the air surrounding it, because in this instance the DALR exceeds the ELR.

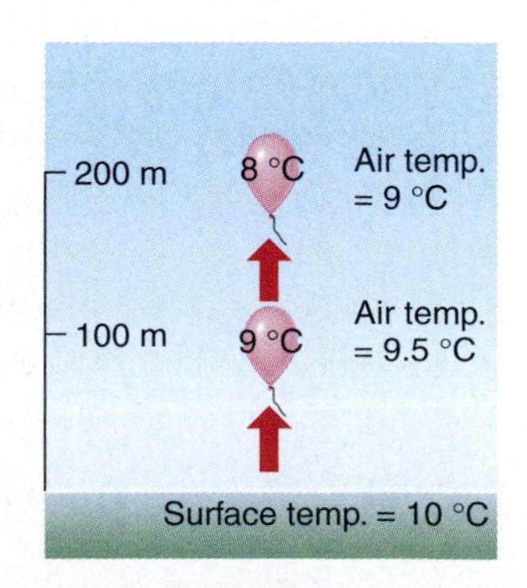

▶ **Figure 5–14**
The adiabatic lapse rates are not to be confused with the environmental (or ambient) lapse rate. A balloon rising through air with an ELR of 0.5 °C/100 m passes through air whose temperature decreases from 10 °C at the surface, to 9.5 °C at 100 m, and 9.0 °C at 200 m. The air within the balloon cools at the dry adiabatic lapse rate of 1.0 °C/100 m, faster in this example than the ELR, and therefore attains a temperature of 8 °C at the 200 m level.

Forms of Condensation

Saturation can lead to the formation of liquid water or ice crystals. The condensation or deposition can occur in the air as cloud or fog, or onto a surface as dew or frost. In this section we discuss the processes that give rise to these various types of condensation. Table 5–4 summarizes the general types of condensation and the processes that form them.

Dew

Dew (Figure 5–15a) is liquid condensation on a surface, often occurring during the early morning hours. At night the loss of longwave radiation can cause the surface to cool diabatically. Air immediately in contact with the cold surface cools by conduction, and if the temperature decreases all the way to the dew point, condensation forms. Dew is most likely to form on clear, windless nights, when the absence of clouds allows much longwave radiation to escape to space and the lack of wind precludes the mixing of warmer air from above. Together these conditions promote rapid cooling within a shallow layer of air immediately adjacent to the surface.

Table 5–4 • General Types of Condensation

Condensation Form	Predominant Processes	Characteristics
Dew	Lowering of temperature to the dew point near the surface. Favored under clear skies and no wind. Diabatic process.	Appears as coating of liquid water on surfaces.
Frost	Lowering of air temperature to saturation point, when the saturation point is below 0 °C (32 °F). Diabatic process.	Appears as large number of small white crystals on surfaces.
Frozen Dew	Formation of dew at temperatures above 0 °C, followed by cooling to temperatures below 0 °C. Diabatic process.	Continuous layer of solid ice on surface.
Fog	Usually by cooling of layer of air with light winds. Sometimes by evaporating water from falling precipitation or by mixing warm, moist air with cold air. Diabatic or adiabatic process.	Large concentration of suspended droplets in layer of air near ground. Under extreme cold, can consist of suspended ice crystals.
Radiation fog	Cooling of air to dew point by longwave radiation loss. Diabatic process.	Same as above.
Advection fog	Cooling of air to dew point as it passes over cool surface. Diabatic process.	Same as above.
Upslope fog	Cooling of air as it flows up slope. Adiabatic process.	Same as above.
Precipitation fog	Increasing the water vapor content of the air by evaporation from falling droplets. Adiabatic process.	Same as above.
Steam fog	Mixing warm, moist air with cold air. Adiabatic process.	Same as above.
Clouds	Usually by lifting of air and adiabatic cooling.	Concentration of suspended droplets and/or ice crystals in air well above the surface.

(a)

(b)

◀ **Figure 5–15**
Dew (a) and frost (b).

Frost

The formation of **frost** is similar to that of dew, except that saturation occurs when the temperature is below 0 °C. When the air is lowered to the frost point, very small ice crystals are deposited onto solid surfaces, giving them a bright white appearance, as shown in Figure 5–15b. This type of deposition, sometimes referred to as *white frost* or *hoar frost*, occurs by the transformation of water vapor directly into ice, without going through the liquid phase.

Because it consists of a huge number of separate ice crystals rather than a solid, continuous coating of ice, frost on the windshield of a car is often easily removed by a swift brushing with a credit card or window scraper. This is in contrast to a more troublesome type of condensation called *frozen dew*.

Frozen Dew

Frozen dew differs from frost in both its structure and its manner of formation. It begins when saturation forms liquid dew at temperatures slightly above 0 °C. When further cooling brings its temperature below the freezing point, the liquid solidifies into a thin, continuous layer of ice. In contrast to frost, frozen dew is neither milky white nor easy to remove but instead bonds tightly to whatever surface on which it forms.

Because it is a continuous coating of solid ice, a mere brushing does not come close to removing frozen dew. In addition to coating your car's windshield, the ice can make it difficult or impossible to get your key into the lock mechanism of your door. And even if you are lucky enough to be able to turn the key, the door can become frozen to the car frame as if it were welded shut. On the other hand, it may be just as well for you that you are not able to get into your car because of the problem of *black ice*, the smooth coating of frozen dew that forms on road surfaces. This is especially likely to form over bridges, causing dangerous, slippery driving conditions.

Fog

Fog is essentially a cloud whose base is at or near ground level (see Figure 5–16). It can be extremely shallow, on the order of a meter or so in depth, or it can extend for tens of meters above the surface. Like any other form of condensation, fog can form by the lowering of the air temperature to the dew point, an increase in the water vapor content, or the mixing of cold air with warm, moist air. We have already described one example of condensation resulting from the addition of water vapor into the air, *precipitation fog*, which results from the evaporation of falling raindrops. Another, although very localized, type of fog occurs from adding water vapor into the air—that which you see right in front of you when you exhale on a cold day. We have also discussed the formation of *steam fogs*, which occur when cold, dry air mixes with warm, moist air above a water surface. All other types of fog result from a cooling of the air to the dew point. They include radiation fogs, advection fogs, and upslope fogs.

Radiation Fog **Radiation fogs** (sometimes called *ground fogs*) develop when the nighttime loss of longwave radiation causes cooling to the dew point. Like dew, a radiation fog is most likely to form on cloudless nights when longwave radiation from the surface easily escapes to space. Unlike dew it is most likely to form with light winds of about 5 km/hr (3 mph) rather than in perfectly still air. Light breezes promote a gentle stirring of the lower atmosphere, which permits condensation to form throughout a layer of air. When the wind speed is much greater than 5 km/hr, the excess turbulence brings warm air down toward the surface and thereby inhibits cooling to the dew point.

Most radiation fogs begin to dissipate within a few hours of sunrise. Although we sometimes talk about a fog "lifting," that is not what really happens. It is probably better (although still somewhat imprecise) to describe it as "burning off." When sunlight penetrates the fog, it warms the surface, which in turn warms the overlying

▶ Figure 5–16
Fog.

air. As the air temperature increases, the fog droplets gradually evaporate. Because the evaporation of the droplets is most rapid near the surface, the fog appears to lift, although it really undergoes no vertical displacement.

When radiation fogs are especially well developed, they can scatter backward the greater part of incoming solar radiation. Because the amount of energy reaching the surface is now reduced, the fog can persist throughout the day, especially in the winter when the days are short and the sun angles are low. Under the most extreme circumstances, fogs can persist for days on end.

A prime location for a persistent radiation fog is the Central Valley of California (Figure 5–17). To the west of the valley the Coast Ranges block the moderating effects of the Pacific Ocean, while the Sierra Nevada isolates the valley from the east. In addition to the clear skies and light winds that often predominate during the winter, this heavily agricultural region has copious amounts of moisture evaporated into the air from the irrigated farmland. When the radiation fog (locally referred to as a *Tule fog*) covers the valley, visibility along the two major north–south highways can be dangerously reduced to near zero. This happened with fatal consequences on February 5, 2002, on California State Highway 99, south of Fresno. As the fog reduced visibility to as little as 50 ft, California Highway Patrol "pace cars" were dispatched to lead traffic at safe speeds. But, as has happened many times in the past, the effort was unsuccessful. Eighty-seven cars were involved in two chain reaction pile-ups that left two persons dead and many others injured. Just one month earlier, a similar set of accidents caused two other fatalities along Interstate 5 near Sacramento airport.

On March 5, 2002—one month after the major pile-up near Fresno—a thick, morning fog led to a major crash involving 125 vehicles on Interstate 75 in northwest Georgia, just south of Chattanooga, Tennessee. Four people died in the accident. This is another region where radiation fogs are common, and that morning the National Weather Service issued numerous advisories about hazardous driving conditions. Despite the warning, the near-zero visibility made any travel along the highway a risky endeavor. According to Sheriff Phil Summers, "It did not appear that there was any fault other than fog."

Radiation fogs—and the risks they impose on travelers—are frequent occurrences wherever air has the opportunity to cool at night with gentle stirring. Various

Figure 5–17
The Central Valley of California often gets radiation fogs in winter that last for days on end. These persistent fogs are visible from space.

state transportation agencies have tried to alleviate the threat of such accidents by painting reflective "fog lines" on at-risk highways and installing automated weather stations to provide real-time weather information, but measures such as these cannot overcome the inherent danger imposed by dense fog.

Radiation fogs are formed by diabatic cooling and are therefore associated with cold air. Because cold air is denser than warm air with otherwise similar characteristics, radiation fogs often settle into local areas of low elevation, where they are called *valley fogs.* It is tempting to attribute the fog's high density to the heavy droplets of liquid water, but they are not the cause. It is the low temperature—not the existence of water droplets—that causes the fog to be dense and settle into valleys. Because all the water droplets replace an equal mass of water vapor from which they formed, their presence does not make the air any heavier.

Advection Fog **Advection fogs** (Figure 5–18) form when relatively warm, moist air moves horizontally over a cooler surface (the term *advection* refers to horizontal movement). As the air passes over the cooler surface, it transfers heat downward, causing it to cool diabatically. If sufficient cooling occurs, a fog forms. Such fogs can be advected for considerable distances and persist well downwind of the area over which they form. One of the most famous examples of this phenomenon occurs during the summer months over the San Francisco Bay area. As relatively warm Pacific air drifts eastward, it passes over the narrow, cold, southward-flowing California ocean current. The cooling of the air offshore forms the fog, which drifts eastward toward San Francisco. It is quite common for the fog to fully engulf San Francisco while Oakland and Berkeley, across the bay, remain warm and sunny.

Advection fogs can also form over water when warm and cold ocean currents are in proximity to each other. Off the coast of New England and the Maritime Provinces, the cold Labrador current flows just to the north of the Gulf Stream. When the moist air from the Gulf Stream region drifts over the Labrador current, it can be cooled to the dew point to form a persistent, dense fog that can last for weeks. Although most common in summer, it can occur any time during the year.

Another example of an advection fog is that which often covers London during the winter. Warm air passing over the warm Gulf Stream is advected over England, where it is chilled by the surface to form a thick fog.

▶ **Figure 5–18**
An advection fog at San Francisco Bay.

The winds associated with advection fogs are often greater than those of radiation fogs. This promotes greater turbulence and allows the droplets to circulate to greater heights. Advection fogs can have thicknesses up to about a half kilometer (1500 ft).

Upslope Fog Of the three types of fog caused by the cooling of air, only **upslope fog** is formed by adiabatic cooling. When air flows along a gently sloping surface, it expands and cools as it moves upward. The western slope of the Great Plains of the United States provides an excellent setting for this type of condensation. Westward from the Mississippi River valley, the elevation gradually increases toward the foothills of the Rocky Mountains. Moist air from the Gulf of Mexico cools adiabatically as it ascends the slope of the plains to create widespread fog. Figure 5–19 presents a highly generalized description of the types of fog most prevalent over regions of the United States and Canada.

Distribution of Fog

Figure 5–20 depicts the number of heavy fog days (defined as limiting visibility to a quarter of a mile or less) across the 48 conterminous United States. The three significant centers of heavy fog are along the Pacific Northwest, New England, and the middle Appalachians. The Pacific Northwest and the coast of British Columbia experience numerous advection fogs, as westerly winds advect moist air over the cold California current. But it is not correct to attribute all of the fog formation to the cooling effect of the cold surface waters. As damp, cool air reaches the shore, fog formation is abetted by the *orographic* effect (the lifting effect caused as air crosses a mountain or similar barrier) as air approaches the steep Coast Ranges. Cape Disappointment, Washington, wins the prize for the foggiest U.S. location, being shrouded in heavy fog nearly one-third of the time. The zone of most persistent fog is confined primarily to the coastal region because the Coast Ranges block the flow of moisture inland.

▲ **Figure 5–19**
The different types of fog commonly found throughout North America.

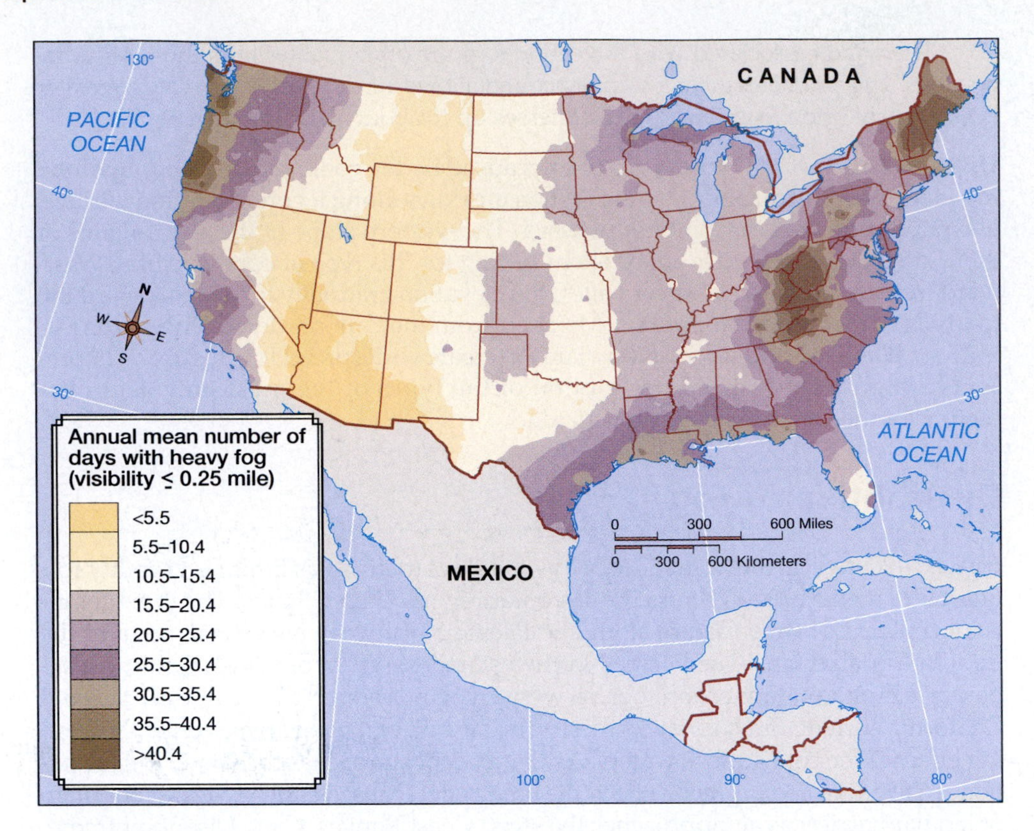

▶ **Figure 5–20**
Average annual number of days with heavy fog in United States.

New England and the Maritime Provinces experience a large number of heavy fog days. Along the coast, advection fogs dominate, with the coast of Maine having the highest incidence of dense cover. The advection fogs are most prevalent in summer. Inland, radiation fogs are very common at some of the higher elevation areas of New Hampshire and Vermont.

The upper reaches of the Appalachians, the third major focus of fog, undergo a large number of radiation fogs, particularly in late summer and fall. Not surprisingly, fog is clearly at a minimum in the desert Southwest.

Formation and Dissipation of Cloud Droplets

Farther from the surface than fog, dew, or frost, clouds are usually the result of the adiabatic cooling associated with rising air parcels. In this section we take a closer look at such parcels as they rise, become saturated, and continue upward above the level at which condensation first occurs.

For reasons we will not discuss here, the dew point decreases as the air rises, at the rate of about 0.2 °C/100 m (1.1 °F/1000 ft). This decrease is called the **dew point lapse rate**. As unsaturated air is lifted, its temperature therefore approaches the dew point by 0.8 °C for every 100 m of ascent (i.e., 1.0 °C minus 0.2 °C). Thus, if the air temperature and dew point start out at 18 °C and 10 °C, respectively, an ascent of 1000 m is necessary to cause the air to be saturated.

Raising an air parcel above the lifting condensation level initially leads to the formation of small cloud droplets. But at about 50 m or so above the LCL, all the condensation nuclei in the air will have attracted water, and further uplift leads only to the growth of existing water droplets. In other words, no new droplets form as the air continues to rise; instead, the existing droplets simply grow larger.

The processes that lead to the formation of a cloud do not continue forever. At some point in time, lifting will cease and there will be no further condensation. But this does not mean that the existing droplets or crystals will immediately disappear. Instead, they will gradually undergo evaporation or sublimation, a process that can take considerable time before the cloud is fully dissipated. In fact, the ice crystals associated with certain clouds may exist days after they were formed. In the eastern United States and Canada, high, wispy clouds of ice sometimes mark the remnants of thunderstorm clouds over the Rocky Mountains.

Now consider what happens if a rising parcel of air reverses its movement and begins to subside. The parcel now warms at the *saturated* adiabatic lapse rate, because the warming by compression is partially offset by the gradual evaporation of the droplets. The evaporation continues until the parcel has descended back to the original level at which condensation occurred. If the air then sinks below the original lifting condensation level, it warms at the DALR, because all the droplets will have evaporated. If brought down to the initial level at which uplift first began, the air will reassume its original temperature and dew point. The net effect of all this is that the cooling of the air and the condensation of liquid water are *reversible processes.* Note that we assume all the condensation products remain in the atmosphere and are thus available for evaporation during descent. Of course, the real atmosphere *loses* moisture by precipitation of rain and snow, which means that these processes are not strictly reversible. But a relatively small portion of condensed water falls out, so the concept remains generally valid.

In thinking about vertical motions and moisture, it's interesting that very small displacements can have such large consequences. After all, we hardly think about the effects of horizontal movements covering 100 km. But in the vertical, movement of just 1 km can make the difference between a fine day and a ruined picnic, as rising air leads to clouds, and clouds give rise to precipitation. Like the formation of clouds, precipitation is no simple matter, as will be seen in Chapter 7.

Summary

Despite the fact that it accounts for only a small proportion of the mass of the air, water in its three phases is extremely important to the atmosphere—and to all life on Earth. To fully understand its impacts on the atmosphere, we need to comprehend the concept of saturation. At the surface of liquid water, molecules are constantly moving about. As they do so, some randomly break free of the surface to become water vapor. In a partially filled jar, this adds vapor to the volume above the water surface and thereby increase the water vapor content. But as the water vapor concentration increases, so does the opposite process—condensation. As long as the rate of evaporation exceeds condensation, the amount of water vapor increases. The condensation rate in the jar eventually becomes equal to the evaporation rate, and saturation occurs.

The foregoing principle applies to the real atmosphere. In meteorology, we are particularly concerned with the exchange of water between a suspended cloud droplet and the air that surrounds it, because the equilibrium between evaporation and condensation is a prerequisite to the persistence of clouds and fog. When evaporation and condensation are in equilibrium, the air is saturated and small droplets can remain in the air without evaporating away.

There is no single best expression of humidity. One way of expressing the amount of water vapor is by the pressure it exerts (vapor pressure). The specific humidity and mixing ratio are similar ratios that relate the mass of water vapor to the air in which it is contained and to the mass of the other gases, respectively. Another measure, absolute humidity, is not widely used but is simply the density of water vapor. In contrast, relative humidity is a quantity we hear about almost daily, even though its partial dependence on temperature presents a serious drawback. Perhaps the most useful of the indices is dew point, expressed as the temperature to which the air must be lowered for saturation to occur.

Although it might seem that condensation should occur at exactly 100 percent relative humidity, the situation in reality is slightly more complex. Condensation is controlled by two factors having opposite effects: curvature and the abundance of hygroscopic nuclei. Curvature retards condensation, so that without the presence of condensation nuclei, condensation would occur only at very high relative humidities. Natural and anthropogenic nuclei are sufficiently numerous that condensation usually occurs at relative humidities between 98% and 100.1%.

Humidity is an easy property to measure. It can be determined by the use of paired thermometers (a psychrometer) that provide dry and wet bulb temperatures. The difference between the two temperatures is the wet bulb depression which, when combined with dry bulb temperature, permits the use of simple tables for the determination of relative humidity and dew point. One of the important features of humidity is its effect on a human's susceptibility to heat-related dangers, which is reflected in the heat index.

For any kind of condensation (such as dew, fog, or clouds) to form, the dew point must equal the air temperature. This can result from raising the vapor content of the air to the saturation level; mixing warm, moist air with cooler, dry air; or by lowering the air temperature to the dew point. The latter process is most important for cloud formation. Lowering the air temperature does not require that heat be removed from the air. In fact, most clouds form by adiabatic cooling—the lowering of the air temperature without the removal of heat. We have shown that adiabatic cooling, which results from the expansion of air that occurs when it is lifted, is a direct application of the first law of thermodynamics.

Other forms of condensation—dew, frost, frozen dew, and fog—are distinguished from clouds by their proximity to or direct contact with the surface. Though one type of fog (upslope fog) results from adiabatic processes, the others result from diabatic (involving the loss of energy) cooling of the air.

We have discussed how rising air leads to adiabatic cooling and to saturation and cloud formation, but we have yet to provide any explanation for these vertical motions. As it happens, there are several mechanisms capable of generating uplift, some of which are quite intricate and closely tied to the cloud form that results. The topic deserves extended discussion, which we provide in the next chapter.

Key Terms

evaporation p. 134
condensation p. 134
saturation p. 134
sublimation p. 135
deposition p. 135
humidity p. 135
vapor pressure p. 135
saturation vapor pressure p. 135
absolute humidity p. 136
specific humidity p. 136
saturation specific humidity p. 136
mixing ratio p. 137
saturation mixing ratio p. 137
relative humidity p. 137
dew point p. 139
frost point p. 141
precipitation fog p. 142
steam fog p. 144
homogeneous nucleation p. 146
hygroscopic p. 146
heterogeneous nucleation p. 146
condensation nucleus p. 146
haze p. 147
supercooled water p. 148
ice nucleus p. 148
sling psychrometer p. 148
wet/dry bulb thermometer p. 148
wet bulb depression p. 149
aspirated psychrometer p. 149
hair hygrometer p. 149
hygrothermograph p. 149
heat index p. 149
diabatic process p. 153
second law of thermodynamics p. 153
adiabatic process p. 153
first law of thermodynamics p. 153
dry adiabatic lapse rate p. 154
lifting condensation level p. 154
saturated (wet) adiabatic lapse rate p. 154
environmental (ambient) lapse rate p. 154
dew p. 156
frost p. 157
frozen dew p. 157
radiation fog p. 158
advection fog p. 160
upslope fog p. 161
dew point lapse rate p. 162

Review Questions

1. Why is it incorrect to refer to the air as "holding" water vapor?
2. What are deposition and sublimation?
3. What is vapor pressure? What units of measurement is it expressed in?
4. Explain the concepts of equilibrium and saturation.
5. What units of measurement are used to describe mixing ratio and specific humidity? Why are the two values nearly equal?
6. Why is absolute humidity seldom used?
7. Define relative humidity.
8. Why is relative humidity a poor indicator of the amount of water vapor in the air?
9. Define dew point. What characteristics make this measure superior to relative humidity?
10. Why can't the dew point temperature exceed the air temperature? What happens if the air temperature is lowered to a value less than the initial dew point?
11. Describe the distribution of average dew point across the United States in summer and winter.
12. What are the three general methods by which the air can become saturated?
13. What are the effects of droplet curvature and solution on the amount of water vapor needed for saturation?
14. Why doesn't homogeneous nucleation form water droplets in the atmosphere?
15. What are condensation nuclei and ice nuclei? Are they typically made of the same materials? Which is more abundant in the atmosphere?
16. What is supercooled water?
17. What are psychrometers? How do they work?
18. Define dry bulb temperature, wet bulb temperature, and wet bulb depression.
19. What is the heat index?
20. What is the first law of thermodynamics and why is it important?
21. Explain the difference between diabatic and adiabatic processes.
22. What are the numerical values of the dry and saturated adiabatic lapse rates? Under what circumstances are they applicable?

23. What does the environmental lapse rate refer to?

24. Describe the various processes that can lead to the formation of dew.

25. What is the difference between frozen dew and frost?

26. Describe the various processes that can lead to the formation of fog.

Critical Thinking

1. When rubbing alcohol is applied to a person's skin, it feels colder than the application of water would. Why?

2. A person sleeps through the night without waking up, but awakes in the morning weighing slightly less than the night before. What happened?

3. A crowded classroom is filled with students. In what way might the presence of the students affect the dew point and relative humidity within the room?

4. A person parks her car in the driveway on a warm afternoon and notices a small puddle of water beneath the car a few minutes later. Explain how the use of air conditioning can account for the puddle.

5. At Wheeling, West Virginia, the evening temperature is 55 °F and the dew point is 48 °F. How would you assess the likelihood of fog forming overnight?

6. A map of North America shows the average distribution of vapor pressure across the continent. Will the distribution shown on the map be only a function of the amount of water vapor in the air or will the distribution be affected by another factor as well?

7. The temperature within a forest is –2 °C (28 °F) and there is frost on the trees but no fog. Outside the woods there is a fog. Why was this fog not encountered in the woods?

8. Is fog more likely to occur downwind or upwind of an oil refinery? Why?

9. All fogs are made of water droplets or ice crystals. Despite the fact that they have the same composition, how would you know if a particular fog is a radiation, advection, or upslope fog?

10. Diesel engines, like four-stroke engines, work because of the burning of fuel, but they do not require a spark plug or similar device for initiating the burning of the fuel. Apply your knowledge of the first law of thermodynamics to explain how the fuel can be forced to burn.

Problems and Exercises

1. Assume that a kilogram of air consists of 995 g of dry air and 5 g of water vapor. Show that the specific humidity and mixing ratio are very nearly equal.

2. Assume that a kitchen measures 4 meters by 5 meters by 3 meters. If the air density is 1 kg/cu. m and the specific humidity is 10 grams of water vapor per kilogram of air, how much water vapor is in the room? If the doors and windows were sealed shut, would boiling 1 kg of water into the air make a substantial change in the humidity of the room?

3. The dry and wet bulb readings in Honolulu are 80 °F and 69 °F. At Charlottesville, Virginia, the readings are 50 °F and 45 °F.
   - **a.** Use Tables 5–1 and 5–2 to determine the relative humidity and the dew point for both locations.
   - **b.** Which of the two locations is more humid?

4. The dry and wet bulb temperatures are 70 °F and 54 °F. Use Tables 5–1 and 5–2 to answer the following questions:
   - **a.** What are the dew point and the relative humidity?
   - **b.** What will the dew point and relative humidity be if the air temperature increases to 80 °F? (*Hint:* Do not assume the same wet bulb depression.)
   - **c.** What will the dew point and relative humidity be if the air temperature drops to 39 °F? (*Hint:* You don't need the tables for this one.)
   - **d.** What will the dew point and relative humidity be if the air temperature drops further, to 35 °F?

5. The numerical value of the specific heat for air in the first law of thermodynamics, c_v, is strictly valid only for air with no water vapor. Water vapor has a specific heat approximately twice as great as that of c_v. Will a humid mass of air undergoing expansion therefore undergo more or less cooling than would a dry mass of air?

6. Assume a parcel of air starts out at the surface with a temperature of 12 °C and a dew point of 9.6 °C is lifted:
   - **a.** What will the air temperature be at the 100 m level?
   - **b.** What will the dew point be at the 100 m level? (*Hint:* Don't forget the dew point lapse rate.)
   - **c.** At what height will condensation occur?
   - **d.** What will the temperature be when condensation occurs?
   - **e.** What will the dew point be when condensation occurs?
   - **f.** What will the temperature be at 500 m above the surface?
   - **g.** If the parcel of air is lowered back to the surface (assuming none of the condensed moisture was removed as rain), what will the temperature and dew point be?

Quantitative Problems

As this chapter has shown, several indices are used to express the moisture content of the air, with each having some advantages and disadvantages. In meteorology we are particularly concerned with how these indices change in response to the lifting of air adiabatically. One excellent way to reinforce your understanding of these measures is by solving the simple problem set given on the Chapter 5 page of this book's Web site **(http://www.prenhall.com/aguado/)**. These questions are particularly valuable as an aid to understanding how saturation is achieved in these rising parcels.

Useful Web Sites

http://www.usatoday.com/weather/whumcalc.htm

Much useful information on humidity, with links to charts and calculaters.

http://www.ems.psu.edu/wx/usstats/dewpstats.html

Map of current dew point distribution across United States.

http://www.wunderground.com/US/Region/US/HeatIndex.html

Map of current heat index across United States.

Media Enrichment

Tutorial

Atmospheric Moisture

This tutorial explains the concepts of saturation, condensation, and various measures of atmospheric moisture. Use it to experiment with arbitrary combinations of temperature and moisture as an aid to understanding the relationship between the various measures.

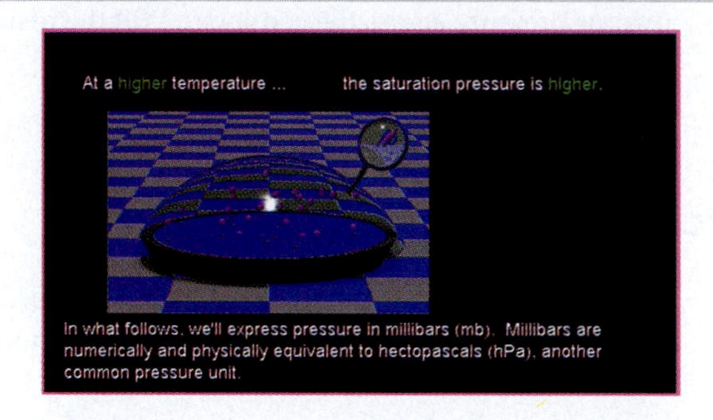

Tutorial

Adiabatic Processes

This tutorial provides details on adiabatic expansion, explaining the intricacies of both the dry and saturated lapse rates. Be sure to complete the moisture tutorial before starting this tutorial.

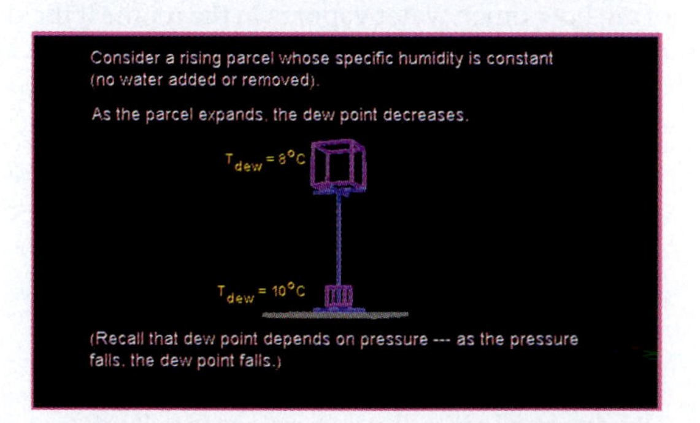

Interactive Exercise

Moisture

As you know, changes in moisture content and temperature can lead to saturation. First, compare the moisture content of air at 10 °C with air at 20 °C, assuming both have a relative humidity of 50%. Which contains more moisture? What is the percentage difference in moisture content? Next change temperature only, noting the effects on relative humidity. As temperature decreases with no water added or removed, what happens to relative humidity? Does a 5 °C change have the same effect regardless of temperature? (*Hint:* Compare a warm parcel and a cold parcel, decreasing the temperature of both by 5 °C.)

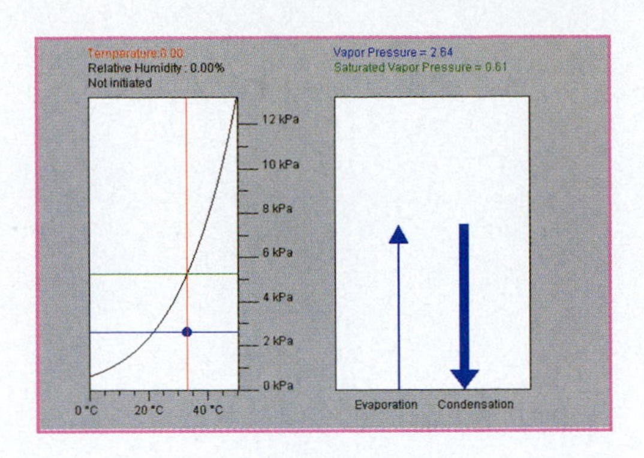

Weather in Motion

January and July Water Vapor Movies

These two movies show primarily the flow of water *vapor*, based on observations at 3-hr intervals. As we will see in the rest of this chapter, there is a very big distinction between water vapor and water in its other two phases. Dark portions of the screen indicate dry air, whereas brighter areas have a higher water vapor content. Very bright areas indicate the presence of high cloud cover, often the tops of thunderstorm clouds.

In January we see a broad movement of moisture, especially in the middle latitudes. Areas of strong convection occur primarily in the Tropics. Moreover, we see bands of moisture at times flowing into the southern United States as parts of air flows called *subtropical jets* (which we describe in Chapter 8). In July the zone of tropical convection shifts northward, in response to the shift in solar declination. Also, thunderstorms are more common in the middle latitudes of the Northern Hemisphere in this movie than they are in the January movie.

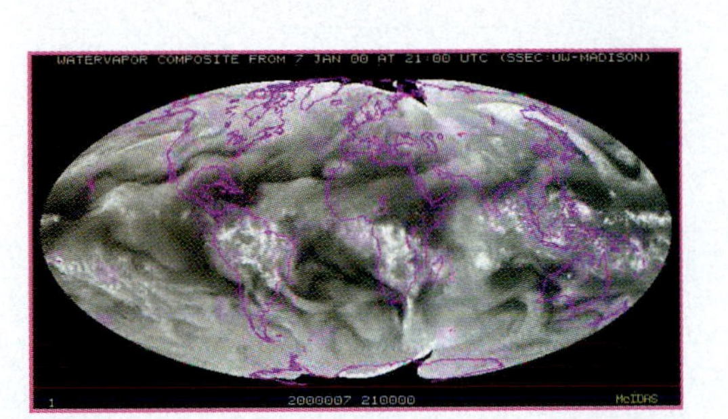

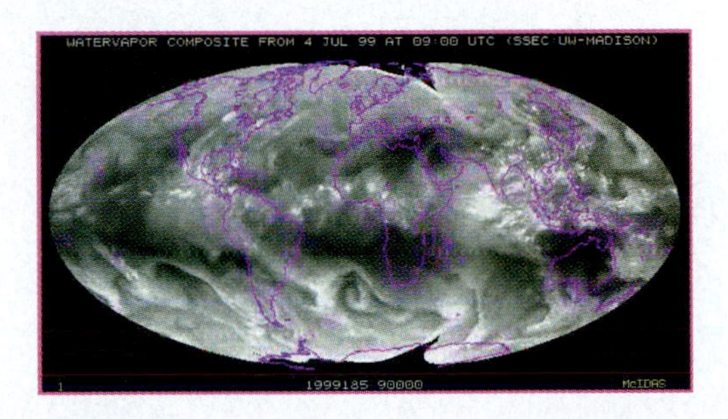

chapter 6

Storm chasers risk life and limb to track violent storms.

Cloud Development and Forms

All of us pay special attention to clouds at one time or another. Some are particularly beautiful, while others portend severe weather. But certain people do a lot more than passively observe certain clouds. These *storm chasers* set out in teams during the severe weather season with the goal of observing the formation and movement of tornadoes. Some storm chasers are nonprofessional weather enthusiasts, while others are trained meteorologists with advanced degrees or students working toward their credentials. The latter group goes out not just for the thrill of witnessing severe weather firsthand (though one can't deny that the excitement accounts for much of the pursuit) but also to acquire information that is important to our overall understanding of these phenomena. Using a combination of tools ranging from advanced technology to sheer intuition, professional storm chasers pursue potentially severe storms to understand the mechanisms within and near clouds that lead to tornado development. These efforts have provided valuable information that has helped meteorologists understand what parts of active thunderstorm clouds are most likely to produce tornadoes. These efforts have also revealed that large areas of rotation within storm clouds often precede tornado formation by about 20 minutes to half an hour—thus providing advanced warning to forecasters and the public.

Storm chasing came of age in the spring of 1979, with the undertaking of the Severe Environmental Storms and Mesoscale Experiment (SESAME) in the central United States. Scientists working with the project assembled information from radar units, weather balloons, and field observations. By a remarkable coincidence, the season in which the project was undertaken coincided with one of the most devastating tornadoes ever to hit the region. During the afternoon of April 10, several teams of storm chasers outside of Seymore, Texas, observed a *wall cloud*, a large protrusion extending below the main base of a thunderstorm where tornadoes commonly develop. Howard Bluestein, now one of the leading experts on the subject of severe storms, photographed the particular wall cloud shown in Figure 6–1. Soon afterward, a tornado

Figure 6–1 Wall cloud associated with severe thunderstorm near Seymore, Texas.

CHAPTER OUTLINE

occurred that the storm chasers were able to observe from its inception to its demise, about 15 minutes later. But the main event was soon to follow. Off to the northeast, another tornado had just formed. Though the observation teams tried to keep pace, the storm outran them as it tracked toward Wichita Falls. When the chasers arrived in town, they witnessed the horrible destruction that had occurred minutes before: 3000 homes were destroyed and 42 people killed. But the tornado was still swirling in full force and moving northeast. As it continued out of Wichita Falls, the chasers saw the back side of the storm, which Bluestein describes as looking like an atomic bomb explosion.

Most clouds are far less interesting than those that spawn tornadoes. Yet even clouds associated with the most violent weather result from the same processes that cause condensation and deposition in fair weather clouds. From Chapter 5 we know that condensation or deposition can occur by adding water vapor to the air; mixing warm, moist air with cold air; or lowering the air temperature to the dew point. Although the first two processes can lead to cloud formation in many situations, lowering of air temperature is the most important (especially for clouds that cause precipitation). This chapter discusses the processes and conditions associated with the formation of clouds due to upward motions. It also describes the cloud types that form as a result of those processes.

Mechanisms That Lift Air

Four mechanisms lift air so that condensation and cloud formation can occur:

1. Orographic lifting, the forcing of air above a mountain barrier
2. Frontal lifting, the displacement of one air mass over another
3. Convergence, the horizontal movement of air into an area at low levels
4. Localized convective lifting due to buoyancy

Orographic Uplift

As shown in Figure 6–2, air flowing toward a hill or mountain will be deflected around and over the barrier. The upward displacement of air that leads to adiabatic cooling is called **orographic uplift** (or the **orographic effect**). The height to

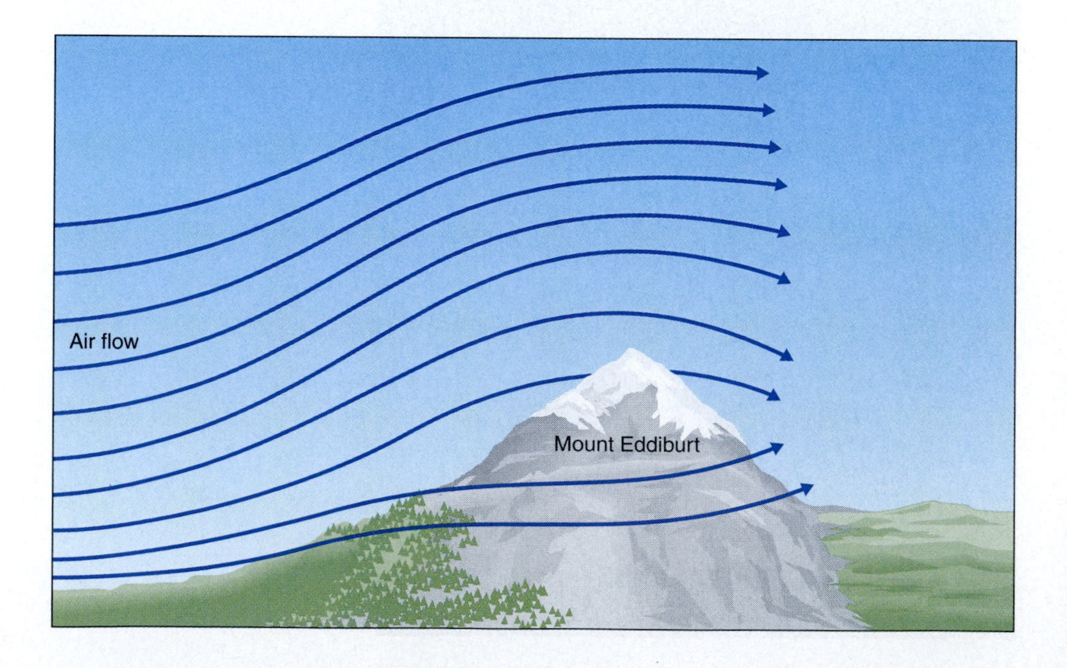

▶ Figure 6–2
Orographic uplift. When air approaches a topographic barrier, it can be lifted upward or deflected around the barrier.

(a)

(b)

▲ **Figure 6–3**
The development of cumulus clouds over the mountains east of San Diego, California. Notice the increased vertical development between (a) and (b). In (c) the cloud has decreased in depth due to precipitation. Photos were taken at about 10-minute intervals.

(c)

which these clouds can rise is not limited to the height of the hill or mountain; the tops of orographic clouds can extend many hundreds of meters higher and even into the lower stratosphere. The height of cloud tops is strongly related to characteristics of the air that vary from day to day, an issue we will describe in more detail later in this chapter. Figure 6–3 shows the development of orographic clouds. Notice that the thickness of the cloud greatly exceeds the height of the orographic barrier that helped create it.

Downwind of a mountain ridge, on its leeward side, air descends the slope and warms by compression to create a **rain shadow** effect, an area of lower precipitation. The Sierra Nevada mountain range (Figure 6–4) provides a dramatic illustration of this effect. The ridge crest of the Sierra runs north-to-south and is essentially perpendicular to the predominantly west-to-east air flow. With much of the range being higher than 3500 m (11,500 ft), precipitation on the western, windward side is greatly enhanced because of orographic lifting; in places, the mean annual precipitation exceeds 250 cm (100 in.). The eastern slope of the range is extremely steep and the valley floor is low, sometimes below sea level. Thus, the descending air on the leeward slope creates one of the strongest rain shadow effects on Earth. Interestingly, Death Valley, one of the driest places in North America, is located just to the east of the range whose windward slopes accumulate the majority of California's usable water. A comparable rain shadow effect exists in South America, where the Andes Mountains form an abrupt barrier to the westerly winds.

Frontal Lifting

Although the temperature normally changes from place to place, experience tells us that such changes are usually quite gradual. In other words, if the temperature is 10 °C (50 °F) in Toronto, Ontario, chances are that the temperature in Buffalo, New York, about 100 km (60 mi) away, won't be very much different. Sometimes, however, transition zones exist in which great temperature differences occur across relatively short distances. These transition zones, called *fronts*, are not like vertical walls separating warm and cold air but rather slope gently, as we discuss in Chapter 9.

(a)

(b)

▲ **Figure 6–4**
The Sierra Nevada forms a major barrier to winds that generally blow from the west. This promotes enhanced precipitation on the windward side (a), and a rain shadow on the leeward side (b).

Air flow along frontal boundaries results in the widespread development of clouds in either of two ways. When cold air advances toward warmer air (a situation called a **cold front**), the denser cold air displaces the lighter warm air ahead of it, as shown in Figure 6–5a. When warm air flows toward a wedge of cold air (a **warm front**), the warm air is forced upward in much the same way that the orographic effect causes air to rise above a mountain barrier (Figure 6–5b).

Convergence

Because the mass of the atmosphere is not uniformly distributed across Earth's surface, large areas of high and low surface pressure exist. These pressure differences set the air in motion in the familiar effect we call the wind. Not surprisingly, the pattern of wind that results is very much related to the pattern of pressure.

In particular, when a low-pressure cell is near the surface, winds in the lower atmosphere tend to converge on the center of the low from all directions. Horizontal movement toward a common location implies an accumulation of mass called **horizontal convergence**, or just **convergence** for short. Does convergence lead to increasing density, with imported air confined to its original altitude? No—instead, vertical motions carry away about as much mass as is carried in. Thus, in the case of low-level convergence, rising air will be the result. This will be explained in more detail later, but for now we can just make the important connection between low-level convergence and rising air with adiabatic cooling.

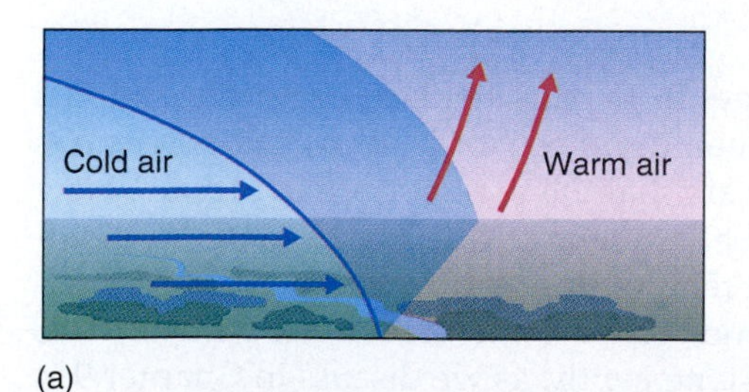

(a)

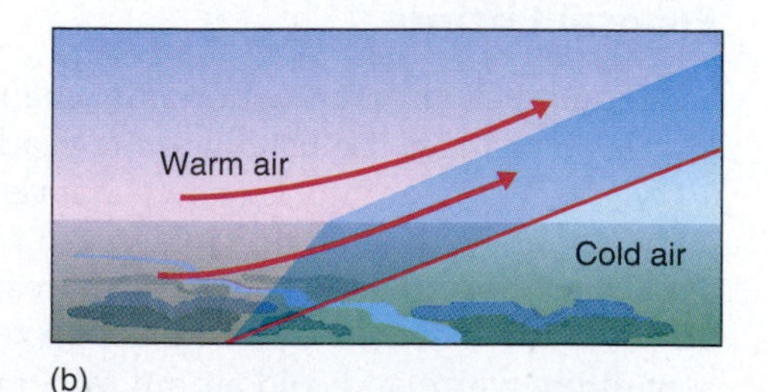

(b)

▶ **Figure 6–5**
Frontal boundaries. Cold fronts (a) cause uplift as cold air advances on warmer, less dense air. Uplift occurs along a warm front (b) when warm air overruns the cold wedge of air ahead of it.

Localized Convection

We saw in Chapter 3 that free convection is lifting that results from heating the air near the surface. It is often accompanied by updrafts strong enough to form clouds and precipitation. During the warm season, heating of Earth's surface can produce free convection over a fairly limited area and create the brief afternoon thunderstorms that disrupt summer picnics. In Canada and the United States east of the Rocky Mountains, high moisture content of the air sometimes allows for tall clouds with bases at relatively low altitudes. Such conditions favor vigorous precipitation over small regions (free convection by its nature does not create updrafts more than several tens of meters in diameter). Even in the deserts of the Southwest, which are usually low in water vapor, intense heating can lead to localized convection intense enough to cause thundershowers.

Global Clouds, 1999

Free convection arises from buoyancy, the tendency for a lighter fluid to float upward through a denser one. By itself, buoyancy can initiate uplift. But buoyancy can also speed or slow the uplift begun by the orographic effect, frontal lifting, and convergence. As we'll see next, meteorology uses the concept of static stability to summarize the effect of buoyancy on uplift.

Static Stability and the Environmental Lapse Rate

Atmospheric Stability
Section 5.1.1

Sometimes the atmosphere is easily displaced and an air parcel given an initial boost upward continues to rise, even after the original lifting process ceases. At other times, the atmosphere resists such lifting. The air's susceptibility to uplift is called its **static stability**. *Statically unstable* air becomes buoyant when lifted and continues to rise if given an initial upward push; statically stable air resists upward displacement and sinks back to its original level when the lifting mechanism ceases. *Statically neutral air* neither rises on its own following an initial lift nor sinks back to its original level; it simply comes to rest at the height to which it was displaced.

Static stability is closely related to buoyancy. When a parcel of air is less dense than the air around it, it has a positive buoyancy and floats upward. (In fact, buoyant parcels of air not only move upward, they increase in speed as they do so—even to the point of creating violent updrafts.) Air that is denser than its surroundings sinks if not subjected to continued lifting forces. In turn, density differences between a parcel and its surroundings are determined by their temperatures. If the parcel is warmer than the surrounding air, it will be less dense and experience a lifting force. If it is colder, it will be more dense and have "negative" buoyancy.

If a rising parcel cools at a rate that makes it colder than the surrounding air, it will become relatively dense. This will tend to suppress uplift. If the lifted air cools more slowly than its surroundings, it will become warm relative to the surroundings and have positive buoyancy. This creates a tendency for a parcel to rise on its own, even in the absence of other lifting mechanisms. Thus, the buoyancy of a rising air parcel depends on its rate of cooling relative to the surrounding air. Temperatures in the parcel are governed by either the dry or saturated adiabatic lapse rate, whereas the surroundings are governed by the **environmental lapse rate** (ELR). (The adiabatic and environmental lapse rates were explained in Chapter 5.)

Consider a parcel of air near the surface that is lifted through the surrounding air. The lifted air cools at one of the adiabatic lapse rates, and the surrounding air maintains its original temperature profile. The relative density of the rising parcel thus depends on two conditions: whether or not the parcel is saturated (which determines the applicable adiabatic lapse rate) and the ELR. These factors combine to produce different types of air with regard to their static stability. These are *absolutely unstable*, *absolutely stable*, and *conditionally unstable*.

Absolutely Unstable Air

Figure 6–6a illustrates what happens when a parcel of unsaturated air is lifted and the ELR is greater than the **dry adiabatic lapse rate** (DALR). In other words, in Figure 6–6a, the rising air is cooling more slowly than its surroundings.

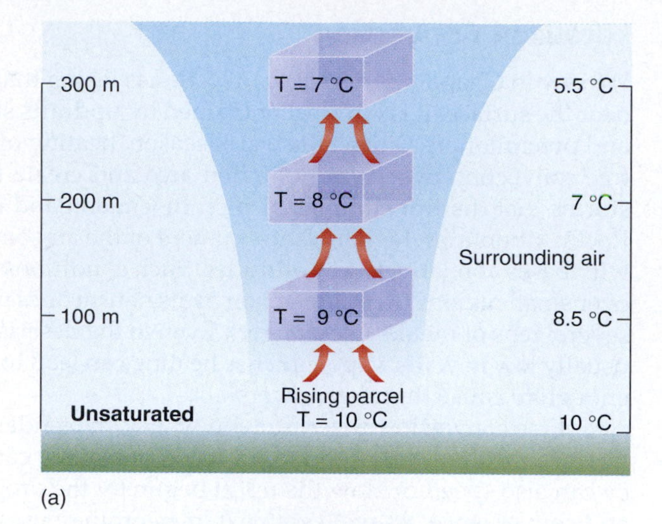

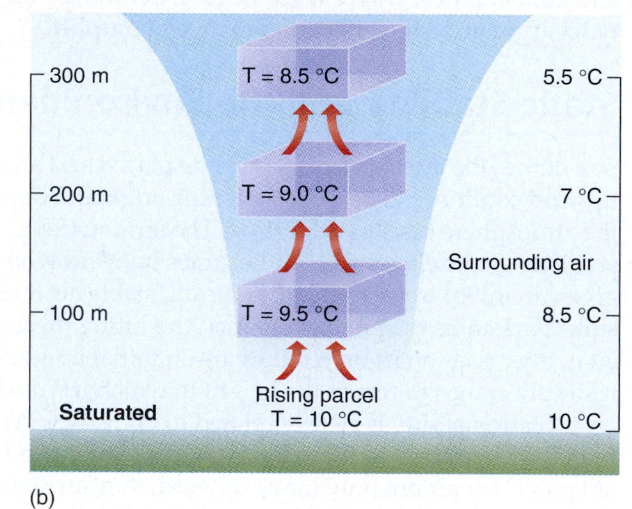

▶ **Figure 6–6**
Absolutely unstable air. In both examples, the ELR of 1.5 °C/100 m exceeds the DALR. Air forced to rise becomes warmer and more buoyant than the surrounding air, whether it is unsaturated (a) or saturated (b).

Let's suppose the air temperature at the surface is 10 °C and the ELR is 1.5 °C/100 m, which means that the air is cooling at the rate of 1.5 °C for every 100 m of height. As our parcel rises, it cools at the DALR (recall that this is 1 °C/100 m). When it is lifted to the 100 m level, the rising parcel cools to 9 °C—half a degree warmer than the surrounding air. This is so because the parcel's temperature has cooled 1 °C during its rise of 100 m and the surrounding atmosphere at that altitude is only 8.5 °C (10°C − 1.5°C = 8.5°C). If the parcel is lifted to the 200 m level, its temperature becomes 8 °C, or one degree Celsius warmer than the surrounding air. Thus, the lifted parcel is becoming progressively warmer and more buoyant than the surrounding air.

The air in this instance is said to be *absolutely unstable* because once a parcel within is lifted, it continues to move upward. Not only does the parcel rise but it does so at an ever-increasing speed. This occurs because the temperature difference between it and the surrounding air continually increases, leading to greater buoyancy, and also because it gathers momentum as it rises.

Figure 6–6b provides a second example of absolutely unstable air. In this case, the ELR is still 1.5 °C/100 m, but the air is now *saturated*. The lifted parcel of air therefore cools more slowly, at the **saturated adiabatic lapse rate** (SALR), and will be warmer than before. Thus, with condensation, the temperature difference between the warm parcel and colder surrounding air is greater, giving rise to a stronger buoyant force. We conclude that the air is again unstable, even more so than in the previous example.

The important conclusion from these two examples is that *whenever the environmental lapse rate exceeds the dry adiabatic lapse rate, the air is absolutely unstable and*

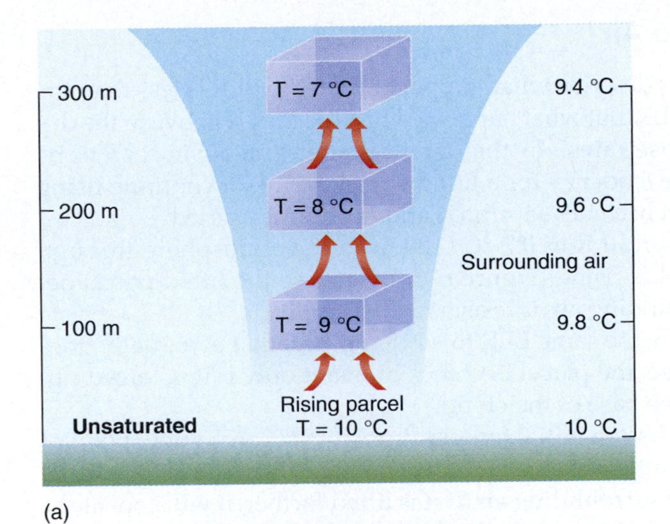

(a)

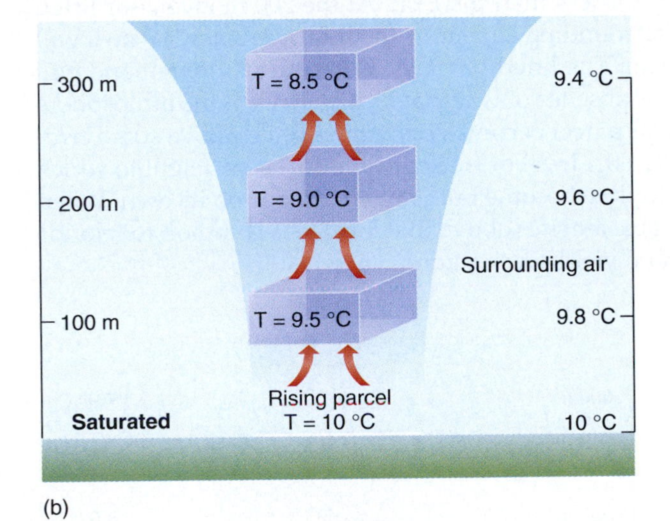

(b)

◀ **Figure 6–7** Absolutely stable air. In both examples, the ELR of 0.2 °C/100 m is less than the SALR. Air forced to rise becomes colder and less buoyant than the surrounding air, whether it is unsaturated (a) or saturated (b).

a parcel contained within it will continue to rise once lifted, regardless of whether or not it is saturated. (Of course, upward motions cannot continue forever. But for now, we will put aside the issue of how far unstable air can rise so that we can focus on the main concept associated with what happens to air in a zone of instability.)

Absolutely Stable Air

Figures 6–7a and 6–7b illustrate what happens when the ELR, in this case 0.2 °C/100 m, is less than the saturated adiabatic lapse rate. As we see in (a), when a parcel of unsaturated air rises, its temperature drops more rapidly than the temperature of the air around it, making the parcel relatively heavier and less buoyant. Because of its negative buoyancy, the lifted air will sink back to its initial level if the lifting mechanism stops. Such air is *absolutely stable*. The same principle applies in part (b) of the figure. The saturated parcel cools at the saturated adiabatic lapse rate and becomes colder than the air around it. Like the unsaturated parcel in (a), it has a tendency to sink back to its original position.

From these two examples we can conclude that *whenever the environmental lapse rate is less than the saturated adiabatic lapse rate, the air will be absolutely stable and will resist lifting, regardless of whether or not it is saturated.* It is important to note that it is possible for the ELR to be such that the temperature does not change at all with height, or even for the temperature to increase with height, as will be discussed later in this chapter. Though we will not provide examples here, the logic described in this section applies to these situations as well. In other words, if the ELR = 0°C/100 m or if the temperature increases with height (a negative ELR), the air will be absolutely stable.

Conditionally Unstable Air

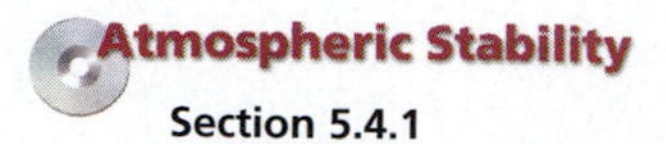

Section 5.4.1

The four preceding examples describe what happens when the ELR is less than the SALR or greater than the DALR. But what happens when the ELR is between the dry and saturated adiabatic lapse rates? In this environment, the air is said to be *conditionally unstable*, and the tendency for a lifted parcel to sink or continue rising depends on whether or not it becomes saturated and how far it is lifted.

Let's suppose there is an ELR of 0.7 °C/100 m in the atmosphere through which an unsaturated parcel is rising (Figure 6–8a). Because the lifted parcel becomes colder than the surrounding air, it resists further uplift.

In Figure 6–8b, we apply the same ELR to a lifted parcel that eventually does become saturated. In this case, the parcel becomes buoyant only if it is forced upward some distance above the base of the cloud.

In this example, the air has an initial temperature of 10 °C and a dew point of 9.2 °C. It cools at the DALR until it reaches saturation at the 100 m level. The lifted parcel is now colder than the surrounding air. If it is lifted farther, it will cool at the saturated adiabatic rate, which is less than the ELR. At the 200 m level, our lifted parcel is still colder than the surrounding air, but by the time it reaches 300 m, it will be warmer than the ambient air. The lifted parcel thus becomes buoyant and will now rise on its own even in the absence of external lifting. Thus, if the atmosphere is conditionally unstable, an air parcel becomes buoyant if lifted above some critical altitude. That altitude, called the **level of free convection**, is the height to which a parcel of air must be lifted for it to become buoyant and to rise on its own. When a parcel of conditionally unstable air rises above that level, it is common for clouds to rapidly increase in depth and yield precipitation.

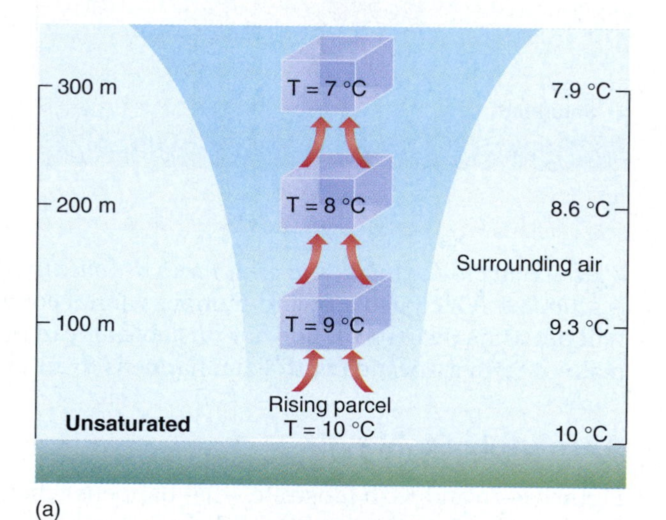

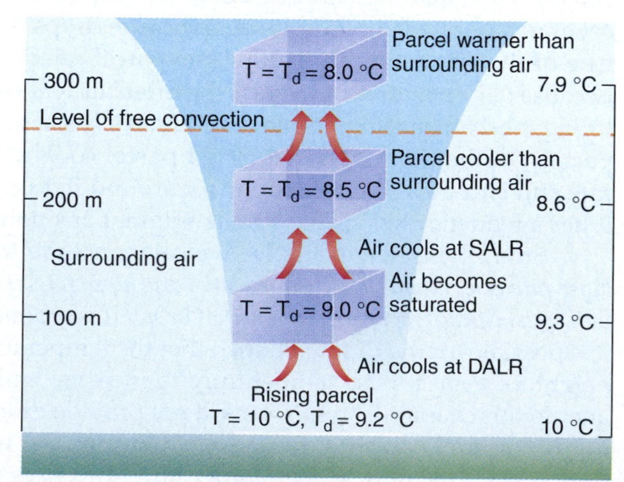

▶ **Figure 6–8**
The atmosphere is conditionally unstable when the ELR is between the dry and saturated adiabatic lapse rates. In (a) the ELR is 0.7 °C/100 m and the air is unsaturated. As a parcel of air is lifted, its temperature is less than that of the surrounding air, so it has negative buoyancy. In (b) a parcel starts off unsaturated but cools to the LCL, where it is cooler than the surrounding air. Further lifting cools the parcel at the SALR. At the 200 m level, it is still cooler than the surrounding air, but if taken to 300 m, it is warmer and buoyant. Thus, if the air is lifted sufficiently, the parcel continues to rise by virtue of its buoyancy.

Factors Influencing the Environmental Lapse Rate

ELRs are highly variable in space and time. Just as the surface air temperature at a location is subject to change, so is the vertical temperature profile. The following three factors can bring about changes in the ELR.

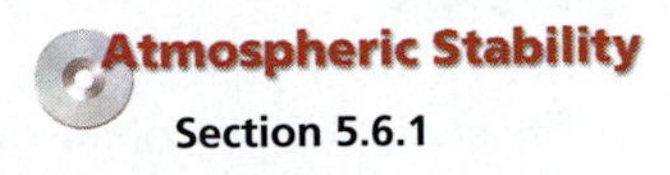

Section 5.6.1

Heating or Cooling of the Lower Atmosphere

During the daytime, solar radiation heats Earth's surface, which in turn warms the atmosphere in contact with it. Because it is heated more rapidly than the air aloft, the lower atmosphere typically has a steeper ELR during the midday, as shown in Figure 6–9. The initial temperature profile indicated by the solid line changes through the course of the day, and steeper profiles (shown by the dashed lines) can result for several hundred meters above the surface. The effect of solar radiation on the lapse rate is (of course) greatest on clear, sunny days, especially above unvegetated land surfaces, where abundant solar radiation is available and little energy is expended on evaporation.

Cooling of the surface, such as occurs at night, chills the lower atmosphere and decreases the ELR. With sufficient cooling, the air near the surface can become colder than the air above and create a situation in which air temperature *increases* with height. (This is known as a temperature inversion, a condition of extremely stable air that we discuss later in the chapter.)

Advection of Cold and Warm Air at Different Levels

Temperature profiles can be influenced by differences between wind direction at low and high levels. In Figure 6–10a, for example, low- and high-level winds are both from the west, where surface and upper-level temperatures are 10 °C and 9.5 °C, respectively. The lapse rate is thus 0.5 °C/100 m. In Figure 6–10b, surface winds are unchanged, but upper-level flow now comes from the colder northeast, so that temperatures aloft are lower, only 9.0 °C. Cold air has been advected above the surface, resulting in a steeper lapse rate. Warm air can be similarly advected, if winds happen to blow from warmer toward colder locations.

Of course, the advection of warm or cold air can occur at any level. For example, if cold air is advected at low levels, the lapse rate declines, producing greater stability. Moreover, advection is not confined to a single altitude; we should not think of a moving slab of air at some height unconnected to the rest of the atmosphere.

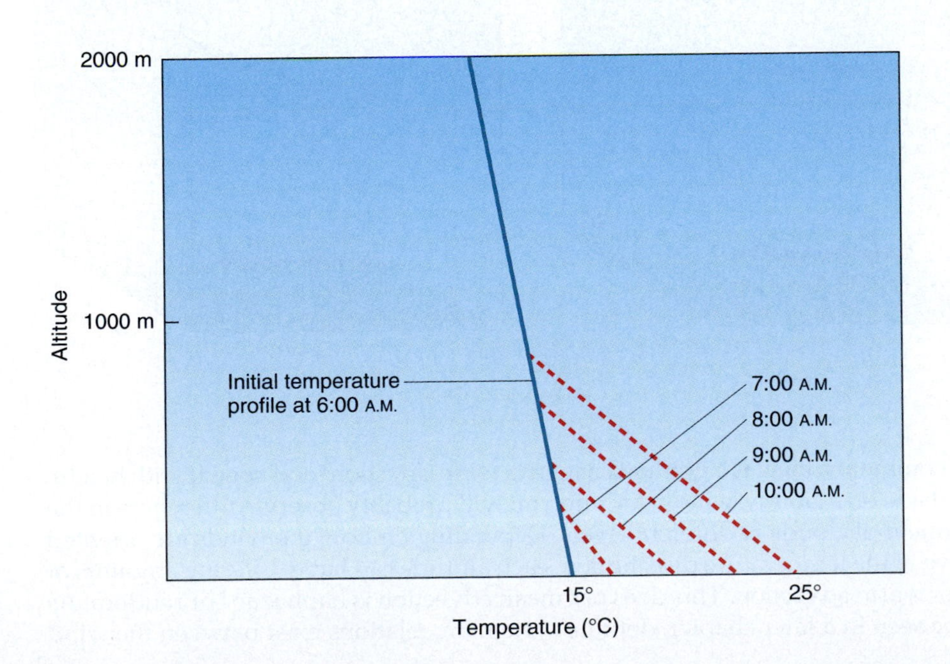

Figure 6–9 The ELR can be changed by heating of the surface, as shown by the sequential changes in this temperature profile.

6–1 Forecasting

Determining Stability from Thermodynamic Diagrams

The static stability of air can be determined numerically by comparing the environmental lapse rate to the saturated and dry adiabatic lapse rates. Thermodynamic diagrams can also be very useful in this regard. Figure 1 provides a simple schematic showing how this is done. The figure compares three hypothetical temperature profiles against the SALR and the DALR on a portion of a simplified thermodynamic diagram. The lines labeled as *moist adiabats* and *dry adiabats* plot the change in temperature a saturated or unsaturated parcel of air would experience if it were lifted or lowered. The three hypothetical temperature profiles, labeled 1, 2, and 3, illustrate examples in which the air is absolutely unstable (the temperature decreases more rapidly than does the DALR or the SALR), conditionally unstable (ELR is between the DALR and SALR), and absolutely stable (ELR < DALR or SALR), respectively.

In the real world, the ELR varies from the surface upward. For example, at one level the air might be absolutely stable, whereas at another level it might be conditionally or absolutely unstable. Thermodynamic diagrams allow the forecaster to observe the resultant changes in stability at different levels visually, rather than by having to compute the ELR repeatedly for comparison to the adiabatic lapse rates. Figure 2 shows a temperature and dew point sounding plotted on a complete thermodynamic diagram that includes dry and moist adiabats. The dry adiabats are shown in green and slope steeply to the left as they extend upward. The dashed blue lines are the moist adiabats. From the surface to 850 mb, the temperature profile is parallel to that of the adjacent dry adiabats, indicating the layer is nearly statically unstable. Above that is a shallow layer, which is statically stable. From 800 mb to about 650 mb, the air is conditionally unstable. And just above that, there is a very shallow inversion.

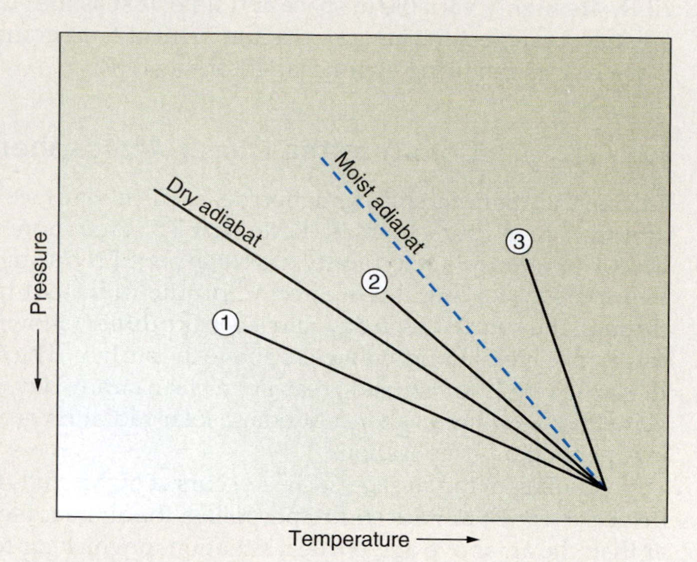

▲ **Figure 1**
Stability can be determined by comparing temperature profiles with the slope of the dry and wet adiabats. In (a) Profile 1 is absolutely unstable, Profile 2 conditionally unstable, and Profile 3 absolutely stable.

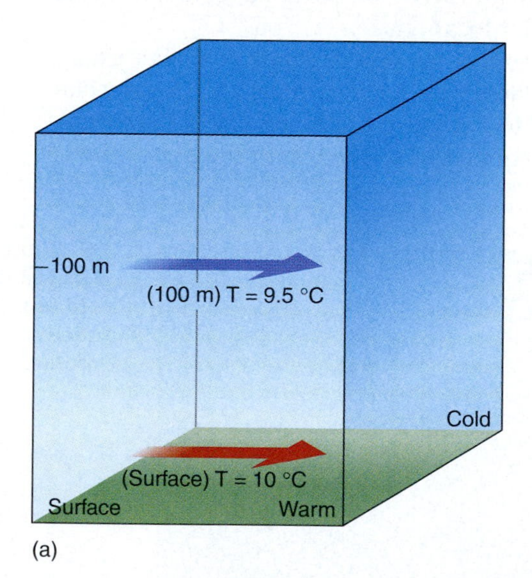

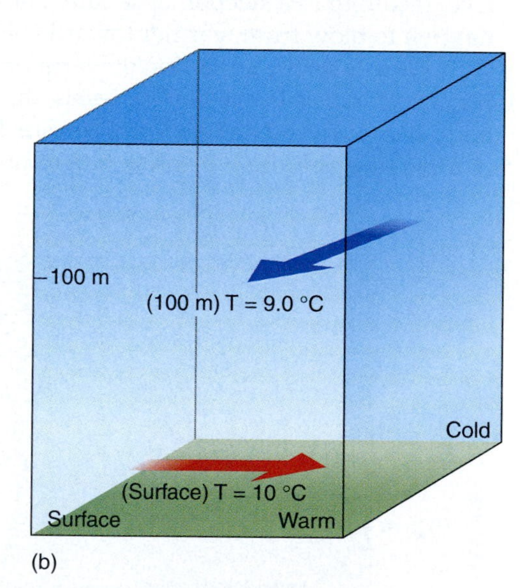

▶ **Figure 6–10**
The ELR can be changed by the advection of air with a different temperature aloft. In (a), the winds at the surface and the 100-m level bring in air with temperatures of 10 °C and 9.5 °C, respectively, yielding an ELR of 0.5 °C/100 m. In (b), the surface winds still bring in air with a temperature of 10 °C. But the wind direction at the 100-m level has shifted to northeasterly, and the advected air has a temperature of 9.0 °C. This yields a steeper ELR of 1.0 °C/100 m.

Most commonly, there is a gradual change in wind direction (and speed) with height. Go outside on a cloudy, windy day and you will probably observe differences in the movement of clouds at different levels. Depending on how the winds are oriented relative to the temperature distribution, each altitude can have differing amounts of cold or warm advection. This does not mean advection is haphazard or random. As will be seen in a later chapter, definite systematic relations exist between the wind

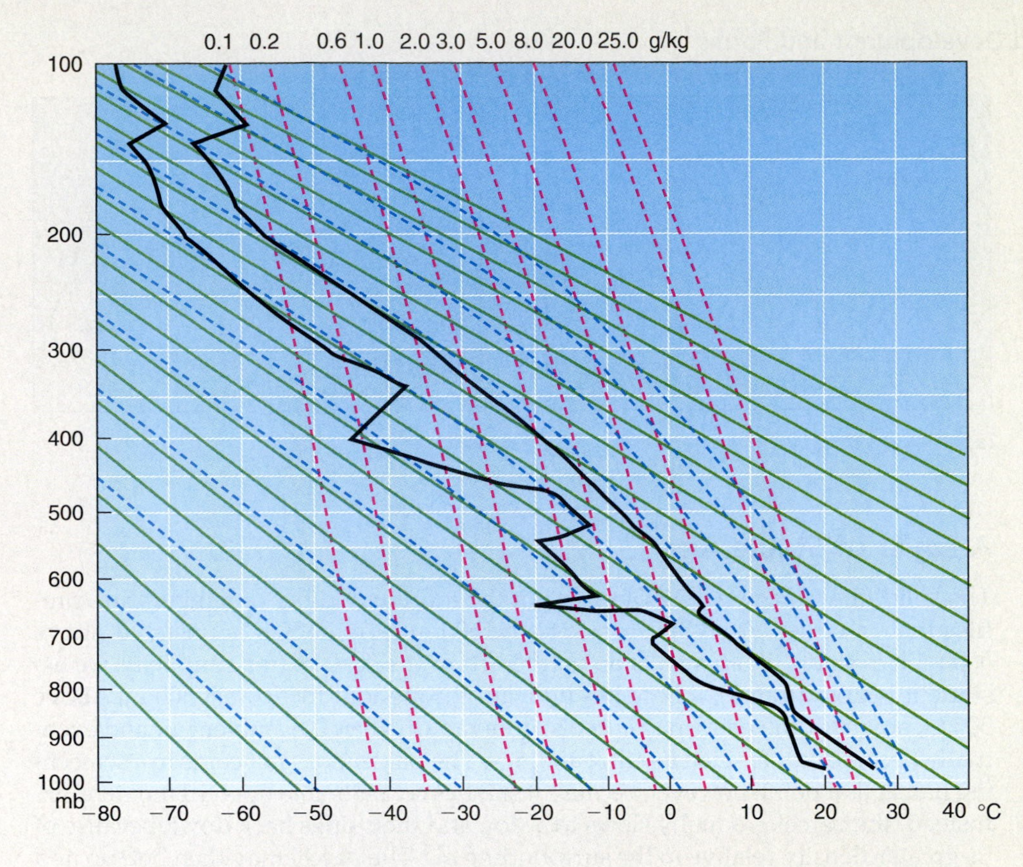

Figure 2
A complete thermodynamic diagram plotting temperature and dew point profiles for Detroit, Michigan, on June 27, 2002.

The changes in stability at different levels may appear to make the use of the thermodynamic diagram a daunting task for the forecaster, but the situation has several remedies. Professional meteorologists have a number of numerical indexes calculated for every sounding. The indexes are based on temperature-dew point combinations at varying levels and are computed automatically when the soundings are plotted. Forecasters refer to these values for initial guidance in their interpretations of how stability conditions will influence the likelihood of cloud cover, precipitation, or violent weather.

and pressure fields. The point, rather, is that the presence of advection is variable, both from day to day and from altitude to altitude within the column, and the effects on atmospheric stability are likewise variable.

Advection of an Air Mass with a Different ELR

The atmosphere has a strong tendency to be arranged into large areas distinguished by small horizontal differences in temperature and humidity. These so-called air masses maintain their temperature and moisture characteristics as they move from one place to another. When an air mass migrates to a particular area and replaces another, the initial ELR at that location gives way to that of the new air mass. In Figure 6–11a, for example, Location *A* has a steeper ELR than does *B*. As the air mass over Location *A* moves over *B*, it brings to that location the new temperature profile.

Limitations on the Lifting of Unstable Air

We now know that once unstable air is lifted, it continues to rise and even increases in speed. This brings up an important question: What causes unstable air to quit rising? If a rising parcel continued to rise forever it would eventually escape Earth, never to be seen again. Given enough time, the continued loss of the unstable air parcels would entirely deplete the atmosphere. Of course, our atmosphere is not exploding out to space, so something must occur to eventually suppress uplift.

The most important braking mechanism for rising parcels is their ascent into a layer of stable air. A second reason is called *entrainment*.

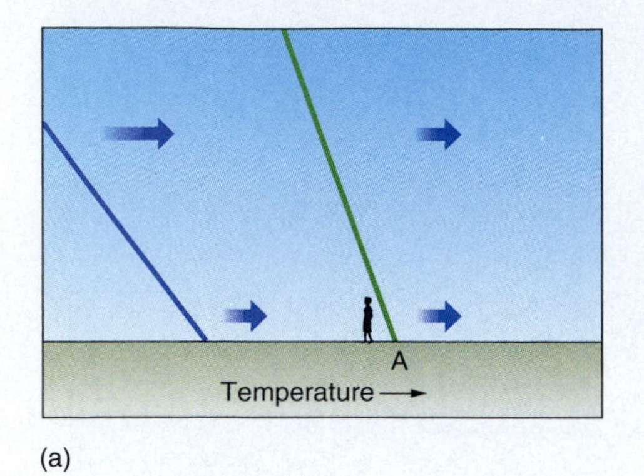

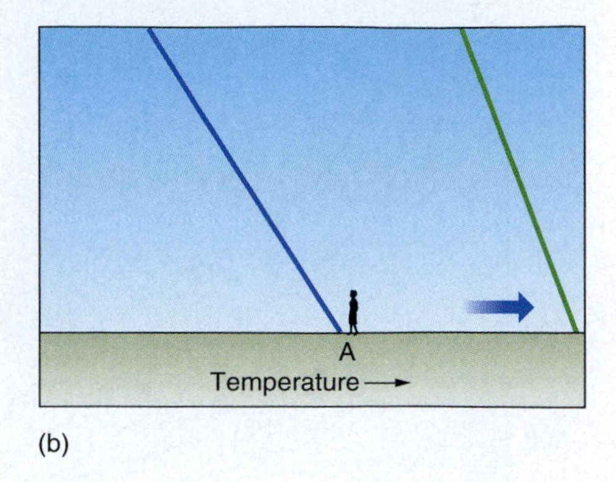

▶ **Figure 6–11**
The ELR changes when a new air mass replaces one that has a different lapse rate. The green line represents the temperature profile originally encountered at position *A*. The blue line depicts the temperature profile replacing the earlier profile.

A Layer of Stable Air

The solid line in Figure 6–12 plots one of the infinite number of temperature profiles that can exist. From the surface to the 500 m level, the air is unstable; above 500 m, it is stable. If a parcel of air is lifted from the surface (for simplicity, we assume it is unsaturated), it becomes buoyant throughout the lowest 500 m. Above 500 m, however, the rising parcel cools more rapidly than the ambient air and eventually becomes cooler than its surroundings. The parcel does not come to a screeching halt at that point, however, because it still has considerable upward momentum. Instead, the parcel gradually slows to a stop and then sinks back down because of its greater density relative to the surrounding air. The parcel may then bob up and down before coming to rest at some equilibrium level.

Is a stable layer always present at some altitude to contain uplift? The answer is yes because, if nothing else, a rising parcel will eventually encounter the stratosphere, which is extremely stable. As a result, even the most rapidly ascending parcels of air must slow down and reach an equilibrium level above the

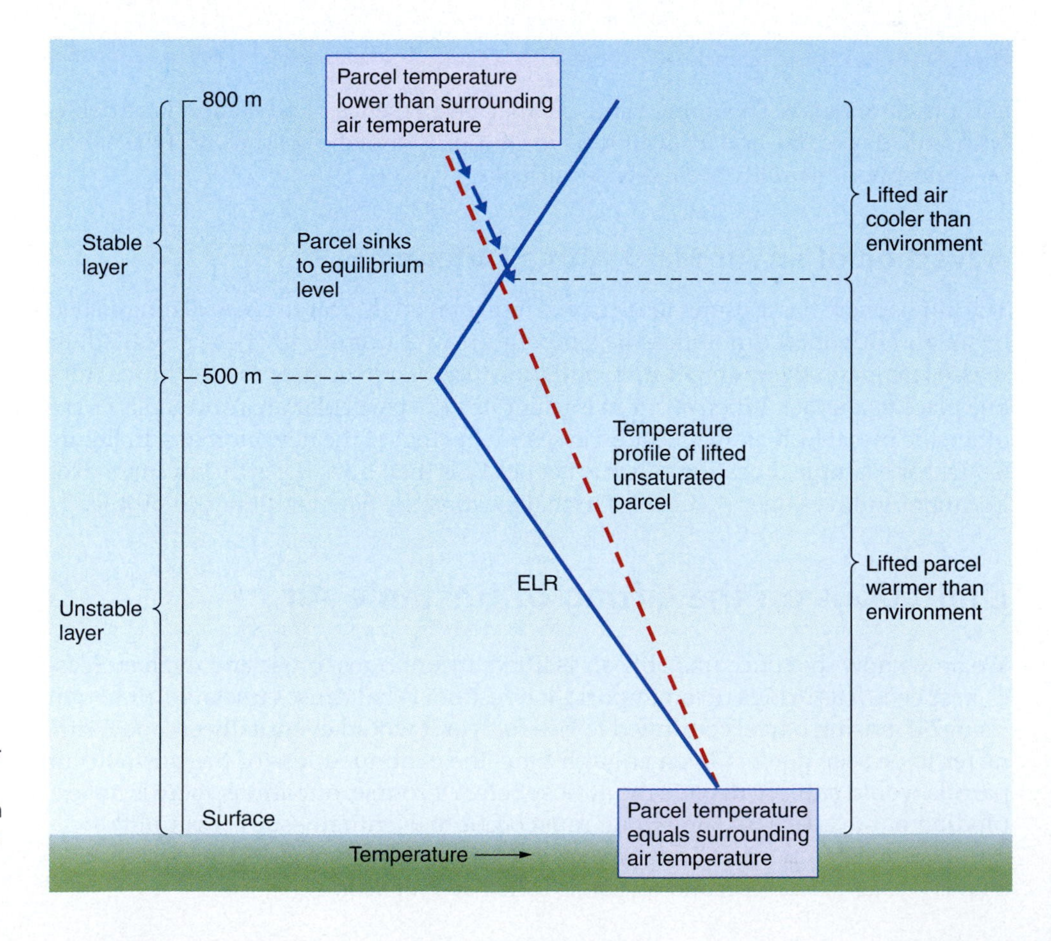

▶ **Figure 6–12**
Air that is unstable at one level may be stable aloft. The solid line depicts a temperature profile that is unstable in the lowest 500 m but capped by an inversion. An unsaturated air parcel displaced upward would cool by the DALR (dashed line), making it initially warm and buoyant relative to the surrounding level. Some distance after penetrating the inversion layer, the rising air is no longer warmer than the surrounding air, and further lifting is suppressed. The parcel continues upward for some distance, however, because it has considerable upward momentum. As it does so, it cools more rapidly than the surrounding air and becomes relatively dense. After coming to a stop, the heavy parcel of air sinks back down and eventually comes to rest at some equilibrium level.

tropopause. Although severe thunderstorms can have updrafts of more than 200 km per hour (120 mph), uplift seldom extends above the lowest kilometer or so of the stratosphere.

Entrainment

When we talk about a rising air parcel, we mean a small mass that undergoes motions distinct from the surrounding atmosphere. To some extent, we can imagine such a parcel as being like the air contained within a balloon. But unlike a balloon, which has a rubber film to isolate the air within, an air parcel has no barrier to prevent it from mixing with its surroundings. In fact, as air rises, considerable turbulence is generated, which causes ambient air to be drawn into the parcel. This process, called **entrainment**, is especially important along the edges of growing clouds. Entrainment suppresses the growth of clouds because it introduces unsaturated air into their margins and thus causes some of the liquid droplets to evaporate. The evaporation consumes latent heat and thereby cools the margin of the cloud, making it less buoyant.

Extremely Stable Air: Inversions

Atmospheric Stability
Section 5.6.3

So far, we have been concerned with the mechanisms that cause air to rise, and the influence of instability on the effectiveness of those mechanisms. Now it is useful to examine the most extreme forms of stable air, those associated with inversions.

Although on average the temperature in the troposphere decreases with elevation, situations do arise in which the inverse of the normal profile exists, and the temperature *increases* with altitude. Layers of the atmosphere in which this situation exists are called **inversions**. Air parcels rising through inversions encounter ever-warmer surrounding air and therefore have strong negative buoyancy. Inversions are thus extremely stable and resist vertical mixing.

Several different processes can cause different types of inversions to develop. One of the most common is a *radiation inversion*, which results from cooling of the surface. On cloud-free nights with little or no wind, longwave radiation emitted by the surface easily escapes to space. This lowers the ground temperature, which in turn chills the air immediately in contact with it. Because the lower air chills more rapidly than that farther from the surface, an inversion develops at ground level.

If cooling is sufficient to lower the temperature to the dew point, a radiation fog forms. Inversions are associated with all radiation fogs, but if the cooling does not lower the temperature to the dew point, a radiation inversion can exist without the appearance of fog.

Radiation inversions occur throughout the world. Though they are usually restricted to fairly shallow depths above the surface, they can have important ramifications for agriculture and other activities. We discuss two examples in *Box 6–2, Special Interest: Radiation Inversions and Human Activities.*

Factors other than diabatic cooling of the surface can also produce inversions. When a cold or warm front is present, for example, a transition zone separates warm and cold air masses. The boundary is not horizontal but rather forms a wedge of cold air that underlies warmer air, as shown in Figure 6–13. The horizontal extent of these *frontal inversions* can be up to several hundred kilometers wide, and the height of the inversion increases with distance from the leading edge of the front. Rain falling into a very cold surface layer can freeze before reaching the ground (resulting in sleet), or on the ground, resulting in the much more dangerous freezing rain.

More extensive and meteorologically important than frontal inversions are the *subsidence inversions* that result from sinking (or *subsiding*) air. Recall that a layer of air is compressed and warmed during descent. As it is compressed, its thickness decreases, which means that the top of the layer descends a greater distance than does the bottom of the layer. The longer descent leads to greater temperature increases at the top of the layer than the bottom, and thus to the formation of an inversion.

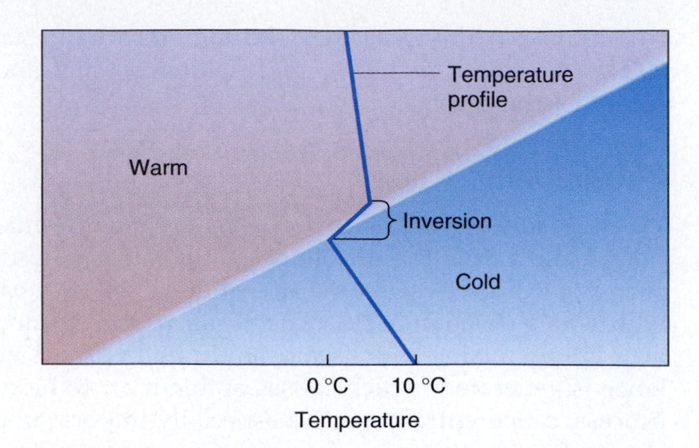

▶ **Figure 6–13**
Frontal inversions. The temperature is plotted with the solid blue line.

Subsidence is common along the eastern margins of large areas of high pressure and downwind of major mountain ranges. Because the subsiding air does not descend all the way to the surface, the base of the inversion can be several hundred meters above the surface. As a result, subsidence inversions are clearly distinguishable from radiation inversions, the bases of which are at ground level. A typical temperature profile for a subsidence inversion is shown in Figure 6–14. Notice that the top of the inversion layer is more than 10 °C (18 °F) warmer than its base. This is a large difference, but not at all uncommon.

A spectacular example of a subsidence inversion can be found every year between the months of April and September in the United States. A large high-pressure system called the Hawaiian High often forms over the middle latitudes of the North Pacific. As upper-level air rotates out of the high-pressure center, subsidence occurs over coastal southern California and forms an inversion. This provides a "cap" for the vertical dispersal of pollutants that helps to give Los Angeles some of the most polluted air in North America.

Cloud Types

Clouds can assume a variety of shapes and sizes and can occur near the surface or at high altitudes. Most cloud types occur in the troposphere, but some appear in the stratosphere and even in the mesosphere. Clouds can contain liquid droplets, ice crystals, or a mixture of the two. They can be thick or thin and have high or low liquid water or ice contents. It is therefore easy to see why meteorologists would want a classification scheme to distinguish the many types of clouds from one another.

▶ **Figure 6–14**
Subsidence inversions occur when air descends toward but not all the way to the surface.

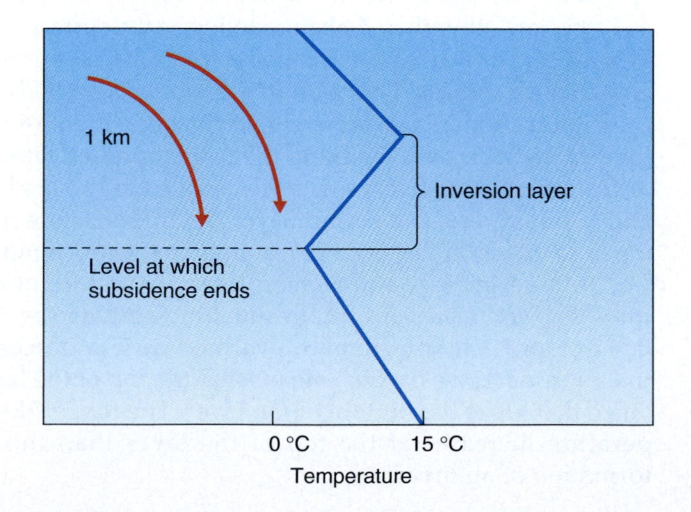

6–2 Special Interest

Radiation Inversions and Human Activities

On clear, still nights, longwave radiation emitted from the surface easily passes through the atmosphere and out to space. This causes the surface to cool rapidly and chill the overlying air diabatically. Of course, air higher up will not be subjected to the same cooling as air in contact with the surface, and temperature differences of several degrees Celsius can be observed over just a few meters in height. Thus, subfreezing temperatures can exist near the ground, while just a short distance above, temperatures may be safely above the freezing level. This has some important effects on agriculture in the southern part of the United States.

When temperatures drop to the freezing point, winter crops are vulnerable to frost damage. To offset this problem, growers often set large "wind machines" atop masts, as shown in Figure 1a. When temperatures near the ground fall dangerously low, the machines are turned on to force the warm air in the upper part of the inversion down toward the surface. Thus, the potentially damaging air near the surface is replaced as warmer air circulates downward.

At even lower temperatures, growers may also activate *smudge pots* that burn heating oil, as shown in Figure 1b. Although the emission of longwave radiation by the smudge pots helps somewhat to protect the crops, their more important effect is to produce free convection. Like the wind machines, the smudge pots produce a continual mixing of air between low and higher levels, so that surface air stays above the freezing point. Yet another tactic is to spray citrus crops with water. As water freezes, latent heat is released, which keeps the fruit within a few degrees of freezing, warm enough to prevent severe damage. Of course, on very cold nights inversions are so deep and cold that none of these remedies is effective. In such a case, frost damage can ruin an entire crop.

Problems associated with radiative cooling and the resultant inversions are not restricted to agriculture. The strong stability of an inversion can suppress the vertical motions that dilute the concentration of pollutants near the surface. As we will see in Chapter 14, radiation inversions are by no means the only kind that can create air quality problems. Nonetheless, it is not uncommon for urban dwellers to notice a low-lying layer of sooty haze near the surface on cold, clear mornings when radiation inversions are most likely to form.

(a)

(b)

▲ **Figure 1**
Agricultural wind machines (a) blow air downward during times of potential frost damage to force warm air in the upper part of a radiation inversion toward the crops. During times of greater stress, smudge pots (b) are also employed.

The first widely accepted system for cloud classification was devised by English naturalist Luke Howard in 1803. It divided clouds into four basic categories:

1. **Cirrus**—thin, wispy clouds of ice
2. **Stratus**—layered clouds
3. **Cumulus**—clouds having vertical development
4. **Nimbus**—rain-producing clouds

Our current classification scheme is a modified version of Howard's typing that retains his four categories and also allows new combinations (for instance, *cirrostratus* clouds have the characteristics of cirrus clouds and stratus clouds). The ten principal types of clouds that result are then grouped according to their height and form:

1. **High clouds**—cirrus, cirrostratus, and cirrocumulus
2. **Middle clouds**—altostratus and altocumulus
3. **Low clouds**—stratus, stratocumulus, and nimbostratus
4. **Clouds with extensive vertical development**—cumulus and cumulonimbus

The ten principal cloud types based on this scheme are outlined in Table 6–1 and Figure 6–15.

Table 6–1 • Ten Principal Cloud Types

Cloud type	Figure
High Clouds (heights greater than 6000 m, or 19,000 ft)	
Cirrus (Ci)	(Figure 6–16)
Cirrostratus (Cs)	(Figure 6–19)
Cirrocumulus (Cc)	(Figure 6–20)
Medium Clouds (heights between 2000 m and 6000 m, or 6000 to 19,000 ft)	
Altostratus (As)	(Figure 6–21)
Altocumulus (Ac)	(Figure 6–22)
Low Clouds (below 2000 m, or 6000 ft)	
Stratus (St)	(Figure 6–23)
Nimbostratus (Ns)	(Figure 6–24)
Stratocumulus (Sc)	(Figure 6–25)
Clouds with Vertical Development (may extend through much of atmosphere)	
Cumulus (Cu)	(Figures 6–26 and 6–28)
Cumulonimbus (Cb)	(Figure 6–29)

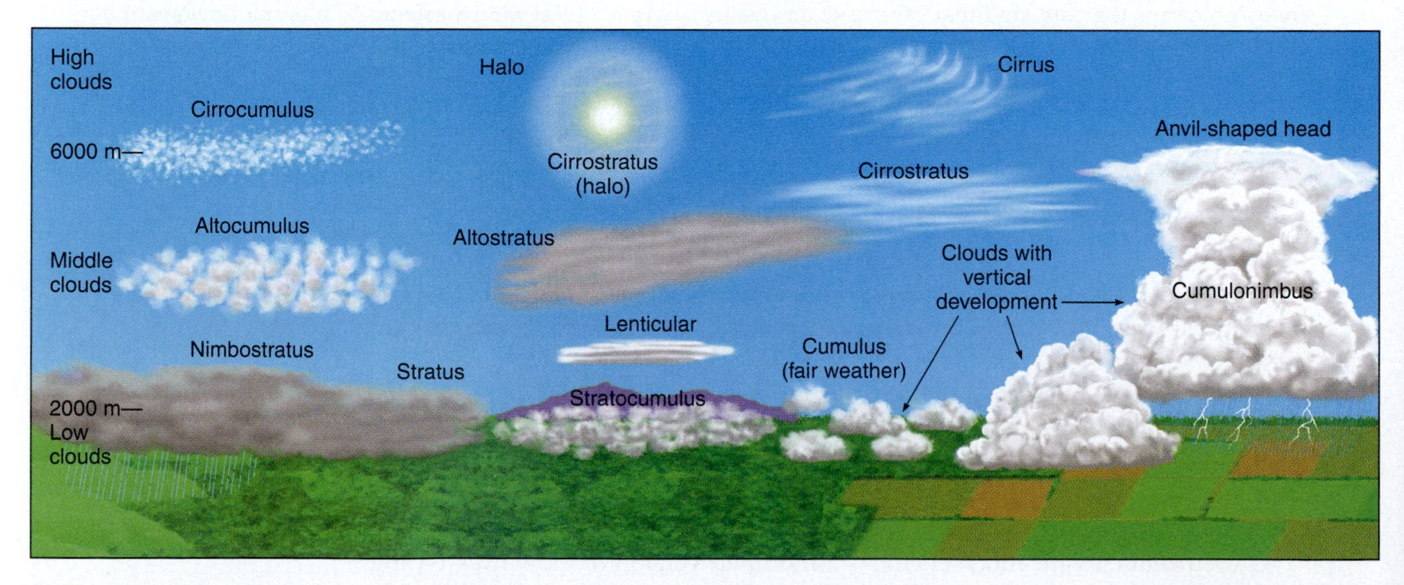

▲ **Figure 6–15**
Generalized cloud chart.

◀ **Figure 6–16**
Cirrus clouds are wispy clouds of ice crystals.

High Clouds

High clouds are generally above 6000 m (19,000 ft). They are almost universally composed of ice crystals instead of liquid droplets. Recall that within the troposphere the average temperature decreases from 15 °C (59 °F) at sea level at a rate of 6.5 °C/1000 m (3.6 °F/1000 ft). As a result, the average temperature for high clouds is usually no higher than –35 °C (–31 °F), which means that cooling to the frost point causes the formation of ice crystals instead of supercooled droplets.

Where surface temperatures are very low, clouds composed exclusively of ice can occur at altitudes as low as 3000 m (10,000 ft). Thus, the definition of a high cloud (usually assigned to those above 6000 m), is somewhat temperature dependent.

The simplest of the high clouds are cirrus (abbreviated Ci), which are wispy aggregations of ice crystals (Figure 6–16). The average thicknesses of cirrus clouds is about 1.5 km (1 mi), but they can be as thick as 8 km (5 mi). Given the very low temperatures at which they exist, they have little water vapor from which to form ice. So although they may be easily visible, the water content of these clouds is extremely low. In fact, the ice content of cirrus clouds is typically only about 0.025 grams per cubic meter, or about one-thousandth of an ounce per cubic yard.

Although the entire mass of ice contained in a cirrus cloud is small, the individual crystals can be as long as 8 mm (0.3 in.). These crystals fall at a speed of about 0.5 meters per second (about 1 mile per hour), which is sufficient for them to overcome updrafts and descend as *fall streaks* (Figure 6–17a).

(a)

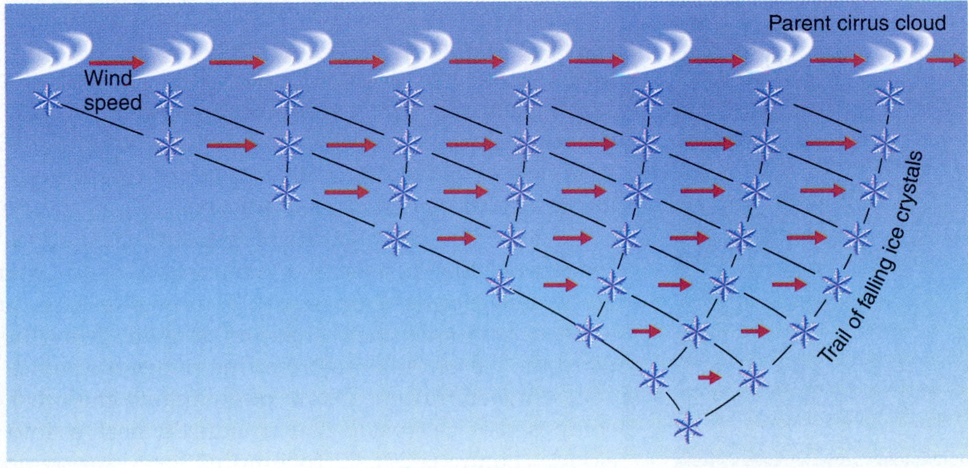

(b)

▲ **Figure 6–17**
Fall streaks (a) result from falling ice crystals. The ice crystals fall through air (b) with gradually decreasing wind speeds and become comma shaped.

▶ **Figure 6–18**
Aircraft contrails.

Fall streaks gradually sublimate (that is, they change to water vapor without first melting) as they fall through the warmer, dry air below and dissipate high above the surface. As shown in Figure 6–17b, the higher, faster-moving crystals (as indicated by the length of the arrows on the figure) are blown horizontally faster than those below, which gives the fall streaks their characteristic curved appearance. Horizontal swirls are also evident in cirrus clouds because of curving winds, leading to mares' tails.

Recent events have taught us a little more about the effect of cirrus clouds on surface temperatures. Immediately following the tragedy of September 11, 2001, the United States put a temporary ban on all commercial aircraft flights, which remained in effect for three days. Because high-altitude aircraft commonly produce **contrails** (Figure 6–18) as water vapor from engine exhaust freezes to form ice crystals, the cessation of flights resulted in a reduced amount of cirrus cloud cover over the nation. Researchers discovered a year later that the reduction in contrails led to a 1.1 °C (1.9 °F) increase in the daily temperature range (Chapter 3) across the 48 conterminous United States relative to the previous 30 years of record for September 11 though 14. This increase in daily temperature range is thought to reflect some combination of higher daytime temperatures due to reduced backscattering of radiation by high clouds and lower nighttime temperatures in response to increased outgoing longwave radiation.

▲ **Figure 6–19**
Cirrostratus clouds. Such clouds often create a halo around the Sun or Moon.

Cirrostratus (Cs) clouds (Figure 6–19), like cirrus, are composed entirely of ice but tend to be more extensive horizontally and have a lower concentration of crystals. Though cirrostratus clouds reduce the amount of solar radiation reaching the surface, enough direct sunlight penetrates to allow objects at the surface to cast shadows. Furthermore, they do not fully obscure the Moon or Sun behind them. Instead, when viewed through a layer of cirrostratus, the Moon or Sun has a whitish, milky appearance but a clear outline.

A characteristic feature of cirrostratus clouds is the *halo*, a circular arc around the Sun or Moon formed by the refraction (bending) of light as it passes through the ice crystals. Sunlight is bent by 22° more than other angles. Because the crystals are aligned randomly, those that are 22° away from the Sun or Moon direct light toward the viewer, making for a ring of brighter light.

Cirrocumulus (Cc) are often among the most beautiful of clouds. They are composed of ice crystals that arrange themselves into long rows of individual, puffy clouds (Figure 6–20). Cirrocumulus form during episodes of *wind shear*, a condition in which the wind speed and/or direction changes with height. Wind shear often occurs ahead of advancing storm systems, so cirrocumulus clouds are often a precursor to precipitation. Because of their resemblance to fish scales, cirrocumulus clouds are associated with the term "mackerel sky."

◀ **Figure 6–20**
Cirrocumulus clouds. These frequently occur arranged in rows of individual, puffy clouds.

Middle Clouds

Middle clouds occur between 2000 and 6000 m (6500 and 19,000 ft) above the surface and are usually composed of liquid droplets. The two major categories in this group are both prefixed by *alto*, which means "middle."

Altostratus (As) clouds (Figure 6–21) are the middle-level counterparts to cirrostratus. They differ from cirrostratus in that they are more extensive and composed primarily of liquid water. Altostratus scatter a large proportion of incoming sunlight back to space, thereby reducing the amount of sunlight that reaches the surface. The insolation that does make its way to the surface consists primarily or exclusively as diffuse radiation, so one way to distinguish the presence of altostratus from cirrostratus is the absence of shadows. Furthermore, when viewing the Sun or Moon behind altostratus, one sees a bright spot behind the clouds instead of a halo.

Altocumulus (Ac) (Figure 6–22) are layered clouds that form long bands or contain a series of puffy clouds arranged in rows. They are often gray in color, although one part of the cloud may be darker than the rest. Consisting mainly of liquid droplets rather than ice crystals, they usually lack the beauty of cirrocumulus.

◀ **Figure 6–21**
Altostratus clouds. These middle-level, layered clouds are composed of water droplets.

▶ **Figure 6–22**
Altocumulus clouds. Layered, mid-level altocumulus clouds are often arranged in bands.

Low Clouds

Low clouds have bases below 2000 m. Stratus (Figure 6–23) are layered clouds that form when extensive areas of stable air are lifted. They are normally between 0.5 and 1 km (0.3 to 0.5 mi) thick, in marked contrast to their horizontal extent, which can exceed that of several states. Usually the rate of uplift producing a stratus cloud is only a few tens of centimeters per second (less than 1 mph), and its water content is low, perhaps just a few tenths of a gram per cubic meter.

Weather in Motion

Time Lapse Movie of Stratus Clouds

Stratus clouds do not necessarily form from the lifting of air but may also result from the turbulence associated with strong winds. Consider a situation in which the air is at rest and there is a moderate decrease in temperature and dew point with height. When the wind begins to blow, it stirs the air by forced convection. This causes a slight decrease in dew point near the surface and an increase in water vapor at the top of the layer, as moisture is redistributed vertically. At the same time, the air temperature decreases more rapidly with altitude after mixing because the uplifted air cools at the DALR rather than the original (more gradual) ELR. The

▶ **Figure 6–23**
Stratus clouds. These are low-level, layered clouds.

6–3 Focus on the Environment

The Unknown Influence of Clouds on Climatic Change

One of the most important scientific issues debated today is whether an increase in certain gases put into the atmosphere by human activity will lead to an increase in global temperatures. To try to answer this question, atmospheric scientists use complex computer programs called **general circulation models** (GCMs).

GCMs estimate the global distribution of temperature and precipitation that would result from hypothesized changes in the atmosphere (such as doubling the carbon dioxide content). The models are first solved using control values representing current conditions, and they are then solved again using perturbed (changed) values corresponding to the altered atmosphere. By comparing the output from the original and the revised runs, researchers can estimate the effect of the change.

Unfortunately, this procedure includes a number of difficulties, not the least of which is our imperfect understanding of the impact of clouds on the atmosphere. Specifically, with regard to global warming, we need to answer two important questions: how might an increase in greenhouse gases affect the global cloud cover, and how would changes in cloud cover affect climate?

One line of reasoning holds that if an increase in these gases was to lead to a global warming, more evaporation would occur from the surface and thereby increase the water vapor content of the atmosphere. An increase in atmospheric water vapor could lead to an increase in global cloud cover. But even if it did, we do not know which type of clouds would increase the most, which is important because each type of cloud can have a very different impact on the climate.

It is easy to see how increased cloud coverage can promote cooling by increasing the percentage of sunlight reflected back to space. However, water droplets are also effective at absorbing longwave energy emitted by Earth's surface. Thus, the albedo effect could lead to global cooling, while the absorption of longwave energy might promote warming (by allowing the atmosphere to retain energy that otherwise would escape to space). The relative magnitude of the two effects depends on the type of cloud.

An increase in the depth of cumulus clouds would not greatly decrease the amount of insolation reaching the surface but could reduce the amount of longwave radiation lost to space by the atmosphere. This process amounts to a positive feedback (self-propagating) mechanism promoting further warming. On the other hand, an increase in the areal extent of *stratiform* (layered) clouds could substantially reduce the amount of insolation at the surface, while having relatively little effect on the amount of longwave radiation lost to space. This is a negative feedback (one that inhibits further change) that would reduce the rate of future warming.

Scientists are also uncertain about whether an increase in cloud cover would occur primarily at low, middle, or high elevations. It is believed that an increase in high cirrus clouds would promote an overall warming of the atmosphere, but that increases in lower and middle clouds would have the opposite effect.

Other important questions remain unanswered. For example, GCMs do not include complex microphysical processes occurring within clouds that influence precipitation. Thus, the issue of how changes in cloud cover might affect the amount and distribution of precipitation across the globe is still subject to much uncertainty. Furthermore, very recent research has indicated that clouds are more effective at absorbing solar radiation than previously believed. These and other unknowns are among the reasons it is so difficult to make predictions about how the atmosphere will respond to changes in greenhouse gas concentrations. Not surprisingly, considerable research activity in atmospheric science is directed at understanding clouds.

reduction in air temperature and increase in moisture content in the upper portion of the layer causes the air to become saturated.

Low, layered clouds that yield precipitation are called **nimbostratus** (Figure 6–24). Because nimbostratus clouds have low liquid water contents and weak updrafts to replenish moisture, they yield only light precipitation. Seen from below, these clouds look very much like stratus, except for the presence of precipitation.

Stratocumulus (Figure 6–25) are low, layered clouds with some vertical development. Their darkness varies when seen from below because their thickness varies across the cloud. Thicker sections appear dark, and thinner areas appear as bright spots.

Clouds with Vertical Development

Cumuliform clouds are those that have substantial vertical development and occur when the air is absolutely or conditionally unstable. Vertical velocities within these clouds are commonly several meters per second, but they can achieve speeds well in excess of 50 meters per second (100 miles per hour). In other words, updrafts in certain cumuliform clouds can be more rapid than the horizontal winds found in weak hurricanes! Liquid water contents are several times greater than those of stratiform clouds.

▶ **Figure 6–24**
Nimbostratus clouds. This type of cloud produces light rainfall.

▶ **Figure 6–25**
Stratocumulus clouds. These are low, layered clouds with some vertical development.

Cumulus clouds fall into several subgroups distinguished by the extent of their vertical development. Fair-weather cumulus (Figure 6–26), called *cumulus humilis*, have a single plume of rising air that often results from localized heating at the surface. They do not yield precipitation (hence the name, "fair-weather cumulus"), and they can evaporate away soon after their formation. Notice in Figure 6–27 that the clouds and the open areas between them form an invisible circulation system. The clouds mark the zone of rising air, and the cloud-free areas occur where the air sinks.

Weather in Motion

Time Lapse Movie of Cumulus Clouds

More intensely developed clouds are *cumulus congestus* (Figure 6–28). Unlike cumulus humilis, they consist of multiple towers, and each tower has several cells of uplift. This gives them a fortress-like appearance with numerous columns of varying heights. Their strong vertical development implies that these clouds form in unstable air.

Figure 6–26
Cumulus humilis, or fair-weather cumulus.

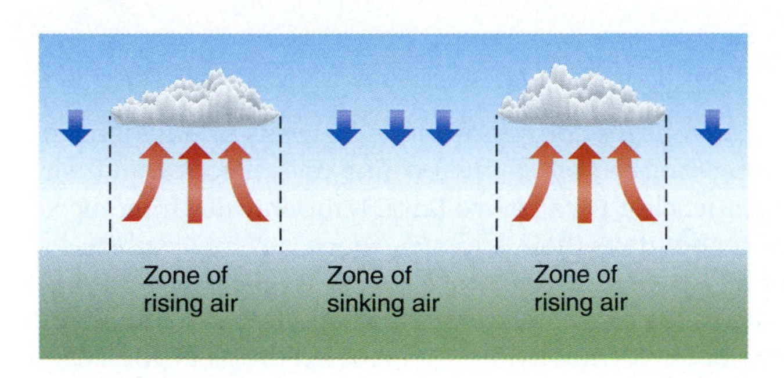

Figure 6–27
Scattered fair-weather cumulus clouds form by rising air parcels, but the area between clouds has weak downdrafts.

Figure 6–28
Cumulus congestus clouds. Note the substantial vertical development.

6–4 Special Interest

Why Clouds Have Clearly Defined Boundaries

The next time you see a cumulus cloud, notice its base and edges. You'll see that the boundaries of the cloud are marked by flat, sharply defined bottoms and edges that are clearly distinct from the surrounding air. This clear definition of the cloud base is partly due to the rapid growth of droplets as they form just above the lifting condensation level.

Recall that liquid water at the base of a cloud initially forms onto condensation nuclei, of which there are a finite (but very large) number. Within a few tens of meters of the lifting condensation level, all the available condensation nuclei have attracted moisture, and further condensation occurs only onto existing droplets. When the droplets are first formed, they are very small, but they quickly attain diameters of about a micrometer, which makes them effective at scattering visible light; hence the clearly visible base. (If instead the droplets grew slowly, there would be only a gradual increase in the number capable of scattering visible light from the base of the cloud. The cloud would therefore have a faintly visible base that would gradually become more discernible with height.)

Another factor is that cloud droplets evaporate within a very short distance of the cloud base. To see why, consider that if a droplet falls into unsaturated air below the cloud, it shrinks by evaporation. Because they are so small, cloud droplets fall slowly to begin with and fall even more slowly as they evaporate. As a result, the distance a cloud droplet can fall without evaporating is minuscule, on the order of a centimeter. Viewed from a great distance, this will obviously look like a flat surface. (By way of contrast, raindrops have survival distances measured in kilometers because they are so much larger.)

The sharp boundaries along the sides of cumulus clouds are the result of entrainment. When unsaturated air just outside the margins of the cloud is drawn into the cloud by turbulence, some of the water droplets evaporate. The rapid evaporation produces the same result as at the base of the cloud—a well-defined, sharp boundary. Droplets at the cloud margins rapidly evaporate, leaving behind a distinct boundary between the unsaturated, ambient air and the large droplets remaining within the cloud.

The individual towers of a cumulus congestus have lifespans of only tens of minutes and are constantly being replaced by newly forming ones. Because of their vertical extent, cumulus congestus clouds can have large temperature differences from top to bottom, and even on hot days their upper portions can have subfreezing temperatures. Liquid droplets do not freeze into ice instantaneously as updrafts carry them into the cold portion of the cloud, however; instead, they can remain in their supercooled state for some time. Eventually the supercooled droplets do freeze, and the liquid cloud becomes dominated by ice crystals. We can actually observe the result of this process from the ground. When a portion of the cloud becomes *glaciated* (composed entirely of ice), it does not exhibit sharply defined edges like the portions consisting of water. Instead, it has a washed-out appearance that makes it readily distinguishable from the liquid portion of the cloud. The upper portion of the cloud in Figure 6–28 provides a good example of glaciation.

Cumulonimbus (Figure 6–29) are the most violent of all clouds and produce the most intense thunderstorms. In warm, humid, and unstable air, they can have bases just a few hundred meters above the surface and tops extending into the lower stratosphere. In other words, these clouds can occupy almost the entire depth of the troposphere and more!

▶ **Figure 6–29**
Cumulonimbus cloud. Extending into the lower stratosphere, they can create violent weather.

A cumulonimbus is distinguished by the presence of an **anvil** (so named for its resemblance to the blacksmith's tool). This feature, composed entirely of ice crystals, is formed by the high winds of the lower stratosphere that extend the cloud forward. The anvil appears as a wedge of ice at the top of the cloud that gradually thins out as it gets farther from the main body of the cloud. Strong winds can propel hailstones toward the anvil, where they are ejected and fall from the cloud. For this reason, airline pilots avoid them.

Although cumulonimbus clouds have extremely strong updrafts, these are by no means uniform across the cloud. The most rapidly rising air is found in about the upper third or so of the cloud, with weaker motions below. Furthermore, updrafts along the margins of the cloud are generally less intense than those in the interior because of entrainment. Cumulonimbus even have regions where air actually descends. Although commercial aircraft can fly through such clouds, the rapid and dramatic shifts in vertical winds cause extreme turbulence that would violently jostle the plane and its occupants. Commercial pilots wisely fly around rather than through cumulonimbus clouds.

Unusual Clouds

Certain cloud types do not neatly fall into the categories mentioned above. **Lenticular clouds** form downwind of mountain barriers and have curved shapes like eyeglass lenses (Figure 6–30). They form when mountain ranges disrupt the flow of air to form a series of waves. As the air rises in each wave, adiabatic cooling leads to condensation; as the air descends, adiabatic warming causes the cloud droplets to gradually evaporate. Although new droplets constantly form on the upwind side of lenticular clouds and old droplets evaporate on the downwind side, the clouds remain in a fixed position. Thus, the flow of moisture into and out of lenticular clouds is similar to the movement of objects on a conveyor belt, with as much mass removed as is added.

Usually no more than two or three lenticular clouds form downwind of the barrier, but if the conditions are just right, as many as six or seven may be observed. **Banner clouds** (Figure 6–31) are similar to lenticular clouds but are individual clouds located immediately above isolated peaks.

Sometimes portions of cumulonimbus clouds hang downward in sac-like shapes called **mammatus** (Figure 6–32). These features occur where downdrafts force water droplets below the cloud base, usually near the anvil. Because of the high liquid water content of these clouds, the droplets contained in the downward-moving air require

◀ **Figure 6–30**
Waves formed by the passage of air over a topographic barrier can lead to the formation of lenticular clouds.

▶ **Figure 6–31**
Banner clouds form atop isolated mountain peaks.

▶ **Figure 6–32**
Mammatus. These clouds are found on cumulonimbus clouds and formed by downdrafts.

substantial descent (and resultant adiabatic warming) before they fully evaporate. Thus, the mammatus extend some distance below the droplets of the surrounding cloud base.

Like most other weather features, the majority of clouds exist in the troposphere. However, two rare cloud types exist at higher levels and can be seen during the twilight hours of winter at high latitudes. **Nacreous clouds** (Figure 6–33) consist of

6–5 Physical Principles

The Surprising Composition of Clouds

We think of clouds as being liquid water and/or ice, but by far the greatest amount of the mass contained in a cloud is air. Although clouds contain a very large number of suspended droplets or particles—typically about 1000 per cubic centimeter—even the largest of these droplets are extremely small. Therefore, despite their large numbers they amount to relatively little mass. This might not be too surprising if you recall that water vapor usually accounts for less than 2 percent of the mass of the atmosphere. Because the water vapor is so limited, it stands to reason that only a small amount of ice or liquid water can be deposited or condensed.

We can apply some simple arithmetic to determine the relative mass of air and liquid water in a cloud. The average cloud droplet has a radius of about 0.001 cm, and the volume of a sphere is equal to $^4/_3 r^3$, where r is the radius. Substituting 0.001 cm for r, we find that the average droplet has a volume of about 4×10^{-9} cubic centimeters.

Because the density of water is about 1 g/cm^3 it follows that the mass of each droplet is about 4×10^{-9} g. Multiplying this value by the 1000 droplets per cubic centimeter normally found in a cloud gives us a liquid water content of 0.000004 g/cm^3.

Now we can compare the 0.000004 g/cm^3 of water to the mass of the air. At an altitude of 5.5 km above sea level, for example, the density of the air is about half that at sea level, or roughly 0.0006 g/cm^3. Compare this to the mass of the liquid, and you will find that about 150 times more air than water is in the cloud!

You can think of this another way: If a cloud has a horizontal area of 1 square kilometer and a height of 1 km, it contains about 4 million kg (or more than 1 million gallons) of water! But this is far less than the approximately 600 million kg of air in the same cloud.

◀ **Figure 6–33** Nacreous clouds. These stratospheric clouds are only observed at high latitudes.

supercooled droplets or ice crystals in the stratosphere at heights of about 30 km (20 mi). They have a soft, whitish appearance and sometimes are called *mother of pearl* clouds. Even higher are the **noctilucent clouds** (Figure 6–34), whose location in the mesosphere allows them to be illuminated after sunset (or before sunrise) when the surface and the lower atmosphere are in Earth's shadow.

Cloud Coverage

In addition to their height and form, another important characteristic of clouds is their breadth or **coverage**. Meteorologists use several terms to describe coverage. When clouds occupy more than nine-tenths of the sky, conditions are said to be *overcast*. When coverage is between six-tenths and nine-tenths, it is called *broken*.

▶ **Figure 6–34**
Noctilucent clouds are in the mesosphere and can illuminate the sky at high latitudes during the twilight hours.

Scattered clouds occupy between one-tenth and one-half of the sky, and less than one-tenth cloud cover is classified as a *clear-sky* condition.

Of course, clouds on any given day are not restricted to a single height above the surface. Clouds can occur simultaneously at several different levels in the atmosphere, and each level can have different cloud types and a different amount of coverage. Thus, for example, a detailed cloud report might describe the sky as having scattered cumulus at 1000 m, broken altostratus at 4000 m, and a layer of overcast cirrostratus at 7000 m.

Summary

In the previous chapter we saw the importance of lifting for the formation of clouds. Here we examine the processes by which this lifting occurs—frontal uplift, convergence, orographic uplift, and convection. The first three processes are enhanced or hindered by the static stability of the atmosphere, whereas free convection necessarily occurs only when the air is unstable. Instability implies that if a parcel is given an initial boost upward, it will become buoyant and continue to rise. On the other hand, if the air is stable, a parcel displaced vertically will tend to return to its original position.

Static stability or instability is determined by the air column's rate of temperature decrease with altitude. When the temperature lapse rate is less than the saturated adiabatic rate, the air is statically stable; when it exceeds the DALR, it is unstable. Conditional instability arises when the lapse rate is between the two adiabatic rates. When the air is conditionally unstable, a lifted parcel will rise on its own accord only if it is lifted above a certain critical point called the level of free convection.

Three processes modify the lapse rate: the inflow of warm and cold air at different altitudes, the advection of a different air mass, and heating or cooling of the surface. ELRs vary not only through time, but also with elevation. Thus, a column of atmosphere might be unstable at one level but stable aloft. In fact, no matter what the condition of the troposphere, the stratosphere is always statically stable and thereby limits the maximum height of updrafts.

Inversions are a special case in which the temperature increases with altitude. Because of their strong static stability, inversions suppress the vertical motions necessary for cloud formation and for the dispersion of air pollution. Inversions are formed by subsidence (sinking air), the emission of longwave radiation from the surface, and the presence of fronts.

Clouds have been categorized into 10 distinct types according to their height and form. Layered high clouds are called *cirrostratus*, for example, while those aligned in rows or clusters with vertical development are called *cirrocumulus*.

Clouds with extensive vertical development are called *cumulus*. Many cumulus clouds are associated with fair weather, but others can lead to precipitation or even extremely violent thunderstorms. The most dramatic of these are the cumulonimbus clouds, whose updrafts can exceed tens of meters per second.

By themselves, clouds are of fundamental interest to meteorology. But of even greater interest, perhaps, is the fact that many clouds precipitate. Why do some clouds cause precipitation and not others? The answer is that a large amount of growth must occur for cloud water droplets and ice crystals to become heavy enough to fall to the surface. The processes by which the droplets grow to precipitation size, and the resultant types of precipitation, are discussed in Chapter 7.

Key Terms

orographic uplift p. 170
rain shadow p. 171
cold front p. 172
warm front p. 172
convergence p. 172
static stability p. 173
environmental lapse rate p. 173
dry adiabatic lapse rate p. 173
saturated adiabatic lapse rate p. 174
level of free convection p. 176
entrainment p. 181
inversions p. 181
cirrus p. 183
stratus p. 183
cumulus p. 183
nimbus p. 183
high, middle, and low clouds p. 184
clouds with vertical development p. 184
contrails p. 186
cirrostratus p. 186
cirrocumulus p. 186
altostratus p. 187
altocumulus p. 187
general circulation models p. 189
nimbostratus p. 189
stratocumulus p. 189
cumuliform p. 189
cumulonimbus p. 192
anvil p. 193
lenticular clouds p. 193
banner clouds p. 193
mammatus p. 193
nacreous clouds p. 194
noctilucent clouds p. 195
coverage p. 195

Review Questions

1. Describe the four mechanisms that lift air and promote cloud formation.
2. Explain how buoyancy affects the air's susceptibility to uplift.
3. Describe the situations that can cause air to be absolutely stable, absolutely unstable, or conditionally unstable.
4. What two factors can ultimately stop rising parcels of air from continuing upward?
5. What will determine whether air that is conditionally unstable will become buoyant?
6. What is the level of free convection?
7. Describe the processes that bring about changes in the environmental lapse rates, and thus the stability of the atmosphere.
8. What is entrainment, and how does it affect the growth of clouds?
9. Define the term *inversion*, and describe the mechanisms that can cause the various types of inversions.
10. Describe the classification scheme for clouds based on their height and form.
11. List the major subtypes of high, middle, and low clouds.
12. What type of cloud produces a characteristic halo?
13. Other than height, what significant difference exists between altocumulus and cirrocumulus clouds?
14. How do cumulus humilus and cumulus congestus clouds differ from each other?
15. What distinctive feature characterizes a cumulonimbus cloud?
16. What conditions cause lenticular and banner clouds to form?
17. What are mammatus?
18. List and describe the types of clouds that exist above the troposphere.

Critical Thinking

1. Orographic uplift can cause cloud or fog to form. How might stability be a factor in determining which develops?
2. Except for the shallow zone near the surface, it is rare that the atmosphere be absolutely unstable. Why is it difficult for a very steep ELR to develop in the middle and upper atmosphere?
3. Localized convection might be more vigorous over the desert in the summer than over a wooded region, but precipitation is more likely in the latter environment. Why?
4. Is the stability of the air more likely to change rapidly near the surface or aloft? At what time of day are major changes in the ELR most likely?
5. What time of year will unstable conditions be most common over the continental United States and Canada?
6. Is it possible for radiation and subsidence inversions to occur simultaneously?
7. Unlike the Sierra Nevada range in the west, the Appalachian Mountains do not exhibit very strong rain shadow effects. Why not?
8. Some of the higher peaks in Hawaii have "cloud forests" at certain levels. What type of environmental situations would favor the presence of ubiquitous cloud cover?
9. In many regions, the orographic effect causes precipitation to increase with elevation. Can you think of any reason why this might not be true all the way up to the top of Mt. Everest?

Problems and Exercises

1. On a daily basis, examine the current surface weather maps and the satellite images depicting cloud cover from any of the Web sites offering those products. Do you notice any relationships between cloud cover and surface pressure distributions? How do clouds tend to appear when associated with cold or warm fronts? Are there any mountainous regions that seem to have a higher incidence of cloud cover?
2. Assume that a parcel of air starts out with a temperature of 12 °C and a dew point of 10.4 °C, and that the ELR = 0.7 °C/100 m. Determine the following:
    a. Is the air stable, unstable, or conditionally unstable?
    b. At what level will the air become saturated?
    c. Where is the level of free convection?
3. Redo problem 1, but assume that the ELR = 0.4 °C. Explain why there will be no level of free convection.

Quantitative Problems

Among other things, this chapter has demonstrated the importance of temperature lapse rates in cloud formation. The change in ambient air temperature with height determines whether the local atmosphere is absolutely stable, absolutely unstable, or conditionally unstable. The quantitative problems on this book's Web site (**http://www.prenhall.com/aguado/**) provide you with the opportunity to assess your knowledge of how these lapse rates determine the susceptibility of the air to upward motions. We suggest that you log on to the site and go to the Chapter 6 page, where you can work out a set of numerical problems to improve your understanding of the concept of stability.

Useful Web Sites

http://asd-www.larc.nasa.gov/SCOOL/cldchart.html
Examples of many types of clouds.

http://www.weather-photography.com/gallery.php?cat=clouds
Extensive library of cloud photos and time lapse movies.

http://www.giss.nasa.gov/research/intro/delgenio_03/
Interesting information on clouds and climate change.

http://www.cnn.com/EARTH/9608/27/cloud.harvest/
A discussion of how people in arid regions "harvest" fog.

Media Enrichment

Tutorial

Atmospheric Stability

This tutorial shows how buoyant forces arise, leading to rising (and sinking) air parcels. It covers all of the stability classes described in the text. Be sure to complete the moisture and adiabatic processes tutorials before starting this one.

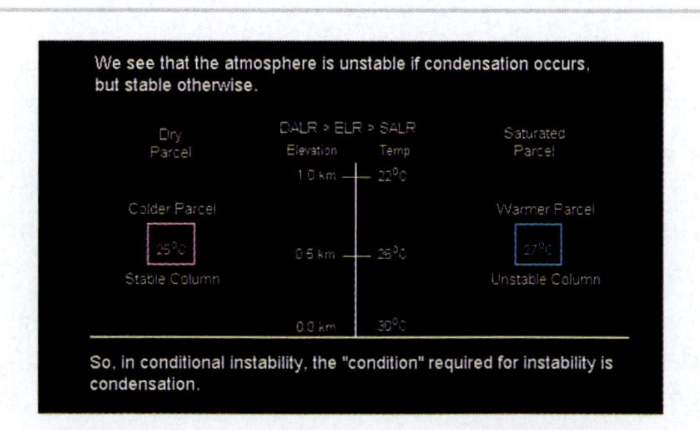

Weather in Motion

Global Clouds, 1999

Although it is sometimes called the "water planet," does Earth actually behave as a big, wet sphere, uniformly covered by ocean? Or instead, do continents affect atmospheric and ocean circulations to the point of influencing global climate patterns? To answer, consider this movie of clouds with land areas suppressed. Obviously, if continents are unimportant, you wouldn't expect their fingerprint to appear in the movie. What do you see? Latitudinal variations are most prominent, with bands of clouds along the equator and in the middle latitudes of both hemispheres. But given the great difference in radiation and energy budgets, we might expect these latitudinal differences even on a uniform Earth.

Are there other, more subtle variations that reveal the role of continents? Yes! First, notice that along the equator some areas are cloud-covered most of the year, whereas other places are far more likely to be cloud-free. If you have a good mental map of the world, you can tie these to particular continents and ocean basins. As another continental fingerprint, compare stormy areas in the middle latitudes of the two hemispheres. You can see that the Southern Hemisphere mid-latitudes (mostly ocean) remain stormy throughout the year. In contrast, the Northern Hemisphere mid-latitudes (with major land masses) have a relatively quiet summer followed by a much stormier winter. Last, look for the role of major mountain ranges. For example, find orographic clouds along the Andes Mountains of South America. Although the details remain to be explained in later chapters, it's clear from this movie that the arrangement of continents is of utmost importance in global climate.

Weather in Motion

Time Lapse Movie of Stratus Clouds

This is the first of two time lapse movies that show examples of a general cloud type. Here we see two instances of stratus clouds, both with minimal vertical development. The clouds maintain a constant thickness as they move across the field of sight because vertical motions are weak.

Weather in Motion

Time Lapse Movie of Cumulus Clouds

Here we have two very different examples of cumulus clouds. In the first, convection is deep, so a thick cloud with extensive vertical development approaches the viewer. Vertical updrafts are strong and continue to high altitudes in this type of situation, which is brought about by a thick layer of unstable air. The second example shows fair weather cumulus. Individual cells of convection are evident as small plumes grow in size as the clouds move downwind. Growth of the cloud tops is limited because the layer of unstable air does not extend to high altitudes.

Weather Image

Altocumulus Over the Mouth of the Amazon River

Sediment carried to the mouth of the Amazon River results in debris plumes clearly visible in this photo taken from space. Mid-level altocumulus clouds are found throughout the area. The striking pattern seen here, known as "cloud streets," is indicative of atmospheric waves.

chapter 7

Aerial view of a rain shaft in a thunderstorm.

Precipitation Processes

We often think of heavy rainfall and its environmental impacts as brief, passing events. A sudden downpour can create major flooding in a matter of minutes, but the danger is usually over in an hour or so. Sometimes, however, preexisting environmental conditions can cause a rainstorm to have an impact far more serious than usual. The widespread flooding in the midwestern United States in 1993 provides an excellent example.

The setting for the severe flooding began in the fall of 1992, when persistent rains saturated the soil across the north-central United States. The rainy autumn was followed by a deep accumulation of winter snow. The combination of deep snowpack atop saturated soil always presents a flooding hazard for the melt season, but 1993 proved to be worse than anybody could have imagined because the melt season was followed by unusually heavy rainfall during June and July.

The magnitude of the flooding was unprecedented. At St. Louis, Missouri, the Mississippi River crested at 6 m (19.5 ft) above flood stage. The river, normally about 800 m (0.5 mi) wide near St. Joseph, Missouri, stretched out to as much as 10 km (6 mi), putting nearly half of St. Charles County under water (Figure 7–1). At Kansas City, Missouri, the Missouri River rose 6.7 m (22 ft) above its banks. Across the Midwest, tens of thousands of homes were damaged or destroyed by the flooding, as entire

CHAPTER OUTLINE

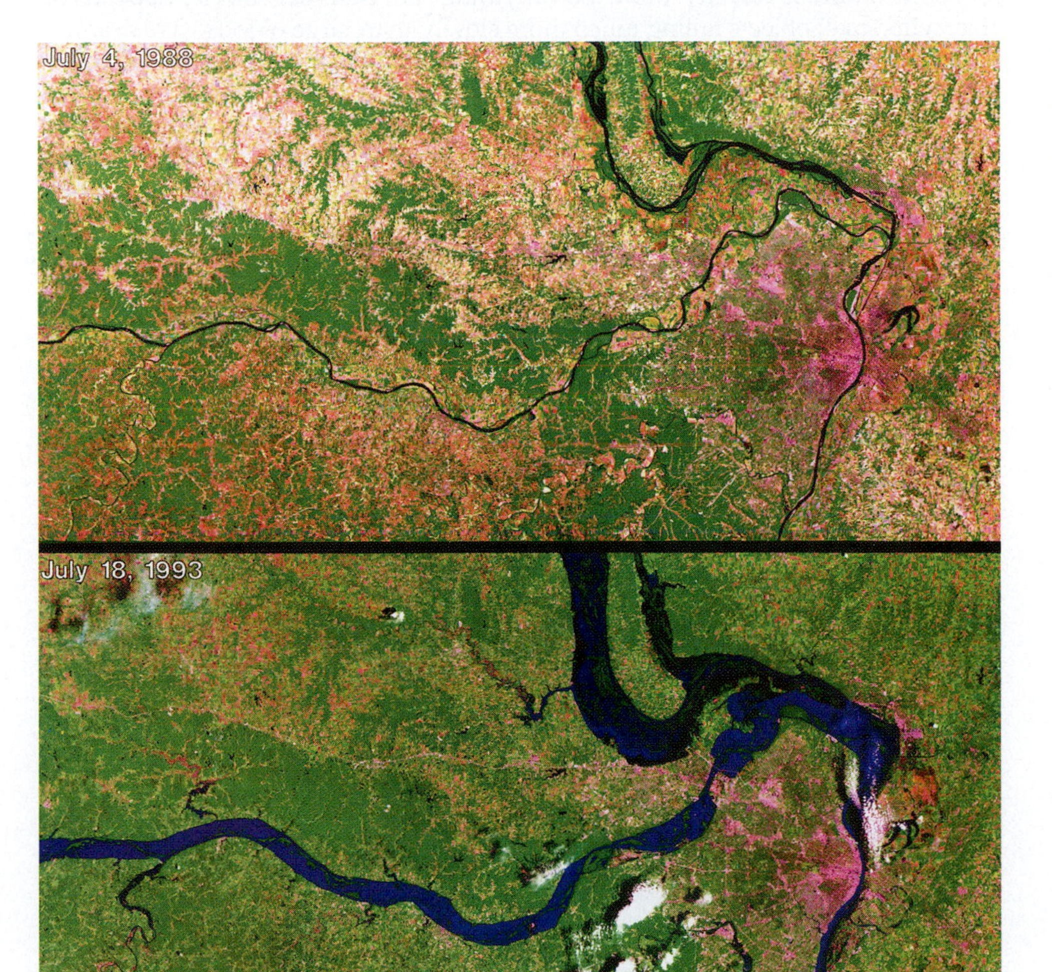

Figure 7–1 Flooding across the Mississippi and Missouri Rivers. The two rivers are confined to their banks in the top image, showing the normal situation. Extreme overbank flooding inundated large areas of several midwestern states during the summer in 1993 (bottom).

neighborhoods and 77 small towns ended up under water. The flooding even brought its share of irony: Des Moines, Iowa, was without potable water for 12 days because of contaminated floodwaters. Forty-year-old Jacki Meek of suburban St. Louis probably spoke for all of the 85,000 people who had to evacuate their homes: "I feel about 65 right now. I see my house on the news, and I just cry."

Rain and other forms of precipitation are a fact of life for everybody, although usually they are of far less consequence than the floods of 1993. In fact, the quest for an answer to the question of what causes precipitation may have been one of the reasons you picked up this book. In Chapter 6 you learned about the processes that led to the formation of all clouds, precipitating or nonprecipitating. In this chapter we explain the processes by which nonprecipitating cloud droplets and ice crystals grow large enough to fall as precipitation.

Growth of Cloud Droplets

Acting alone, gravity would accelerate all objects toward the surface. But gravity is not the only force acting on a falling object; at the same time, the air exerts an opposing resistance or **drag**. As speed increases, so does resistance, until its force equals that of gravity and the acceleration ceases. The object falls, but at a constant speed, its **terminal velocity**. More than anything else, terminal velocity depends on size, with small objects falling much more slowly than large objects. (We examine details of the relationship between size and terminal velocity in *Box 7–1, Physical Principles: Why Cloud Droplets Don't Fall.*) As we have implied, cloud droplets fall slowly because they are so tiny. Their small size is largely explained by the fact that condensation nuclei are very abundant; thus, cloud water is spread across numerous small droplets rather than being concentrated in a fewer number of large drops. With their small size, cloud droplets initially have extremely low terminal velocities, making it impossible for them to reach the surface.

This effect is apparent in Figure 7–2, which shows terminal velocities for various cloud constituents. The smallest are the condensation nuclei, on which liquid droplets form (for the sake of simplicity, the figure applies only to clouds consisting of liquid water alone, without ice crystals). Condensation nuclei are so small that they fall at an imperceptibly slow rate. Larger cloud droplets (but not falling as precipitation) typically range from about 10 m to about 50 m in radius (recall that 1 m is one-millionth of a meter). These have fall speeds ranging from about 1 cm/sec (0.02 mph) to about 25 cm/sec (0.5 mph). By way of contrast, the much larger raindrops shown in the figure fall at 650 cm/sec, about 25 times faster.

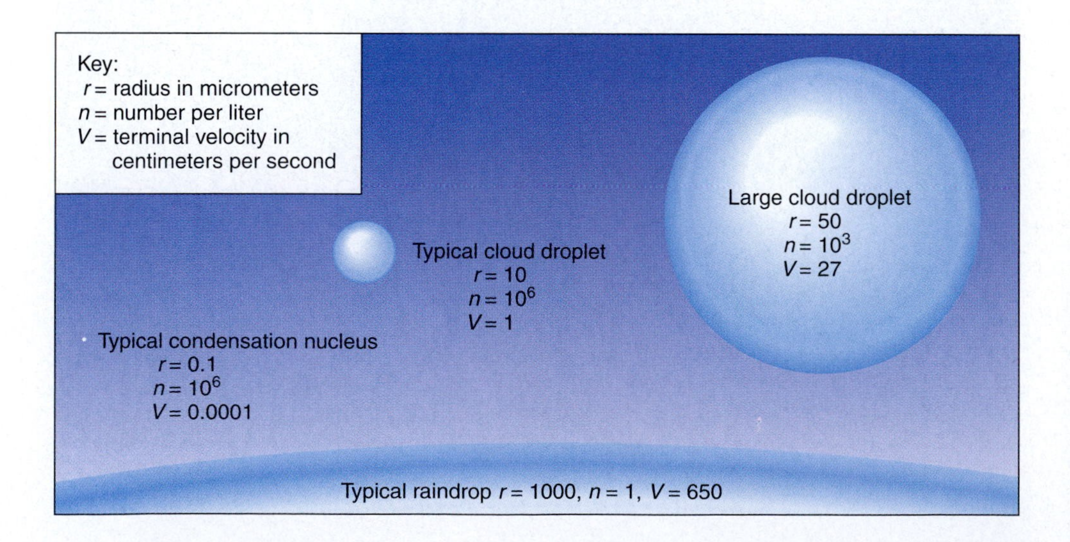

▶ **Figure 7–2**
The average characteristics of cloud constituents.

7–1 Physical Principles

Why Cloud Droplets Don't Fall

You are likely familiar with the legendary, late-sixteenth-century experiment of Galileo Galilei, who dropped two objects—a light one and a heavy one—off the Leaning Tower of Pisa. Both objects, being subjected to the same gravitational acceleration, hit the ground at nearly the same time. Galileo's demonstration may seem inconsistent with our everyday experience, as an ant would surely take longer to fall from the top of a tall building than would a golfball. It is also at odds with our claim that small droplets fall slowly. The solution must be that there is another force acting on falling objects besides gravity, and indeed there is: wind resistance, or drag. By examining how these two forces work together, we will gain some insight into why cloud droplets don't fall. To keep the discussion simple, we will assume spherical droplets throughout—using more realistic shapes would not change our conclusions.

Newton's second law tells us that if a net force is applied to a mass, it will undergo an acceleration (or change in velocity through time). For a given mass, the acceleration is directly proportional to the net force. In equation form, the law is given as

$$\text{net force} = \text{mass} \times \text{acceleration}$$

Notice that Newton's second law says that we must consider the net force, the result of all the forces acting on the object. As far as a falling droplet is concerned, there is the downward gravitational force, which is opposed by the force of wind resistance (drag). A droplet suddenly released in the atmosphere falls at increasing speed, but not indefinitely. Eventually the force of drag (F_d) balances the force of gravity (F_g), resulting in no net force:

$$\text{net force} = F_g - F_d = 0$$

With no net force, there is no acceleration, and the droplet falls at its terminal velocity. How fast does it fall? To answer that, we need to know something about the magnitude of the two forces.

Gravitational Force

The force of gravity is directly proportional to mass. Whenever we step on a scale, we measure this force. For a droplet containing nothing but water molecules, mass is density times volume. We therefore have F_g as

$$F_g = \rho \frac{4}{3}\pi r^3$$

where is the density of water and r is droplet radius. Notice that gravity is proportional to the cube of radius. (You do remember that the volume of a sphere is $\frac{4}{3}\pi r^3$, don't you?)

Force of Drag

Drag between the droplet and surrounding air depends on the rate of fall and on the size of the droplet. Just like an automobile on a highway, a faster-moving droplet experiences greater resistance as it moves through the air. In fact, to a good approximation, the drag force increases with the square of wind speed (v^2). So how does size influence drag? For a sphere, the effect of droplet size is largely one of surface area. As surface area increases, the force of drag increases proportionally. Remembering that the surface area of a sphere is $4\pi r^2$, we can combine the effects of speed and surface area as

$$F_d = kv^2 4\pi r^2$$

where k is some constant. The value of k is not important here; what matters is that F_d is proportional to the square of both fall rate (v^2) and radius (r^2).

Terminal Velocity

For a droplet falling at terminal velocity, we've said that gravity and drag are equal. If we use v_t for terminal velocity, we get

$$\underset{(F_g)}{\rho \frac{4}{3}\pi r^3} = \underset{(F_d)}{kv_t^2 4\pi r^2}$$

To find the terminal velocity, we rearrange and solve for v_t:

$$v_t^2 = \rho\frac{4}{3}\pi r^3 / k4\pi r^2$$
$$= \left(\frac{4}{3}\rho/4k\right) r^3/r^2$$
$$= \left(\frac{1}{3}\rho/k\right) r$$

or

$$v_t = c\sqrt{r}$$

where c is another constant $\sqrt{\frac{1}{3}\rho/k}$. From this equation, we obtain the important result that as droplet radius increases, there is an increase in terminal velocity. Equivalently, large droplets fall faster than small droplets. What happens physically is that both F_g and F_d increase with radius, but gravity increases more than drag, and a higher fall rate is therefore required to cancel F_g. Notice that as far as the droplet is concerned, falling through a still atmosphere at v_t is the same as remaining stationary in an updraft of speed v_t. Thus, the equation says that a strong updraft is needed to hold a large droplet aloft, whereas a small droplet is easily suspended.

Going back to the Leaning Tower of Pisa situation described at the beginning of this box, we can now understand why Galileo's objects fell at nearly the same speed. With such large objects, the gravitational forces far exceeded the drag forces throughout their short fall. With negligible drag, gravity accelerated both at nearly the same rate. If he had used objects of greatly different size, or if the objects had fallen far enough to reach their terminal velocities, differences in v_t would have emerged. The old, familiar story would be about wind resistance, and books like this would have no need for a feature on the topic.

Raindrops fall to the surface when they become large enough that gravity overcomes the effect of updrafts. How large is large enough? In terms of radius, raindrops are about 100 times bigger than typical cloud droplets. At this size, most of the molecules are in the interior of the drop, with relatively few exposed to the atmosphere. With a smaller proportion susceptible to drag, the terminal velocity is high enough for precipitation.

We should note that in terms of volume or mass of water, raindrops are larger than cloud drops by a factor of a million, rather than just a hundred. The difference arises because volume for a sphere is proportional to the cube of radius. If the radius is 100 times larger, the volume is $100 \times 100 \times 100$ (1 million) times larger. Raindrops are not truly spherical, but the principle holds: precipitation particles are vastly more massive than cloud drops. Although we don't think of clouds yielding massive falling objects, they certainly do, at least from the point of view of a cloud droplet. In the paragraphs that follow, we outline the processes that give rise to these "massive" falling objects.

Growth by Condensation

When cloud droplets begin to form by the adiabatic cooling of ascending air, they do so on condensation nuclei. But within a few dozen meters above the lifting condensation level, all the available condensation nuclei have attracted water, and any further condensation can only occur on existing droplets.

Condensation can lead to rapid growth for very small water droplets, but only until they achieve radii up to about 20 µm, far smaller than necessary to fall as precipitation. Beyond this point, further growth by condensation is minimal. To understand why, recall that relatively little water vapor is available for condensation. With so many droplets competing for a limited amount of water, none can grow very large. It's clear that if growth by condensation were the only process operating, we would experience little, if any, precipitation on Earth. We should therefore think of condensation as only the starting point for rain and snow, not the sole cause. Two other processes are responsible for further growth, the relative importance of which depends on the temperature characteristics of the clouds.

Growth in Warm Clouds

Most precipitating clouds in the Tropics, and many in the middle latitudes, are **warm clouds**, those having temperatures greater than 0 °C throughout. In warm clouds, the **collision–coalescence process** causes precipitation, which depends on the differing fall speeds of different-sized droplets.

Cloud droplets come in different sizes, and therefore attain different terminal velocities. Refer to Figure 7–3 and consider what will happen when the largest droplet (the **collector drop**) falls through a warm cloud. As the collector drop falls, it overtakes some of the smaller droplets in its path because of its greater terminal velocity. This provides the opportunity for collisions and coalescence.

▲ **Figure 7–3** Because of their greater mass, collector drops have greater terminal velocities (indicated by the length of the downward-pointing arrows) than do the smaller droplets in their path. Collector drops overtake and collide with the smaller ones.

Collision As it falls, a collector drop collides with only some of the droplets in its path. The likelihood of a **collision** depends on both the absolute size of the collector and its size relative to the droplets below. If the collector drop is much larger than those below, the percentage of collisions (the *collision efficiency*) will be low. Figure 7–4 illustrates why. As the collector drop falls, it compresses the air in its path. The compressed air creates a small gust of wind that pushes the smaller droplets out of the way. The small gust of wind cannot push aside larger droplets, however, and the collector is able to collide with them. As a result, the collision efficiency is greater for the droplets that are not very much smaller than the collector drops.

You have probably witnessed a similar phenomenon on a larger scale while driving down a country road in summer, with your windshield turning relatively large flying insects into mere "bug juice." Too heavy to get swept aside by the compressed air immediately ahead of the windshield, they follow their own paths until the fateful moment of impact. Smaller bugs, in contrast, get blown out of harm's way.

Collision efficiencies are also low for droplets nearly equal in size to the collector drop, because their terminal velocities are so close to the collector's velocity that it is difficult for the collector to catch up to and collide with them. Continuing with the car analogy, collisions between vehicles are unlikely as long as all move at the same speed and direction.

Under certain situations, collision efficiencies can actually exceed 100 percent, and the collector can collide with more droplets than are in its path. A falling drop creates turbulence that can entrain small droplets outside its path and carry them back toward the top of the collector, where collision occurs.

Recent research using mathematical models shows that turbulence in the form of whirling vortices greatly enhances collision efficiency. The vortices function like small centrifuges, separating droplets according to size as they spin around the center. The resulting variations in concentration significantly increase the average collision rate. In addition, rapid spinning causes jets of droplets to detach from the air flow like a rock thrown from a sling. The ejected droplets have a high probability of colliding with other droplets, so this process also enchances collision efficiency. Calculations show that only mild turbulence is required for the centrifuge and sling effects, which implies that the processes operate in most clouds.

(a)

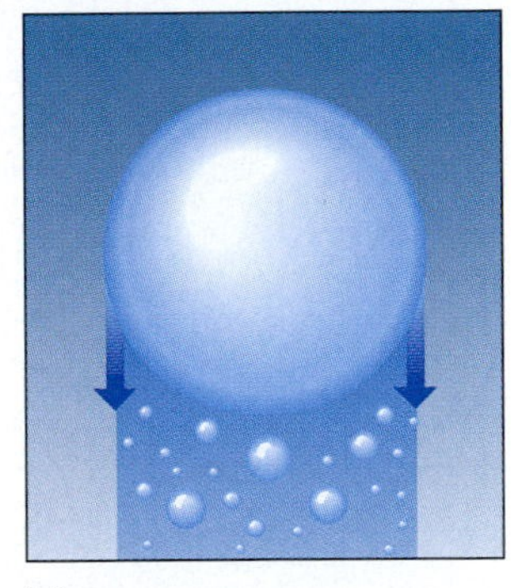

(b)

(c)

▲ **Figure 7–4**
As a collector drop falls (a), it compresses the air beneath it (b). This causes a pressure gradient to develop that pushes very small droplets out of its path (c). The small droplets get swept aside and avoid impact.

Coalescence When a collector drop and a smaller drop collide, they can either combine to form a single, larger droplet or bounce apart. Most often the colliding droplets stick together. This process is called **coalescence**, and the percentage of colliding droplets that join together is the *coalescence efficiency*. Because most collisions result in coalescence, coalescence efficiencies are often assumed to be near 100 percent. In other words, we can safely ignore the coalescence efficiency of two similar-sized drops, because they are unlikely to collide in the first place.

Collision and coalescence together form the primary mechanism for precipitation in the Tropics, where warm clouds predominate. In the middle latitudes, most precipitating clouds have freezing temperatures, at least in their upper portions. This favors the growth of precipitation by another mechanism involving the coexistence of ice crystals and supercooled water droplets, the **Bergeron process** (also known as the Bergeron-Findeisen or ice crystal process) described in the next section.

Growth in Cool and Cold Clouds

Unlike their counterparts in the Tropics, at least a portion of most mid-latitude clouds have temperatures below the melting point of ice. Some, such as the one in Figure 7–5a, have temperatures below 0 °C throughout and consist entirely of ice crystals, supercooled droplets, or a mixture of the two. These are referred to as **cold clouds**.

Cool clouds (Figure 7–5b), on the other hand, have temperatures above 0 °C in the lower reaches and subfreezing conditions above. As we discussed in Chapter 5, saturation at temperatures between about –4 °C (25 °F) and –40 °C (–40 °F) can lead to the formation of ice crystals if ice nuclei are present or supercooled liquid droplets if they are absent. Thus, a well-developed cumulus cloud might be composed entirely of water droplets in its lower portion, a combination of supercooled droplets and ice crystals in its middle section, and exclusively ice crystals in its upper reaches (Figure 7–6). The processes described in this section operate in cold and cool clouds having a mixture of ice and liquid water.

As we will now see, the coexistence of ice and supercooled water droplets is essential to the development of most precipitation outside the tropics. The process by which droplets and crystals in mid-latitude clouds grow to precipitation size was first described by one of the preeminent figures of modern meteorology, Tor Bergeron. This process is therefore often referred to as the Bergeron process.

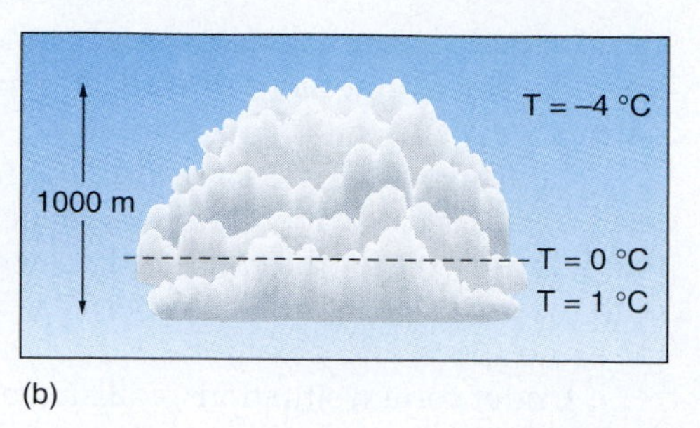

▶ **Figure 7–5**
Cold clouds (a) have temperatures below 0 °C from their base to their top. Cool clouds (b) have temperatures above 0 °C in their lower portions with subfreezing temperatures above.

The principle underlying the Bergeron process is that the saturation vapor pressure over ice (the amount of water vapor needed to keep it in equilibrium) is less than that over supercooled water at the same temperature.* In other words, if there is just enough water vapor in the air to keep a supercooled droplet from evaporating away, then there is more than enough water vapor to maintain an ice crystal. Let us see how that leads to precipitation.

Refer to Figure 7–7 and consider the situation in which ice crystals and supercooled droplets coexist, and the vapor pressure is equal to that needed to keep the droplets in equilibrium. In (a) the rate of condensation onto the liquid droplet equals the rate of evaporation. But while vapor pressure in the cloud equals the saturation vapor pressure for the droplet, it exceeds that for the ice. This causes some of the water vapor in the air to be deposited directly on the ice. The vapor content of the air then falls, which in turn causes the liquid droplet to evaporate as it gives up water to restore equilibrium (b).

But the process does not end there, because evaporation from the droplet increases the water vapor content of the air, which causes further deposition onto the ice crystals (c). This leads to a continuous transfer in which the liquid droplets surrender water vapor, which is subsequently deposited onto the ice crystals. In other words, the ice crystals continually grow at the expense of the supercooled droplets. Although Figure 7–7 suggests this process involves distinct steps, evaporation and deposition actually occur simultaneously.

The growth of ice crystals by the deposition of water vapor is usually not enough to create more than a very light precipitation. As the ice crystals grow, their increasing mass enables them to fall through the cloud and collide with droplets and other ice crystals. The collisions cause two other important processes to occur that greatly accelerate the growth rate of the ice crystals: riming and aggregation.

▲ **Figure 7–6**
A cumulonimbus cloud. The lower portion consists entirely of liquid droplets, the middle a mixture of ice and liquid, and the upper portion entirely of ice. Note the less sharply defined margins of the glaciated portion composed of ice.

Riming and Aggregation We have seen that the formation of ice crystals in the atmosphere usually requires the presence of ice nuclei, or particles that initiate freezing. It so happens that ice itself is a very effective ice nucleus. Thus, when ice crystals fall through a cloud and collide with supercooled droplets, the liquid water freezes onto them. This process, called **riming** (or **accretion**), causes rapid growth of the ice crystals, which further increases their fall speeds and promotes even further riming.

Another important process in the development of precipitation is **aggregation**, the joining of two ice crystals to form a single, larger one. Aggregation occurs most easily when the ice crystals have a thin coating of liquid water to make them more "adhesive." The presence of such water is more likely when the cloud temperature is not much below 0 °C, so adhesion is more important at the warmer end of cold clouds. (Perhaps you've noticed that very large snowflakes are more common during warm, early season snows, as opposed to those that come in the dead of winter.)

---

*This is because molecules in an ice crystal bond to each other more tightly than do those of liquid water. Interestingly, Bergeron did not discover this. That discovery was made years earlier by Alfred Wegener, who is better known for his contributions to the theory of plate tectonics (sometimes called *continental drift*). Bergeron's contribution was to apply this principle to the growth of cloud constituents and the formation of precipitation.

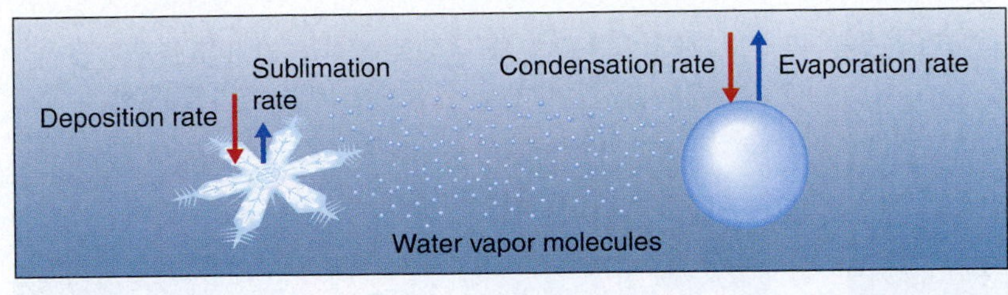

Figure 7–7
The Bergeron process. If exactly enough water vapor is in the air to keep a supercooled water droplet in equilibrium, then more than enough moisture is present to keep an ice crystal in equilibrium. This causes deposition (i.e., the transfer of water vapor to ice) to exceed sublimation (i.e., the transfer of ice to water vapor), and the crystal grows in size (a). This, in turn, draws water vapor out of the air, causing the water droplet to undergo net evaporation (b). Evaporation from the droplet puts more water vapor into the air and facilitates further growth of the ice crystal (c). Although this is shown here as a sequence of discrete steps, the processes occur simultaneously.

The combination of riming and aggregation allows ice crystals to grow much faster than by the deposition of water vapor to ice alone. In fact, growth rates from the three processes combined allow the formation of precipitation-sized crystals within about half an hour from the initial formation of the ice. When the ice crystals begin to fall, precipitation begins. What happens to these crystals as they fall determines the type of precipitation that occurs.

Forms of Precipitation

In the Tropics, precipitation occurs primarily by the collision–coalescence process, and it can therefore occur only as rain. In the middle latitudes, where ice crystal processes dominate, precipitation occurs as a solid or a liquid, depending on the temperature profile of the air through which it falls. If precipitation reaches the surface without ever having melted, we recognize it as snow. If it melts on the way down, it might reach the surface as rain. But raindrops sometimes freeze again before, or immediately after, reaching the surface, and then a different type of precipitation results. We now discuss the various types of precipitation.

Snow

Snow results from the growth of ice crystals through deposition, riming, and aggregation. Ice crystals in clouds can have a wide variety of shapes, including six-sided plates, columns, solid or hollow needles, and complex dendrites with numerous long, narrow extensions (Figure 7–8). The structure depends on the temperature and moisture conditions that exist when the crystal is formed.

If all of a crystal's growth occurs under similar conditions, its structure can be quite simple. If, on the other hand, the temperature and moisture conditions change during growth, a complex mixture of plate, needle, and dendrite can develop. Consider, for example, a crystal that originates in the cold, upper reaches of a cloud and gradually falls through a warmer environment. Because each combination of

(a)

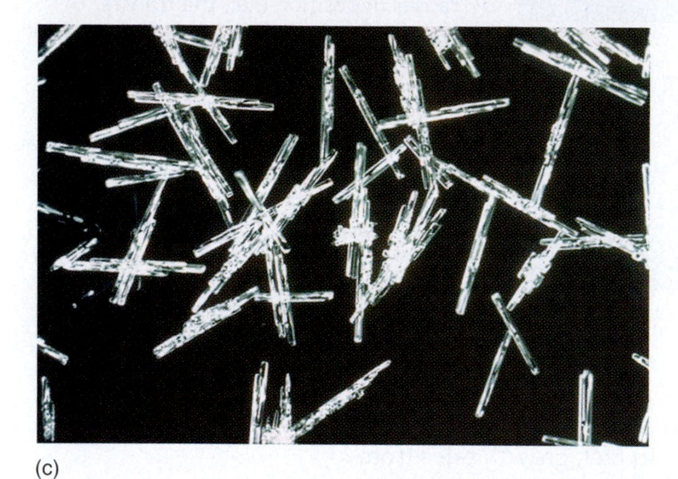

(c)

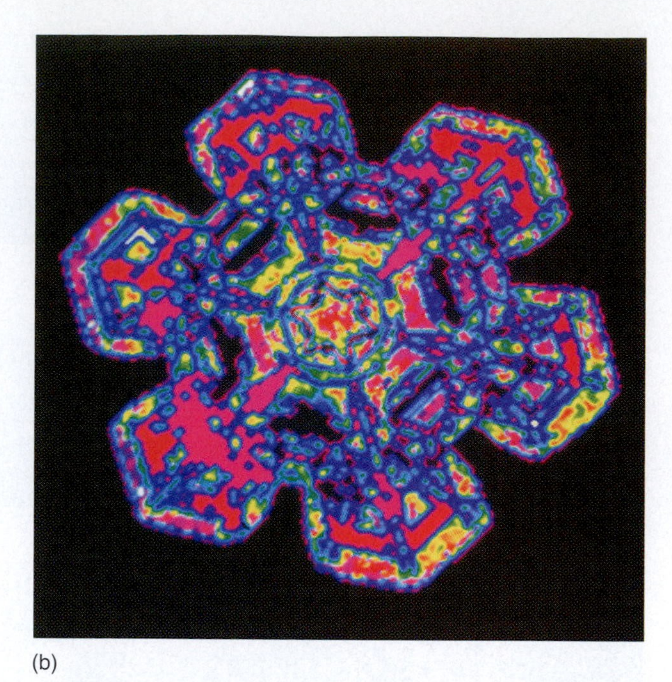

(b)

▲ **Figure 7–8**
Ice crystals can assume several general shapes, including dendrites (a), plates (b), and columns (c). Each is favored under certain conditions of moisture content and temperature.

moisture and temperature tends to favor a different type of structure, the crystal can have a particular form at its nucleus, with other forms superimposed.

Snowflakes exist in a wide range of shapes and sizes. They can be as small as about 50 μm or as large as 5 mm. Where riming is the dominant growth process (which is the case in relatively warm clouds), the crystals tend to form a dense "wet" snowpack, ideal for snowball fights but no friend to snowblowers. In contrast, very cold snow typically forms small snowflakes that accumulate on the ground with a lower density. Because of their low temperature, these crystals have less adhesion and are difficult to pack. Skiers know this type of snow as *powder*.

Figure 7–9a maps the distribution of mean annual snowfall across Canada and the United States. In the western portion of North America, the distribution of snowfall is governed largely by the presence of north–south mountain ranges (the Coast Ranges, the Sierra Nevada, the Cascades, and the Rockies) that provide orographic uplift and enhance the precipitation from passing storm systems. At high elevations, these ranges have winter temperatures low enough so that most precipitation occurs as snow. Over the eastern two-thirds of the continent, there is an increase in mean snowfall with latitude, mostly because the lower temperatures at higher latitudes favor snow rather than rain.

The distribution of annual snow (a) is in marked contrast to the distribution of annual precipitation—rain plus the water equivalent of snow (b). Precipitation over the eastern two-thirds of North America decreases with latitude rather than increases, largely due to the fact that there is less water vapor in the air with increasing distance from the Gulf of Mexico. Furthermore, lower temperatures typically found at higher latitudes reduce the amount of water vapor that can exist in the air. Note too the decrease in precipitation westward across the Great Plains, revealing a rain shadow in the lee of the Rockies.

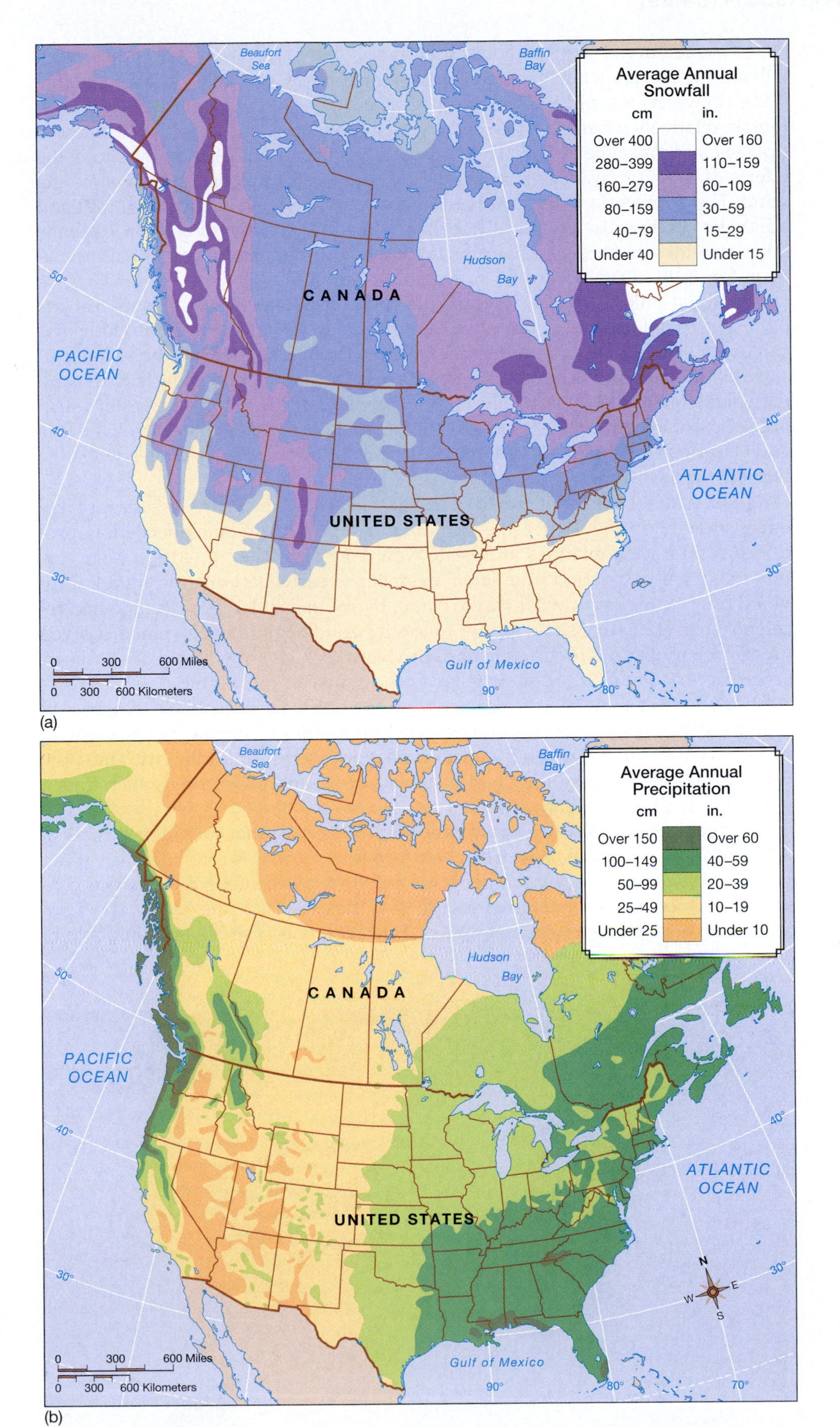

◀ **Figure 7–9** Average annual snowfall in Canada and the United States (a), and average annual precipitation (b).

One feature of the snow distribution not shown in Figure 7–9 is the strong enhancement of snowfall that occurs downwind of the Great Lakes (and other large bodies of water). As cold air from the north or northwest flows over the lakes, which remain relatively warm even into early winter, heat and water vapor are transferred upward and make the air moist and unstable. As the air passes over the shore, the effects of topography, vegetation, and other features of the land surface slow the wind. The decrease in wind speed causes convergence, a mechanism for uplift and adiabatic cooling discussed in Chapter 6. Thus, the passage of cold air over the lakes provides the three mechanisms favorable for precipitation: an initial mechanism for uplift, unstable air, and sufficient moisture.

This so-called *lake effect* (Figure 7–10) often produces snow showers restricted to a strip of land that can be anywhere from 1 to 80 km (0.6 to 50 mi) long and can extend more than 100 km (60 mi) inland. (It can also increase the amount of snowfall from storm systems passing over the lakes.) Lake-effect snow is most common along the northern part of the Upper Peninsula of Michigan, the western strip of the Lower Peninsula along Lake Michigan, and the southern shores of Lake Erie and Ontario. It is most likely to occur in the late autumn and early winter, when the lakes are still warm.

The winter of 1976–77 provided one of the most noteworthy seasons for lake-effect snowfall in the Great Lakes region, when 51 days of lake-effect snow produced record accumulations in upstate New York. The 103 cm (40.5 in.) of snow that fell in a 4-day period from late November to early December in Buffalo provided only a preview of what was to come. By the end of January, Buffalo had received 3.6 m (12.5 ft) of snow for the 3-month period beginning November 1. Even greater accumulations occurred in northern New York, downwind of Lake Ontario, where up to 9.5 m (33 ft) of snow fell over the course of the winter.

Lake-effect snowfall can also occur on the northern side of Lake Ontario. In January 1999, for instance, Toronto experienced a series of storms that brought almost 120 cm (50 in.) of snowfall—an all-time monthly record for the city. The snow was concentrated during the first two weeks of the month, during which time a normal year's worth of snow fell.

There is one widely held misconception about snow that should be dispelled. Although some people believe it can be too cold to snow, this is not the case. It can be too cold to snow *a lot*, but it is never too cold to snow *at all*. Because mass is conserved, any ice crystals that form must do so at the expense of the water vapor content of the air. At very low temperatures, however, only a small amount of water

Figure 7–10
Heavy lake-effect snow in Buffalo, New York.

vapor can exist in the air. And without an ample supply of water vapor, cooling of the air can cause the deposition of only a limited supply of ice. Nonetheless, it is still possible for some snow to occur, no matter how low the temperature.

Rain

As we have already seen, most precipitation in the Tropics comes from warm clouds whose temperatures are somewhat above the melting point of ice. Furthermore, the air temperature below the clouds is well above freezing, so the rain does not freeze after leaving the base of the cloud. Thus, virtually all precipitation in this part of the world occurs as **rain**, except in high mountains such as Kilimanjaro in Tanzania. In the middle latitudes, precipitation is usually initiated by the Bergeron process, so most rainfall results from the melting of falling snow.

Rain Showers Convection can lead to the development of cumuliform clouds and precipitation within a few minutes. The episodic precipitation from these rapidly developing clouds is called **showers**. Showers can occur as either rain or snow, but because convection is usually most vigorous in the warm season, showers are more likely to occur as rain.

During a steady rain, droplets occur in a wide variety of sizes. In a shower, the first droplets all tend to be large and widely spaced, but within a short period of time the large droplets give way to a greater number of smaller ones. What happens is quite simple: large and small droplets fall from the base of the cloud together, but the larger raindrops have greater terminal velocities and reach the surface while the smaller ones are still falling through the air.

Another factor favors the occurrence of large drops at the beginning of a rain shower. Because they take longer to fall through the unsaturated air, small raindrops are more likely to evaporate before reaching the surface. (Evaporation does decrease after a few minutes, however, when the first drops have sufficiently increased the moisture content of the air.)

Raindrop Shape One of the prevailing myths about weather is that raindrops are teardrop-shaped. In reality, raindrops are initially spherical (Figure 7–11a). As they grow by collision and coalescence, their velocities increase, and the greater wind resistance flattens them along the bottom to give them a parachute or mushroom shape (b). As the bottom of the drops flattens out, the greater surface area increases the wind resistance and leads to even further flattening (c). Eventually, wind resistance exceeds the surface tension that holds the droplets together and they break apart (d). The resulting small droplets then begin to grow again by collision and coalescence.

The breakup of falling drops explains why collision and coalescence do not produce enormously large droplets. If the drops were to grow continually on their way down, they could conceivably attain the size of basketballs! Under special conditions, droplets can have diameters of up to 5 mm or more, but they are seldom larger.

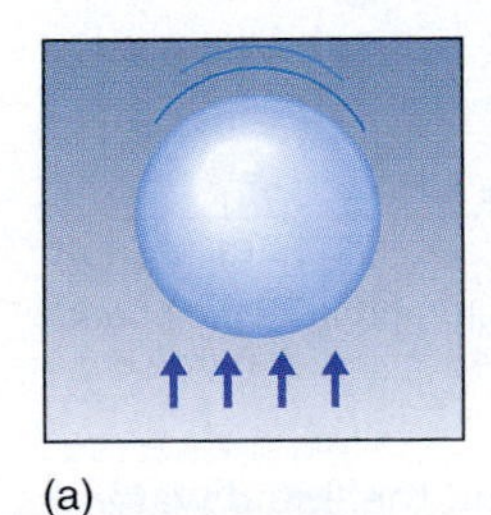
(a)

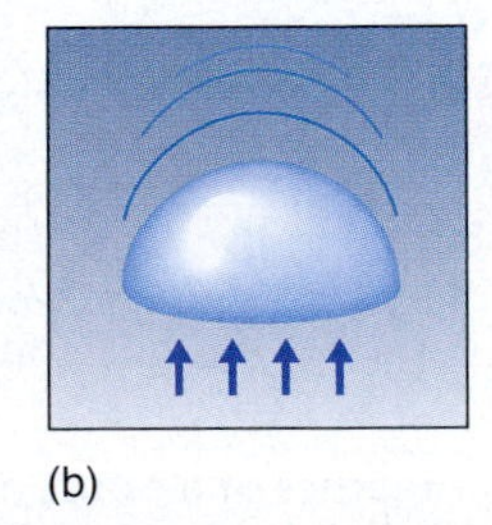
(b)

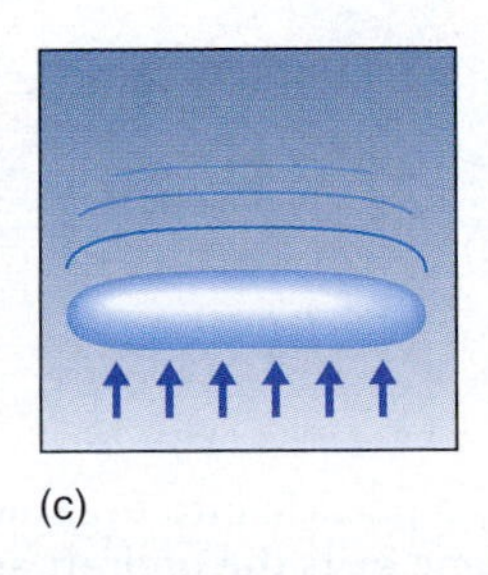
(c)

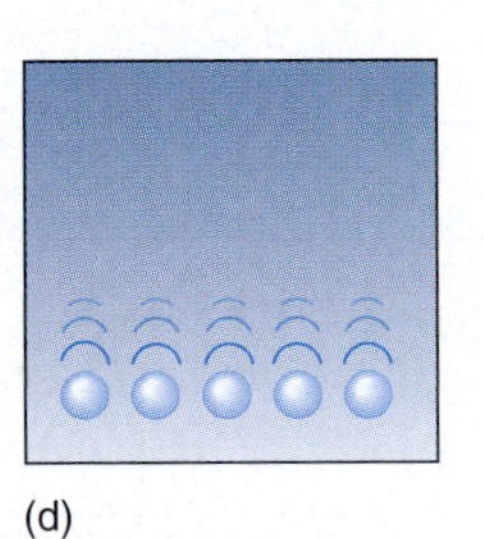
(d)

▲ **Figure 7–11**
Raindrops are not teardrop-shaped. They are initially spherical (a) but flatten out on the bottom as they fall (b). As they flatten, the greater surface area on the bottom causes greater resistance and further flattening (c). Eventually, the droplet breaks apart (d).

▲ Figure 7–12
Large hailstones. Each of the concentric rings was formed during an upward passage above the freezing level in the cloud.

Graupel and Hail

When an ice crystal takes on additional mass by riming, its original six-sided structure becomes obscured and its sharp edges are smoothed out. The new ice may contain very small air bubbles that give it a spongy texture and milky-white appearance. This type of modified ice crystal is called **graupel** (pronounced GRAU-pull). Graupel pellets attain diameters up to 5 mm or so, giving them terminal velocities of about 2.5 m/sec (5 mph). Graupel pellets can fall to the ground as precipitation, but under other circumstances they can remain in the cloud and provide the nucleus upon which hail forms.

Hail consists of ice pellets formed in roughly concentric layers (Figure 7–12). Figure 7–13 illustrates the processes that shape hail in a cumulonimbus cloud. Initially, an updraft carries a graupel pellet or water droplet above the freezing level to form the core of a hailstone. At some point, the pellet falls from the updraft and collides with liquid droplets that coat it with a film of liquid water. If the pellet is again lifted above the freezing level by an updraft, the liquid water freezes to form a second layer of ice. The sequence can repeat itself numerous times, with each passage above the freezing level resulting in another shell of ice.

The ultimate size of the hailstone is determined by the intensity of the updrafts—stronger vertical motions are capable of lifting larger hailstones. Although most hailstones are pea-sized, they can become as large as marbles, golfballs, or under the most violent of conditions, even softballs!

Hailstones consist mainly of ice with only a small amount of air mixed in. Because ice is relatively dense—90 percent as dense as liquid water—hailstones can become fairly heavy. Compare hailstones to snowflakes, whose volume is occupied mostly by air. Snowflakes have relatively little mass and low terminal velocities, so they flutter to the ground and make barely a sound. Hailstones, on the other hand, sound like a barrage of falling marbles as they hit the surface. Table 7–1 lists the terminal velocities of various sizes of hailstones.

Weather Image

Large Hailstone

Hailstones

Hailstones the size of baseballs (radius = 3.5 cm) contain about 160 g of ice (weighing about a third of a pound) and fall at about 40 m/sec (88 mph)! No wonder they are capable of producing tremendous damage. Hailstorms present a major threat to the Great Plains of the United States (Figure 7–14), where they are known to destroy entire fields of crops in a matter of minutes. Because they are most common in the spring and summer, it is often too late to replant the acreage with new seed.

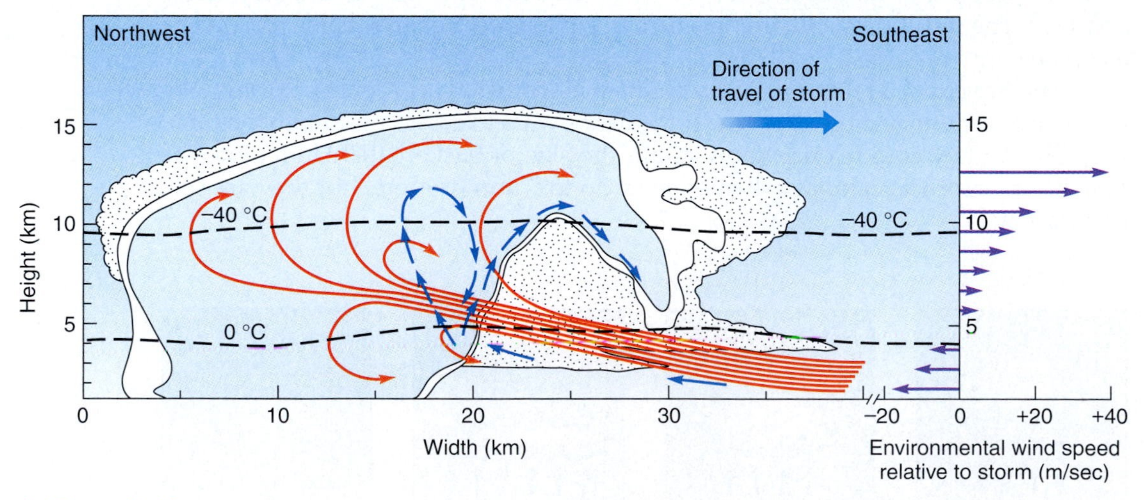

▲ Figure 7–13
The formation of hail. Updrafts carry a particle into the colder reaches of a cloud, and the liquid water coating the ice freezes. When the stone exits the updraft and falls, it becomes wet from its collisions with liquid droplets. The hailstone can be captured once again by an updraft, and the coating of water freezes. This process, when it occurs repeatedly, forms large hail.

Table 7–1 • Terminal Velocities of Hailstones

$v_t = \sqrt{\frac{1}{3}\rho/k}\,\sqrt{r} = 20\sqrt{r}$ (**r in cm, v_t in m/sec**)

Radius (cm)	Terminal Velocity (meters per second)
0.1	6 (13 mph)
1.0	20 (44 mph)
2.0	28 (62 mph)
3.0	35 (77 mph)

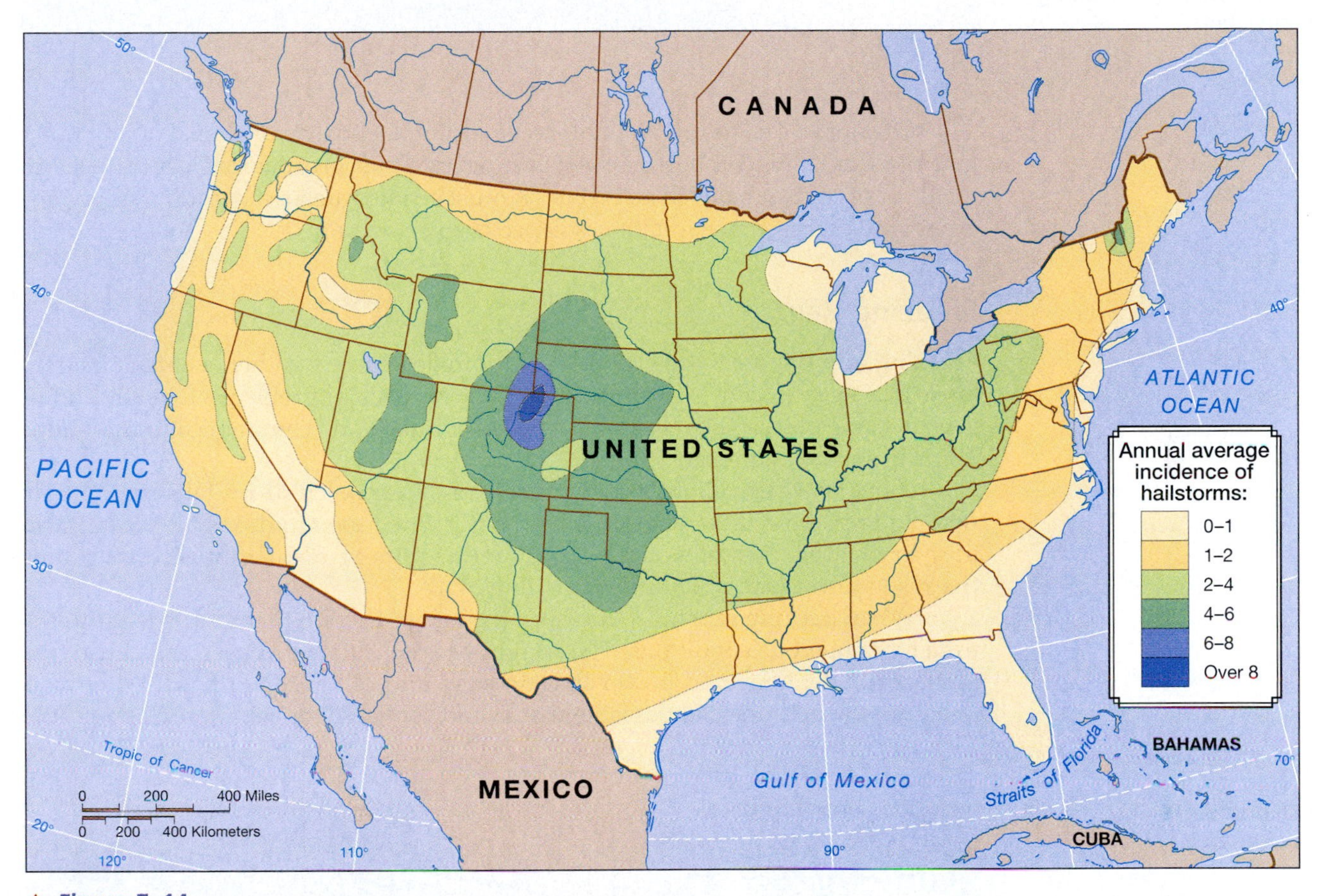

▲ **Figure 7–14**
The annual average number of hailstorms over the United States.

Sleet

Sleet forms when raindrops freeze in the air while falling to the surface. Because most rain outside the Tropics originates from the ice crystal process, sleet begins as falling ice crystals or snowflakes. As the ice falls through the air, it encounters warmer air and melts to form a raindrop. If the falling raindrop then encounters a lower layer of air whose temperature is below 0 °C, it can refreeze to form sleet. This process, shown in Figure 7–15, results in semi-transparent pellets smaller than about 5 mm (0.2 in.) in diameter. Because the formation of sleet requires that a droplet fall through air that is cooler near the surface than aloft, it necessarily requires an inversion, usually one associated with the presence of a warm front (which we describe in Chapter 9).

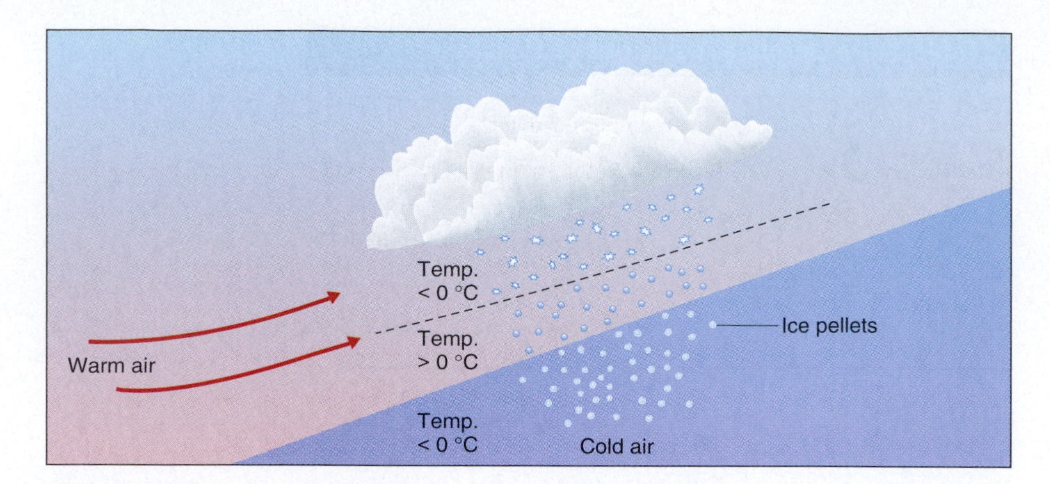

▶ **Figure 7–15**
Sleet occurs as rain, falling from a cloud, passes through a cold layer and freezes into ice pellets. This is most common along warm fronts.

Of course, a raindrop will not freeze instantaneously; sufficient cooling must take place as it falls through the surrounding air. Thus, in order for sleet to develop, the layer of cold air beneath an inversion must be fairly deep. If it is too shallow, another type of frozen precipitation will occur, freezing rain.

Freezing Rain

Freezing rain (Figure 7–16) is one of the more deceptive weather events. It usually looks like a gentle rain—certainly nothing to cause major problems. In reality, however, freezing rain can literally paralyze transportation and communications for hundreds of square kilometers.

It begins when a light rain or drizzle of supercooled drops falls through air with a temperature at or slightly below 0 °C. When the raindrops hit the surface, they form a thin film of water, but only for a moment. Soon afterward the water freezes to form a slick, continuous coating of ice.

When freezing rain hits roadways, the loss of friction that results leads to extremely dangerous conditions. The weight of accumulated ice also can cause tree limbs and telephone and power lines to snap and fall to the ground. When you imagine downed lines, impassable roads, and broken debris scattered about, it is

▶ **Figure 7–16**
Freezing rain.

7–2 Physical Principles

The Effect of Hail Size on Damage

You might wonder why large hailstones are more damaging than small ones. After all, isn't the issue how much ice falls to the ground, and aren't many small stones roughly equivalent to a smaller number of large stones? As it happens, this is far from true. Damage done by hail increases very rapidly in a nonlinear fashion with increasing hailstone size. What matters is the amount of kinetic energy hail carries as it falls to the surface.

Kinetic energy depends on mass (m) and speed (v) according to

$$KE = \frac{1}{2}mv^2$$

The mass of a stone is its density (ρ) times its volume

$$\left(\frac{4}{3}\pi r^3\right).$$

Velocity can be found using the formula in Table 7–1. We can thus write

$$KE = \frac{1}{2}\left(\rho\frac{4}{3}\pi r^3\right)(20^2 r)$$

We see that the kinetic energy of a hailstone is proportional to the fourth power of its radius. As a result, a hailstone with a radius of 1 cm packs not twice as much punch as a 0.5 cm hailstone, but 16 times as much!

easy to see why freezing rain can be so disruptive to human life. To make matters worse, freezing rain is often associated with slowly moving storm systems and may therefore persist for several days.

Measuring Precipitation

Given the importance of precipitation to everyday activities, it is no surprise that it is measured at a large number of locations. Just as precipitation occurs in several different forms, different types of gauges exist for observing precipitation amounts. Each method has its own advantages and disadvantages. Figure 7–17 shows the distribution of mean annual precipitation based on worldwide weather records.

Global Precipitation

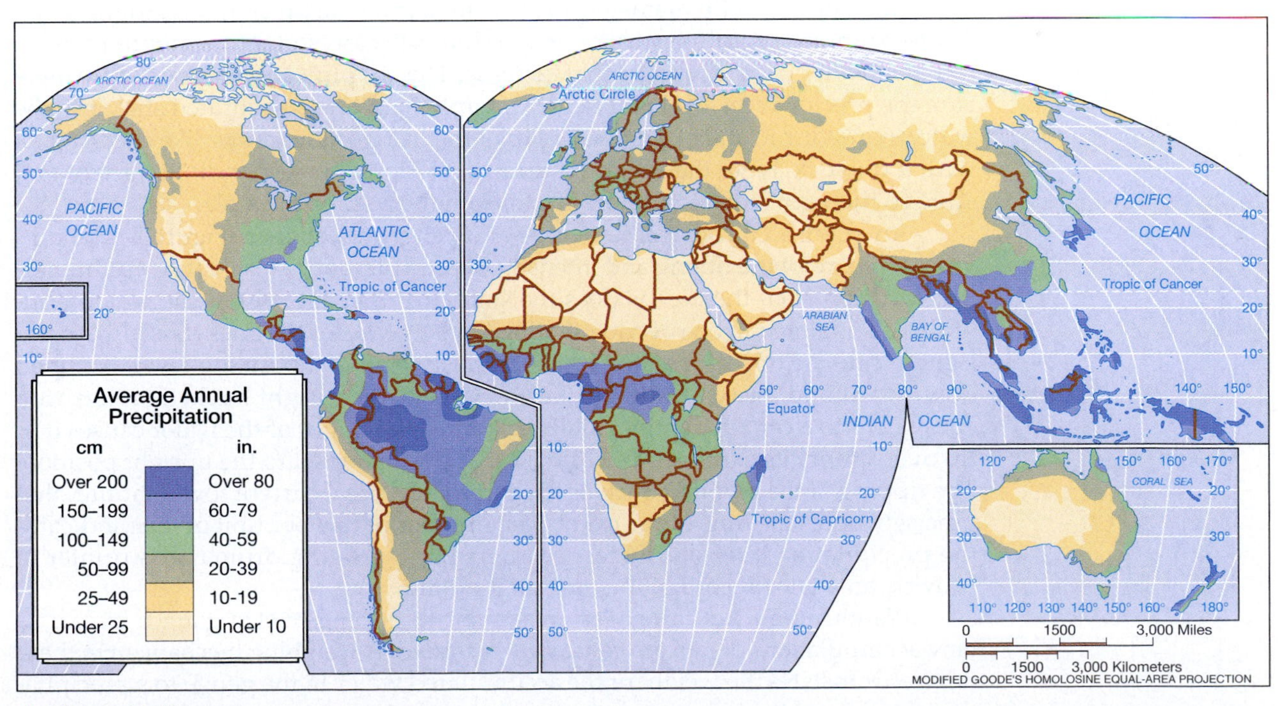

▲ Figure 7–17
Average annual world precipitation.

(a)

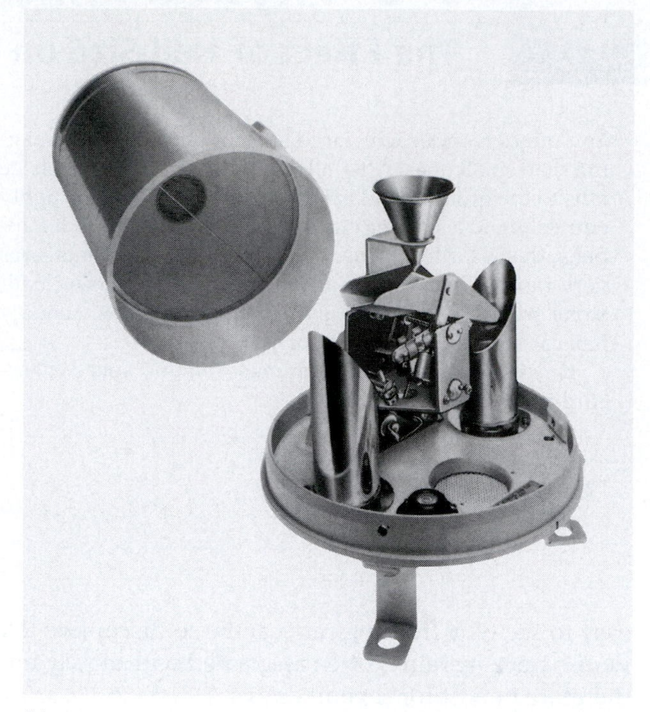

(b)

▲ **Figure 7–18**
A standard raingage with its component parts (a). Rain captured by the collector (bottom left) is funneled into the narrow tube (right). The calibrating stick is inserted into the collection tube, and the length of the wetted portion indicates the precipitation accumulation. The interior workings of a tipping-bucket gage are shown in (b).

Raingages

Rainfall is usually measured by a **raingage** (Figure 7–18a). Standard gauges have collecting surfaces with diameters of 20.3 cm (8 in.). The precipitation funnels into a tube with one-tenth the surface area of the collector, so that the depth of accumulated water undergoes a tenfold increase. This amplification permits us to measure the precipitation level precisely by simply inserting a calibrated stick into the water, removing it, and noting the depth of the wet portion, rather like checking oil with a dipstick. Note that the measuring stick has a correspondingly graduated scale so the 1-cm mark is actually 10 cm from the base of the scale.

An automated collector known as a **tipping-bucket gage** (Figure 7–18b) provides a record of the timing and intensity of precipitation. This instrument funnels precipitation from the top, as does the standard raingage, but as the water accumulates it is stored in one of two pivoting buckets. One of the buckets is always upright, while the other, mounted on the opposite end of a pivoting lever, is tipped downward and away from the collector. When the upright bucket gathers rain equivalent to a certain depth (usually 0.01 in.), the weight of the water causes it to tip over, empty its contents, and bring the opposite bucket to the upright position. The tipping of the pivoting buckets triggers an electrical current to a computer that precisely notes the time of the event. The number of tips per unit of time indicates the precipitation intensity. Older recorders use a rotating drum and a printer to provide an analog record of the precipitation rate.

Weighing-bucket raingages are similar to tipping-bucket gauges insofar as new accumulations of rain are constantly recorded. A weighing mechanism in these devices translates the weight of the accumulated water in the gauge to a precipitation depth, and the information is stored automatically.

Raingages are found at virtually all weather-recording stations, which makes precipitation data plentiful in economically developed countries. Data are scarce in much of the rest of the world, especially over the more than 70 percent of Earth's surface covered by ocean. Furthermore, measurement accuracy is a concern (even with modern instruments), due to problems such as evaporation from the gage, winds that can prevent rain from entering the gage, and other factors.

Raingage Measurement Errors We tend to put unquestioned confidence in the readings we get from precipitation gages, but unfortunately they have several inherent sources of error. First of all, they are *point measurements*, meaning that they represent the amount of precipitation that has fallen at only a single point or location—not across the street, 100 m away, or down the block. Compared to air temperature, precipitation shows wide variations across different locations, so we face some difficulty in trying to generalize from gage readings. This not only frustrates attempts to estimate precipitation at points where measurements are unavailable, but it also greatly complicates the mapping of precipitation patterns.

Precipitation gages have other flaws. First, wind-generated turbulence near the top of a gage deflects precipitation away from the collecting surface, leading to an underestimate of the true value (especially for snow). Of the precipitation that does find its way into the gage, a certain amount goes unrecorded—some because the water splashes out upon impact, and some because it is retained as a thin film that does not accumulate at the base of the collector. On hot, windy days some of the collected water evaporates from the collector prior to measurement.

Other measurement errors have the opposite effect and cause an overestimate of precipitation. Just as water can splash out of a gage, some can bounce off the surrounding ground and into the collector. Similarly, wind can cause snow to drift from nearby surfaces into the gage. These errors can affect gage catches even under the most carefully monitored conditions. Other things can cause errors as well, such as failing to completely empty the gage after the last measurement, placing it on a nonlevel surface, or spilling some of the accumulated water. At well-run weather stations where conditions can be controlled, errors such as these are usually very small. Imagine, however, the difficulty in making reliable observations of precipitation on a ship or buoy. The surface of the gage can be kept only as horizontal as a pitching and rolling ship deck will allow, and it is extremely difficult to prevent windblown seawater from entering the gage. Thus, the precipitation data for the three-quarters of Earth's surface covered by ocean are not only sparse (confined mainly to shipping lanes) but also of questionable quality.

The moral here is that the distribution of precipitation is not as well recorded as we might suppose, nor as well recorded as befits its extreme importance to agriculture and the welfare of humankind. Figure 7–19 shows estimated gage errors for the world, based on a recent study that developed techniques for error estimation and correction. The most severe errors, which occur at high latitudes and over water, sometimes exceed 80 percent of the true values. Clearly, uncorrected gage measurements must be used cautiously.

Precipitation Measurement by Weather Radar During recent years, the network of raingages has been augmented by radar measurements. Though radar will be discussed in more detail in Chapter 11, for now we can say that radar can measure the intensity of precipitation by emitting microwave radiation with wavelengths of several centimeters. Precipitating droplets, ice crystals, and hailstones can scatter the emitted radiation back to the radar unit, which records the intensity of the backscattered radiation. In general, the more intense the backscattered radiation, the more intense the precipitation. Meteorologists have developed schemes that relate the intensity of radiation to the rate of precipitation.

Radiation is not emitted continuously by the transmitter, but just for very brief periods that are interspersed between momentary pauses. Sufficient time is allowed for each pulse of radiation to echo back to the transmitter/receiver unit before the

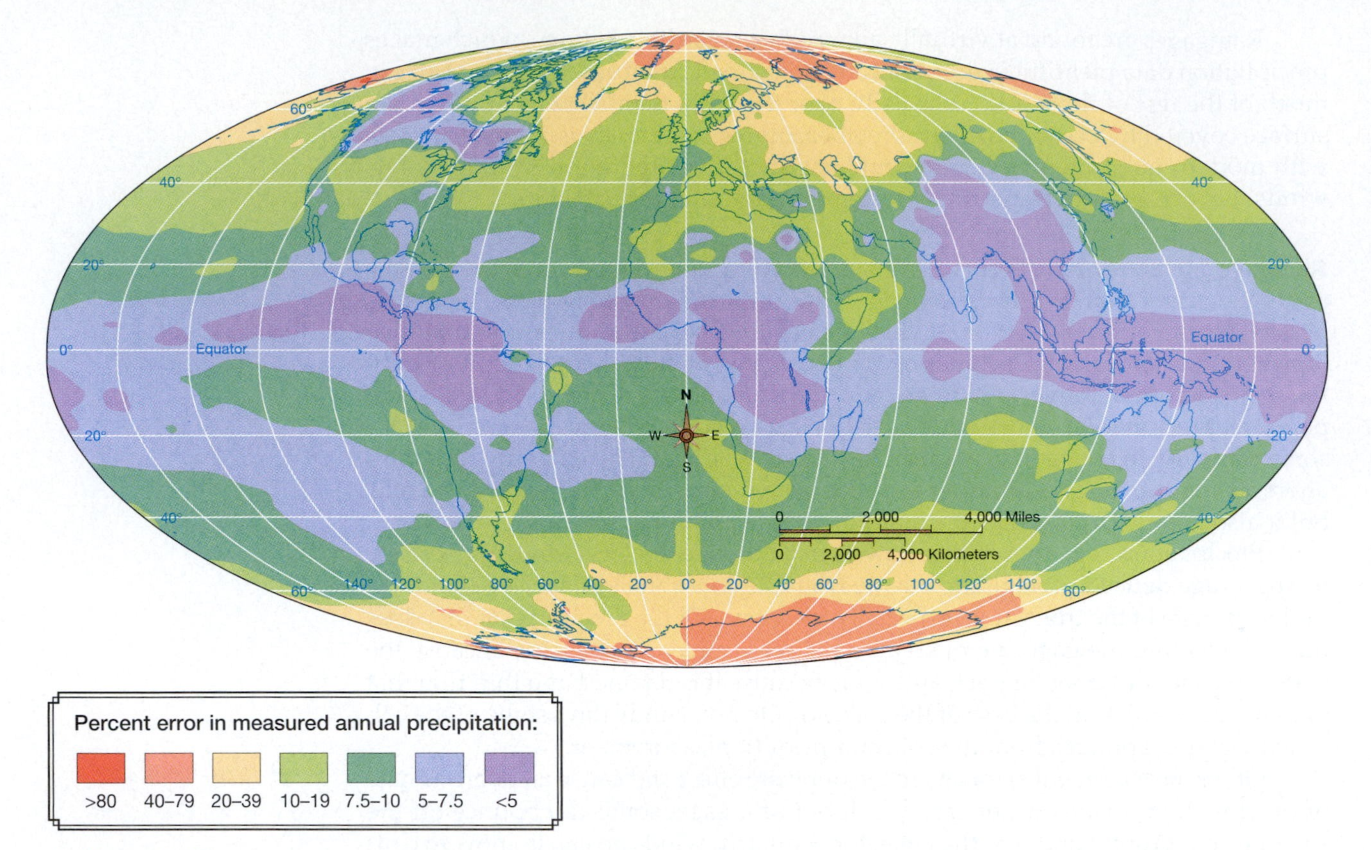

▲ **Figure 7–19**
The global distribution of the percent error in measured annual precipitation. Darker areas have greater percent errors, with intervals drawn at 2.5, 5, 10, 20, 40, and 80 percent.

next beam is emitted. The closer the precipitation is to the radar, the quicker the pulse will return to the unit. By measuring the strength of the return radiation and the time taken for it to return to the unit, a profile can be taken showing how much precipitation is occurring and how far from the radar it is.

The transmitter/receiver slowly rotates as it emits and receives the radar beams, thus allowing a two-dimensional depiction of the precipitation for several hundred kilometers from the radar. The information can be continually monitored and stored, so that the total amount of precipitation occurring over a fairly extensive region can be estimated (Figure 7–20). Such measurements have proven to be particularly useful in providing short-term forecasting for potential floods.

Snow Measurement

Raingages are particularly unreliable when precipitation occurs as snow, because captured snow can obstruct the inlet to the collecting tube. To get an estimate of the precipitation in these environments, we measure the depth of accumulated snow. The *water equivalent* of the snow, which is the depth of water that would result if all the snow were melted, can then be roughly estimated using a conversion ratio of 10:1. Although the 10-to-1 ratio is commonly used, in reality the ratio of snow depth to water equivalent can vary greatly—from about 4:1 to even 50:1.

In remote mountainous areas, particularly in the western United States and Canada, observations of snow cover have been made for decades at hundreds of *snow courses*. Usually about ten observations are made at each snow course by pushing a collection tube into the snow, extracting the tube and its contents, and weighing them on a spring balance. The weight of the snow-filled tube is directly proportional to the water equivalent of the snow cover, and the average of the ten or so readings is used as the representative value.

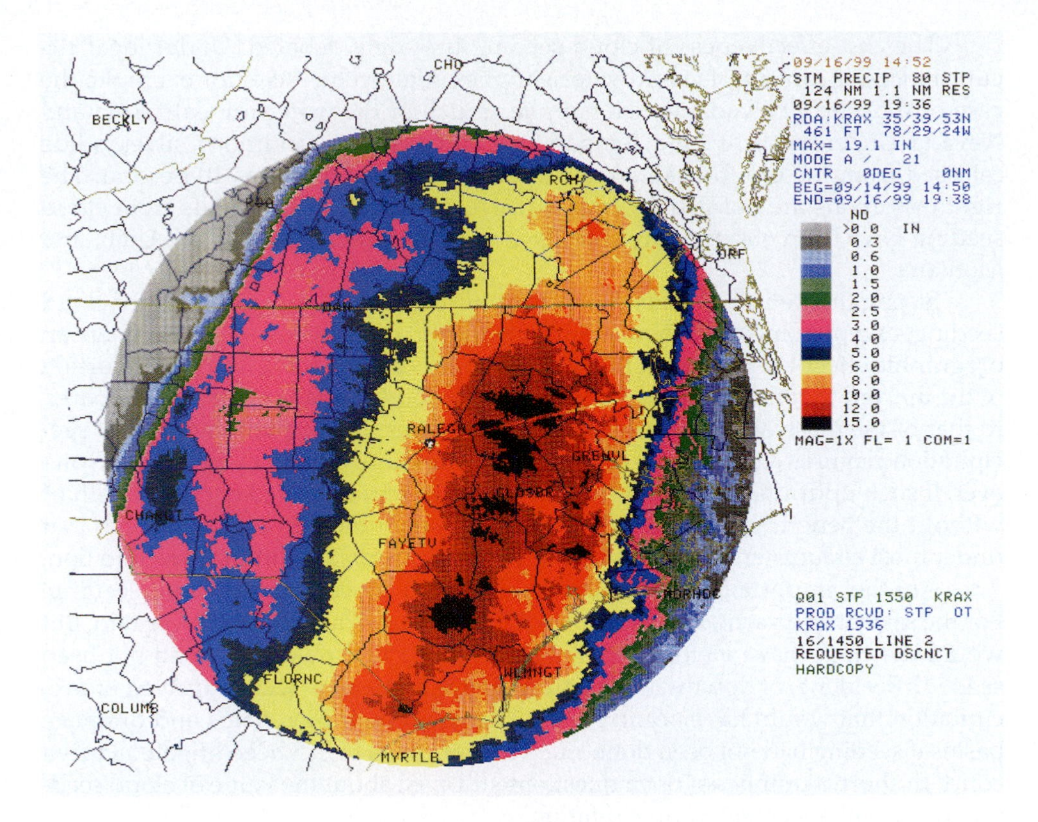

Figure 7–20 Precipitation estimates depicted by Doppler radar.

Over the last few years, manual snow course measurements have given way to the use of automated snow pillows. **Snow pillows** are large air mattresses filled with an antifreeze liquid and connected to pressure recorders. As snow accumulates on a pillow, the increased weight is recorded and converted to a water equivalent. These instruments have radio devices that transmit the data to a centralized receiving station. Interestingly, the transmitted radio signals are reflected back to the receiving station by meteorites in the upper atmosphere.

Cloud Seeding

Since the late 1940s, people have tried to induce precipitation from clouds, most often to alleviate droughts. This process, called **cloud seeding**, involves injecting one of two materials into nonprecipitating clouds. The objective is to convert some of the supercooled droplets in a cool cloud to ice and cause precipitation by the Bergeron process.

One of the materials, *dry ice* (frozen carbon dioxide), promotes freezing because of its very low temperature (below –78 °C or –108 °F). When introduced into a cloud, dry ice lowers the temperature of the droplets so that freezing can occur by homogeneous nucleation. (Recall that at temperatures below about –40 °C, water droplets require no ice nuclei in order to freeze.) Dry ice can be introduced by ejecting very small shavings of the material from a plane flying through the cloud.

The second agent for cloud seeding, *silver iodide*, initiates the Bergeron process by acting as an ice nucleus at temperatures as high as –5 °C (23 °F). Silver iodide owes its effectiveness as an ice nucleus to its inherent six-sided structure. Like dry ice, silver iodide can be introduced directly into a cloud from aircraft. More often, it is mixed with a material that produces smoke when ignited in ground-based burners. Updrafts then carry the smoke and the silver iodide into the cloud. If a portion of the seeded cloud is cold enough, some of the supercooled droplets will freeze and begin the Bergeron process.

The cost-effectiveness of cloud seeding is widely debated. Under ideal circumstances, it can supplement water supplies somewhat. Take, for example, the case of the Sierra Nevada, which supplies much of the water for California and Nevada. Under the right wind, temperature, and moisture conditions, silver iodide released from the ground can enhance precipitation by perhaps 10 percent. The right conditions are not usually met, however. Similarly poor results from cloud seeding have been observed from trials in the mountains of Colorado, Utah, and Montana.

Strong theoretical reasons provide grounds to doubt the usefulness of cloud seeding, except in regions where a continued uplift of air occurs (such as where an orographic effect exists). Recall that water vapor accounts for only a small portion of the air, and that the liquid water content of a cloud is relatively small compared to that of the mass of the air contained within. Thus, the formation of heavy precipitation requires a constant resupply of moisture into clouds by updrafts. However, if such updrafts are already occurring, precipitation will likely occur with or without the benefit of seeding. Consequently, many meteorologists believe that under most circumstances cloud seeding yields little or no additional precipitation.

Another important question raises an ethical concern about cloud seeding. For the sake of argument, let's assume that the seeding of a cloud produces rain. But would the cloud have yielded precipitation farther downwind if it had not been seeded? Residents of downwind locations might argue they were deprived of precipitation that would have occurred naturally over their own fields and drainage basins if seeding had not been done. Such matters have in fact been litigated in civil court. In short, a number of open questions still exist about the value of cloud seeding for the enhancement of precipitation.

Interestingly, cloud seeding has also been attempted for the reduction of hail intensity. It was once believed that by seeding hail-producing clouds one could increase the number of growing ice pellets. Because clouds only contain so much available water that can freeze onto growing hailstones, an increase in the number of hailstones would theoretically reduce their average size. The decrease in size would reduce the kinetic energy of the falling hailstones and thereby lower the likelihood of damage near the ground. However, a multiyear experiment in northeastern Colorado in the 1970s was unable to support the usefulness of seeding as a hail suppression measure, and such attempts have been discontinued in the United States and Canada.

Summary

When you first picked up this book, no doubt you thought you would learn how precipitation occurs. That has been the focus of this chapter.

We can summarize the start of precipitation as the result of cloud droplets growing beyond a size that can remain suspended in the air. But the amount of growth necessary for each droplet is tremendous. Some growth occurs through condensation onto existing droplets, but further growth depends on other processes. In the Tropics, the primary mechanism for droplet growth is collision and coalescence. Outside the Tropics, collision and coalescence are still important, but the Bergeron process dominates, wherein ice crystals grow at the expense of supercooled droplets. Once the Bergeron process has been set in motion, riming and aggregation promote even further growth of ice.

Precipitation occurs in several different forms. Outside the Tropics, rain usually results from the melting of ice crystals or snowflakes as they fall to the surface. Ice crystals that do not melt prior to reaching the surface form snow. Graupel and hail form when supercooled water attaches to ice crystals and freezes. In the case of hail, growth occurs when water repeatedly freezes onto existing ice pellets as they pass above the freezing level. Sleet and freezing rain both involve the freezing of raindrops. For sleet, freezing occurs before the drop reaches the surface; for freezing rain, it takes place upon contact with the surface.

The standard instrument for measuring precipitation is the simple raingage, a collecting device that funnels precipitation into a narrow tube for measurement with a calibrated stick. The timing and intensity of precipitation can be recorded with a modified precipitation gage called a tipping-bucket gage. As direct and uncomplicated as precipitation measurement may seem, it is subject to a host of potential errors. As a result, our knowledge about the worldwide distribution of this important meteorological variable is subject to much uncertainty.

When precipitation is not sufficient to meet human needs, people sometimes resort to cloud seeding, which is the introduction of materials into clouds to stimulate precipitation by the ice crystal process. This can be done with dry ice, which causes supercooled droplets to freeze by homogeneous nucleation, or with silver iodide, which serves as an ice nucleus. At present, strong reasons exist for doubting the efficacy of cloud seeding other than in certain limited circumstances.

Key Terms

drag p. 202
terminal velocity p. 202
warm cloud p. 204
collision–coalescence process p. 204
collector drop p. 204
collision p. 204
coalescence p. 205
Bergeron process p. 205
cold cloud p. 205
cool cloud p. 205
riming p. 206
aggregation p. 206
snow p. 207
rain p. 211
shower p. 211
graupel p. 212
hail p. 212
sleet p. 213
freezing rain p. 214
raingage p. 216
tipping-bucket gage p. 216
weighing-bucket raingage p. 216
snow pillow p. 219
cloud seeding p. 219

Review Questions

1. What determines the terminal velocity of falling droplets and raindrops?
2. Describe the characteristics that distinguish warm, cool, and cold clouds.
3. How do the growth processes of droplets in warm and cold clouds differ from each other?
4. Why isn't growth by condensation able to create precipitation size droplets on its own?
5. How do collision and coalescence work to increase the size of cloud droplets?
6. Explain how variations in the saturation vapor pressure for ice crystals and supercooled water droplets affect the development of precipitation.
7. Why can't the Bergeron process take place in warm clouds?
8. What are riming and aggregation?
9. Why is precipitation greater in Mississippi than in Michigan?
10. How do lakes enhance precipitation downwind?
11. Why do rain showers initially occur only as large drops?
12. Explain why the formation of sleet requires an inversion.
13. It is never too cold for snow to occur. Is that also true of sleet?
14. Why does hail consist of multiple layers of ice?
15. What are some of the inherent sources of error in raingages?
16. How do weighing-bucket and tipping-bucket gages measure rainfall?
17. Explain how snow pillows measure snow accumulation.
18. What are the materials used in cloud seeding, and how do they stimulate (or inhibit) precipitation?

Critical Thinking

1. Industrial activity can increase the number of cloud condensation nuclei. Would an increase in the number of such nuclei tend to promote the formation of rainfall or inhibit it?
2. How might a warming of the atmosphere alter the formation of rainfall in the middle latitudes?
3. It is not possible for a cloud to precipitate all of its ice crystals or water droplets. Why not?
4. Precipitable water refers to the depth of water that would precipitate if all the water in a column of air above the surface were to be rained out. Typically precipitable water is on the order of about 2.5 cm (1 in.), but precipitation amounts can greatly exceed this amount. How is this possible?
5. Weather radar is capable of examining vertical profiles of clouds and precipitation shafts. It is easy to determine from the profiles the height at which the temperature is 0 °C (32 °F). Why is that true?
6. Aircraft icing is a serious threat to aviation at temperatures of about –4 °C to 0 °C (25 °F to 32 °F). Why is it less of a problem at lower temperatures?

7. Typical raindrops fall at a speed of about 6 m/sec—roughly 25 km/hr (15 mph). Falling snowflakes obviously fall more slowly. What does this imply about the depth of cloud required to yield precipitation through the collision–coalescence process compared to through the Bergeron process?
8. A shampoo company once advertised that its product was "pure as rainwater." Do you think this is true, and if so does this speak well of the shampoo?
9. Hail is sometimes observed to propel out of a cumulonimbus near the anvil. How can large hailstones be found in this relatively shallow portion of the cloud?
10. Ice in the upper reaches of a cumulonimbus over Colorado may be observed 2 days later over the eastern United States. How can these ice crystals manage to survive without having been precipitated out of the cloud or sublimated away?

Problems and Exercises

1. On a regular basis, examine the weather radar map at **http://weather.noaa.gov/radar/mosaic/DS.p19r0/ar.us.conus.shtml**. Then click on any region for a closer, regional view of the precipitation. Would you describe the precipitation intensity as uniform or spotty? This pattern is likely to vary by season and region. How might season, type of precipitation, and geography influence the spatial variability of precipitation within a storm?
2. A particular collector drop has a fall speed of 0.26 m/sec while a smaller, 10 μm droplet directly below falls at a rate of 0.01 m/sec. How long will it take for the two to collide? How far will each of them have fallen prior to collision?
3. The terminal velocity of spherical falling droplets and raindrops is proportional to the square root of their radii.
   a. If a cloud droplet with a radius of 10 μm falls at 0.1 m/sec, how fast would a large droplet with a radius of 100 μm fall?
   b. If both droplets are within a cloud, 100 m above the cloud base, how long would it take the two of them to fall to the bottom of the cloud, assuming no growth or diminution in size?

Quantitative Problems

The Web site for this book (**http://www.prenhall.com/aguado/**) offers several problems to help you obtain a quantitative understanding of precipitation. We suggest you enhance your comprehension of precipitation by spending a few minutes working on those problems.

Useful Web Sites

http://weather.noaa.gov/radar/mosaic/DS.p19r0/ar.us.conus.shtml

Composite map showing radar returns across the country. Click on any region to get a more detailed view of the local precipitation distribution.

http://weatheroffice.ec.gc.ca/radar/index_e.html

Canadian radar maps.

http://www.intellicast.com/

Excellent source of information, including map of radar-derived precipitation amounts for last 24-hour or weekly period. To obtain map, from the main page go to U.S. Weather and click on RADAR. On the next page click on the pull-down window called HISTORIC, and select DAILY or WEEKLY PRECIPITATION.

http://lwf.ncdc.noaa.gov/oa/climate/severeweather/rainfall.html

Maps of rainfall records for the 50 states, as well as a number of reports on significant precipitation.

http://www.weather.unisys.com/surface/snow_cover.html

Map of U.S. snow cover.

http://www.weather.unisys.com/surface/prec_day.html

Daily U.S. precipitation totals.

Media Enrichment

Weather in Motion

Global Precipitation

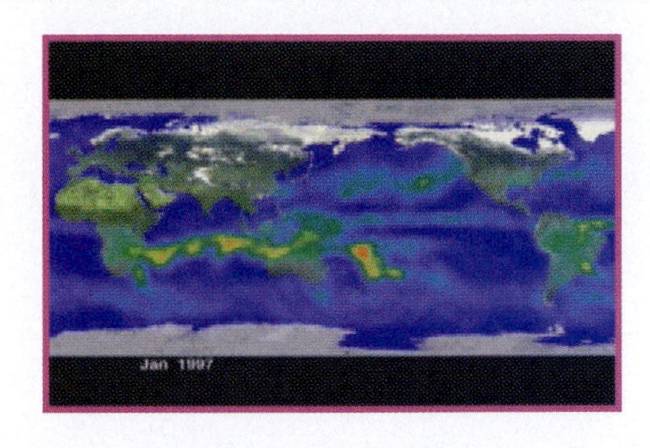

This movie shows the distribution of heavy precipitation from January 1997 to May 1998. The Tropics clearly have a greater amount of precipitation than do the higher latitudes, and the zone of maximum tropical rainfall can be seen to shift north and south in response to the movement of the solar declination. Pay close attention to the rapid development of intense precipitation off the west coast of India in June; it is associated with the Asian monsoon, described in Chapter 8. Precipitation over North America is most intense over the southern part of the continent, especially in spring and summer.

Weather Image

Large Hailstone

Some hailstones grow to be extremely large, such as the one shown in this photo. Large hailstones fall at high speeds and can be extremely damaging.

Weather Image

Hailstones

These hailstones have been cut open to reveal their concentric nature. A new layer of ice forms each time a hailstone is lifted above the freezing level.

PART THREE

Distribution and Movement of Air

High winds in a typhoon.

chapter 8

A blizzard complicates life on Wall Street, New York City.

Atmospheric Circulation and Pressure Distributions

During the summer of 1999, vast portions of the eastern United States and Canada experienced several weeks of searing heat, with temperatures persistently well above 38 °C (100 °F). These high temperatures, combined with minimal rainfall, exacerbated a water shortage that had begun during the dry summer of 1998 (Figure 8–1). By mid-1999, much of the region, especially in the mid-Atlantic states, was in the midst of its worst drought in decades. River flows and reservoir levels were critically low, forcing several eastern governors to declare drought emergencies outlawing such nonessential water use as watering lawns and washing cars.

Agricultural losses were estimated to be $800 million. The heat and drought severely reduced crop harvests for many farmers. Others endured a different problem: unusually early ripening of produce that caused an early glut and reduced prices. These low prices kept farmers such as Dale Benson of Delaware from selling his produce at local auctions.

The drought also brought about some unexpected animal behavior. Black bears, for example, wandered into suburban areas outside major eastern cities in search of food. One bear was spotted a mere 15 km (10 mi) from downtown Baltimore, and another broke into a family's kitchen near Stillwater, New Jersey.

Events such as these typically result from pressure patterns that, once established, persist for unusually long periods of time. Relief comes only when the pressure pattern evolves to permit wetter conditions. In Chapter 4 we saw that atmospheric pressure varies from one place to another, but its distribution is not haphazard. Instead, well-defined patterns dominate the distribution of pressure and winds across the global surface. The largest-scale patterns, called the **general circulation,** can be considered the background against which unusual events occur, such as the drought described above. Likewise, even mundane daily wind and pressure variations can be thought of as departures from the general circulation.

Our first goal in this chapter is to describe the dominant planetary wind motions and look at the processes that generate them. In particular, we examine the interrelationships between the winds of the upper and lower atmosphere and the important connections that occur at the boundary between the surface of the oceans and

CHAPTER OUTLINE

Figure 8–1 Crops wither from prolonged drought.

the lower atmosphere. We then consider wind and pressure patterns at sequentially smaller spatial and temporal scales. The chapter concludes with a discussion of air–sea interactions.

Single-Cell Model

Scientists have sought to describe general circulation patterns for centuries. As early as 1735 a British physicist, George Hadley (1685–1768), proposed a simple circulation pattern called the **single-cell model** to describe the general movement of the atmosphere. One of his primary goals was to explain why sailors so often found winds blowing east to west in the lower latitudes. (Winds blowing east-to-west or west-to-east are referred to as **zonal winds**; those moving north-to-south or south-to-north are called **meridional***). Hadley's idealized scheme, shown in Figure 8–2a, assumed a planet covered by a single ocean and warmed by a fixed Sun that remained overhead at the equator. Hadley suggested that the strong heating at the equator caused a circulation pattern in which air expanded vertically into the upper atmosphere, diverged toward both poles, sank back to the surface, and returned to the equator. Hadley did not think winds would simply move north and south, however. He believed instead that the rotation of Earth would deflect air to the right in the Northern Hemisphere and to the left in the Southern Hemisphere, leading to the east–west surface winds shown in the figure.†

Hadley's main contributions were to show that differences in heating give rise to persistent large-scale motions (called *thermally direct circulations*) and that zonal winds can result from deflection of meridional winds. His idea of a single huge cell in each hemisphere was not so helpful, however.

A somewhat more elaborate model does a better—though still simplified—job of describing the general circulation. This **three-cell model** (Figure 8–2b) was proposed by U.S. meteorologist William Ferrel (1817–1891) in 1865.

The Three-Cell Model

The three-cell model divides the circulation of each hemisphere into three distinct cells: the heat-driven **Hadley cell** that circulates air between the Tropics and subtropics, a **Ferrel cell** in the middle latitudes, and a **polar cell**.** Though more realistic than the single-cell model, the three-cell model is so general that only fragments of it actually appear in the real world. Nonetheless, the names for many of its wind and pressure belts have become well established in our modern terminology, and it is important that we understand where these hypothesized belts are located.

The Hadley Cell

Along the equator, strong solar heating causes air to expand upward and diverge toward the poles. This creates a zone of low pressure at the equator called the **equatorial low**, or the **Intertropical Convergence Zone** (ITCZ). The upward motions that dominate the region favor the formation of heavy rain showers, particularly in the afternoon. Heavy precipitation associated with the ITCZ is observable on weather maps and satellite images (Figure 8–3). Notice in the figure that the equatorial low exists not as a band of uniform cloud cover, but rather as a zone

---

*The wind is seldom purely zonal or purely meridional, but instead usually moves in some intermediate direction. In that case we will think of the wind as having two parts both a zonal and a meridional component. It is possible for the two components to be equal (as in a southwesterly wind), but in general one component will be larger than the other.

†Hadley lived between the years 1685–1768, before Gaspard de Coriolis (1792–1843) quantitatively described the acceleration due to Earth's rotation. Nonetheless, Hadley had a qualitative knowledge of the Coriolis force and incorporated it into his model.

**You might wonder, "Why three cells, rather than some other number?" These wind belts arise because of interplay between Earth's rotation rate and the energy gradient between the equator and the poles. If Earth rotated faster, we would expect more belts. Thus, Jupiter, which turns on its axis every 11 hours, has many belts, not just three in each hemisphere.

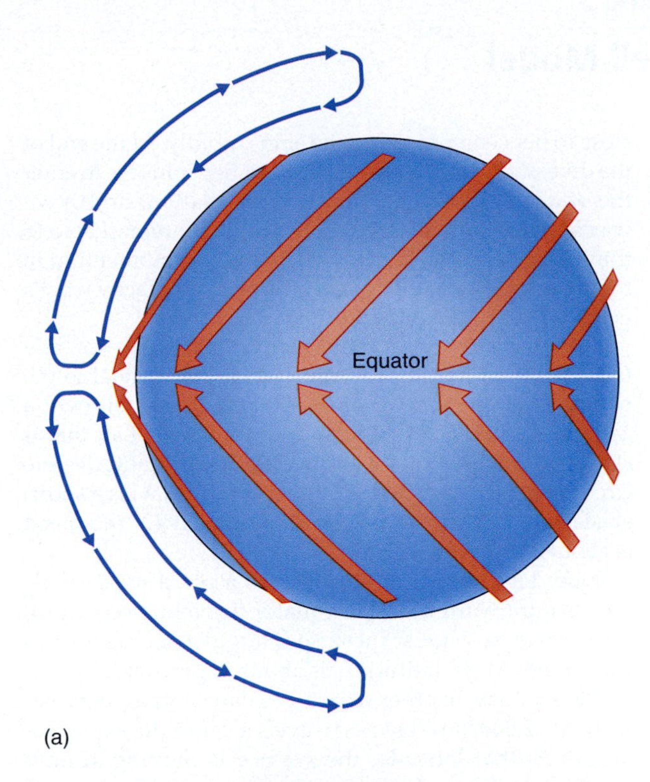

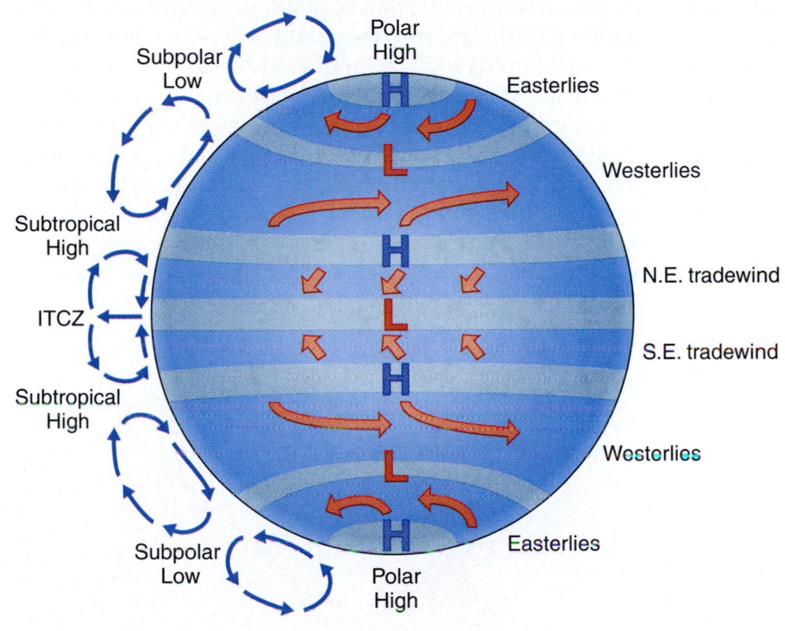

Figure 8–2
The single-cell (a) and three-cell (b) models of atmospheric pressure and wind. In the single-cell model, air expands upward, diverges toward the poles, descends, and flows back toward the equator near the surface. In the three-cell model, thermally driven Hadley circulation is confined only to the lower latitudes. Two other cells (more theoretical than real) exist in each hemisphere, the Ferrel and polar cells.

containing many clusters of convectional storms. The ITCZ is the rainiest latitude zone in the entire world, with many locations accumulating more than 200 days of rain each year. Imagine how listless you might feel in such an environment, with hot, humid afternoons giving way to heavy rain showers all year long. It is for this reason that the ITCZ is sometimes called the *doldrums*.

Within the Hadley cell, air in the upper troposphere moves poleward to the subtropics, to about 20° to 30° latitude. As it travels, it acquires increasing west-to-east motion, primarily because of the conservation of angular momentum (see *Box 8–1, Physical Principles: Problems with the Single-Cell Model*). This westerly component is so strong that the air circles Earth a couple of times before reaching its ultimate destination in the subtropics. In other words, the upper-level air follows a great spiraling path out of the Tropics, with zonal motion much stronger than the meridional

8–1 Physical Principles

Problems with the Single-Cell Model

Hadley's single-cell circulation model is a simple one based on the differential heating of Earth's surface. Contrary to Hadley's description, this thermally driven circulation occupies only the range of latitudes between about 30° N and 30° S. Why doesn't the single-cell model cover the entirety of each hemisphere instead of just a zone near the Tropics? The answer boils down to the *conservation of angular momentum.*

Just as an object moving in a straight line has "linear" momentum, an object following a curved path has "angular" momentum. If we start a weight twirling at the end of a rope, as in Figure 1, we see the rope sweeping out a growing pie-shaped angle, hence the term *angular momentum.* The amount of angular momentum obviously increases with increasing mass (imagine trying to stop a very heavy weight). Angular momentum also increases with speed, because a larger area is swept out each second. Lastly, angular momentum increases with increasing radius, again because a larger area is swept out per unit time.

Putting this together, we can say that angular momentum is the product of mass, speed, and radius, or

$$A = mvr$$

where A is angular momentum, m is mass, v is speed, and r is the radius of rotation.

If angular momentum in conserved, then A is constant. This simple statement has some important consequences. Suppose, for example, that we decrease the radius by pulling on the rope. As r decreases, the speed must increase, even though we do not twirl harder. The same phenomenon allows an Olympic diver to twist and flip in the air. After leaving the diving board, the diver's angular momentum is fixed. When she tucks, bringing legs and arms close to her center of mass, she spins rapidly. At the end of the dive, she straightens out, slowing her spin rate to enter the water vertically without much rotation, and is rewarded with a high score (hopefully). It's remarkable to think that divers can't adjust their angular momentum in the air, but rather must leave the board with exactly what's needed for a particular dive.

Now consider the rotating Earth. At the equator, any fixed point on the surface travels a distance equal to the planetary circumference of 40,000 km (24,000 mi) over a 24-hour period. The same applies to a parcel of air that is stationary relative to the surface. At higher latitudes the circumference is smaller, so a parcel at rest travels a shorter distance in a day's time. For example, at 40° the speed is about 31,000 km per day.

Now let's see what happens as a fixed mass of air moves northward from the equator. Traveling poleward, r decreases, so v must increase if angular momentum is conserved. At 40° latitude, v is about 30 percent larger, or 52,000 km/day. In other words, the parcel is moving eastward at 52,000 km/day as it travels around the axis of rotation. At that latitude, the surface is moving at only 31,000 km/day, so the parcel is moving over the surface. Standing on the ground, we would observe a wind speed of 21,000 km/day (52,000 minus 31,000). This is a huge value, equivalent to a wind speed of 243 m/sec, or 547 mph!

According to the single-cell model, upper-level air should flow all the way from the equatorial region to the poles. But if that were the case, the conservation of angular momentum would send the poleward-moving air eastward at unimaginably high speeds—far greater than what we experience in even the most severe tornadoes. Physical

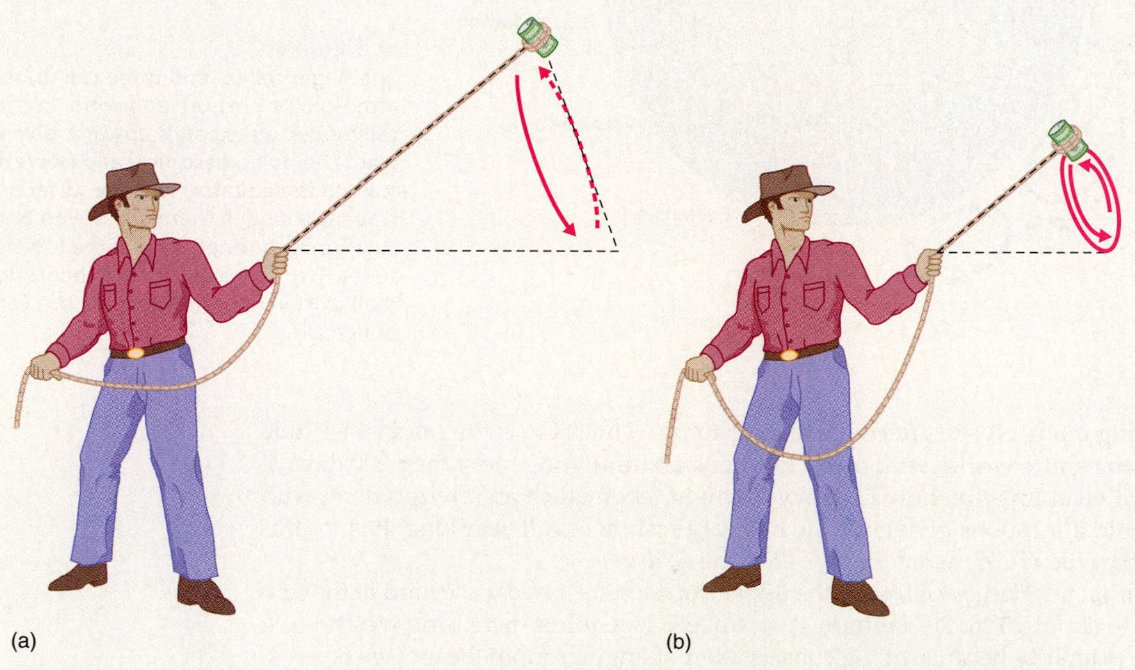

▲ **Figure 1**
The twirling of a weight at the end of a rope (a) illustrates the conservation of angular momentum. As the rope is pulled closer to the man's body (b), its speed of rotation increases in order to maintain its angular momentum.

considerations prevent the air from maintaining such extreme flows, because even a slight random disruption would cause the wind to break down into numerous eddies with strong north–south motions.

The single-cell model involves another difficulty, also related to conservation of angular momentum. To a large degree, the momentum possessed by the entire Earth–atmosphere system is unchanging—nothing stands outside the planet giving it a shove to increase its spin rate or to slow it down. (We are ignoring gravitational variations caused by roving planets and other small factors.) But angular momentum is transferred between Earth and atmosphere by friction whenever the air moves past the surface. Where there is westerly wind, the atmosphere transfers momentum to the surface as it "pushes" in the direction of rotation. Similarly, easterly (from the east) winds imply a transfer of momentum from the surface to the atmosphere.

The single-cell model calls for easterly surface winds everywhere. If this were to occur, the planet would everywhere supply westerly momentum to the atmosphere, which would bring the atmosphere to rest within a week or two. To avoid this problem, easterly winds must be balanced by westerly winds somewhere else. More precisely, averaged over the whole Earth, momentum transferred from the surface to the atmosphere must equal momentum transferred from the atmosphere to the surface. Thus, a single-cell circulation would not be possible on Earth—or on any other rotating planet.

◀ **Figure 8–3**
The Intertropical Convergence Zone (ITCZ) is observable as the band of convective clouds and showers extending from northern South America into the Pacific on this satellite image.

component. Among other things, this explains why material ejected by tropical volcanic eruptions spreads quickly over a wide range of longitudes.

Upon reaching about 20° to 30° latitude, air in the Hadley cell sinks toward the surface to form the **subtropical highs**, large bands of high surface pressure. Because descending air warms adiabatically, cloud formation is greatly suppressed and desert conditions are common in the subtropics. The subtropical highs generally have weak pressure gradients and light winds. Such conditions exert minimal impact on long-distance travel today. But in preindustrial days when oceangoing vessels depended on the wind, its prolonged absence could be catastrophic. Ships crossing the Atlantic from Europe to the New World risked getting stranded in mid-ocean while crossing the subtropics. Often among the cargo were cattle and horses to be brought to the New World, and legend has it that the crews of stalled ships threw their horses overboard. The jettisoned cargo has lent its name to what we colloquially call the **horse latitudes**.

In the Northern Hemisphere, as the pressure gradient force directs surface air from the subtropical highs to the ITCZ, the weak Coriolis force deflects the air slightly to the right to form the **northeast trade winds** (or simply the *northeast trades*). In the Southern Hemisphere, the northward-moving air from the subtropical high is deflected to the left to create the **southeast trade winds**. Notice that the trade winds are fairly shallow. Moving upward through the troposphere, the easterly motions weaken and are eventually replaced by westerly motion aloft. Together, the subtropical highs, the equatorial low, the trade winds, and upper-level westerly motions form the Hadley cells. Because it is produced thermally, Hadley circulation is strongest in the winter season, when temperature gradients are strongest.

The Ferrel and Polar Cells

According to the three-cell model, the Hadley cell accounts for the movement and distribution of air over about half of Earth's surface. Immediately flanking the Hadley cell in each hemisphere is the Ferrel cell, which circulates air between the subtropical highs and the **subpolar lows**, or areas of low pressure. On the equatorial side of the Ferrel cell, air flowing poleward away from the Northern Hemisphere subtropical high undergoes a substantial deflection to the right, creating a wind belt called the **westerlies**. In the Southern Hemisphere, the pressure gradient force propels the air southward, but the Coriolis force deflects it to the left—thus producing a zone of westerlies in that hemisphere as well. Unlike the Hadley cell, the Ferrel cell is envisioned as an indirect cell, meaning that it does not arise from differences in heating, but is instead caused by turning of the two adjacent cells. Imagine three logs placed side-by-side, touching one another. If the two outer logs are turned in the same direction, friction will set the middle one in motion. Thus, the Ferrel cell shows the same kind of overturning as the others, but for different reasons.

In the polar cells of the three-cell model, surface air moves from the **polar highs** to the subpolar lows. Like the Hadley cells, the polar cells are considered thermally direct circulations. Compared to the poles, air at subpolar locations is slightly warmer, resulting in low surface pressure and rising air. Very cold conditions at the poles create high surface pressure and low-level motion toward the equator. In both hemispheres, the Coriolis force turns the air to form a zone of **polar easterlies** in the lower atmosphere.

The Three-Cell Model vs. Reality: The Bottom Line

Do the wind and pressure belts of the three-cell model adequately describe real-world patterns? The answer is: sort of. We have already seen that the ITCZ is real enough to be observed from space and that many deserts exist in their predicted locations. Furthermore, the trade winds are the most persistent on Earth. We would have to say that the Hadley circulation provides a good account of low-latitude motions. On the other hand, the Ferrel and polar cells are not quite as well represented in reality, though they do have some manifestation in the actual climate.

With regard to surface winds, much of the middle latitudes experience the strong westerly winds depicted by the model, especially in the Southern Hemisphere. Of course, local conditions often override this tendency (in fact, much of the central United States is dominated by a southerly flow during the summer). It is even more difficult to observe a persistent pattern of polar easterlies in the overall wind regime. They emerge in long-term averages, but are not a prevailing wind belt as the trades are.

With regard to upper-level motions, the three-cell model is not realistic at all. For example, where the Ferrel cell implies easterly motion in the upper troposphere, there is overwhelming westerly wind. Moreover, large overturning cells do not exist outside of the Hadley zones. Thus, the three-cell model mainly provides a starting point for a more detailed account. But perhaps its failures aren't surprising, given that it doesn't consider land–ocean contrasts or the influence of surface topography, two factors that surely ought to influence planetary winds and pressure.

Semipermanent Pressure Cells

The three-cell model provides a good beginning for describing the general distribution of wind and pressure, but the real world is not covered by a series of belts that completely encircle the globe. Instead, we find a number of alternating **semipermanent cells** of high and low pressure, as shown in Figure 8–4. They are called *semipermanent* because they undergo seasonal changes in position and intensity over

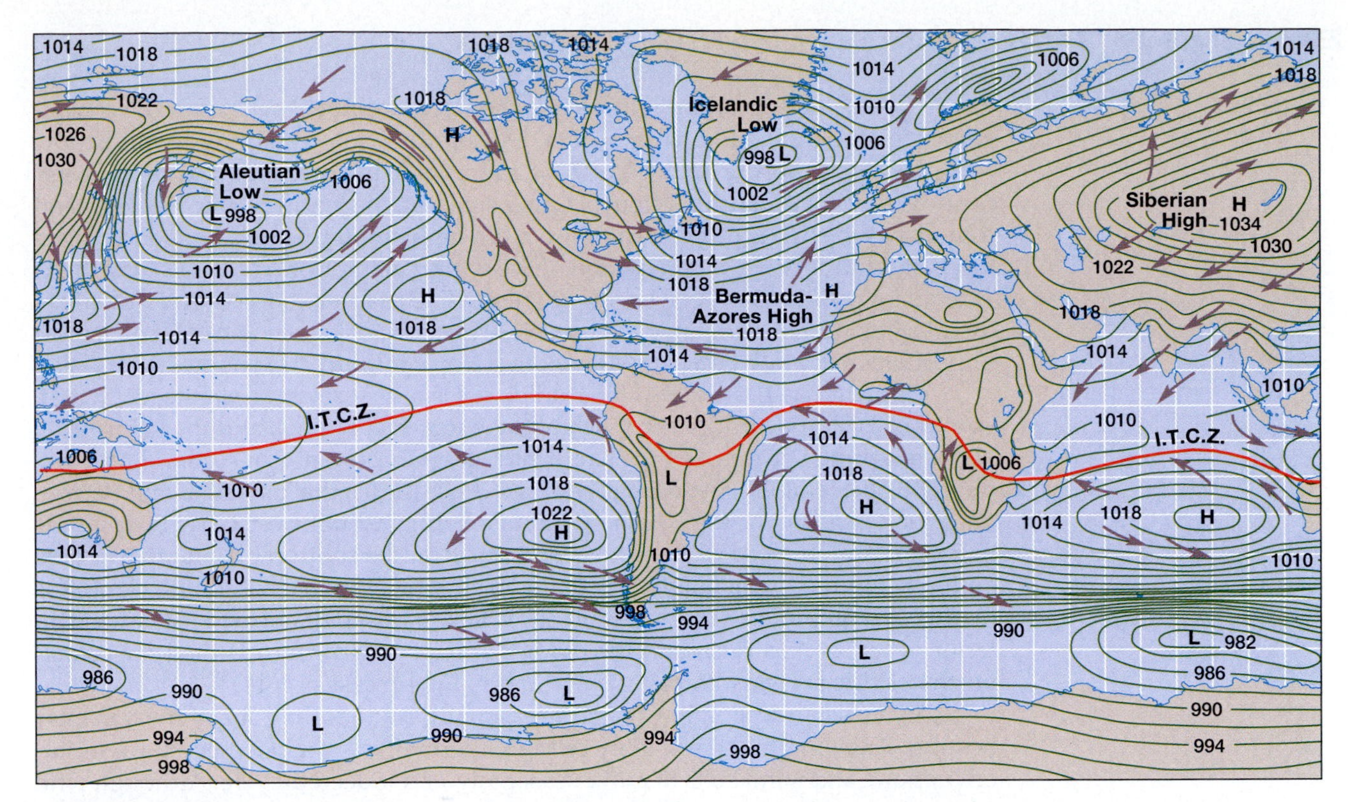

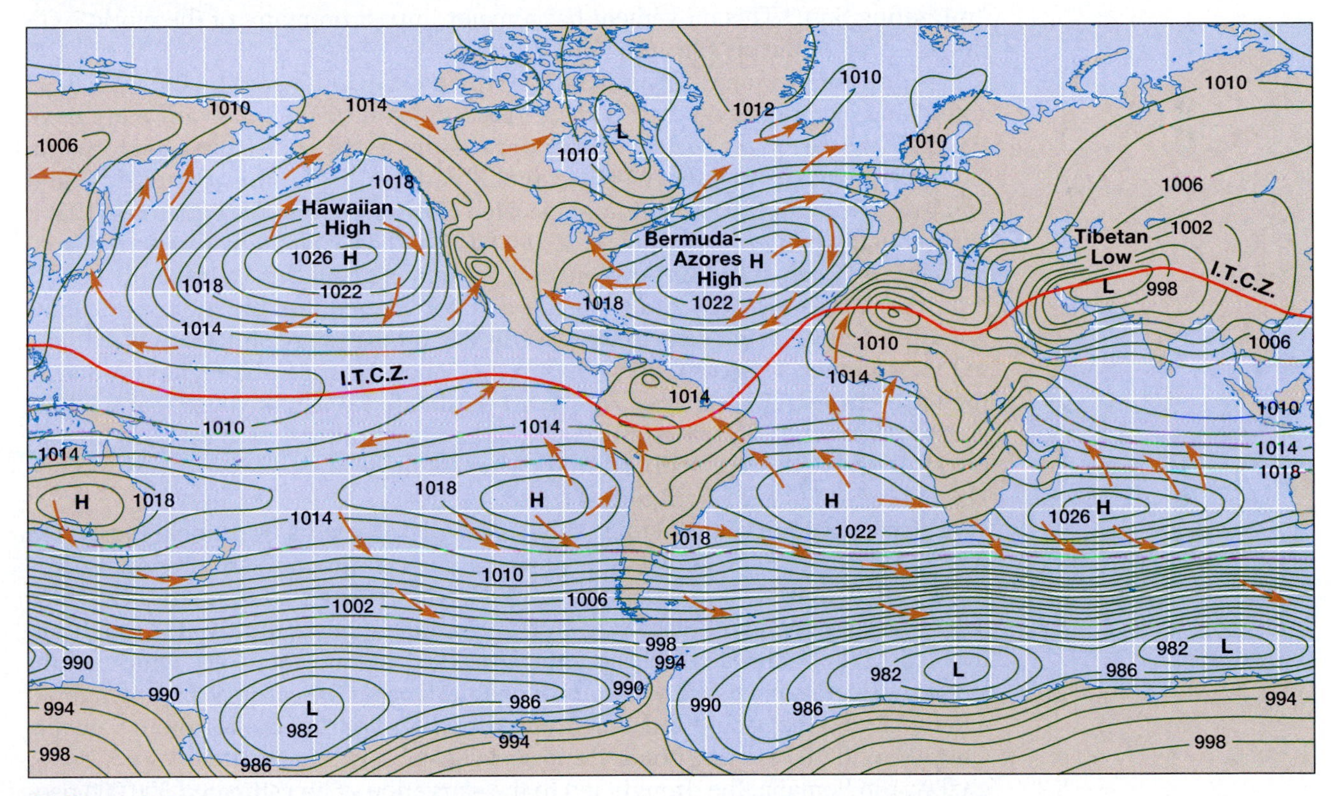

Figure 8–4
Sea level pressure for January (a) and July (b).

the course of the year. Some of these cells result from temperature differences, and others from dynamical processes (meaning that they are related to the motions of the atmosphere). Among the most prominent features in the Northern Hemisphere during winter (a) are the **Aleutian** and **Icelandic lows**—over the Pacific and Atlantic Oceans, respectively—and the **Siberian high** over central Asia. In summer (b), the best-developed cells are the **Hawaiian** and **Bermuda-Azores highs** of the Pacific and Atlantic Oceans and the **Tibetan low** of southern Asia.

The size, strength, and locations of the semipermanent cells undergo considerable change from summer to winter. During the winter, a strong Icelandic low occupies a large portion of the North Atlantic, while the Bermuda-Azores high appears as a small, weak anticyclone. During summer, the Icelandic low weakens and diminishes in size, and the Bermuda-Azores high strengthens and expands. Even more dramatically, the Siberian high of the winter in interior Asia gives way to the Tibetan low of summer. As we will see later in this chapter, the seasonal shift in the distribution of semipermanent cells plays an important role in one of Earth's most important circulation patterns—the monsoon of southern and southeastern Asia.

We mentioned earlier that the Hadley cell is fairly easy to see in the real world, with the ITCZ appearing over much of the equatorial regions and the subtropical highs existing over much of the subtropics. But as you can also see in Figure 8–4, the subtropical highs exist primarily over the oceans (as the Hawaiian and Bermuda-Azores highs over the Northern Hemisphere) and not over land. Despite the absence of pronounced high sea level pressure over the subtropical land masses, the air in the middle troposphere does undergo sinking motions that inhibit cloud formation and promote desert conditions. The Sahara Desert of northern Africa, the interior desert of Australia, and the deserts of the southwestern United States and northwest Mexico clearly reflect that this sinking process occurs.

As the solar declination changes seasonally, so does the zone of most intense heating. Knowing that the Hadley cells are thermal, we might expect the associated pressure and wind belts to move seasonally, and indeed they do. Although with a lag of several weeks, the ITCZ, subtropical highs, and trade winds all follow the "migrating Sun." This movement has a major impact on many of the world's climates—and on the people who inhabit them.

For example, many areas along the equator are dominated by the ITCZ year round and experience no dry season. Areas located near the poleward margins of the ITCZ, however, are subject to brief dry seasons as the zone shifts equatorward. Compare, for example, the average rainfall patterns for Iquitos, Peru (3° S), and San Jose, Costa Rica (9° N). Iquitos is located close enough to the equator so that it is perennially influenced by the ITCZ, but San Jose has a relatively dry period from January to March, when the low-pressure system is displaced to the south.

Similarly, some areas located on the equatorward edge of the subtropical highs are dry for most of the year, except briefly when the system shifts poleward during the summer. This condition exists in the Sahel of Africa, the region bordering the southern margin of the Sahara Desert (Figure 8–5a). Unlike the Sahara, which is dry all year, the Sahel normally experiences a brief rainy period each summer as the ITCZ enters the region (b). During the rest of the year, the descending air of the Hadley cell leads to dry conditions (c).

The migration of the Hadley cell has long supported a traditional lifestyle in which African herders followed the northward and southward shifting rains. During the 1960s and 1970s, the population of the region increased dramatically, which led to overgrazing and set the stage for catastrophe when a multiyear drought hit the area. Millions of head of livestock died from lack of food and water, which in turn led to the deaths of tens of thousands of people. During the early 1990s, low rainfall again plagued the Sahel—this time along the eastern part of the region in Somalia. Coupled with the existing political and social instability that eventually gave rise to civil war in Somalia, the drought led to the starvation of an estimated 300,000 people. Thus, the existence of these cells—and variations in the way they develop during unusual years—are more than mere abstractions. They have real-world ramifications for millions of people.

The Upper Troposphere

In Chapter 4 we saw that pressure decreases more rapidly with altitude where the air is cold. We also know that temperature in the lower troposphere generally decreases from the subtropics to the polar regions. These two principles are critically important in understanding the distribution of wind and pressure in the upper troposphere.

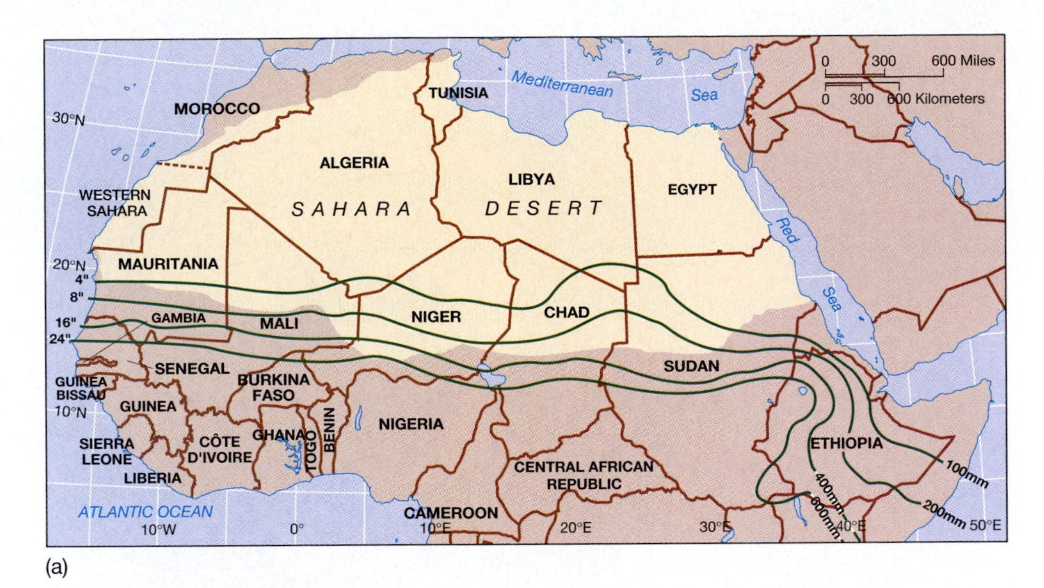

(a)

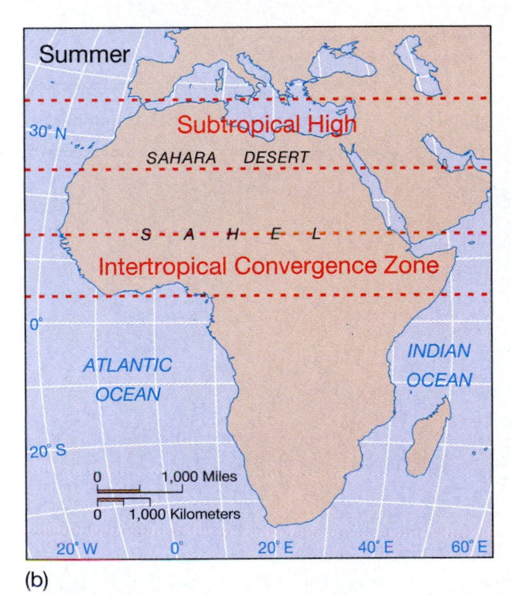

(b)

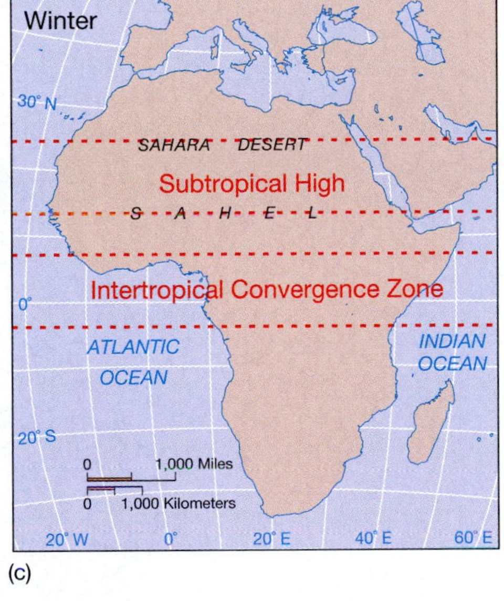

(c)

Figure 8–5 The Sahel is a region of Africa bordering the southern Sahara Desert (a). During the summer (b), the ITCZ usually shifts northward and brings rain to the region. For much of the year, the ITCZ is located south of the Sahel, and the region receives little or no precipitation (c).

Figure 8–6 maps the global distributions of the mean height of the 500 mb surface (a convenient level representing conditions in the middle troposphere) for January (a) and July (b). In both months, the height of the 500 mb level exhibits a strong tendency to decrease toward the polar regions, due to the lower temperatures found at higher latitudes. In January, the average height of the 500 mb surface over much of the southern United States is about 5670 m (18,600 ft), while over northern Canada it decreases to less than 5300 m (17,400 ft). A similar but less extreme change occurs in July as well.

Three important features stand out from the maps in Figure 8–6. First, for both January and July the 500 mb heights are greatest over the Tropics and decrease with latitude. Second, the gradient in height is greater in the hemisphere experiencing winter (the Southern Hemisphere in July and Northern Hemisphere in January). Third, at all latitudes the height of the 500 mb level is greater in the summer than during the winter. All three of these features result from the general distribution of temperature in the lower-middle atmosphere; areas of warm air have greater 500 mb heights.

Westerly Winds in the Upper Atmosphere

Section 8.2.2

Recall from Chapter 4 that height differences correspond to pressure differences, and that when the 500 mb surface slopes steeply, a strong pressure gradient force exists. We can therefore infer from the 500 mb maps that there is always a pressure

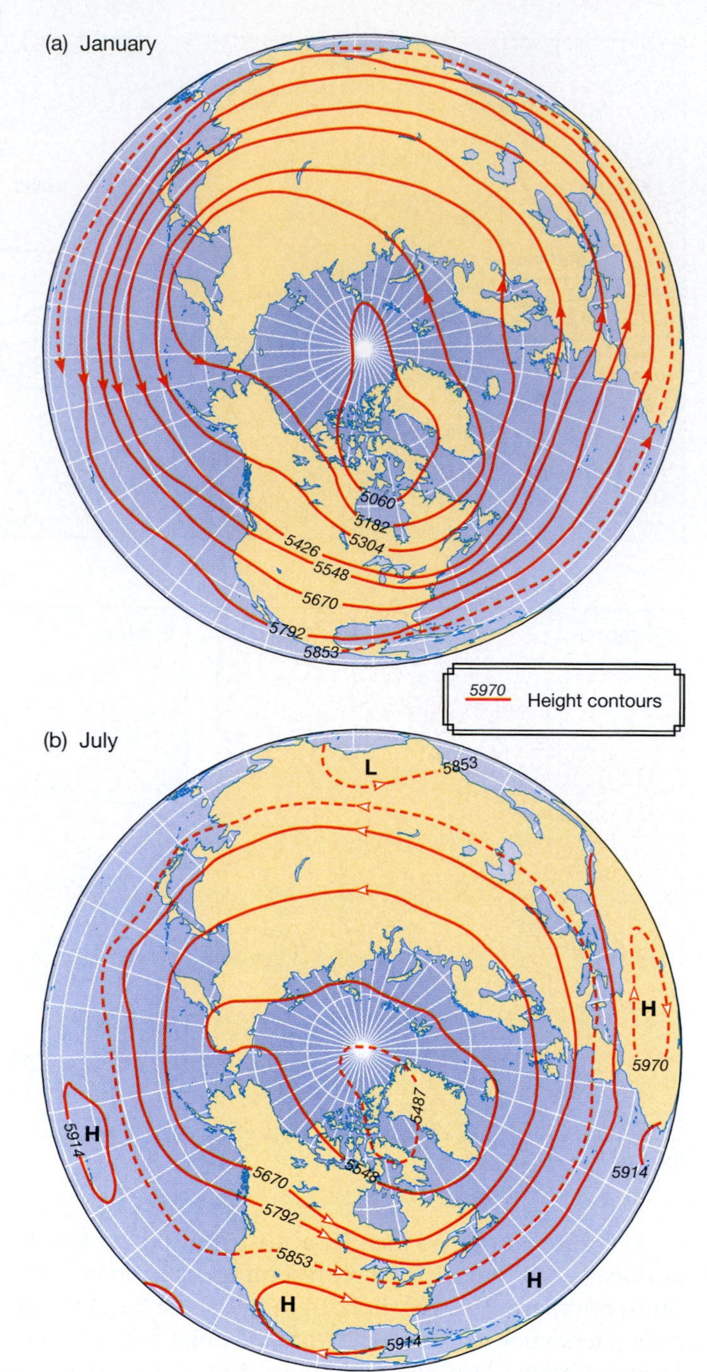

Figure 8–6
The mean heights of the 500 mb levels for January (a) and July (b). The pattern is mostly zonal with decreasing heights toward the poles.

gradient force across the middle latitudes trying to push the air toward the poles. Of course, in the absence of friction, the winds do not blow poleward, but rather blow parallel to the height contours, from west to east. The pressure gradient force is strongest in winter (the height contours are closely spaced), so the upper-level westerlies are strongest in winter. What does this mean to you? For one thing, it explains why most mid-latitude weather systems migrate from west to east. In other words, it tells us why a storm over Chicago might find its way over the East Coast a day or two later, but seldom if ever does such a storm make the reverse trip.

The predominance of westerly winds in the upper troposphere also affects aviation. For example, a commercial aircraft going from Chicago to London has an estimated flight time of about 7.5 hours, while the return trip normally takes an hour longer because it must overcome headwinds. The difference in flight time would be even greater were it not for the fact that airlines route their planes to take advantage of tailwinds and avoid headwinds.

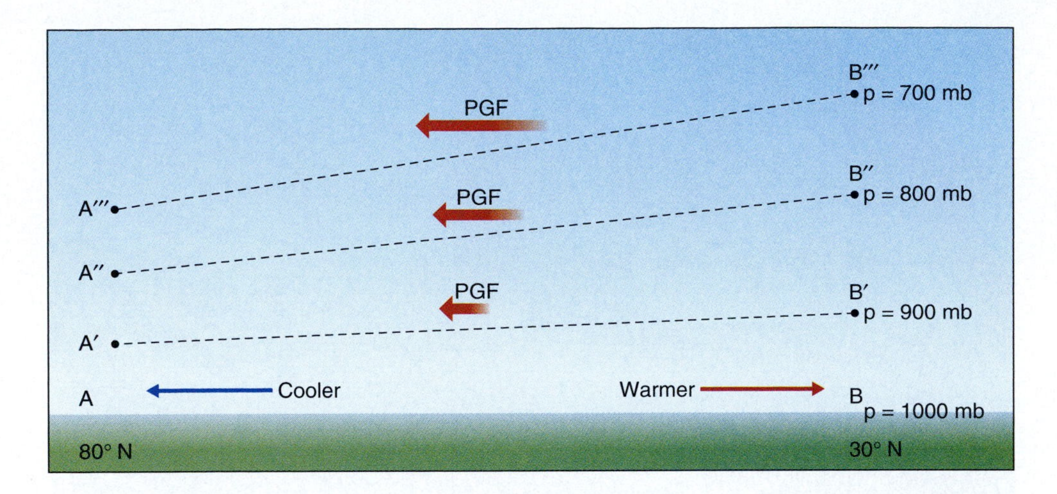

Figure 8–7 Latitudinal temperature gradients cause pressure surfaces to slant poleward. In this example, we assume a constant gradual decrease in temperature with latitude. The intensity of the pressure gradient force remains constant from one latitude to another, but increases with altitude.

Wind speeds generally increase with height between the surface and the tropopause. Partly this is because of decreasing friction, but more importantly, the pressure gradient force is typically stronger at high altitudes. As illustrated in Figure 8–7, the surfaces representing the 900, 800, and 700 mb levels all slant downward to the north, but not by the same amount. Higher surfaces slope more steeply, which means that the pressure gradient force is greater. You may recall from Chapter 4 that the pressure gradient force is directly proportional to slope, without regard to density.

But why do those higher surfaces slope more steeply? The air is warmer at point *B*, so the layer of air from 900 mb to 1000 mb is thicker at point *B* than point *A*. Similarly, the thickness from 800 mb to 900 mb is greater at point *B*. In other words, the height change from *B*′ to *B*″ is greater than the change from *A*′ to *A*″. It must be, therefore, that the 800 mb surface slopes downward more than the 900 mb surface (*B*″–*A*″ is greater than *B*′–*A*′). The difference in heights between successive surfaces continues to increase upward, leading to stronger winds.

The Polar Front and Jet Streams

The gradual change in temperature with latitude depicted in Figure 8–7 does not always occur in reality. Instead, areas of gradual temperature change often give way to narrow, strongly sloping boundaries between warm and cold air. One such boundary, the **polar front**, is shown in Figure 8–8.

Outside of the frontal zone, the changes in temperature with latitude are gradual (as they were in Figure 8–7), and the slopes of the 900, 800, and 700 mb levels respond accordingly. But within the front, the slope of the pressure surfaces increases greatly because of the abrupt horizontal change in temperature. With steeply sloping pressure surfaces there is a strong pressure gradient force, resulting in the **polar jet stream**. Thus, we see the jet stream as a consequence of the polar front, arising because of the strong temperature gradient. But at the same time, the jet stream also reinforces the polar front. In Chapter 10, we will see that a jet stream is necessary to maintain the temperature contrast across the front.

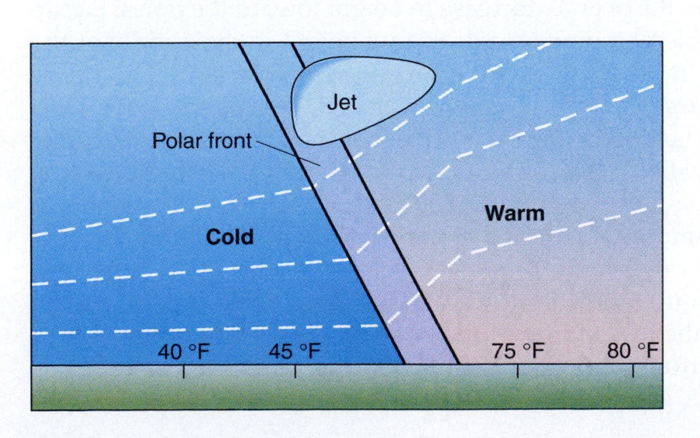

Figure 8–8 The polar jet stream is situated above the polar front near the tropopause.

▶ **Figure 8–9**
The subtropical jet stream appears in this infrared satellite image as the band of cloudiness extending from Mexico through Florida.

Jet streams can be thought of as meandering "rivers" of air, usually 9 to 12 km (30,000 to 40,000 ft) above sea level. Their wind speeds average about 180 km/hr (110 mph) in winter and about half that in summer, though peak winds can exceed twice these values. Like rivers on land, jet streams are highly turbulent, and their speeds vary considerably as they flow. But unlike rivers on land, they have no precisely defined banks. Furthermore, a single jet stream will often diverge at some point and fork off into two distinct jets. Thus, the locations and boundaries of these features on weather maps are often difficult to accurately pinpoint.

The polar jet stream greatly affects daily weather in the middle latitudes. Nearer the equator is another prominent jet, the **subtropical jet stream**, associated with the Hadley cell. As the upper-level air flows away from the ITCZ, the conservation of angular momentum imparts ever-growing westerly motion. When moving toward the northeast, the subtropical jet stream can bring with it warm, humid conditions. Figure 8–9 shows the flow of moisture associated with a subtropical jet stream.

Troughs and Ridges

Section 8.4.1

Although on average 500 mb heights decrease toward the poles, at any given time significant departures from the general trend will exist. Typically undulations, or waves, are superimposed on the overall decrease in height toward the poles. Figure 8–10 is a cartoon view of this, showing an axis of low height in the middle of the United States, flanked by "mountains" of high heights on either side. The valley of low heights is called a *trough*, and the upward bulges are called *ridges*. Also shown on the diagram are height contours—notice that they too have a wavelike character. The air flows parallel to the contours, so there is wavelike motion to the air stream as well.

Figure 8–11 shows simplified contour maps depicting the relationship between 500 mb heights and ridges and troughs. No east–west height changes (no trough, no ridge) occur in (a), and the flow is completely zonal. In (b), on the other hand, a trough is in the midsection of the country. Going from point 1 to 2, there is a height decrease from 5610 m to 5580 m. Going from 2 to 3, heights increase again. In other words, going from 1 to 3 requires us to cross a valley (trough)

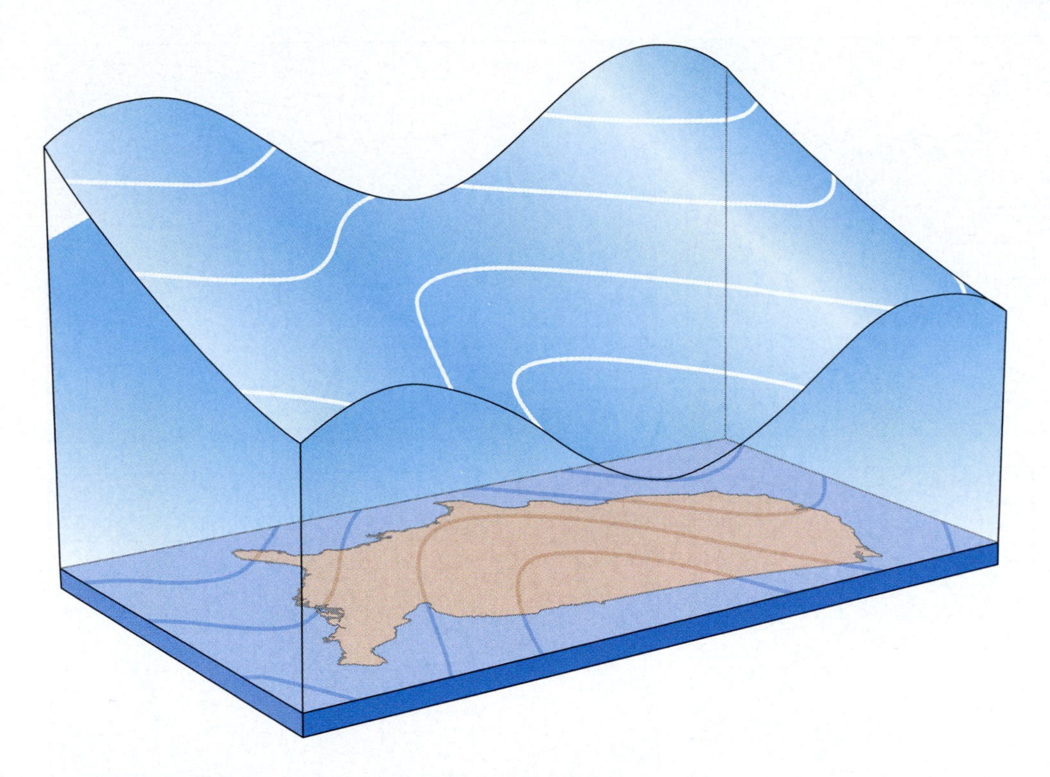

◀ **Figure 8–10**
A hypothetical drawing of the 500 mb surface. Heights decrease from south to north but also rise and fall through the ridges and trough. Vertical changes are highly exaggerated in the figure. Actual height changes are very small compared to the size of the continent.

of low heights. We see that height contours are displaced toward the equator in troughs and toward the pole in ridges. We also see that air winding its way poleward around ridges and equatorward around troughs will have a meridional component as well as a zonal component. In fact, when pronounced waves are present, we say the flow is "meridional," as opposed to "zonal" when the flow is nearly all westerly.

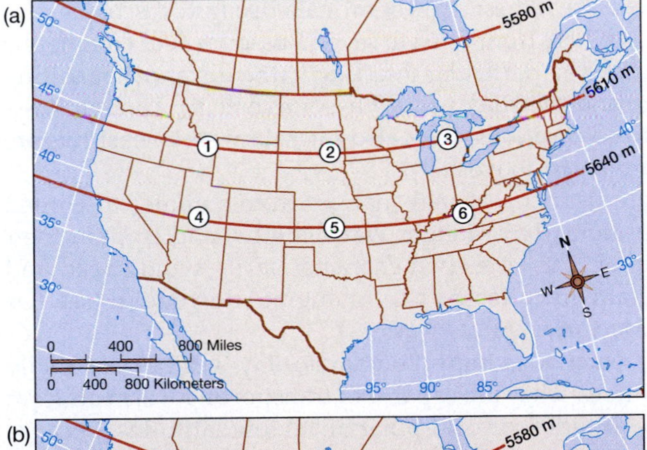

◀ **Figure 8–11**
Troughs occur in the middle troposphere where the 500 mb height contours dip equatorward. In (a) positions 1–3 have the 500 mb level at 5610 m. Farther to the south, at positions 4–6, the 500 mb level is at 5640 m. In (b) the contour lines are in the same position over the East and West Coasts as they were in (a), but they shift equatorward over the central portion of the continent. Thus, positions 2 and 5 have lower pressure than the areas east and west of them. The zone of lower pressure over the central part of the continent is a trough.

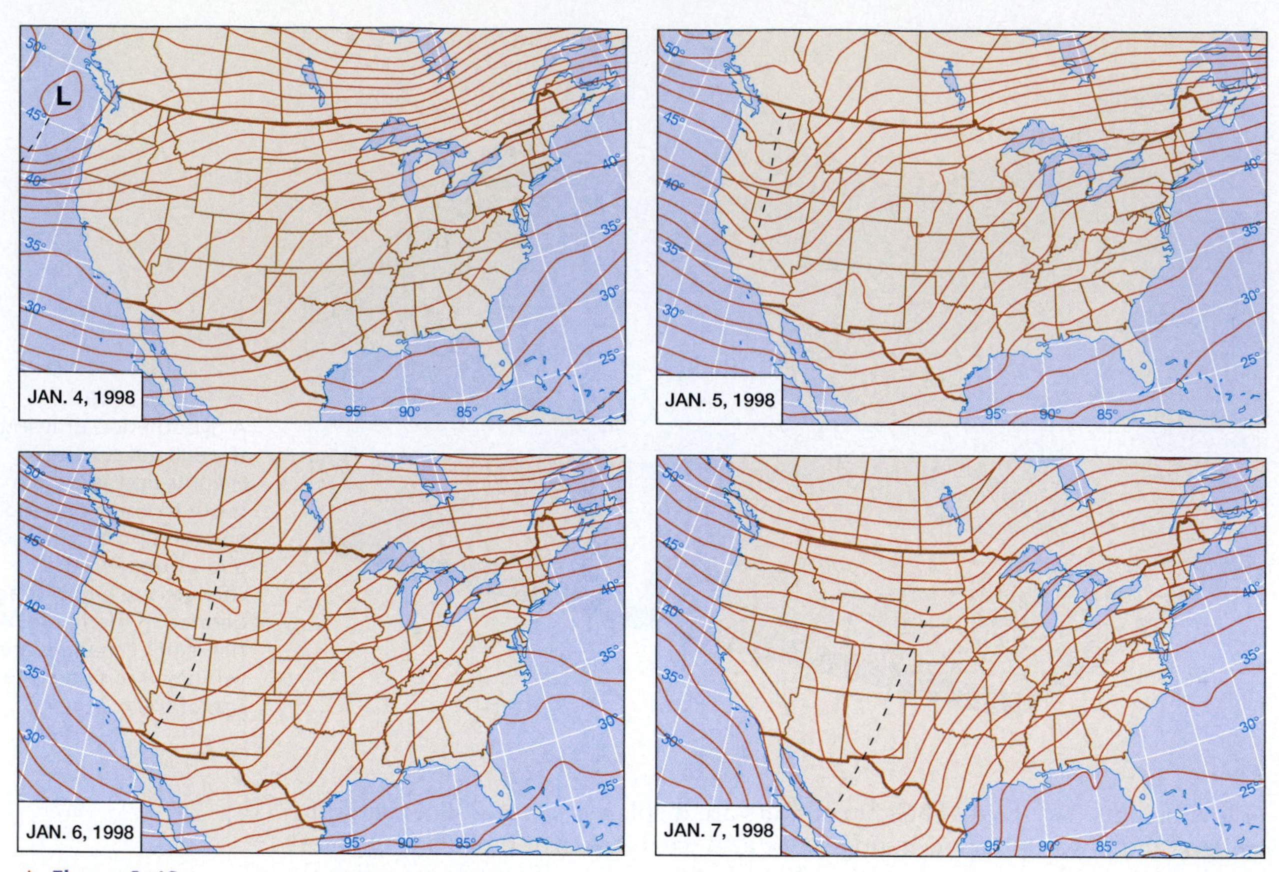

▲ **Figure 8–12**
A sequence of 300 mb maps showing the migration of Rossby waves at 24-hour intervals.

Rossby Waves

We've seen that ridges and troughs give rise to wavelike flow in the upper atmosphere of the middle latitudes. The largest of these are called *long waves*, or **Rossby waves**.* Usually, there are anywhere from three to seven Rossby waves circling the globe. Like other waves, each has a particular *wavelength* (the distance separating successive ridges or troughs) and *amplitude* (its north–south extent). Though Rossby waves often remain in fixed positions, they also migrate west to east (Figure 8–12), or on rare occasions from east to west.

Rossby waves undergo distinct seasonal changes from summer to winter. They tend to be fewer in number, have longer wavelengths, and contain their strongest winds during winter. The latter two characteristics—wind speed and wavelength—affect the rate at which Rossby waves migrate downwind (see *Box 8–2, Physical Principles: The Movement of Rossby Waves*).

Upper-Level Winds and Pressure
Section 8.6.1

Rossby waves exert a tremendous impact on day-to-day weather, especially when they have large amplitudes. They are capable of transporting warm air from subtropical regions to high latitudes, or cold polar air to low latitudes. Because upper-level air tends to change temperature only slowly in the absence of strong vertical motions, Rossby waves can bring anomalous temperatures to just about any place within the middle to high latitudes. This is illustrated by Figure 8–13, which shows a strong Rossby wave over North America on September 22, 1995. Record-breaking low temperatures for the date were observed over much of the central United States as the wave brought cold air from the far north. Farther upwind, a southwesterly flow brought mild air to the extreme northwest of North America, with Fairbanks, Alaska, basking in temperatures in the mid-20s Celsius (mid-70s Fahrenheit).

*Named for Carl Gustov Rossby, who contributed much of the pioneering research on upper-level air flow in the early 1900s.

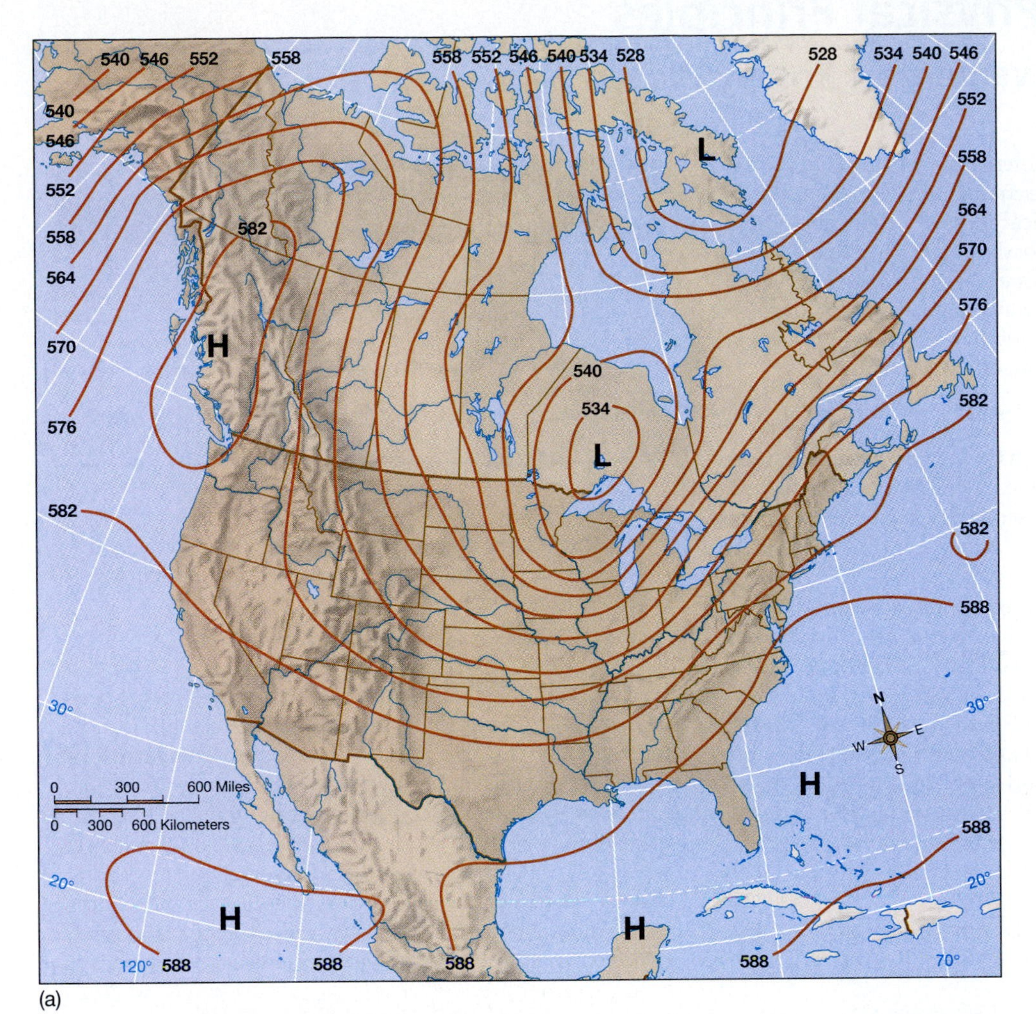

(a)

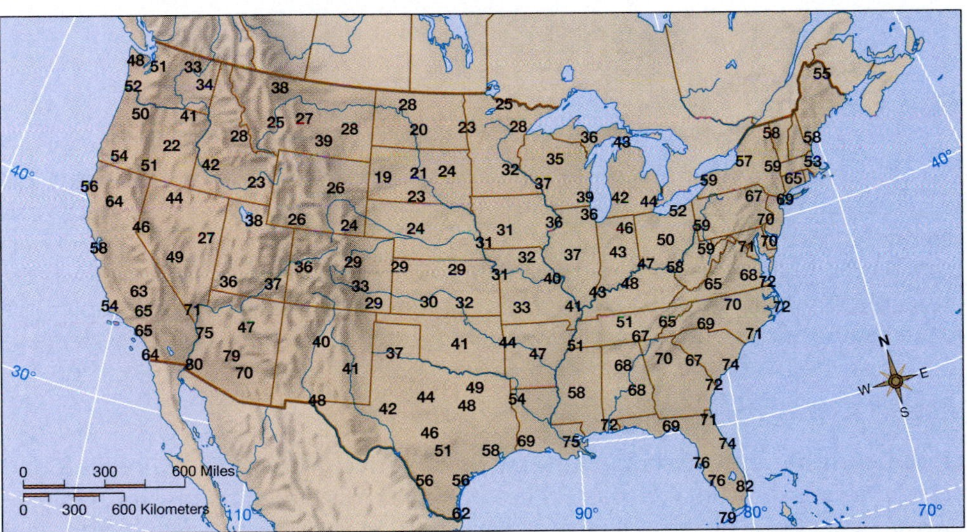

(b) Minimum temperatures (°F)

◀ **Figure 8–13**
Rossby waves in the upper atmosphere can advect cold or warm air from one location to another. On September 22, 1995, such a wave brought very mild conditions to interior Alaska, while southward-moving air brought record-breaking low temperatures to much of the central United States.

In addition to the redistribution of cold or warm air, Rossby wave patterns have an equally important but less obvious impact on local weather. Changes in the flow along the wave lead to *divergence* and *convergence*. When air in the upper atmosphere diverges (or spreads out), it draws air upward from below, causing adiabatic cooling. Thus, divergence in the upper atmosphere can serve as an important mechanism for cloud development and precipitation. Convergence in the upper atmosphere has the opposite effect of forcing air downward and inhibiting cloud formation. We will return to this important concept in Chapter 10.

8–2 Physical Principles

The Movement of Rossby Waves

Like all atmospheric phenomena, the movement of Rossby waves results not from mere chance, but from the combined actions of numerous physical forces. Three factors determine the rate at which a Rossby wave propagates: (1) the westerly component of its internal wind speed, (2) its latitudinal position, and (3) its wavelength. In particular, those Rossby waves with vigorous winds and shorter wavelengths move most rapidly, as indicated by the formula

$$C = u - bL^2$$

where C is the speed at which the wave propagates downwind (m/sec); u is the westerly component of the wind speed within the wave (m/sec); b is a function of latitude equal to 1.6×10^{-11} m^{-1} sec^{-1} at 45°, and L is the wavelength (m).*

The rates of downwind migration for Rossby waves at 45° latitude with various wavelengths and wind speeds are presented in Table 1. As you can see, a wave having a westerly wind speed of 20 m/sec (45 mph) and a wavelength of 3000 km advances 60 percent faster than does one with the same speed but a wavelength of 5000 km.

Note that for a given wind speed and latitude, there is some particular wavelength, L_{crit}, at which the waves do not migrate at all. Waves longer than this critical value actually migrate from east to west and are said to be *retrograding waves*. By rearranging the equation and setting $C = 0$ (that is, by assuming the waves are stationary), we can determine the critical wavelength:

$$L_{crit} = \sqrt{u/b}$$

For example, at 45° a Rossby wave with a jet-stream speed of 40 m/sec will migrate upwind if its wavelength is more than about 10,000 km. This, however, is a particularly long wavelength. Thus, at high wind speeds, retrograding motion occurs only for exceptionally long waves, which are quite rare. For lower wind speeds, the critical wavelength is less (only 5000 km for wind at 10 m/sec). As it happens, shorter waves (with lower wind speeds) are more common. Thus, when westward movement appears, it tends to occur when upper-level winds are weak, not strong.

*More specifically, b is given as 5.8×10^{-13} times the cosine of the latitude. At latitude 60°, it equals 2.9×10^{-13} m^{-1} sec^{-1}.

Table 1 • Rates of Downwind Rossby Wave Migration at Latitude 45° N or 45° S

Wavelength	20 m/sec Wind Speed	40 m/sec Wind Speed
3000 km	16 m/sec wave speed	36 m/sec wave speed
5000 km	10 m/sec wave speed	30 m/sec wave speed

The Oceans

As we have seen, the movement of the atmosphere is strongly influenced by the input of heat from the surface. We have also seen that land and water undergo differential rates of heating and cooling; in part this is because of the vertical and horizontal motions of the water's surface. In this section, we take a close look at how the atmosphere affects the movement of the oceanic waters. Later in this chapter, we look at some of the mutual interactions between the ocean–atmosphere systems.

Ocean Currents

Ocean currents are horizontal movements of surface water that are often found along the rims of the major basins. These currents, discussed briefly in Chapter 3, have a great impact on the exchange of energy and moisture between the oceans and the lower atmosphere. In many instances their effect on climate is conspicuous, with warm ocean currents, for example, favoring the existence of warm, humid air.

Ocean currents are driven by winds in the lower atmosphere that exert a drag on the water. Contrary to what you might expect, the surface water moves not in the same direction as the wind, but at an angle of 45° to the right of the air flow in the Northern Hemisphere (to the left in the Southern Hemisphere). Furthermore, neither the direction nor the speed of the current is uniform with depth. The current turns increasingly to the right (in the Northern Hemisphere) and decreases in speed at greater depths. At about 100 m below the surface, the direction of the current approaches 180° to the direction of the wind, and the current dies out. Figure 8–14 illustrates this pattern, called the **Ekman spiral**.

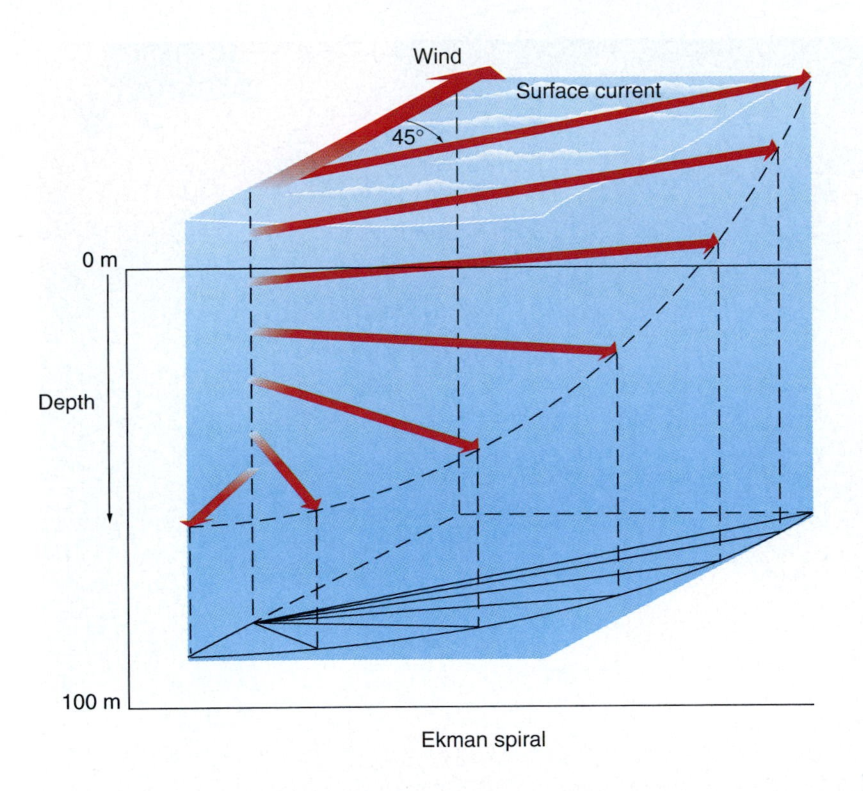

◀ **Figure 8–14**
The Ekman spiral. Surface currents flow at an angle 45° to the right (in the Northern Hemisphere) of the winds that drive them and continue to shift clockwise as their speed decreases. At a depth of about 100 m, the current approaches the opposite direction of the surface current and begins to die out.

The movement of the current at the surface corresponds well to the global wind patterns described earlier in this chapter, as shown in Figure 8–15. Let's have a look at the surface currents of the North Atlantic, whose pattern is quite typical of those in the other major oceans. Just north of the equator, the easterly trade winds drag the surface water westward as the **North Equatorial Current**. Upon reaching South America, most of the westward moving water turns north, but some is deflected to the south toward the equator. A similar pattern appears in the Southern Hemisphere, where a portion of the **South Equatorial Current** is deflected northward to the equator. The water from the North and South Equatorial Currents converges and piles up in the western equatorial Atlantic and creates the eastward-moving **Equatorial Countercurrent**.

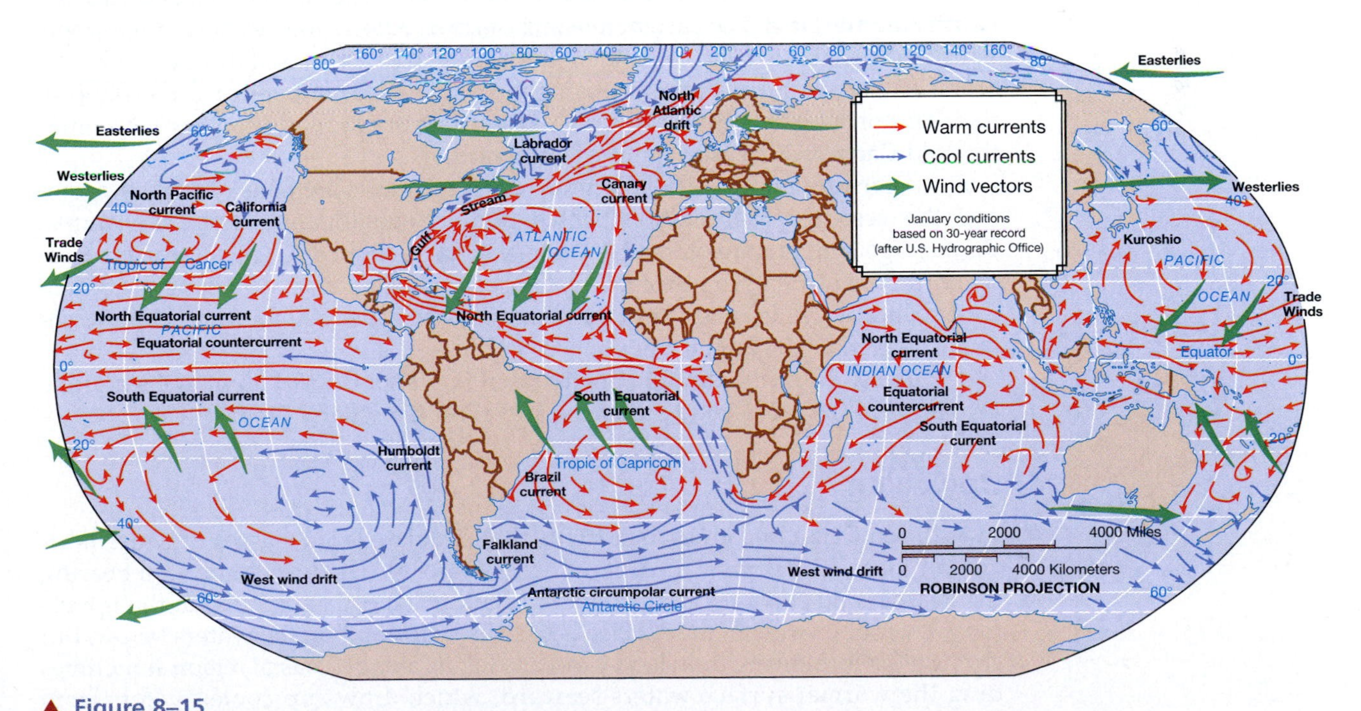

▲ **Figure 8–15**
The ocean currents of the world with the approximate location of some of the major global wind systems.

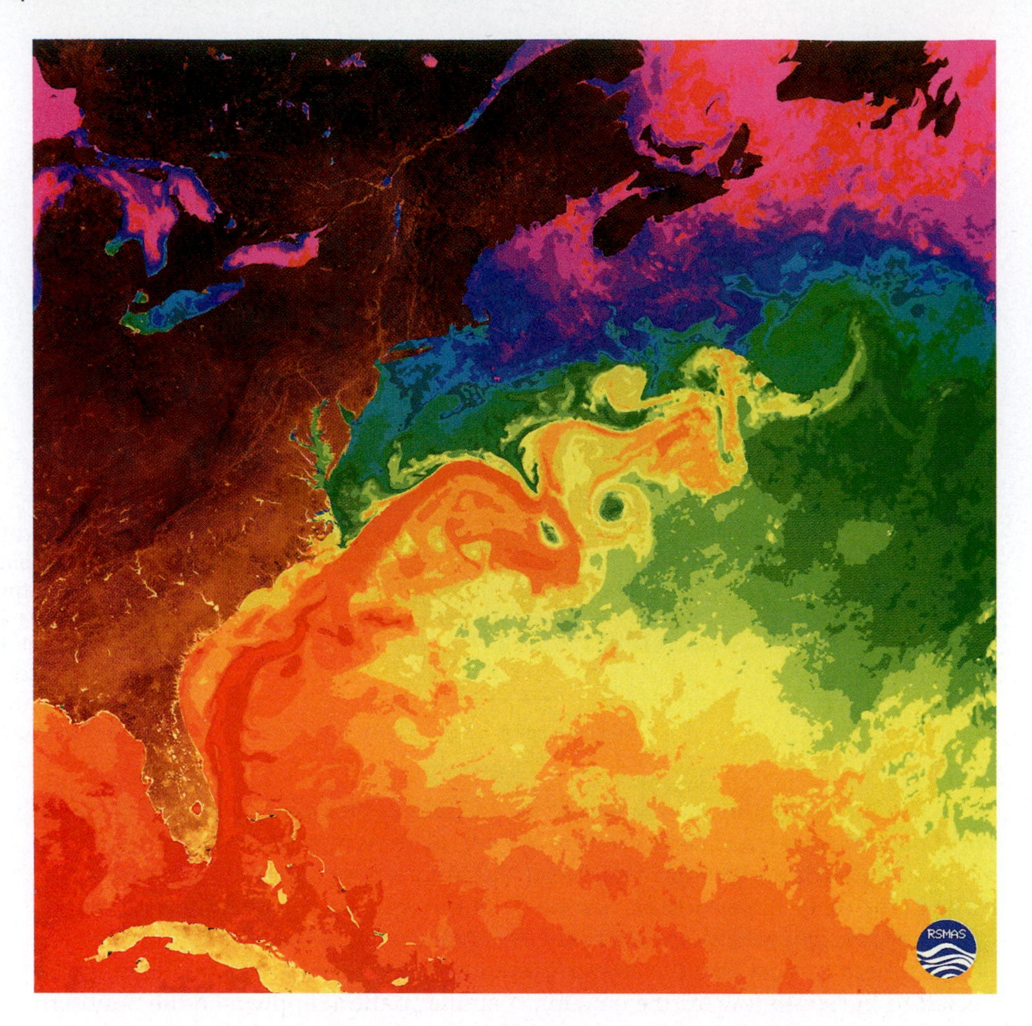

▶ **Figure 8–16**
The Gulf Stream flows in a complex pattern with eddies of different size superimposed, as shown by the infrared satellite image.

In the Northern Hemisphere, most of the North Equatorial Current reaching the South American coast turns northward to form the warm **Gulf Stream** (Figure 8–16). Near 40° N, the westerlies force the current to the east, where it becomes the **North Atlantic Drift**. The current remains warm as it flows toward northern Europe, which makes winter conditions there unusually mild for those latitudes. Even the Scandinavian countries have surprisingly warm winter temperatures, considering their far northerly position. The North Atlantic Drift gradually cools and becomes the cold **Canary Current** as it turns southward.

At corresponding middle latitudes, temperatures are considerably warmer over the western part of the Atlantic than they are over the east. Thus, for example, the average temperature off the New Jersey shore is about 8 °C (14 °F) warmer than off northern Portugal. The effect of a cold ocean current is also seen along the beaches of California, where even during the summer the water temperature seldom rises much above 22 °C (72 °F).

Finally, the cold **Labrador Current** that flows southward along the Maritime Provinces of Canada is fed by the **East** and **West Greenland Drift**.

Upwelling

In addition to the major ocean currents that circulate large masses of water horizontally, localized vertical motions of ocean water also have an important bearing on weather and climate. Because solar radiation is mostly absorbed in the uppermost layer of the ocean, the surface is usually warmer than the waters below. But strong *offshore* (blowing from land to ocean) winds along a coastal region sometimes drag the warmer surface waters seaward, which draws up cooler waters from below to take their place. This process, called **upwelling**, greatly influences sea

8–3 Physical Principles

The Dishpan Experiment

Although Rossby waves are the largest of the atmospheric waves, other swirling motions of varying sizes likewise exist. Why this complexity? At the simplest level, the behavior of the upper atmosphere is the inevitable result of three factors: (1) the unequal heating of the atmosphere from the equator to the poles, (2) the rotation of the planet, and (3) the inherently turbulent nature of the atmosphere.

To illustrate the interaction of these three, we can reproduce the migrating waves and eddy motions of the upper atmosphere with a relatively simple piece of hardware—a pan of water that rotates at a constant speed with a cooling of the fluid near the center and warming along the edge (Figure 1). The "dishpan experiment" simulates the rotating Earth with a surplus of net incoming radiation at low latitudes, and a net deficit closer to the poles. Even this very simple exercise yields motions of the fluid that in many ways resemble those of the upper troposphere.

Long waves form in the pan, resembling atmospheric Rossby waves. Superimposed on the long waves are smaller-scale eddies similar to smaller flows on Earth. Changes in the speed of rotation or the differential heating between the edge and center of the pan cause observable changes in the waves and eddies, with more extreme differences in heating and slower rotation rates leading to an increase in the amplitude of large waves at the expense of smaller-scale eddies. This implies that the oscillations in the atmosphere represent an inherent characteristic of any fluid (liquid or gaseous) on a rotating surface with spatially varying inputs of heat. Such observations are not restricted to simple dishpan experiments; the elaborate computer models that simulate the motions of the atmosphere reveal similar patterns.

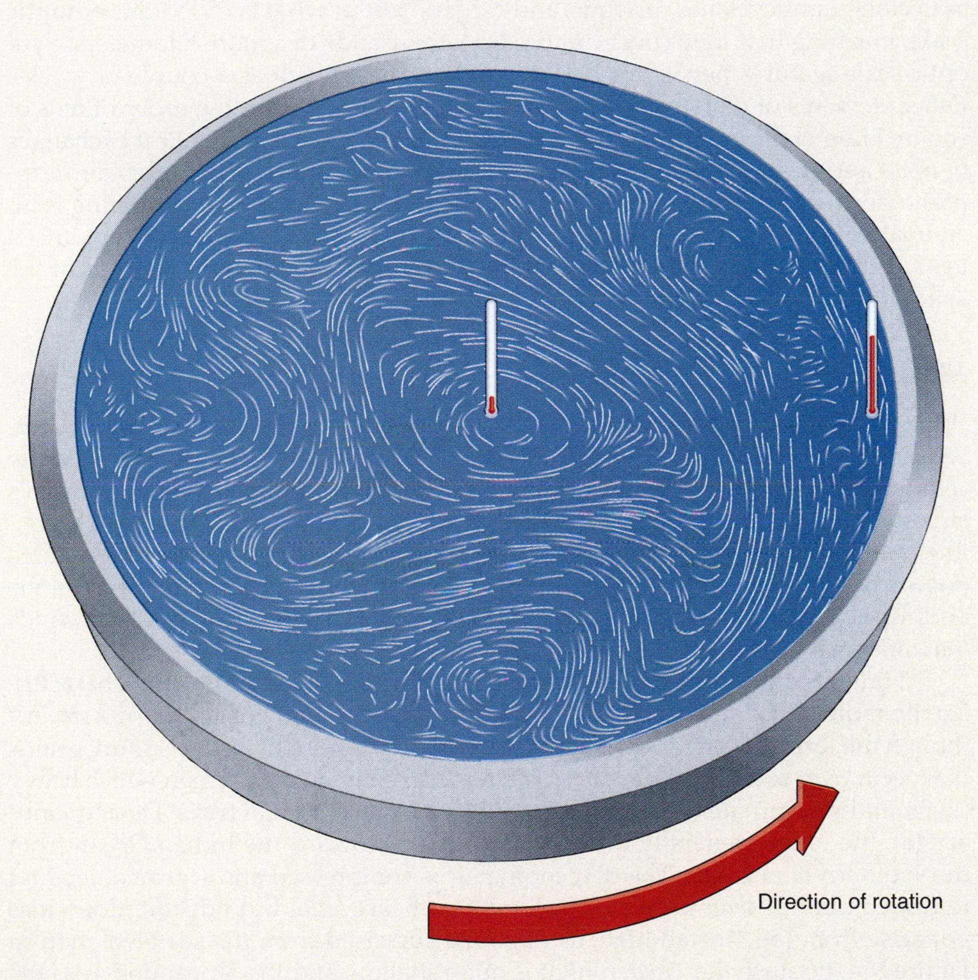

Figure 1 The pattern of eddies of different size of a "dishpan experiment."

surface temperatures (and thus weather and climate) over the eastern portions of the major oceans.

Nowhere is upwelling better illustrated than along the western coast of South America. The equatorial coast of Colombia is one of the rainiest places in the world, with average annual precipitation approaching 700 cm (290 in.). But nearby, between the latitudes of about 7° and 30° S, lies the world's driest desert—the Atacama. At the heart of this desert is Arica, Chile (18° S), which does not have a single month in the year that averages as much as 1 mm (0.04 in.) of precipitation—and most years receive no precipitation at all! This dryness is due largely to the upwelling of cold water along the coast, because cold water chills the lower atmosphere and favors the development of stable air.

The same process occurs along the central California coast during the summer. Air circulating out of the Hawaiian high sets up an Ekman spiral, with a net flow of water away from the coast. Upwelling is particularly strong just north of the San Francisco Bay region, making the water there particularly cool. Beachgoers in southern California are also familiar with the occasional effects of upwelling. Episodes of hot, dry Santa Ana winds (discussed later in this chapter) bring huge crowds to the beaches. But although the high air temperatures and clear skies are great for sunbathing, the offshore winds lead to strong upwelling that creates low water temperatures and keeps many people out of the water.

Major Wind Systems

Earlier in this chapter we covered **global scale** features such as the Intertropical Convergence Zone, the westerlies, and large Rossby waves. Smaller features, such as cyclones, anticyclones, troughs, and ridges, exist at what is called the **synoptic scale**, meaning that they cover hundreds or thousands of square kilometers. Synoptic scale features persist for periods of days to as much as a couple of weeks. Other elements of daily weather operate at the **mesoscale**, on the order of tens of square kilometers and for periods as brief as half an hour. The smallest exchanges of mass and energy operate at the **microscale**, such as those that might cause ripples to form on snow or a sandy beach—or the swirling of smoke emanating from an unattended fry pan. Microscale features are usually covered in more advanced texts. The remainder of this chapter deals with pressure and wind patterns at the synoptic and mesoscale levels.

Monsoons

We know that several large regions of the world undergo changes in mean pressure between summer and winter. The seasonal oscillation between high- and low-pressure cells is nowhere more evident—or more dramatic in its effects—than over Earth's largest continent, Asia. The great size of this continent by itself would foster strong continentality (described in Chapter 3), but the presence of the Himalaya Mountains (Figure 8–17) enhances the effect in two ways: it imposes a barrier that blocks the northward and southward flow of moisture and it alters the flow of upper-level winds that influence surface conditions.

Figure 8–18 illustrates the seasonal reversal in surface winds that characterize the **monsoon** (from the Arabic word meaning "season") of southern Asia. Although the term is often erroneously associated only with the heavy summer rains that occur over southern Asia, *monsoon* refers to the climatic pattern in which heavy precipitation alternates with hot, dry conditions on an annual basis. During January (a), the winds generally flow southwestward toward the Indian Ocean from the southern Himalayas. The descending air is compressed and warmed, leading to dry conditions over most of India and Southeast Asia. But offshore flow does not arise from the Tibetan high, as might be surmised from the sea level map in Figure 8–4, because the Tibetan high is quite shallow and the air cannot cross the Himalayas into southern India. Instead, the offshore flow is maintained by the subtropical jet stream, which branches south of the Himalayas. Jet-stream winds there are convergent, which leads to subsidence, as we discussed previously.

The situation changes abruptly during late spring or early summer, when heating of the continent contributes to a reversal in the wind direction at both low and high levels. Aloft is an easterly jet stream, with often divergent motion promoting uplift. At low levels, onshore flow occurs, bringing warm, moist, and unstable air from the Indian Ocean to the southern part of the continent, where it rises

(a)

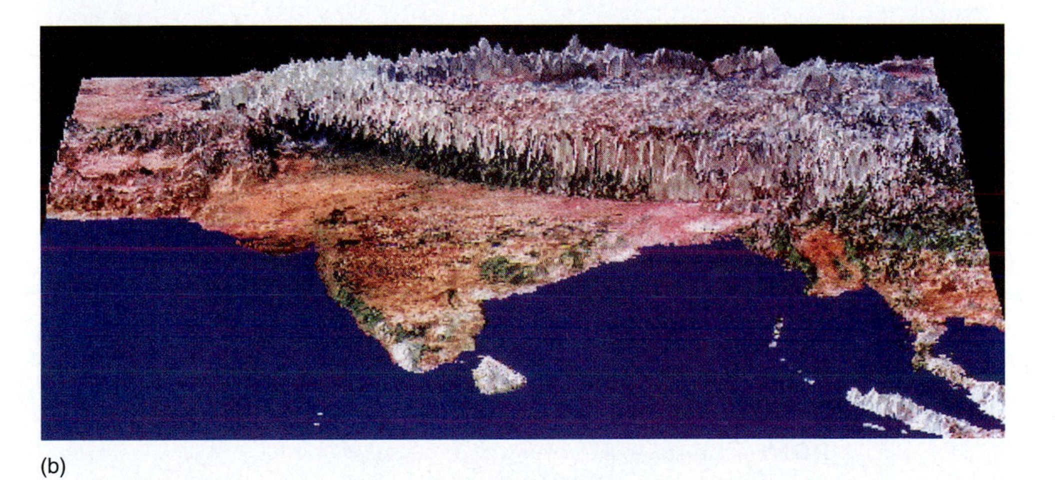

(b)

Figure 8–17
Computer-generated images of the Himalayas and Asia.

orographically and by convection as it passes over the hot surface. Cloud formation is further enhanced by a much stronger orographic effect as the air ascends the southern slopes of the Himalayas.

The combination of moist air and strong uplift produces precipitation in amounts unimaginable to inhabitants of more moderate climates. The heavy rainfall in northeastern India is enhanced even further by the movement of **monsoon depressions** (or monsoon lows), areas of low pressure superimposed in the southeasterly air flow out of the Bay of Bengal. Consider, for example, the average distribution of monthly precipitation at Cherapunji, India (Figure 8–19). Rainfall amounts are generally low during the winter months, but by early summer mean values approach 300 cm (125 in.) *per month!* Keep in mind that this extreme climate exists in a very heavily populated area of the world, affecting the lives of millions of people.

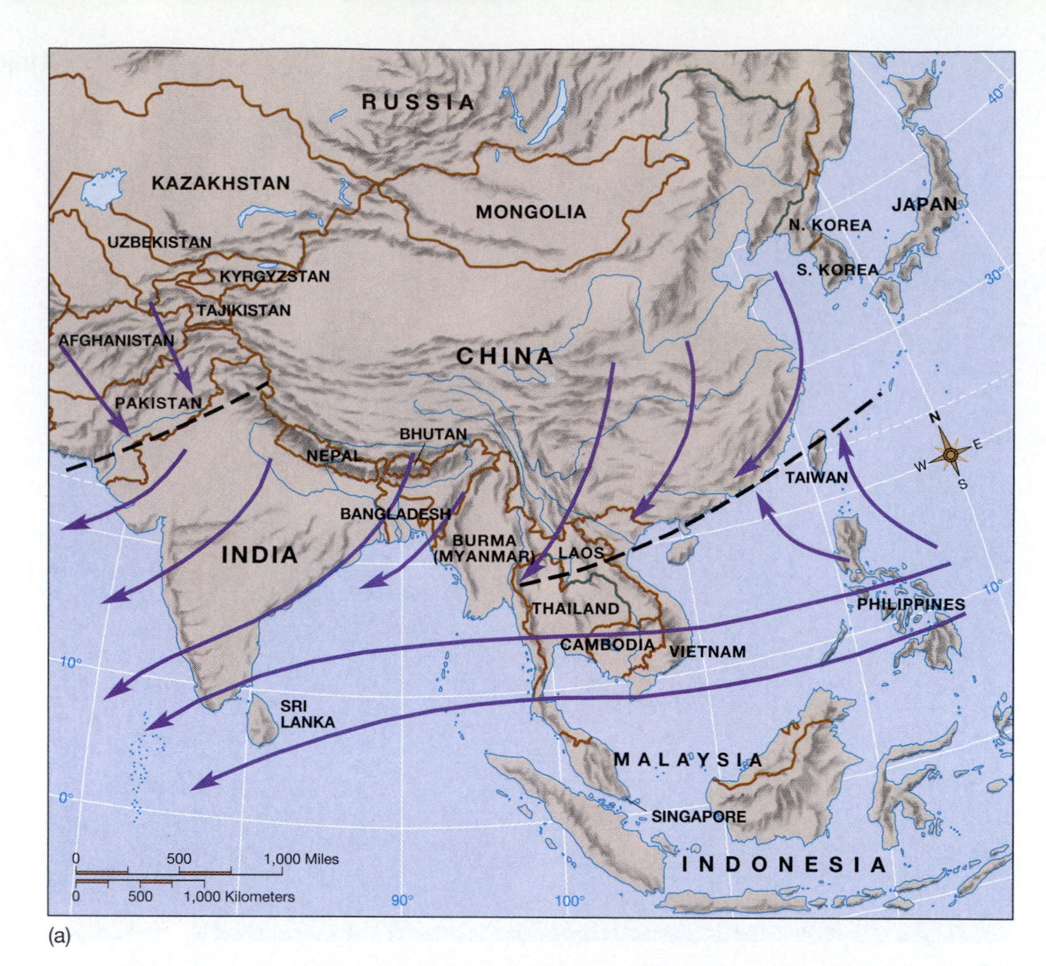

(a)

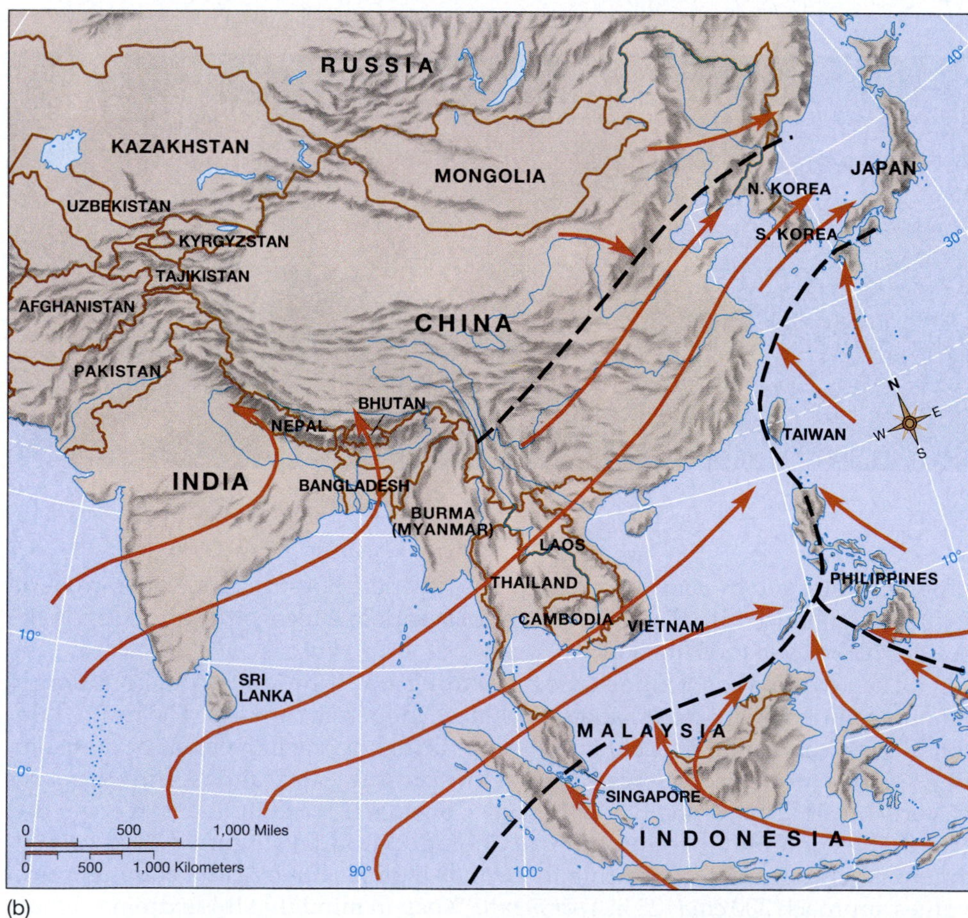

(b)

Figure 8–18
The monsoon of Asia results from a reversal of the winds between the winter and summer. During winter (a), dry air flows southward from the Himalayas. When summer arrives (b), moist air is drawn northward from the equatorial oceans. Surface heating, convergence, and a strong orographic effect cause heavy rains over the southern part of the continent.

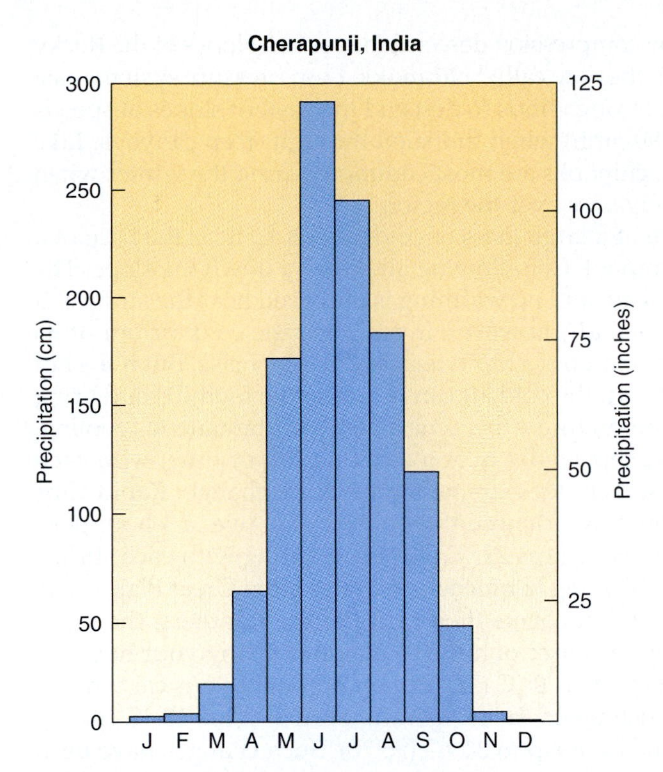

◀ Figure 8–19
Monthly mean precipitation at Cherapunji, India, highlights the sudden increase in precipitation that occurs when the south winds of the monsoon begin. Over much of the monsoon region, there is an abrupt increase in precipitation in May or June.

The southeastern part of North America is geographically similar to Asia. Like Asia, it lies to the north of a warm water body, the Gulf of Mexico, and it undergoes a seasonal reversal of the surface winds due to warming and cooling of the land mass. But despite this similarity, the southern United States does not have a strong monsoon climate. Part of the reason for this is that the smaller size of North America results in weaker oscillations in pressure than those of Asia. More important, however, is the fact that the southern United States does not have a major east–west mountain chain comparable to the Himalayas. In fact, the area is essentially flat, except for the relatively low southern Appalachians that extend southwest-to-northeast. The Appalachians do not create a strong orographic barrier to the southerly wind flow of summer, and therefore they do not promote the extreme rainfall seen in south Asia. They also do not block the passage of winter storms from the north, as do the Himalayas, nor do they promote persistent upper-level convergence. With all of that, it's easy to see why winter is hardly a dry season in eastern North America.

The desert of the southwestern United States experiences what residents often call the *Arizona monsoon*.* In spite of its name, the Arizona monsoon bears little resemblance to the real thing. It occurs each summer when warm, moist air from the south combines with strong surface heating to trigger scattered thundershowers. Unlike the intense precipitation of the Asian monsoon, that of the Arizona monsoon does little to alleviate the desert conditions of the American Southwest.

Foehn, Chinook, and Santa Ana Winds

Foehn (pronounced like the common houseplant, "fern") is the generic name for synoptic scale winds that flow down mountain slopes, warm by compression, and introduce hot, dry, and clear conditions to the adjacent lowlands. Although the term *foehn* strictly applies to winds coming from the Alps of Europe, we generally use it to describe this type of wind anywhere in the world. In Europe, foehns develop when mid-latitude cyclones approach the Alps from the southwest. The air rotates counterclockwise toward the center of low pressure and descends the northern slopes. These winds bring unseasonably warm conditions to much of northern Europe during the winter, when they are most prevalent.

*Also called the *Southwest monsoon* or the *Mexican monsoon*.

When winds warmed by compression descend the eastern slopes of the Rocky Mountains in North America, they are called **chinooks**. Low-pressure systems east of the mountains cause these strong winds to descend the eastern slopes at speeds that can exceed 150 km/hr (90 mph) when funneled through steep canyons. Like their European counterparts, chinooks are most common during the winter when mid-latitude cyclones routinely pass over the region.

Sometimes the presence of a large mass of cold, dense air near the base of a mountain range prevents a chinook from flowing all the way down the slope. The hot air then overrides the cold air, and no warming is observed near the surface. If the chinook strengthens sufficiently, however, it can push the cold air out of its path, and the foothill region undergoes a rapid temperature increase. But if the hot winds weaken even momentarily, the cold air can return to the foothills and bring another sudden change in temperature—this time a nearly instantaneous cooling. Such reversals can take place repeatedly over a short period of time, with each bringing another rapid and sometimes extreme temperature change. Rapid City, South Dakota, for example, once experienced three such cycles over a 3-hour period, with temperature changes as large as 22 °C (40 °F) occurring with each shift.

Chinook winds can be a blessing to ranchers in the western Great Plains who rely on them to melt the snow that covers their rangelands. To others, the rapid temperature oscillations can be a source of misery. Imagine leaving your house in the morning when the temperature is 0 °C (32 °F), getting out of your car when it is 38 °C (100 °F), going for lunch when the temperature is down to –10 °C (14 °F), and returning from lunch when it is up to 35 °C (95 °F). Such changes have been known to occur from chinooks, and there is anecdotal evidence (although nothing has been proven) suggesting an increase in violent crime, depression, and suicide during such episodes. Other problems not related to human discomfort can also arise from chinooks. During the 1988 Winter Olympics at Calgary (Canada), chinook winds melted the snow cover and forced a postponement of the ski competition for several days.

In parts of the western Great Plains, chinooks are frequent enough and strong enough to increase the average winter temperatures, as exemplified by Rapid City and Sioux Falls, South Dakota. Rapid City is situated in the foothills of the Black Hills and commonly experiences chinooks, whereas Sioux Falls is several hundred kilometers to the east and well out of their range. But despite the fact that its elevation is 500 m (1650 ft) greater, Rapid City has an average January temperature 4 °C (7 °F) greater than that of Sioux Falls, thanks to frequent chinook winds.

The **Santa Ana winds*** of California are similar to foehns and chinooks, but they arise from a somewhat different synoptic pattern. These winds, common in the fall and to a lesser extent in the spring, occur when high pressure develops over the Rocky Mountains (Figure 8–20). Air flowing away from the high pressure descends the western slopes and warms by compression, just as air flowing along the eastern slopes does for the chinooks. The difference is that the Santa Ana winds occur in response to a large area of high pressure causing air to flow out of the Rockies, whereas the chinook forms in response to air flowing across the range.

During Santa Ana conditions, the sinking air can warm by 30 °C (54 °F) or more and attain temperatures in excess of 40 °C (104 °F) near the coast. Contrary to what some people believe, Santa Anas are *not* warm because they pass over hot desert surfaces—it is compression, and compression only, that causes their high temperatures. In fact, during a well-developed Santa Ana, the coastal areas of California are usually hotter than interior desert locations such as Las Vegas, Nevada.

Weather in Motion
San Diego Fire

Santa Ana winds often contribute to the spread of tremendously destructive fires in California. The natural vegetation of the region is dominated by an assemblage of species collectively referred to as *chaparral* (Figure 8–21a), which is dry and highly flammable. When Santa Anas develop, the combination of hot,

*Several explanations, some rather implausible, have been offered for the origin of the name *Santa Ana*. The most widely accepted explanation is that during the early part of the twentieth century a local newspaper in Orange County reported on such a wind blowing out of the nearby Santa Ana Canyon. Eventually the name was used to describe the type of wind rather than its location.

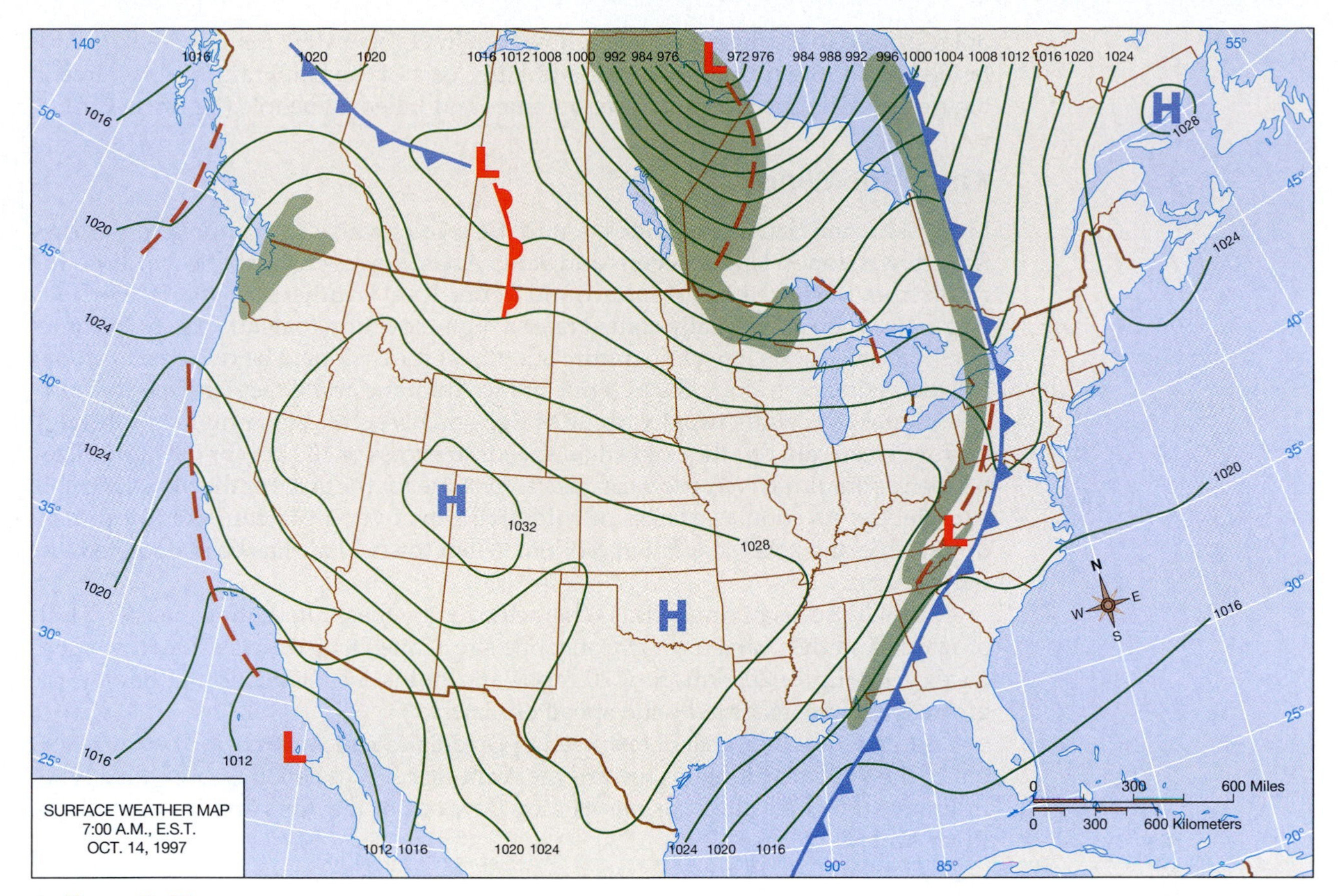

▲ **Figure 8–20**
A weather map showing a high-pressure system over the Rockies, causing a Santa Ana wind over southern California.

dry winds, low humidity, and an abundant source of fuel can set the stage for a major conflagration. Until a few years ago, the most notable of these was the 1961 fire in the exclusive Bel Air neighborhood of Los Angeles that destroyed nearly 500 homes. This was surpassed in 1993 when a series of fires over a 2-week period consumed more than a thousand coastal homes, from Laguna Beach in Orange County to Malibu and Santa Barbara farther north and west. Such fires are

(a)

(b)

▲ **Figure 8–21**
Much of coastal California is covered by chaparral, a vegetation type adapted to the summer-drought conditions of the region (a). The Oakland Hills fire (b) was spread by hot, dry winds and destroyed hundreds of homes in northern California.

not restricted to southern California, however. In October 1991, Santa Ana-like winds fanned a wildfire through the Oakland Hills, east of the San Francisco Bay area. It too destroyed more than a thousand homes and killed 24 people (Figure 8–21b).

Katabatic Winds

Like foehn and Santa Ana winds, **katabatic winds** warm by compression as they flow down slopes. Unlike foehns and Santa Anas, however, katabatic winds do not result from the migration of surface and upper-level weather systems. Rather, they originate when air is locally chilled over a high-elevation plateau. The air becomes dense because of its low temperature and flows downslope. The two best locations for such winds are along the margins of the Antarctic and Greenland ice sheets.

Katabatic winds usually occur as light breezes, but when funneled through narrow, steep canyons they can attain speeds in excess of 100 km/hr (60 mph). They happen sporadically because it can take some time for the air over the plateau to chill sufficiently. As soon as a mass of cold air forms over a plateau, gravity pulls it downslope and sets the wind in motion. When the cold air has been depleted, the wind ceases.

Much of coastal Antarctica is characterized by these alternating gusts and lulls of wind. And they can be very strong; one site, Cape Denison, occasionally experiences gusts up to 200 km/hr (120 mph) and holds the distinction of having the greatest average recorded wind speed on Earth.

Katabatic winds are not restricted to Greenland and Antarctica. They also flow out of the Balkan Mountains toward the Adriatic coast, where they are called *boras.* In France they are called *mistrals* as they flow out of the Alps and into the Rhone River Valley.

Sea and Land Breeze

Near coastal regions or along the shores of large lakes, the differential heating and cooling rates for land and water form a diurnal (daily) pattern of reversing winds. During the daytime (especially in summer), land surfaces warm more rapidly than the adjacent water, which causes the air column overlying the land to expand and rise upward (Figure 8–22). At a height of about 1 km, the rising air spreads outward, which causes an overall reduction in the surface air pressure. Over the adjacent water less warming takes place, so the air pressure is greater than that over land. The air over the water moves toward the low-pressure area over the land, which sets up the daytime **sea breeze** (remember, winds are always named for the direction *from* which they blow). (See Figure 8–23.)

As a sea breeze encroaches landward, a distinct boundary exists between the cooler maritime air and the continental air it displaces. This boundary, called the **sea breeze front**, usually produces a small but abrupt drop in temperature as it passes. This does not mean that the temperatures will not continue to rise after the sea breeze front moves on—only that there is a temporary lull in the rate of warming. Farther inland more heating can occur before the sea breeze passes, and temperatures become higher than those along the coastal strip.

Note that though the rising air in the heated column creates low pressure at the surface, it also creates higher pressure in the middle atmosphere. But we know that pressure always decreases with height. So how can this be? Remember that the terms *high pressure* and *low pressure* are relative—that is, they mean that the pressure is higher or lower than surrounding air *at the same level.* Thus, in this example the surface pressure over land is less than that over the adjacent ocean because the outflow of air above the 1-km level reduces the total amount of air over the surface. In the middle atmosphere, however, the rising of air from below increases the amount of mass above a particular level, and thereby increases the pressure relative to that of the surrounding air (though the pressure is still less than that below).

At night, when the land surface cools more rapidly than the water, the air over the land becomes dense and generates a **land breeze**. That is, lower land temperatures

(a)

(b)

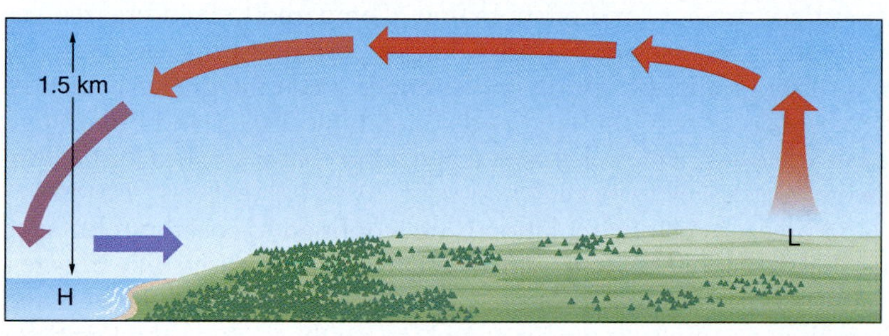

(c)

◀ **Figure 8–22**
The development of a sea breeze.

◀ **Figure 8–23**
This image of Hawaii shows the effect of a sea breeze. Heating of the land causes the air to expand upward. Coastal air flowing toward the interior is lifted as it passes over the mountains, causing orographic cloud cover.

Table 8–1 • Average Wind Speed and Direction—Los Angeles, CA

	Winter		Spring		Summer		Fall	
Time (PST)	Dir	Speed (m/s)	Dir	Speed (m/s)	Dir	Speed (m/s)	Dir	Speed (m/s)
4 A.M.	ENE	1.0	E	0.5	WSW	0.4	ENE	0.6
10 A.M.	ENE	1.4	WSW	1.9	WSW	3.2	WSW	1.0
1 P.M.	WSW	2.5	WSW	5.3	WSW	5.5	WSW	4.5
4 P.M.	WSW	3.5	WSW	5.5	WSW	5.8	W	5.1
10 P.M.	NE	0.5	W	1.6	WSW	2.4	W	0.7
1 A.M.	ENE	1.0	—	0	WSW	1.0	NNE	0.2

Source: California Air Resources Board.

make for higher surface pressure and offshore flow. Compared to land breezes, sea breezes are usually more intense and last for a longer period of time each day. Sea breezes tend to be strongest in the spring and summer when the greatest daytime temperature contrasts occur between land and sea. A typical sea/land breeze pattern is described in Table 8–1, which shows average wind characteristics for Los Angeles, California.

The sea/land breeze type of circulation is not confined to coastlines but also occurs along the shores of large lakes, in which case it creates daytime **lake breezes**. Such systems occur along the margins of the Great Lakes in the United States and Canada, but there they occupy a narrower zone than do those along the Pacific or Atlantic coasts.

Valley and Mountain Breeze

A diurnal pattern of reversing winds similar to the land/breeze system also exists among mountains and valleys. During the day, mountain slopes oriented toward the sun heat most intensely. The air over these sunny slopes warms, expands upward, and diverges outward at higher altitudes in much the same way as it does over inland areas when a sea breeze develops. The **valley breeze** occurs when air flows up from the valleys to replace it (Figure 8–24a).

At night, the mountains cool more rapidly than do low-lying areas, so the air becomes denser and sinks toward the valleys to produce a **mountain breeze** (Figure 8–24b). Mountain breezes are usually just about as intense as valley breezes but tend to be somewhat gustier. Whereas valley breezes blow fairly continuously at speeds below 15 km/hr (10 mph), the nighttime air may be still for several minutes and then suddenly flow downslope.

Air–Sea Interactions

Earlier in this chapter we described how atmospheric circulations propel ocean currents. In a similar manner, the oceans exert an important influence on the input of heat and moisture to the atmosphere. Warm surface waters heat the overlying atmosphere by the transfer of sensible and latent heat. The addition of this heat, in turn, affects atmospheric pressure. Thus, the atmosphere and the ocean are linked as a complex system.

Compared to the atmosphere, oceanic motions and temperature changes are exceedingly slow. Temperatures in the lower troposphere can change tens of degrees in a matter of hours, while oceanic temperatures are very stable. Because the ocean surface changes so slowly, information about the current distribution of temperature

(a)

(b)

Figure 8–24
A valley breeze (a) forms when daytime heating causes the mountain surface to become warmer than nearby air at the same altitude. The air expands upward and the air flows from the valley to replace it. Nocturnal cooling makes the air dense over the mountain and initiates a mountain breeze (b).

can be a useful tool for atmospheric scientists in making long-term weather forecasts. In this section we examine some of the important interrelationships between the ocean and the atmosphere.

El Niño, La Niña, and the Walker Circulation

Before its dramatic reappearance in 1983, the phenomenon known as **El Niño** was largely unknown to the public. But that changed in the early part of the year when the unusually warm waters in the eastern Pacific Ocean that mark an El Niño helped spawn a series of powerful storms in southern California. The storms not only caused severe flooding but also generated heavy surf that caused extensive coastal property damage and completely washed the sand away from many beaches. Another recent (1997–98) El Niño was also unusually strong, again leading to episodes of heavy surf, landslides, and flooding in southern California. In addition, precipitation across the southern tier of states was well above normal and severe storms were more frequent, some of which spawned very damaging tornadoes. Residents of the northern United States and eastern Canada also experienced anomalous weather conditions with unusually mild temperatures during the winter. During the fall and winter of 2002–2003, another weaker El Niño developed over the eastern Pacific.

High water temperatures promote two conditions favorable for major storm activity: increased evaporation into the air and reduced air pressure. There is therefore little doubt that the unusual episodes of 1983 and 1997–98, in which the surface waters were as much as 6 °C (11 °F) warmer than normal, played a role in the formation and passage of the storms. So what exactly is El Niño and how does it form?

At 2- to 5-year intervals (every 40 months on average), the surface waters of the eastern Pacific, especially near the coast of Peru, become unusually warm. (The

8–4 Focus on the Environment

Wildfires

Weather in Motion

Arizona Fire

In the fall and summer of 2002, the landscape of the western United States was particularly dry due to an extended drought. That situation, coupled with the accumulation of potential fuel that had built up over the years since the previous fires occurred, set the potential for devastating wildfires. Many ecologists believe that fire suppression, though it may save homes and large areas of woodland in the short run, creates the potential for uncontrollable fires. According to this argument, allowing fires to burn simply allows the ecosystem to maintain its normal conditions and precludes the buildup of highly volatile material that ultimately sets off massive conflagrations.

In 2002 the worst fears were indeed realized. By the first day of summer, hundreds of thousands of acres of land had already burned across the West. Unusually high temperatures and strong winds set the stage for the first of the most destructive fires in June in western Colorado. A couple weeks later, two major fires in Arizona joined together to form a massive wall of flames. And before summer was over, enormous, long-lived fires had also broken out over Oregon. All three fires were the worst ever in the history of the respective states.

Part of what makes massive fires such as these so devastating is that they create their own weather in a way that fosters more rapid spreading. Intense flames create strong thermal updrafts that lower the surface air pressure. This sets up strong pressure gradients, and the resulting strong winds help to transport burning embers over great distances. Little wonder that such fires are so difficult to contain and why they are often able to jump fire lines.

In one regard, fires sometimes set up weather conditions that actually assist fire crews. The same convection that can set up strong pressure gradients can also trigger rain showers that help extinguish the flames.

It is ironic, but not surprising, that while the major fires of the spring and summer of 2002 were burning in the West, other regions were coping with severe flooding. Rivers topped their banks in the upper Midwest in the spring. In the summer it was Texas that endured extremely high floodwaters, especially near San Antonio. Why is it not surprising that floods occurred in the central United States while the West was coping with drought? The answer is related to the Rossby waves, discussed earlier in this chapter. When these waves assume favored positions for extended periods of time, some portions of the wave favor high precipitation while others promote dry conditions. This is discussed in Chapter 10.

tendency for this warming to occur near the Christmas season led to the name El Niño, a reference to the Christ child.) But El Niño is much more than a simple oceanic warming; it results from a complex interaction with the atmosphere in what is called the **Walker circulation** (Figure 8–25).

Under normal conditions, the trade winds move warm surface waters near the equator westward, causing higher temperatures and even an increase in sea level about half a meter greater in the western Pacific than in the eastern Pacific. At the same time, the upwelling of colder water from below replaces the warm water migrating westward in the eastern Pacific. Warmer water in the western Pacific leads

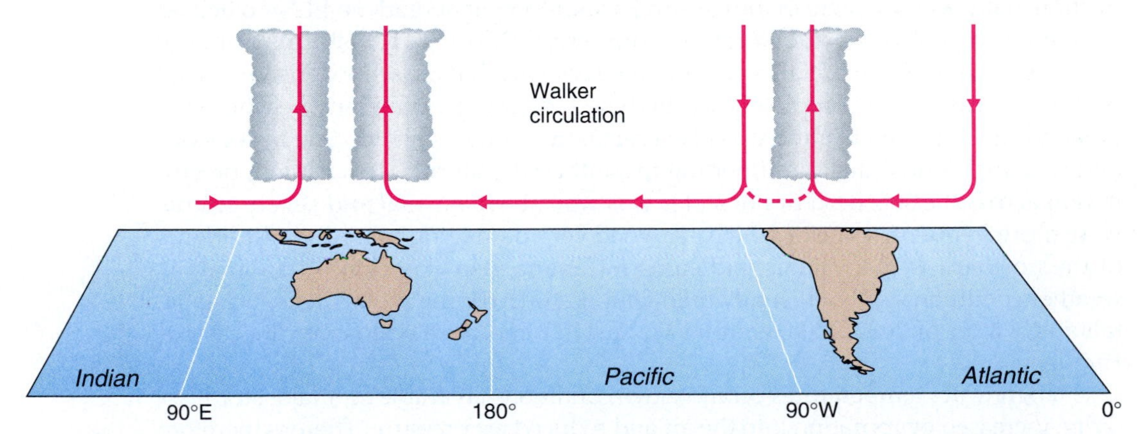

▲ **Figure 8–25**
The Walker circulation involves a westward flow of surface air over the equatorial Pacific. (Although many texts show return flow in closed loops, recent evidence indicates this is wrong. Sinking air is supplied by convergence in the upper atmosphere, not eastward flow in the upper atmosphere across the Pacific basin.) The surface flow drags warm surface waters into the western Pacific. When the westerly flow weakens or reverses, the warm waters to the west migrate eastward and cause an El Niño.

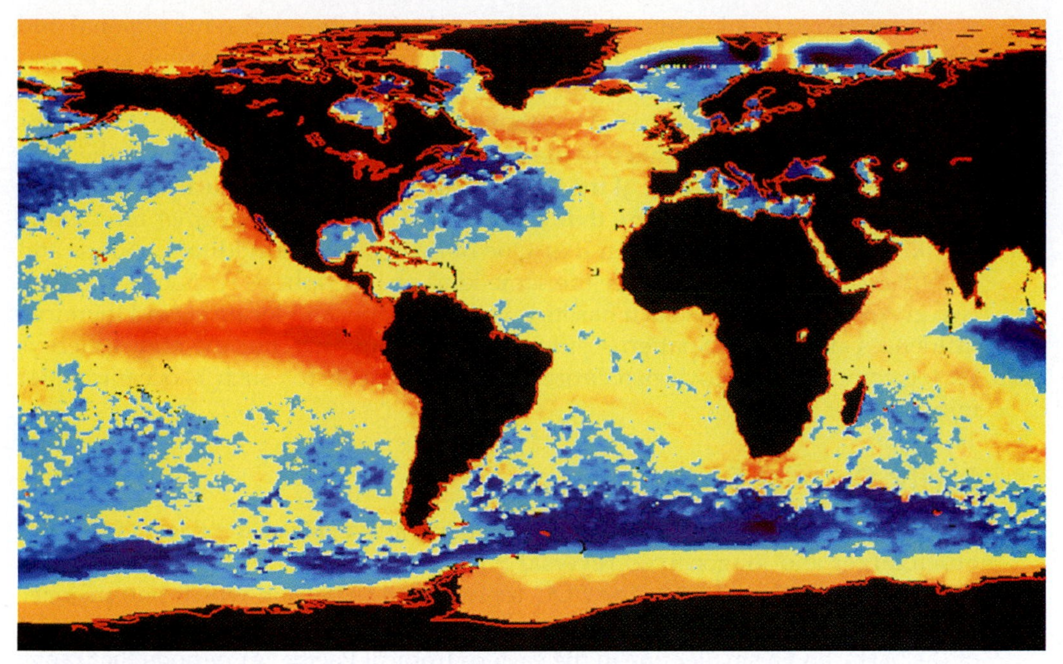

▲ Figure 8–26
The 1997–98 El Niño event contained a very large area of much above-normal sea surface temperatures over the tropical east Pacific (shown in bright red). The image is based on satellite data obtained on November 3, 1997.

to higher air temperatures, lower surface air pressure, and an increase in convectional precipitation.

El Niño develops when the trade winds weaken or even reverse and flow eastward (see Figure 8–26). The warm water normally found in the western Pacific gradually "sloshes" eastward, in part due to the higher sea level, and eventually makes its way to the coast of North and South America. This situation, called the **Southern Oscillation**, is so closely intertwined with El Niño that atmospheric scientists refer to their combined occurrences as **ENSO events** (from **E**l **N**iño and **S**outhern **O**scillation).

El Niño

La Niña

Some useful observations have been made regarding the impact of El Niño events on seasonal weather. The central coast of California, for example, appears to have a greater likelihood of unusually high amounts of precipitation when El Niño is present, while the northwestern United States and Canadian Pacific coast regions tend to be unusually dry. The effects of El Niño are not restricted to the western coast of North America, however. As you have seen, the large-scale patterns of the atmosphere are strongly influenced by the position of Rossby waves. When high-pressure or low-pressure systems exist in some locations, they not only impact local weather conditions, but also influence the overall size, shape, and position of the entire Rossby wave pattern. The establishment of an upper-level trough over the Pacific Coast, for example, will promote the development of a ridge farther to the east. For this reason, certain weather conditions in the eastern United States are set up through these **teleconnections**. El Niños favor the formation of ridge-trough patterns that often bring rainy conditions to the southeastern United States and mild, dry conditions to the northeastern United States and eastern Canada.

These teleconnections are not limited to North America; El Niños tend to promote enhanced precipitation over coastal Ecuador and Peru, the central Pacific, parts of the Indian Ocean, and eastern equatorial Africa. The western tropical Pacific, Australia, India, southeastern Africa, and northeastern South America usually undergo decreased precipitation with the occurrence of El Niño events.

When an El Niño dissipates, it can be followed either by a return to normal sea surface conditions, or by further cooling of the tropical eastern Pacific waters. The resultant pattern, the reverse of an El Niño, is called a **La Niña**. Like its warm water counterpart, a La Niña tends to favor distinct (but different) regional and teleconnection patterns, including dry conditions along the California coast and enhanced precipitation in the Pacific Northwest.

The fact that certain types of weather tend to occur in the presence of El Niños or La Niñas should not be taken to mean that some particular type of weather will always accompany the event. If the relationship between sea surface temperatures and atmospheric patterns were that well defined, long-range weather forecasting would be very much easier than it is. One of the reasons individual El Niños and La Niñas do not always produce the same kind of atmospheric response is that no two are exactly the same. Each has its own size, shape, and location, and therefore influences atmospheric patterns differently. Not only that, but the atmosphere is far too complex to be governed entirely by any single factor, such as oceanic water temperatures.

Pacific Decadal Oscillation

The ENSO is not the only important oscillation pattern across the Pacific Ocean. A much larger and longer-lived reversal pattern also exists, known as the **Pacific Decadal Oscillation** (PDO). The PDO represents a pattern in which two primary nodes of sea surface temperature exist, a large one in the northern and western part of the basin, and a smaller one in the eastern tropical Pacific. At periods that range from about 20 to 30 years, the sea surface temperatures in the two zones undergo fairly abrupt shifts. For example, the period from about 1947 to 1976–77 was marked by generally low temperatures in the northern and western part of the ocean and high temperatures in the eastern tropical Pacific. Then an abrupt transition occurred and the reverse temperature pattern became established until the late 1990s.

Information recently obtained by the TOPEX/Poseidon and Jason 1 satellites suggests that another reversal has apparently occurred (Figure 8–27). The image shows a pattern of warm water extending off the southern coast of Japan. A wedge of warmer waters also lies in the eastern tropical Pacific. Mostly cooler waters exist off the west coast of North America. Because of the enormous heat content of the oceans, the PDO may exert a direct major impact on the pressure distribution of the atmosphere.

Recent research has also suggested that the phase of the PDO influences the impact that El Niño events have on climate. When the PDO is in a "warm phase" (high temperatures in the eastern tropical Pacific), El Niño's impacts on weather are more pronounced than when the PDO is in a cold phase (low temperatures in the eastern tropical Pacific).

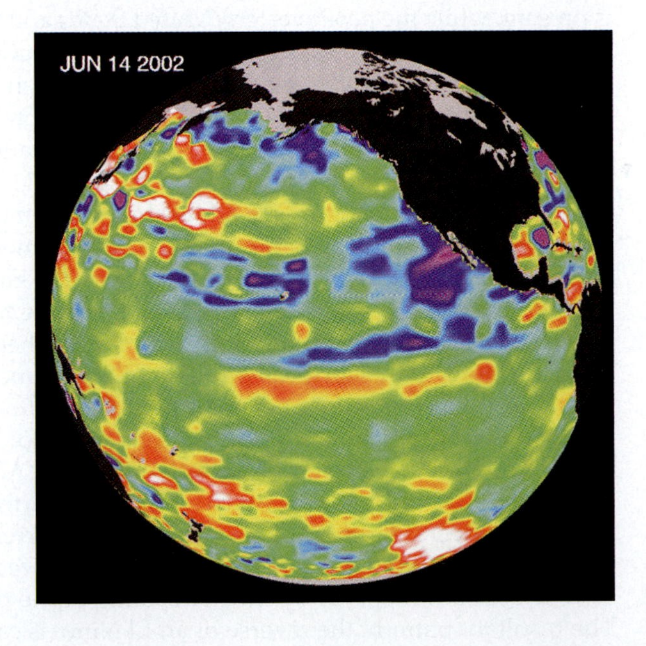

Figure 8–27
Pacific sea surface temperatures obtained in June 2002 suggest that a reversal in the Pacific Decadal Oscillation has begun. If so, this represents a major shift from the pattern in existence between 1976 and the late 1990s.

Arctic Oscillation

Multiyear oscillations are not restricted to the Pacific Ocean. An important one exists in the Atlantic. It is called the **Arctic Oscillation** (AO), and is closely related to another observed pattern called the **North Atlantic Oscillation** (NAO). Like the PDO, the AO alternates between a warm and a cold phase in sea surface temperatures. During the warm phase, there is relatively low pressure over polar locations and higher-than-normal pressure at lower latitudes. A weaker polar high means cold winter air masses do not penetrate as far south in the eastern United States, making for warm winters there. At the same time, cold air masses are more frequent in Newfoundland and Greenland, so winters are colder. The higher-than-normal pressure over the central Atlantic during this phase results in stronger westerly winds, steering warm (and moist) air to northern Europe. During the cold phase, higher-than-normal pressure dominates over the pole, with lower pressure in the central Atlantic. The eastern United States experiences cold winters, as do northern Europe and Asia, while Newfoundland and Greenland are relatively warm. We see the AO includes both a north–south oscillation in pressure and an east–west oscillation in temperature, with Greenland and Europe experiencing opposite conditions.

The Greenland–Europe seesaw in temperature has been known for centuries, and climatologists have studied it for decades as part of the NAO. What's new is a realization that the temperature oscillation is part of a hemisphere-wide phenomenon and that it is closely linked to changes in the Arctic Ocean. In a warm phase, Atlantic water pushes farther into the Arctic, and as a result sea ice thins (by more than 1 m during the most recent warm phase). During a cold phase, low-level winds are strong, which effectively isolates the Arctic Ocean from warmer, saltier water to the south and promotes a thicker layer of sea ice.

The AO switches phase irregularly, roughly on a time scale of decades. This tends to produce a run of similar winter conditions. For example. Hitler's invasion of Russia during World War II bogged down in the extreme cold-phase winter of 1941–42, which itself followed two severely cold winters. (Ironically, in planning the campaign, he relied on a forecast of normal or mild conditions, which was largely based on the mistaken idea that three cold winters in succession would never occur.) It's interesting that the last 20 years or so have seen an unusually warm phase, exceeding anything observed in the last century. This might be further evidence of human-induced warming, or it might simply be the latest excursion in a long-observed natural cycle.

Summary

Although air pressure is not the first thing that comes to mind when we think of weather, it is the driving force for atmospheric motions. Horizontal changes in air pressure distribution not only initiate the wind but also influence the vertical motions that promote or inhibit cloud formation.

While the three-cell model of wind and pressure belts, or bands, provides a good foundation for understanding the general circulation, in reality the atmosphere is dominated by a number of semipermanent cells of pressure.

The circulation of the upper troposphere is largely the result of latitudinal differences in temperature, with a strong tendency for westerly winds outside of equatorial latitudes. Despite the dominance of the westerlies, other atmospheric movements also occur in the upper troposphere. The largest and most prominent are the Rossby waves, the large, looping motions of air associated with the jet streams. Adding further complexity to the upper-level winds are short waves and other eddying motions.

Although this chapter (and indeed this book) deals primarily with the atmosphere, weather and climate phenomena are closely intertwined with the oceans. The large-scale winds of the atmosphere initiate the large, slow-moving currents of the ocean that likewise influence the input of energy and water vapor into the air. The most famous interaction between the oceans and the atmosphere is the ENSO event (El Niño–Southern Oscillation).

One of the most significant patterns of wind flow is the monsoon of Asia, the seasonal reversal of winds that are dry in the winter and extremely wet in the summer. Smaller-scale winds can also affect weather and climate. Foehn winds occur near many mountain ranges and bring hot, dry conditions. In North America, foehn winds called *chinooks* and *Santa Anas* warm the Great Plains and the West Coast, respectively. Daily sea/land breeze patterns are prominent features of coastal zones, while mountain and valley breezes produce similar patterns inland.

The circulation of the atmosphere is strongly intertwined with sea surface conditions across the oceans. We discussed three important patterns of atmospheric/oceanic behavior: the ENSO phenomenon, the Pacific Decadal Oscillation, and the Arctic Oscillation.

Key Terms

general circulation p. 227
single-cell model p. 228
zonal wind p. 228
meridional winds p. 228
three-cell model p. 228
Hadley cell p. 228
Ferrel cell p. 228
polar cell p. 228
Intertropical Convergence Zone p. 228
subtropical high p. 231
horse latitudes p. 231
trade winds p. 231
subpolar low p. 232
westerlies p. 232
polar high p. 232
polar easterlies p. 232
semipermanent cells p. 232
Aleutian low p. 233
Icelandic low p. 233
Siberian high p. 233
Hawaiian high p. 233
Bermuda-Azores high p. 233
Tibetan low p. 233
polar front p. 237
polar jet stream p. 237
subtropical jet stream p. 238
Rossby wave p. 240
ocean currents p. 242
Ekman spiral p. 242
North Equatorial Current p. 243
South Equatorial Current p. 243
Equatorial Countercurrent p. 243
Gulf Stream p. 244
North Atlantic Drift p. 244
Canary Current p. 244
Labrador Current p. 244
East Greenland Drift p. 244
West Greenland Drift p. 244
upwelling p. 244
global scale p. 246
synoptic scale p. 246
mesoscale p. 246
microscale p. 246
monsoon p. 246
monsoon depressions p. 247
foehn wind p. 249
chinook wind p. 250
Santa Ana wind p. 250
katabatic wind p. 252
sea breeze p. 252
sea breeze front p. 252
land breeze p. 252
lake breeze p. 254
valley breeze p. 254
mountain breeze p. 254
El Niño p. 255
Walker circulation p. 256
Southern Oscillation p. 257
ENSO p. 257
teleconnections p. 257
La Niña p. 257
Pacific Decadal Oscillation p. 258
Arctic Oscillation p. 259
North Atlantic Oscillation p. 259

Review Questions

1. Describe the single-cell and three-cell models of the general circulation.
2. What is the Hadley cell and where is it found?
3. Of the pressure and wind belts described in the three-cell model, which have the strongest basis in reality?
4. Why do the trade winds flow from the northeast and southeast instead of directly from the east?
5. What are the Ferrel and Polar cells?
6. Describe the distribution of semipermanent cells and their seasonal changes in location and size.
7. What is the Sahel? Describe its seasonal cycle of rainfall and explain its origin.
8. Describe the average patterns of the 500 mb level for January and July. What causes the patterns?
9. Explain why upper-atmospheric winds outside the Tropics have a strong westerly component on average.
10. Why is it that the equatorward bending of height contours for the 500 mb level implies the presence of a trough?
11. Explain how temperature patterns lead to the development of the polar jet stream.
12. Describe the distribution of Rossby waves and their impact on daily weather.
13. What is the Ekman spiral?
14. What is upwelling? How is it caused?
15. Describe the scope of global, synoptic, mesoscale, and microscale wind systems.
16. Describe the wind patterns associated with the monsoon of southern Asia.
17. Describe foehn winds.
18. How do katabatic winds differ in origin from foehn winds?
19. What causes sea/land and mountain/valley breezes to develop?
20. What is an El Niño and how is it related to the Walker circulation?
21. Describe the Pacific Decadal Oscillation and the Arctic Oscillation.

Critical Thinking

1. Figure 8–2a depicts the classic description of the one-cell model. Can you think of any reason why the arrows should not be directed in a straight line direction toward the equator?
2. The three-cell model of circulation places the center of the equatorial low right along the equator. Do you think that the varying solar declination through the course of the year would be able to shift the center of the low all the way to the Tropics of Cancer and Capricorn?
3. Which of the belts depicted in the three-cell model is likely to exhibit the greatest temperature gradients?
4. Examine Figure 8–4 and determine which of the semi-permanent cells have the highest and lowest surface air pressures. How do the strength of the winter and summer cells in the Northern and Southern Hemispheres compare to each other?
5. Figure 8–4 shows that pressure gradients, and therefore wind speeds, are strong in the middle-high latitudes of the Southern Hemisphere (e.g., at the southern tip of South America). Why doesn't a similar feature exist in the Northern Hemisphere?
6. If the western Great Plains were to have very low temperatures and the East Coast relatively warm conditions, what inferences would you make about the position and amplitude of the Rossby wave pattern?
7. Weather forecasters often use a type of weather map depicting the thickness of the 1000–500 layer of the atmosphere. What information would this map convey to the meteorologist?
8. Why isn't upwelling an important process of the Gulf Stream?
9. Why do ocean surface temperature patterns change so slowly, especially when compared to atmospheric patterns?
10. El Niño and La Niña conditions help to make seasonal forecasts for the southeastern United States—thousands of kilometers away. What atmospheric mechanism permits such extrapolations?

Problems and Exercises

1. Go to **http://weather.noaa.gov/fax/nwsfax.html**. In the first group of maps (labeled "standard barotropic levels"), select 500 mb. Then click on the link under *4a*, called *Height/temp*. The map that will appear is a map of the 500 mb pattern, with solid lines depicting the height of the 300 mb level (in tens of meters).
   a. Locate the position of the major troughs and ridges.
   b. Is the current pattern a zonal or meridional one?
   c. Where is the jet stream most prominent?
   d. Does today's jet stream appear across the entire region mapped?
   e. How does today's compare to the mean distributions shown in Figure 8–6? Why is the map for today (in all likelihood) less zonal than those in Figure 8–6?
2. If you have a highspeed connection to the web, go to **http://grads.iges.org/pix/movie.html** to see an animation depicting the change in the 500 mb pattern as predicted by a numerical model.
   a. Describe the predicted movement of the Rossby waves.
   b. How does the movement of the Rossby waves respond to their wavelength and internal wind speeds (as inferred by the spacing of the contours)?

Quantitative Problems

Rossby wave patterns and other phenomena discussed in this chapter are best understood when actual numbers are applied to them. We suggest you log on to this book's Web site (**http://www.prenhall.com/aguado/**) and work out the quantitative problems for Chapter 8.

Useful Web Sites

http://weather.noaa.gov/fax/nwsfax.html

National Weather Service site offers many different current weather maps. The first group includes standard maps for various pressure levels, including maps for North America and the entire Northern Hemisphere.

http://grads.iges.org/pix/pix.html

Numerous maps of current conditions, animations, computer model output, and El Niño forecasts.

http://www.ssec.wisc.edu/data/sst/latest_sst.gif

Satellite image depicting current ocean temperatures.

http://www.cpc.ncep.noaa.gov/products/analysis_monitoring/enso_update/index.html

Weekly update on ENSO with several animations showing sea surface temperature distributions.

http://topex-www.jpl.nasa.gov/science/el-nino.html

Information on El Niño and La Niña, along with numerous links (including one on the PDO).

http://www.enn.com/news/enn-stories/2001/07/07102001/arctic_44224.asp

Describes recent research on Arctic Oscillation.

http://virga.sfsu.edu/gif/jetsat_00.gif

Satellite image of North America with superimposed arrows depicting strength and position of current jet stream.

Media Enrichment

Tutorial

Atmospheric Forces and Motion

This is an extensive review of the equation of motion, showing the joint effects of the various forces involved in horizontal motion. It covers the development of geostrophic flow (and its close relative, gradient flow) and shows wind flow around cyclones and anticyclones. In addition, an interactive exercise lets you experiment with various combinations of friction and pressure gradient to see the resulting effects on wind speed and direction. Animations are used throughout to present many issues difficult to depict with static diagrams.

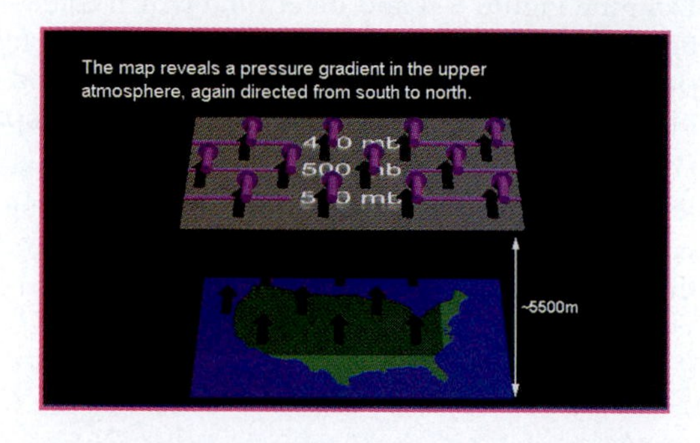

Tutorial

Upper-level Winds and Pressure

This tutorial develops the relationship between upper-level pressure and height fields and shows how ridges and troughs are reflected on contour maps. Rossby waves are presented as common features of the mid-latitudes, with animations used to depict simultaneous movement of the wave and prevailing westerly flow through the wave.

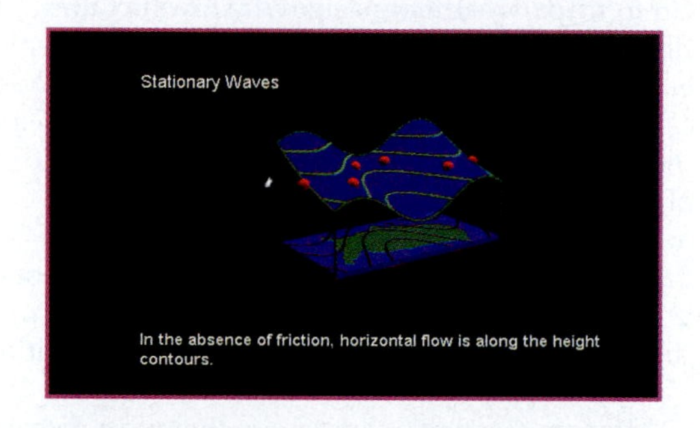

Interactive Exercise

Seasonal Migration of the ITCZ in Africa

This tutorial illustrates the movement of the ITCZ across Africa through the course of the year. Click on the PLAY button to watch the migration of the ITCZ over the year, or click on an individual month to observe the mean position for that month. You can view histograms (charts depicting the average monthly temperature and precipitation) for a number of locations by rolling your cursor over the city names.

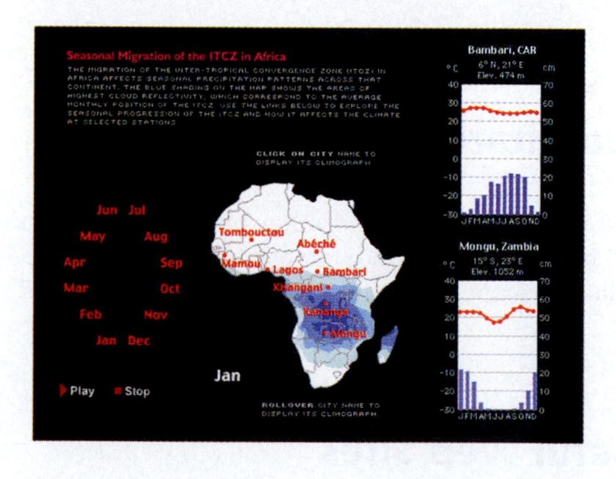

Weather in Motion

El Niño

This movie shows the development and recession of an El Niño event from early May 1997 to late June 1998. High sea surface temperatures (SSTs), indicated by red, begin to appear in May 1997, off the west coast of tropical South America. The pool of warm water becomes large by late June and reaches a maximum size by approximately late August–early September 1998. The zone of high SSTs changes shape and shifts position for the next few months, and begins to recede early in 1998. This El Niño was followed by the opposite situation, a La Niña. The La Niña SSTs are shown in an accompanying movie.

Weather in Motion

La Niña

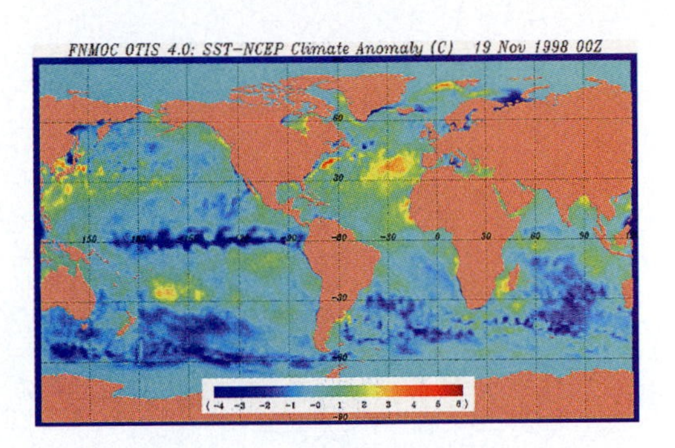

Here we see a continuation of the SST pattern from the previous movie (with about four months' overlap). Warm SSTs exist in the tropical east Pacific in the spring of 1998, but a small pool of cool water appears along the equator at 120° W longitude in late May. The zone of low SSTs expands eastward for a while and covers much of the equatorial region by mid-summer 1998. Notice that a small zone of cool water appears off the South American coast at this time. This smaller area of low SSTs does not remain continuously over the area but comes and goes over time, eventually merging with the larger area of cool water to the west. By mid-February 1999, the low temperatures have disappeared over the easternmost tropical Atlantic. The SST patterns of the El Niño and La Niña years at the latter part of the 1990s are believed to have contributed to anomalous weather conditions across much of the world.

Weather Image

Algerian Dust Storm

Viewed from the space shuttle, this North African dust storm appears as a brownish blob. The margin of the dust storm is capped in places by cumulus clouds, formed by the uplift of air ahead of the wind. These storms can continue into the Atlantic Ocean, spreading dust as far away as North and South America.

Weather Image

Southern California Sea Breeze

This satellite view of the Los Angeles area shows the development of orographic clouds along Catalina Island and the Palos Verdes Peninsula. These clouds formed when the sea breeze was lifted as it approached land.

chapter 9

A cold, damp air mass over Ketchikan, Alaska.

Air Masses and Fronts

Nobody ever expects warm winters in Alaska, but sometimes conditions are even more severe than usual. This was particularly true in late January and early February of 1999, when a prolonged period of extreme cold produced the worst cold snap in a decade. Galena, in the northern part of the state, just missed an all-time low temperature when the reading dipped to –53 °C (–64 °F) in February. Conditions were so extreme that the city government officially closed down, maintaining only emergency services. If that wasn't cold enough, consider the –75 °C (–103 °F) wind chill factor at Kotzebue along the northwestern coast! Things weren't exactly balmy in Fairbanks, either, when on February 15 the low temperature dipped below –37 °C (–35 °F) for a record-breaking nineteenth consecutive day. Needless to say, the air was also extremely dry, as is always the case under cold conditions.

Contrast those conditions with those experienced in Texas half a year earlier. College Station, by way of example, had its warmest May, June, and July average temperatures ever, as maximum temperatures exceeded 38 °C (100 °F) 51 times between May 31 and September 3. The heat was further compounded by high humidities and unusually sparse rainfall throughout the period.

Have you ever wondered about episodes such as the two above, in which large areas experience more or less similar weather? At times like these, broadcasters use phrases such as "throughout the Midwest," or "across the eastern seaboard," or "Today the Pacific Northwest experienced. . . ." By way of contrast, it often happens that places within an hour's drive of each other have very different weather, with essentially nothing in common. What's the explanation for this behavior? Why does the atmosphere sometimes organize itself into broad uniform patches, and other times show extreme variation over short distances? This chapter addresses these and related questions using some very simple, but nevertheless powerful concepts.

The Alaska and Texas situations cited above represent two extreme instances in which large regions are covered by a body of air having more or less uniform temperature and moisture. These large volumes of air are called **air masses**. Often an area the size of North America will be covered by several air masses at the same time so that, for example, the northeastern United States and southeastern Canada may experience cold, dry conditions, while the southern United States is dominated by warm, moist air. Consequently, a person might board an airplane in Nashville feeling perfectly comfortable in shirt sleeves, only to end up shivering in Boston. Moreover, these air masses are commonly separated from each other by fairly narrow boundary regions, called **fronts,** across which conditions change rapidly. The passage of these fronts are significant weather events because they often bring abrupt changes in temperature, humidity, and wind. They also provide a lifting mechanism that can lead to the formation of clouds and precipitation.

In this chapter we describe the formation and nature of air masses, the fronts that separate them, and their influences on local weather.

CHAPTER OUTLINE

Formation of Air Masses

The temperature, pressure, and moisture characteristics of the atmosphere arise in large part from the continuous exchange of energy and water vapor near the surface. When energy inputs exceed energy losses, the temperature of the air increases.

In the same way, when more evaporation than precipitation takes place, the moisture content of the atmosphere increases. But because heat and water are not uniformly distributed across the globe, the cooling and warming of the atmosphere vary from place to place, as does the net input of water vapor. Thus, air over the tropical Pacific, for example, takes on different characteristics from air over northern Canada.

Source Regions

The areas of the globe where air masses form are called **source regions**. The heating or cooling of large bodies of air requires many days, as do changes in the moisture content, so air must remain over a source region for a substantial length of time in order for an air mass to form. Air mass source regions occur only in the high or low latitudes; middle latitudes are too variable, and thus they do not have the quiet periods necessary for an air mass to take on the characteristics of the underlying surface. Also, an area must be quite large—many tens of thousands of square kilometers—to act as a source region. Iceland is too small, for example, to allow the formation of air masses, despite its northerly location.

Although air masses have fairly uniform temperature and moisture content in a horizontal direction, they are not uniform from the surface to the upper atmosphere. In fact, large vertical gradients in temperature can easily occur in an air mass. These vertical differences in temperature affect the stability of the atmosphere (Chapter 6), which has important ramifications with regard to the likelihood of precipitation. Thus, some air masses, by their very nature, are more likely than others to yield precipitation.

Air masses are classified according to the temperature and moisture characteristics of their source regions. Based on moisture content, air masses can be considered either *continental* (dry) or *maritime** (moist). According to their temperature, they are either *tropical* (warm), *polar* (cold), or *arctic* (extremely cold). Meteorologists use a two-letter shorthand scheme for categorizing air masses. A small letter *c* or *m* indicates the moisture conditions, followed by a capital letter *T*, *P*, or *A* to represent temperature. Thus, continental polar air, for example, is designated cP, and maritime tropical air is mT. While this combination of letters could theoretically yield six possible types of air masses, maritime arctic (mA) air masses do not occur in nature because water bodies do not get cold enough to foster arctic air (they freeze at arctic temperatures, which largely removes their maritime character). Thus, there are five types of air masses. The major source regions of North America are shown in Figure 9–1.

Given the wide continuum of temperature and moisture contents that can exist, the arbitrary categorization of air masses into only five types might seem somewhat limiting. How, for example, would you classify a day with a temperature of 20 °C (68 °F) and a dew point of 10 °C (50 °F)? There really is no answer in that case, as it would hardly feel either tropical or polar. Or looking at the question another way, why even try to classify such bodies of air? The answer to the latter question is that the air mass concept is useful when we wish to identify the air on either side of frontal boundaries, or when a very simple description of the air will suffice.

Air masses are not permanently confined to their source regions; they are able to migrate to regions marked by less extreme weather conditions. The movement of an air mass away from its source causes two things to happen: (1) the region to which the air mass has moved undergoes a major change in temperature and humidity and (2) the air mass becomes more moderate. We will now examine the various types of air masses, the weather they bring, and the transformations they experience as they travel.

---

*Sometimes called *marine*.

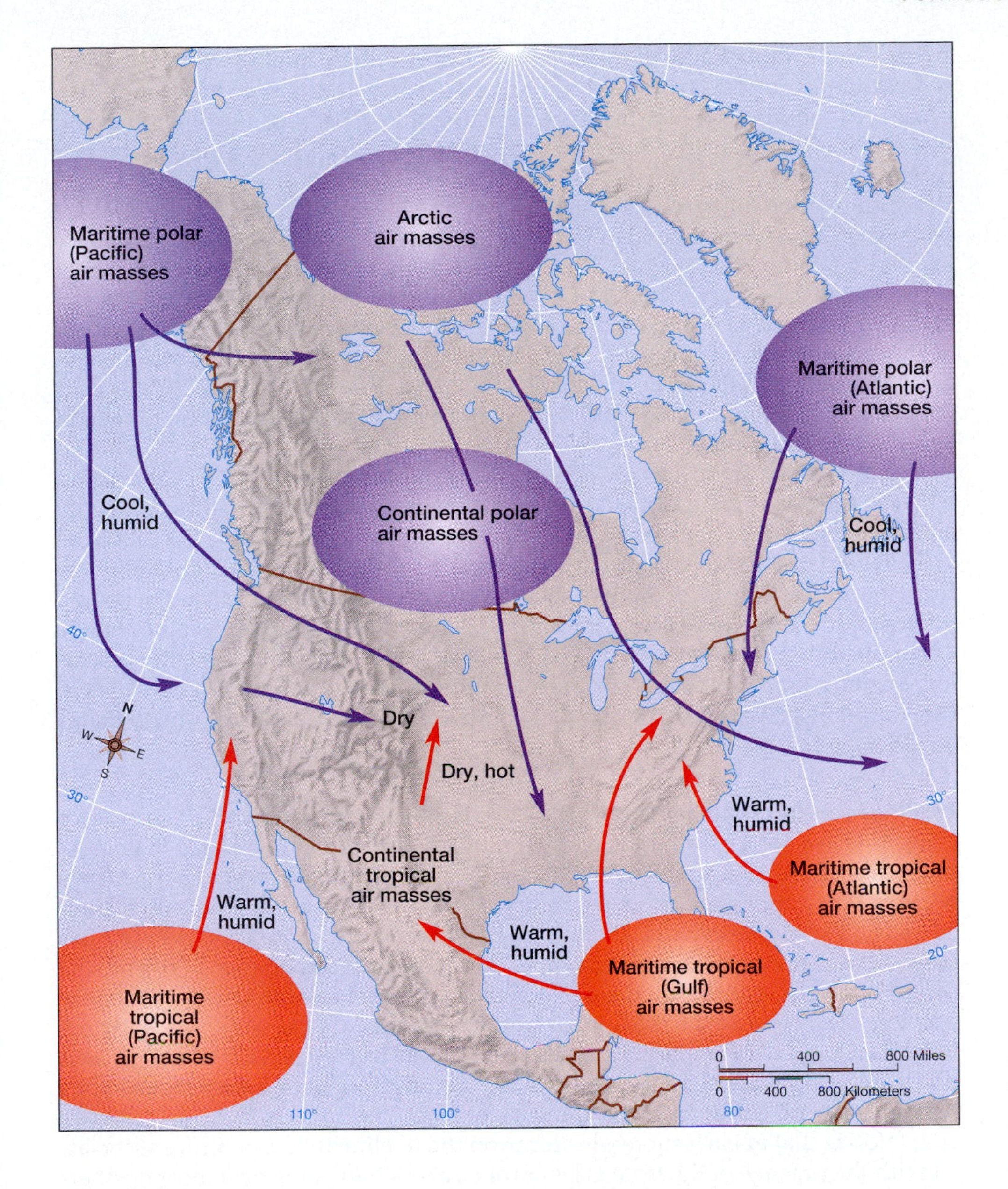

◀ **Figure 9–1**
The source regions for North American air masses.

Continental Polar (cP) and Continental Arctic (cA) Air Masses

Continental polar (cP) air masses form over large, high-latitude land masses, such as northern Canada or Siberia. In winter, these regions have short days and low solar angles. They also are usually snow-covered during the winter and therefore reflect much of whatever solar radiation does reach the surface. This combination of circumstances virtually guarantees that the air will lose more radiant energy in the winter than it receives. The cooling of the air from below leads not only to low temperatures but also to radiation inversions and highly stable conditions.

In addition to having very low temperatures, winter cP air masses are extremely dry. Recall from Chapter 5 that little water vapor can exist in cold air and that the dew point (or frost point) temperature is always equal to or less than the air temperature. Thus, if the air temperature is very low—say –30 °C (–22 °F)— a kilogram of air at sea level can contain a maximum of only 0.24 g of water vapor. If the air is not saturated, the actual amount of moisture will be even lower.

The combination of dry air and stable conditions ensures that few if any clouds will form over a cP source region. Furthermore, the lack of water vapor reduces the absorption of incoming solar radiation by the atmosphere. Thus, despite their low temperatures, cP air masses over their source regions are usually bright and sunny. On the other hand, the stability of the atmosphere inhibits vertical mixing, so pollutants introduced at the surface remain concentrated near the ground. Because cold conditions lead to increased consumption of heating fuel (often coal and fuel oil), it isn't surprising that cP air is often associated with poor air quality over urban areas.

Summer cP air masses are similar, but much less extreme. That is, they are both warmer and more humid than the winter version. They tend to remain at higher latitudes compared to winter cP air masses, so they do not influence as much of the globe as their winter counterparts. Temperature inversions do not form, because the air mass develops over a snow-free surface when the days are long. In fact, it's not uncommon for some convective uplift to occur, resulting in "fair weather cumulus" clouds (scattered puffy clouds in an otherwise blue sky). Still, these air masses form over a continent, and thus do not contain enough moisture to sustain significant precipitation.

Continental arctic (cA) air is colder than continental polar, but the distinction between the two is more than just a matter of degree. More importantly, cA and cP air are separated by a transition zone similar to the polar front (Chapter 8), called the *arctic front*. Unlike the polar front, which can extend upward several kilometers from the surface, the arctic front is shallow and does not usually extend beyond a kilometer (0.6 mi) or two above the surface (Figure 9–2). Thus, we feel the change in temperature associated with the passage of an arctic front, but the relatively shallow front does not produce a great deal of uplift that might promote snowfall. On rare occasions, cA air can extend as far southward as the Canadian–United States border region.

Weather Image

The Perfect Storm

Modification of cP Air Masses When leaving their source region, cP air masses bring cold weather to more temperate latitudes. Figure 9–3 illustrates the movement of a hypothetical mass of continental polar air. In (a) the frontal boundary is north of Minneapolis, and moderate temperatures exist throughout the central United States. Twenty-four hours later (b), the boundary of the continental polar air has passed over Minneapolis, causing a 20 °C (36 °F) drop in temperature. Farther to the south, St. Louis and Birmingham experience little temperature change from the previous day, because the cold air has not yet extended that far southward. By the third day (c), however, the cold air has penetrated to the Gulf Coast and caused noticeable temperature declines at St. Louis and Birmingham, while Minneapolis remains under the influence of extremely cold air.

Notice that at each successive location the decline in temperature associated with the passage of the front is less pronounced than at the next most northerly site. In other words, Minneapolis experienced the most severe drop in temperature, while St. Louis and Birmingham underwent less extreme cooling. This occurs because of the gradual moderation of the cP air as it leaves its source region.

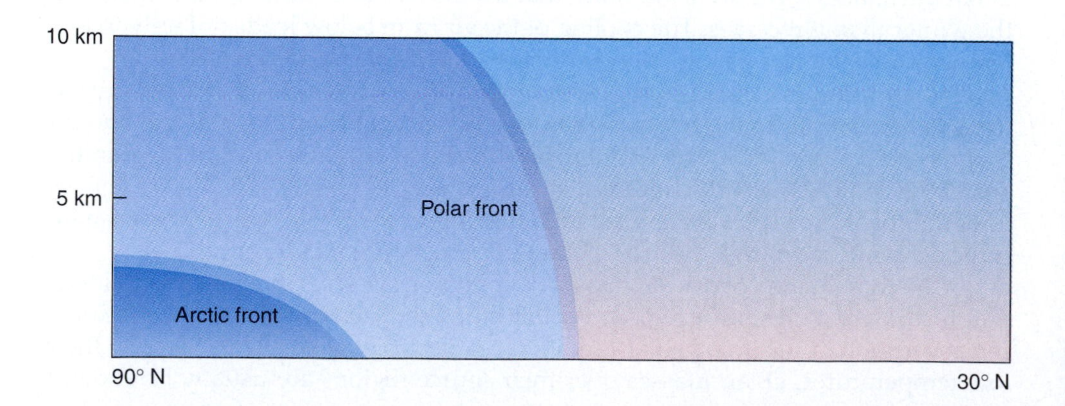

▶ **Figure 9–2** The arctic front separates a shallow layer of extremely cold arctic air from cold polar air. It differs from the polar front by virtue of its much more limited vertical extent.

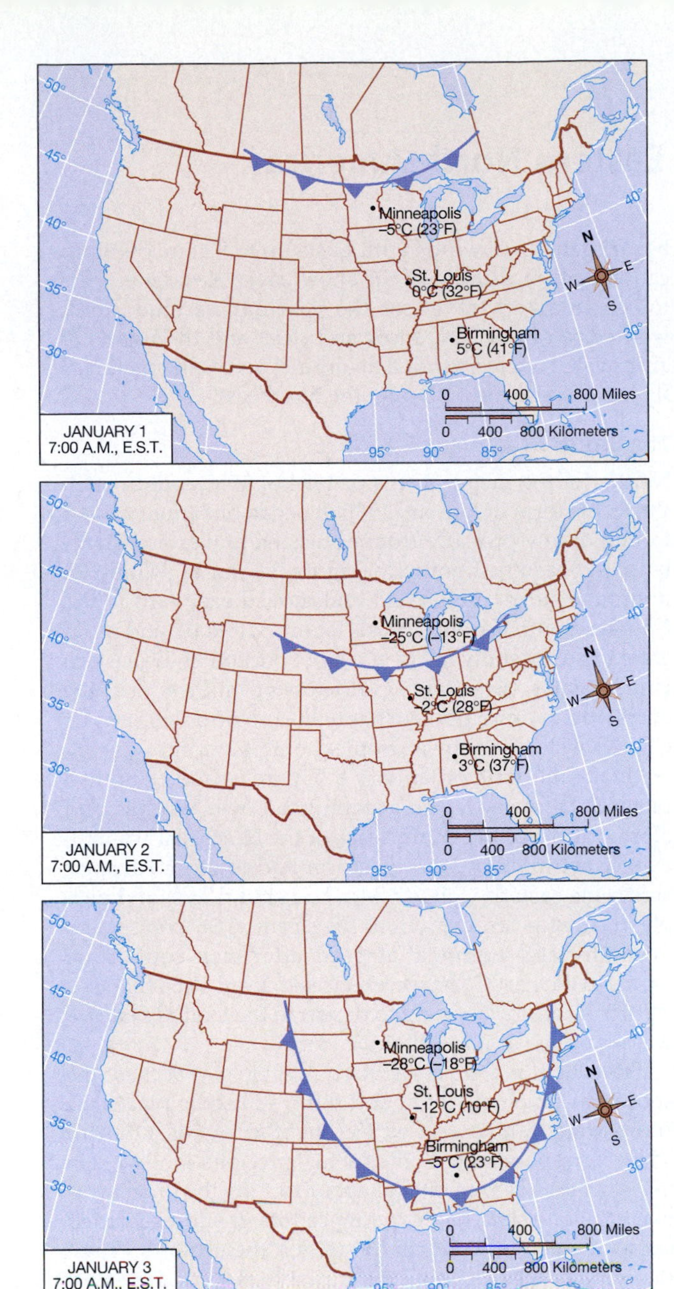

◀ **Figure 9–3**
A sequence of surface weather maps showing the southward movement of continental polar air.

Maritime Polar (mP) Air Masses

Weather Image
Pacific Storm Water Vapor

Maritime polar (mP) air masses are similar to continental polar air masses but are more moderate in both temperature and dryness. Maritime polar air forms over the North Pacific as cP air moves out from the interior of Asia. The warm Japan current adds heat and moisture to the cold, dry air and converts it from cP to mP. The developing air masses migrate eastward across the Pacific, with most of them passing across the Gulf of Alaska before reaching the West Coast of North America. They approach the northern west coast of North America throughout the year but influence the California coast mainly in the winter.

Maritime polar air also affects much of the East Coast, but the manner in which it approaches is different. This type of air is associated with the circulation of air around mid-latitude cyclones *after* they pass over a region. As the air rotates cyclonically (counterclockwise in the Northern Hemisphere), it sweeps around the low-pressure system and approaches the coast from the northeast. The resultant winds are the famous **northeasters** (or *nor'easters*) that can bring cold winds and heavy snowfall (see *Box 9–1, Special Interest: Maritime Air Masses Invade Eastern North America*).

9–1 Special Interest

Maritime Air Masses Invade Eastern North America

The movement of air masses produces many abrupt and important changes in the weather. Examples of such transitions are almost limitless and new ones occur regularly. Three recent examples in the northeastern United States and eastern Canada illustrate how such air mass transitions can have an enormous impact on the local environment: the major storms of the winters of 1994, 1995, and 1996.

The Storm of March 2–4, 1994

Residents of the East Coast vividly recall the winter of 1993–94 as one of the snowiest of all time. On March 3, the fifteenth major snowstorm of the season hit the northeastern United States with what the *New York Times* referred to as a "classic northeaster." The storm originated over the southeastern United States on the morning of March 2 (Figure 1a and b). A cold front extended southward from the mid-latitude cyclone into the Gulf of Mexico, and a warm front extended along the coast to Cape Hatteras, North Carolina. The counterclockwise circulation over the southeastern United States brought with it maritime polar air from the Atlantic, along with rain from Maryland to Georgia.

By the second day (Figure 1c and d), the system had intensified and moved to the New Jersey coast. Strong northeasterly winds brought warm, moist air from offshore, which was subject to strong convergence as it entered the center of low pressure. The combination of ample moisture and strong convergence brought heavy snow, with Martinsburg, West Virginia, accumulating 34 cm (14 in.). The storm intensified even further as it moved northward into New England and the Maritime Provinces (Figure 1e and f), bringing the total snowfall for the season to more than 215 cm (90 in.) in the Boston area. Ultimately, the system moved offshore into the Atlantic.

The Storm of February 5, 1995

Unlike the previous winter, the first half of 1994–95 had been unusually mild in the northeastern United States, with generally balmy temperatures and little snow. This situation came to an abrupt end on February 5 (Figure 2). Like the March 1994 storm, the system migrated northward along the Atlantic Coast and pulled in maritime polar air from offshore. It was centered near Chesapeake Bay on the morning of the 4th (Figure 2a and b) and intensified greatly as it moved over New England and the Maritime Provinces on the 5th (Figure 2c and d). Over that 24-hour period, the center of low pressure had decreased from about 995 mb to 962 mb.

In the New York City–New Jersey area, snow began falling just after midnight, and by morning a total of 38 cm (16 in.) had accumulated at Princeton, New Jersey. Although the heavy snowfall subsided just hours after it began, gale-force winds with gusts up to 80 km (50 mph) caused severe drifting of the snow cover. Residents were forced to avoid travel for the next day as road crews worked to remove the snow and sleet, and thousands of air travelers were stranded at airports across the region as flights were delayed all over the Northeast.

The Storm of January 6–8, 1996

Neither of the major events of 1994 or 1995 could match the great storm of January 1996. It began on January 6 as a trough of low pressure from Louisiana to western Maryland and eastern Kentucky (Figure 3a and b). Within 24 hours the storm intensified and moved eastward to the Georgia–South Carolina coast (Figure 3c and d), and moderate to heavy snow fell over a large portion of the eastern United States. (See "Media Enrichment" at the end of the chapter for a description of a satellite movie on this book's CD-ROM showing the movement of this storm.)

By the morning of January 8 (Figure 3e and f), up to a meter (3 ft) of snow had accumulated over Virginia and North Carolina. But a large area of the Northeast was still experiencing the worst the storm had to offer. Airports across the eastern United States had shut down, and people in cities as far away from the storm as St. Louis spent uncomfortable nights at airports after their connecting flights to the East Coast were canceled. In all, 7600 flights—a third of those scheduled to depart in the United States—were canceled as a result of the storm.

Snow depths were particularly great in part because the snow was unusually light and fluffy—in some places the snow–water ratio exceeded 18:1. In other words, 18 cm of snow corresponded to 1 cm of actual precipitation—well in excess of the average ratio of about 10:1. By the time it was over, Philadelphia had experienced its greatest recorded snowfall ever, with 73 cm (30 in.) on the ground. Providence, Rhode Island, suffered through its second-greatest snowfall, and Washington and New York experienced their third-deepest accumulations.

At least 86 people died as a result of the storm. But despite the human toll, things could have been much worse. Fortunately, the National Weather Service forecast for the region had been extremely accurate in predicting the storm's timing, location, and intensity. This gave local citizens and disaster authorities ample time to prepare for the onslaught of maritime air.

Notice that air mass characteristics were important in all these storms but were not the complete story. In each case, there were characteristic circulations around a low-pressure cell (cyclone) that delivered those air masses in systematic ways. The relations between air masses, fronts, and cyclones is too big a topic for discussion here, but don't despair—it will be covered in the next chapter.

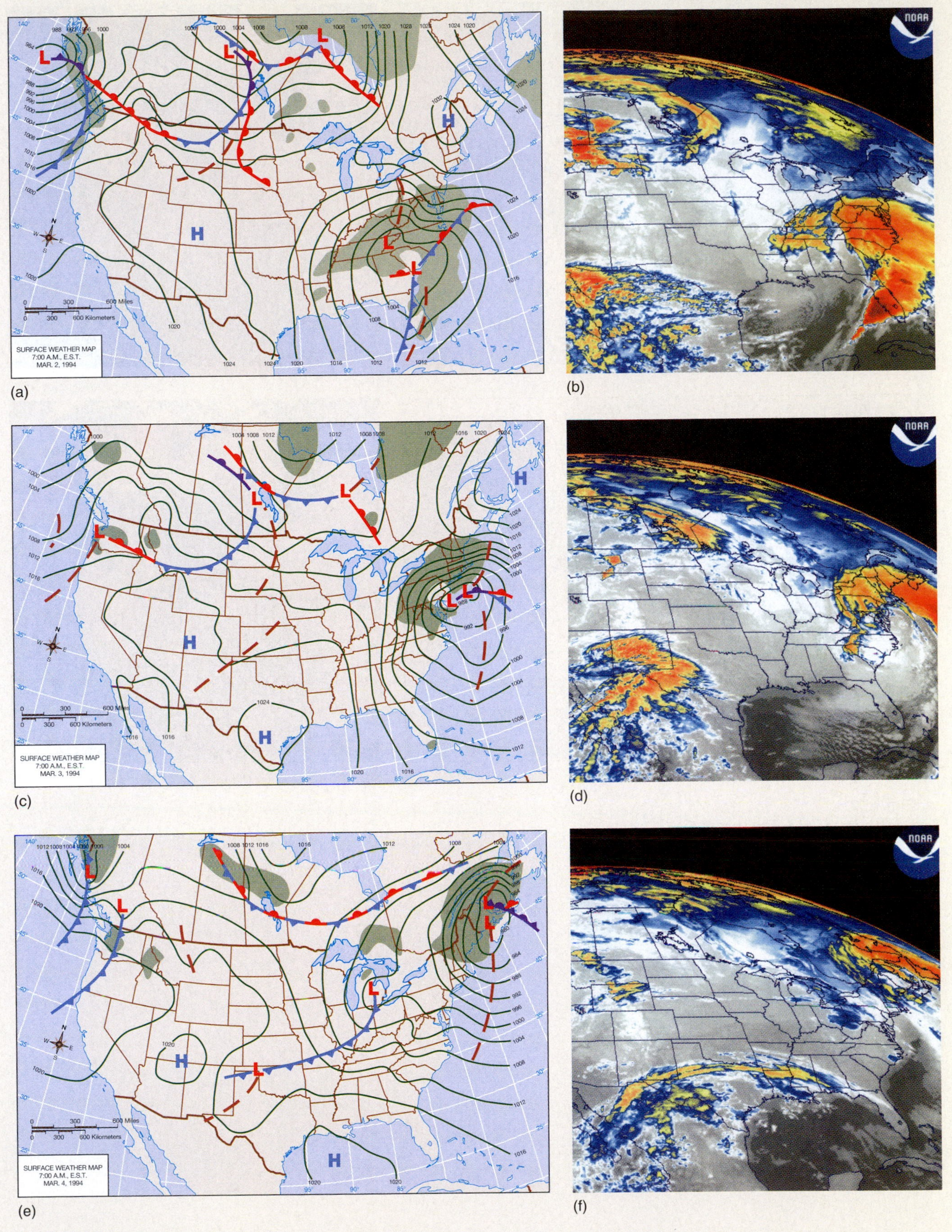

▲ **Figure 1**
The northeaster of March 1994. The maps on the left panels and the satellite images on the right show the position of the storm at 7 a.m. EST.

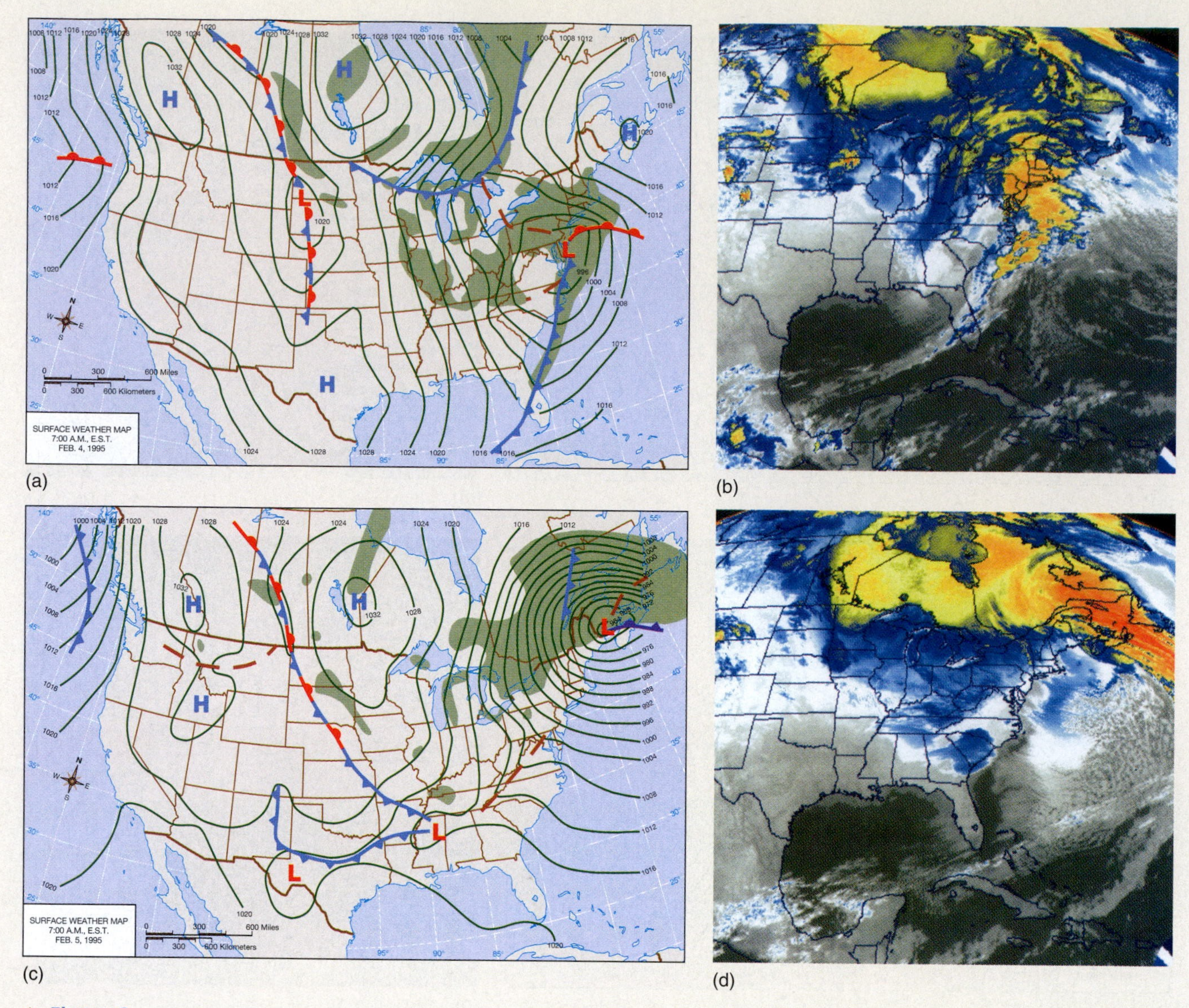

▲ **Figure 2**
The northeaster of February 1995.

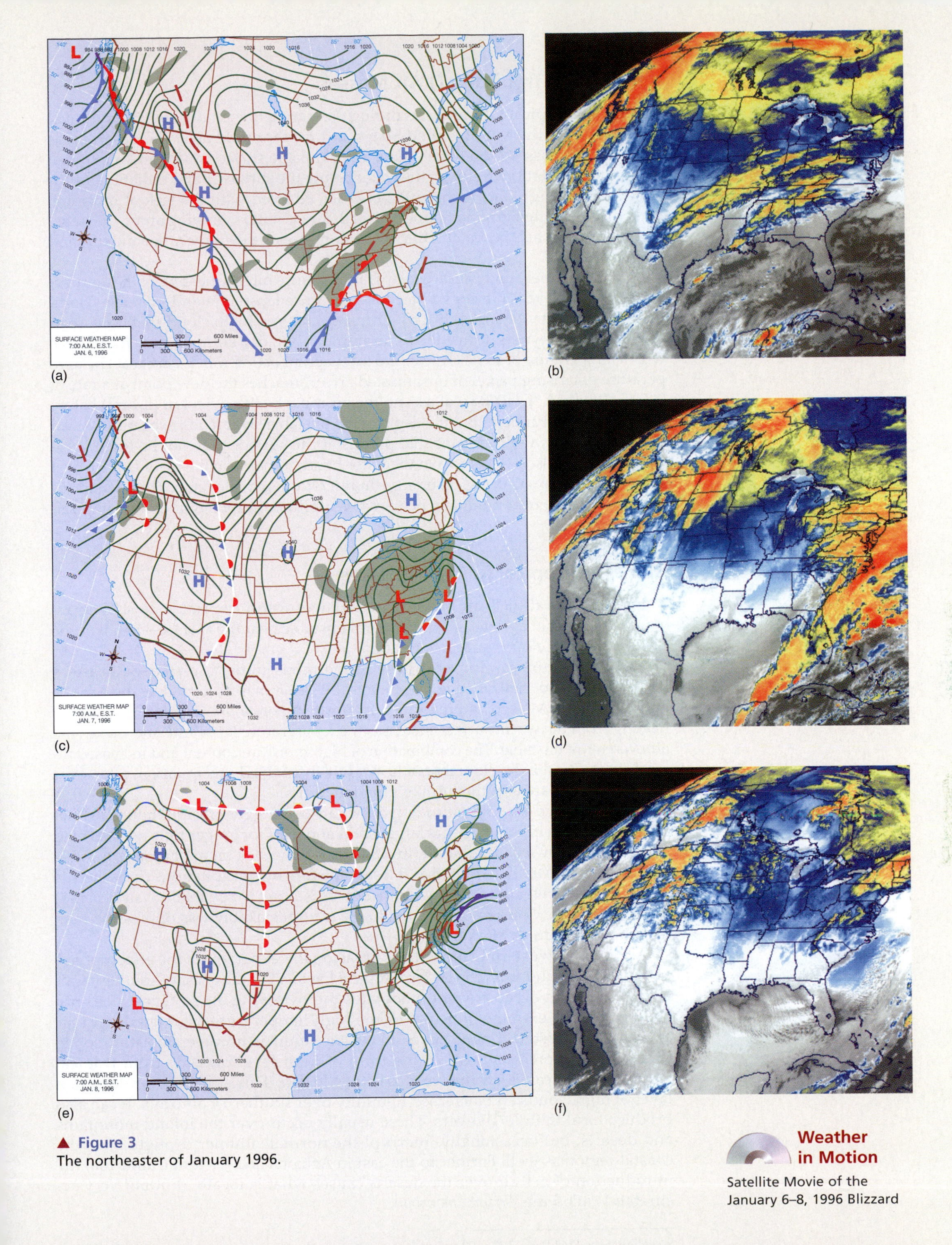

▲ **Figure 3**
The northeaster of January 1996.

Weather in Motion
Satellite Movie of the January 6–8, 1996 Blizzard

Continental Tropical (cT) Air Masses

Continental Tropical Air Mass

Continental tropical (cT) air forms during the summer over hot, low-latitude areas, such as the southwestern United States and northern Mexico. The desert areas where cT air masses form have little if any available surface water and a minimal amount of vegetation to extract the water below the surface. Solar radiation inputs are extremely high during the summer. That, coupled with the lack of moisture, makes for very high ground temperatures that warm the overlying air by the transfer of sensible heat. The result is that these air masses are extremely hot and dry, and often cloud-free.

The extremely high surface temperatures cause intense heating of the air nearest the ground, which brings a steep temperature lapse rate and unstable conditions. But despite the instability, cT air masses often remain cloud-free because of their inherent dryness. Consider, for example, a hot, dry air mass near the surface with a temperature of 45 °C (113 °F) and a dew point of 0 °C (32 °F). Because the temperature of a rising parcel of unsaturated air approaches the dew point at a rate of 0.8 °C/100 m (5 °F/1000 ft) of ascent, the air must be lifted 5.6 km (3.4 mi) before it can become saturated.* But the depth of the unstable layer tends to be much lower than this height, which means that parcels at the surface may not be lifted sufficiently for clouds to develop. On the other hand, if the unstable layer is deep or if some moisture exists in the air, intense thunderstorms can develop. Thunderstorms can also develop near mountain peaks where uplift is promoted by converging valley breezes.

Maritime Tropical (mT) Air Masses

Maritime tropical (mT) air masses develop over warm tropical waters. They are warm (though not as hot as cT), moist, and unstable near the surface—ideal for the development of clouds and precipitation.

Maritime tropical air masses have an enormous influence on the southeastern United States, especially during the summer. These air masses form over the Atlantic and the Gulf of Mexico and migrate into North America. As the air flows inland, heating from the warm surface increases the temperature lapse rate and makes the air even more unstable. The combination of high moisture content and increased instability favors the development of heavy but short-lived precipitation, often over relatively small areas in the form of thunderstorms.

The passage of a mid-latitude cyclone can also trigger precipitation in mT air. This kind of precipitation covers larger areas and lasts for longer time periods than those initiated by localized uplift. In either case, however, an mT air mass moving poleward is gradually modified as it loses moisture by precipitation. Thus, there is a shift toward lower dew points northward. This is not to say that the air becomes dry as it reaches Chicago or Toronto, but it is unmistakably less oppressive than over New Orleans or Miami.

The southwestern United States occasionally experiences the effects of mT air advected inland from the eastern tropical Pacific. Off the coast of southern California, the cold ocean current moderates the overlying air so that it does not have either the extreme moisture or the high temperature associated with air over the southeastern United States. In the late summer, however, moisture sometimes moves northeastward in the form of high-level outflow from tropical storms or hurricanes off the west coast of Mexico. This situation can lead to a layer of high clouds and increased humidity over southern California. It can also produce local thundershowers. These usually occur over the inland mountains and deserts, but occasionally interrupt the normal summer drought over the coastal region as well. Farther to the east in Arizona and eastern California, the introduction of mT air over the deserts causes what is locally (but not very accurately) called the *Arizona monsoon.*

---

*Recall that the DALR = 1.0 °C/100 m, but at the same time the dew point lapse rate = 0.2 °C/100 m. Thus, a rising parcel of unsaturated air approaches the dew point at a rate of 0.8 °C/100 m.

As we emphasized at the beginning of this chapter, it is often difficult to categorize air masses. The concept of an air mass is most meaningful when applied to large bodies of air separated by boundary zones with horizontal extents of tens (or perhaps a couple hundred) of kilometers. These boundary zones are categorized into four types of fronts.

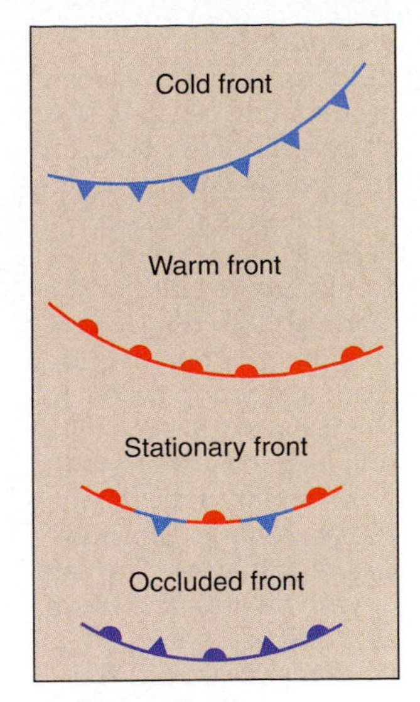

▲ Figure 9–4
The symbols used to represent the four types of fronts on weather maps.

Fronts

Fronts are boundaries that separate air masses with differing temperature and other characteristics. Often they represent the boundaries between polar and tropical air. They are important not only for the temperature changes they bring but also for the uplift they cause. Cold air is typically more dense than warm air, so when one air mass encroaches on another, the two do not mix together. Instead, the denser air remains near the surface and forces the warmer air upward. These upward motions lead to adiabatic cooling and sometimes to the formation of clouds and precipitation.

A **cold front** occurs when a wedge of cold air advances toward the warm air ahead of it. A **warm front**, on the other hand, represents the boundary of a warm air mass moving toward a cold one. A **stationary front** is usually similar to a cold front in structure, but it differs in that neither air mass has recently undergone substantial movement (in other words, it remains stationary). Unlike the other three types of fronts, **occluded fronts** do not separate tropical from polar air masses. Instead, they appear at the surface as the boundary between two polar air masses, with a colder polar air mass usually advancing on a slightly warmer air mass ahead of it. Figure 9–4 shows the symbols used to show the location of these fronts on surface weather maps.

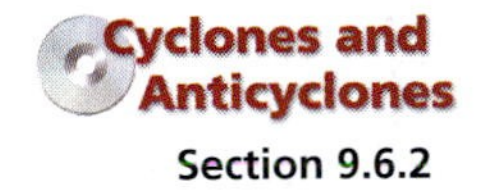

Section 9.6.2

Though fronts are named for the temperature contrasts associated with them, distinct changes in other weather elements also occur along these boundaries. People experiencing the passage of a front not only notice a rapid change in temperature; they may also note a shift in wind speed and direction, a change in moisture content, and an increase in cloud cover.

Figure 9–5 illustrates the overall structure of typical mid-latitude cyclones, of which cold, warm, and (sometimes) occluded fronts are an important part (we discuss mid-latitude cyclones in detail in Chapter 10). The pressure distribution is somewhat circular in the larger cold sector on the poleward side of the cyclone, leading to a counterclockwise rotation of the air (in the Northern Hemisphere). The cold front separates the cold air that typically comes out of the northwest from the warmer air ahead. Between the cold front and the warm front ahead of it, pressure decreases toward the apex where the two fronts join together, and the wind is typically southwesterly. Let's now look at the structure of these fronts in more detail.

Cold Fronts

We are all familiar with the dramatic shifts in weather that can occur over very short time periods with the approach and passage of a cold front. During the winter, they can disrupt life for millions of people as they cause intense flooding from heavy rain or white-out conditions from blowing snow.

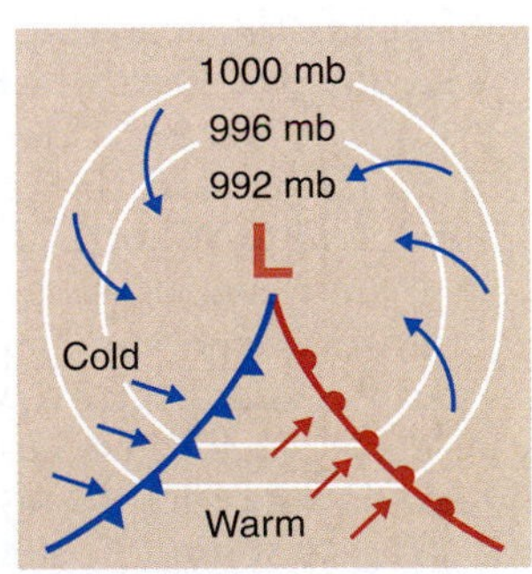

▲ Figure 9–5
A typical mid-latitude cyclone. Cold and warm fronts separated by a wedge of warm air meet at the center of low pressure. Cold air dominates the larger segment on the north side of the system.

Figure 9–6 shows the structure of a typical cold front, with its upper boundary sloping back in the direction of the cold air. Note that the vertical scale in this figure is highly exaggerated and makes the slope of the front appear far steeper than it really is. In reality, the typical cold front has a surface slope of about 1:100, meaning that its surface rises only 1 m for every 100 m of horizontal extent.

Observe also that the forward edge of the front is not flat but curves backward, due to the decrease in friction with increased height. As the front advances, friction from the surface slows the lowest portion of the cold air. But the effect diminishes with distance from the surface, so the air at higher levels advances more rapidly. This gives the cold front its characteristic profile, with a steeper slope near the leading edge at the surface. Cold fronts move at widely varying speeds, ranging from a virtual standstill to about 50 km/hr (30 mph).

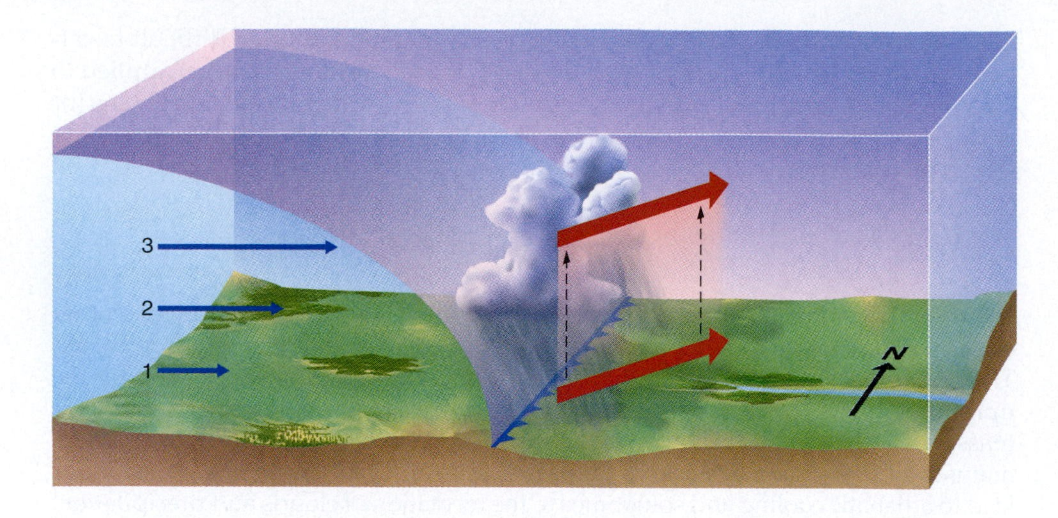

▶ **Figure 9–6**
Cold fronts typically move more rapidly and in a slightly different direction from the warm air ahead of them. This causes convergence ahead of the front and the uplift of the warm air that can lead to cumuliform cloud development and precipitation. In this example, the cold air (in blue) advances from west to east (notice that the wind speed depicted by the thin arrows increases with height). The warm air (in red) is blowing toward the northeast. The cold air wedges beneath the warm air and lifts it upward.

The cloud cover we usually observe with the passage of a cold front results from the convergence of the two opposing air masses. Differences in wind speed and direction allow the cold northwesterly wind to converge on the warm air ahead of it and displace it upward. Furthermore, the air ahead of a cold front tends to be unstable and therefore easily lifted. For this reason, cumuliform clouds are likely to develop along these boundaries. With their large vertical extent, cumuliform clouds are often capable of producing intense precipitation. However, because of the limited horizontal extent and rapid movement of the frontal wedge, such precipitation is often of short duration.

The production of surface weather maps is now almost entirely automated and performed by computer. But meteorologists still plot the location of fronts manually, because identifying them is often a subjective process. Though the position of cold fronts are plotted as a line on surface maps, they are three-dimensional and usually extend upward past the 500 mb level. Nonetheless, they are not plotted on upper-level weather maps, in part because there are not enough data available aloft

9–2 Special Interest

The Pineapple Express

Most precipitation along the West Coast of North America results from storms that cross the Pacific Ocean between November and April, the majority of them passing through the Gulf of Alaska. Unlike the winter storms that affect the central part of Canada and the United States, the Pacific storms are dominated by maritime polar air, which seldom brings extremely low temperatures to the region. Thus, the precipitation associated with them almost always occurs as rain along the immediate coastal region, and as rain or snow in inland (especially high-elevation) locations.

The snow line (the elevation marking the change from snow to rain) for these storms depends on the temperature. It usually varies between several hundred and a few thousand meters above sea level. Obviously, the colder the storm, the lower the snow line. When air temperatures remain above 0 °C long enough, the snow begins to melt. In southern California, this can be a matter of hours or days after snowfall. But in the upper reaches of the Sierra Nevada, snow can remain on the ground for months before it begins to melt in the spring. In fact, snowmelt is the main input to the many reservoirs that supply water for agricultural and urban consumption.

In some years, most Pacific storms do not approach the West Coast from the northwest but instead travel eastward from the vicinity of the Hawaiian Islands. This storm track is referred to locally as the *Pineapple Express*. When storms take this path repeatedly, as they did during the El Niño winters of 1982–83 and 1992–93, the greatest amounts of precipitation are concentrated farther to the south than during more normal years. Instead of a general pattern of increasing precipitation with latitude, the southern part of the coast might receive greater amounts than the northern coast. Furthermore, because the Pineapple Express passes over warmer waters than do storms from the Gulf of Alaska, the air tends to be warmer and more humid, and the height of the snow line increases considerably. As a result, even though this type of storm track can lead to heavy precipitation in the mountains, a smaller percentage of it accumulates as snow—much to the chagrin of local residents who depend on the snow for skiing and other recreational activities.

to adequately determine their location, and in part because their location is most useful in analyzing surface weather.

When seen from space, the cloud bands ahead of cold fronts often appear to contain a fairly uniform distribution of cloud cover. Closer examination, however, reveals that the cloud bands often contain pockets of thicker cloud cover and more intense precipitation. For example, Figure 9–7 shows a frontal zone extending across the central United States on March 31, 1998. The portion of the front that extends from northeastern Missouri to central Texas is the cold front. The satellite view (Figure 9–7a) provides a compelling example of how the clouds are aligned parallel to the front, with a sharp transition between the warm, moist air ahead of the front and much drier, colder air behind. But as is often the case, this image gives a false impression of uniform precipitation.

The radar composite map (Figure 9–7b) gives us a much better description of the intensity of precipitation along the front. Weather radar imagery is obtained by emitting energy with wavelengths of about 10 cm. Clouds composed of large droplets scatter some of the electromagnetic energy back toward the radar unit, which displays the position and other characteristics of the cloud. The color on the

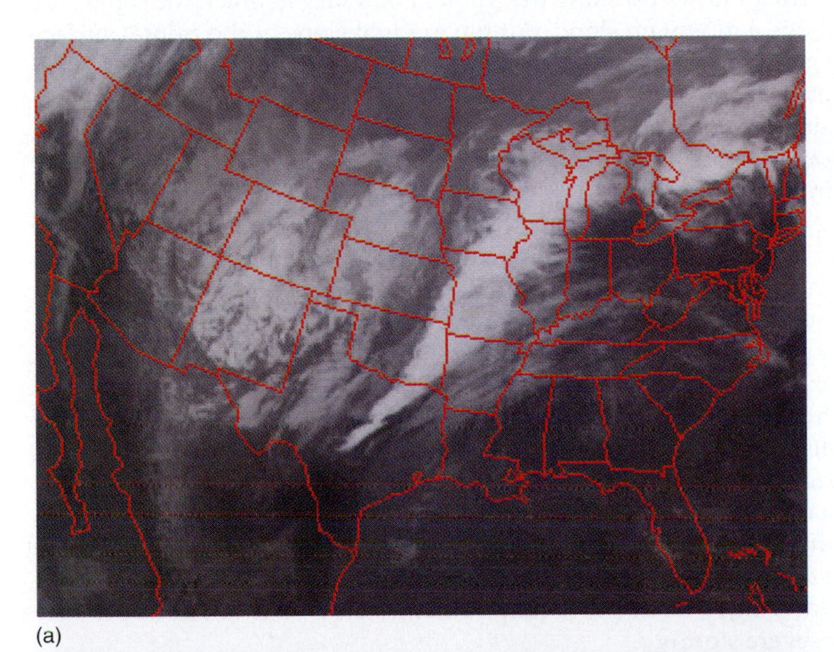
(a)

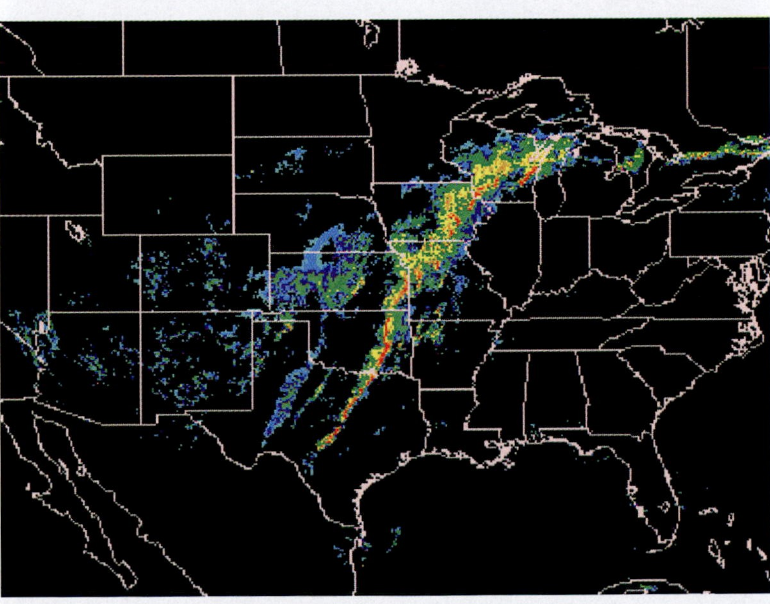
(b)

Figure 9–7
A mid-latitude cyclone. The cloud cover in the satellite image (a) appears to be a continuous, uniform band. The radar composite map (b) reveals that the cloud cover is, in fact, marked by areas of varying precipitation intensities.

radar display indicates the intensity of the returned energy, with bright red portions indicating large droplets and heavy precipitation.

In this instance, three pockets of very heavy precipitation are located in central Missouri, northeastern Oklahoma, and central Texas. Notice that at the time of the map, no precipitation is indicated along the front in northern Texas and Oklahoma. However, those areas were hardly dry for the duration of the frontal passage. In fact, central Oklahoma received more rain than any other location, with recorded amounts above 4 cm (1.6 in.). This tendency for spotty, short-lived precipitation is typical of strong cold fronts, a feature that places them in distinct contrast to warm fronts, which we discuss next.

Warm Fronts

Warm fronts separate advancing masses of warm air from the colder air ahead. As is the case with cold fronts, the differing densities of the two air masses discourage mixing, so the warm air flows upward along the boundary. This process is called **overrunning**.

Figure 9–8 illustrates the typical sequence of clouds that form along the frontal surface. The warm air flows up along the frontal boundary in much the same way that air rises above a mountain slope. But the gradual slope of the frontal surface leads to a very gradual progression of cloud types. As the air rises along the frontal boundary, adiabatic cooling first leads to the formation of low-level stratus clouds. As the air continues to rise along the front, a sequence of higher clouds develops, with nimbostratus, altostratus, cirrostratus, and, finally, cirrus occurring in that order. As the front moves eastward or northeastward, the leading segment of the cloud sequence (the cirrus) is seen first, followed by the continually thickening and lowering cloud cover. Thus, even an amateur forecaster can predict an episode of continuous light rain from a warm front a day or two in advance, simply by observing the sequence of clouds as they pass overhead.

Warm fronts are about half as steep as cold fronts (their slopes are about 1:200), which causes the lifting of the warm air to extend for greater horizontal distances than for cold fronts (Figure 9–9). Compare the horizontal extent and intensity of the radar results for the warm and cold fronts in Figure 9–7. Although the clouds above the warm front cover a greater horizontal extent than do those of the cold front, they usually consist of smaller droplets and have a lower liquid water content. This results in less intense precipitation than is usually associated with cold fronts. For this reason, warm fronts in the summer are considered a friend to farmers; they provide a light, steady rain that nourishes crops but does not produce flash floods or severe storms.

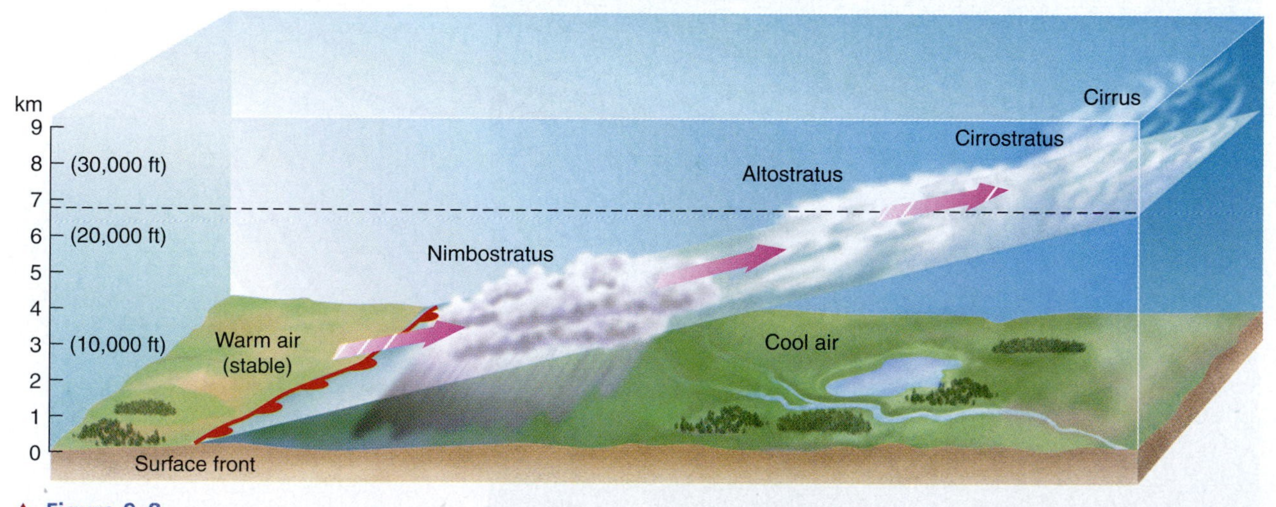

▲ **Figure 9–8**
A warm front. Overrunning leads to extensive cloud cover along the gently sloping surface of cold air.

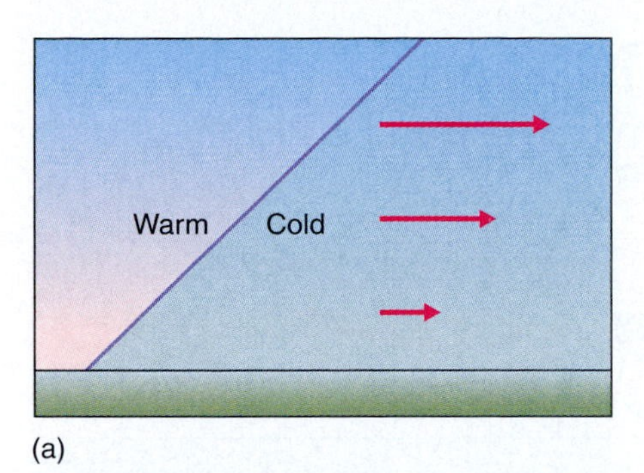

(a)

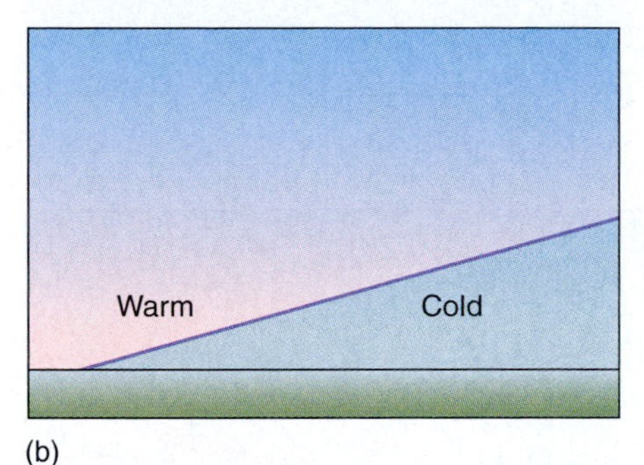

(b)

◀ **Figure 9–9**
Warm fronts have gentler sloping surfaces and do not have the convex-upward profile of cold fronts. Surface friction decreases with distance from the ground, as indicated by the longer wind vectors away from the surface (a). This causes the surface of the front to become less steep through time (b).

The overrunning air above a warm front is usually stable, which normally leads to the formation of wide bands of stratiform cloud cover. Though the rising air may sometimes be unstable and lead to the formation of cumuliform clouds, this is the exception rather than the rule. Precipitation along the front tends to be light, but the wide horizontal extent and typically slow movement of warm fronts (usually about 20 km/hr—12 mph) allows rain to persist over an area for up to several days.

The clouds along a warm front exist in the warm air above the wedge of dense, cold air. Thus, when rain falls from the base of these clouds, the falling droplets pass through the cold air below on their way to the surface. If the falling droplets are substantially warmer than the air they fall through, they can evaporate rapidly and form a frontal fog.

If the air is cold enough during the passage of winter warm fronts, the drops freeze on their way down to form sleet, or they solidify on contact with the surface to form freezing rain. Thus, unlike summer warm fronts, which are generally beneficial, those in the winter can cause widespread problems.

Stationary Fronts

Eastern U.S. Water Vapor

More often than not, the boundaries separating air masses move gradually across the landscape; at other times, they stall and remain in place for extended time periods. Nonmoving boundaries are called *stationary fronts*. Although they do not move as rapidly as cold or warm fronts, they are identical to them in terms of the relationship between their air masses. As always, the frontal surface is inclined, sloping over the cold air.

Determining whether a front is truly stationary is somewhat subjective. For example, if the front advances at a rate of 1 km/hr (0.6 mph), does that constitute enough movement for it to be nonstationary? If not, what about 2 km/hr? In practice, a meteorologist will make the designation by looking at the previous one or two surface weather maps (compiled at 3-hour intervals). If there has been no movement during that period, the front is considered stationary.

If this sounds simple, keep in mind that the exact position of a front at any point in time is difficult to pinpoint. This is in part because fronts are zones of transition rather than abrupt boundaries; thus, they are located at no precise line. Furthermore, the maps are compiled from only a finite number of weather stations. Thus, a front may have undergone some movement that is not detected because it did not move across an observing station.

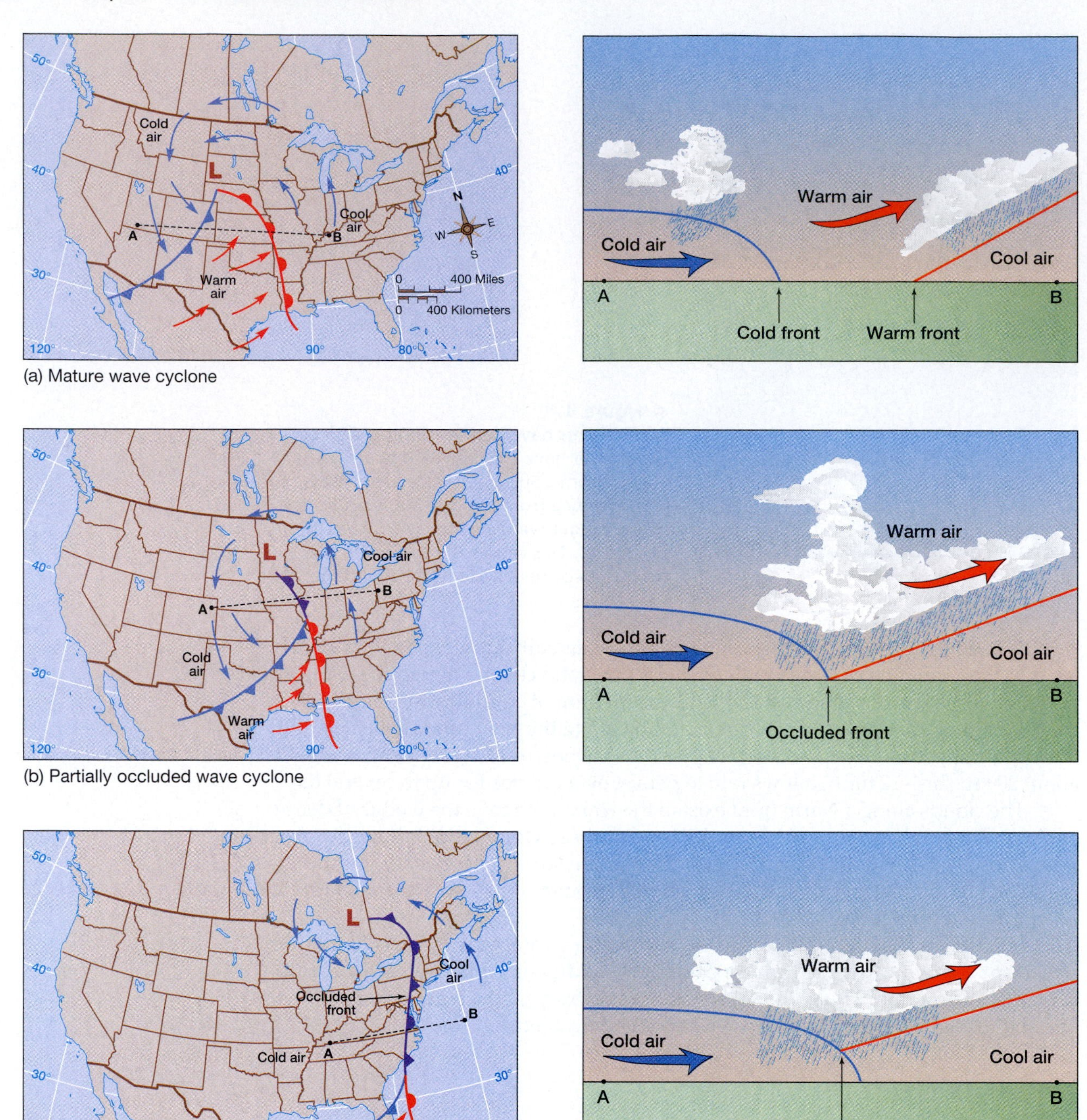

(a) Mature wave cyclone

(b) Partially occluded wave cyclone

(c) Occluded wave cyclone

▲ **Figure 9–10**
The traditional explanation of the occlusion process, with a cold front overtaking a warm front.

Occluded Fronts

The most complex type of front is an occluded front (sometimes called an *occlusion*). The term *occlusion* refers to closure—in this case, the cutting off of a warm air mass from the surface by the meeting of two fronts. According to the original model of cyclone evolution, occluded fronts form when a faster-moving cold front "catches up" to the warm front ahead. Occluded fronts separate air masses of different temperature, as do warm, cold, and stationary fronts. But at the surface, one mass of cold air merges with another; thus, there is a smaller difference in temperature from one side of an occluded front to the other. A warm air mass is present, but it is aloft, pinched off from the surface, and therefore not reflected in surface temperature, as shown in Figure 9–10c.

Figure 9–10a shows a typical mid-latitude cyclone prior to occlusion, with the cold and warm fronts intersecting at the center of low pressure. When the cold front meets the warm front ahead of it, that segment becomes occluded, as shown in (b). The warm air does not disappear, of course, but gets lifted upward, away from the surface. The occluded front becomes longer as more and more of the cold front converges with the warm front. Eventually, the cold front completely overtakes the warm front (c), and the entire system is occluded.

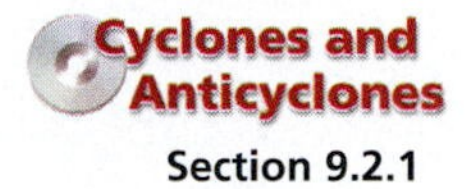

Section 9.2.1

In this occlusion, the air behind the original cold front was colder than that ahead of the warm front. This is an example of a *cold-type occlusion*, originally considered to be the more common type because air in the western part of the storm is imported from colder, higher latitudes. In British Columbia and the Pacific Northwest of the United States, *warm-type occlusions* are more common. In these, relatively mild mP air behind a Pacific cold front moves onshore and meets up with more frigid cP air.

This description of the occlusion process is essentially what was proposed by the leading meteorologists of the early twentieth century—and it probably does apply to many mid-latitude cyclones. However, in many cases, perhaps most, structures resembling an occluded front form in ways unrelated to the traditional explanation. For example, occlusions sometimes occur when the circular core of low pressure near the junction of the cold and warm fronts changes shape and stretches backward, away from its original position (Figure 9–11). In (a), the cold and warm fronts are joined at the dashed line. At some later time period (b), the cold and warm fronts have the same orientation with respect to each other as they did in (a), but both have been pulled back beyond the dashed line. The circular isobar pattern of (a) becomes elongated to form a trough over the occluded region. Thus, the structure of the occluded front is the same as that proposed in the traditional model, but the process responsible for its formation is somewhat different. In other cases, the cold front moves eastward relative to the warm front, so that the point where the fronts intersect moves progressively farther east (Figure 9–12). The occluded front appears as a relic portion of the warm front, situated to the west of the new intersection of the cold and warm fronts.

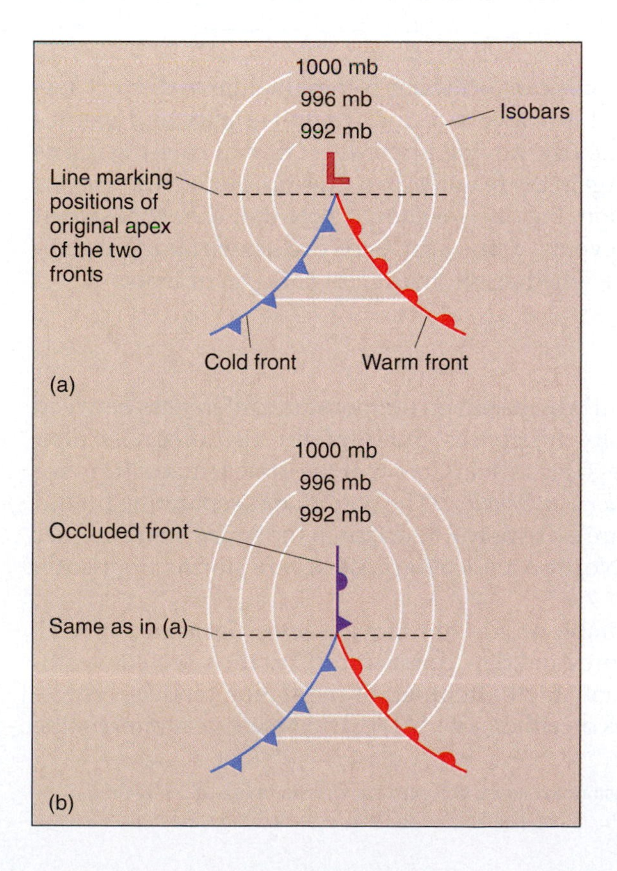

Figure 9–11 Some occluded fronts form when the surface low elongates and moves away from the junction of the cold and warm fronts.

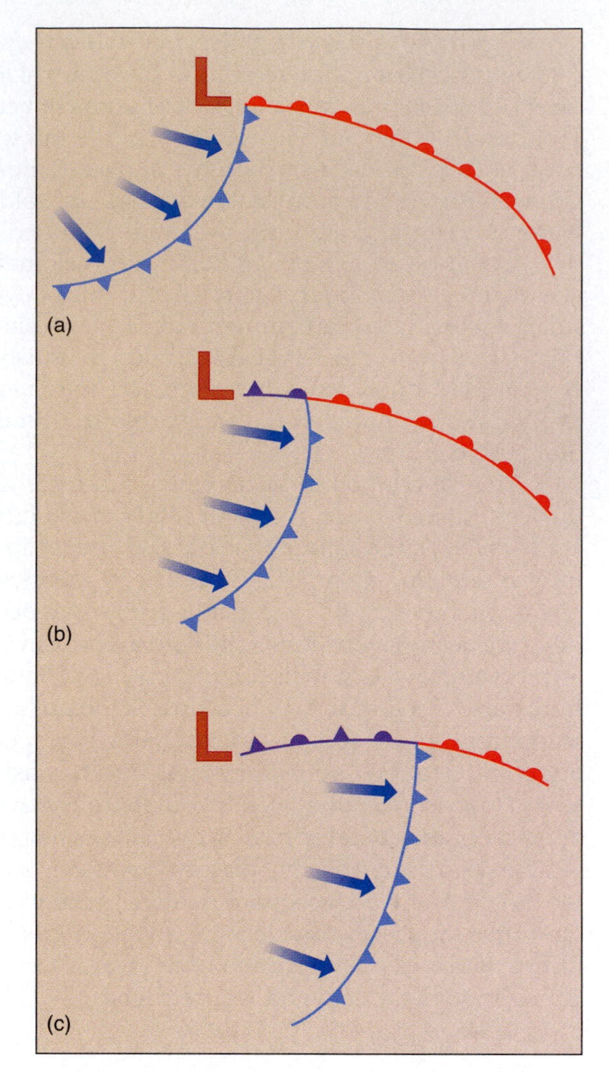

▶ **Figure 9–12**
Some occlusions occur when the intersection of the cold and warm fronts slides along the warm front.

Yet another way occluded fronts can form is by a combination of distinct surface and upper-level processes. At low levels, a cold front overtakes and merges with a warm front, while aloft a rapidly moving upper-level frontal zone overtakes and joins with the surface front. Again the result is an occlusion, but not for reasons outlined in the classical description. Finally, we should note that cold-type occlusions are now thought to be rare events, with warm-type occlusions far more common for all mid-latitude cyclones, whether originating over ocean or land.

Drylines

The fronts described so far have all been based on temperature differences between air masses and the resultant density differences. But humidity also affects the density of air, with humid air being less dense than dry air at the same temperature (see *Box 4–2, Physical Principles: Variations in Density*). The boundaries separating humid air from dry air, called **drylines**, are extremely important in the spring and summer in the southern Great Plains of North America because severe storms frequently form along them.

Figure 9–13 illustrates an example of a dryline moving across Texas at noon CST on March 30, 2002. The temperatures and dew points (in °F)* at the selected weather stations are shown on the top left and bottom left of the station models, respectively. Although the temperatures on either side of the dryline are very similar, the

*We are using °F rather than °C in this instance because U.S. surface weather maps still rely on the Fahrenheit scale.

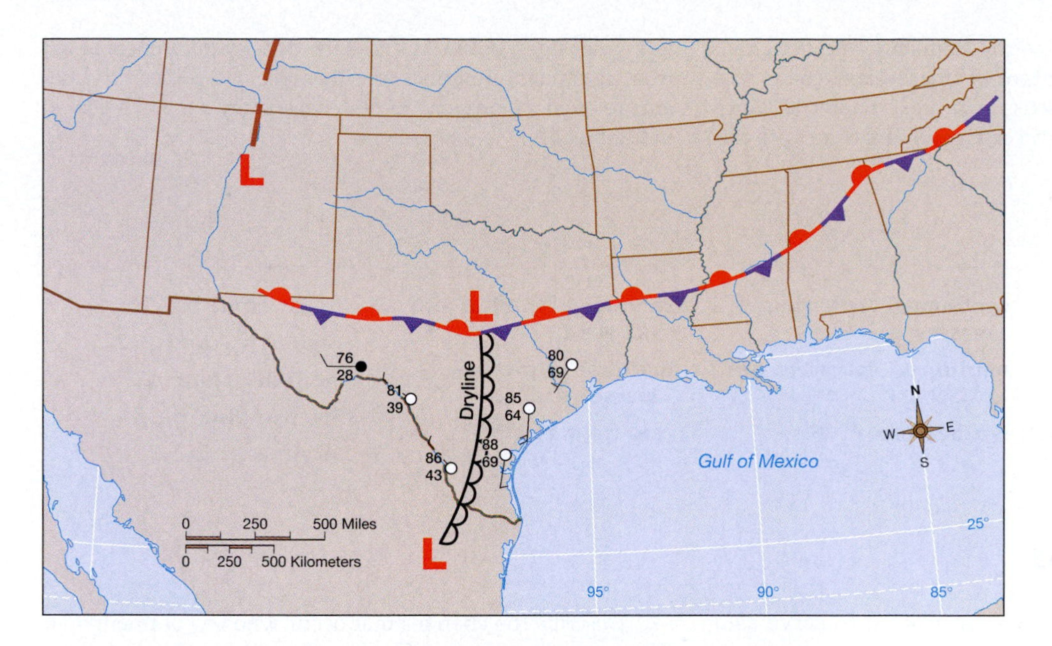

Figure 9–13
A dryline over Texas. West of the line the humidity is low, dew points ranging from the upper 20s to the 40s (°F). East of the line the air humidity is greater, owing to the flow of air from the Gulf of Mexico.

Table 9–1 • Temperatures and Dew Points Prior to and Following the Passage of a Dryline

Time (CST)	Temperature °F (°C)	Dew Point °F (°C)
10 A.M.	66 (19)	64 (18)
11 A.M.	66 (19)	62 (17)
12 P.M.	71 (22)	64 (18)
1 P.M.	77 (25)	66 (19)
2 P.M.	78 (26)	62 (17)
3 P.M.	78 (26)	46 (8)
4 P.M.	80 (27)	46 (8)
5 P.M.	75 (24)	39 (4)

dew points on the east side of the line are significantly greater than those on the west. The dryline clearly demarcates the moist air from the Gulf and the drier air to the west. As the dryline moves eastward, the denser dry air can lift the warm, moist air ahead of it—in much the same way that convergence along a cold front initiates uplift. When the uplift is coupled with the presence of moist, unstable air to the east of the line, tornadoes and severe thunderstorms are a real possibility.

Table 9–1 shows the change in temperature and dew point observed at an automated weather-observing system near Austin. A substantial drop in the humidity occurred as the dryline passed between 2 and 3 P.M., though only a minor increase in temperature occurred. It should also be noted that thunderstorms had been going on in the area during the morning hours as a result of the uplift along the dryline.

Summary

We have seen that the atmosphere tends to arrange itself into large masses with relatively little horizontal change in temperature and humidity. These air masses form over particular source regions—large areas of high or low latitudes of either continental or maritime locations. Five major types of air masses were described: continental polar, continental arctic, maritime polar, continental tropical, and maritime tropical.

The boundaries of these masses, called *fronts*, are particularly important to meteorology—not only because their passage causes abrupt changes in the temperature and humidity, but also because they promote uplift and cloud formation. The four types of fronts are cold, warm, stationary, and occluded. They are important components of weather systems called *mid-latitude cyclones*, which we discuss in the next chapter. Drylines are similar to fronts in that they separate bodies of air with differing densities, but in this case the density differences result from dissimilar humidities on either side of the lines. Drylines are preferred regions for severe storm activity, a topic discussed in Chapter 11.

Key Terms

air masses p. 265
fronts p. 265
source regions p. 266
continental polar air masses p. 267
continental arctic air masses p. 268
maritime polar air masses p. 269
northeasters p. 269
continental tropical air masses p. 274
maritime tropical air masses p. 274
cold front p. 275
warm front p. 275
stationary front p. 275
occluded front p. 275
overrunning p. 278
drylines p. 282

Review Questions

1. What are the requirements for an area to serve as a source region?
2. Where are the primary air mass source regions in North America located?
3. Describe the characteristics of cA, cP, cT, mT, and mP air masses.
4. Of the five types of air masses, which are the hottest, driest, coldest, and most damp?
5. What is the primary difference between arctic and polar air masses?
6. Which of the air mass types are likely to be stable or unstable?
7. Describe the changes that occur when a continental air mass migrates out of its source region.
8. Describe of the structure of cold, warm, stationary, and occluded fronts.
9. What is overrunning?
10. Why do cold fronts have steeper slopes than warm fronts?
11. How do the alternative models of the occlusion process differ from the traditional model?
12. What is the difference between warm-type and cold-type occlusions?
13. What are drylines and why are they important?

Critical Thinking

1. Contrast the formation of air masses in the Northern and Southern Hemispheres.
2. Explain the limitations and benefits of classifying air into distinct masses.
3. Continental polar air masses can migrate into Florida during the winter but not into northern India. Why not?
4. What parts of North America are likely to experience the most frequent changes in air mass during the summer and winter?
5. Distinct temperature changes can be detected across narrow regions that are not associated with the presence of fronts. What can cause such conditions to exist?
6. The southwestern United States experiences what is locally referred to as a monsoon. Can we say the same thing about Florida or Texas?
7. Warm fronts are extremely rare over southern California. Why?
8. What does the presence of a continental polar air mass tell you about the height of the 500 or 300 mb levels?
9. Where in North America might you expect to see the collision of mT and cT air masses? What time of year would this be most likely?
10. Which types of fronts are most or least likely to have inversions?

Problems and Exercises

1. The following temperatures and dew points are observed. What are the likely types of air masses present for each?
   a. T = 29 °C (85 °F) Dew Pt = 19 °C (66 °F)
   b. T = –18 °C (0 °F) Dew Pt = –21 °C (–5 °F)
   c. T = 3 °C (38 °F) Dew Pt = 0 °C (32 °F)
   d. T = 38 °C (100 °F) Dew Pt = –4 °C (25 °F)
2. Assume that a cold front has a slope of 1:100, and that the height of the 700 mb level is 1500 meters. How far behind the surface position of the front will the 700 mb position be?

Quantitative Problems

The Web site for this book (**http://www.prenhall.com/aguado/**) offers some quantitative problems that can help reinforce the concept of air masses and fronts. We encourage you to go to the Chapter 9 page of the Web site and answer the problems.

Useful Web Sites

http://www.udel.edu/dgs/Publications/pubsonline/OFR40/1998damagingstorms.htm

Some background on the effects of northeasters.

http://weather.unisys.com/surface/sfc_front.html

Map of current position of surface fronts across North America.

Media Enrichment

Weather in Motion

Satellite Movie of the January 6–8, 1996 Blizzard

This movie shows the blizzard of 1996 as seen by color infrared satellite imagery. The bright red portions show the regions of most intense precipitation. As you can see, the zone of heaviest storm activity moves northward along the Atlantic Coast. As that portion of the storm approaches New England, a distinct flow of moisture rotates westward toward the Great Lakes, feeding moisture into the system and promoting exceptionally strong blizzard conditions. By the end of the movie, a large sweeping arc of heavy moisture associated with maritime polar air extends well into the continental interior. As we will see in Chapter 10, this sweeping zone clearly illustrates one of the three distinct flows associated with the conveyor belt model of mid-latitude cyclones.

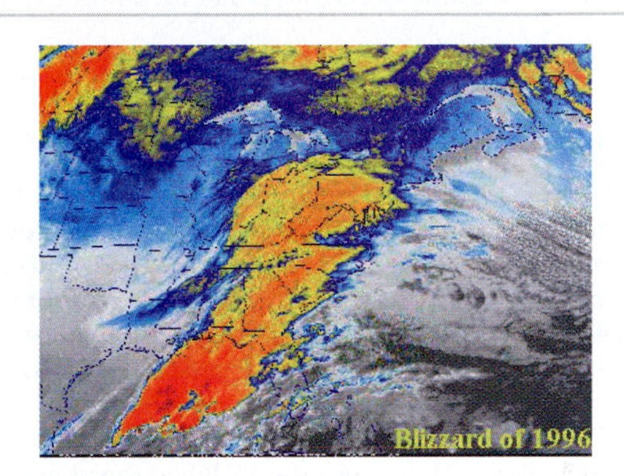

Weather Image

The Perfect Storm

This GOES 7 color-enhanced infrared image (taken 1200 UTC October 30, 1991) shows an enormous mid-latitude cyclone that wreaked havoc along the entire Atlantic Coast. This storm was called "the Perfect Storm" by the National Weather Service and was the subject of a best-selling book and popular movie. This storm is an excellent example of a "nor'easter," which is an extremely strong extratropical cyclone that may develop during the autumn along the East Coast of the United States due, in part, to the contrast between continental polar (cP) air masses from Canada and milder, maritime air masses along the Atlantic Coast. This storm developed along a cold front located off the East Coast of the United States. A strong upper air low and remnant moisture from Hurricane Grace caused the storm to explosively intensify. During the storm, hurricane-force winds were reported on Cape Cod, Massachusetts, and flooding and record high tides occurred all along the mid-Atlantic and New England coastlines.

Weather Image

Pacific Storm Water Vapor

This GOES satellite water vapor image shows a mid-latitude cyclone located near the coast of the western United States. The image was taken on December 31, 1996. This storm brought maritime polar (mP) air into the Pacific Northwest.

Weather Image

Continental Tropical Air Mass

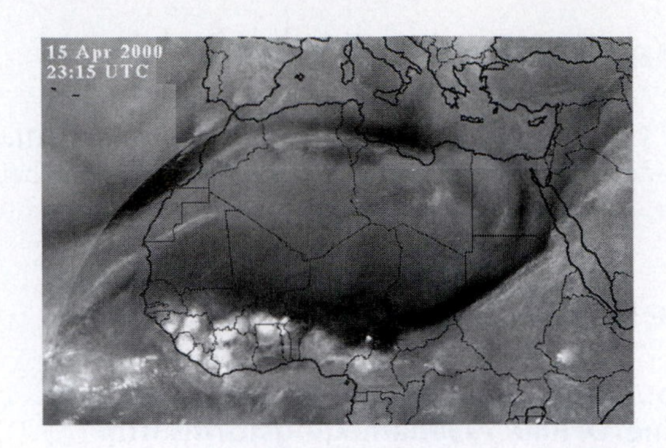

This image shows atmospheric water vapor content over northern Africa at 23:15Z on April 15, 2000. Notice that much of the atmosphere over the Sahara Desert is very dry in comparison to other regions in the image. Like the southwestern United States and northern Mexico, northern Africa is a source region for continental tropical (cT) air masses. In this region there is very little available surface water and minimal vegetation. In addition, surface temperatures are very high due to northern Africa's subtropical location. Consequently, air masses over northern Africa—and other continental tropical source regions—are extremely hot, dry, and often cloud-free. In this image, there are some clouds (most likely due to thunderstorms) present along the Ivory Coast where moisture from the Atlantic Ocean has penetrated northward.

Weather Image

Eastern U.S. Water Vapor

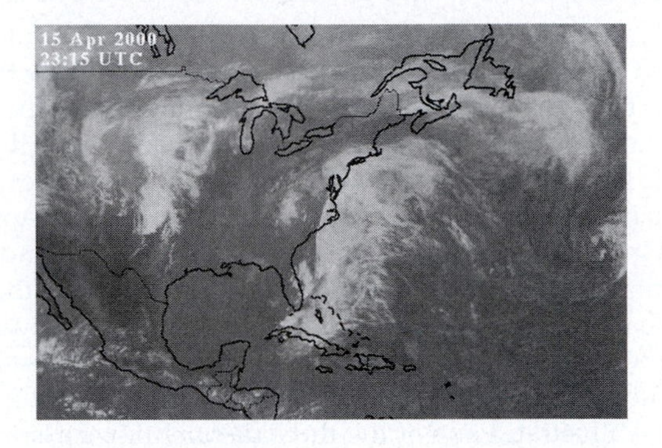

This image shows atmospheric water vapor content over the eastern United States at 23:15Z on April 15, 2000. This image comprises data from the water vapor channels of four weather satellites in geosynchronous orbit (GOES satellites). Light areas correspond to relatively wet regions (often clouds), and dark areas correspond to relatively dry regions. Notice the region of high water vapor content along the eastern coast of the United States. This cloudiness is due to the presence of a stationary front located along the East Coast.

PART FOUR

Disturbances

Winter storm hits Denver, Colorado.

chapter 10

Severe flooding in the Midwest.

Mid-latitude Cyclones

Less than a week before the end of 1999, as people everywhere fretted about the Y2K computer bug, much of Europe was shaken by a natural event unrelated to the turning of the calendar. The most powerful European storm in 50 years hit Europe with winds that topped 190 km/hr (110 mph). It brought power outages to nearly two million households, shut down three nuclear power plants, and crippled much of France's air and land transportation systems. Worse still, 97 people from across western Europe died from the storm, mostly from falling or flying debris. When asked about the situation, one of the city workers cleaning up the debris, David Chézeaud, said, "Since I was born, I've never seen anything like that storm. All over Paris there are problems like this."

Among the European countries, France was the worst hit—both in terms of loss of life and material damage. Many of Paris's most famous landmarks were badly damaged, and 10,000 trees were toppled in Versailles. Germany, Switzerland, and Great Britain also sustained major damage and numerous fatalities.

Although European forecasters saw the storm approaching from the Atlantic and predicted that it would reach the continent, they did not expect its winds to be as severe as they were by the time it made landfall. As is the case with the west coast of North America, weather forecasting in Europe is complicated by the paucity of weather stations over the expanse of ocean to the west of the continent—a fact sometimes overlooked by the public. No sooner had the criticism started to mount over the understated forecast than another, even stronger, storm arrived 2 days later, killing at least 22 more people. Thus, one very rare event was followed very shortly by a second. This rare combination of events should remind us that regardless of the technological sophistication of the twenty-first century, we are still subject to the random whims of nature.

The two storm systems brought hurricane-force winds but were distinctly different from hurricanes. Unlike hurricanes, which originate over tropical waters, these *mid-latitude cyclones* originate in the middle or high latitudes and are marked by well-defined fronts separating two dissimilar air masses. This chapter describes the mid-latitude cyclones that are a common and important feature of the weather outside the Tropics.

CHAPTER OUTLINE

Polar Front Theory

During the years from 1914 to 1918, the world experienced one of history's great catastrophes, World War I. The advent of new weapons, including the machine gun and mustard gas, made it nearly impossible for opposing armies to gain large tracts of ground from their opponents. Until then, an army attacking with rifles and bayonets could charge its opponents with some reasonable chance of success. However, against a foe dug into trenches and armed with the latest weapons, such maneuvers were almost doomed to failure. Thus, the war zone remained stagnant for long periods, since neither army could advance across the *front*, or battle line.

While the war was going on in western Europe, Vilhelm Bjerknes (pronounced *bee-YURK-ness*) established the Norwegian Geophysical Institute in the city of Bergen. With several colleagues,* including his son Jacob, Bjerknes developed a modern theory of the formation, growth, and dissipation of mid-latitude cyclones. Recall that these storms, neither tropical nor hurricanes, form along a front in middle and high latitudes. Bjerknes observed the systems forming along a boundary

*Tor Bergeron, who discovered the ice crystal process for the formation of precipitation (Chapter 7), was among the scientists in this group.

separating polar air from warmer air to the south. Comparing that boundary to the one separating the opposing armies in western Europe, he called his model the **polar front theory** (also called the *Norwegian cyclone model*). This theory has stood the test of time remarkably well, and though we now have far more observational information available than did Bjerknes (especially for the middle and upper troposphere), we still describe the life cycle in much the same way that the scientists of the Bergen school did decades ago.

Mid-latitude cyclones are large systems that travel great distances and often bring precipitation—and sometimes severe weather—to wide areas. Lasting a week or more, and covering large portions of a continent, they are familiar as the systems that bring abrupt changes in wind, temperature, and sky conditions. Indeed, all of us who live outside the Tropics are well acquainted with the effects of these common events.

The Life Cycle of a Mid-latitude Cyclone

The Bergen meteorologists were perfectly placed to witness the atmosphere along the polar front, and they used their observations to describe the formation of mid-latitude cyclones, a process called **cyclogenesis**, along this boundary. Although many cyclones do originate along the polar front, they also form in other areas, especially downwind of major mountain barriers. We first discuss the formation of mid-latitude cyclones at the polar front, as described in Bjerknes' classical model. Later in this chapter we incorporate more recent insights into cyclogenesis.

Cyclogenesis

Figure 10–1 illustrates the classical description of the life cycle of a typical mid-latitude cyclone. Initially, the polar front separates the cold easterlies and the warmer westerlies (a). As cyclogenesis begins, a minor "kink" (b) develops along the boundary. The cold air north of the front begins to push southward behind the cold front, and air behind the warm front advances northward. This creates a counterclockwise rotation (in the Northern Hemisphere) around a weakly developed low-pressure system. In other words, the mid-latitude cyclone has begun to take form. With further intensification (c), the low pressure deepens even further and distinct warm and cold fronts emerge from the original polar front. Convergence associated with the low pressure can lead to uplift and cloud formation, while linear bands of deeper cloud cover develop along the frontal boundaries, as described in Chapter 9. Cyclogenesis along the polar front may seem to be an abstraction to people in lower latitudes, but it represents the initial stage in the development of a system that may affect millions of people thousands of kilometers away several days later.

The Bergen scientists could not explain *why* cyclogenesis occurs, but they did observe that it commonly happened near zones of thermal contrasts (such as along coastal regions or at the boundaries between warm and cold ocean currents) or where topographic features (such as major mountain chains) disrupt the normal air flow.

Mature Cyclones

A Mid-latitude Cyclone Passes over the Southeast

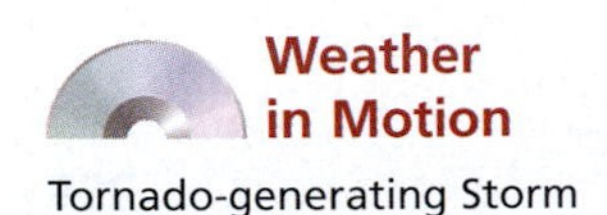

Tornado-generating Storm

Figure 10–2a illustrates the cloud patterns, wind, uplift processes, and precipitation patterns associated with a **mature cyclone**. (The precipitation probabilities, listed as percentages in the figure, should not be interpreted too literally; they merely give a general picture.) A band of mostly cumuliform cloud cover runs along and ahead of the cold front, caused by the displacement of the warm air by the denser cold air. The likelihood of precipitation along the front increases toward the center of the low pressure, where large-scale convergence adds to the uplift caused by the meeting of the two air masses. Because of the high moisture content and generally unstable conditions typically found ahead of a cold front, precipitation—in the form of rain, snow, or even sleet or hail—can be intense. But the band of

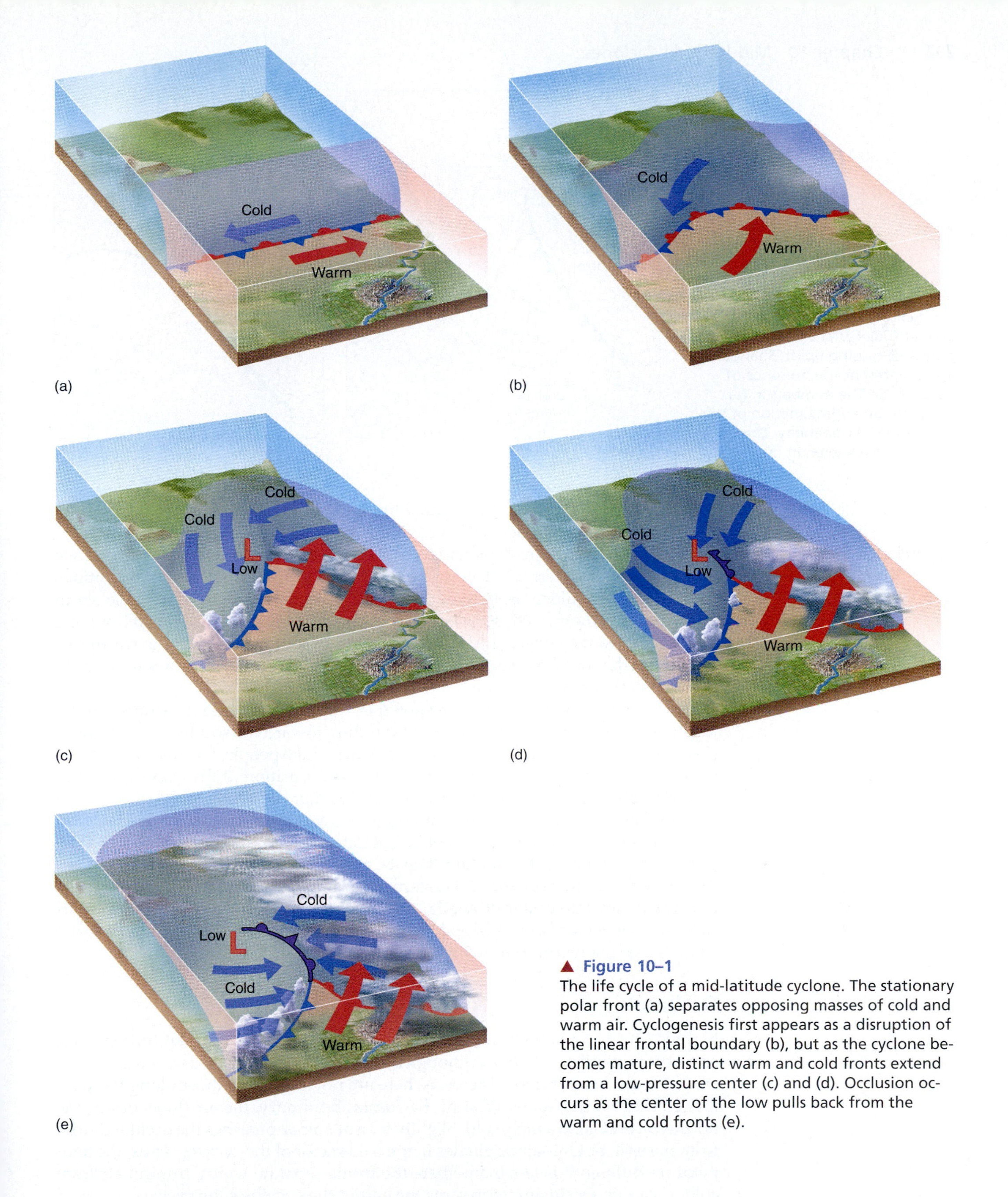

▲ **Figure 10–1**
The life cycle of a mid-latitude cyclone. The stationary polar front (a) separates opposing masses of cold and warm air. Cyclogenesis first appears as a disruption of the linear frontal boundary (b), but as the cyclone becomes mature, distinct warm and cold fronts extend from a low-pressure center (c) and (d). Occlusion occurs as the center of the low pulls back from the warm and cold fronts (e).

cloud cover and precipitation is relatively narrow, so precipitation may last for only a brief period of time before the frontal zone moves on.

A wider band of mostly stratiform clouds lies ahead of the warm front. As we found with the cold front, the likelihood of precipitation increases toward the center of low pressure. Precipitation tends to be light along the warm front because its more gradual slope leads to slower uplift. But the warm front's greater horizontal extent and generally slower forward motion allow clouds and precipitation to last longer. Clear skies characteristically occur over the warm sector between the

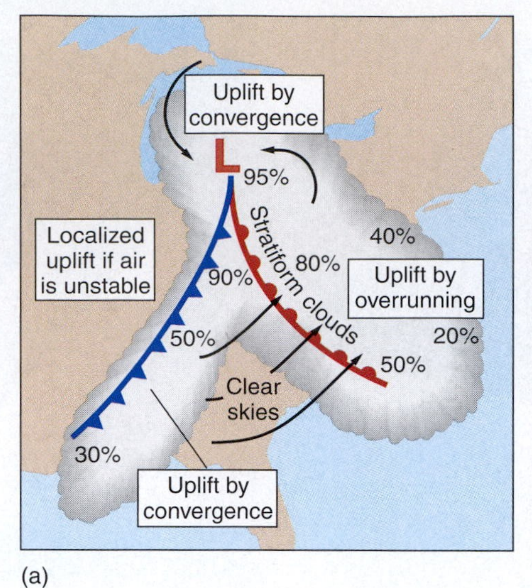

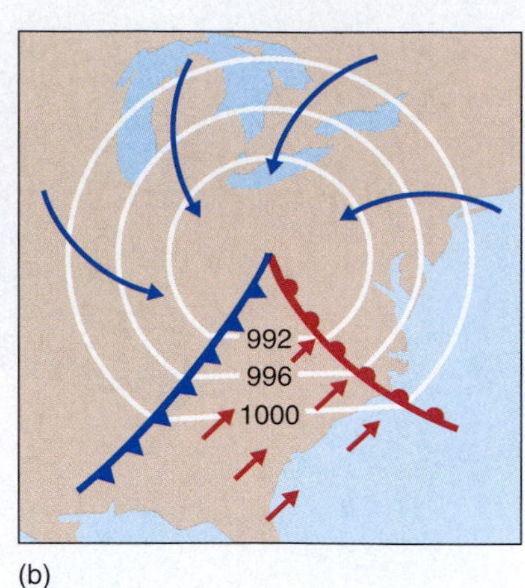

(a) (b)

▶ **Figure 10–2**
The typical structure of a mature mid-latitude cyclone and the processes causing uplift. Shaded areas represent the presence of cloud cover. The numbers in (a) represent an approximation of the precipitation probability. The isobar pattern is shown in (b).

cold and warm fronts, although squall lines and other disturbances (discussed in Chapter 11) develop under certain conditions.

Section 9.6.2

The isobar pattern, depicting the distribution of pressure within the cyclone (Figure 10–2b), is interrupted along the two fronts, which causes abrupt transitions in wind direction along the boundaries. The isobars are nearly straight in the warm sector but become curved in the larger, cold region. Looking at the warm front, the winds shift from southeasterly on the cold side to southwesterly in the warm sector. Across the cold front, the winds shift from southwesterly in the warm sector to northwesterly on the cold side.

Though common, the pattern shown in Figure 10–2—with the fronts coming together in an inverted "V" shape and extending toward the southwest and southeast—does not apply to all mid-latitude cyclones.* Like people, few mid-latitude cyclones look exactly alike, and the orientation and position of the fronts can differ considerably. Figure 10–3 shows two examples that depart from the generalized picture. The mid-latitude cyclone in (a) has a warm front that stretches eastward from the center of the low and a cold front extending to the south. In (b) the low-pressure system over the southern Great Lakes area has a well-defined cold front that sweeps southwestward into eastern Texas and a stationary front that extends northeastward into eastern Canada. Although the exact orientation of the fronts can vary between storms, the one consistent characteristic mid-latitude cyclones share is that the warm front will be located ahead of the cold front.

Occlusion

Figures 10–1d and e show the later stages in the life cycle of a mid-latitude cyclone as it becomes fully occluded. Although a temperature contrast exists across the occluded front, temperature differences here are not as great as those along the original cold or warm fronts. West of the frontal boundary, the air flows out of the northwest and is extremely cold. Slightly warmer air approaches the occluded front from the east, but this air originates in the cold sector of the cyclone. Thus, the temperature difference is less than where the fronts separate warm, tropical air from cold, polar air. **Occlusion** represents the end of the cyclone's life cycle.

The transitions from cyclogenesis to maturity, and from the mature phase to occlusion, are gradual, so no obviously identifiable points in time exist when the cyclone changes from one stage to another. Also, the evolution of the system coincides with a generally eastward migration of the mid-latitude cyclone, though it may also have a northward or southward component.

*In the Southern Hemisphere, typical mid-latitude cyclones have a similar shape, but the "V" opens to the north. Thus, in both hemispheres warmer air is on the equatorial side of the storm.

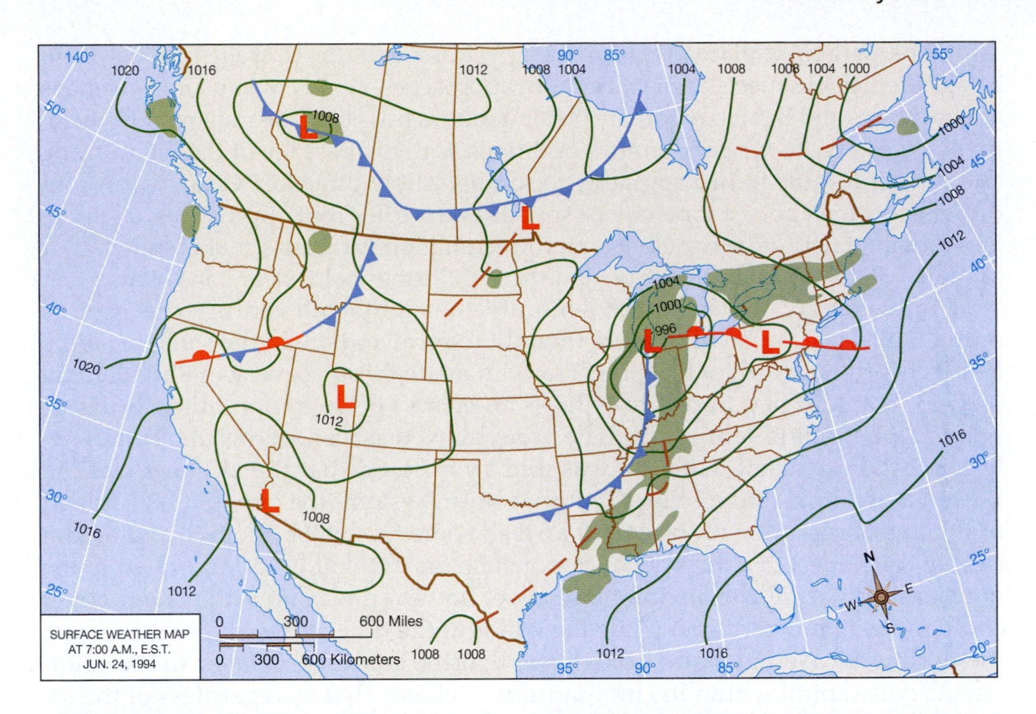

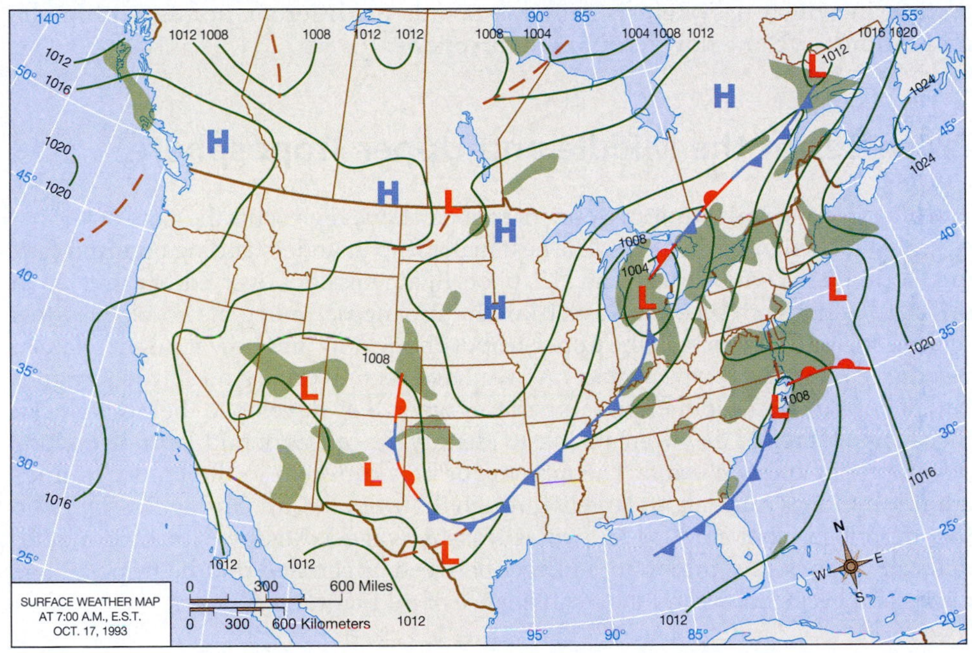

Figure 10–3 Two examples of mid-latitude cyclones. On June 24, 1994 (a), a fairly typical system was centered just south of the Great Lakes. A cold front extends southwestward into north Texas, and a warm front stretches eastward to the Atlantic Coast. On October 17, 1993 (b), another low is centered south of the Great Lakes, with a stationary front extending northeastward and a cold front oriented toward the southwest.

Evolution and Movement of Cyclones

Let's look at a hypothetical but realistic scenario to illustrate how the development and movement of the cyclone impacts the weather. A weak disturbance in the airflow may have little noticeable effect as cyclogenesis begins off the east coast of Japan. But when the system develops into maturity and moves eastward, it can bring rain to the coastal portions of western North America and snow to the coastal mountains. If this storm occurs in the winter, upper-level winds may guide the storm southward into central and southern California and then eastward into the Rocky Mountain states. Passing over the lee of the mountains, the mid-latitude cyclone may intensify and then barrel northeastward to bring blizzard conditions to the northeastern United States and southeastern Canada. As it moves offshore into the western Atlantic, the storm may undergo complete occlusion a week or two after its formation in the western Pacific.

At a particular place (say, Kansas City, Missouri), the passage of the system brings predictable effects. As a warm front approaches, cloud cover usually deepens and increases and light to moderate precipitation is possible. The rain or snow eventually gives way to warmer, sunny conditions with the passage of the warm front, and the wind shifts from a southerly to southwesterly direction. Clear, warm conditions may then persist for a day or so. But as the cold front approaches, a strong, fast-moving band of heavy clouds and precipitation can cause major snowfall or rainfall. Afterward, the cold air behind the front brings cold, clear conditions.

Interestingly, it was not known until the eighteenth century that storms move at all—previously they were thought to form and die in the same location. By an unfortunate turn of luck, Ben Franklin made this discovery in Philadelphia in 1743. Franklin had hoped to witness an eclipse of the moon but was disappointed when a mid-latitude cyclone brought overcast conditions that completely obscured the event. Later he was told by a friend that the sky was clear in Boston at the time of the eclipse, but that storm system (the one that had ruined Franklin's observation of the event) arrived some time later. From this, Franklin rightly concluded that the system responsible for the clouds had migrated to the northeast. What he was unable to resolve, however, was how the cloud cover could move to the northeast while the winds at the time of the eclipse were from the northwest. We now know, of course, that the answer is related to the counterclockwise spiral within the mid-latitude cyclone. That is, regardless of the direction in which a system is moving, winds at different points within the mid-latitude cyclone flow in different directions.

Processes of the Middle and Upper Troposphere

Upper-Level Winds and Pressure
Section 8.6.1

The life cycle of mid-latitude cyclones described above represents the state of knowledge that existed up to the 1940s. The next major leap in understanding occurred during World War II, when British and U.S. pilots flying missions over Europe and Japan observed winds with speeds up to 400 km/hr (250 mph). Among meteorologists, this finding stirred an interest in the upper-tropospheric flow and how it might relate to weather conditions on the surface. As we have seen, Bjerknes and his colleagues had no information about upper-air patterns when they developed their polar front theory, and therefore they were unable to identify the causes of mid-latitude cyclone development and occlusion. The next major breakthrough in the theory of mid-latitude cyclones came about largely through the work of Carl Gustav Rossby (who first described what are now known as **Rossby waves**). Rossby explained mathematically many of the important linkages between upper- and middle-tropospheric winds and cyclogenesis and the maintenance of mid-latitude cyclones.

Rossby Waves and Vorticity

In Chapter 8 we described the large Rossby waves of the upper troposphere. Figure 10–4 illustrates how the air turns left and right as it flows through these waves. Moving from points 1 to 3, the air rotates counterclockwise (as indicated at the bottom left of the figure). Between points 3 and 5, it rotates clockwise. The rotation of a fluid (such as air) is referred to as its **vorticity**.* The figure shows vorticity changes in the moving air, relative to the surface. Viewed from space, there is an additional component of vorticity arising from Earth's rotation on its axis. The overall rotation of air, or its **absolute vorticity**, thus has two components: **relative vorticity**, or the vorticity relative to Earth's surface, and **Earth vorticity**, which is due to Earth's daily rotation about its axis.

Relative vorticity depends on air motions with respect to Earth's surface, while Earth vorticity is a function solely of latitude—the higher the latitude, the greater

*For this discussion, we are only concerned with rotation with respect to the local vertical (e.g., a merry-go-round). *Vorticity* can also refer to rotation around horizontal axis (e.g., a roll of paper towels).

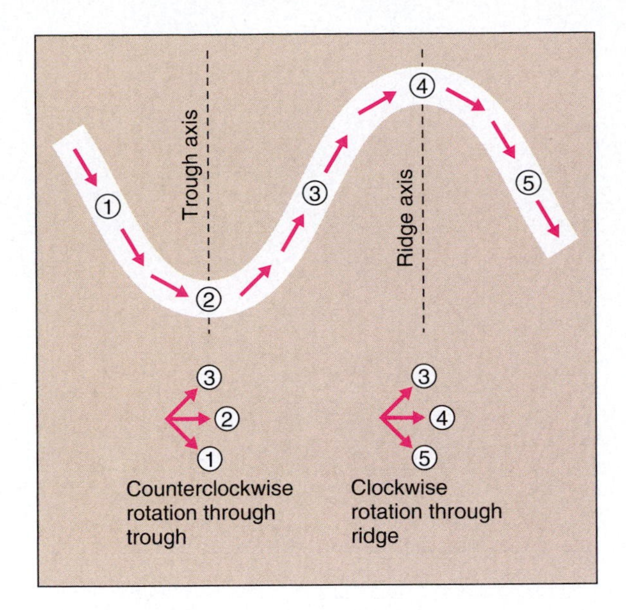

Figure 10–4
Vorticity around a Rossby wave. As air flows from positions 1 to 3, it undergoes a counterclockwise rotation. Along the ridge, the air turns clockwise from positions 3 to 5. The bottom of the figure shows the wind vectors representing the flow at the five positions.

the vorticity—with zero vorticity at the equator.* If the flow of air relative to the surface is in the same direction as the rotation of Earth itself (counterclockwise in the Northern Hemisphere), relative and Earth vorticity complement each other and increase the total or absolute vorticity (Figure 10–5). For this reason, counterclockwise rotation in the Northern Hemisphere is said to have *positive vorticity*, to be consistent with the convention used for the Coriolis force. Air rotating clockwise possesses *negative vorticity*.

Figure 10–6 shows the trough from Figure 10–4 in greater detail so we can examine the air's vorticity. In segment 1, the air flows southeastward with no change in direction or speed. Because it undergoes no rotation, the air has zero relative vorticity. In segment 2, the air continuously turns to its left to yield positive relative vorticity. In segment 3, the air flows continuously toward the northeast and has no relative vorticity. Thus, the trough has three distinct regions: two with zero relative vorticity and one with positive relative vorticity. Two transition zones separate the regions of maximum and minimum (zero) relative vorticity. Across transition zone *A*, vorticity increases as the air flows, whereas across transition zone *B* the relative vorticity decreases. (The Rossby wave shown here covers a relatively limited range of latitude. As a result, Earth vorticity changes only slightly in

*Earth vorticity is proportional to the sine of the latitude. Thus, Earth vorticity is only 16 percent greater at latitude 55° (sin 55° = 0.819), than at 45° (sin 45° = 0.707). In other words, over the mid-latitudes where Rossby waves are most likely to occur, the Earth vorticity changes with latitude are fairly small.

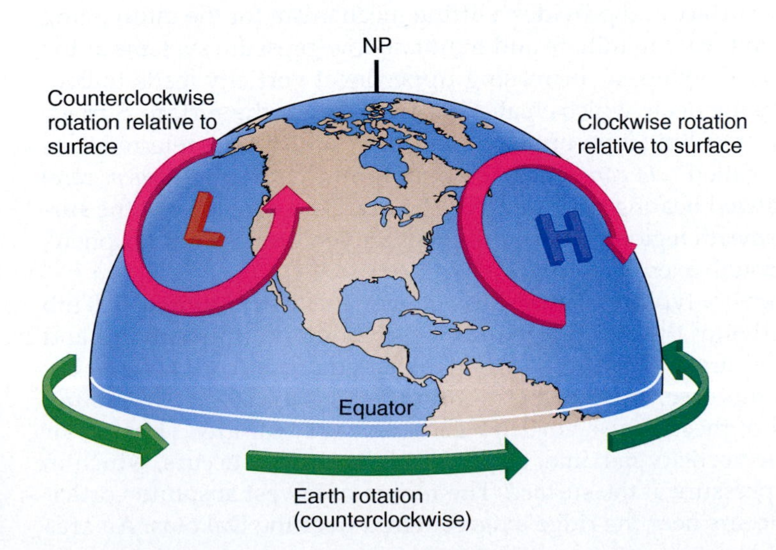

Figure 10–5
Earth vorticity and relative vorticity of air. As the Northern Hemisphere rotates counterclockwise, it generates Earth vorticity. Relative vorticity is the rotation of air relative to the surface, without regard to the planet's rotation. Absolute vorticity is the sum of the two.

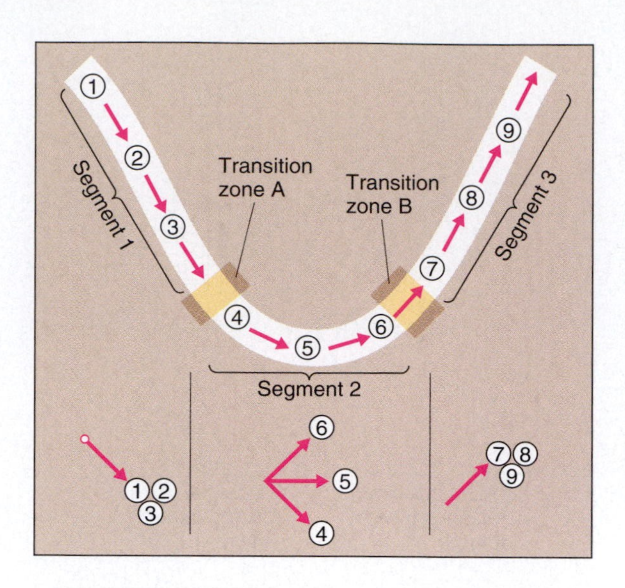

▶ **Figure 10–6**
The change in vorticity along a Rossby wave trough. As the air flows from positions 1 to 3, it undergoes little change in direction and thus has no relative vorticity. From positions 4–6, it turns counterclockwise and thus has positive relative vorticity. The air flows in a constant direction from positions 7–9. Thus, the trough has three segments based on vorticity, separated by two transition zones.

Figure 10–6, and changes in absolute vorticity correspond closely to changes in relative vorticity.)

At this point you might logically ask, "So what?" The answer is that vorticity changes in the upper troposphere lead to pressure changes near the surface. Let us see how. As you learned in Chapter 8, angular momentum is conserved in the absence of any outside forces. As a cowboy's twirling rope is pulled in, the reduction in the area swept by the rope causes it to twirl faster. The same thing happens when the vorticity or rotation of a parcel of air changes. That is, as the horizontal area occupied by an air parcel decreases by convergence, its vorticity or spin must increase, as in transition zone *A*. Decreasing vorticity, as in transition zone *B*, likewise leads to divergence. This very important relationship can be summarized in the simple equation

$$-\frac{1}{\zeta}\frac{\Delta\zeta}{\Delta t} = div$$

where $-\frac{1}{\zeta}\frac{\Delta\zeta}{\Delta t}$ is the change (here, the decrease) in absolute vorticity with respect to time, and *div* = divergence. Conversely, an increase in absolute vorticity with respect to time leads to convergence. (For simplicity, we limit our discussion of divergence and convergence to horizontal changes in area.)

Divergence in the upper atmosphere, caused by decreasing vorticity, draws air upward from the surface and provides a lifting mechanism for the intervening column of air. This, in turn, can initiate and maintain low-pressure systems at the surface (Figure 10–7). Conversely, increasing upper-level vorticity leads to convergence and the sinking of air, which creates high pressure at the surface. Surface low-pressure systems resulting from upper-tropospheric motions are referred to as **dynamic lows** (also called *cold core lows*)—distinct from the **thermal** (*warm core*) **lows** caused by localized heating of the air from below. Cold core lows at the surface typically exist beneath regions of decreasing vorticity in the upper atmosphere, just downwind of trough axes.

Figure 10–8 shows a typical relationship between the distribution of 500 mb heights (representative of the pressure pattern in the middle troposphere) and absolute vorticity. The areas of greatest vorticity (shaded purple) occur along the two trough axes (in this case, over northern California and the lower Mississippi Valley). Downwind of these zones, vorticity decreases very rapidly. Thus, as air flows away from the vorticity maxima, upper-level divergence occurs, which in turn promotes low pressure at the surface. The region of lowest absolute vorticity (shaded in red) occurs near the ridge axis, centered over the Dakotas. An area

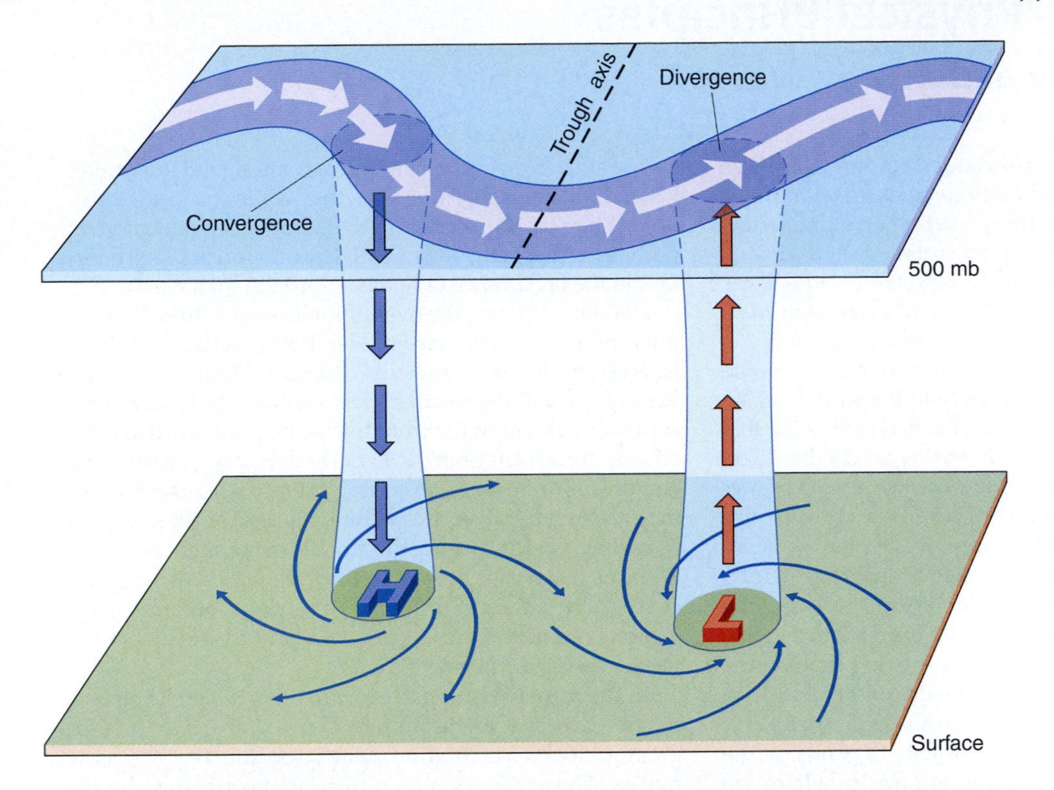

◀ **Figure 10–7**
Upper-level convergence and divergence along favored positions on a Rossby wave create high and low pressure at the surface.

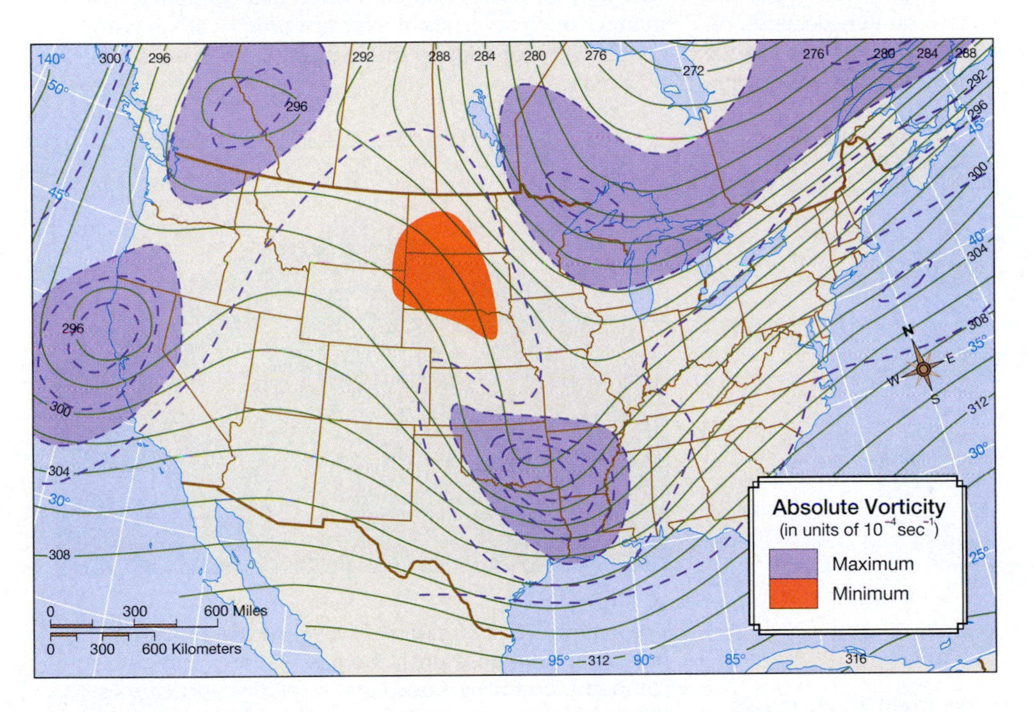

◀ **Figure 10–8**
Values of absolute vorticity (shown in dashed lines) on a hypothetical 500 mb map. Notice that the greatest values appear near trough axes (vorticity in units of 10^{-4} sec^{-1}).

of increasing vorticity that is a likely center of high pressure at the surface exists just downwind of this region. (See *Box 10–1, Physical Principles: Vorticity and Rossby Waves* for further information on vorticity patterns.)

Surface Fronts and Upper-Level Patterns

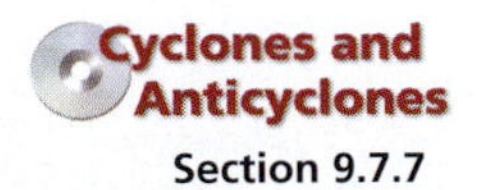

Section 9.7.7

At this point, we know that upper-level divergence causes the formation and intensification of surface mid-latitude cyclones, whereas upper-level convergence causes high pressure at the surface. We have also seen that the airflow along Rossby waves can generate upper-level divergence and convergence. Thus, the airflow in the middle and upper troposphere has important impacts on surface patterns. But

10–1 Physical Principles

Vorticity and Rossby Waves

We've seen that the vorticity associated with airflow has two components. The first, called *Earth vorticity*, arises from the planet's 24-hour rotation. A person sitting on a chair at the North Pole undergoes one complete rotation each day and thus has maximum Earth vorticity. Like the Coriolis force, Earth vorticity increases with latitude so that it is at its maximum at either pole and is nonexistent at the equator.

We define the second source of rotation, *relative vorticity*, in terms of motions of air relative to the surface. Relative vorticity itself has two sources: the first is the *shear* that occurs when the speed of a fluid varies across the direction of flow. Figure 1a illustrates this process as water flows through a channel (the same applies to air, but we use water as an example simply because it is easier to visualize). The flow is faster along the right bank. If a paddlewheel was fixed in the middle of this stream, the faster-moving water to the right would exert a greater force on the wheel than would the slower-moving flow on the left. This would cause the wheel to rotate counterclockwise and thereby undergo vorticity due to shear.

The second source of relative vorticity depends on the *curvature* of the flow, as shown in Figure 1b, where the curved channel forces a fluid to turn counterclockwise. As air flows through a Rossby wave, it undergoes this type of curvature. As we said, relative vorticity is the sum of curvature and shear, but remember, they are signed quantities. When they have the same sign, they act in the same direction. When the signs are different, they offset one another, perhaps completely. In general, shear is much less important in the occurrence of relative vorticity than curvature.

Absolute vorticity, the overall rotation of a fluid, like angular momentum, is conserved—that is, in the absence of intervening forces, it remains constant. Figure 2 shows a Rossby wave in the Northern Hemisphere. As the air flows southward, west of the trough axis, its Earth vorticity decreases (recall that Earth vorticity decreases toward the equator). But because absolute vorticity is conserved, an increase in relative vorticity compensates for the decrease in Earth vorticity, causing the air to turn counterclockwise. Then, as it starts to flow poleward, the Earth vorticity increases. Thus, the air turns back to its right and once again exhibits negative relative vorticity. (For the sake of clarity, we are assuming no shear exists.)

In short, there is a constant trade-off between Earth and relative vorticity. As the air moves poleward, it assumes a greater clockwise rotation relative to the surface; as it moves toward the equator, it turns in a counterclockwise manner. Such reversals in relative vorticity, along with the conservation of angular momentum, help maintain Rossby waves.

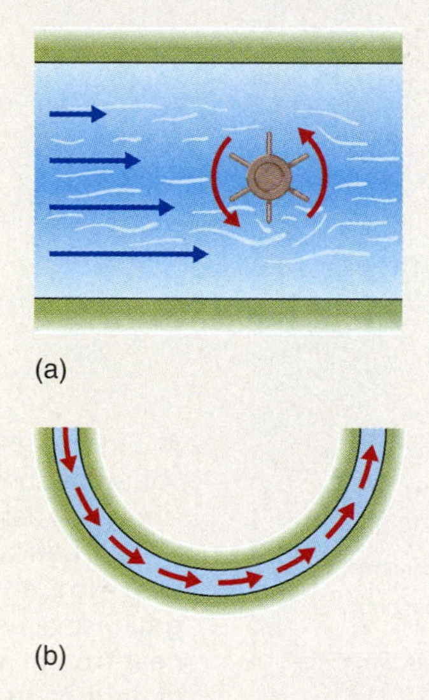

▲ **Figure 1**
Relative vorticity by shear (a) occurs when a fluid moves at a differential speed across the direction of flow. A paddle wheel fixed across the fluid rotates counterclockwise as a greater forward stress is exerted to the right of the direction of flow. Curvature in the direction of flow can also produce relative vorticity (b).

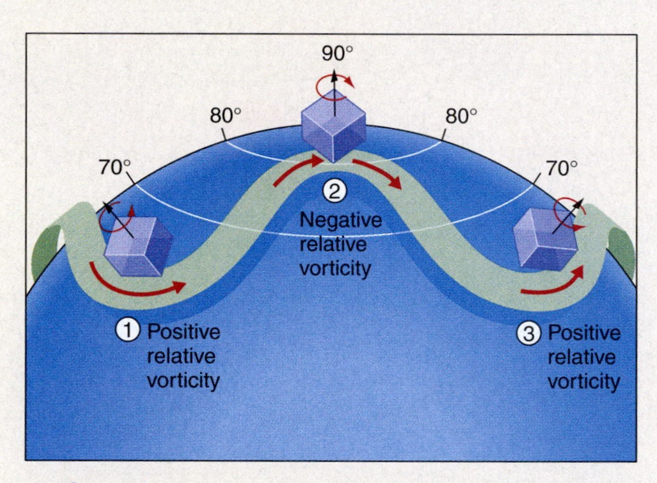

▲ **Figure 2**
The maintenance of a Rossby wave. As the air moves poleward from position 1, the gain in Earth vorticity is compensated for by a decrease in relative vorticity. As the parcel approaches position 2, the relative vorticity becomes negative and the parcel turns back equatorward. At position 3, the reduction of Earth vorticity causes an increase in relative vorticity, and the air again turns poleward.

the causality is not one way, because patterns at the surface, in particular the presence of cold and warm fronts, also exert their own direct impacts on the middle and upper troposphere.

As you have seen in Chapter 4, the hydrostatic equation states that the decrease in pressure with altitude (the vertical pressure gradient) is determined by the density of the air—cold, dense air has a greater vertical pressure gradient than does warm, light air. It follows that in a cold air column, the greater decrease in pressure with altitude should lead to lower pressure aloft compared to warm air. Thus, the differences in temperature on either side of a cold front must lead to significant differences in upper-level pressure there (the same reasoning applies to warm fronts). We develop this idea more fully in the following section.

Cold Fronts and the Formation of Upper-Level Troughs

Rossby waves consist of large, alternating troughs and ridges that establish patterns of upper-level divergence and convergence. The troughs in the waves normally develop behind the position of surface cold fronts, not by some grand coincidence but in response to the presence of the fronts. Figure 10–9 illustrates how this happens by showing the temperature and pressure changes in a 1 km-thick layer of air on either side of a cold front. The air above point *A* lies entirely within the warm sector ahead of the front. The frontal boundary slopes backward so that at *B* the air is cold in the lowest 500 m and warm in the upper 500 m. Cold air occupies the entire kilometer-thick layer above *C*.

Now let's assume (for simplicity's sake) that the surface pressure is 1000 mb at all three locations and compare the vertical pressure distributions. Above *A*, the pressure drops 55 mb in the lowest 500 m and another 53 mb in the next 500 m, to yield a pressure of 892 mb at the 1 km level. Above *B*, the pressure drops 58 mb in the cold, lowest 500 m (3 mb more than that above *A*). But, as in the upper 500 m above *A*, the pressure drops 53 mb to the 1 km level. Thus, the pressure at that height is 889 mb—3 mb less than at *A*. Over *C*, the pressure drops 58 and 56 mb, respectively, in the lower and upper 500 m layers, so that the pressure at 1 km is 886 mb. Thus, the pressure at 1 km decreases from 892 to 889 to 886 mb across the frontal boundary. In short, the differing temperature characteristics cause a horizontal pressure gradient in the middle atmosphere. The sloping boundary of a warm front exerts the same sort of effect.

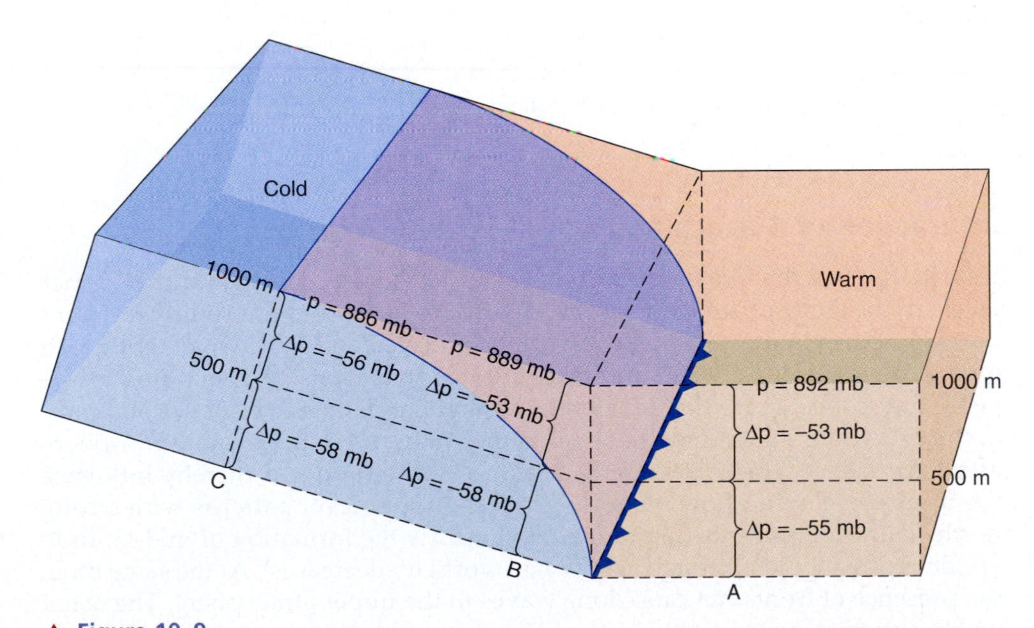

▲ **Figure 10–9**
Temperature distributions in the lower atmosphere lead to variations in upper-level pressure. Above *A* the entire column of air in the lower atmosphere is warm, so the pressure drops relatively slowly with height. At *B* cold air occupies the lowest 500 m, with warmer air aloft. This leads to a slightly lower pressure at the 1 km level. At *C* cold air occupying the lowest 1000 km causes a greater rate of pressure decrease with altitude and, as a result, a lower pressure at the 1 km level. Thus, the existence of sloping frontal boundaries establishes horizontal pressure gradients in the upper and middle atmosphere.

10–2 Physical Principles

A Closer Look at Divergence and Convergence

Upper-level divergence and convergence are changes in the horizontal area occupied by an air parcel, but their real importance lies in their effects on vertical motions. In its simplest form, divergence is nothing more than air spreading out over a greater horizontal area (convergence, of course, is the exact opposite). Divergence and convergence can occur in two ways: the first is by an increase or a decrease in the speed of air as it flows. The second is by a stretching out or pinching inward of the air, in a direction perpendicular to the direction in which it is moving. The divergence and convergence described earlier in this chapter can take either form.

Speed Divergence and Speed Convergence

Speed divergence and **speed convergence** occur when air moving in a constant direction either speeds up or slows down. Consider the two parcels of air, *A* and *B*, in Figure 1. Both parcels are moving in the same direction, but parcel *B* moves faster, as indicated by the length of the arrows. Because the leading parcel has greater speed than the one behind it, the distance between the two increases with time (b). This is an example of speed divergence.

This form of divergence is analogous to what might happen in a race with many entrants at the starting line. Initially, the runners cluster together, with little space between them. When the starting gun goes off, the people at the front of the pack dash away from those farther back, who shuffle along as they wait for the crowd to move forward. The cluster of people gradually thins out as the faster runners pull away from the slower ones, and the same number of people now occupy a greater area.

Because wind speed on an upper-level weather map is directly proportional to the spacing of height contours, we

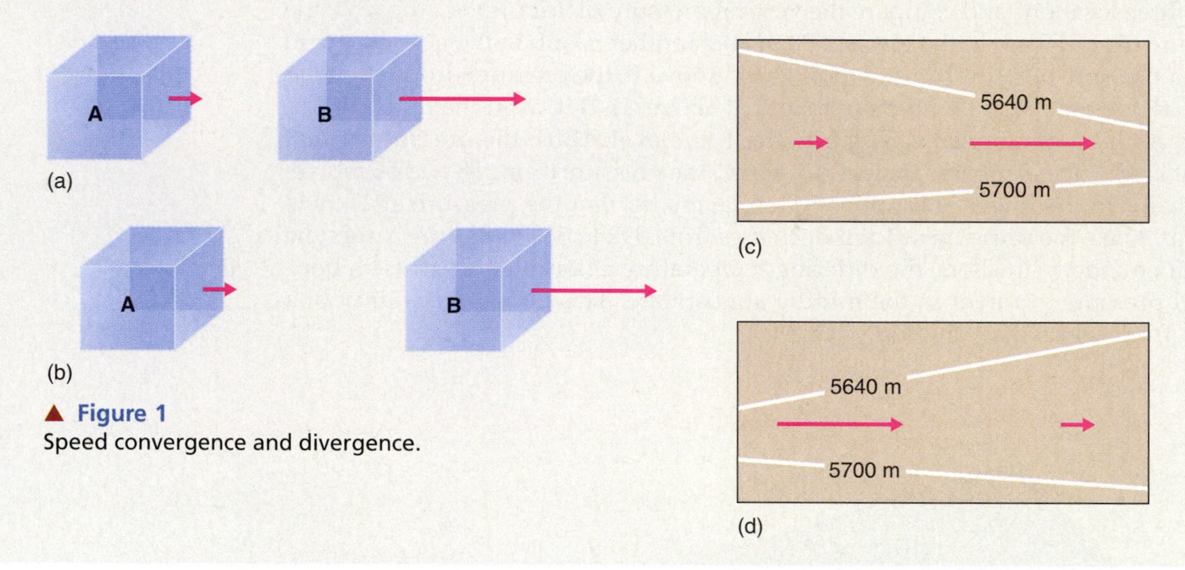

▲ **Figure 1**
Speed convergence and divergence.

Interaction of Surface and Upper-Level Conditions

Despite the fact that we commonly refer to the "surface level" and the "upper level" of the atmosphere, they are *not* separate entities; they are simply different parts of a single atmosphere that are fully connected and intertwined with each other. The upper-level winds influence surface conditions by generating divergence and convergence that lead to the formation of surface cyclones and anticyclones. At the same time, the temperature patterns in the lower atmosphere affect the rate at which pressure decreases with altitude and thereby influence the upper-level wind flow. More specifically, upper-level patterns with strong north–south components (meridional flow) cause the formation of mid-latitude cyclones downwind of trough axes (where vorticity decreases). At the same time, the presence of fronts can cause long waves in the upper atmosphere. The usual juxtaposition of surface cold fronts and upper-level troughs and ridges is shown in Figure 10–10.

The bottom line is that the interconnectedness between surface patterns and those aloft provides the true foundation for understanding the life cycle of mid-latitude cyclones. While the model of cyclogenesis, maturity, and occlusion described by Bjerknes and his colleagues provided an excellent *description* of the life cycle of mid-latitude cyclones, the *explanation* behind the processes eluded the early can use these maps to identify regions of speed divergence. Specifically, speed divergence occurs where contour lines come closer together in the downwind direction. In Figure 1c, the wind speed, indicated by the length of the arrows, increases in the direction of flow and causes speed divergence.

Speed convergence occurs when faster-moving air approaches the slower-moving air ahead. In the example of runners in the race, convergence might occur behind a muddy part of the track that slows the runners. The fastest runners, who have pulled ahead of the others, are the first to encounter the muddy spot. As they slog through the muck, the trailing runners have the opportunity to catch up. The entire pack bunches up in a smaller area and convergence occurs. A similar phenomenon occurs in Figure 1d, where speed convergence occurs.

Diffluence and Confluence

A second type of divergence and convergence, **diffluence** and **confluence**, occurs when air stretches out or converges horizontally due to variations in wind direction. In Figure 2a, a certain amount of air is contained in the shaded area between points 1 and 3. As it passes to the region between points 2 and 4, the same amount of air occupies a greater horizontal area. This is diffluence, a pattern that commonly appears wherever vorticity changes cause divergence to occur. Confluence is shown in Figure 2b.

Close inspection of Figure 2 reveals an interesting relationship between the different types of convergence and divergence. Recall that diffluence occurs where height contours on an upper-level map spread apart in the upwind direction (confluence occurs where they converge). But as the height contours spread apart, the horizontal pressure gradient decreases downwind and the air slows down. This, of course, creates *speed convergence*. Stated another way, one type of divergence (diffluence) is accompanied by another type of convergence (speed convergence). How does divergence actually occur downwind of a trough axis when diffluence and speed convergence occur simultaneously? The answer is that in most instances the diffluence downwind of trough axes is slightly greater in magnitude than the speed convergence, so the sum of the two yields a net divergence.

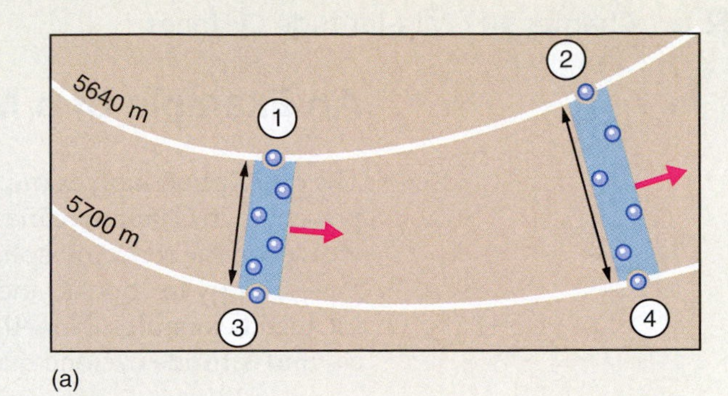

(a)

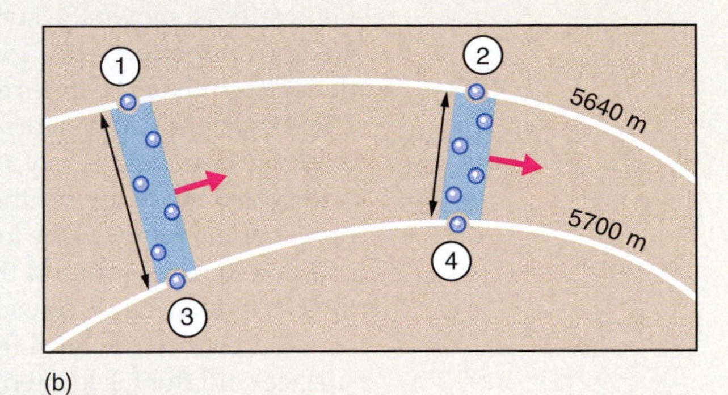

(b)

▲ **Figure 2**
Diffluence and confluence.

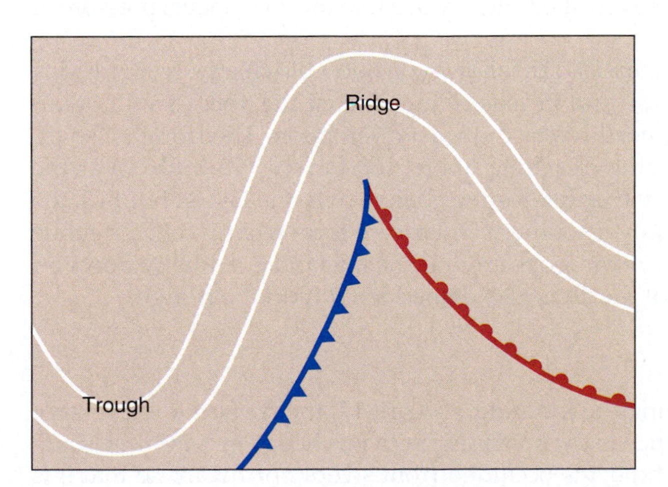

◀ **Figure 10–10**
The effect of differing vertical pressure gradients on either side of warm and cold fronts leads to upper-tropospheric troughs and ridges.

scientists. Today we know that upper-level patterns and their associated divergence and convergence affect pressure distributions (and hence the changes in mid-latitude cyclones) at the surface. Furthermore, we know that the upper-level patterns are influenced in turn by temperature conditions near the surface.

An Example of a Mid-latitude Cyclone

The descriptions and examples of troughs, ridges, cyclones, anticyclones, and fronts presented in Chapters 4 and 9, as well as in this chapter, represent idealized patterns. However, the real atmosphere seems not to have learned the lessons presented in meteorology textbooks, and real-life conditions often depart markedly from these idealized examples. Nonetheless, the following example from 1994 is a fairly typical mid-latitude cyclone making its trek across North America.

April 15

Figure 10–11 shows the surface and 500 mb weather maps and a satellite image of North America on April 15. Clear skies, low humidities, and light winds dominate the western United States and southwest Canada. In contrast, the mid-latitude cyclone over the north-central part of the United States has brought strong winds, overcast skies, and heavy rain showers. The surface wind in the warm sector flows northward out of the southern states and turns somewhat to the northwest as it approaches the center of the low pressure. North and west of the system, the air rotates counterclockwise around the low. Temperatures in the warm sector are typically in the 60s to low 70s °F, considerably greater than those to the west of the cold front.*

At the 500 mb level (b), a well-defined trough exists to the west of where the surface cold front is located. Strong westerly winds in excess of 100 km/hr (60 mph) occur over the western portion of the Canadian–United States border and then become northwesterly as they enter the trough. Downwind of the trough axis, the air again flows eastward toward the North Atlantic.

Just downwind of the trough axis, the vorticity decreases over northern Missouri, Iowa, and Minnesota. As we would expect, diffluence in this region leads to net divergence aloft and low pressure at the surface. Out west, downwind from the ridge axis, upper-level convergence leads to high pressure at the surface centered over central Idaho.

April 16

Over the next 24 hours (Figure 10–12), the mid-latitude cyclone migrated some 800 km (500 mi) to the northeast. It also intensified, with about a 5 mb decrease in sea level pressure from the previous morning. For the first time, a portion of the cyclone becomes occluded north of the Great Lakes and rain showers cover southeast Canada and the northeastern United States. The anticyclone over the western United States expanded southeastward and intensified slightly to a maximum sea level pressure of about 1027 mb.

The 500 mb pattern (Figure 10–12b) shows considerable change from the day before. The trough moved east, and its axis changed from the north–south orientation of the day before to a northwest–southeast orientation. The trough also intensified, with peak winds now reaching about 150 km/hr (90 mph) over the east-central United States. Another interesting change over the last 24 hours is the closing of the 5280-m contour over southern Ontario, to form what is called a **cutoff low**. Although the main flow of air loops around the low and eventually flows off the map, a circular rotation of the air is also embedded within the trough.

April 17

During the 24 hours preceding the morning of April 17, the surface low-pressure center migrated a short distance to the northeast, with no change in central pressure (Figure 10–13). At the same time, the occluded front swept northward so that it is

*Observed temperatures, dew points, wind velocities, and sea level pressures are depicted on the surface map for a few stations. Note that the temperatures and dew points (° F) are shown on the upper and bottom left, respectively, of the so-called *station models*. A line coming out of the circle shows the direction from which the wind is blowing, and the number of short and long tick-marks attached near the end of the line represents the wind speed. *Appendix C: Weather Map Symbols* provides more detailed information.

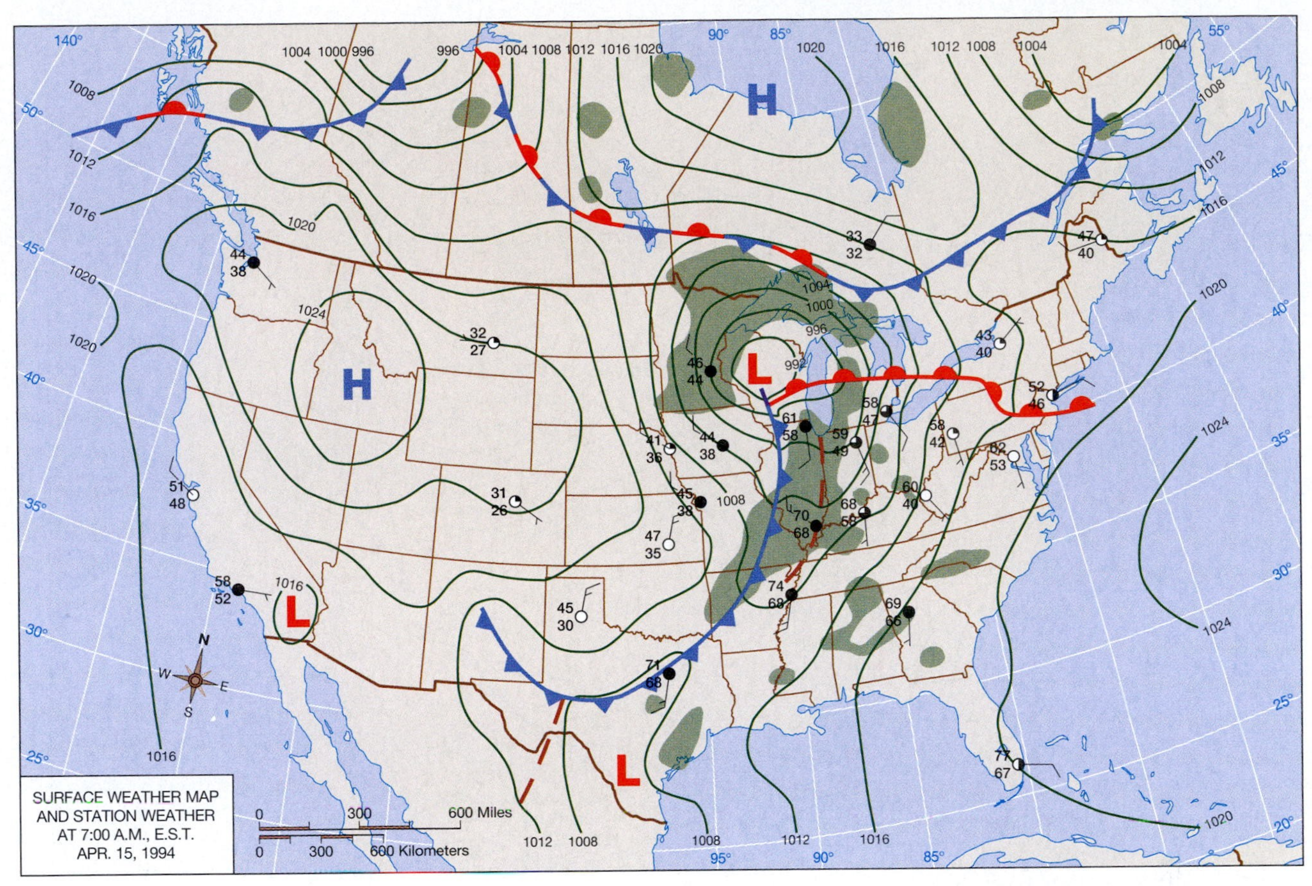

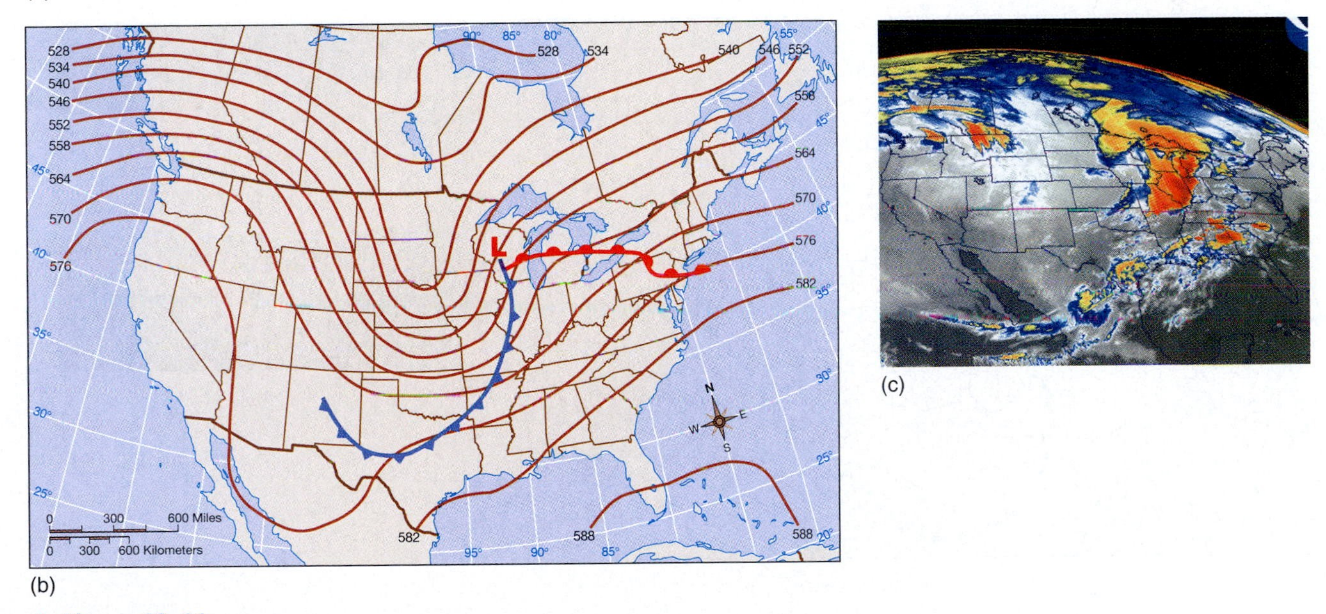

▲ **Figure 10–11**
Weather maps of the surface (a) and 500 mb level (b), and a satellite image (c) for April 15, 7 A.M. EST. Note that the positions of the surface fronts have been superimposed on the 500 mb map.

situated in a nearly west-to-east direction over southeast Canada. Precipitation, mainly as snow, is scattered along the frontal boundary and concentrated near the low-pressure center east of Hudson Bay.

The upper-level low (Figure 10–13b) has continued to deepen from the day before, as indicated by the lower heights of the 500 mb level. Not only has the height of the 500 mb level at the center of the cutoff low decreased by about 90 m, but the number of closed height contours increased to four (from the previous one).

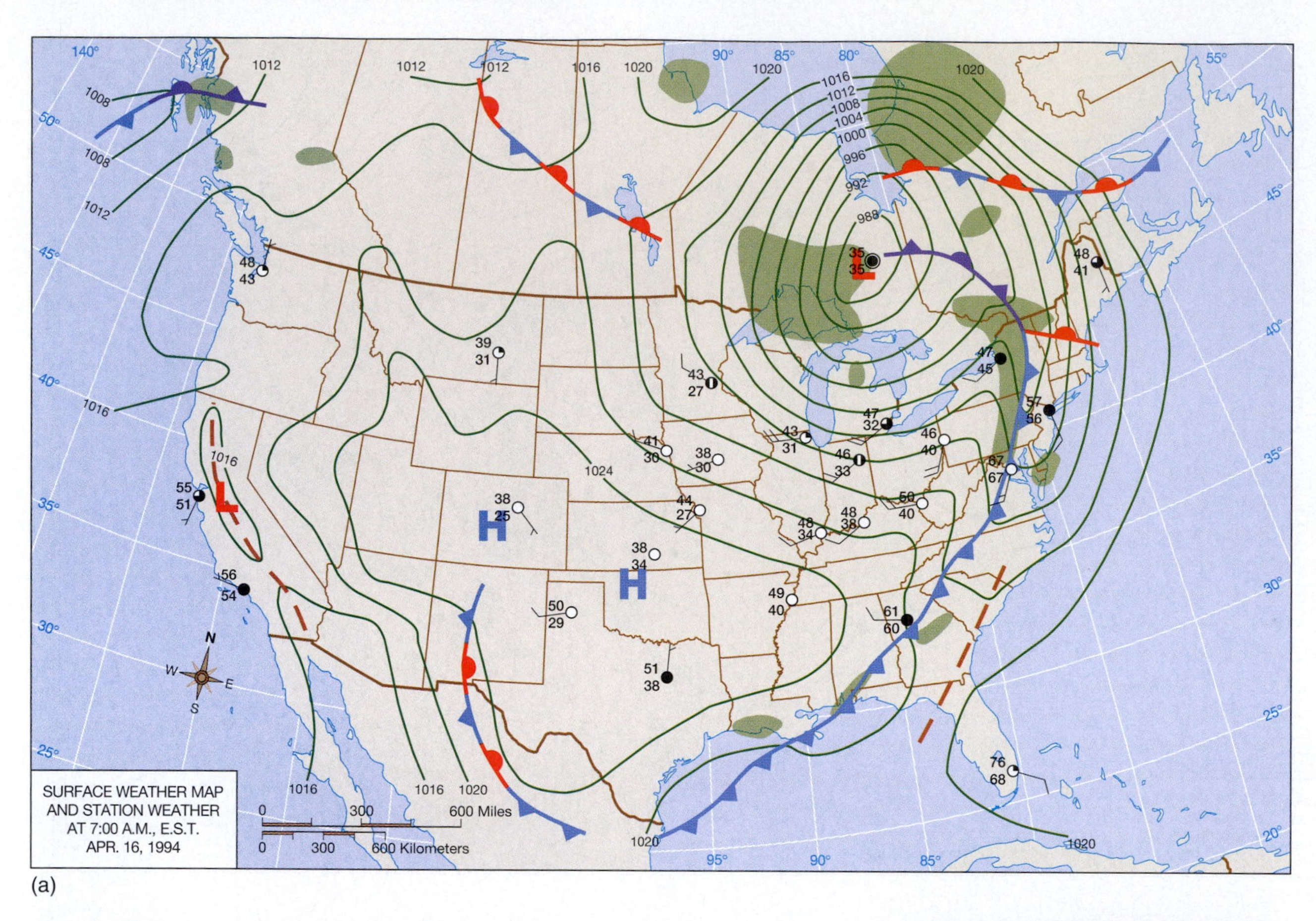

(a)

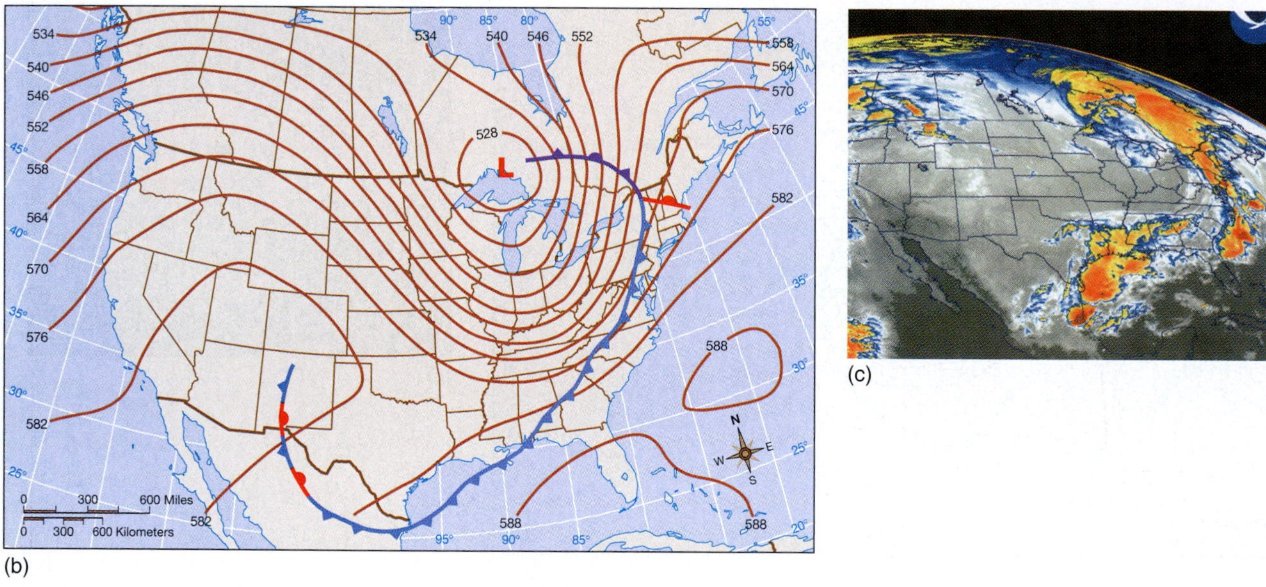

(b)

(c)

▲ **Figure 10–12**
Weather maps and a satellite image as in Figure 10–11, for April 16.

April 18

By April 18 (Figure 10–14), the center of the low-pressure system and most of the frontal boundaries have migrated off the surface map, but the system still is evident on the 500 mb map as the large trough extending to the southeast. As far as most of the population of eastern Canada and the United States is concerned, however, the mid-latitude cyclone no longer exerts any direct influence on the weather.

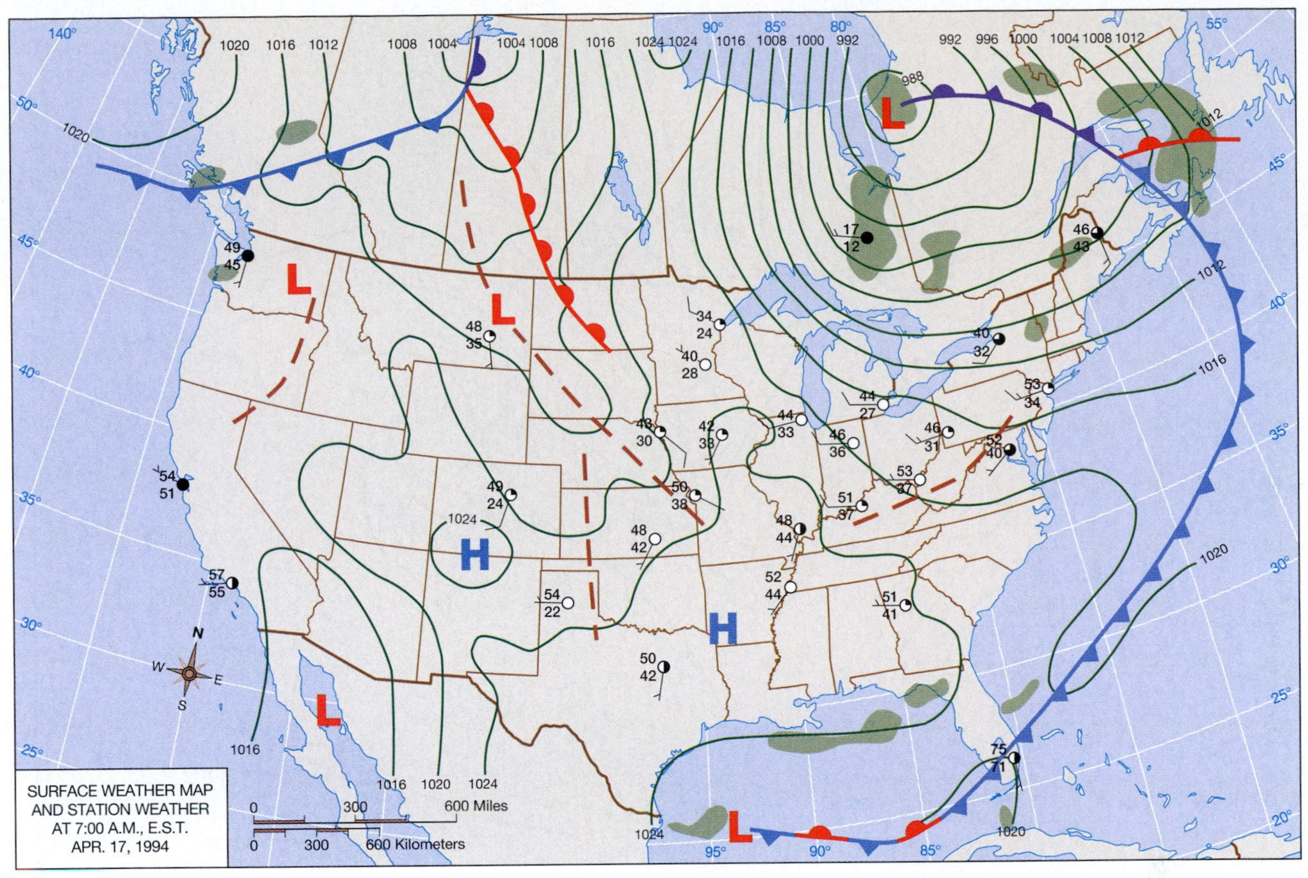

(a)

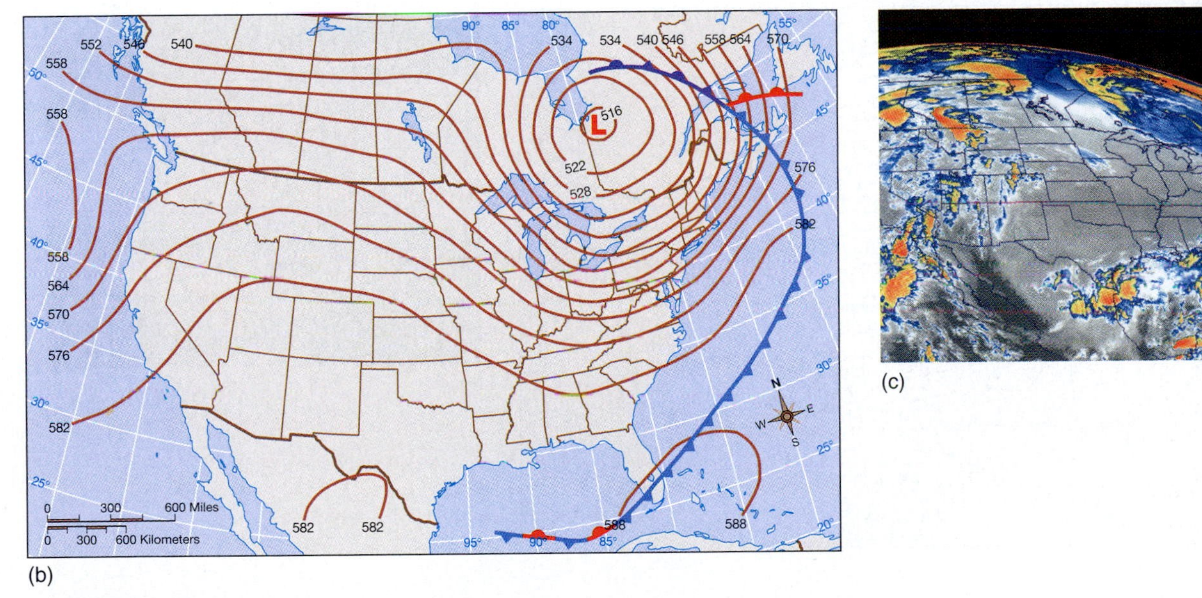

(b)

(c)

▲ **Figure 10–13**
Weather maps and a satellite image as in Figure 10–11, for April 17.

Flow Patterns and Large-scale Weather

We have seen that changes in upper-level vorticity create divergence and convergence patterns that influence the formation, intensification, and dissipation of surface cyclones and anticyclones. Thus, strong looping motions of the upper-level air are likely to create distinct regions of high and low pressure at the surface. On the other hand, air over a large area such as North America flowing in a

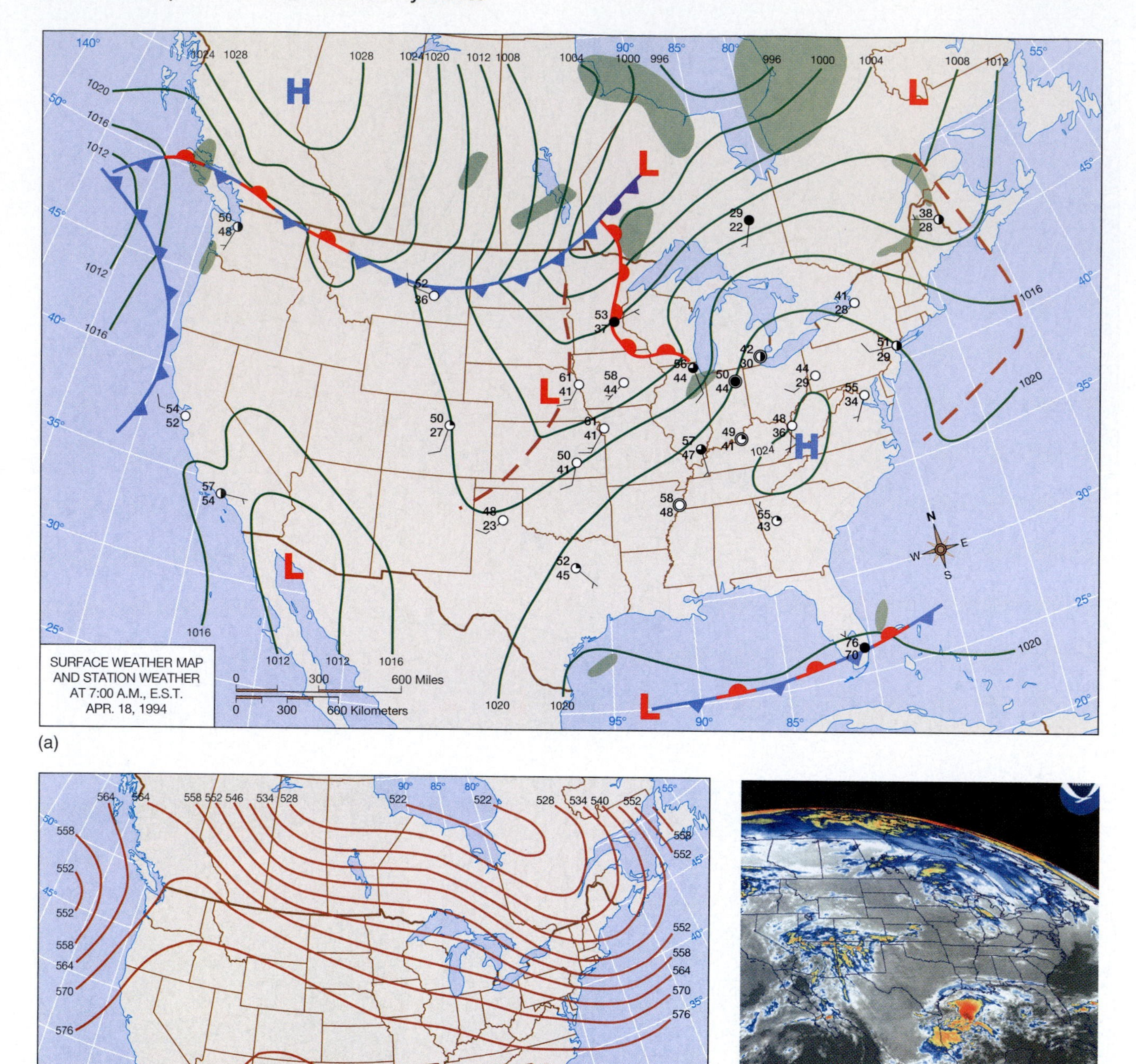

▲ **Figure 10–14**
Weather maps and a satellite image as in Figure 10–11, for April 18.

straight westerly direction will have uniform vorticity, with little chance for significant upper-level divergence or convergence.

Compare the 500 mb maps in Figure 10–15. In (a) the height contours exhibit a zonal pattern with a minimum of north–south displacement. In contrast, the pattern in (b) shows a strong meridional component.

Because they have no pronounced vorticity changes, zonal patterns hamper the development of intense cyclones and anticyclones. They are therefore more often associated with a large-scale pattern of light winds, calm conditions, and no

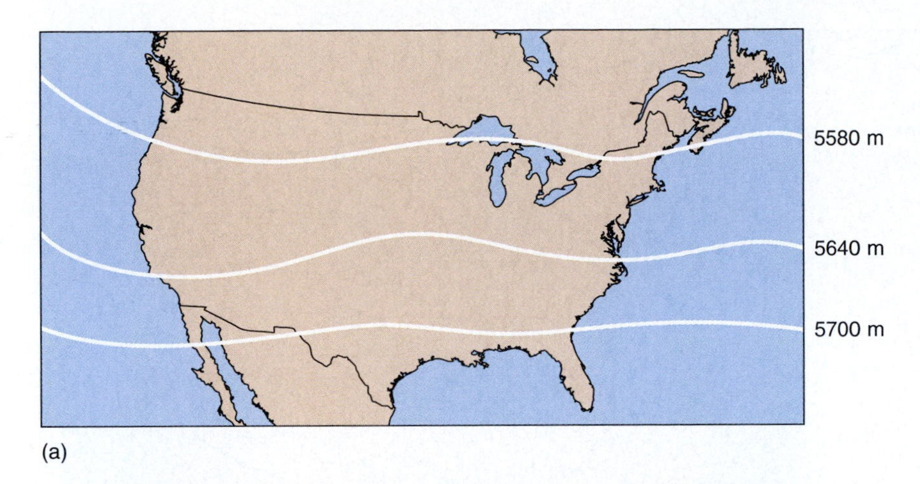

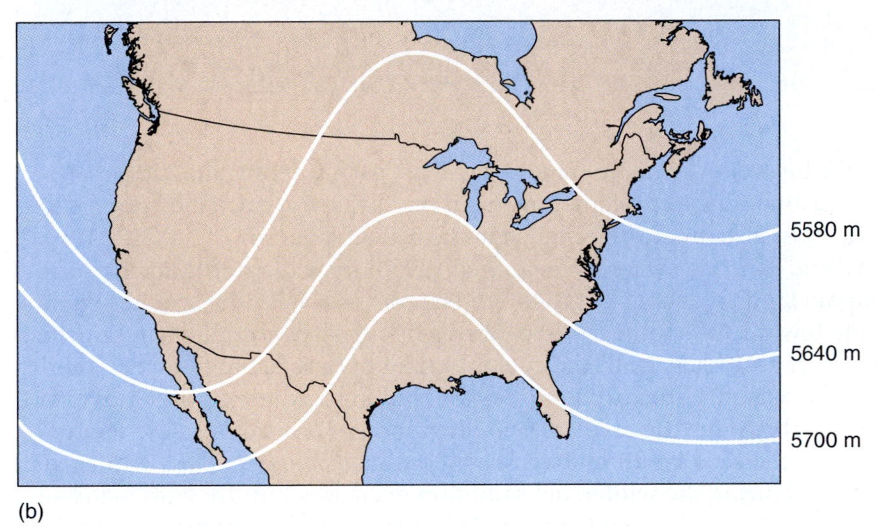

◀ **Figure 10–15**
Zonal (a) and meridional (b) flow patterns.

areas of widespread precipitation. Certainly there may be areas of localized precipitation—and the precipitation may even be quite heavy, as when orographically produced—but this activity will be spotty and widely scattered. Meridional flow, in contrast, can lead to the formation of major cyclones and anticyclones. If you look at an upper-level weather map and see strongly meridional flow, you can expect that some areas are experiencing cloudy and wet conditions while others are calm and dry. If you see a zonal pattern, it is less likely that large temperature contrasts exist from place to place or that there are large areas of heavy precipitation.

Experience shows that large-scale wind patterns in the upper atmosphere often persist, with one general type of pattern dominating for weeks or longer at a time. Such persistence of a zonal or meridional pattern can lead to droughts or episodes of heavy precipitation. A persistent zonal pattern can cause very widespread droughts to occur due to the lack of vorticity. Regional droughts can also occur if a meridional pattern remains in place, with the zone of upper-level convergence downwind of a ridge axis persisting over a particular region.

The Steering of Mid-latitude Cyclones

Upper-level winds have another important effect on surface conditions by governing the direction and speed at which the surface systems move. Outside of the Tropics, the upper atmosphere includes a strong component of west-to-east flow. Likewise, experience tells us that both cyclones and anticyclones outside of the Tropics typically migrate eastward. These two facts are not mere coincidence. In fact, the movement of surface systems can be predicted by the 500 mb pattern, with the surface systems moving in about the same direction as the 500 mb flow, at about one half the speed. Keep in mind, however, that the 500 mb level wind pattern changes through time, so

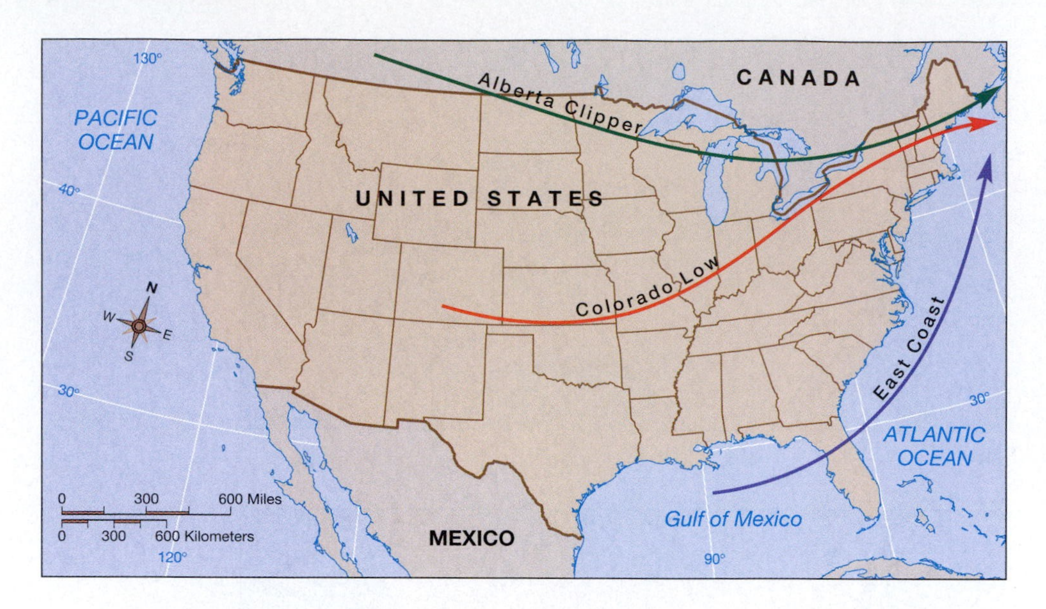

Figure 10–16
Typical winter storm paths. Alberta Clippers migrate in a west-to-east pattern along the United States–Canada border. Colorado lows move toward the northeast from the southern plains. Some storms have their origin near the Gulf of Mexico and move northward along the eastern seaboard.

predicting the track of a cyclone involves more than just examining the current upper-level flow and assuming a constant movement parallel to the current height contour pattern; one must also predict the change in the 500 mb pattern.

Many mid-latitude cyclones have their origin over the north Pacific off the coast of Japan. Upon reaching the Aleutian Islands of Alaska, the systems can die out, migrate toward the southeast, or continue on an eastward path across British Columbia. The most likely path the cyclones take upon reaching North America varies with the season, with northern treks favored in the summer, and movement toward the southeast having an increased likelihood of occurring in the winter.

Upper-level winds are about twice as vigorous on average in the winter than in the summer. During the winter, net radiation decreases rapidly with increasing latitude, giving rise to a stronger latitudinal temperature gradient than one finds in summer (Chapter 3). This results in greater pressure gradients (and winds) in the upper atmosphere. Thus, it is no surprise that mid-latitude cyclones generally move faster in the winter.

Though a winter mid-latitude cyclone can take many different paths across North America, two are particularly common: the Alberta Clipper and the Colorado low (Figure 10–16). The Alberta Clipper is associated with zonal flow and a polar jet stream that sweeps across southern Canada and the northern United States. Though it can bring frigid conditions, snowfall is usually light. In contrast, some mid-latitude cyclones passing farther to the south over western North America spawn new centers of low pressure as they pass over the central Rocky Mountains. They then follow a path from the southern Plains toward the northeastern United States and eastern Canada. These storms, usually warmer and containing greater amounts of water vapor in the air, often produce extremely heavy snowfall.

Storms occasionally have their genesis well to the south of the polar front (in contrast to the original model of the early Norwegian meteorologists) and can track northward along the eastern United States. Such storms often have strong uplift and high water vapor contents—conditions favorable for the development of extremely heavy snowfalls. Such conditions produced the "storm of the century" in March of 1993, which produced strong winds and record-breaking snow accumulations over the eastern third of the United States.

Migration of Surface Cyclones Relative to Rossby Waves

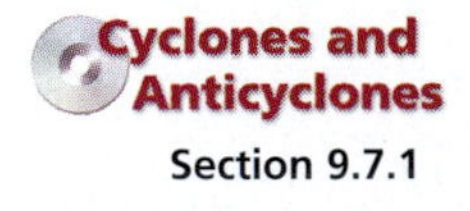

Section 9.7.1

For a mid-latitude cyclone to form, there must be upper-level divergence. If there is more divergence aloft than convergence near the surface, the surface low deepens and a cyclone forms. If the convergence at the surface exceeds the divergence aloft, the low gradually fills until it ceases to exist.

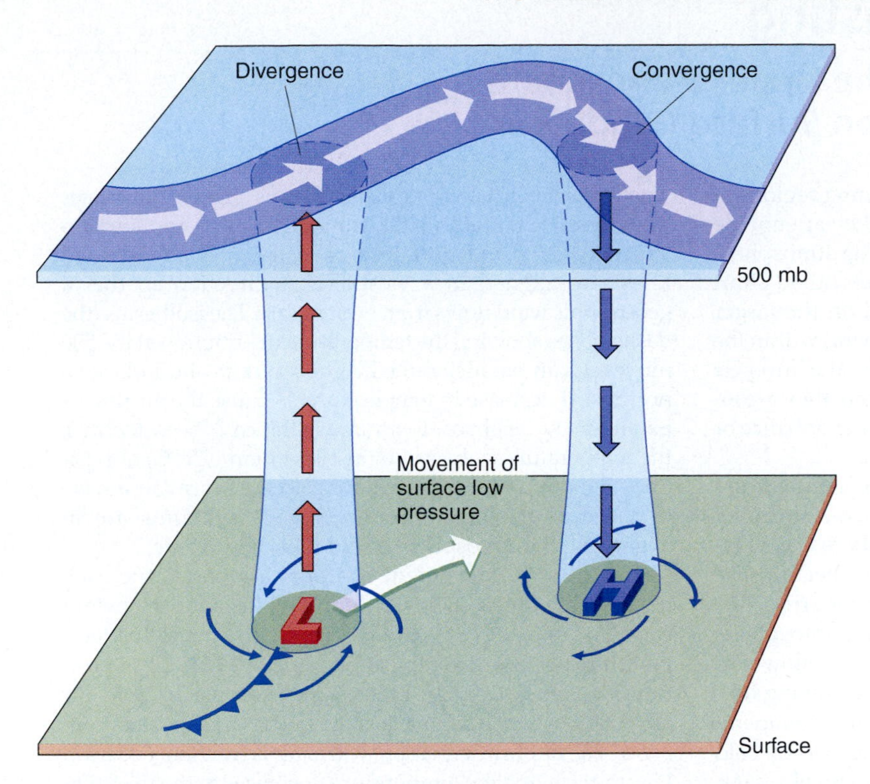

◀ **Figure 10–17**
Surface high- and low-pressure centers can migrate relative to the Rossby wave aloft.

Although the optimal place for mid-latitude cyclones to develop is just beneath the zone of decreasing vorticity aloft, they don't usually remain in a fixed position relative to the upper-level trough. Instead, they are usually pushed along so that they migrate in the same direction (and at about half the speed) as the winds at the 700 mb level—typically about 3 km (2 mi) above the surface. Figure 10–17 shows a center of low surface pressure gradually moving northeastward relative to a Rossby wave. In doing so, it moves away from the region of maximum divergence aloft and evolves as it travels. It goes through the various stages of its life cycle, typically occluding and dissipating as it approaches the upper-level ridge.

The Modern View—Mid-latitude Cyclones and Conveyor Belts

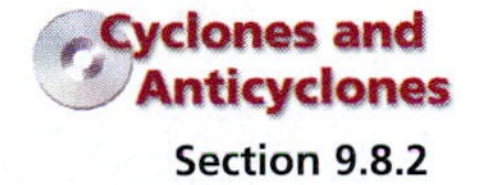

Section 9.8.2

We have examined the structure of mid-latitude cyclones at the surface and in the middle troposphere, but we should not overlook the fact that cyclones are three-dimensional entities. In other words, the 500 mb level (or any other level) is not separate from the surface portion—it is simply a different part of the same system. The **conveyor belt model** (Figure 10–18a), which we look at next, provides a better depiction of the three-dimensional nature of mid-latitude cyclones and accounts for their characteristic comma-shaped cloud pattern shown from above in Figure 10–18b.

January 1999 Blizzard

This model describes the mid-latitude cyclone in terms of three major flows. The first flow, the *warm conveyor belt*, originates near the surface in the warm sector and flows toward and over the wedge of the warm front. As the warm belt flows up the frontal surface, adiabatic cooling leads to condensation and precipitation. Moreover, as the air rises into the middle troposphere, it begins to turn to its right and become incorporated into the general westerly flow downwind of the upper-level trough. The cloud cover associated with the rising warm conveyor belt appears prominently as the bright, wide band extending from south to north in (b).

The *cold conveyor belt* lies ahead (north) of the warm front. It enters the storm at low levels as an easterly belt flowing westward toward the surface cyclone. But like the warm conveyor belt, it too ascends as it flows, turns anticyclonically (clockwise in the Northern Hemisphere), and becomes incorporated into the general westerly flow aloft. Although it originates as cold (and therefore relatively dry) air, the

10–3 Forecasting

Short Waves in the Upper Atmosphere and Their Effect on Surface Conditions

While Rossby waves play a role in establishing regions of upper-level divergence and convergence, they are not the only waves in the atmosphere that do so. The atmosphere also contains smaller eddies. Some of these, called **short waves**, are smaller ripples superimposed on the larger Rossby waves. These eddies migrate downwind within the Rossby waves and exert their own impact on the life cycle of mid-latitude cyclones. Depending on where they are located within the Rossby waves, they can either enhance or reduce the local divergence or convergence.

The formation of short waves depends on **temperature advection**, the horizontal transport of warm or cold air by the wind. Because air in the upper troposphere is well removed from the direct source of atmospheric heating (the surface), the temperature changes we experience from day to night are barely perceptible in the upper atmosphere. Thus, upper-level air changes temperature very slowly as it moves from one region to another, and air flowing horizontally from a warm region can retain its high temperature as it moves into a region otherwise occupied by cold air. We refer to the horizontal movement of relatively warm air as **warm air advection**. The opposite, of course, is called **cold air advection**. Both types of temperature advection appear on Rossby waves and, as we will see, affect the development of surface cyclones.

A useful method for detecting warm and cold air advection on a map of the upper atmosphere is to compare the orientation of height contours and isotherms. Figure 1 illustrates three possible patterns of 500 mb height levels and temperature advection. In Figure 1a, parallel height contours (solid lines) are aligned in a west-to-east direction so that a geostrophic wind flows from west to east. The isotherms (the dashed lines showing the temperature distribution at the 500 mb level) run parallel to the height contours and indicate a northward decrease in temperature. Because the air flow is parallel to the height contours (and in this case the isotherms), the temperature is the same (just less than –25 °C) at positions 1 and 2, and there is neither cold nor warm air advection. When the height contours and isotherms are in alignment, the atmosphere is said to be **barotropic**.

In Figure 1b, the height contours are parallel to each other, as are the isotherms. But in this instance the height contours and isotherms intersect each other, and the temperature increases from position 3 (below –25 °C) to position 4 (above –20 °C). This is an example of cold air advection, wherein a parcel of colder air is transported from 3 to 4. The opposite situation, warm air advection, occurs in Figure 1c, where the temperature decreases in the direction of airflow. When the height contours and the isotherms intersect, as in both (b) and (c), the atmosphere is said to be **baroclinic**.

Refer to Figure 2 and observe the two baroclinic zones at positions 1 (cold advection) and 2 (warm advection). Where cold advection exists, the entering air is denser than

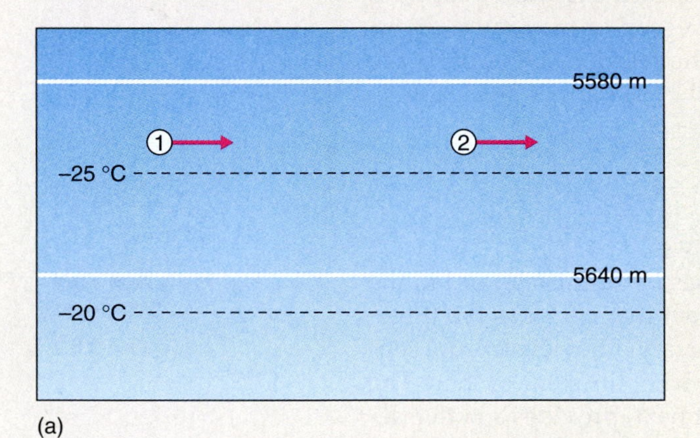

(a)

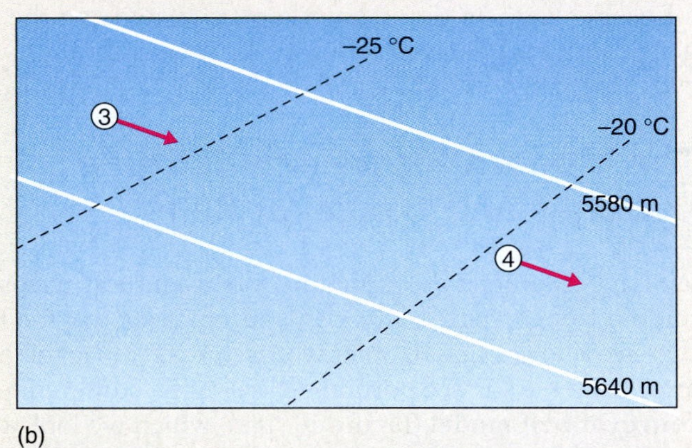

(b)

▲ Figure 1
A barotropic atmosphere (a) exists where the isotherms (the dashed lines showing the temperature distribution) and height contours (solid lines) are aligned in the same direction. No temperature advection occurs when the atmosphere is barotropic. A baroclinic atmosphere occurs where the isotherms intersect the height contours. Cold air advection is occurring in (b), warm air advection in (c).

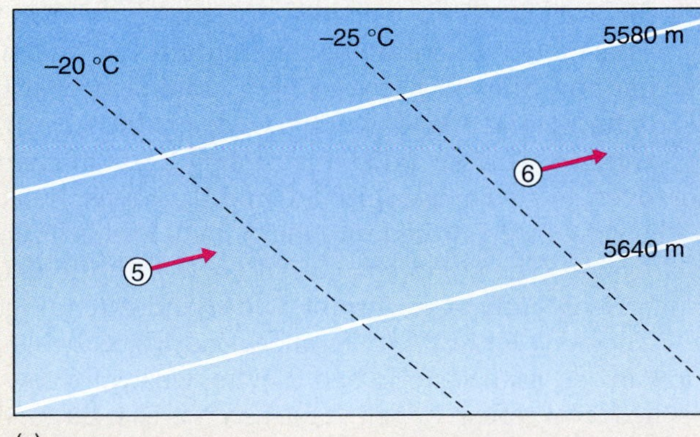

(c)

the air ahead of it because of its lower temperature. This gives it a negative buoyancy that causes it to sink downward, bringing cold air toward the surface. Cold air advection typically occurs behind a cold front, thereby enhancing the temperature contrast found on either side of the front. Where warm advection occurs, entering air is warmer and more buoyant than the air ahead of it and therefore rises. The warm and cold air advection thus cause vertical motions similar to those associated with static instability. This situation is called **baroclinic instability**.

In addition to undergoing rising or sinking motions, the air in areas of warm or cold air advection also undergo a slight turning—to the right in areas of cold air advection and to the left in regions of warm air advection. These motions are what cause the ripples (short waves) to form on the Rossby waves. When a short wave is located downwind of a Rossby wave trough axis, the divergence is enhanced and surface cyclones intensify.

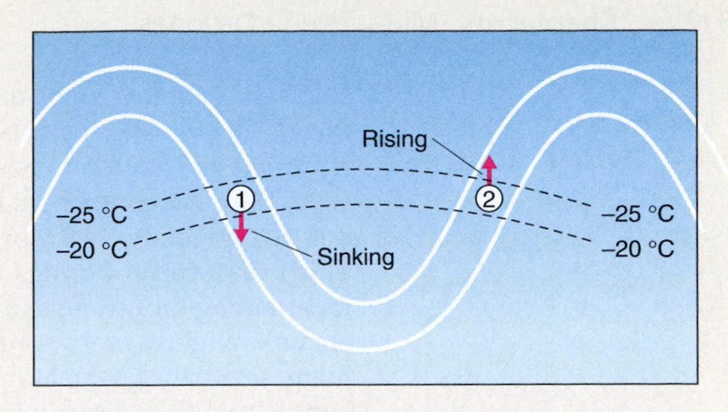

▲ **Figure 2**
Warm and cold air advection around a Rossby wave. The air at position 1 flows from colder to warmer air, resulting in cold air advection. Along a zone of cold air advection, sinking motions and a turning of the air to the right tend to take place. Warm air advection occurs at position 2 along with a rising of the air.

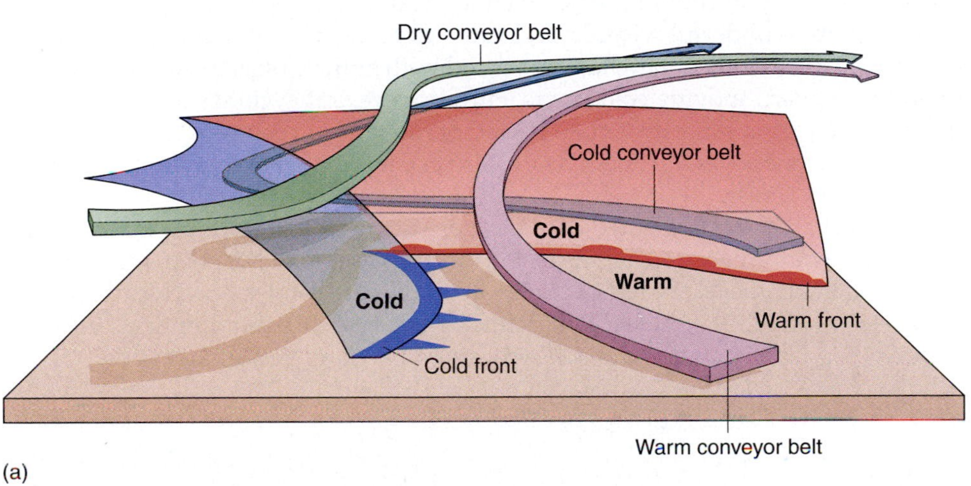

(a)

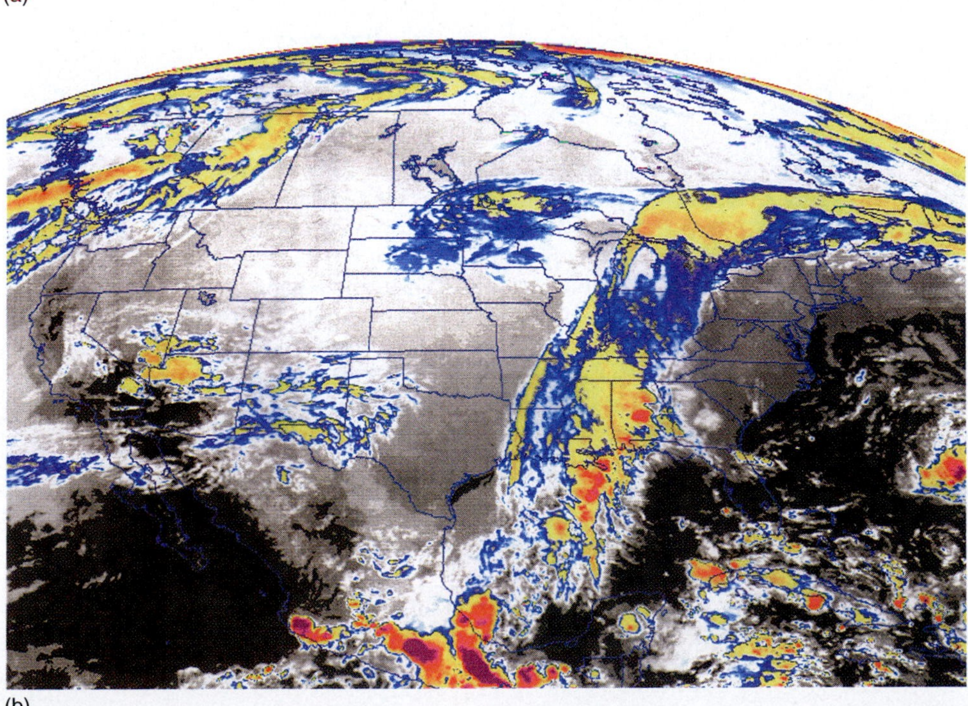

(b)

◀ **Figure 10–18**
The conveyor belt model of mid-latitude cyclones.

cold conveyor belt gains moisture from the evaporation of raindrops falling from the warm conveyor belt above. The cold conveyor belt extends in (b) from northern Michigan to eastern South Dakota.

The final component of the three-dimensional circulation is the *dry conveyor belt* that originates in the upper troposphere as part of the generally westerly flow. This broad current brings the coldest air into the cyclone, and it is important in maintaining the strong temperature contrast across the cold front. The upper-level air sinks slightly as it approaches the cold front from the west, but then it rises and merges with the general upper-level flow. The dry conveyor belt separates the cloud bands from the warm and cold conveyor belts, which helps to give the cloud distribution its distinctive comma-shaped appearance.

Anticyclones

So far we have said little about anticyclones, but they are as much influenced by upper-level conditions as are cyclones, and they exert an important impact on weather. (Recall from Chapter 8 that these phenomena are areas of high pressure around which the wind blows clockwise in the Northern Hemisphere.) While cyclones can bring heavy precipitation and strong winds, anticyclones foster clear skies and calm conditions because the cool air within them slowly sinks toward the surface. But that should not be taken to mean that anticyclones are always associated with wonderful weather. Indeed, outbreaks of continental polar (cP) air over the eastern United States are associated with anticyclones behind southward- or southeastward-moving cold fronts. Furthermore, anticyclones often tend to remain over a region for an extended period of time, which can lead to droughts. Finally, anticyclones over the Rocky Mountains can lead to Santa Ana wind conditions over the West Coast.

This chapter has presented the characteristics of mid-latitude cyclones and anticyclones that influence weather outside the Tropics. Yet the atmosphere often undergoes violent types of weather, usually on smaller time and space scales than those associated with mid-latitude cyclones. In Chapter 11 we'll meet thunderstorms and tornadoes, phenomena that can cause considerable damage and loss of life.

Summary

Although much of our knowledge of mid-latitude cyclones comes directly from the work of the Norwegian meteorologists in the early twentieth century, recent insights into upper-level winds have greatly increased our understanding. Mid-latitude cyclones and anticyclones both depend on a close interaction between processes occurring in the upper and lower troposphere. Counterclockwise rotation in the Northern Hemisphere has positive vorticity, while clockwise rotation has negative vorticity. The greatest positive vorticity occurs around the trough axis of a Rossby wave. The decreasing vorticity immediately downwind of the axis causes divergence, which leads to the formation of cyclones at the surface. At the same time, the fronts associated with mid-latitude cyclones help form the Rossby waves that create upper-level convergence and divergence. Thus, a constant interaction constantly takes place between the upper and lower atmosphere.

Traditionally, the distribution of clouds along frontal boundaries has been linked to convergence and overrunning on either side of the fronts. The modern approach explains the cloud distributions as the result of three separate air flows, or conveyor belts. A cold conveyor belt flows toward the center of low pressure ahead of the warm front. As it approaches the low pressure, it rises, and adiabatic cooling produces the wide band of cloud cover. Similarly, a rising conveyor belt of warm air flows ahead of the cold front to provide another band of cloud cover. Both the warm and cold conveyor belts turn anticyclonically near the center of low pressure and join the upper-level westerly flow. A dry conveyor belt in the middle atmosphere flows above cold fronts.

Surface cyclones and anticyclones migrate in the direction of the mid-tropospheric (700 mb) winds of Rossby waves. Cyclones thus move from regions of upper-level divergence (which help maintain or intensify the surface low pressure) to regions of upper-level convergence (which weaken the cyclones).

Key Terms

polar front theory p. 290
cyclogenesis p. 290
mature cyclone p. 290
occlusion p. 292
Rossby wave p. 294
vorticity p. 294
absolute vorticity p. 294
relative vorticity p. 294
Earth vorticity p. 294
dynamic low p. 296
thermal low p. 296
speed divergence p. 300
speed convergence p. 300
diffluence p. 301
confluence p. 301
cutoff low p. 302
conveyor belt model p. 309
short waves p. 310
temperature advection p. 310
warm air advection p. 310
cold air advection p. 310
barotropic p. 310
baroclinic p. 310
baroclinic instability p. 311

Review Questions

1. Define *cyclogenesis*. Where does it most commonly occur according to the original polar front theory?
2. Describe the isobar and wind patterns associated with mature mid-latitude cyclones.
3. Where is precipitation most likely to be found within mid-latitude cyclones?
4. Where within a mid-latitude cyclone are overrunning, convergence, and instability likely to serve as precipitation inducing processes?
5. What are Earth, relative, and absolute vorticity?
6. Why is counterclockwise rotation in the Northern Hemisphere said to have positive vorticity?
7. Where are the zones of positive, negative, and zero vorticity in a typical ridge and trough pattern?
8. In what part of a ridge and trough system do you find the areas of decreasing and increasing vorticity? Why is their existence important?
9. How do dynamic and thermal lows differ from each other?
10. Where are upper-level divergence and convergence most likely to occur?
11. What type of upper-level condition typically lies above and behind a cold front?
12. Where are upper-level ridges generally located relative to mid-latitude cyclones?
13. What are diffluence and speed divergence? How do they differ from confluence and speed divergence?
14. If the upper-level air flow over North America is zonal, what can you infer with regard to widespread precipitation conditions?
15. Which type of general upper-level air flow is more likely occur if minimal temperature variations exist across North America?
16. Describe the three conveyor belts of the conveyor belt model of mid-latitude cyclones. How does the conveyor belt model differ from the description of air flow presented in the original polar front model?
17. Explain how the movement of mid-latitude cyclones relative to the upper-level air flow contributes to the demise of cyclones.

Critical Thinking

1. Why is the term *polar front theory* probably a misnomer in view of current knowledge about cyclogenesis?
2. A commercial aircraft is flying at the 300 mb level and goes across a mid-latitude cyclone over both the cold and warm fronts. What kind of weather changes might the aircraft encounter?
3. After a front is fully occluded, its demise is imminent. Why can't the occluded front persist for several more weeks?
4. Why can't systems similar to mid-latitude cyclones develop over the Tropics?
5. Vorticity is usually discussed with reference to a vertical axis. What types of Earth, relative, and absolute vorticity conditions would you expect to exist with regard to a horizontal axis?
6. Why don't thermal lows migrate like dynamic lows do?
7. Why do forecasters take particular interest in the distribution of vorticity on 500 mb weather maps?
8. Clear skies often portend warm conditions during the summer but are often associated with very cold conditions in the winter. Explain why this is true.
9. Are Rossby waves likely to have greater representation in the Southern Hemisphere or the Northern Hemisphere?

Problems and Exercises

1. Go to **http://weather.uwyo.edu/upperair/uamap.html** and view the current 500 mb map for North America. Without referring to a surface map, make an educated guess about the position of surface mid-latitude cyclones, cold fronts, and warm fronts. Then go to **http://weather.uwyo.edu/surface/front.html** to see how well your educated guess worked out (be sure to highlight analysis for image type).
2. Using the same Web site as for problem 1, print the maps of the surface, 300 mb, 500 mb, 700 mb, and 850 mb levels. What patterns emerge as you move upward?
3. Visit **http://weather.uwyo.edu/surface/front.html** on a daily basis and keep track of existing mid-latitude cyclones and their associated fronts. Do the systems correspond well to the life cycle described in this chapter?

Quantitative Problems

This chapter has discussed the importance of temperature differences in the establishment of upper-level troughs and the critical roles of vorticity and divergence in the life cycle of mid-latitude cyclones. The Chapter 10 page of this book's Web site (**http://www.prenhall.com/aguado/**) gives you the opportunity to perform some calculations to help you better understand the material. We recommend you access the site and perform some of the simple but revealing problems.

Useful Web Sites

http://www.giss.nasa.gov/data/stormtracks/

An online atlas of storm tracks from 1961 through 1998. For each year, maps are available of seasonal or monthly storm track frequency, intensity, or individual paths.

http://www.hpc.ncep.noaa.gov/sfc/namfntsfcwbg.gif

Simplified, current surface weather map highlighting frontal systems.

http://weather.unisys.com/archive/index.html

An excellent source of archived surface and upper air weather maps and satellite images.

Media Enrichment

Tutorial

Cyclones and Anticyclones

Beginning with the formation of cyclones and anticyclones, this tutorial demonstrates how surface pressure is related to upper-level divergence and convergence. It puts particular emphasis on the structure of mid-latitude cyclones, with numerous three-dimensional diagrams and animations used to show frontal surfaces, conveyor belts, and other important features of these powerful storms.

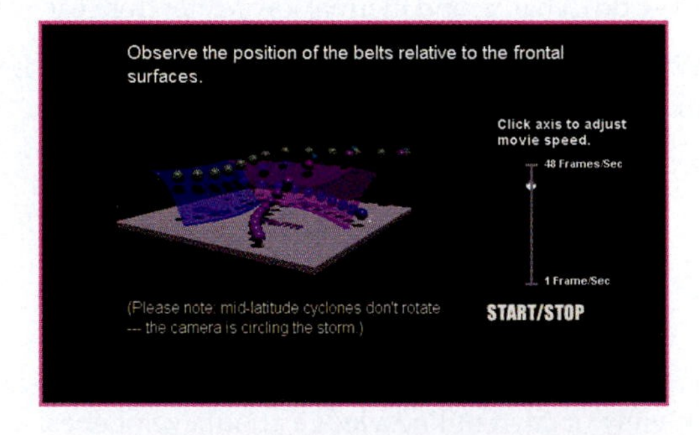

Tutorial

Upper-level Winds and Pressure

This tutorial develops the relationship between upper-level pressure and height fields and shows how ridges and troughs are reflected on contour maps. Rossby waves are presented as common features of the mid-latitudes, with animations used to depict simultaneous movement of the wave and prevailing westerly flow through the wave.

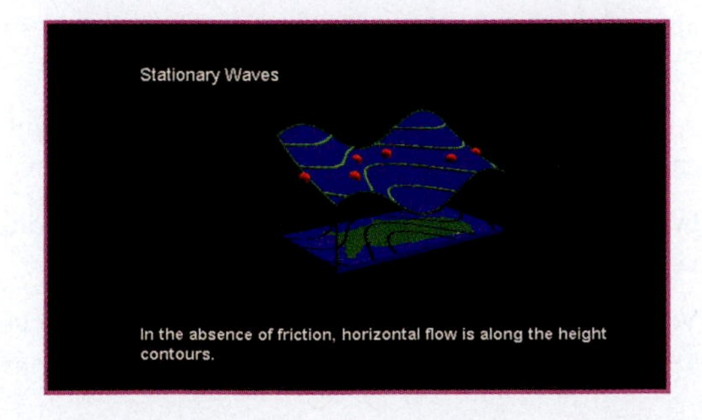

Weather in Motion

A Mid-latitude Cyclone Passes over the Southeast

This movie shows the movement of a storm over slightly more than 2 days, beginning at about 1000 UTC (5 A.M. EST) on January 26, 1998. Initially, several bands of clouds approach Mississippi, Tennessee, and Kentucky from the west with a minimal amount of counterclockwise rotation. By 1500 UTC the individual bands merge into a large area of continuous cloud cover that begins to show more discernible rotation. By 0000 UTC on the 27th, clouds with lower tops can be seen over the Ohio River Valley, moving northeast away from the center of rotation. By about noon UTC, the region of deepest cloud cover extends from South Carolina to Virginia within a broad area of precipitation covering the East Coast. As the movie comes to an end, the rotation into the low pressure center over Kentucky and West Virginia remains strong.

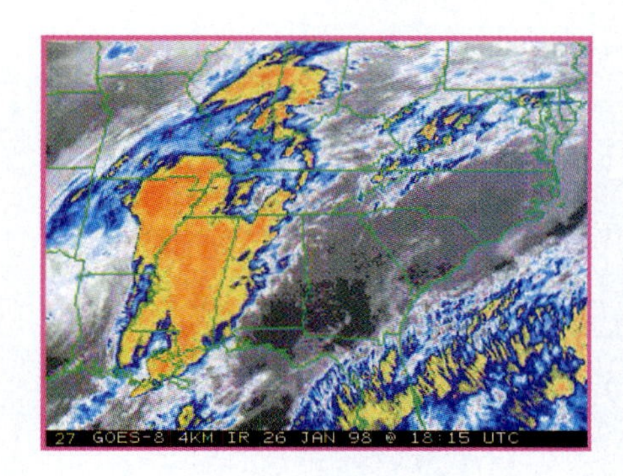

This movie reinforces the idea that although the original polar front theory nicely approximates the structure and evolution of mid-latitude cyclones, the actual movement of air and moisture within the system is more complex.

Weather in Motion

Tornado-generating Storm

This mid-latitude cyclone on March 20, 1998, produced tornadoes over northern Georgia and western North Carolina. The most violent weather took place along the leading edge of the southern portion of the cold front. This highlights the fact that storm activity is not usually uniform along the length of a frontal boundary.

Weather in Motion

January 1999 Blizzard

This mid-latitude cyclone brought blizzard conditions to the north-central United States and southern California. A strong counterclockwise rotation appears over the western part of the system, associated with a strong low-pressure system. The warm conveyor belt appears prominently during the last half of the movie. It is the band of intense cloud cover that flows south to north until it approaches the core of low pressure. The airflow then bends toward the east, where it is incorporated into the westerly airflow of the middle and upper troposphere.

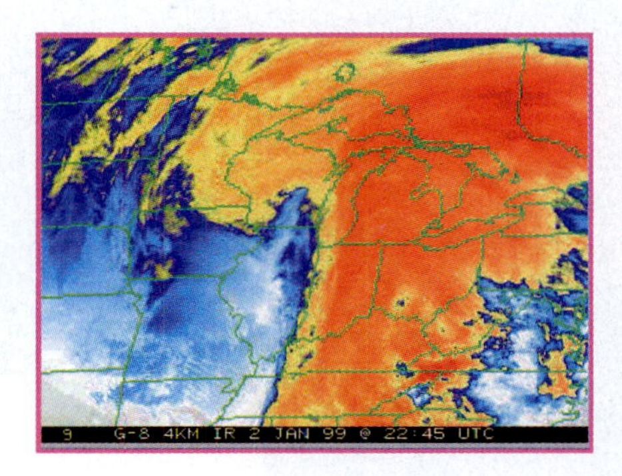

Weather Image

Cyclones Near the Martian Poles

Over 20 years ago NASA's *Viking* orbiter allowed astronomers to detect spiral storms on the surface of Mars. These storms occur near the poles during the Martian summer. In 1999 the Hubble Space Telescope led scientists to the discovery of very large cyclones near the Martian North Pole, three times larger than those previously seen on the planet. Unlike the spiral storms previously observed on Mars, or the mid-latitude cyclones of Earth, these storms feature three or more cloud bands surrounding a cloud-free eye, similar to tropical storms on our planet. Such storms, though rare, have also been observed at high latitudes on Earth, sometimes generating extreme winds.

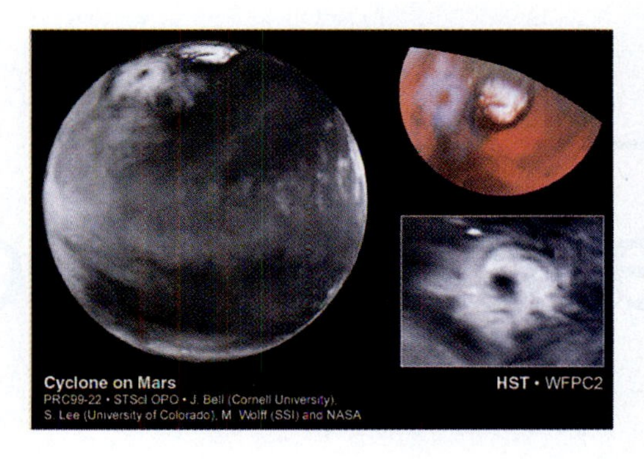

chapter 11

Lightning hitting Half Dome, Yosemite National Park in California.

Lightning, Thunder, and Tornadoes

It is hard to imagine anybody failing to be impressed by the beauty—as well as the danger—brought about by a thunderstorm. Spectacular though they may be, such storms are not rare. In fact, they are common throughout the world and occur about 40,000 times each day. Although their frequency varies greatly from place to place, virtually every location on Earth is vulnerable to thunder and lightning from time to time.

Lightning can create inconveniences—such as blowing out all the electrical appliances in a house. It can also do considerable damage, such as starting forest fires. And, of course, it can kill; during an average year, about 200 people are killed by lightning in the United States and Canada. But considering that the population of these two countries approaches 300 million people, it is easy to see that your chances of being struck are extremely remote.

Yet people do get hit by lightning. Consider the experience of the McQuilken family on their trip to Sequoia National Park, California, in August 1975. As the sky began to darken, Sean, Michael, and their sister Mary noticed their hair standing on end. Recognizing the apparent comedy of the situation, the boys posed for the photograph shown in Figure 11–1. Hail followed almost immediately. Then lightning struck—literally—and Sean was knocked unconscious. Michael quickly administered artificial respiration, which probably saved Sean's life. Another victim was less fortunate, however. The lightning had apparently forked off, with another branch hitting two nearby people, one of whom was killed.

But the effects of lightning and thunder are eclipsed by an even greater menace—tornadoes. We will now examine how, where, and why violent weather occurs, and we'll look at the situations that cause some storms to be weak and others to become destructive and deadly.

CHAPTER OUTLINE

Figure 11–1 Sean and Michael McQuilken in a strong electric field just prior to lightning.

Processes of Lightning Formation

About 80 percent of all lightning results from the discharge of electricity *within* clouds, as opposed to discharge from cloud to surface. This *cloud-to-cloud* lightning occurs when the voltage gradient within a cloud, or between clouds, overcomes the electrical resistance of the air. The result is a very large and powerful spark that partially equalizes the charge separation. Cloud-to-cloud **lightning** causes the sky to light up more or less uniformly. Because the flash is obscured by the cloud itself, it is commonly called **sheet lightning.**

The remaining 20 percent of lightning strokes are the more dramatic events in which the electrical discharge travels between the base of the cloud and the surface. Most of this *cloud-to-ground* lightning occurs when the negative charges accumulate in the lower portions of the cloud. Positive charges are attracted to a relatively small area in the ground directly beneath the cloud. This establishes a large voltage difference between the ground and the cloud base. The positive charge at the surface is a local phenomenon; it arises because the negative charge at the base of the cloud repels electrons on the ground below. Farther away, the surface maintains its normal negative charge relative to the atmosphere. Although the term *cloud-to-ground* is used, the same effect occurs in water—and lightning often strikes lakes, rivers, and oceans.

Although a stroke of lightning may come and go in just a few moments, a regular sequence of events must occur first in order for the event to take place. Electrification of a cloud is the initial stage in all lightning. After that, a path must develop through which electrons can flow. Only then is electricity actually discharged to produce a lightning stroke.

Charge Separation

All lightning requires the initial separation of positive and negative charges into different regions of a cloud. Most often the positive charges accumulate in the upper reaches of the cloud, negative charges in lower portions. Small pockets of positive charges may also gather near the cloud base (Figure 11–2a). So now the question is: How does this **charge separation** occur in the first place? Nobody knows for sure, because clouds that produce lightning and thunder happen to be particularly inhospitable laboratories. But we do know several important facts from which we can get some idea of how charges separate. First, lightning occurs only in clouds that extend above the freezing level and is also restricted to precipitating clouds. Thus, the ice crystal processes responsible for precipitation must also influence charge separation. Over the years, numerous theories were proposed and hotly debated. None provided a definitive answer until a major breakthrough in 1998, when a model involving charge transfer across thin films of water present on ice crystals and hailstones was confirmed by experiment.

Though we normally don't notice it, solids are often coated with a liquid surface layer just a few molecules thick. This layer consists of molecules only weakly bound to the solid below and is present even at temperatures well below the freezing point. (Among other things, the presence of this layer explains why ice is so slippery at temperatures well below zero degrees Celsius.*) In a cloud, when an ice crystal and a hailstone collide, some of the liquid-water molecules on the hailstone's surface migrate to the ice. In fact, evidence exists suggesting that the collision actually increases the tendency for liquification and thereby increases the mass transfer. Along with the water molecules, there is a transfer of positive charge from the hailstone to the ice crystal or, equivalently, a transfer of negative charge from the crystal to the hailstone. In this way ice crystals surrender negative ions to the much larger hailstones, which then fall downward toward the cloud base.

*You may have heard that pressure from an ice skate's blade melts enough ice to create a slippery film of water, but that's not correct. After all, the ice is just as slippery whether you press hard or not.

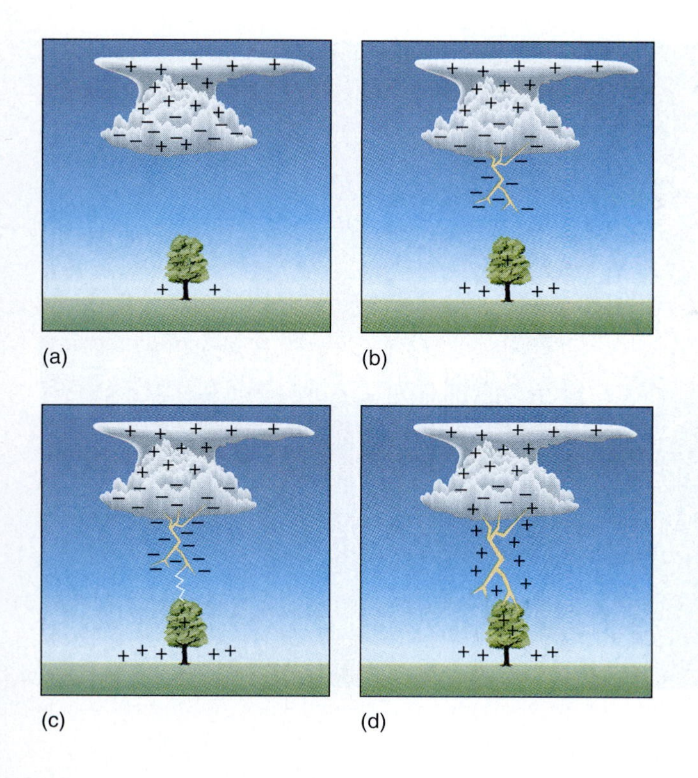

Figure 11–2
The downward movement of a stepped leader occurs after charge separation has taken place within a cloud (a). The leader approaches the surface as a very rapid sequence of steps (b) and (c), until contact is made with an object at the ground. The flow of electricity produces the lightning stroke (d).

Leaders, Strokes, and Flashes

In cloud-to-ground lightning, the actual lightning event is preceded by the rapid and staggered advance of a shaft of negatively charged air, called a **stepped leader** (Figure 11–2b), from the base of the cloud. The leader is not a single column of ionized air; it branches off from a main trunk in several places. Only about 10 cm (4 in.) in diameter, each section of the column first surges downward about 50 m (165 ft) from the base of the cloud in about a millionth of a second (or microsecond). This invisible leader pauses for about 50 μm, then surges downward another 50 or so meters. The downward movement in a rapid sequence of individual steps gives the stepped leader its name.

When the leader approaches the ground (Figure 11–2c), a spark surges upward from the ground toward the leader. When the leader and the spark connect, they create a pathway for the flow of electrons that initiates the first in a sequence of brightly illuminated **strokes,** or **return strokes.** The electrical current, flowing at about 20,000 amperes (A), appears to work its way downward from the base of the cloud, but the stroke actually propagates upward (Figure 11–2d). The conducting path is completed at the surface, from which there is a surge of positive charge upward toward the cloud. The current heats the air in the conducting channel to temperatures up to 30,000 K (54,000 °F), or five times that of the surface of the Sun! (See Figure 11–3.)

Weather in Motion
Lightning

The electrical discharge of the first stroke neutralizes some, but not all, of the negatively charged ions near the base of the cloud. As a result, another leader (called a **dart leader**) forms within about a tenth of a second, and a subsequent stroke emerges from it. This sequence of dart leaders and strokes may repeat itself four or five times.* Because the individual strokes occur in such rapid succession, they appear to be a single stroke that flickers and dances about. We call the combination of strokes a lightning **flash,** the net effect of which is to transfer electrons from the cloud to the ground.

The total transfer of electrons is not large, only about as many as we use in burning a 100-watt lightbulb for a minute or so. So how is lightning able to split trees and perform other dramatic work? For one thing, in lightning the charge transfer

*On occasion, a flash can consist of more than 20 individual strokes.

▶ Figure 11–3
Lightning.

is rapid, and so the electric current is about 100 times that of household current. To understand why, think of the lightbulb example as having low current flowing for a relatively long time. The total charge transferred is the same as in lightning, where a huge current flows for just an instant. Another factor is that the voltage gradient is much larger than in a household circuit, so the energy release is much larger for each electron transferred. Taken together, these facts mean that a large amount of energy gets released over a very brief period of time, making each stroke extremely powerful, far more than the 100-watt lightbulb.

Positive charges found at the top of thunderstorm clouds can also lead to lightning. When high-level winds are strong, thunderstorm clouds become tilted with positive charges carried ahead of the storm (Figure 11–4). These positive charges induce negative charges at the surface, resulting in a lightning strike that lowers positive charge to the surface. As a result, it often happens that the first of a storm's lightning strikes are positive, and measurements reveal that they can be twice as strong as the negative strokes that follow. This positive form of lightning is therefore particularly dangerous. It can occur several miles away from the storm, where people do not feel threatened; it tends to have larger peak electrical currents; and it typically lasts longer, making fires more likely.

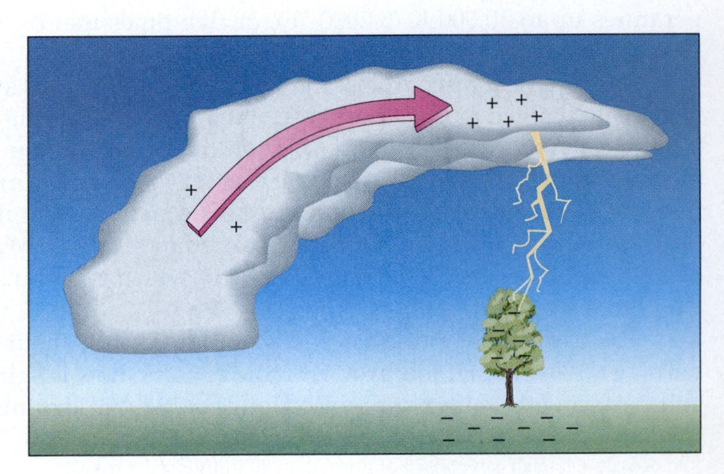

▶ Figure 11–4
Positive lightning stroke.

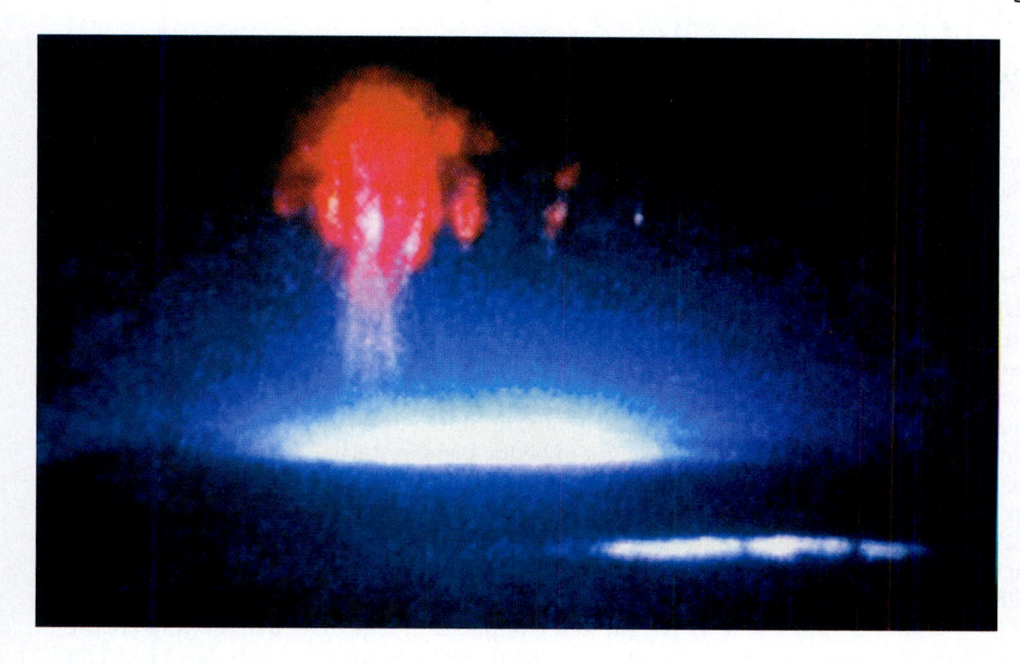

◀ **Figure 11–5**
A sprite.

Types of Lightning

Far less common than strokes and leaders is the bizarre type of electrification called **ball lightning.** Ball lightning appears as a round, glowing mass of electrified air, up to the size of a basketball, that seems to roll through the air or along a surface for 15 seconds or so before either dissipating or exploding. One form is a free-floating, reddish mass that tends to avoid good electrical conductors and flows into closed spaces or through doorways and windows. Another form is considerably brighter and is attracted to electrical conductors (including people). Various explanations for ball lightning have been offered for at least 150 years, but until recently none could account for all aspects of the phenomenon, and most had glaring weaknesses. The situation improved significantly early in 2000, with the report of experiments involving artificial lightning strikes on soil. It was seen that lightning reduces silicon compounds in the soil to tiny nanoparticles of silicon carbide (SiC), silicon monoxide (SiO), and metallic silicon (Si). Unlike the original silicon compounds, these contain significant chemical energy and are unstable in an oxygen environment. They are ejected into the air, where they cool rapidly and condense into filmy chains and networks. The networks are light, so they float easily in the atmosphere. And most important, they burn brightly as they oxidize, releasing the stored energy in the form of visible light.

St. Elmo's fire is yet another rare and peculiar type of electrical event. Ionization in the air—often just before the formation of cloud-to-ground lightning—can cause tall objects such as church steeples or ships' masts to glow as they emit a continuous barrage of sparks. This often produces a blue-green tint to the air, accompanied by a hissing sound.

Recent observations and photographs from space shuttle missions have revealed the existence of previously unknown electrical phenomena at the top of thunderstorms. **Sprites** (Figure 11–5) are very large but short-lived electrical bursts that rise from cloud tops as lightning occurs below. A sprite looks somewhat like a giant red jellyfish, extending up to 95 km above the clouds, with blue or green tentacles dangling from the reddish blob. Sprites accompany only about 1 percent of all lightning events. (Interestingly, military and commercial pilots now admit having seen sprites before they were observed from shuttle missions, but they did not often report them earlier lest they be accused of having hallucinations.)

Blue jets (Figure 11–6) are upward-moving electrical ejections from the tops of the most active regions of thunderstorms. They shoot upward at about 100 km/sec and attain heights of up to 50 km above the surface. In all likelihood, other types of electrical activity above thunderstorms remain to be discovered.

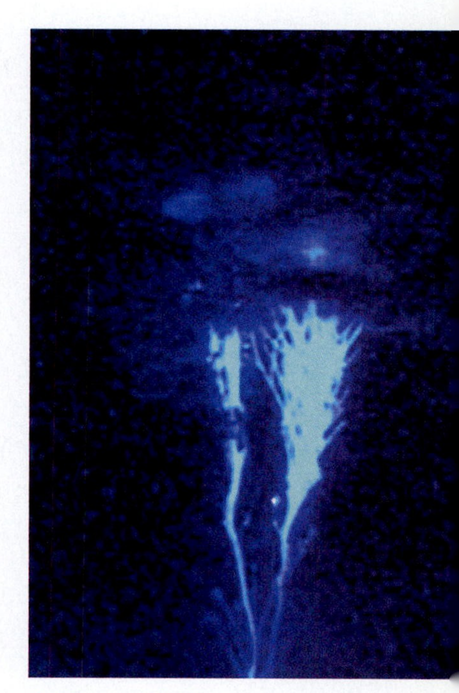

▲ **Figure 11–6**
A blue jet.

11–1 Physical Principles

Electricity in the Atmosphere

Lightning is, of course, an electrical disturbance, much of which can be explained by the basic principles of atmospheric electricity. You know from Chapter 1 that ions (charged particles) are most abundant high in the atmosphere, with the ionosphere being found from about 90 to 480 km. The upper atmosphere has a positive charge, just as we find near the positive pole of a battery. In the same way that a battery stores energy, electrical charges in the atmosphere represent stored energy and have the potential to do work. For both batteries and the atmosphere, this electrical potential is expressed by voltage, which is simply the energy per unit charge. For example, if a battery is rated at 1.5 volts (V), it means that 1.5 joules are available per coulomb of charge (1.5 J/C). A coulomb (C) is equivalent to the charge carried by about 6×10^{19} electrons. The higher the voltage, the greater the energy release for each coulomb transferred.

In the case of Earth, a huge voltage difference exists between the surface and the ionosphere—about 400,000 volts! This voltage gradient sets up what we call the **fair weather electric field.** The fair-weather field is always present, even in bad weather, so a better name might be the **mean electric field.** The fair weather field can be thought of as the background situation, on which extreme events such as lightning are superimposed.

Does electricity flow in response to the voltage gradient of the fair weather field? Yes, but because air is a good insulator, the current is weak, about 2000 coulombs per second for the entire planet (2000 A). In North America, individual houses are typically wired for 200 A service, so we see that the atmospheric current is truly very small. Nevertheless, it does represent a continuous leakage, whereby electrons are transferred from the surface, or (equivalently) positive charges are transferred from the atmosphere. This implies that for the mean electric field to be maintained, it must be continuously replenished. As a matter of fact, lightning discharges in thunderstorms are thought to be the primary recharge mechanism. In other words, cloud-to-ground lightning discharges transfer electrons to the surface, maintaining the voltage difference and the resulting electric field.

In the lower atmosphere, the fair weather electric field gradient is on the order of 100 V per meter. (Although this might sound impressive, remember that few ions are present, so the total available energy is very low.) For lightning to occur, the field strength must be greatly intensified above the background value. Only if the voltage gradient reaches millions of volts per meter will the resistance of the atmosphere be broken down, resulting in the "spark" we call *lightning* and the accompanying clap of thunder.

The question of how lightning forms is thus the question of how storms become sufficiently electrified. Even today only a partial answer can be given, as this chapter describes.

Thunder

The tremendous increase in temperature during a lightning stroke causes the air to expand explosively and produce the familiar sound of **thunder.** Although sound travels rapidly—about 0.3 km (0.2 mi) per second—it is much slower than the speed of light (300,000 km, or 186,000 mi, per second). This difference creates a lag between the flash of light and the sound of thunder; the farther away the lightning, the longer the time lag. You probably know the familiar rule of thumb for estimating the distance of a lightning stroke: simply count the number of seconds between the stroke and the thunder and divide by three to determine the distance in kilometers (divide by five for the distance in miles).

This method can fail for very distant strokes, those more than about 20 km (12 mi) away. The decrease in the density of air with height causes sound waves to be bent upward. At relatively short distances, the amount of bending is negligible. But at about 20 km, it is sufficient to displace the sound waves so that they cannot be heard at ground level. Lightning that seems to occur without thunder is sometimes called **heat lightning,** though this term is misleading in that it implies there is something unusual about it. The only oddity is that the sound of thunder does not reach the listener.

You have probably noticed that nearby thunder sounds like a loud, brief clap, while more distant thunder often occurs as a continuous rumble. A lightning stroke producing thunder may be several kilometers in length, so one part of it may be significantly farther from a listener than other parts. Thus, thunder makes a continuous sound as it takes longer for the sounds of more distant parts of the stroke to reach the listener. At greater distances, the echoing of sound waves off of buildings and hills can cause the thunder to make a rumbling sound.

11–2 Special Interest

A Personal Account of Ball Lightning*

I saw ball lightning during a thunderstorm in the summer of 1960. I was 16 years old. It was about 9:00 P.M., very dark, and I was sitting with my girlfriend at a picnic table in a pavilion at a public park in upstate New York. The structure was open on three sides and we were sitting with our backs to the closed side. It was raining quite hard. A whitish-yellowish ball, about the size of a tennis ball, appeared on our left, 30 yards away, and its appearance was not directly associated with a lightning strike. The wind was light. The ball was 8 feet off the ground and drifting slowly towards the pavilion. As it entered, it dropped abruptly to the wet wood plank floor, passing within 3 feet of our heads on the way down. It skittered along the floor with a jerky motion (stick-slip), passed out of the structure on the right, rose to a height of 6 feet, drifted 10 yards further, dropped to the ground and extinguished non-explosively. As it passed my head, I felt no heat. Its acoustic mission I liken to that of a freshly struck match. As it skittered on the floor it displayed elastic properties (a physicist would call them resonant vibrating modes). Its luminosity was such that it was not blinding. I estimate it was like staring at a less than 10-watt lightbulb. The whole encounter lasted for about 15 seconds. I remember it vividly even today, as all eyewitnesses do, because it was so extraordinary. Not until 10 years later, at a seminar on ball lightning, did I realize what I had witnessed.

*Source: Graham K. Hubler. Reprinted by permission from *Nature*. Copyright 2000, Macmillan Magazines Ltd.

Lightning Safety

Despite its splendor, we must not forget that lightning can be lethal, killing an average of 69 people a year in the United States and 7 in Canada. Fortunately, our current understanding of lightning suggests some important safety rules.

First and foremost, in the presence of lightning, always take cover in a building, being careful not to make contact with any electrical appliances or telephones. And of course, don't watch the lightning storm from a pool or hot tub! Automobiles (other than convertibles) are relatively safe, but not because the rubber tires provide insulation against grounding (as believed by many people). The real reason is that if a car is hit, the electricity will flow around the car body rather than through the interior (or its occupants). The same fact explains why lightning seldom brings down airplanes, even though any particular commercial aircraft is hit on an average of once a year.

We often associate deadly lightning strikes with golfing during a thunderstorm and other foolish behaviors, but the danger is not always easily avoided. On January 1, 2000, for instance, a single lighting bolt killed a family of six near Mount Darwin, Zimbabwe, in a tragedy eerily similar to one that occurred a few months earlier in Zimbabwe, when a single strike killed six persons near the city of Gokwe. Lightning deaths are becoming ever more frequent throughout that region as forests are cleared, leaving villages more exposed in open areas. The problem is greatly exacerbated by the use of dry thatch as a roofing material. Soot from cooking fires impregnates the thatch with carbon, making the roof highly conductive and attractive for lightning.

While deforestation has been increasing the danger of lightning deaths in some parts of the world, in the United States a substantial decline (greater than 50 percent) has occurred in the number of lightning fatalities since the 1920s. This fact is even more impressive when one considers that since that time the country's population has increased from about 150 million to 270 million. The decline in fatalities is believed to be due in part to a decline in the population in rural areas, where the risk of being struck is greater. Better public education, improved warnings, advances in medical aid, and modernized electrical systems in homes and other buildings are also believed to have contributed to the decline.

Thunderstorms: Self-Extinguishing vs. Self-Propagating

Fortunately for us, the majority of lightning events are associated with localized, short-lived storms that dissipate within tens of minutes after forming. These storms, called **air mass thunderstorms,** actually extinguish themselves by creating downdrafts that cut off the supply of moisture into the precipitating clouds. For that reason, they do not normally produce severe weather. On other occasions, however, downdrafts from heavy precipitation actually intensify the storms that generate them. These storms are classified as **severe** and produce the greatest damage and loss of life.

Air Mass Thunderstorms

Air mass thunderstorms are the most common and least destructive of thunderstorms. They also have very limited life spans, usually lasting for less than an hour. Despite the name, which implies that these thunderstorms might occupy entire air masses (which are very large), air mass thunderstorms are very localized. But the term does make sense when you consider that air mass thunderstorms occur within individual air masses and are well removed from frontal boundaries. Think of it this way: air mass thunderstorms are contained *within* uniform air masses, but they do not occupy the *entire* air mass.

Our current understanding of air mass thunderstorms is based on the Thunderstorm Project, which examined such events in Ohio and Florida during the late 1940s. An air mass thunderstorm normally consists of a number of individual cells, each undergoing a sequence of three distinct stages—cumulus, mature, and dissipative (Figure 11–7).

Cumulus Stage The first stage of an air mass thunderstorm begins when unstable air begins to rise, often by the localized convection that occurs as some surfaces undergo more rapid heating than others. Because air mass thunderstorms frequently occur at night when the air is cooling, we know that other lifting processes can also trigger uplift. Regardless of which process causes uplift, the rising air cools adiabatically to form fair weather cumulus clouds. These initial clouds may exist for just a matter of minutes before evaporating. Although they do not directly lead to any precipitation, the initial clouds play an important role in thunderstorm development by moving water vapor from the surface to the middle troposphere. Ultimately the atmosphere becomes humid enough that newly formed clouds do not evaporate but instead undergo considerable vertical growth. This growth represents the **cumulus stage** in the air mass thunderstorm.

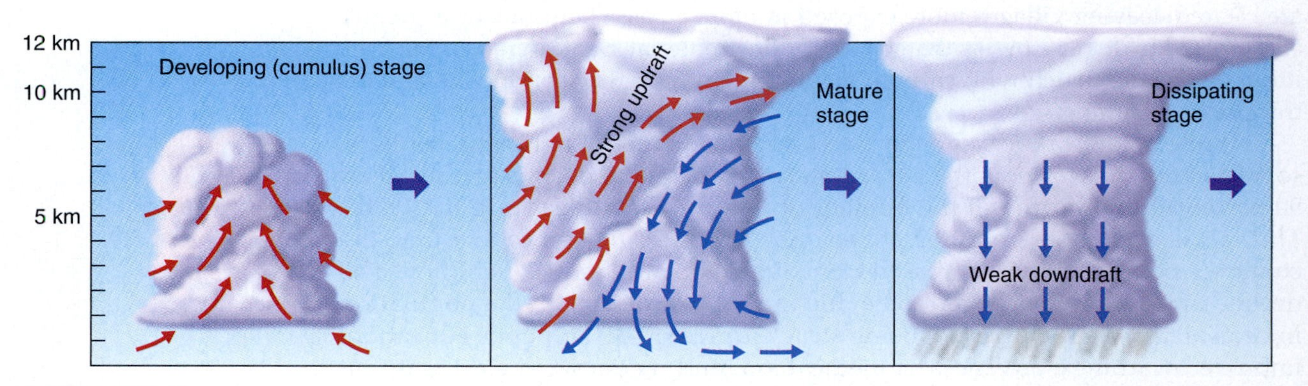

▲ **Figure 11–7**
The cumulus (a), mature (b), and dissipative (c) stages of an air mass thunderstorm.

Clouds in the cumulus stage grow upward at 5 to 20 m/sec (10 to 45 mph). Within the growing clouds, the temperature decreases with height at roughly the saturated adiabatic lapse rate, and a portion of the cloud extends above the freezing level. Ice crystals begin to form and grow by the Bergeron process. The sky rapidly darkens under the thickening cloud; when precipitation begins to fall, the storm enters its next stage of development.

Mature Stage The **mature stage** of the air mass thunderstorm begins when precipitation—as heavy rain or possibly graupel—starts to fall. As the falling rain or graupel drags air toward the surface, downdrafts form in the areas of most intense precipitation. You can observe this process in your own yard. Simply turn on a garden hose full blast and put your hand just outside the stream of water; you will notice a breeze in the direction in which the hose is pointed. The downdrafts are strengthened by the cooling of the air—by as much as 10 °C (18 °F)—that occurs as the precipitation evaporates.

The mature stage marks the most vigorous episode of the thunderstorm, when precipitation, lightning, and thunder are most intense. Air mass thunderstorms usually consist of multiple cells, located in different parts of the cloud and formed at different times. The top of the cloud extends to an altitude where stable conditions suppress further uplift. Strong winds at the top of the cloud push ice crystals forward and create the familiar anvil shape extending outward from the main part of the cloud.

During the cumulus and mature phases of the storm, an abrupt transition exists between the edge of the cloud and the surrounding unsaturated air. Updrafts dominate the interior of the cloud, while downdrafts occur just outside it. This sets up a highly turbulent situation that encourages entrainment (Chapter 6). The entrainment of unsaturated air causes the droplets along the cloud margin to shrink and cool the cloud by evaporation. The outer part of the cloud becomes more dense and less buoyant, thus suppressing further uplift.

Dissipative Stage As more and more of the cloud yields heavy precipitation, downdrafts occupy an increasing portion of the cloud base. When they occupy the entire base, the supply of additional water vapor is cut off and the storm enters its **dissipative stage**. Precipitation diminishes and the sky begins to clear as the remaining droplets evaporate. Only a small portion—perhaps 20 percent—of the moisture that condenses within an air mass thunderstorm actually falls as precipitation. The greatest amount simply evaporates from the cloud.

Cloud and 3-D Flow Tracers

Cloud and Vertical Velocity

Vertical Motion Cross-Section

Cloud and Horizontal Flow

Figure 11–8 shows an air mass thunderstorm. As is typical for thunderstorms in the mature phase, each tower consists of an individual cell and is in a different part of its life cycle. Notice in particular that some of the storm cloud appears washed out and less well defined than the rest. Such areas consist entirely of ice crystals, with no liquid droplets, and are said to be *glaciated*. They are not necessarily colder than other parts of the cloud; they are merely old enough so that all the supercooled droplets have had a chance to freeze.

Severe Thunderstorms

By definition, severe thunderstorms are those whose wind speeds exceed 93 km/hr (58 mph)*, have hailstones larger than 1.9 cm (0.75 in.) in diameter, or spawn tornadoes. Unlike those associated with air mass thunderstorms, the downdrafts and updrafts in severe storms reinforce one another and thereby intensify the storm. This reinforcement usually requires suitable conditions over an area from 10 to 1000 km across, a size referred to as *mesoscale*. In other words, most severe thunderstorms get a boost from a mesoscale atmospheric pattern that allows the wind, temperature, and moisture fields to "cooperate" and thereby create very strong storms.

*This seemingly odd value was originally designated as 50 nautical miles per hour, or knots.

▶ Figure 11–8
An air mass thunderstorm. The part of the cloud that has a washed-out appearance has become glaciated.

Because the conditions favoring strong thunderstorms exist over a rather large area, they typically appear in groups, with several individual storms clustered together. Such clusters of thunderstorms are generally referred to as **mesoscale convective systems (MCSs)**. In some cases, MCSs occur as linear bands called **squall lines**. At other times, they appear as oval or roughly circular clusters called **mesoscale convective complexes (MCCs)**. Regardless of the way in which they are arranged, the individual storm cells of an MCS form as part of a single system. In other words, MCSs are more than just a grouping of individual storms that happen to be near each other; instead, the storm cells develop from a common origin or exist in a situation in which some cells directly lead to the formation of others.

MCSs can bring intense weather conditions to areas covering several counties. They often have life spans of up to 12 hours, but in some cases they can exist for as long as several days. They are fairly common in North America; and in some parts of the central United States and Canada, they account for as much as 60 percent of the annual rainfall. Because the surrounding circulation supports an MCS, they lead to much stronger winds and heavier precipitation than is normally found in an air mass thunderstorm. When one or more of the defining characteristics is present, a storm within an MCS is classed as severe. Severe thunderstorms can also arise from **supercells**, intensely powerful storms that contain a single updraft zone. Though supercells often appear in isolation, they can also occur as a part of an MCS.

Certain conditions are necessary for the development of all severe thunderstorms. Among these are wind shear, high water vapor content in the lower troposphere, some mechanism to trigger uplift, and a situation called *potential instability*, described in *Box 11–3, Forecasting: Potential Instability*.

We will now briefly describe squall lines, MCCs, and supercells. Keep in mind that the descriptions of these storm types are quite general, and individual systems might not be easily categorized. It is also common for storm systems to evolve from one type of system to another.

11–3 Forecasting

Potential Instability

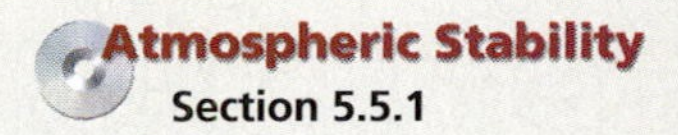

In Chapter 6, we discussed the concept of static stability, a function of the temperature lapse rate. Another type of instability that influences vertical air motions, called **potential instability**, arises when a layer of dry air rests above one that is warm and humid. If the air is potentially unstable, lifting of an entire layer of air can cause its temperature lapse rate to increase, thus making it statically unstable.

Consider the inversion situation shown in Figure 1. Just below the base of the inversion (Parcel 1), the temperature (T) equals 26 °C and the dew point (T_d) is 22 °C. Just above the base of the inversion (Parcel 2), the temperature and dew point are 27 °C and 19 °C, respectively. Now consider what happens between Parcels 1 and 2 if some lifting process lifts the entire air mass. Both parcels are unsaturated, so they cool at the dry adiabatic lapse rate of 1 °C per 100 m, and their dew points decrease at 0.2 °C per 100 m. After 500 m of ascent, the temperature of both parcels has fallen by the same amount, so the temperature difference between them is unchanged. However, Parcel 1 is now saturated, so further lifting will cause its temperature to decline at the saturated adiabatic lapse rate (SALR). Meanwhile, Parcel 2 is still unsaturated, so further lifting leads it to cool at the dry adiabatic lapse rate (DALR).

Now let's lift the two parcels another 500 m. Assuming an SALR of 0.5 °C per 100 m, the lower parcel cools 2.5 °C to 18.5 °C, while the upper parcel cools at the DALR to 17 °C. We can now see how uplift of the layer has affected its stability. Initially, the upper parcel was warmer than the one below, which meant the layer containing the two parcels was statically stable. After lifting both parcels 1000 m, however, the upper parcel became 1.5 °C cooler than the one below. This yields a temperature lapse rate between the two parcels of 1.5 °C per 100 m, making the air statically unstable. Thus, the air that is statically stable has the potential to become statically unstable, given sufficient uplift—hence the term *potential instability.*

Both theory and experience show that potential instability is an important factor in the development of severe thunderstorms. During spring and summer, the southern Great Plains region often has warm, humid air near the surface advected from the Gulf of Mexico. In the middle troposphere above the region, westerly winds bring dry air from the southern Rockies. This air in the middle troposphere sinks somewhat after passing the Rockies to form a subsidence inversion, which inhibits the development of air mass thunderstorms. But given sufficient uplift, the surface layer of air can become statically unstable and severe thunderstorms can develop.

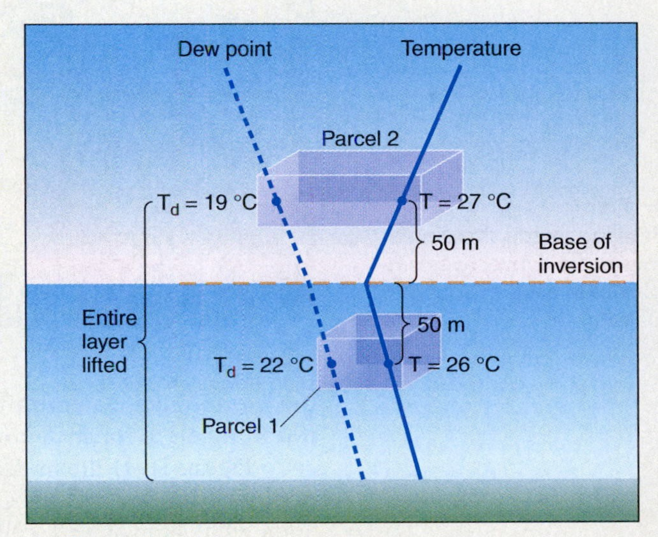

▲ **Figure 1**
Potential instability occurs when a layer of warm, dry air overlies moist air. The temperature (solid line) and dew point (dashed line) profiles show such a situation. Examine what happens to the parcels of air 50 m above and below the base of the inversion. Because the lower parcel is nearly saturated (i.e., the dew point is close to the air temperature), lifting causes it to become saturated after 500 m of uplift. The parcel above the inversion does not become saturated until 1000 m of uplift. Thus, after both have been lifted 1000 m, the upper parcel has undergone more cooling than the lower one, and the temperature lapse rate becomes steep enough to produce statically unstable air.

Mesoscale Convective Complexes In the United States and Canada, severe weather arises most often from mesoscale convective complexes (MCCs) (Figure 11–9). In the most general sense, they are defined as oval or roughly circular, organized systems containing several thunderstorms.*

Although not all MCCs create severe weather, they are self-propagating in that their individual cells often create downdrafts, leading to the formation of new, powerful cells nearby. To see how this occurs, imagine a large cluster of thunderstorms. At the surface a flow of warm, humid air comes from the south, and in the middle troposphere the wind flows from the southwest. This setting provides the wind shear necessary for a severe thunderstorm. As we have already seen, the precipitation from each thunderstorm cell creates its own downdraft, which is enhanced by the cooling of the air as the rain evaporates and consumes latent heat. Upon hitting

*Meteorologists have some precise criteria for classifying a system an MCC, based on its signature on satellite imagery. For our purposes, we will simply apply the term to organized systems of thunderstorms clustered in a pattern that is closer to circular than linear.

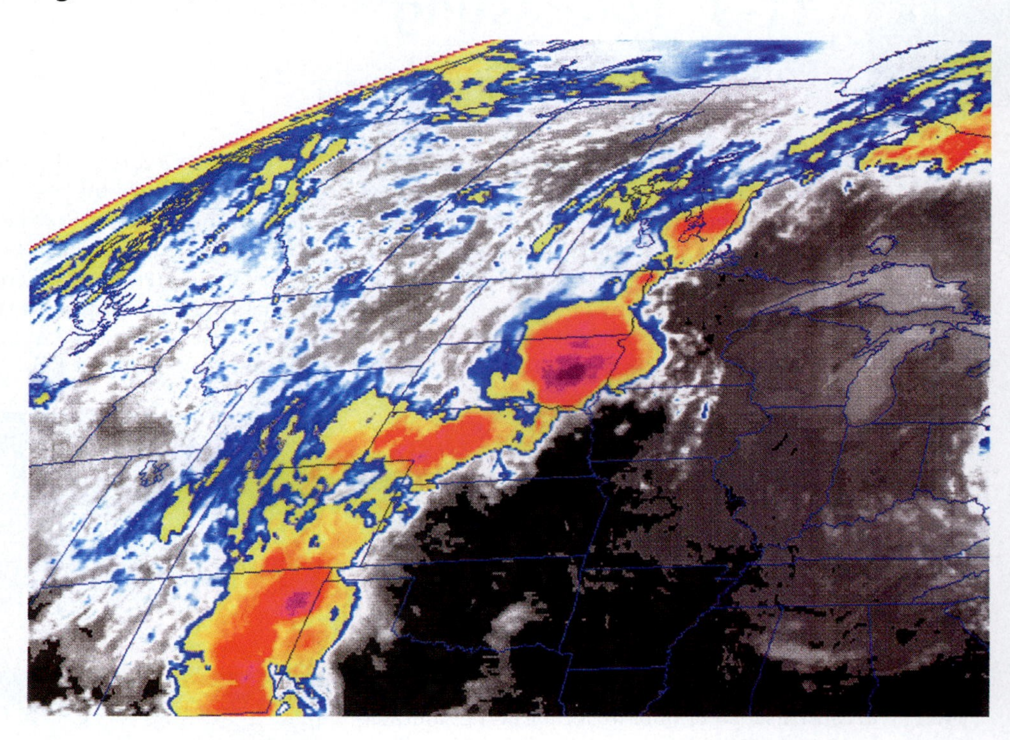

▶ **Figure 11–9**
Satellite image showing a mesoscale convective complex over eastern South Dakota.

the ground, the downdrafts spread outward and converge with the warmer surrounding air to form an **outflow boundary** (Figure 11–10).

Figure 11–11 illustrates the progressive movement of cells in an MCC. Initially, at time $t = 0$, five cells labeled *A, B, C, D,* and *E* are moving toward the northeast. Somewhat later, at $t = 1$, cells *B* through *E* have migrated to the northeast. Cell *A*, located farthest to the north, has died out, while along the southern margin of the complex a new cell, *F*, has formed. At $t = 2$, cell *B* has died out, to be replaced by cell *G* along the south. In this manner, the complex of thunderstorms moves eastward, though each individual cell moves to the northeast. Near the southern side of the MCC, the cold outflow collides with the large-scale southerly surface flow and lifts it upward. The warm, humid air is drawn into the southern edge of the MCC, where it forms new cells. At the same time, the older cells on the northern side of the MCC dissipate because they lack the updrafts needed for replenishment. Overall, the entire storm seems to move to the east even though each individual cell moves to the northeast.

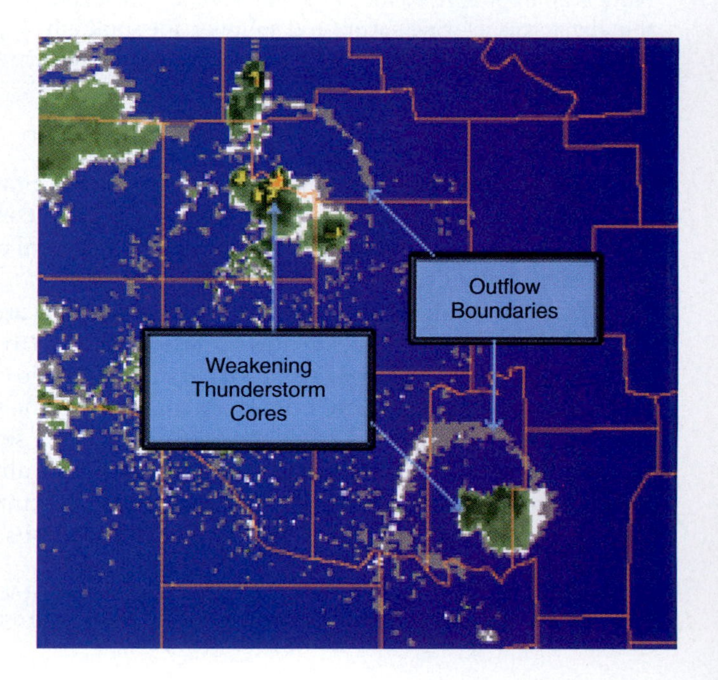

▶ **Figure 11–10**
A radar image highlighting two outflow boundaries.

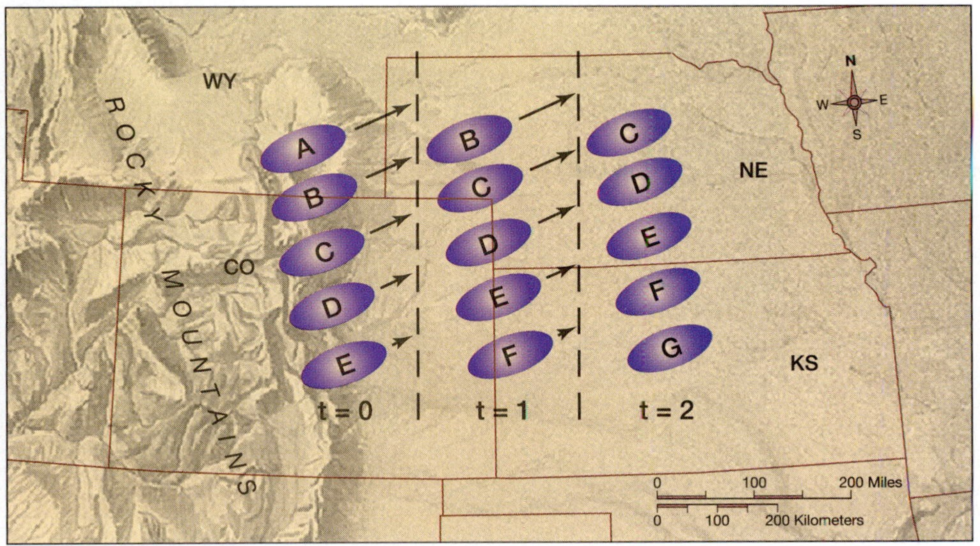

Figure 11–11 The movement of thunderstorm cells in a mesoscale convective complex. Initially (at $t = 0$) all the cells are moving toward the northeast. The cells in row *A* are the oldest, those in *E* the most recently formed. Later ($t = 1$), the cells in row *A* have dissipated, but a new row, *F*, has formed along the southern margin of the complex. At $t = 2$, row *B* has dissipated while a new row (*G*) has formed.

Squall Line Thunderstorms Squall line thunderstorms consist of a large number of individual storm cells arranged in a linear band, about 500 km (300 mi) in length (Figure 11–12). They tend to form parallel to and about 300 to 500 km (180 to 300 mi) ahead of cold fronts, especially over the southern United States during spring and summer. The average squall line has a life span on the order of 10 hours, though some have lasted up to four days.

Squall Line Tornado over Northwestern Louisiana

As is the case for all intense thunderstorms, strong vertical wind shear is an important component of squall line thunderstorms. As shown in Figure 11–13, wind velocities in the direction of storm movement typically increase with height. The strong winds aloft push the updrafts ahead of the downdrafts and allow the rising air to feed additional moisture into the storm. As the downdrafts reach the ground, they surge forward as a wedge of cold, dense air, called a **gust front**. Gust fronts act in much the same way as advancing cold fronts by displacing warm air upward (see Figure 1–7).

Supercell Storms Few weather systems are as awesome as a supercell storm (Figure 11–14). With diameters that range from about 20 to 50 km (12 to 30 mi), they are smaller than either squall lines or MCCs. On the other hand, they are usually more violent and provide the setting for the majority of very large tornadoes. Unlike MCCs and squall lines, a supercell storm consists of a single, extremely powerful cell rather than a number of individual cells.* They also undergo a large-scale rotation absent from squall lines and MCCs. The typical life span of a supercell is 2 to 4 hours.

*Supercells usually occur as isolated storms, though "squall lines" of supercells have been observed.

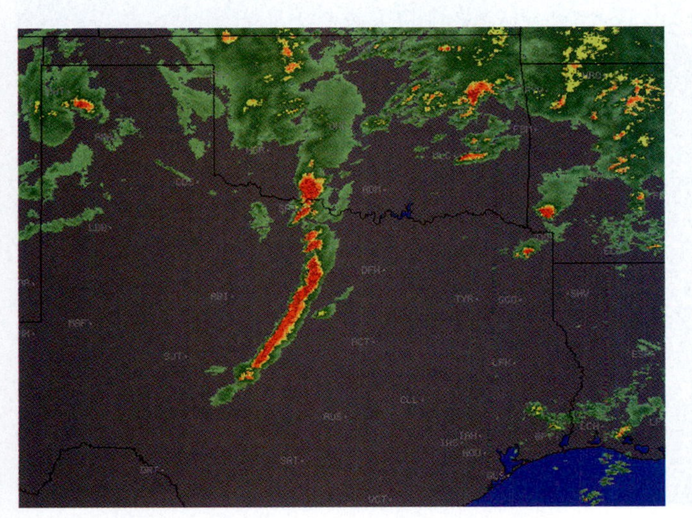

Figure 11–12 Radar image of a squall line.

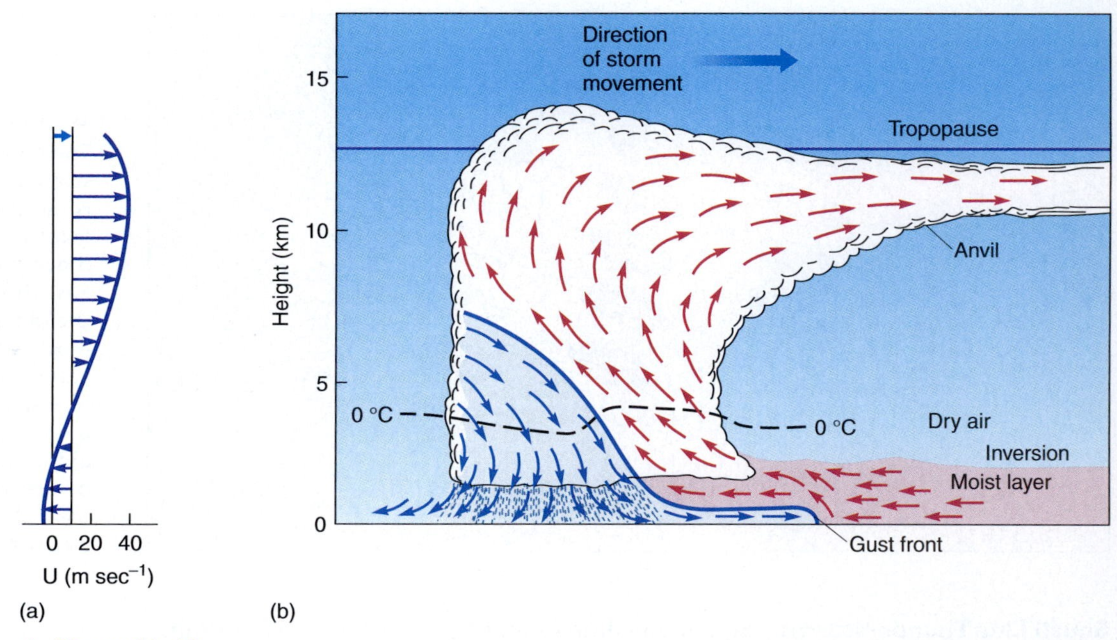

▲ **Figure 11–13**
Squall line thunderstorms require the presence of wind shear. The arrows in (a) represent the wind speeds with respect to the movement of the storm. The movement of the air within the cumulonimbus is shown in (b). The upper part of the cloud is pushed forward more rapidly than the lower part, which helps to draw in warm, moist air. Note the gust front near the ground ahead of the rain shaft.

Despite their single-cell structure, supercell storms are remarkably complex, with the updraft and downdraft bending and wrapping around each other due to strong wind shear (Figure 11–15). As in any other weather system that spawns severe weather, the downdrafts serve to amplify the adjacent updrafts.

Meteorologists follow supercells with tremendous interest. Fortunately, they have at their disposal an extremely useful tool in the form of weather radar. One particular type, called **Doppler radar**, is described in *Box 11–4, Forecasting: Doppler Radar*. Radar can reveal one of the most noteworthy features of a supercell, called a **hook**, which looks like a small appendage attached to the main body of the storm on the radar image (Figure 11–16).* Hooks are significant because their appearance usually means tornado formation is imminent.

*A hook is sometimes called a *hook echo* in reference to the way radar waves reflect off it, as sound waves echo off canyon walls.

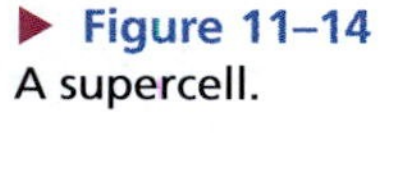

▶ **Figure 11–14**
A supercell.

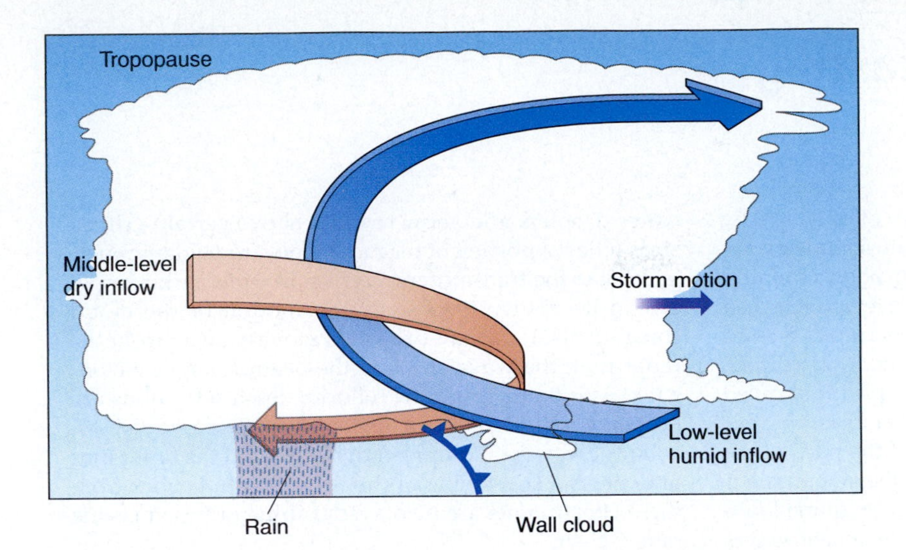

Figure 11–15
The internal structure of a supercell.

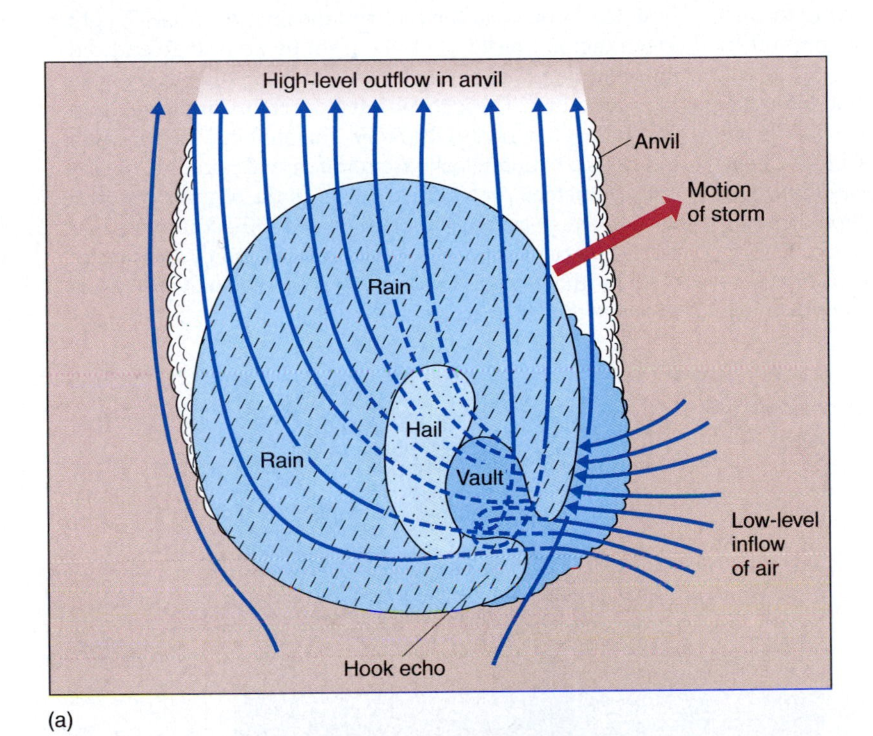

(a)

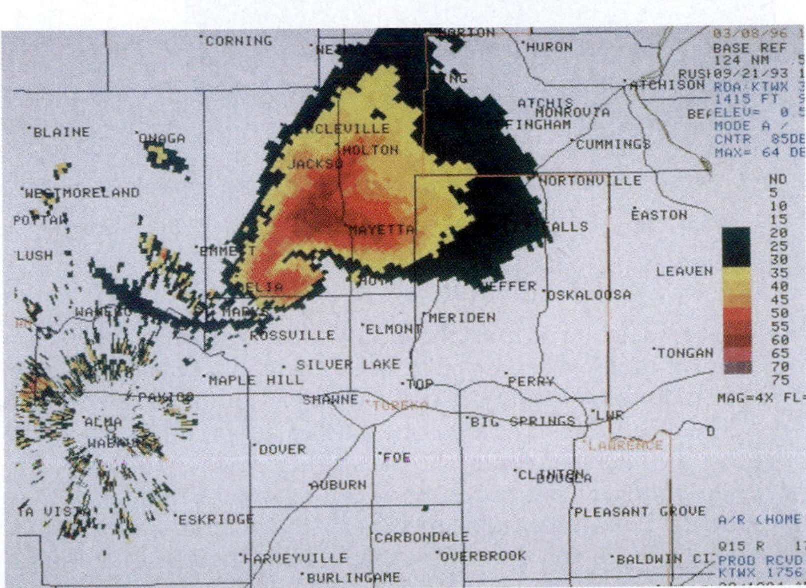

(b)

Figure 11–16
The typical organization of a supercell as seen on a radar image (a). The zone of no radar return in the southeast is the vault, which is flanked by a hook echo to the south. An actual radar image showing a hook is seen in (b).

11–4 Forecasting

Doppler Radar

Just as we are able to distinguish different colors of light by their wavelengths, so can we differentiate sounds by the length of their sound waves. If an object making a sound is moving away from a listener, the sound waves are stretched out and assume a lower pitch. Sound waves are compressed when an object moves toward the listener, making them higher pitched. Unconsciously, we use this principle, called the **Doppler effect**, to determine whether an ambulance siren is coming closer or moving away. If the pitch of the siren seems to become higher, we know the ambulance is getting nearer (of course, the siren would also sound louder). A similar process occurs when electromagnetic waves are reflected by a moving object: the light shifts to shorter wavelengths when reflected by an object moving toward the receiver, and to longer wavelengths as it bounces off an object moving away from the receiver.

Doppler radar is a type of radar system that takes advantage of this principle. It allows the user to observe the movement of raindrops and ice particles (and thus determine the wind speed and direction) from the shift in wavelength of the radar waves. Like any other type of radar, Doppler radar has a transmitter that emits pulses of electromagnetic energy with wavelengths on the order of several centimeters. Depending on the wavelength used, water droplets and snow crystals above certain critical sizes reflect a portion of the radar's electromagnetic energy back to the transmitter/receiver. Doppler radar is special in its ability to observe the motion of the cloud constituents. If a cloud droplet is moving away from the radar unit, the wavelength of the beam is slightly elongated as it bounces off the reflector. Such reflections are normally indicated on the display monitor as reddish-to-yellow. Likewise, a droplet moving toward the radar unit undergoes a shortening of the wavelength. Echoes from these constituents are displayed as blue or green on the radar screen.

Figure 1 shows a Doppler radar display during a severe thunderstorm. Near the center of the image is a small, light green region bordered on the right by an orange and red area. The warm colors represent winds blowing away from the transmitter. Those to the left (cool colors) indicate rapid winds directed toward the radar. This shift in wind direction indicates a counterclockwise rotation with a radius of several kilometers. As we discuss later in the chapter, this pattern, called a *mesocyclone*, often precedes the formation of a tornado. After the onset of rotation, it takes only 30 minutes or so for the tornado to form, which allows meteorologists to give warning in advance.

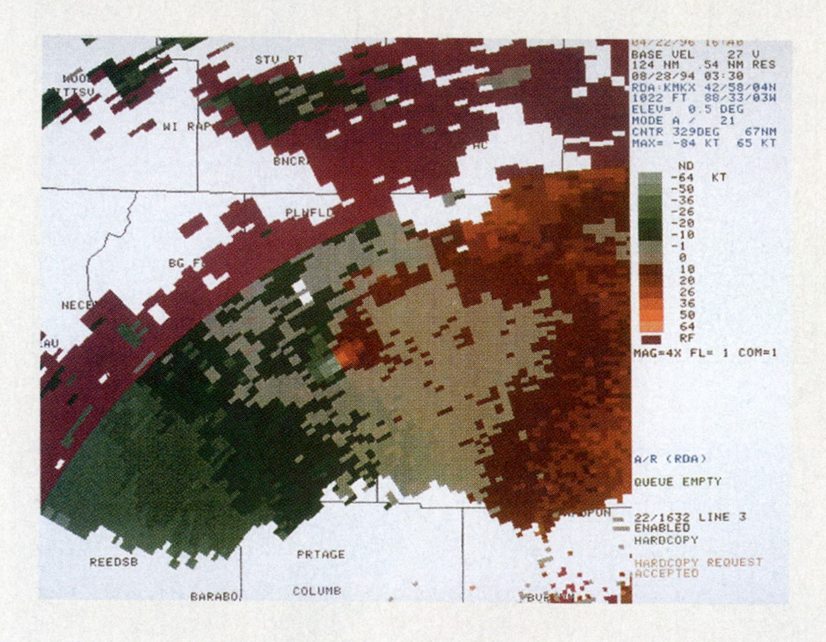

▶ **Figure 1**
A Doppler radar image.

When we view it by satellite or from the ground, the supercell appears as a large, continuous mass of cloud. When it is displayed on a radar screen, however, we see that a large portion of the storm—usually along its southeast quadrant—seems to be missing (see Figure 11–16). This zone, known as a **vault**, is where the inflow of warm surface air enters the supercell. The air entering the vault rises, and water vapor condenses to form a dense concentration of water droplets. But the newly formed droplets in the vault are too small to effectively reflect radar waves. Thus, this zone does not show up on the radar image despite the dense concentration of water droplets.

As recently as the 1980s, the U.S. National Weather Service used a network of radar units made in the 1950s. The equipment was so antiquated that it relied on vacuum tubes rather than transistors! During the early 1990s, the Weather Service began replacing the old system of conventional radar with a modern network called NEXRAD (for NEXt Generation Weather RADar). The first unit, installed at Norman, Oklahoma, became operational in early 1991 and on its very first day in service tracked a tornado that destroyed two houses. Fortunately, the radar allowed forecasters to issue a warning that may have contributed to no one being killed or injured.

Today about 160 Doppler sites are scattered across the United States (Figure 2). The National Weather Service operates 113 of these sites; the rest are owned by the Federal Aviation Administration and the Department of Defense. In part because of budgetary cutbacks, the Atmospheric Environment Service of Canada has just a handful of Doppler radar installations. Because both sides of the border area tend to be heavily populated, Doppler radar from the United States provides extensive coverage of severe storms that could affect many large Canadian urban centers.

NEXRAD is also useful for flood forecasting, providing continual precipitation estimates over large areas. Doppler radar can sometimes observe wind movements even when no clouds exist, as large clusters of flying bugs or heavy dust concentration scatter radar waves back toward the transmitter. The resultant echoes are called *clear air echoes.*

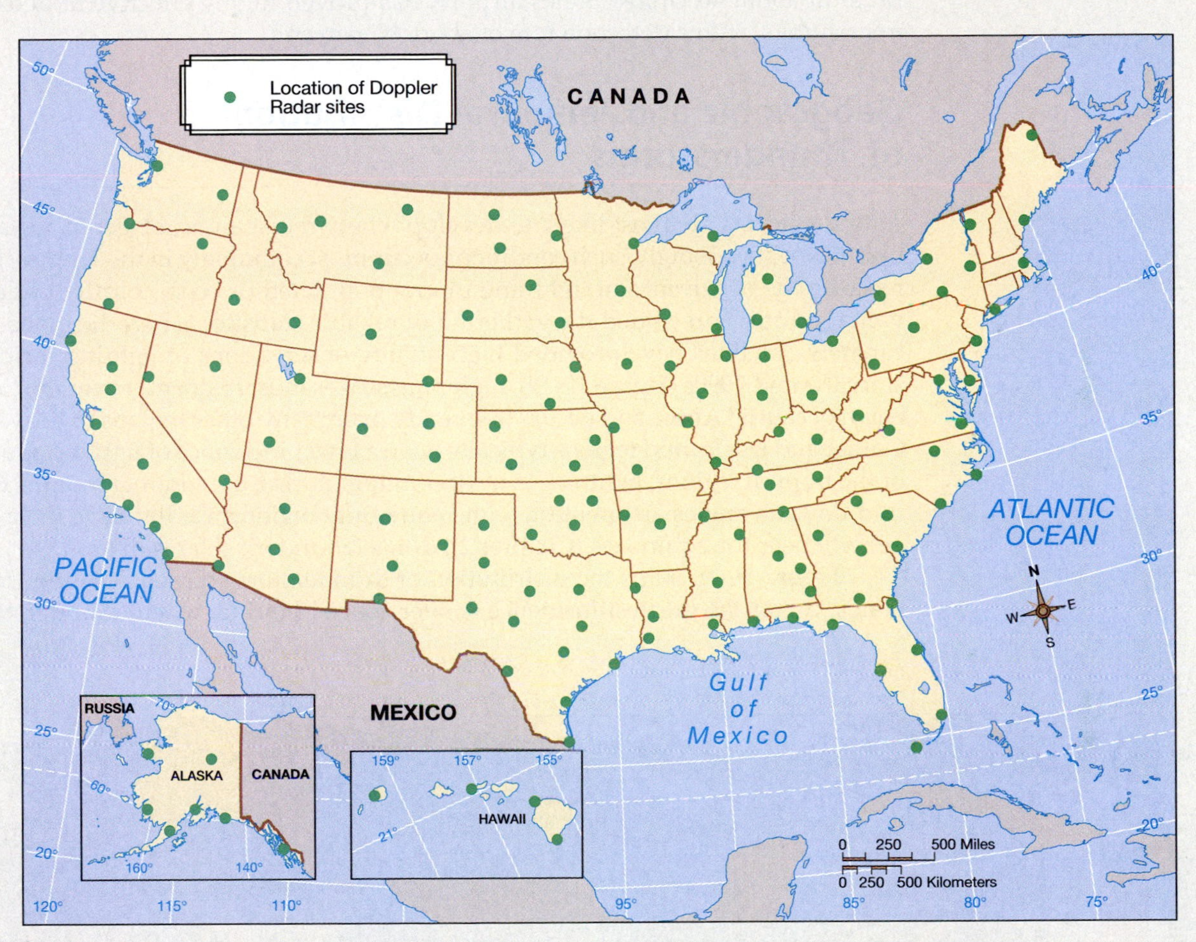

▲ **Figure 2**
Doppler radar sites in the United States.

Downbursts and Microbursts

We have seen how downdrafts are an important feature of thunderstorms, especially in the maintenance of severe thunderstorms. Strong downdrafts may also create **downbursts**, potentially deadly gusts of wind that can reach speeds in excess of 270 km/hr (165 mph). When strong downdrafts reach the surface, they can spread outward in all directions to form intense horizontal winds capable of causing severe damage at the surface. In fact, damage sometimes mistakenly attributed to tornadoes may in some cases be the result of downbursts.

Downbursts with diameters of less than 4 km (2.5 mi), called **microbursts**, can produce a particularly dangerous problem when they occur near airports. The horizontal spreading of a microburst creates strong wind shear when it reaches the surface. For example, air may flow westward on one side of the microburst while spreading eastward on the opposite side. Imagine what this might do to an aircraft attempting to land in a microburst. As the plane enters the microburst, a headwind provides lift, to which the pilot might respond by turning the aircraft downward. As soon as the plane passes the core of the downdraft, however, the headwind not only disappears, it is replaced by a tailwind, decreasing lift. Coming after the pilot's earlier downward adjustment, this causes the plane to abruptly drop in altitude. Because the plane is not far above the ground when these events occur, the pilot may not have time to compensate before a deadly crash occurs. Fortunately, such disasters are rare. They are also becoming less likely because the installation of Doppler radar at about 40 United States airports has proven highly effective at detecting microbursts, with a detection rate of about 95 percent.

Geographic and Temporal Distribution of Thunderstorms

Thunderstorms are most likely to develop where moist air is subject to sustained uplift. Not surprisingly, such conditions occur most commonly in the Tropics. Until recently the occurrence of lightning in low-populated or economically underadvantaged areas precluded the gathering of reliable statistics across the globe. Fortunately, satellites have provided high-quality observations of lighting incidence since the mid-1990s (Figure 11–17). Lightning strikes most frequently over the Congo basin of central Africa and occurs frequently over many other regions of the world. Colder and less humid regions typically have a lower incidence of lighting. Outside of the tropical regions, lightning is most common during the summer months when mid-day sun angles are greatest, while equatorial regions usually have their peak activity near the equinoxes (Chapter 2), due to abundant solar radiation.

Figure 11–18 maps the distribution of thunderstorm days across the United States. By far the greatest incidence is over central Florida, where thunderstorms

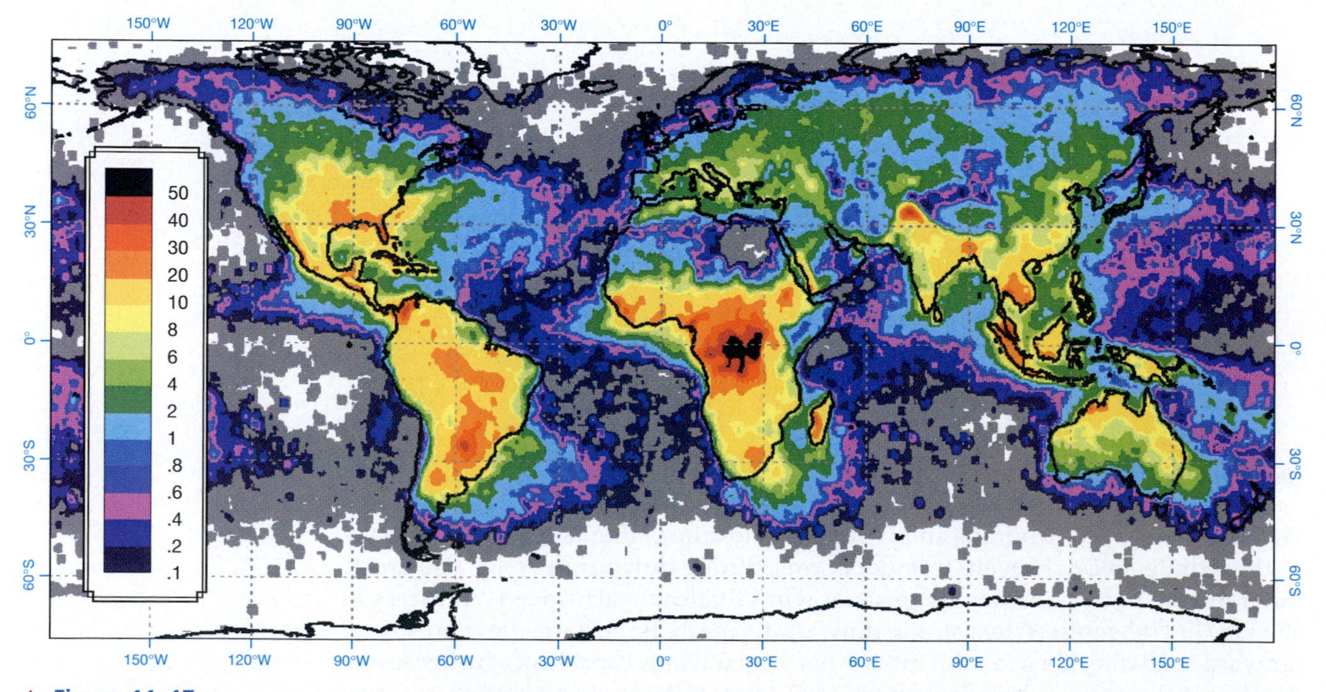

▲ **Figure 11–17**
Data from space-based optical sensors reveal the uneven distribution of worldwide lightning strikes. Units: flashes/km^2/yr.

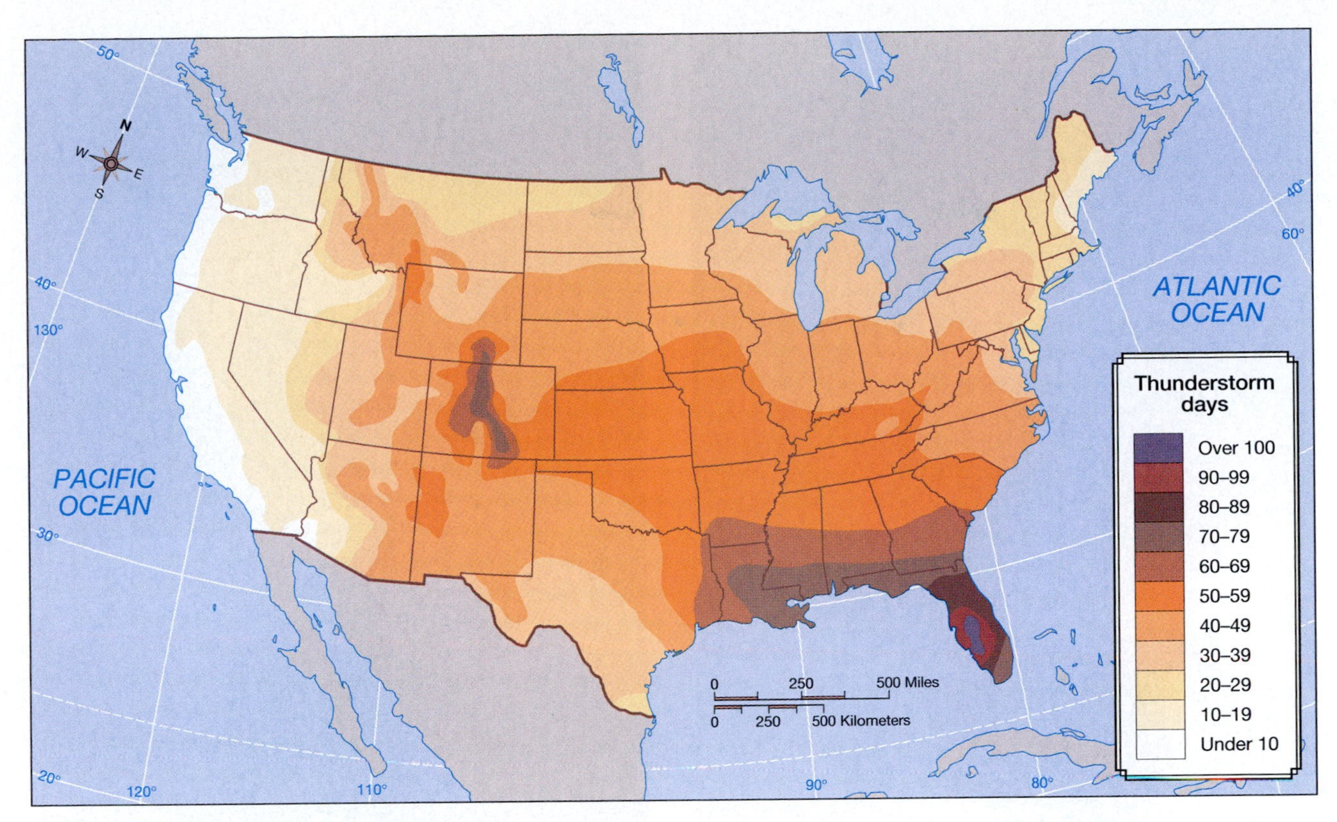

▲ **Figure 11–18**
The mean distribution of thunderstorms across the 48 conterminous United States.

occur on average more than 100 days per year. Much of the necessary lifting of air results from strong solar heating of the surface. But the situation over Florida is unique within the continental United States, because the state is almost completely surrounded by warm water. Thus, air that flows into the interior to replace lifted air has a very high moisture content, which in turn supports heavy precipitation and the development of thunderstorms.

Lightning Across the Globe

Tornadoes

The large hail and strong winds of a severe thunderstorm can bring widespread destruction, but even hailstorms are relatively tame compared to **tornadoes** (Figure 11–19). Tornadoes are zones of extremely rapid, rotating winds beneath the base of cumulonimbus clouds. Though the overwhelming majority of tornadoes rotate cyclonically (counterclockwise in the Northern Hemisphere), a few spin in the opposite direction. Some appear as very thin, rope-shaped columns, while others have the characteristic funnel shape that narrows from the cloud base to the ground. Regardless of their shape or spin, tornadoes are extremely dangerous.

Strong tornadic winds result from extraordinarily large differences in atmospheric pressure over short distances. Over just a few tenths of a kilometer, the pressure difference between the core of a tornado and the area immediately outside the funnel can be as great as 100 mb. To put this in perspective, on a typical day the highest and lowest sea level pressure across all of North America may differ by only about 35 mb—and this difference exists over horizontal distances of up to thousands of kilometers.

Tornado Characteristics and Dimensions

Tornado

It is difficult to generalize about tornadoes because they occur in a wide variety of shapes and sizes. While the majority have diameters about the length of a football field (100 yards or so), some are 15 times as large. Usually they last no longer than

(a) (b)

▲ **Figure 11–19**
Tornadoes come in a wide range of shapes and sizes.

a few minutes, but some have lasted for several hours. Tornadoes normally move across the surface at speeds comparable to a car driving down a city street—about 50 km/hr (30 mph). A typical tornado covers about 3 or 4 km (2 to 2.5 mi) from the time it touches the ground to when it dies out.

Estimates of wind speeds within tornadoes are based primarily on the damage they have produced. The weakest have wind speeds as low as 65 km/hr (40 mph); the most severe are in excess of about 450 km/hr (280 mph).

Tornado Formation

Tornadoes can develop in any situation that produces severe weather—frontal boundaries, squall lines, mesoscale convective complexes, supercells, and tropical cyclones (Chapter 12). The processes that lead to their formation are not very well understood. Typically, the most intense and destructive tornadoes are those that arise from supercells.

Supercell Tornado Development In a supercell storm, the first observable step in tornado formation is the slow, horizontal rotation of a large segment of the cloud (up to 10 km—6 mi—in diameter). Such rotation begins deep within the cloud interior, several kilometers above the surface. The resulting large vortices, called **mesocyclones,** often precede the formation of the actual tornado by some 30 minutes or so.

The formation of a mesocyclone depends on the presence of vertical wind shear. Moving upward from the surface, the wind direction in the storm shifts from a southerly to westerly direction as the wind speed increases. This wind shear causes a rolling motion about a horizontal axis, as shown in Figure 11–20a. Under the right conditions, strong updrafts in the storm tilt the horizontally rotating air so that the axis of rotation becomes approximately vertical (Figure 11–20b). This provides the initial rotation within the cloud interior.

Intensification of the mesocyclone requires that the area of rotation decrease, which leads to an increase in wind speed.* The narrowing column of rotating air stretches downward, and a portion of the cloud base protrudes downward to form a **wall cloud** (Figure 11–21). Wall clouds form where cool, humid air from zones of precipitation is drawn into the updraft feeding the main cloud. The cool, humid air condenses at a lower height than does the air feeding into the rest of the cloud. Wall clouds most often occur on the southern or southwestern portions of supercells, near areas of large hail and heavy rainfall (Figure 11–22). **Funnel clouds** form when a narrow, rapidly rotating vortex emerges from the base of the wall cloud. A funnel cloud has all the characteristics and intensity of a true tornado; the only difference between the two is that a funnel cloud has yet to touch the ground.

*This is another application of the conservation of angular momentum, initially described in Chapter 8.

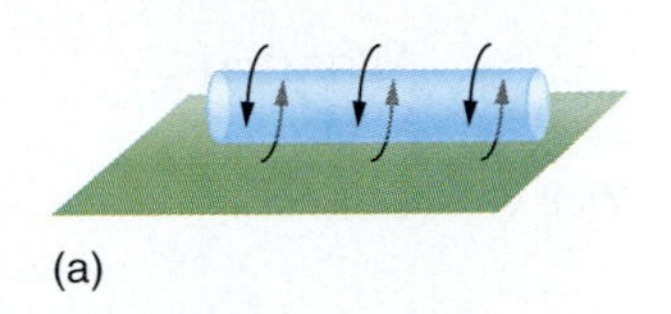

(a)

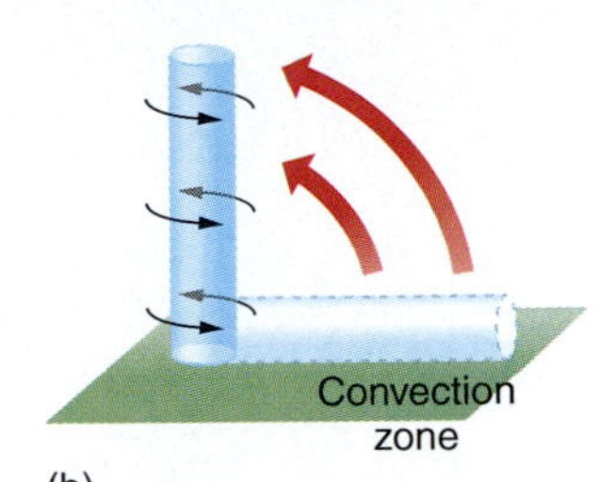

(b)

▲ **Figure 11–20**
Mesocyclones can form when a horizontal vortex of air (a) becomes tilted upward (b).

◄ **Figure 11–21**
A wall cloud protrudes below the main body of a supercell.

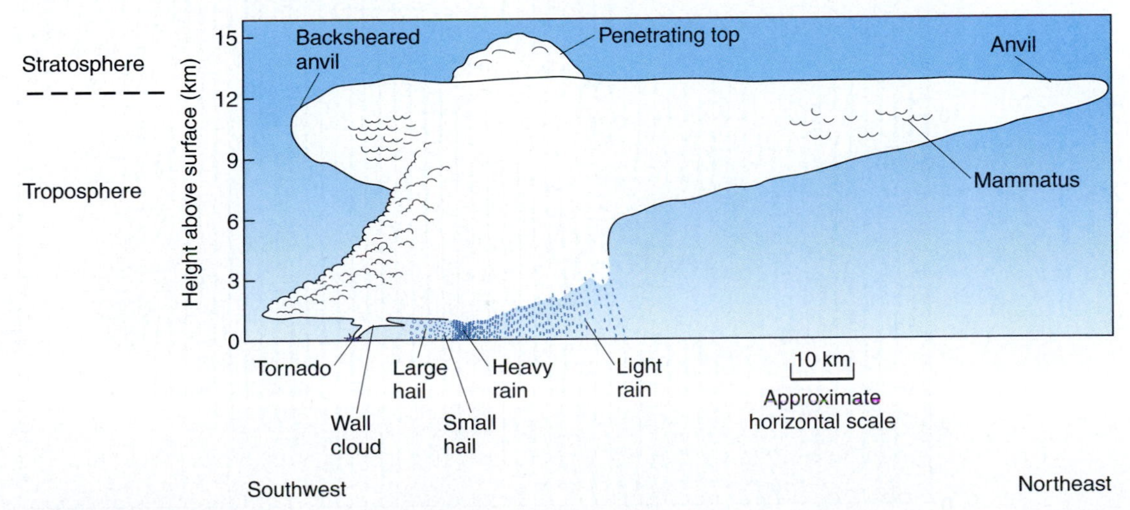

▲ **Figure 11–22**
An idealized supercell, showing a wall cloud on the southwestern portion.

Wall Clouds I and II

Because Doppler radar enables forecasters to observe the rotating winds of mesocyclones, the new network of these radars should greatly increase lead times in the issuance of tornado warnings (refer again to *Box 11–4, Forecasting: Doppler Radar*). Only about 50 percent of all mesocyclones actually spawn tornadoes, however. Exactly why some mesocyclones produce tornadoes and others do not is unknown, and, for that reason, forecasters cannot tell in advance which mesocyclones will produce tornadoes. Despite these uncertainties, Doppler radar has already proven itself to be the most important tool yet developed in "nowcasting" (making short-term predictions about) tornadoes.

Nonsupercell Tornado Development The exact mechanisms that lead to nonsupercell tornadoes are also poorly understood, though recent research indicates that these tornadoes have their origins nearer to the surface than do those that begin as mesocyclones. Figure 11–23 illustrates one situation that may lead to nonsupercell tornadoes. The arrows show the outflow of air from two thunderstorm regions: at the top left and bottom right of the figure. From the bottom left to the top right of the figure is a zone of convergence between the two masses of air. At certain areas along the convergence zone (the circled areas), strong rotation develops. Another possible mechanism is shown in Figure 11–24, where strong convection along the convergence zone causes uplift and the formation of a cumulus cloud. In Figure 11–24c, the cloud develops into a cumulonimbus, and the strong rotation stretches down from the cloud base to form a funnel cloud.

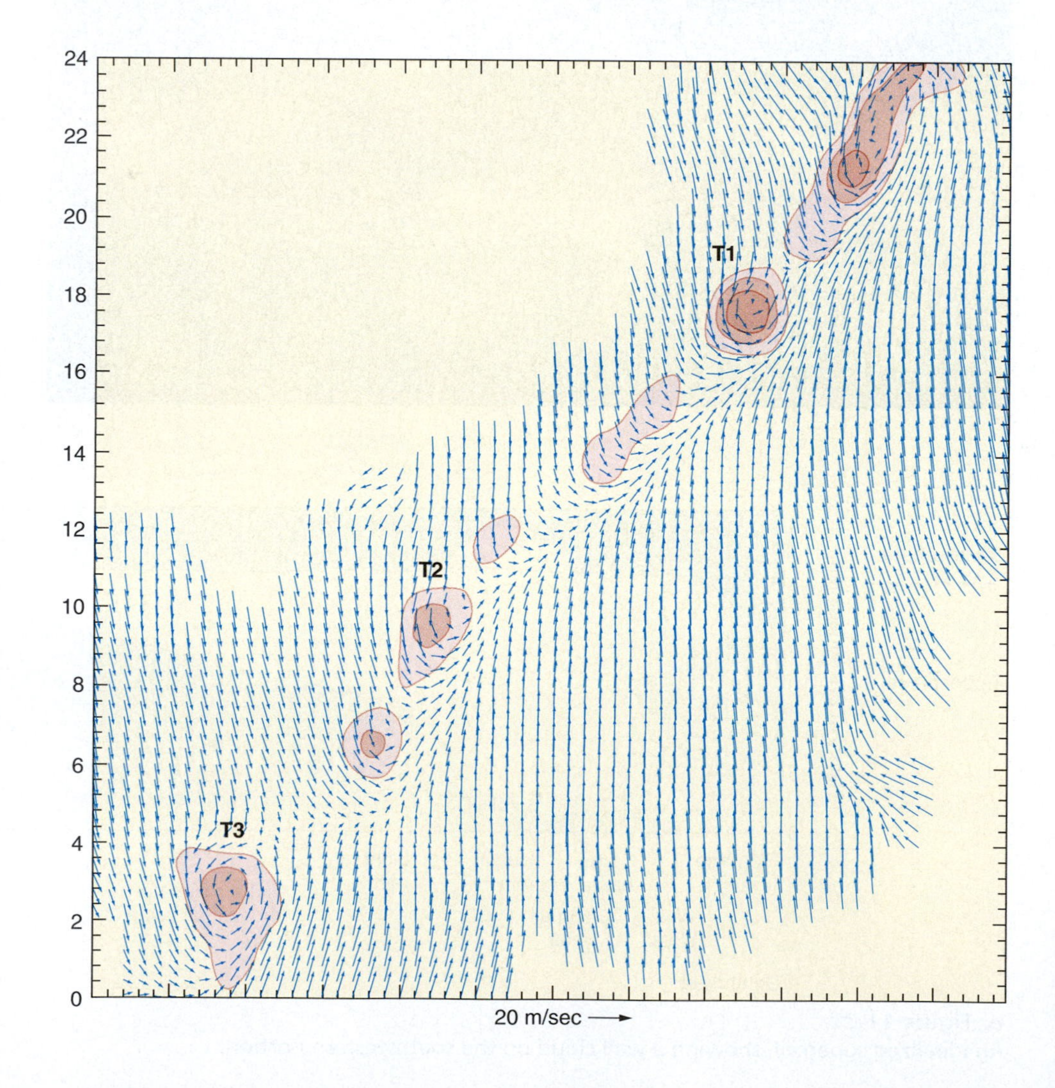

▶ **Figure 11–23**
Some nonsupercell tornadoes appear to develop where the outflow from separate storm downdrafts causes convergence. The arrows depict the wind flow over a region of about 400 sq km. The circled areas, organized from the bottom left to upper right, represent zones of rapid rotation where tornadoes appeared.

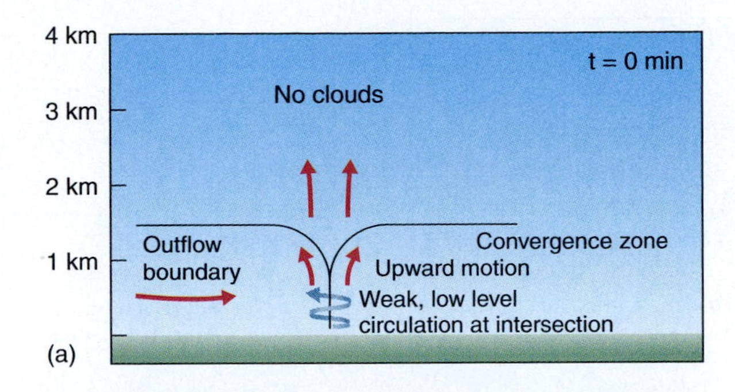

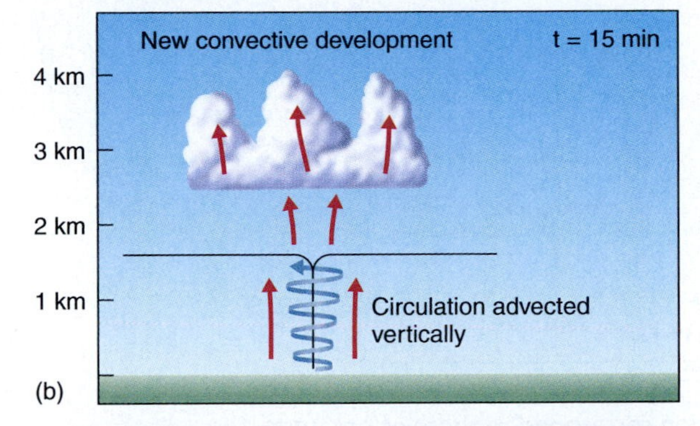

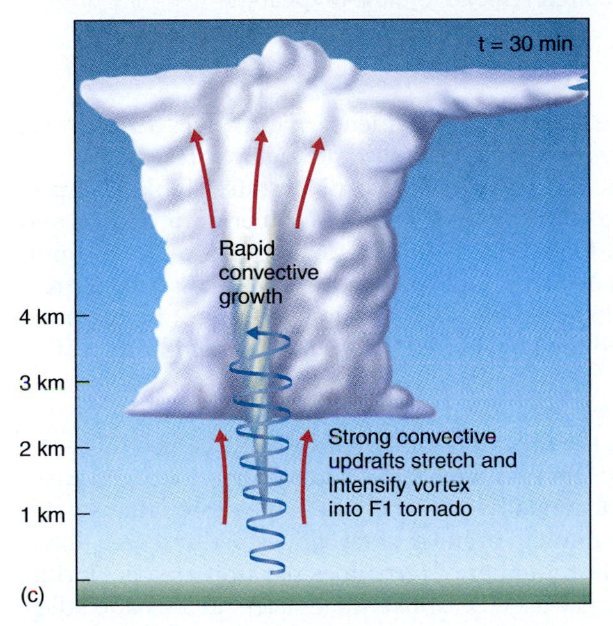

◀ **Figure 11–24**
The evolution of a tornado along a convergent boundary. Spinning motions along the boundary (a) can be carried upward if there is sufficient convection (b). Once the cumulonimbus develops (c), the downward movement of the strong rotation can lead to tornadoes.

The Location and Timing of Tornadoes

Tornadoes are a very American phenomenon; no other country in the world has nearly as many as does the United States (Figure 11–25). Several factors combine to make North America a haven for tornadoes. The continent covers a wide range of latitudes; its southeastern portion borders the warm Gulf of Mexico, while the northernmost portion extends into the Arctic. Furthermore, much of the eastern portion of the continent is relatively flat and, in particular, no major mountain range extends in an east–west direction. Together, these features allow for a collision of northward-moving maritime tropical air from the Gulf of Mexico with southward-moving continental polar air along the polar front. This setting, coupled with the frequent presence of potential instability (refer again to *Box 11–3, Forecasting: Potential Instability*), provides a favorable situation for tornado development. The frequent occurrence of

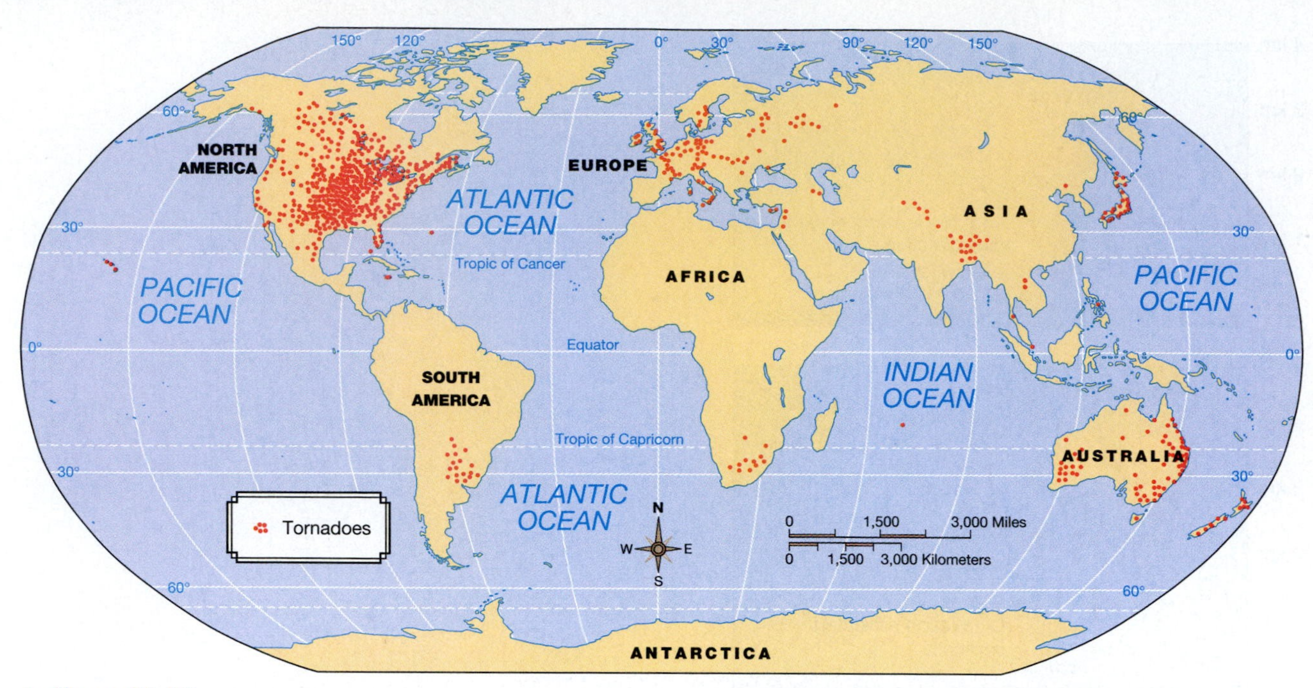

▲ **Figure 11–25**
Tornadoes around the globe. The areas of greatest dot concentration correspond to those of greatest tornado frequency.

drylines (Chapter 9) also contributes to the high incidence of tornadoes across much of the southern Great Plains states (such as Oklahoma and much of Texas).

Tornadoes occur at least occasionally in almost all 50 states. Table 11–1 presents the rank-ordering by state of the average annual number of tornadoes and tornado fatalities for the period 1950–1994. Figure 11–26 maps the average annual number of tornadoes per 10,000 sq km (26,000 sq mi) for each state. A great many tornadoes touch down along a wide strip running southwest to northeast between the southern Plains and the lower Great Lakes region, commonly called *Tornado Alley.*

Ever since Dorothy was swept by a tornado into the Land of Oz, many people have believed that Kansas leads the nation in tornado incidence. In fact, that state ranks only third. Texas easily leads the rest of the states in total number of tornadoes, with considerably more than 100 annually. But when the large size of that state is taken into account, it ranks only ninth in number of tornadoes per unit area. Interestingly, the state with the greatest number of tornadoes per unit area is Florida, which lies altogether outside Tornado Alley. Unlike the twisters in Tornado Alley, many of Florida's tornadoes are embedded in passing hurricanes and tropical storms, and many occur offshore as relatively weak waterspouts (described later in this chapter). Oklahoma is a close second to Florida in concentration of tornadoes.

Tornadoes can occur at any time of the year. But, as shown in Figure 11–27, a strong concentration of tornadoes occurs during the spring, when air mass contrasts are especially strong. May has the greatest number of tornadoes, with June a close second. The peak tornado season is not uniform throughout the country (Figure 11–28), because spring does not arrive everywhere at once. Louisiana and Arkansas have the greatest number of tornadoes in April. Illinois, Wisconsin, and Nebraska are most apt to have them in June. Farther to the west, the Plains states of Texas, Oklahoma, and Kansas all share May as the peak month.

Tornadoes are far less common in Canada, with an annual average of only about 100. The greatest concentration is in the extreme southern part of Ontario, between Lake Huron and Lake Erie. The majority of Canadian tornadoes outside Ontario occur in the southern region of the Prairie provinces and southwestern Quebec. The Canadian tornado season extends from April through October, with the greatest frequency in

Table 11–1 • Tornadoes in the United States, 1950–1994

	Average Annual Number of Tornadoes		Average Annual Number of Tornadoes per 10,000 sq km		Average Annual Number of Fatalities	
1	TX	122.0	FL	19.8	TX	10.6
2	OK	51.1	OK	19.0	MS	8.6
3	KS	46.9	KS	14.8	AR	6.2
4	FL	44.6	IN	14.1	AL	6.1
5	NE	37.2	IA	14.0	MI	5.3
6	IA	30.5	LA	13.1	IN	4.8
7	MO	25.9	MS	12.6	OK	4.8
8	SD	25.3	NE	12.5	KS	4.4
9	IL	25.3	TX	11.9	IL	4
10	CO	24.7	IL	11.7	TN	4
11	LA	24.1	AL	9.9	OH	3.8
12	MS	23.1	MO	9.7	MO	3.4
13	GA	19.7	AR	9.3	LA	3
14	AL	19.7	OH	9.1	GA	2.5
15	IN	19.7	GA	8.7	KY	2.3
16	AR	19	WI	8.7	MA	2.2
17	WI	18.8	SD	8.5	WI	2.1
18	MN	18.5	NJ	8.3	MN	1.9
19	ND	17.8	SC	7.9	FL	1.8
20	MI	15.8	CT	7.3	NC	1.8
21	OH	14.4	MI	7.0	PA	1.7
22	NC	13.1	TN	6.9	IA	1.4
23	TN	11.2	ND	6.5	NE	1.1
24	PA	10	NC	6.5	SC	1
25	WY	9.6	CO	6.2	VA	0.6
26	SC	9.4	MN	5.7	ND	0.5
27	NM	8.7	PA	5.7	NY	0.5
28	KY	8.3	KY	5.3	SD	0.2
29	VA	6.2	NH	4.5	WA	0.1
30	NY	5.5	RI	4.3	CT	0.1
31	MT	5.3	VA	4.0	AZ	0.1
32	CA	4.8	MD	3.0	NM	0.1
33	AZ	3.4	NY	2.9	AK	0
34	MD	3.2	WY	2.6	CA	0
35	MA	3	HI	2.4	CO	0
36	ID	2.6	MT	2.0	DC	0
37	NJ	2.5	VT	2.0	DE	0
38	WV	1.8	NM	1.9	HI	0
39	ME	1.8	WV	1.9	ID	0
40	UT	1.7	DE	1.5	MD	0
41	NH	1.6	ME	1.4	ME	0
42	CT	1.4	MA	0.9	MT	0
43	WA	1.2	AZ	0.8	NH	0
44	DE	1.2	CA	0.8	NJ	0
45	NV	1.1	ID	0.8	NV	0
46	OR	1	WA	0.5	OR	0
47	VT	0.7	NV	0.3	RI	0
48	HI	0.6	OR	0.2	UT	0
49	RI	0.2	UT	0.2	VT	0
50	AK	0	AK	0	WV	0
51	DC	0	DC	0	WY	0

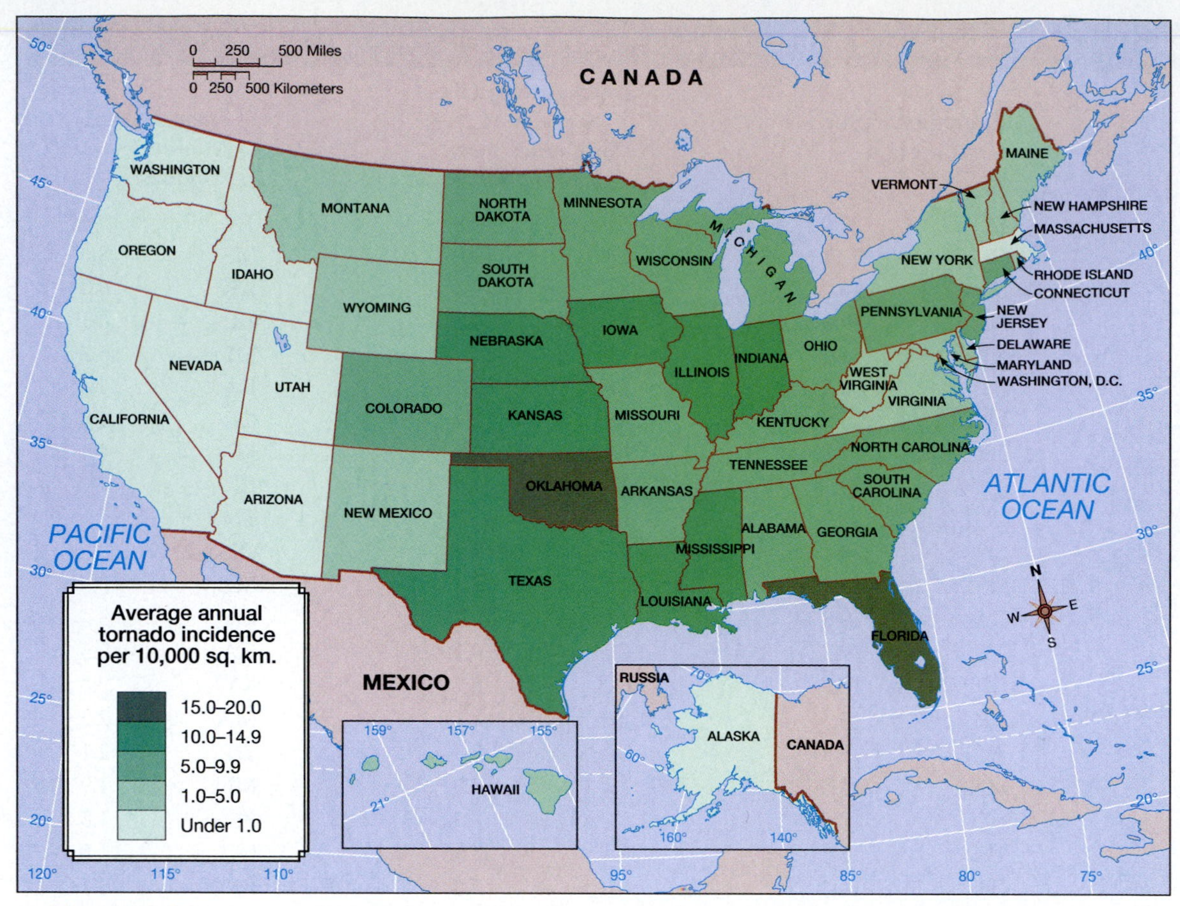

▲ **Figure 11–26**
Average number of tornadoes per unit area by state.

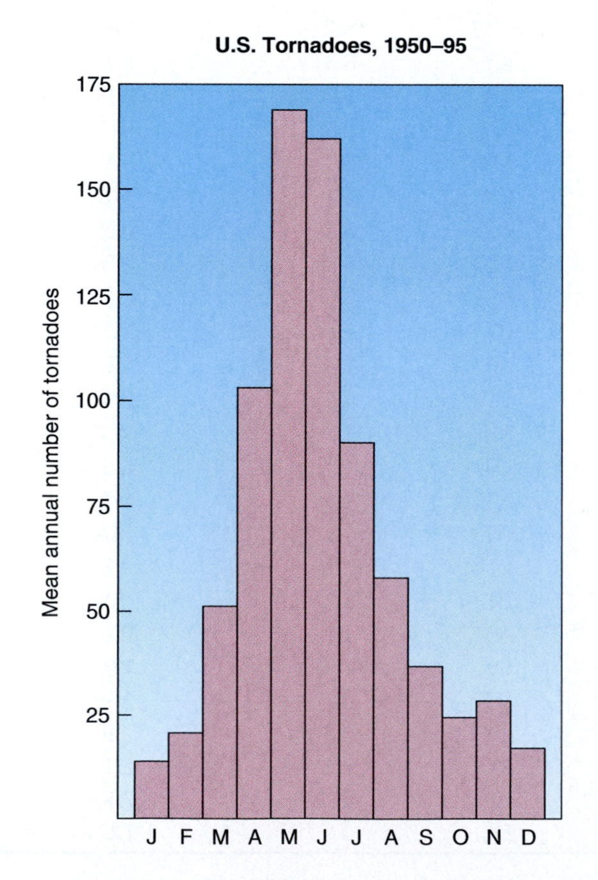

▶ **Figure 11–27**
U.S. tornadoes occur with greatest frequency in May and June. Values represent monthly averages for the years 1950 to 1995.

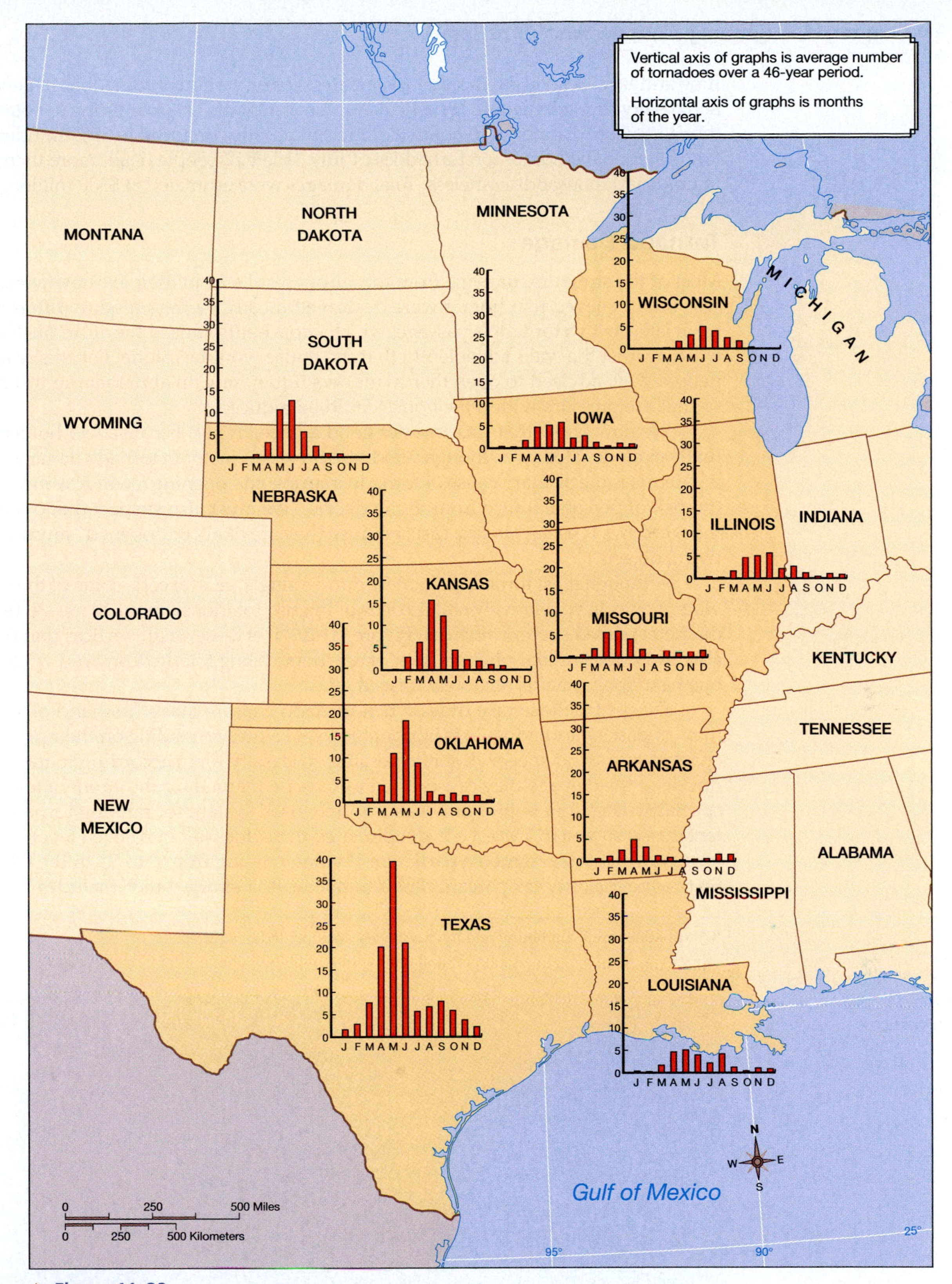

▲ **Figure 11–28**
The distribution of tornadoes during the year by state. Tornadoes tend to occur later in the year along the Great Lakes than along the Gulf Coast. The southern Plains states from Texas to Kansas have the greatest monthly totals in May. Farther north they are more prevalent in June.

June and July. Despite the fact that the greatest concentration of tornadoes in general and of very strong tornadoes in particular occurs in Ontario, it is interesting to note that the worst Canadian tornado outbreak in recent decades occurred in 1987 in Edmonton, Alberta. The Edmonton tornadoes of July 31 left 27 people dead, more than 300 injured, and thousands homeless. Total damages were estimated at $300 million.

Tornado Damage

Most of the structural damage from tornadoes results from their extreme winds. It was once believed that homes were destroyed mostly by the pressure differences associated with a tornado's passage, which supposedly caused the air to push outward against the walls so violently that the house would explode. For this reason, people were advised to open their windows if they saw an approaching tornado, so that the pressure within the house could be reduced.

We now know that this was not good advice, in part because few homes actually explode. Moreover, though winds are the major factor in tornado damage, flying debris is the primary cause of tornado injuries, and opening a window increases the risk of personal injury from flying debris. (We must also suspect that opening the windows is useless in any case, because they are likely be "opened" anyway by flying objects.)

Although most tornadoes rotate around a single, central core, some of the most violent ones have relatively small zones of intense rotations (about 10 m—30 ft—in diameter) called **suction vortices** (Figure 11–29). It is these small vortices that probably cause the familiar phenomenon of one home being totally destroyed while the one next door remains relatively unscathed.

Except for those rare times when tornado chasers make firsthand observations of passing tornadoes, it is impossible to get a precise reading on their pressure changes and wind speeds. But it is possible to classify them according to the magnitude of the damage they cause. The **Fujita scale** (named for the eminent tornado specialist Theodore Fujita) provides a widely used system for ranking tornado intensity. As shown in Table 11–2, documented tornadoes fall into seven levels of intensity, with each assigned a particular F-value ranging from 0 to 5.* In the United States, the majority (69 percent) fall into the *weak* category, which includes F0 and

*F6 tornadoes are hypothetical and have not been documented in nature.

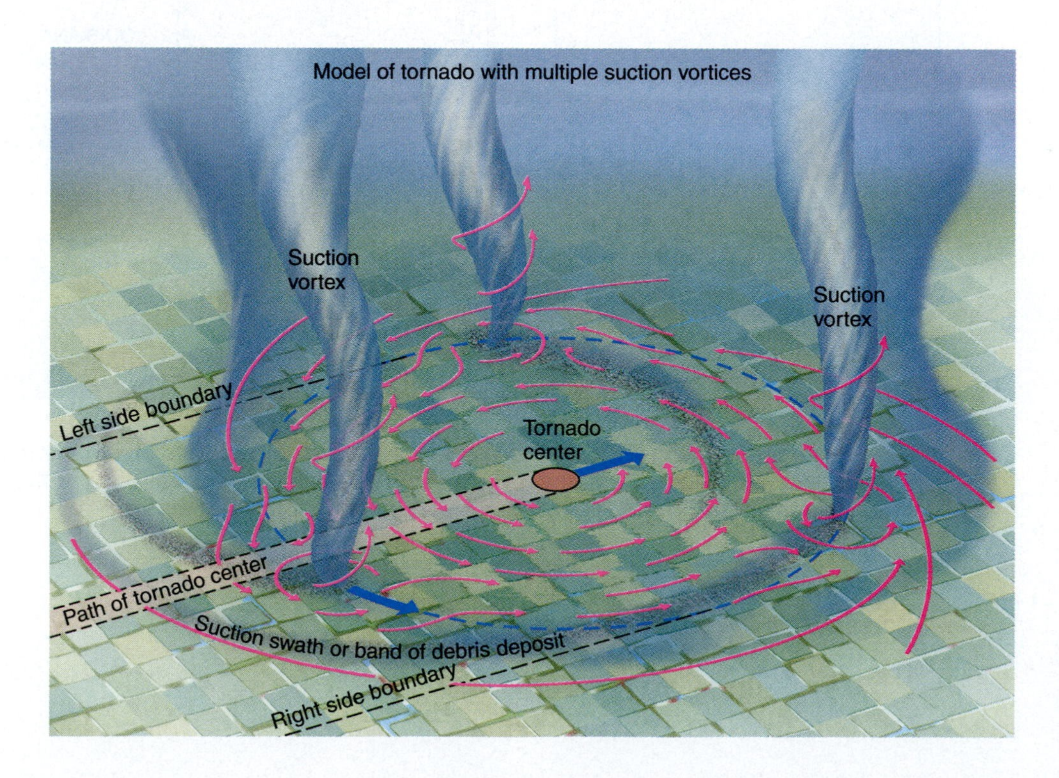

Figure 11–29 Suction vortices.

Table 11-2 • Fujita Intensity Scale

Intensity	Wind Speed (km/hr)	Wind Speed (mph)	Typical Amount of Damage
F0	< 116	< 72	Light: Broken branches, shallow trees uprooted, damaged signs and chimneys.
F1	116–180	72–112	Moderate: Damage to roofs, moving autos swept off road, mobile homes overturned.
F2	181–253	113–157	Considerable: Roofs torn off homes, mobile homes completely destroyed, large trees uprooted.
F3	254–332	158–206	Severe: Trains overturned, roofs and walls torn off well-constructed houses.
F4	333–419	207–260	Devastating: Frame houses completely destroyed, cars picked up and blown downwind.
F5	420–512	261–318	Incredible: Steel-reinforced concrete structures badly damaged.
F6	>513	>319	Inconceivable: Might possibly occur in small part of an F4 or F5 tornado. It would be difficult to identify the damage done specifically by these winds, as it would be indistinguishable from that of the main body of the tornado.

Note: F0 and F1 tornadoes are collectively called weak, F2 and F3 strong, and F4 and F5 violent.

F1 tornadoes. Twenty-nine percent of tornadoes are classified as *strong* (F2 and F3), which makes them capable of causing major structural damage even to well-constructed homes. Fortunately, only 2 percent of tornadoes are *violent* (F4 and F5). Those tornadoes are capable of wreaking incredible destruction. Cars can be picked up and carried tens of meters, pieces of straw can be driven into wooden beams, and freight cars can be carried off their tracks. Indeed, these storms are the true stars in all the movies and videos about tornadoes.

Between 1950 and 1999, 51 F5 tornadoes occurred in the United States and none in Canada. Thus, they happen about once a year in all of North America. Texas holds the lead for the greatest number of F5 tornadoes (six) during that period, while Alabama, Iowa, Kansas, and Oklahoma each had five.

Fatalities

Because they are small and last for such a short time, the overwhelming majority of tornadoes kill no one. Consider the fact that between 1950 and 1994, an average of 91 people died in 760 tornadoes each year. This means that at least 88 percent of all tornadoes kill no one. In fact, the actual proportion of fatality-free tornadoes is even higher than 88 percent, for at least two reasons. First, tornadoes in which no one dies are preferentially undercounted, because a storm that kills is almost certain to be reported. Second, most fatalities result from a few very large storms that kill up to dozens of people. According to the National Severe Storms Laboratory (NSSL), fewer than 5 percent of all U.S. tornado deaths are associated with weak (F0 or F1) tornadoes, nearly 30 percent with strong tornadoes (F2 and F3), and about 70 percent with violent tornadoes (F4 and F5). Thus, only 2 percent of all tornadoes are responsible for more than two-thirds of all fatalities.

If it seems that news reports of killer tornadoes tend to focus on mobile home parks, it's for a reason. A disproportionately large percentage of tornado-related deaths do in fact occur in mobile homes. Trailers offer little protection to their occupants because they are easily blown off their foundations and tossed about by strong winds.

A large percentage of tornado fatalities also occur among passengers and drivers of trucks and automobiles. These victims often panic and make the fatal mistake of leaving the relative safety of their homes to outrun the storm in their vehicles. By far the safest place you can be when a tornado threatens is inside a well-constructed building, preferably in the basement and away from the windows.

A famous video shot in 1991 by a television news crew has led many people to believe that highway overpasses provide good shelter from a tornado. The video shows a film crew along a Kansas highway fleeing an approaching tornado. The reporters stopped at an overpass, where they advised a terrified father and his daughter to take shelter under the girders. Huddled with the man and child, the news crew continued to film the tornado as it passed directly over them—with nobody being hurt. The video then proceeded to show a truck driver who luckily survived after his rig had been blown off the highway, and reported the death of another nearby motorist. The film provided a remarkable view of a tornado from the inside and naturally was repeatedly played on television stations across the country. Unfortunately, it did not issue a disclaimer about the relative lack of safety provided by overpasses in most instances. Though the people in the video survived their ordeal and may have saved their lives by moving under the overpass, this is not usually the best course of action. In this case, the center of the tornado passed just to the south of the overpass so that the road above happened to provide considerable sheleter. However, had the tornado taken a slightly different path, the outcome would not have been so fortuitous. Also, this particular tornado was a weak one—either an F0 or F1—and the overpass was a strong one with a narrow opening that allowed the people to hold on tightly as they were buffeted by winds. And finally, the fact that the incident was in a very rural area led to a small amount of debris being thrown toward the people—a situation that would be far different in a more heavily populated area.

People in the midst of a tornado should always obey the following safety rules when a tornado threatens:

1. Stay indoors and seek shelter in a basement.
2. If you are in a building with no basement, move to an interior portion of the lowest floor and crouch to the floor. If possible, cover yourself with a mattress or some other form of padding to protect you from falling or flying debris.
3. If you are in a mobile home, evacuate. If no fixed building is nearby, move away from all mobile homes and lie flat on low ground.
4. If driving a car or truck, you might be able to avoid the path of a tornado by driving at right angles to its path. Remember that most tornadoes move from southwest to northeast. If you cannot avoid the tornado's path, pull off the road and seek shelter. If none is available, run to low ground away from the road.

Watches and Warnings

Without question one of the most important responsibilities of the National Weather Service is to issue severe weather advisories. These take two forms, watches and warnings, either of which can be issued for severe storms or tornadoes.* The

*Warning and watches can be issued for other types of threatening weather, such as hurricanes and flash floods.

declaration of a watch does not mean that severe weather has developed or is imminent; it simply tells the public that the weather situation is conducive to the formation of such activity. Most watches are issued for a period of 4 to 6 hours, for an area that normally encompasses several counties—about 50,000 to 100,000 sq km (20,000 to 40,000 sq mi).

The Storm Prediction Center (SPC) of the U.S. Weather Service in Norman, Oklahoma, has responsibility for putting out severe storm and tornado watches for the entire country. Operating 24 hours a day, every day, the center constantly monitors surface weather station data, information from weather balloons and wide-bodied commercial aircraft, and satellite data for all of the United States. If any particular part of the country appears vulnerable to impending severe storm activity, SPC issues a severe storm or tornado watch. The advisory then goes to the local office of the National Weather Service, which notifies local television and radio stations. The broadcast media then relay the information to the public. Watches for the entire country are also available from a number of sources, many of which are on the World Wide Web. Here is the text of a sample severe thunderstorm watch:

BULLETIN - IMMEDIATE BROADCAST REQUESTED

SEVERE THUNDERSTORM WATCH NUMBER 875

STORM PREDICTION CENTER NORMAN OK

319 PM CDT FRI JUL 28 1995

THE STORM PREDICTION CENTER HAS ISSUED A SEVERE THUNDERSTORM WATCH FOR A LARGE PART OF MISSISSIPPI EFFECTIVE THIS FRIDAY AFTERNOON AND EVENING UNTIL
1000 PM CDT.

LARGE HAIL..DANGEROUS LIGHTNING AND DAMAGING THUNDERSTORM WINDS ARE POSSIBLE IN THESE AREAS.

THE SEVERE THUNDERSTORM WATCH AREA IS ALONG AND 75 STATUTE MILES EAST AND WEST OF A LINE FROM 15 MILES SOUTH SOUTHEAST OF MC COMB MISSISSIPPI TO 5 MILES EAST OF OXFORD MISSISSIPPI.

REMEMBER..A SEVERE THUNDERSTORM WATCH MEANS CONDITIONS ARE FAVORABLE FOR SEVERE THUNDERSTORMS IN AND CLOSE TO THE WATCH AREA. PERSONS IN THESE AREAS SHOULD BE ON THE LOOKOUT FOR THREATENING WEATHER CONDITIONS AND LISTEN FOR LATER STATEMENTS AND POSSIBLE WARNINGS.

OTHER WATCH INFORMATION. THIS SEVERE THUNDERSTORM WATCH REPLACES SEVERE THUNDERSTORM WATCH NUMBER 874. WATCH NUMBER 874 WILL NOT BE IN EFFECT AFTER 345 PM CDT.

If a severe thunderstorm has already developed, the public is notified by a **severe thunderstorm warning.** Likewise, **tornado warnings** alert the public to the observation of an actual tornado (usually by a trained weather spotter) or the detection of tornado precursors on Doppler radar. Unlike watches, which are issued by the SPC, warnings are given by the local weather forecast office. They warn the public to take immediate safety precautions, such as finding shelter in a basement. The information is broadcast immediately by television and radio stations, and civil defense sirens are sounded.

Sometimes Doppler radar enables meteorologists to give warning of an impending tornado about half an hour before it actually forms. At other times, warnings are issued only after a funnel cloud has been seen. In this case, the observer

11–5 Special Interest

Tornado Research

Despite the inherent difficulty of studying something as complex, violent, and short-lived as a tornado, scientists can peer into their structure using a number of research tools. For many years, much of the information was obtained by films and pictures of tornadoes. Unfortunately, dust and debris kicked up by a tornado, as well as screening by trees and buildings, often make it impossible to observe the part of the tornado nearest to the ground (and of most concern to people).

In 1980 scientists developed the Totable Tornado Observatory, better known by its acronym, "Toto" (an intentional reference to the dog in *The Wizard of Oz*). Toto was a lightweight package that could be carried in the back of a pickup truck and set up by storm chasers in 30 seconds or less. Ideally, the instrument would be placed in the path of a tornado, where it could make observations of temperature, wind, and pressure. Unfortunately, getting Toto in the path of a tornado proved exceedingly difficult. Moreover, wind tunnel tests have shown that it tends to tip over at wind speeds near 50 m/sec (110 mph), well below those that can occur in a tornado. For these reasons storm chasers no longer use Toto.

More recently, scientists from the University of Oklahoma have begun placing a network of smaller instruments called "turtles" in the paths of tornadoes. Though they only record temperature and pressure, turtles have an advantage in that they can be laid out in series across a road, crossing the path of a tornado. They remain a potentially useful tool for examining the environment of a tornado.

Other methods have been attempted for making direct observations within a tornado or mesocyclone. One is to release portable weather balloons with radiosonde instruments in the near-storm environment. If they are released from the right location, the wind can carry the balloons right into the heart of a severe storm. Less successfully, researchers have attempted to fire light instrument-carrying rockets (on the order of 1 kg) into severe storms from small aircraft. Unfortunately, the measurement devices have proven to be too fragile to survive the turbulence of the storm and are no longer used.

By far the most useful instrument for studying the winds of tornado-producing storms has been Doppler radar. Although the NEXRAD network of Doppler radar units has only recently been completed, a marked improvement in the issuance of tornado warnings has already resulted. A recent study has shown that installation of the first six Doppler radar units in the heart of Tornado Alley has improved forecasters' ability to predict tornadoes. Prior to the radar deployment, 33 percent of the tornadoes in the study area occurred without a tornado warning; afterward, the figure dropped to 13 percent.

The biggest limitation of the Doppler radar network is that the farther a storm is from the transmitter/receiver unit, the less detailed is the information about its internal winds. Because tornadoes are usually quite small, they must pass within a few kilometers of the radar if precise wind information is to be obtained. For this reason, tornado researchers also use portable Doppler radar units carried on research vehicles and aircraft.

must report his or her findings to the local Weather Service office. The meteorologist in charge then passes the information on to the broadcast media. The procedure normally takes about 7 minutes. But most tornadoes have life spans of about 3 to 4 minutes. In other words, in many instances the warning is issued to the public only *after* the tornado has died out. Though this might make the procedure seem pointless, that really is not the case. First of all, the formation of one tornado is often followed by others within the same storm system. Consequently, the precautions taken for the defunct tornado might save lives when subsequent ones appear. Second, rare killer tornadoes that represent the greatest threat to life tend to have considerably longer lifetimes and can remain in existence long after the warning has been issued.

Despite the excellent warning system, disasters do indeed happen. At other times, sheer luck is all that stands in the way of major loss of life. On September 2, 2002, at 4:20 P.M. CDT, an F3 tornado hit the town of Ladysmith, Wisconsin, head on, destroying 26 businesses and 17 homes. This time no tornado warning had been issued until 17 minutes *after* the tornado hit the town, although a severe thunderstorm warning had been in effect for more than half an hour. (The fact that the nearby Doppler radar unit at La Crosse, Wisconsin, had been temporarily out of order may have contributed to the absence of a more timely warning.) Although most residents received no warning about the impending tornado, somehow the town escaped with no fatalities.

The Tornado Outbreak of May 3–6, 1999

In May 1999, a devastating series of large tornadoes reminded the world of just how devastating tornadoes can be. Hardest hit was central Oklahoma, where a series of tornadoes on the first day of the outbreak killed 44 people and destroyed 2600 homes and businesses (Figures 11–30 and 11–31). On the same day, tornadoes killed another 5 people and destroyed 1100 buildings in the Wichita, Kansas, area. And on May 5 and 6 the storms claimed another 5 lives in Texas and Tennessee. This was the deadliest outbreak of tornadoes since the record-breaking "Xenia Tornado Outbreak" of April 3–4, 1974, which spawned a total of 148 tornadoes (7 of which were rated as F5) from northern Alabama and Georgia all the way north to Windsor, Ontario, in Canada, and killed more than 300 people.

Many factors contibute to the loss of life inflicted by tornadoes, including their maximum wind speed and their paths relative to large populations. In this case, extremely powerful winds descended upon large population centers. But the effectiveness of the warning system also impacts the death total and, fortunately, the system was at its very best during the 1999 outbreak. Field crews observed and reported the early development of conditions favorable for tornadoes, which complemented the forecasting efforts of meteorologists at the respective Weather Service

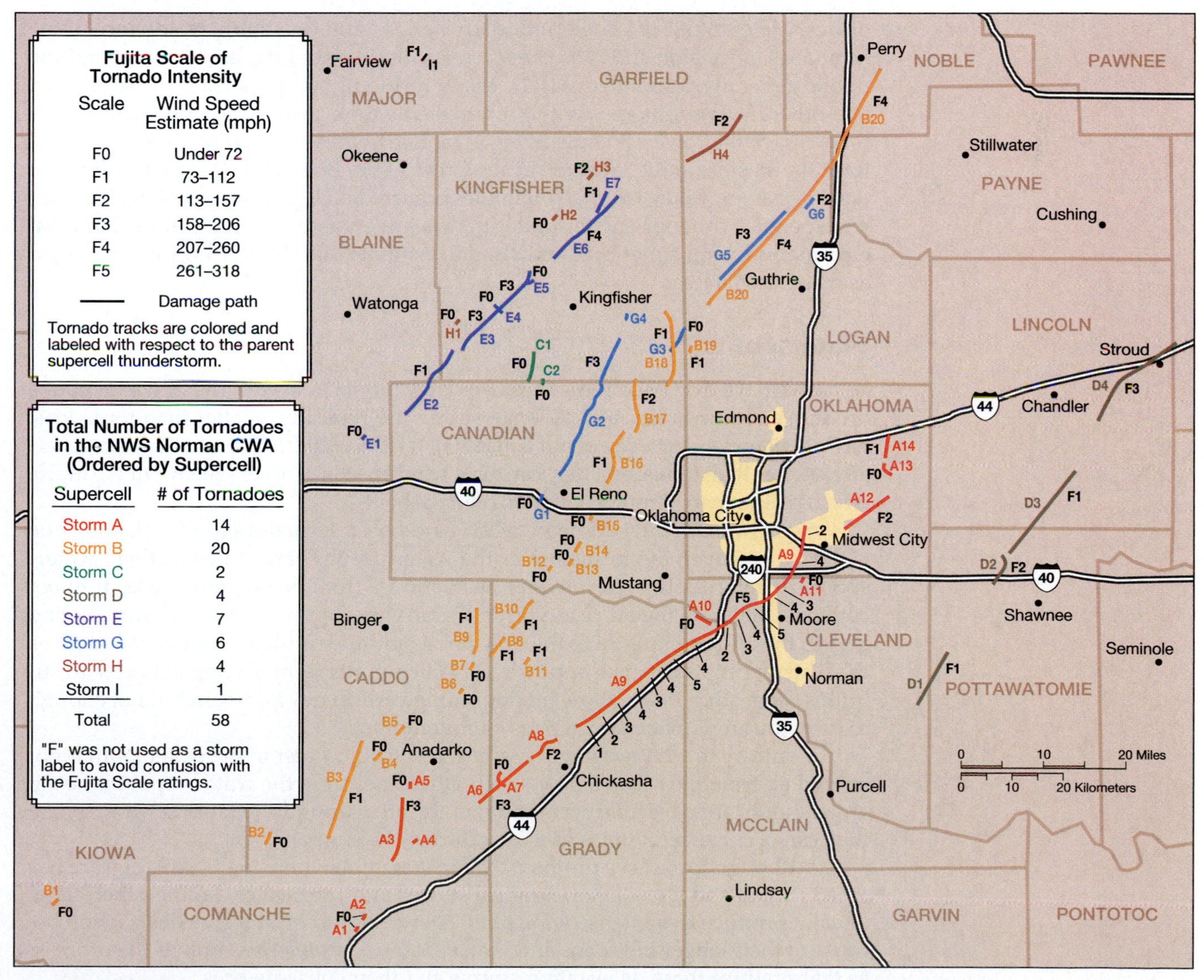

▲ **Figure 11–30**
Map of May 3, 1999, tornadoes.

▶ **Figure 11–31**
Tornado damage in central Oklahoma.

offices. Meteorologists issued public advisories a full two hours in advance of the tornadoes, television and radio news crews diseminated the information, and the public generally responded wisely. Many residents sought shelter in basements, and others living in homes without basements took cover in interior hallways in time to save their lives. (People living in mobile homes received no protection at all from the storms; in the town of Bridge Creek every one of the 11 people who died were in trailers. Likewise, many fatalities occurred on highways—3 of the fatalities were people who took shelter under freeway overpasses!) Experts at the National Oceanic and Atmospheric Administration estimate that the warning system saved as many as 700 lives.

Waterspouts

So far we have discussed tornadoes over land. Similar features, called **waterspouts,** occur over warm-water bodies. Waterspouts are typically smaller than tornadoes, having diameters between about 5 and 100 m (17 to 330 ft). Though they are generally weaker than tornadoes, they can have wind speeds of up to 150 km/hr (90 mph), which makes them strong enough to damage boats.

Some waterspouts originate when land-based tornadoes move offshore. The majority, however, are formed over the water itself. These "fair weather" waterspouts develop as the warm water heats the air from below and causes it to become unstable. As the air rises within the unstable atmosphere, adiabatic cooling lowers the air temperature to the dew point, and the resultant condensation gives the waterspout its ropelike appearance. Waterspouts form in conjunction with cumulus congestus clouds, those having strong vertical development but not enough to form the anvil that characterizes cumulonimbus.

Contrary to what we might assume, the visible water in the waterspout is not sucked up from the ocean below; it actually comes from the water vapor in the air. Waterspouts are particularly common in the area around the Florida Keys, where they can occur several times each day during the summer.

Although tornadoes produce the most extreme winds on Earth, they are relatively small and short-lived. Tropical storms and hurricanes, on the other hand, usually produce less intense winds but can wreak more extensive destruction because of their longer life spans and larger extent. Chapter 12 examines the processes that produce these devastating storms and their consequences.

Summary

Some of the most dramatic of all storms are those that produce lightning and thunder. Lightning begins when negative electrical charges build up near the base of a cloud and positive charges gather at the top. In the case of cloud-to-ground lightning, rapidly growing leaders extend downward from the base of the cloud. When they connect with some object at the surface, a visible stroke develops. Most often a rapid sequence of multiple strokes occurs following the initial one to produce a lightning flash. The extreme heating of the air within the stroke causes the air to expand explosively and create the sound of thunder.

Thunderstorms create downdrafts that can cause the storm to die out or intensify. In air mass thunderstorms, downdrafts eventually destroy the storm. They undergo a sequence from the cumulus to the mature to the dissipative stages in tens of minutes. Severe storms—those that produce large hail, damaging winds, or tornadoes—develop when downdrafts reinforce the storm. Such storms occur within squall lines, in multicellular mesoscale convective complexes, or as supercells.

Tornadoes, among the most fearsome of all natural phenomena, occur more often in the United States than in any other country of the world—and nearly all the states and Canadian provinces experience them sooner or later. They are most likely to form in the spring or early summer, but they can occur at any time of the year. Many tornadoes (especially those that emerge from supercells) follow the formation of large rotating areas within storm clouds called *mesocyclones,* but other unexplained processes can also lead to their formation.

The majority of tornadoes are classified as weak and cause no fatalities, but relatively rare, extremely large tornadoes can cause tremendous devastation and loss of life. Fortunately, the U.S. Weather Service has a system of tornado watches and warnings that alert the public to threatening weather. In recent years, Doppler radar has shown tremendous value as a research and forecast tool. This tool has also become valuable in the tracking of another important weather phenomenon—hurricanes—which we will examine in Chapter 12.

Key Terms

lightning p. 318
sheet lightning p. 318
charge separation p. 318
stepped leader p. 319
return stroke p. 319
dart leader p. 319
strokes p. 319
flashes p. 319
ball lightning p. 321
St. Elmo's fire p. 321
sprites p. 321
blue jets p. 321
fair weather electric field p. 322
mean electric field p. 322
thunder p. 322
heat lightning p. 322
air mass thunderstorms p. 324
severe thunderstorms p. 324
cumulus stage p. 324
mature stage p. 325
dissipative stage p. 325
mesoscale convective systems p. 326
squall lines p. 326
mesoscale convective complexes p. 326
supercells p. 326
potential instability p. 327
outflow boundary p. 328
gust front p. 329
Doppler radar p. 330
hook p. 330
Doppler effect p. 332
vault p. 332
downbursts p. 333
microbursts p. 334
tornadoes p. 335
mesocyclones p. 336
wall clouds p. 337
funnel clouds p. 337
suction vortices p. 344
Fujita scale p. 344
severe thunderstorm warning p. 347
tornado warning p. 347
waterspouts p. 350

Review Questions

1. How common is cloud-to-ground lightning relative to cloud-to-cloud lightning?
2. Describe the current theories regarding the formation of charge separation.
3. What is the difference between a lightning stroke and a lightning flash?
4. Describe the sequence by which electrical imbalances lead to lightning strokes.
5. Briefly describe the following phenomena:
   a. ball lightning
   b. St. Elmo's fire
   c. sprites
   d. blue jets
6. What causes thunder to occur?
7. Why is the term *heat lightning* misleading?
8. What are the three stages of an air mass thunderstorm?
9. How big are air mass thunderstorms and how long do they usually persist?
10. Explain why some thunderstorms are self-extinguishing and others are able to develop into severe thunderstorms.
11. Describe the following types of storm systems:
    a. mesoscale convective systems
    b. squall lines
    c. mesoscale covective complexes
    d. supercells
12. How are outflow boundaries formed, and what effect do they have?
13. Describe the processes that lead to tornado development in supercell and nonsupercell storms.

14. What features of Doppler radar make it an effective tool for severe storm forecasting?
15. What are hook echoes and vaults, and why are they important?
16. Explain how microbursts form, and why they present a serious threat to aviation?
17. Describe the location and timing of tornadoes in North America.
18. Describe the process of tornado formation from supercell storms.
19. What are wall clouds, and why is their appearance a cause for concern?
20. What is the leading threat to human safety when tornadoes hit?
21. Describe the Fujita scale for the classification of tornadoes. Which category is most common? What is the highest F-value that can actually occur in nature?
22. What is the difference between a tornado watch and a tornado warning?
23. How do waterspouts compare to tornadoes on average, in terms of intensity?

Critical Thinking

1. Is charge separation necessary for sheet lightning and/or ball lightning?
2. Why does the environmental lapse rate affect the distance at which "heat lightning" can be observed?
3. You are outside on a sunny afternoon and observe a thunderstorm far to the west. An hour later, the storm passes over you. Is this more likely to have been an air mass thunderstorm or some sort of mesoscale convective system?
4. In what fundamental ways are gust fronts different from passing cold fronts?
5. Why is the incidence of thunderstorms much less near the Pacific Coast than at the Atlantic Coast?
6. What conditions east of the Rocky Mountains promote a much greater incidence of tornadoes than exists in western North America?
7. Why is it not possible for a mesocyclone to occur within an air mass thunderstorm?
8. Why is it extremely unlikely that a tornado will move from east to west?
9. Other than their location with respect to land vs. water, how do waterspouts differ from tornadoes?

Problems and Exercises

1. Compare the map of annual hailstorms (Figure 7–14) to the map depicting the incidence of thunderstorms (Figure 11–18). Identify the regions where both are frequent, and those (if any) where only hail or thunderstorms frequently occur.
2. A tornado has a ring of uniform winds of 200 km/hr (120 mph) around the vortex 20 m away from its center. If the tornado moves to the northeast at 50 km/hr (30 mph), what is the effective wind speed on the northwestern and southeastern portions of the tornado?
3. On a regular basis, go to the Storm Prediction Center (SPC) website at **http://www.spc.noaa.gov/** and note the location of all current severe thunderstorm or tornado watches. Then go to one of the Web sites mentioned in previous chapters to observe surface weather maps, satellite images, and radar return maps. Describe the position of the watch area relative to what the maps and images depict.

Quantitative Problems

The Web site for this book, **http://www.prenhall.com/aguado**, offers quantitative problems that can help reinforce your understanding of several of the concepts discussed in this chapter. The problems deal with the timing of the arrival of thunder following a lightning strike, the characteristics of tornadic winds and pressure gradients, and potential instability. We suggest you go to the Web page for Chapter 11 and work out these straightforward problems.

Useful Web Sites

http://www.lightningstorm.com/

Click on the lightning map icon to see where lightning has occurred over the United States during the last 30 minutes. A similar map for Canada exists at http://weatheroffice.ec.gc.ca/lightning/index_e.html.

http://www.nssl.noaa.gov/edu/ltg/

This site provides interesting lightning information from the National Severe Storms Laboratory.

http://www.nssl.noaa.gov/headlines/outbreak.shtml and http://www.srh.noaa.gov/oun/storms/19990503/index.html

At these two sites, you'll find comprehensive information on the deadly tornado outbreak of May 1999.

http://www.lightningsafety.noaa.gov/

Comprehensive information from the National Weather Service includes medical effects of being struck by lightning, myths, and indoor and outdoor safety tips.

http://www.srh.noaa.gov/oun/papers/overpass.html

This is an outstanding presentation on the dangerous use of highway overpasses as tornado shelters.

Media Enrichment

Weather in Motion

Lightning

This brief movie shows cloud-to-ground lightning, bolts of cloud-to-cloud lightning, and sheet lightning.

Weather in Motion

Cloud and 3-D Flow Tracers

This is the first of four computer simulations illustrating the typical development of thunderstorm clouds. The white arrows show the movement of air. Initially uplift at the surface causes the development of cumulus clouds. The clouds deepen as rising motions continue well beyond the lifting condensation level. The largest of the thunderstorm clouds penetrates into the stratosphere, where differential wind velocities shear the top of the cloud to produce the characteristic anvil. The scene then rotates 360 degrees to offer a three-dimensional perspective.

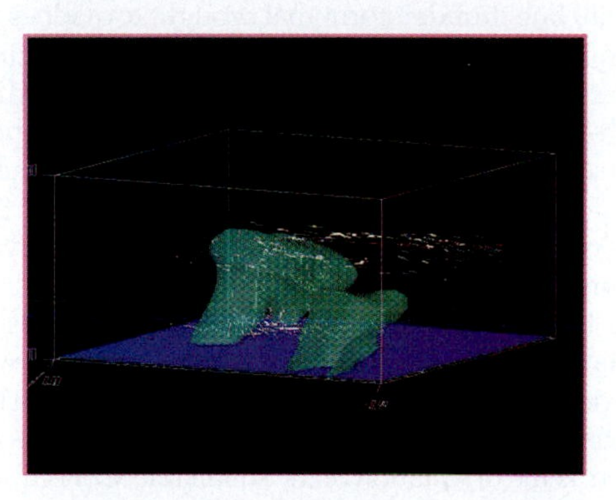

Weather in Motion

Cloud and Vertical Velocity

This movie presents the same scene as above but uses color to highlight the speed and direction of vertical motions. Red portions of the clouds indicate rapid updrafts up to 38 m/sec (84 mph), while blue portions represent areas of downdraft. Clearly the clouds are dominated by zones of rapid uplift, with descending winds found only in limited regions, primarily in the downwind margins of the clouds.

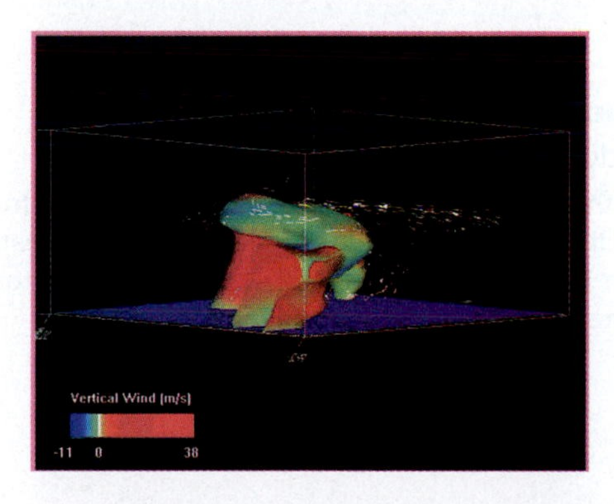

Weather in Motion

Vertical Motion Cross Section

This movie, the third in the series, displays a vertical cross section of the winds in the region. Initially, the airflow is primarily horizontal at all but the very lowest level near the surface. Eventually air moves toward the region of convection and rapidly turns upward, where updrafts upset the general pattern of left-to-right winds. Notice that substantial areas of descending air exist outside the margins of the clouds.

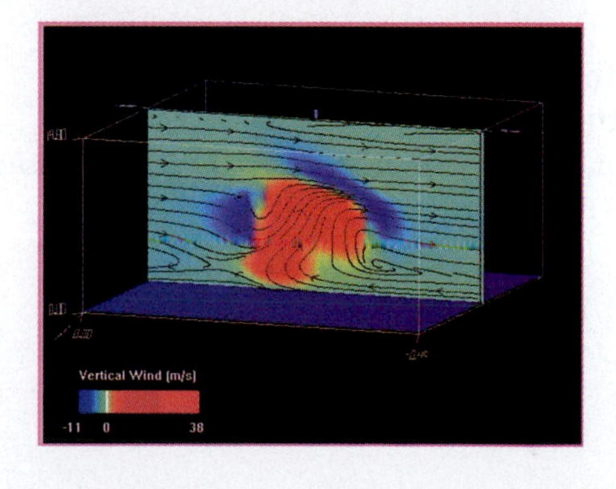

Weather in Motion

Cloud and Horizontal Flow

This is the final movie in the sequence showing thunderstorm development. In this case, we see the horizontal wind flow pattern through a progression of altitudes after the thunderstorm is well developed. The airflow consistently turns clockwise with increasing altitude, yielding a pattern of *veering* winds associated with advection of warm air. Also observe that the wind flow is deflected around and over the upper portions of the main storm cell.

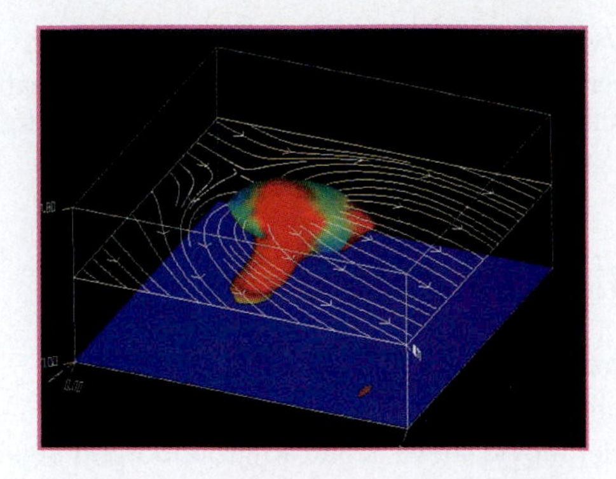

Weather in Motion

Squall Line Tornado over Northwestern Louisiana

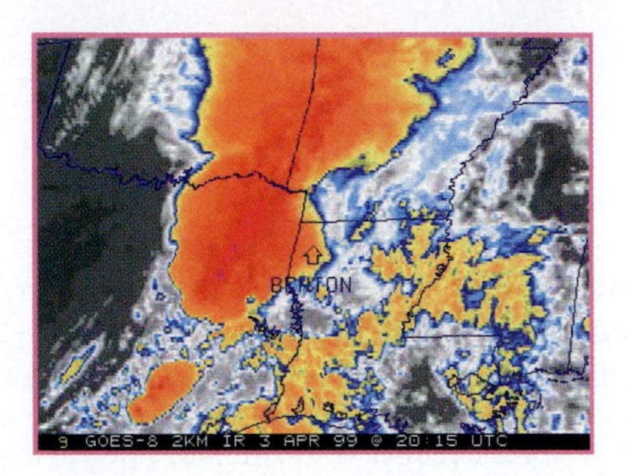

Severe thunderstorms can develop as part of mesoscale convective complexes, supercells, and squall lines. This movie uses color-enhanced, infrared images to depict the movement of a squall line thunderstorm that brought tornadoes to northwestern Louisiana on April 3, 1999. The movie begins with a wide band of cloud cover and storm activity extending nearly north–south from eastern Texas to eastern Oklahoma and western Arkansas. Within the band is a narrow zone of more severe storm activity, indicated by a deep red color. Scattered regions of somewhat less intense convection exist over northern Louisiana.

As the band approaches northwestern Louisiana, the squall line containing the most intense activity grows longer and wider, and tornadoes occur around Benton. The storm moves from the southwest to the northeast, which is the same direction as the orientation of the squall line. Consequently, the squall line lingers more than an hour before leaving Benton.

Notice that throughout the movie, the leading (eastern) edge of the squall line is not sharply defined. The zone of most intense activity merges with a region mosaicked by cloud cover of varying intensity. In contrast, the western boundary of the storm is very sharp.

Weather in Motion

Tornado

One of the most recognizable forms of all weather, tornadoes are capable of wreaking widespread death and destruction. Though this is a large and destructive example, tornadoes are sometimes even larger than this one.

Weather Image

Lightning Across the Globe

This image maps the distribution of lightning flashes over a 1-year period.

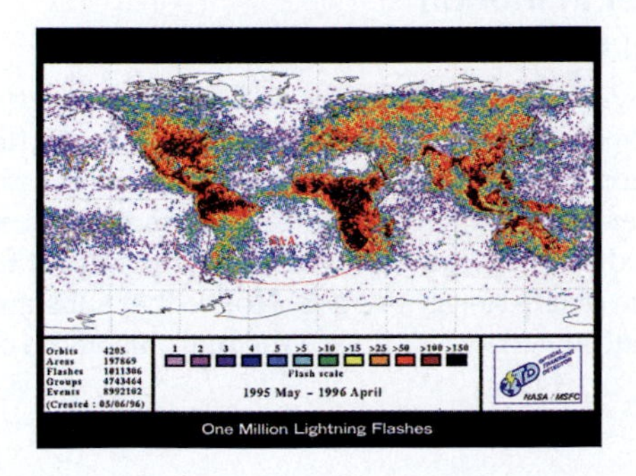

Weather Image

Shelf Cloud over a Gust Front

Strong storms can produce rapid downdrafts. Precipitation causing the downdraft evaporates partially, consuming latent heat and thereby cooling the downdraft. This cooling increases the density of the air in the downdraft. The downdraft spreads outward upon hitting the ground to create a gust front. Because the gust front is denser than the air it displaces (due to evaporative cooling), it lifts the warm air and can produce shelf clouds, as seen in this image.

Weather Image

Mammatus I, II, and III

These pouch-like protrusions are sometimes formed near the anvil of cumulonimbus clouds. Each protrusion forms where air is sinking along the margins of the cloud. The water droplets and ice crystals extend out from the margin of the cloud and evaporate into the surrounding, unsaturated air.

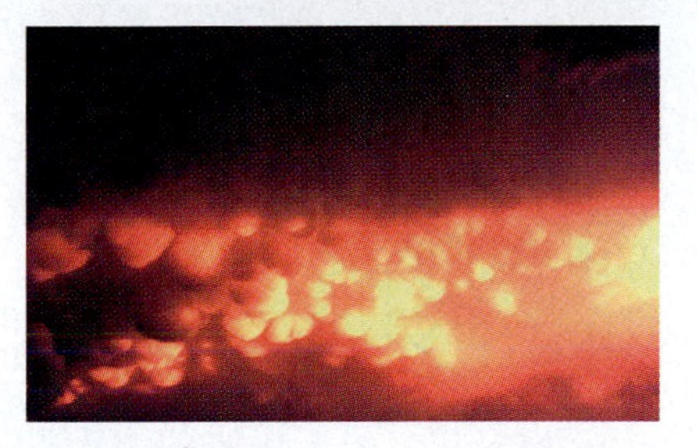

Weather Image

Wall Clouds I and II

Wall clouds are important features of supercell storms. Moist air is drawn into this portion of the supercell, causing a lower lifting condensation level and cloud base. Wall clouds often form the location for subsequent tornado development.

Weather Image

Tornado

Tornadoes assume a wide continuum of shapes and sizes. This is one example of a ropelike tornado. Contrast this one to the tornado shown in the *Weather in Motion: Tornado* movie for this chapter.

Bangladesh after a cyclone.

chapter 12

Tropical Storms and Hurricanes

Nobody who lived in southern Florida during the August 1992 passage of Hurricane Andrew (Figure 12–1) will ever forget the experience. Andrew was fairly small but remarkably powerful, with wind gusts of up to 280 km/hr (175 mph) that moved rapidly across the peninsula. In one regard Andrew was not as destructive as many other hurricanes; although it did cause some local flooding, it did not yield extremely heavy rainfall. This contrasts strongly with Florida's experience with Hurricane Irene in October 1999.

Irene was typical of the strong tropical storms that form in the latter part of hurricane season: large and slow-moving. Although these storms lack the wind speeds of storms such as Andrew, they can bring heavy rainfall for days and cause extensive flooding. In the case of Irene, much of southern Florida received up to 27 cm (17 in.) of rainfall that produced widespread flooding. Nadia Gorriz of Miami-Dade County was among the many victims. The floodwaters that created a swamp around her house made it uninhabitable for humans but perfectly fine for the snakes and fish that took up residence. She put the cleanup in perspective: "We made it through Andrew. You just threw everything out and there wasn't that much water. This is the worst cleanup. We've gone through two gallons of Clorox II. So far."

Hurricanes do not restrict their fury to coastal and inland regions; they have been the nemesis of mariners for centuries. They have sunk an untold number of ships and even played an important role in World War II when a single typhoon (the equivalent of a hurricane over the western Pacific) sank or heavily damaged several U.S. ships, destroyed hundreds of carrier-based aircraft, and killed more than 800 sailors. The death toll exceeded that of most naval battles during the war.

In this chapter we first describe the setting for hurricanes and tropical storms. We then describe their general characteristics, stages in development, and typical patterns of movement, concluding with hurricane monitoring and warning systems.

CHAPTER OUTLINE

Figure 12–1 Modified satellite image showing the progression of Hurricane Andrew as it crossed Florida and entered the Gulf of Mexico in August 1992.

Hurricanes Around the Globe

Extremely strong tropical storms go by a number of different names, depending on where they occur. Over the Atlantic and the eastern Pacific they are known as **hurricanes**. Those over the extreme western Pacific are called **typhoons**; those over the Indian Ocean and Australia, simply **cyclones**. In structure, the three kinds of storms are essentially the same, although typhoons tend to be larger and stronger than the others. We will use the term *hurricane* for the general class of storm, regardless of its location.

Most U.S. residents associate hurricanes with storms that form in the Atlantic Ocean or the Gulf of Mexico. Yet, other parts of the world have a much greater incidence of hurricanes (Table 12–1 and Figure 12–2). The Atlantic and Gulf of Mexico receive an average of 5.4 hurricanes each year, while the eastern North Pacific off the coast of Mexico has an average of 8.9. Most tropical storms in the east Pacific move westward, away from population centers, and so they receive little public attention. Sometimes, however, they do migrate to the northeast and bring severe flooding and loss of life to western Mexico.

The region having the greatest number by far of these events is the western part of the North Pacific. In a typical year, nearly 16 typhoons hit the region. Even during the least active season between 1968 and 1989, there were 11 typhoons. At the other extreme, no hurricanes form in the Southern Hemisphere Atlantic, even at tropical latitudes. As will be seen later, hurricanes depend on a large pool of warm water, a condition that does not arise in the relatively small South Atlantic basin.

The Tropical Setting

In Chapter 8 we saw that during much of the year air spirals out of massive high-pressure cells that occupy large parts of the Atlantic and Pacific Oceans. Middle- and upper-level air along the eastern side of these anticyclones sinks as it approaches the west coasts of the adjacent continents. Because the air does not descend all the way to the surface, a subsidence inversion (see Chapter 6) forms above the surface. This particular subsidence inversion is called the **trade wind inversion**. The air below the inversion, called the **marine layer**, is cool and relatively moist.

The thickness of the marine layer and the height of the inversion base vary across the tropical oceans. The inversion is lowest along the eastern margins of the oceans, where upwelling and cold ocean currents maintain a relatively cool marine layer. Here the inversion may be only a few hundred meters above the surface. Farther to the west, the warmer surface waters heat the marine layer and cause

Table 12–1 • Maximum, Minimum, and Average Number of Hurricanes (and Their Counterparts) per Year over Various Parts of the World's Oceans, 1968 to 1989 (1968 to 1990 for the Southern Hemisphere)

Basin	Maximum	Minimum	Average
Atlantic	12	2	5.4
Eastern Pacific	14	4	8.9
Western Pacific	24	11	16.0
Northern Indian Ocean	6	0	2.5
Southwestern Indian Ocean	10	0	4.4
Southeastern Indian Ocean/Australia	7	0	3.4
Australia/Southwestern Pacific	11	2	4.3
Global	65	34	44.9

Source: Colorado State University

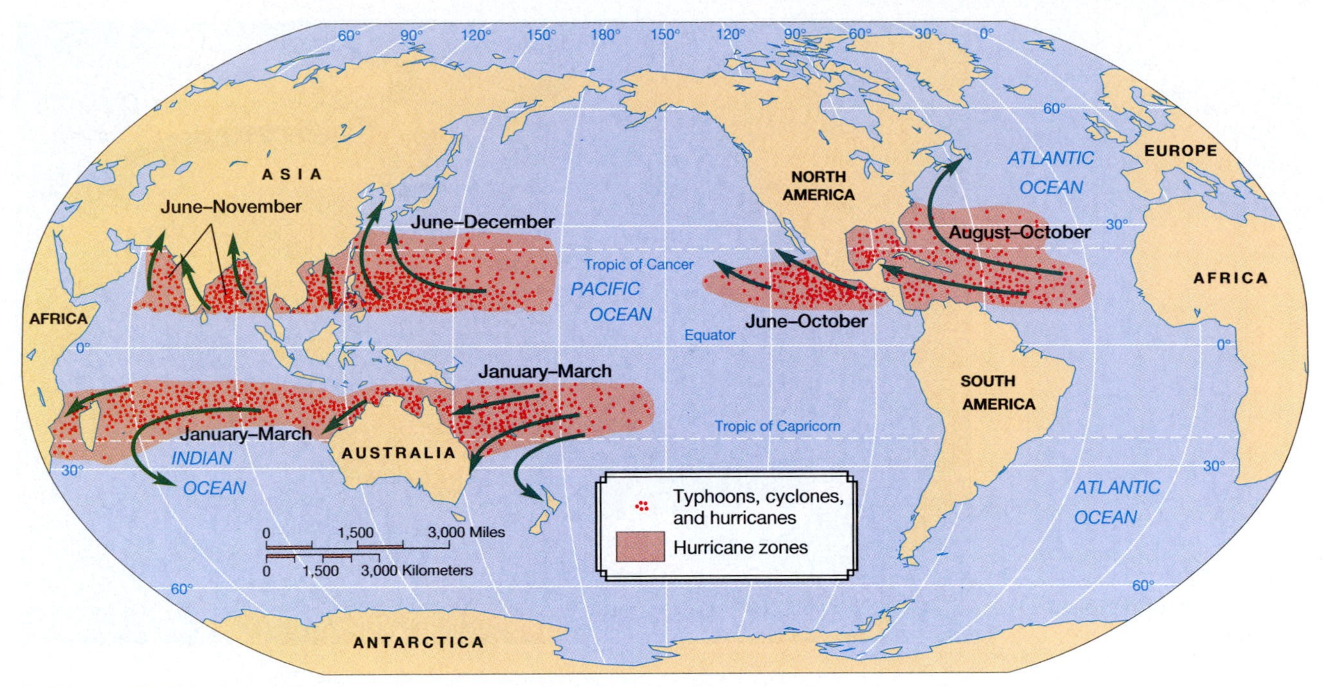

▲ **Figure 12–2**
Hurricanes around the globe.

it to expand to a greater height. Over the eastern part of the oceans, the low inversion inhibits vertical cloud growth, and low stratus clouds often occupy the region. Farther to the west, the greater height of the inversion (or even its total disappearance) allows for more convection, and deep cumulus clouds are more likely to form. For this reason, more hurricanes occur along the western portion of the ocean basins.

Hurricane Characteristics

Hurricanes are the most powerful of all storms. The energy unleashed by just a single hurricane can exceed the annual electrical consumption of the United States and Canada. By definition, they have sustained wind speeds of 120 km/hr (74 mph) or greater. Though their wind speeds are less than those of tornadoes, hurricanes are very much larger and have far longer life spans. Sea level pressure near the center of a typical hurricane is around 950 mb, but pressures as low as 870 mb have been observed for extremely powerful hurricanes. The weakest hurricanes have central pressures of about 990 mb.

In contrast to tornadoes, whose diameters are typically measured in tens of meters, hurricanes are typically about 600 *kilo*meters (350 mi) wide. Thus, the typical hurricane has a diameter thousands of times greater than that of a tornado. Remembering that the area of a circle is proportional to the square of its radius, and knowing that tornadoes and hurricanes are roughly circular, we see that the area covered by a hurricane is likely to be millions of times greater than the area covered by a tornado. Furthermore, a tornado exists only for a couple of hours at most, while a hurricane can have a lifetime of several days or even a week or more.

Because hurricanes obtain most of their energy from the latent heat released by condensation, they are most common where a deep layer of warm water fuels them. Given that tropical oceans have their highest surface temperatures and evaporation rates in late summer and early fall, it is not surprising that August and September are the prime hurricane months in the Northern Hemisphere, with January to March the main season in the Southern Hemisphere.*

*The United States National Hurricane Center defines the hurricane season as the period from June 1 to November 30. Tropical storms in other months are rare events—from 1871 to 1996, only six storms formed in December.

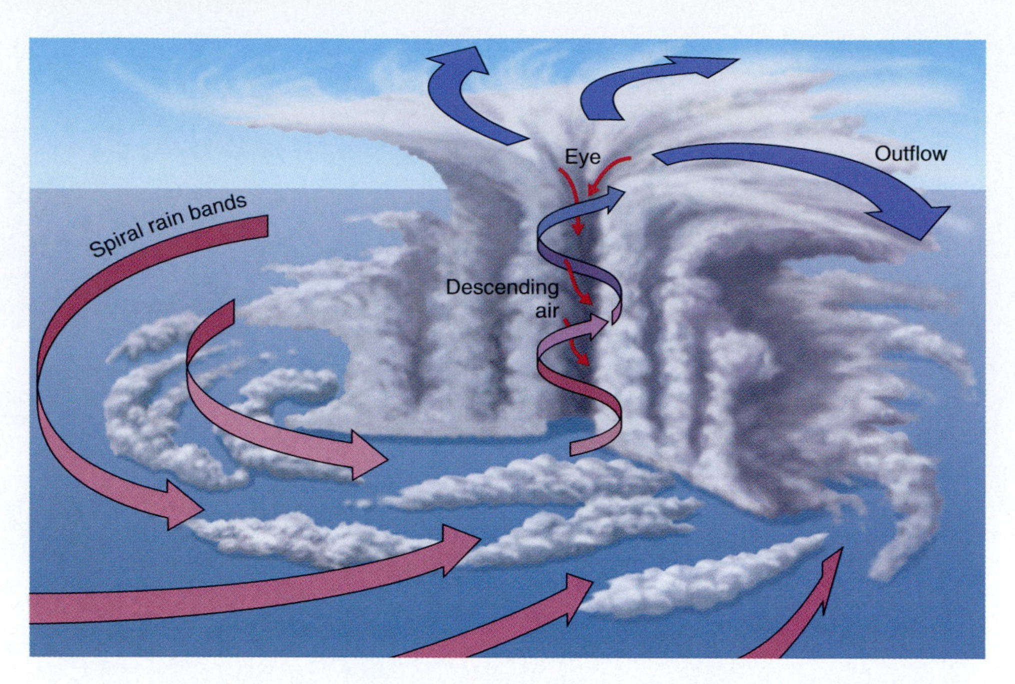

Figure 12–3
A cross section of a typical hurricane.

Hurricanes do not consist of only one uniform convective cell. Instead they contain a large number of thunderstorms arranged in a pinwheel formation, with bands of thick clouds and heavy thundershowers spiraling counterclockwise (in the Northern Hemisphere) around the storm center (Figure 12–3). The bands of heavy convection are separated by areas of weaker uplift and less intense precipitation. The wind speed and the intensity of precipitation both increase toward the center of the system, reaching a maximum 10 to 20 km away from the center, at what is called the *eye wall* (described in a later section).

Though hurricanes are usually about one-third the size of mid-latitude cyclones, the pressure difference across a hurricane is about twice as great. They therefore have extreme horizontal pressure gradients that generate powerful winds: average hurricanes have peak winds of about 150 km/hr (90 mph), and the most intense hurricanes have winds up to 350 km/hr (210 mph). In addition to being smaller and more powerful than mid-latitude cyclones, hurricanes also differ by not having the fronts that characterize cyclones outside the Tropics.

Hurricanes also differ from mid-latitude cyclones in that they are warm-core cyclones. As air flows inward toward lower pressure, the warm ocean surface supplies large amounts of latent and sensible heat to the overlying air. Because pressure within the moving air decreases as it flows toward the low, adiabatic expansion keeps the temperature from increasing dramatically, with the result that there is little temperature difference across the base of the storm. Nevertheless, much thermal energy is added, resulting in a "warm" central core. Aloft, after condensation and the release of latent heat, the warmth is reflected in temperature, so that temperatures near the center are much higher than those of the surrounding air (Figure 12–4).

As a warm-core low, pressure within a hurricane decreases relatively slowly with increasing altitude (Chapter 10). Thus, the horizontal pressure gradient within the storm gradually decreases with altitude. At about 7.5 km (25,000 ft)—about the 400 mb level—the air pressure is the same as that immediately outside the storm. From this height to the lower stratosphere, the hurricane has relatively high pressure. So unlike the lower part of the hurricane in which the air rotates cyclonically, the air in its upper portion spirals anticyclonically from its center (clockwise in the Northern Hemisphere).

In the upper reaches of the storm, the low temperatures cause water droplets to freeze into ice crystals. As the ice crystals spiral out of the storm center, they create a

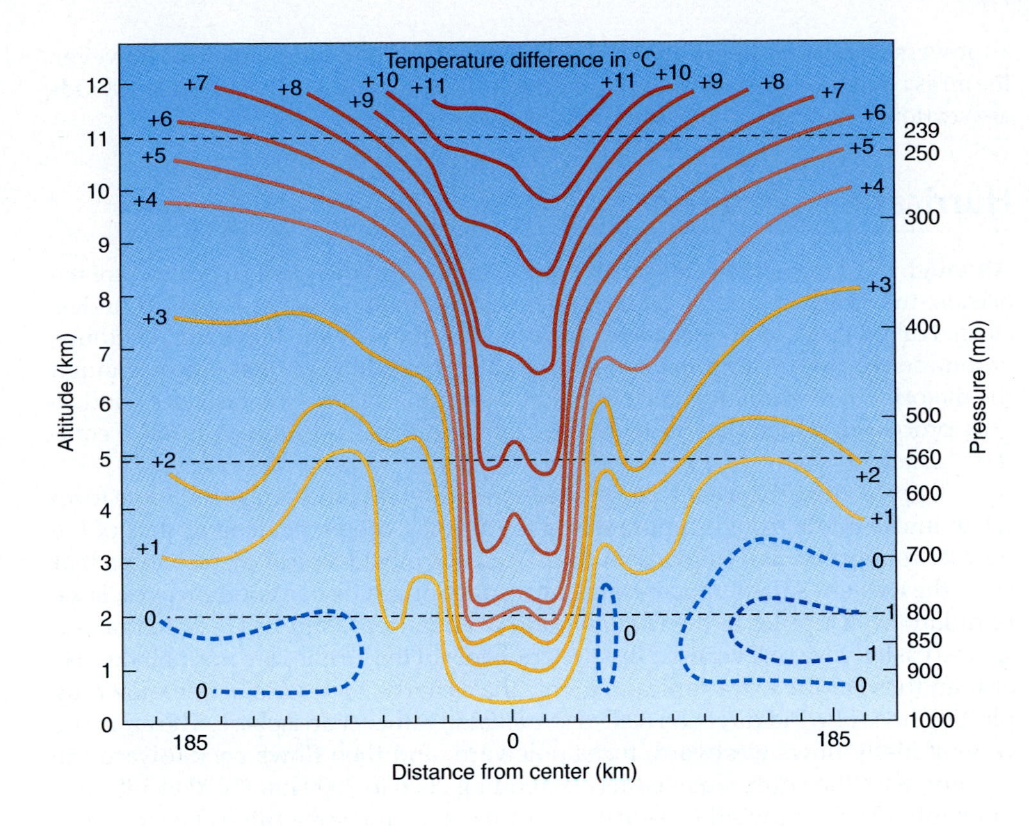

◀ Figure 12–4
Temperature differences across a hurricane relative to the surrounding air (°C). Near the surface, temperatures increase only slightly toward the eye. Aloft, however, temperatures exceed those of the surroundings by about 10 °C (18 °F).

blanket of cirrostratus clouds that overlies and obscures the pinwheel-like structure of the storm. That explains why hurricanes on satellite images often appear to have a uniform thickness and intensity, when in fact they are strongly banded.

The Eye and the Eye Wall

One of the most distinctive characteristics of a hurricane is its **eye**, a region of relatively clear skies, slowly descending air, and light winds. Hurricane eyes average about 25 km (15 mi) in diameter, with most ranging from 20 to 50 km (15 to 30 mi). Eye diameters vary considerably among individual storms, with some as small as 6 km (3.5 mi) and others almost as large as 100 km (60 mi). The change in the size of an eye through time gives some indication of whether the hurricane is intensifying or weakening. Generally, a shrinking eye indicates an intensifying hurricane.

Along the margin of the eye lies the **eye wall**, the zone of most intense storm activity. The eye wall contains the strongest winds, thickest cloud cover, and most intense precipitation of the entire hurricane. Directly beneath the eye wall, rainfall rates of 2500 mm/day (100 in./day) are not uncommon. The abrupt transition from an eye wall to the eye causes a strong and rapid change in weather. Imagine a hurricane about to make a direct hit on a small island. As the hurricane center approaches the island, the intensity of the wind and rain steadily increase, becoming most intense as the eye wall arrives. But when the eye reaches the island, the storm seems to suddenly dissipate, as blue skies and calm conditions take hold. Of course, the storm has not dissipated at all. Rather, there is just a brief lull until the opposite side of the eye wall covers the island and storm conditions resume. Because the average hurricane eye is about 20 km in diameter and travels at about 20 km/hr, the calm associated with passage of the eye lasts about an hour. Clearly, if the eye just grazes the island, the break in the storm will be even shorter.

The air temperature at the storm's surface within an eye is several degrees warmer than outside the eye because compression of the sinking air causes it to warm adiabatically. The air is also drier, because the warming of the unsaturated air lowers its relative humidity. Contrary to what is commonly believed, however, the air is not entirely cloud-free within the eye; instead, fair weather cumulus clouds are scattered throughout the otherwise blue sky.

Hurricane Formation

Although most tropical storms attain hurricane status in the western portions of the oceans, their earliest origins lie far to the east as small clusters of small thunderstorms called **tropical disturbances**. Tropical disturbances are disorganized groups of thunderstorms having weak pressure gradients and little or no rotation. Though the majority of disturbances die out before becoming major systems, a few undergo a process in which they grow in size, join together, and rotate around a common center.

Tropical disturbances can form in several different environments. Some form when mid-latitude troughs migrate into the Tropics; others develop as part of the normal convection associated with the ITCZ. But most tropical disturbances that enter the western Atlantic and become hurricanes originate in **easterly waves**, large undulations or ripples in the normal trade wind pattern. Figure 12–5 illustrates a typical easterly wave. Because pressure gradients in the Tropics are normally weaker than those of the extratropical regions, the easterly waves are better shown by plotting lines of wind direction (called *streamlines*) rather than isobars. The air in the wave initially flows westward, turns poleward, and then flows back toward the equator, with the entire wave pattern extending 2000 to 3000 km (1200 to 1800 mi) in length. On the upwind (eastern) side of the axis, the streamlines become progressively closer together, indicating that surface motions are convergent. With convergence there is rising motion (Chapter 6), thus the tropical disturbance is located upwind of the wave axis (the dashed line) of the easterly wave. Surface divergence downwind of the wave axis leads to clear skies. (An explanation for why the streamlines are convergent and divergent is somewhat complicated; the most important

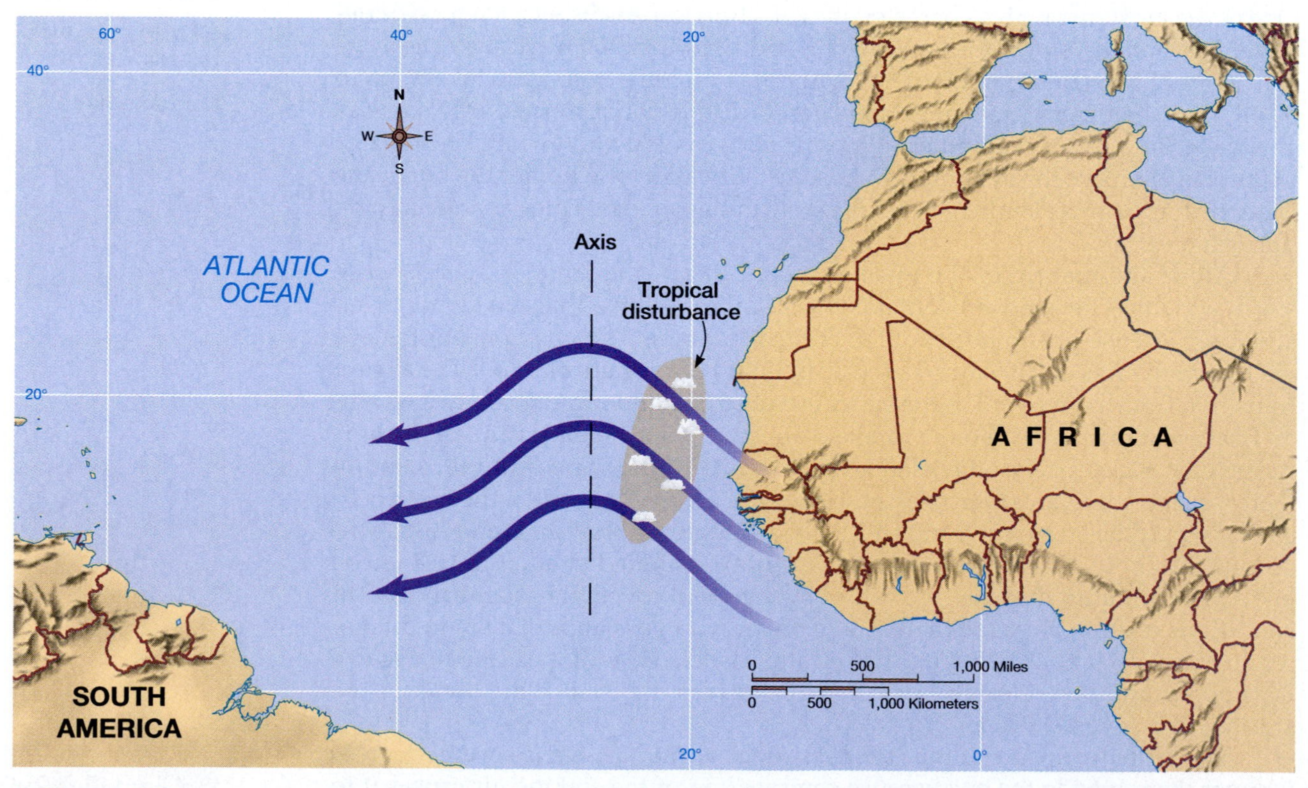

▲ **Figure 12–5**
Easterly waves have surface convergence and cloud cover east of the axis and divergence to the west.

factor involves changes in relative vorticity that occur as the air moves poleward and equatorward.)

The tropical disturbances that affect the Atlantic Ocean, Caribbean, and the Americas mostly form over western Africa, south of the Saharan desert. Being in the zone of the trade winds, these storms tend to migrate westward. When they reach the west coast of Africa, they weaken as they pass over the cold Canary current over the eastern Atlantic. There the low water temperatures chill the air near the water surface and cause the air to become statically stable. If the disturbances migrate beyond the coastal zone of surface upwelling, however, the warmer waters farther offshore raise the temperature and humidity of the lower atmosphere and cause the air to become unstable. Then, as the storms continue westward, a small percentage develop into more intense and organized thunderstorm systems. Easterly waves move westward at about 15 to 35 km/hr (10 to 20 mph), and so it typically takes about a week to 10 days for an embedded tropical disturbance to migrate across the Atlantic.

The vast majority (probably more than 90 percent) of tropical disturbances die out without ever organizing into more powerful systems. But some undergo a lowering of pressure and begin to rotate cyclonically. When a tropical disturbance develops to the point where there is at least one closed isobar on a weather map, the disturbance is classified as a **tropical depression**. If the depression intensifies further and maintains wind speeds above 60 km/hr (37 mph), it becomes a **tropical storm**. A further increase in sustained wind speeds to 120 km/hr (74 mph) creates a true hurricane. While only a small fraction of tropical disturbances ever become tropical depressions, a larger proportion of depressions become tropical storms, and an even greater percentage of tropical storms ultimately become hurricanes.

The location at which hurricanes are most likely to form varies seasonally. Early in the Atlantic season, weak fronts in the western ocean extend southward over warm tropical water. Wind shear across the fronts provides the circulation necessary for cyclone development. Later in the season, fronts are confined to higher latitudes and no longer play a role in cyclogenesis. Instead, warm waters are found progressively farther to the east, so that disturbances leaving the African continent can grow into full-scale cyclones. The net effect is that the birthplace of tropical cyclones moves from west to east across the tropical ocean during the first half of the season. In the late fall, the breeding ground moves westward as frontal activity again emerges as a primary agent of cyclone genesis.

As with their Atlantic counterparts, Pacific hurricanes move westward during their formative stages. Many come near Hawaii, but most bypass the islands or die out before reaching them. Unfortunately, this is not always the case. In September 1992, Hurricane Iniki battered the island of Kauai with wind gusts up to 258 km/hr (160 mph) and brought heavy flooding to the beach resort areas. The hurricane destroyed or severely damaged half of the homes on the island and devastated most of the tourist industry.

Conditions Necessary for Hurricane Formation

Although the dynamics of hurricanes are extremely complex, meteorologists have long recognized the conditions that favor their development. Great amounts of heat are needed to fuel hurricanes, and the primary source of this energy is the release of latent heat supplied by evaporation from the ocean surface. Because high evaporation rates depend on the presence of warm water, hurricanes form only where the ocean has a deep surface layer (several tens of meters in depth) with temperatures above 27 °C (81 °F). The need for warm water precludes hurricane formation poleward of about 20 degrees because sea surface temperatures are usually too low there. This also explains why hurricanes develop most often in late summer and early fall, when tropical waters are warmest.

Hurricane formation also depends on the Coriolis force, which must be strong enough to prevent filling of the central low pressure. The absence of a Coriolis effect at the equator prohibits hurricane formation between 0° and 5° latitude. This factor and the need for high water temperatures explain the pattern shown in Figure 12–2, in which tropical storms attain hurricane status between the latitudes of 5° and 20°.

Stability is also important in hurricane development, with unstable conditions throughout the troposphere an absolute necessity. Along the eastern margins of the oceans, cold currents and upwelling cause the lower troposphere to be statically stable, inhibiting uplift. Moreover, the presence of the trade wind inversion puts a cap on any mixing that might otherwise occur. Moving westward, water temperatures typically increase and the trade wind inversion increases with height or disappears altogether, and so hurricanes become more prevalent. Finally, hurricane formation requires an absence of strong vertical wind shear, which disrupts the vertical transport of latent heat.

Once formed, hurricanes are self-propagating (just as severe storms outside the Tropics are self-maintaining). That is, the release of latent heat within the cumulus clouds causes the air to warm and expand upward. The expansion of the air supports upper-level divergence, which draws air upward and promotes low pressure and convergence at the surface. This leads to continued uplift, condensation, and the release of latent heat. So, if hurricanes are self-propagating, can they intensify indefinitely, until they attain supersonic speeds? No, because they are ultimately limited by the supply of latent heat, which in turn is constrained by the temperature of the ocean below and by the processes underlying evaporation and convection. The importance of ocean temperature suggests that if the oceans were to become warmer, hurricanes would theoretically become more intense. This topic has received considerable attention lately because of the possibility of climatic warming, which could be accompanied by higher ocean temperatures.

Hurricane Movement and Dissipation

The movement of tropical systems is related to the stage in their development. Tropical disturbances and depressions are guided mainly by the trade winds, and therefore tend to migrate westward. The influence of the trade winds often diminishes after the depressions intensify into tropical storms. Then the upper-level winds and the spatial distribution of water temperature become more important in determining their speed and direction (with the storms tending to move toward warmer seas). Once fully developed, tropical storms become more likely to move poleward, as shown in Figures 12–2, 12–6, and 12–7.

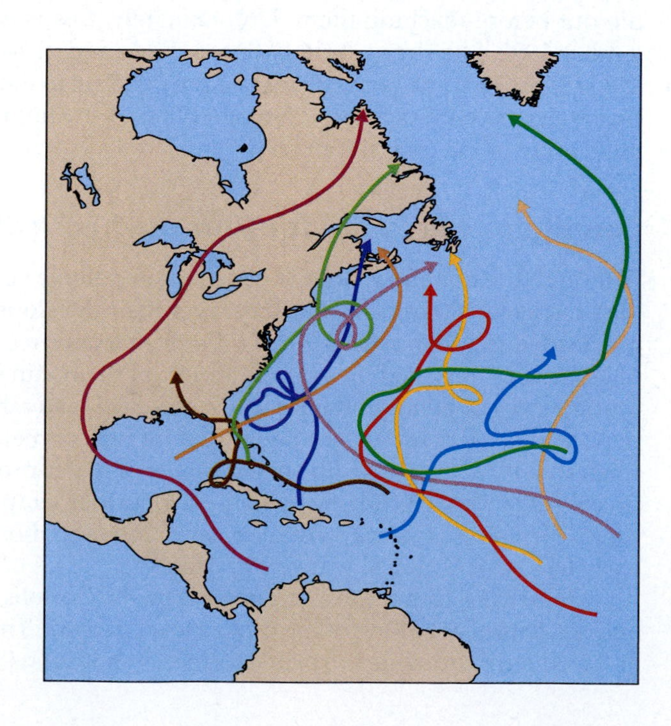

Figure 12–6
Tropical storms and hurricanes have a tendency to move north or northeast out of the Tropics along the southeast coast of North America. Their paths are often erratic, as evidenced by these examples.

Hurricanes and tropical storms often move in wildly erratic ways. One might, for example, move in a constant direction for a time, then suddenly change speed and direction, and even backtrack along its previous path. Figure 12–6 illustrates the variability of hurricane movement by plotting some particularly erratic paths along the east coast of North America.

Although Atlantic tropical storms and hurricanes can travel great distances along the North American east coast, they usually weaken considerably as they approach the northeastern United States and the Maritime Provinces of Canada. These storms usually lack the strong winds that characterize hurricanes in the low latitudes, but can still bring intense rainfall and flooding. On rare occasions, however, the storms can maintain their strong winds even as they move considerable distances away from the subtropics. An intense September 1938 hurricane, for example, brought 200 km/hr (120 mph) winds to Long Island, New York, as it moved toward New England. Its estimated 600 fatalities made it the fourth deadliest of all U.S. hurricanes. More recently, in September 1985, Hurricane Gloria brought considerable wind and flood damage to Long Island and Connecticut (Figure 12–7).

Although hurricanes and tropical storms can move into the northeastern United States, along the West Coast they do not migrate nearly as far north without weakening to tropical depressions. The reason for this is the difference in water temperatures along the two coasts. The Pacific Coast is dominated by upwelling and the cold California current, while the warm Gulf Stream flowing along the East Coast provides a greater supply of latent heat. Sometimes, however, storms off the coast of Mexico move to the northeast across Baja California and into southern California. These storms lose their supply of latent heat and lose their intensity as they

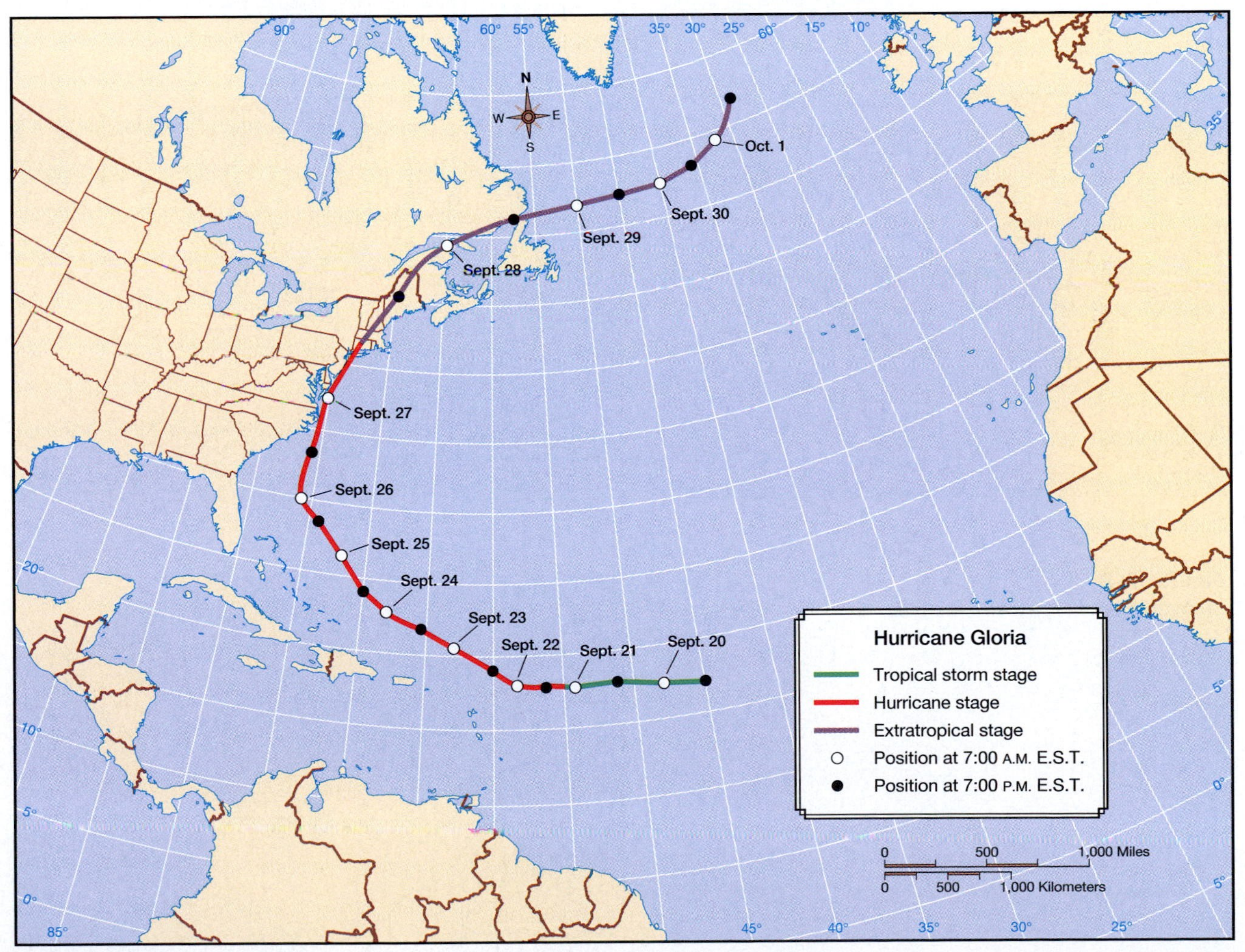

▲ **Figure 12–7**
Path of Hurricane Gloria.

move inland, but they can still bring heavy rains and flooding. In 1976 Hurricane Kathleen caused massive flooding in the desert of southern California that wiped out part of Interstate Highway 8.

After making landfall, a tropical storm may die out completely within a few days. Even as the storm weakens, though, it can still import huge amounts of water vapor and bring very heavy rainfall hundreds of kilometers inland. This is especially true when the remnant of a hurricane moving poleward joins with a mid-latitude cyclone moving eastward. Exactly this happened in 1969 when one of North America's most notorious hurricanes, Camille, moved northward from the coast of Mississippi (Figure 12–8). After its high winds and tidal flooding brought extreme

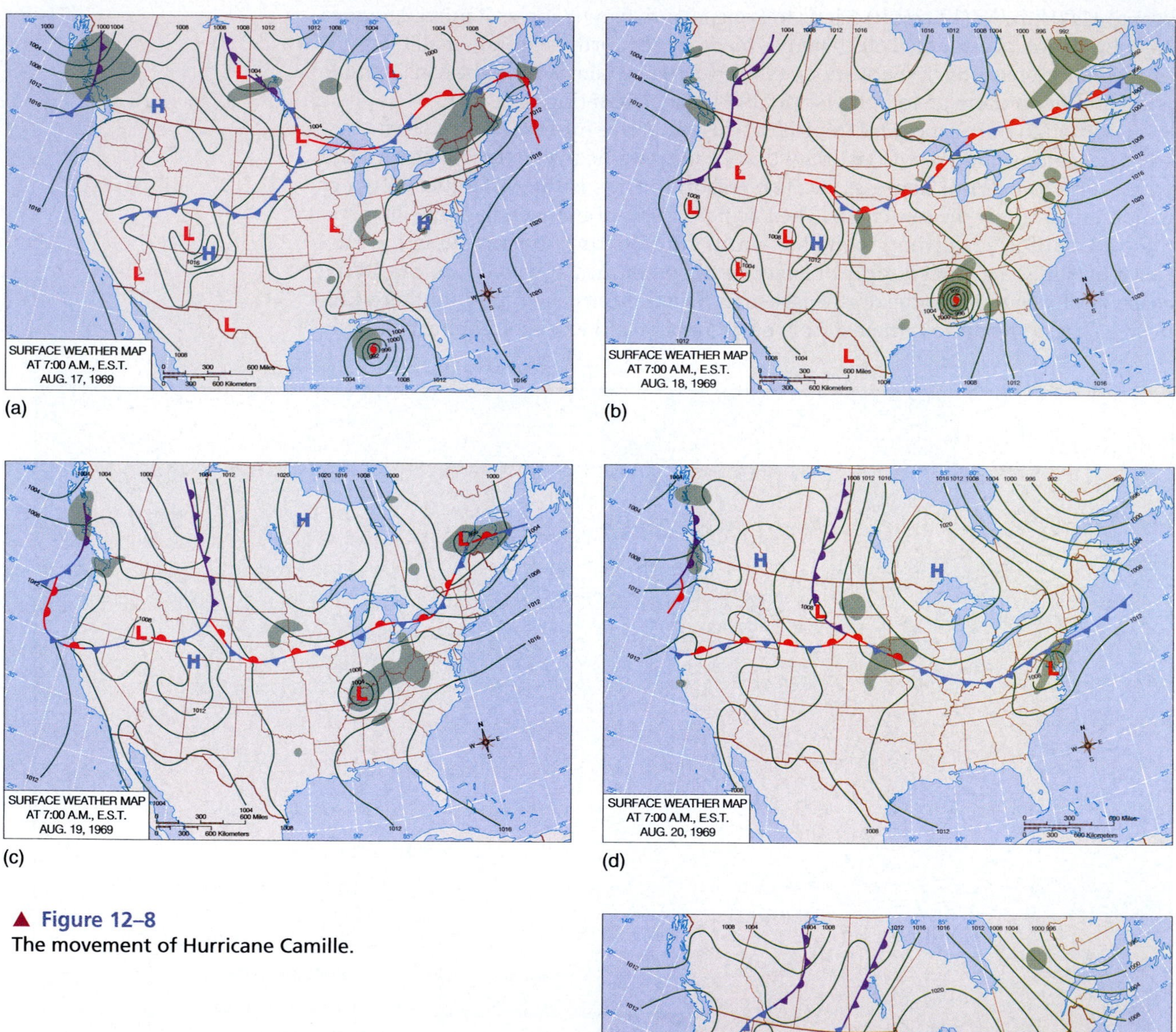

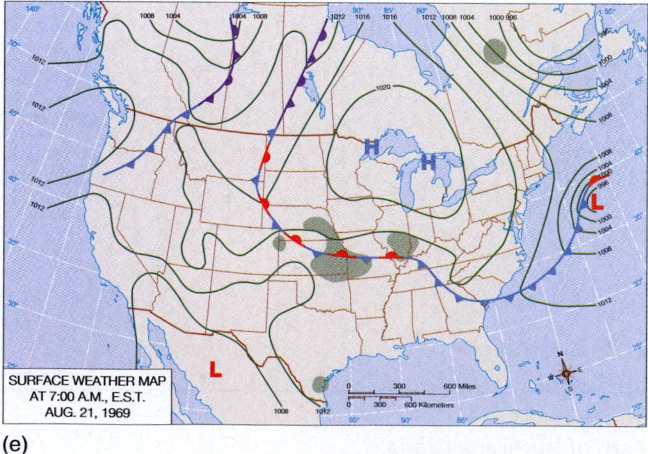

▲ **Figure 12–8**
The movement of Hurricane Camille.

damage to the Gulf Coast, the storm moved northeastward toward the western slopes of the Appalachians. There, orographic uplift coupled with low pressure and high water vapor content of the remnant hurricane could easily have produced serious flooding. But to make matters worse, an eastward-moving cold front reached the mountains at the same time as the former hurricane. The combination of moist air, low pressure, an approaching front, and the orographic effect set the stage for intense rains and flash flooding that killed more than 150 persons.

Hurricane Destruction and Fatalities

By definition, hurricane winds exceed 120 km/hr, and many are much faster. It is not surprising, then, that hurricane-force winds can cause extensive damage and even destroy well-built homes. As we mentioned, hurricanes also produce staggeringly intense rainfall, with rates on the order of meters per day found beneath the eye wall. The rate for a location that remains stationary beneath a passing storm is smaller, but still huge—on the order of 25 cm/day (10 in./day).

Many hurricanes also contain clusters of tornadoes, most often in the right-forward quadrant (Figure 12–9). They usually occur far enough away from the center that they are surrounded by relatively weak winds. It appears that slowing of the wind by friction at landfall contributes to tornado formation. Hurricane-spawned tornadoes tend to have shorter life spans than tornadoes in the central United States.

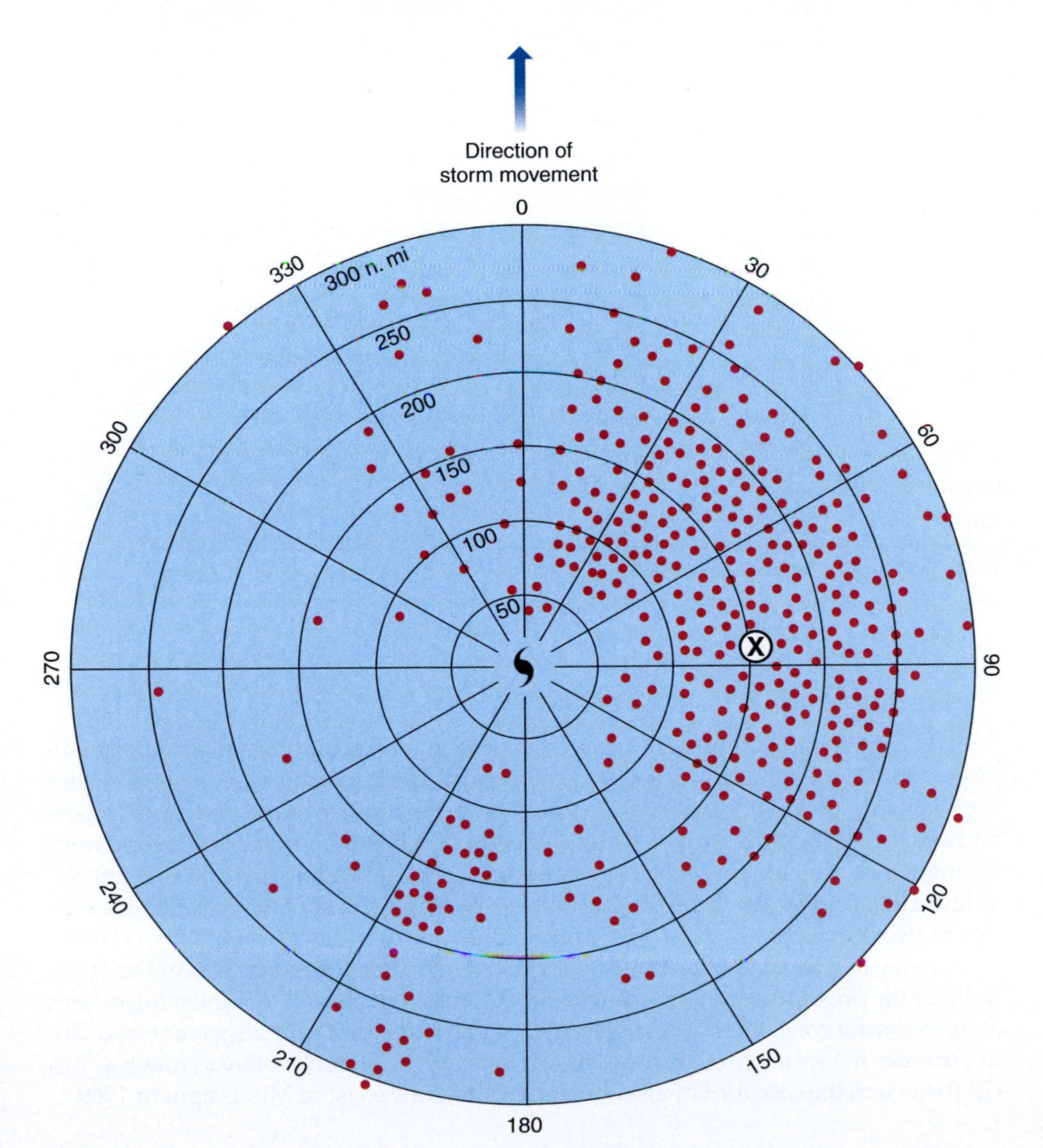

Figure 12–9 Tornadoes most often form in the right-forward quadrant of hurricanes (based on the direction in which the storm is moving). The figure is based on data from 373 hurricane-embedded tornadoes between the years 1948 and 1972. Each dot represents the occurrence of a tornado. The circled X indicates the mean position of tornadoes relative to the storm center.

12–1 Special Interest

The Naming of Hurricanes

During hurricane season, several tropical storms or hurricanes can arise simultaneously over the various oceans. Meteorologists identify these systems by assigning names when they reach tropical storm status. The World Meteorological Organization (WMO) has created several lists of names for tropical storms over each of the oceans. The names on each list are ordered alphabetically, starting with the letter *A* and continuing up to the letter *W*. When a depression attains tropical storm status, it is assigned the next unused name on that year's list. At the beginning of the following season, names are taken from the next list, regardless of how many names were unused in the previous season. Six lists have been compiled for the Atlantic Ocean, and the names on each list are used again at the end of each 6-year cycle. English, Spanish, and French names are used for Atlantic hurricanes.

Table 1 • Western Atlantic Tropical Storm and Hurricane Names

2003	2004	2005	2006	2007	2008*
Ana	Alex	Arlene	Alberto	Andrea	Arthur
Bill	Bonnie	Bret	Beryl	Barry	Bertha
Claudette	Charley	Cindy	Chris	Chantal	Cristobal
Danny	Danielle	Dennis	Debby	Dean	Dolly
Erika	Earl	Emily	Ernesto	Erin	Edouard
Fabian	Frances	Franklin	Florence	Felix	Fay
Grace	Gaston	Gert	Gordon	Gabrielle	Gustav
Henri	Hermine	Harvey	Helene	Humberto	Hanna
Isabel	Ivan	Irene	Isaac	Ingrid	Isidore
Juan	Jeanne	Jose	Joyce	Jerry	Josephine
Kate	Karl	Katrina	Kirk	Karen	Kyle
Larry	Lisa	Lee	Leslie	Lorenzo	Lili
Mindy	Matthew	Maria	Michael	Melissa	Marco
Nicholas	Nicole	Nate	Nadine	Noel	Nana
Odette	Otto	Ophelia	Oscar	Olga	Omar
Peter	Paula	Philippe	Patty	Pablo	Paloma
Rose	Richard	Rita	Rafael	Rebekah	Rene
Sam	Shary	Stan	Sandy	Sebastien	Sally
Teresa	Tomas	Tammy	Tony	Tanya	Teddy
Victor	Virginie	Vince	Valerie	Van	Vicky
Wanda	Walter	Wilma	William	Wendy	Wilfred

Some of the scheduled 2008 storm names are subject to retirement.

In addition to the threat of heavy rain, strong winds, and tornadoes, coastal regions are also vulnerable to a special problem called the **storm surge**, a rise in water level induced by the hurricane. Two processes create a storm surge, the more important one being the piling up of water as heavy winds drag surface waters forward. Strong winds blowing toward a coast force surface waters landward and thereby elevate sea level, while also bringing heavy surf. The low atmospheric pressure in a hurricane also contributes to the storm surge, in the same way that the height of a column of mercury in a barometer responds to variations in atmospheric pressure. For every millibar the pressure decreases, the water level rises 1 cm (0.4 in.). For most hurricanes along a coastal zone, the storm surge elevates the water level only a meter or two. But in extreme circumstances, storm surges can increase the water level by as much as 7 m (23 ft), as was the case for Hurricane Camille along the coast of Mississippi in 1969.

Particularly notable hurricanes can have their names "retired" by the WMO if an affected nation requests the removal of that name from the list. All replacements are made with names of the same gender and language. As of 2002, 48 names have been retired from the Atlantic hurricane list.

The practice of naming hurricanes appears to have begun during World War II when meteorologists in the Pacific assigned female names (possibly after wives and girlfriends) to tropical storms and typhoons. This practice was adopted by the U.S. National Weather Service (then called the Weather Bureau) in 1953 and maintained until 1979, when male names were added to the lists. The names for the Atlantic and east Pacific hurricanes for the years 2003–2008 are presented in Tables 1 and 2.

Table 2 • Eastern Pacific Tropical Storm and Hurricane Names

2003	2004	2005	2006	2007	2008*
Andres	Agatha	Adrian	Aletta	Alvin	Alma
Blanca	Blas	Beatriz	Bud	Barbara	Boris
Carlos	Celia	Calvin	Carlotta	Cosme	Cristina
Dolores	Darby	Dora	Daniel	Dalila	Douglas
Enrique	Estelle	Eugene	Emilia	Erick	Elida
Felicia	Frank	Fernanda	Fabio	Flossie	Fausto
Guillermo	Georgette	Greg	Gilma	Gil	Genevieve
Hilda	Howard	Hilary	Hector	Henriette	Hernan
Ignacio	Isis	Irwin	Ileana	Ivo	Iselle
Jimena	Javier	Jova	John	Juliette	Julio
Kevin	Kay	Kenneth	Kristy	Kiko	Kenna
Linda	Lester	Lidia	Lane	Lorena	Lowell
Marty	Madeline	Max	Miriam	Manuel	Marie
Nora	Newton	Norma	Norman	Narda	Norbert
Olaf	Orlene	Otis	Olivia	Octave	Odile
Patricia	Paine	Pilar	Paul	Priscilla	Polo
Rick	Roslyn	Ramon	Rosa	Raymond	Rachel
Sandra	Seymour	Selma	Sergio	Sonia	Simon
Terry	Tina	Todd	Tara	Tico	Trudy
Vivian	Virgil	Veronica	Vicente	Velma	Vance
Waldo	Winifred	Wiley	Willa	Wallis	Winnie
Xina	Xavier	Xina	Xavier	Xina	Xavier
York	Yolanda	York	Yolanda	York	Yolanda
Zelda	Zeke	Zelda	Zeke	Zelda	Zeke

Some of the scheduled 2008 storm names subject to retirement.

Storm surges along low-lying coastal plains can be extremely devastating where the rise in sea level brings waters far inland. Furthermore, the heavy waves generated by the strong winds pound away at structures, with debris carried by the waves adding to the problem. In the case of Hurricane Camille, the storm completely destroyed the Richelieu Apartments along the Mississippi coast (Figure 12–10). Despite previous warnings to evacuate, about a dozen residents decided to have a "hurricane party," figuring that if the storm surge got too high they would just move up to the third floor of the building. Unfortunately for them, the storm surge undermined the apartment foundation, and more than two dozen died when the building completely collapsed. Storm surges are most destructive when they coincide with high tides, especially over bays and inlets that have an extreme range of height between high and low tide.

Storm Surge

(a)

(b)

▲ **Figure 12–10**
The Richelieu Apartments before (a) and after (b) the passage of Hurricane Camille in 1969.

Weather in Motion
Hurricane Damage

Hurricane winds and storm surges are most intense on the right-hand side of the storm relative to the direction it is moving. To understand why, imagine that a hypothetical hurricane in the Gulf of Mexico is moving northward at 50 km/hr, with a uniform eye wall having 200 km/hr—120 mph—winds (Figure 12–11). If this hurricane makes landfall along the coast of Mississippi and Alabama, the winds to the east of the eye move at 250 km/hr (the sum of wind speed and storm speed). Along the western part of the eye wall, the winds are only 150 km/hr (200 km/hr minus 50 km/hr). In addition to having greater winds, the right-hand side of the storm will also have a greater storm surge because the winds push the water onshore rather than offshore.

Though storm surges present the greatest potential for catastrophic destruction and have claimed thousands of American lives over the last few centuries, over the last few decades there has been a major decrease in the incidence of storm surge fatalities. A recent study published by a researcher at the National Hurricane Center revealed that between 1970 and 1999, more than half the fatalities from tropical storms and hurricanes in North America resulted not from storm surges, but rather from freshwater flooding. Only about one quarter of the fatalities associated with tropical storms and hurricanes (or their remnants) occurred in coastal counties. These figures do not imply that storm surges are no longer extremely dangerous; that is not the case at all. The reduction in deaths associated with them is partially the result of a random decrease in the number of strong hurricanes hitting populated coastal regions during the 30-year period, along with a better ability to predict the movement of hurricanes and improved evacuation procedures.

Hurricane specialists are concerned that there may be a significant increase in the number of hurricanes and intense hurricanes making landfall in the United States. Though the period from the early 1970s to the mid-1990s was marked by a lower than normal incidence of Atlantic hurricanes, the 8-year period from 1995 and 2002 has been exceptionally active. Scientists believe that a change in a multiyear cycle of sea surface temperatures in the North Atlantic may have provided increased fuel for hurricane development that has led to the increased storm activity, and these conditions may persist over the next few decades.

We are more vulnerable than ever to the destructive potential of hurricanes for the simple reason that there has been enormous population growth along the Atlantic and Gulf Coasts in recent decades, swelling the number of residents to 139 million. As a result, a huge number of people in the region may underestimate the likelihood of being hit by a major hurricane and live in homes not proven to withstand the full effects of a hurricane's fury. To make matters worse, the improvements in the highway system necessary for a full evacuation have not kept pace with the rate of growth in the area.

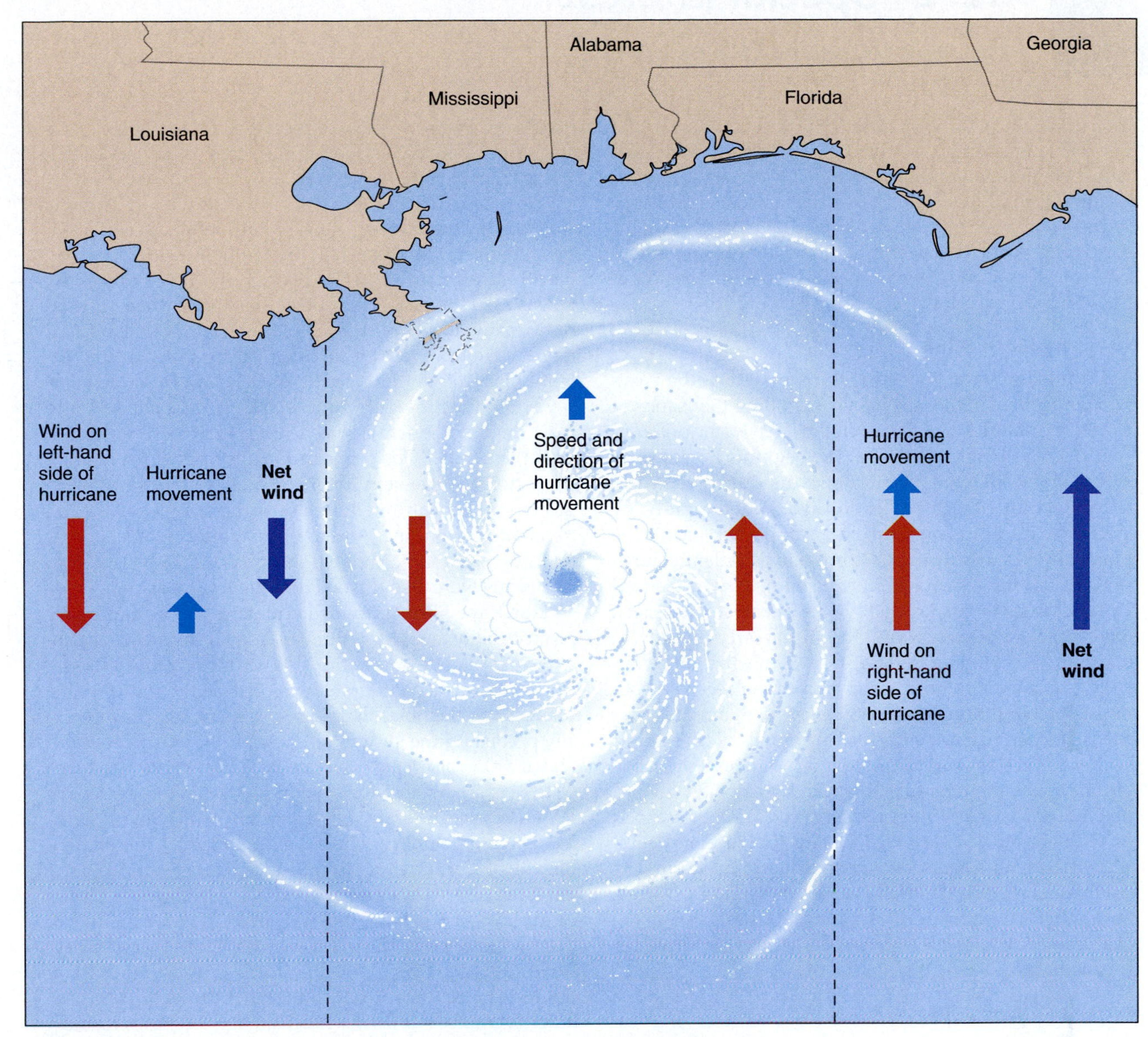

▲ **Figure 12–11**
The varying intensities of the wind on the left and right sides of hurricanes. The direction and speed of the wind components are signified by the direction and length of the arrows, respectively. A hypothetical hurricane moves northward at 50 km/hr. Along the right-hand side, the 200 km/hr winds are in the same direction as the movement of the storm, so there is a net wind speed of 250 km/hr. On the left side, the internal winds blow opposite the direction of storm movement, so the net winds are southward at 150 km/hr.

Hurricane Forecasts and Advisories

Responsibility for tracking and predicting Atlantic and east Pacific hurricanes lies with the National Hurricane Center (NHC) in Miami, Florida. During hurricane season, this office of the National Weather Service obtains constantly updated surface reports and satellite data to determine current storm conditions. Sophisticated numerical models on a supercomputer predict the formation, growth, and movement of tropical storms and hurricanes. When active hurricanes approach land, specially equipped aircraft fly into the storms and provide additional reconnaissance data from airborne radar and *dropsondes*, packages containing temperature, pressure, and moisture sensors and transmitters released from the plane into the storm.

12–2 Special Interest

Recent Notable Hurricanes

The period between 1995 and 2002 has been marked by an unusually high number of Atlantic tropical storms and hurricanes, with many making landfall. Of the eight seasons during this time span, all but two had at least eight hurricanes—well above the annual average of 5.8. Several of the storms and hurricanes over this period have been particularly noteworthy and deserve special mention (the paths of the hurricanes are plotted in Figure 1).

Hurricane Mitch, 1998

Hurricane Mitch was the worst hurricane to hit the Western Hemisphere in the last 200 years, killing thousands of people in Central America in October 1998 (some of the effects of Mitch were described at the beginning of Chapter 3). The storm attained hurricane status during the early morning hours of October 24 and intensified rapidly. By October 26 its central pressure had dropped to 905 mb, the fourth lowest pressure ever recorded in an Atlantic hurricane. Mitch then swept a curved path toward Central America as its winds gradually lost intensity. Tragically, however, once the storm hit land, it moved very slowly, lingering and bringing intense rains that lasted for several days. Parts of Honduras and Nicaragua received estimated precipitation totals of 85 cm (35 in.), causing extensive flooding and mudslides. Although Mitch's winds weakened substantially, the remnant system brought heavy rainfall across the region as it tracked northward toward the Gulf of Mexico. It then reintensified when it reached the Gulf and turned eastward toward the Yucatan Peninsula, where it once again weakened to a tropical storm. The system made its final landfall over southern Florida, where it brought 100 km/hr (60 mph) winds. It dissipated in the Atlantic on November 5, two weeks after having formed in the Caribbean.

Hurricane Floyd, 1999

The hurricane season of 1999 produced eight hurricanes, of which five were Category 4 (the largest number of Category 4 hurricanes to have occurred in a season since accurate records began in the 1940s). The first of the season's storms to have a major impact was Hurricane Dennis. Dennis made landfall at the Cape Lookout National Seashore on the afternoon of September 5, but fortunately its eye never hit land, keeping the stronger right side of the hurricane offshore. One of the major effects of Hurricane Dennis was its heavy rains that saturated the soil and finally put an end to a prolonged drought. While the end of the drought was good news, the elevated water table set the stage for catastrophic flooding less than two weeks later, when Hurricane Floyd arrived.

Much of Hurricane Floyd's path in the western Atlantic was roughly parallel to Dennis's. Unlike the case with Dennis, however, Floyd's eye made landfall near Cape Fear, North Carolina, shortly after midnight on September 16. It was now a Category 2 hurricane. Floyd then tracked northeastward, passing over Norfolk, Virginia, later that day. Weakened to a tropical storm, it continued up the Atlantic Coast along the Maryland, Delaware, and New Jersey shorelines toward Long Island. It then crossed into New England, where it was downgraded to a depression on September 17. The primary damage from Floyd was not from storm surge activity. Instead, abundant rainfall on ground earlier saturated by Hurricane Dennis resulted in tremendous flooding. And to make matters worse, it wasn't just water that inundated the area. Tons of waste from hog farms were also flushed into the floodwaters.

Hurricane Floyd provided an interesting lesson on our present ability to forecast hurricanes. While hurricane watches were in effect, errors in the 24-hour tracking forecast were about average according to the record for the last 10 years. There was a clear bias in the error, with the forecasts generally predicting the system to track farther west than it did and at a slower rate of movement. The forecasts also tended to overpredict the storm's winds—sometimes by tens of kilometers per hour. Despite these errors, however, warnings did not go unheeded, and an estimated 3 million people—the largest evacuation in U.S. history—took to the interstate highways and headed west. In the end, Hurricane Floyd was a major disaster, although it was nowhere near as deadly and destructive as had been feared. Fifty-seven people died from the storm (more than half of them in North Carolina), making it the deadliest U.S. hurricane since Agnes in 1972. Damage estimates reached $6 billion.

Tropical Storm Allison, 2001

Tropical storm Allison demonstrated in 2002 that a tropical storm need not attain hurricane status to become a major disaster. Allison hit the south Texas coast on the fifth of June and hovered over the area for nearly a week. Its heavy rainfall (as much as 96 cm, or 40 inches) caused major flooding across Texas and Louisiana, killing 24 people and flooding more than 46,000 homes and businesses. Even after weakening to a subtropical depression, Allison

The NHC uses the standard computer models for conventional weather forecasting (discussed in Chapter 13), as well as others developed specifically for hurricanes. The latter fall into three categories: *statistical*, *dynamical*, and *hybrid*. Statistical models apply information on past hurricane tracks and use them as predictors for current storms. Dynamical models take information on current atmospheric and sea surface conditions and apply the governing laws of physics to the current data. Hybrid models combine elements of statistical and dynamical models. The models repeatedly forecast the movement and internal changes of hurricanes for short time continued to bring heavy rain to the southeast as it moved toward North Carolina and then northward. Even parts of the northeastern United States found themselves victim to Allison's floodwaters.

Hurricane Lili, 2002

Lili attained tropical storm status on September 23 as it arced westward in the western Atlantic. The system advanced over water for about a week until it made landfall over Jamaica, where it brought four days of flooding. Lili became a hurricane on October 1 just before hitting western Cuba, and intensified rapidly as it approached the Louisiana coast. On October 3 wind speeds reached 235 km/hr (145 mph), making it a Category 4 hurricane. Fortunately, at that point Lili began to weaken rapidly so that when it hit the Louisiana coast some 13 hours later, it had downgraded to a Category 1 with wind speed of about 145 km/hr (90 mph). Although the strong winds, storm surge, and heavy rainfall caused nearly half a billion dollars in damages in Louisiana and Mississippi and was responsible for the death of an elderly woman, it could have been very much worse. Lili made landfall in a relatively low populated region of Louisiana and had weakened considerably from its maximum strength a mere half day earlier. Had it continued as an intense hurricane (Category 3 or higher) and made a direct hit on New Orleans, Lili would have wreaked far greater destruction. In all, at least 13 people died as a result of Hurricane Lili, all but two of them in Haiti and Jamaica.

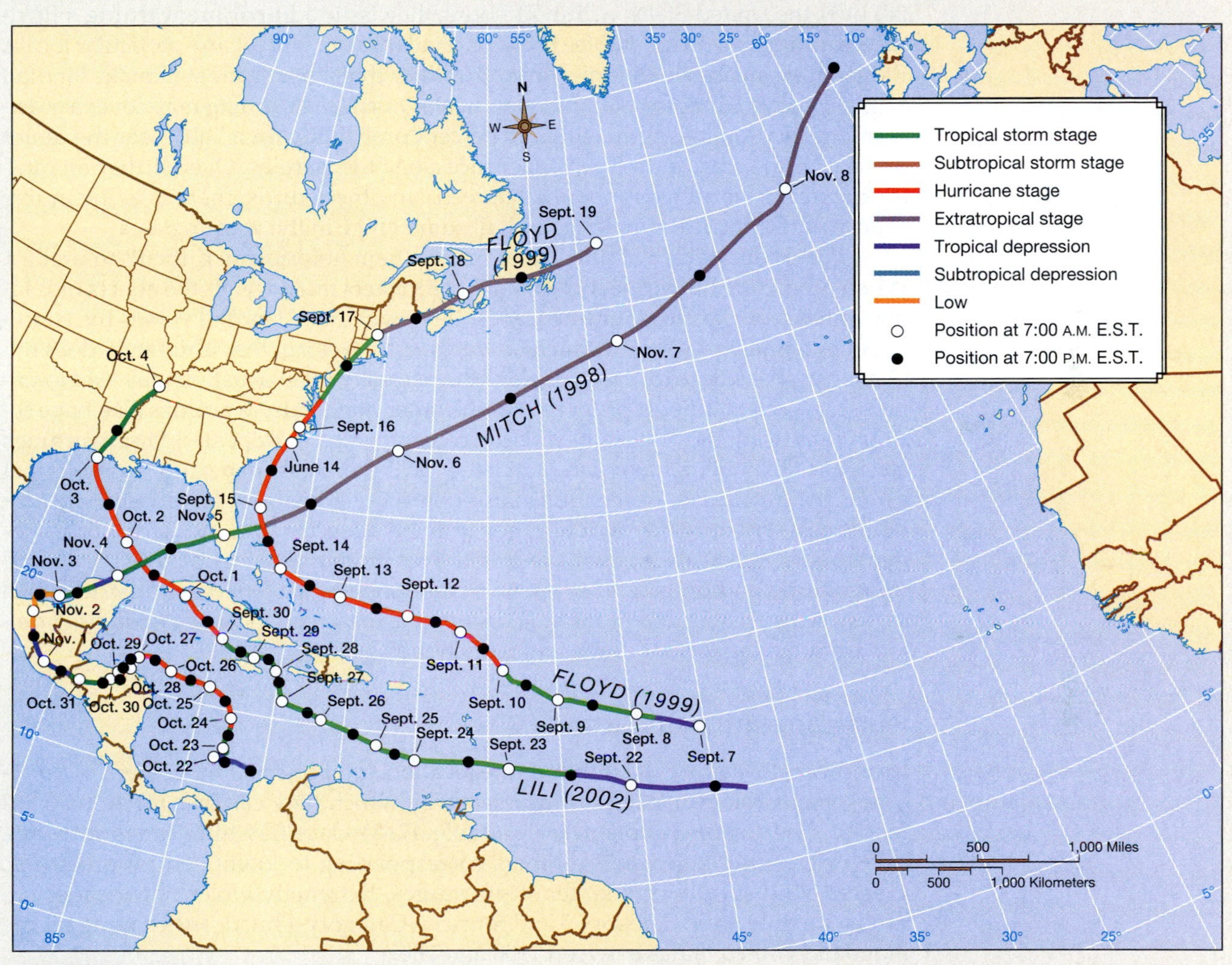

▲ **Figure 1**
The paths of Hurricanes Mitch, Floyd, and Lili.

increments and then print information on projected storm positions, air pressure, and wind at 6-hour intervals. Not surprisingly, model forecasts become less accurate as lead time increases and are unreliable for more than about 72 hours.

Hurricane forecasting requires a tremendous amount of data and places a great demand on computer hardware. For both these reasons, the NHC has recently undergone substantial modernization. The *National Oceanic and Atmospheric Administration (NOAA)* has recently deployed new geostationary satellites (meaning they remain over fixed locations on Earth), GOES 10 and 12, to provide improved data

Interview with Chase Plane Pilot

acquisition. NOAA has also purchased a new jet airplane that can fly higher into a hurricane than could its older propeller airplanes. Furthermore, a new supercomputer was installed in 1995 that can run increasingly complex models with improved accuracy. The improvement in weather forecasting in recent years has been substantial. In 1965 the average error in the 24-hour forecast position of a hurricane was about 250 km (155 mi). By the end of the century, that figure had been cut to about 160 km (100 mi). Still, the movement of hurricanes is particularly difficult to predict.

Hurricane Watches and Warnings

When forecasters at the NHC predict that an approaching hurricane will reach land in more than 24 hours, they issue a **hurricane watch**. If it is expected to make landfall over the United States within 24 hours, they issue a **hurricane warning**. Hurricane warnings do not guarantee that a particular storm will hit any particular locale. Rather, they are issued in terms of probabilities that a hurricane will make landfall at given points. Thus, for example, a particular hurricane warning may cover a coastline some 500 km in length, with a 25 percent probability for a "hit" near the center of the warning area, and a 5 percent chance along the margins. Overall, the "hit rate" for hurricane warnings is about 20 percent. In other words, there is a one in five chance of the hurricane making landfall somewhere in the warning area.

The erratic nature of hurricanes makes them notoriously difficult to predict. When predicting hurricane movements, forecasters must weigh the effects of issuing watches or warnings for hurricanes that never make landfall versus the consequences of failing to issue a watch or warning for a storm that ultimately does hit. Obviously, the failure to warn people of the necessity for evacuation may lead to unnecessary loss of life and property. On the other hand, false warnings also have serious ramifications, especially if they occur repeatedly. Repeated false warnings can make the public so complacent that they will eventually disregard warnings that prove accurate. Also, evacuations based on these advisories have immense economic costs for the general public, government agencies, and industry. Local residents and small businesses will have their lives thoroughly disrupted as they board up windows and prepare to evacuate or take shelter, while large industries (such as petroleum mining and processing) incur costs measured in tens of millions of dollars from having to shut down and reopen their plants.

Hurricane Intensity Scale

In addition to alerting the public to the location and projected movement of hurricanes, meteorologists need a simple scale by which they can categorize their intensity. The **Saffir-Simpson scale** (Table 12–2) classifies hurricanes into five categories, with increasing numbers corresponding to lower central pressures, greater wind speeds, and larger storm surges. Extremely violent hurricanes are rare, with only three Category 5 and fourteen Category 4 hurricanes having hit the mainland United States between 1900 and 2002.

Table 12–2 • The Saffir-Simpson Scale

Category	Pressure mb	Wind Speed km/hr	Wind Speed mph	Storm Surge m	Storm Surge ft	Damage
1	≥ 980	119–154	74–95	1–2	4–5	Minimal
2	965–979	155–178	96–110	2–3	6–8	Moderate
3	945–964	179–210	111–130	3–4	9–12	Extensive
4	920–944	211–250	131–155	4–6	13–18	Extreme
5	< 920	> 250	> 155	> 6	> 18	Catastrophic

12–3 Forecasting

Seasonal and Long-Term Hurricane Forecasts

Scientists are currently developing techniques for predicting the number and intensity of hurricanes and tropical storms on a seasonal basis. The best-known forecasts are those put out three times each year by the research group at Colorado State University (headed by Professor William Gray). Issued in November, June, and August, the forecasts predict the number of tropical storms and hurricanes, as well as unusually intense hurricanes, that will occur each season in the western Atlantic, Caribbean Sea, and Gulf of Mexico. The forecasts do not predict characteristics of individual storms, such as date or landfall.

The seasonal forecasts are based on a number of meteorological and oceanographic conditions around the globe, including the presence or absence of El Niño conditions, seasonal rainfall in western Africa, temperature, stratospheric and upper-tropospheric wind directions, and air pressure over the Caribbean Sea. The theoretical underpinnings that make these climatological conditions good hurricane predictors are supported by empirical observations. For example, a recent study showed that during El Niño years, an average of 1.04 hurricanes hit the U.S. mainland, in contrast to the 2.23 that hit during La Niña years and the 1.61 that hit in years that are neither El Niño nor La Niña years.

The Colorado State group has had improved success in recent years, as shown in Table 1. In 1997 the team significantly overpredicted the number of named storms and hurricanes, the number of days in which they would occur, the number of intense hurricanes, and the net tropical cyclone activity (an overall indicator of such events). In 1998 they underpredicted the same indicators. Between 1999 and 2001, however, their forecasts were largely on target for all categories. Their forecast for the 2002 season underpredicted the number of named storms and storm days, but accurately predicted the number of hurricanes and intense hurricanes.

Researchers are trying to determine whether there has been a trend in the number of hurricanes over the last several decades. As far as the Atlantic Basin is concerned, there has been no long-term trend in hurricane frequency, though the quarter century from the 1970s through the early 1990s had relatively few intense hurricanes (notwithstanding Hurricane Andrew in 1992). Overall, the 1990s were very erratic: During the period from 1991 through 1994, there were fewer Atlantic hurricanes than in any other 4-year period on record. Then came an abrupt transition in the seasons from 1995 to 2001—an extremely active period of Atlantic hurricane activity, despite the relatively quiescent 1997 season. Because of poorer monitoring of tropical storms outside of the Atlantic, it is difficult to determine whether there has been any trend in their worldwide incidence.

The prospect of human-induced climatic warming has led to much debate about whether such a change would affect the number or severity of tropical storms. Most of the speculation has centered on the possibility that a warmer climate would promote increased hurricane activity, though some researchers have presented arguments suggesting that higher temperatures would lead to decreasing numbers of hurricanes. Not surprisingly, there is no definitive answer to this question.

Table 1

	1997		1998		1999		2000		2001		2002	
	Predicted	Observed	Predicted	Observed	Predicted	Observed	Predicted	Observed	Predicted	Observed	Predicted	Observed
Number of Tropical Storms	11	7	10	14	14	12	11	14	12	15	9	12
Number of Tropical Storm Days	45	28	50	80	75	77	55	66	60	62	35	54
Number of Hurricanes	6	3	6	9	9	8	7	8	7	9	4	4
Number of Hurricane Days	20	10	25	47	40	43	30	32	30	27	12	11
Number of Intense Hurricanes	2	1	2	3	4	5	3	3	3	4	1	2
Number of Intense Hurricane Days	4	2	5	9	10	15	6	5	5	5	2	2
Net Tropical Cyclone Activity*	100%	54%	110%	172%	160%	193%	130%	134%	120%	142%	45%	80%

**An overall indicator of tropical storm and hurricane activity. It is defined as the average seasonal percentage of the total number of tropical storms, tropical storm days, hurricanes, hurricane days, intense hurricanes, and intense hurricane days in the year, relative to their long-term averages.*
The United States National Hurricane Center defines the hurricane season as the period from June 1 to November 30. 'Tropical storms in other months are rare events—from 1871 to 1996, only six storms formed in December.

12–4 Special Interest

The Galveston Hurricane of 1900

Certain natural disasters are so embedded in our folklore that virtually everybody knows about them. We have all heard about the San Francisco earthquake of 1906 and the Great Chicago Fire of 1871. Yet the single deadliest natural disaster in U.S. history, the Galveston (Texas) hurricane of 1900, seems to have been lost from the national memory. In just a few hours, rising sea waters and heavy surf drowned 6000 persons on Galveston Island—a narrow strip of land that peaks at less than 3 m (9 ft) above sea level (Figure 1).

The loss of life resulted not from lack of warning but rather from a failure to take the threat seriously. Two days earlier, a strong storm was reported moving westward into the Gulf of Mexico off Cuba, and ships returning from the Gulf of Mexico reported encountering the storm offshore the day before it made landfall. Furthermore, the local weather forecaster, Isaac Cline, observed the combination of winds and heavy surf that the storm would move onshore. But evidence of the impending landfall seems to have been largely unheeded, in part because some meteorologists erroneously believed it was virtually impossible for a storm in the Caribbean to track across the Gulf. Scientists (including Cline) were also erroneously convinced that the gently sloping seafloor offshore would protect Galveston from major flooding in the event of a hurricane.

There is some uncertainty as to when the hurricane was taken seriously. According to Cline's account of the disaster, he rode through Galveston Island urging residents to evacuate, but recent research casts doubt about the degree to which he actually warned the populace. Regardless of how urgent Cline's warnings were, however, few people evacuated, and some residents even rode to the beach to watch the heavy waves crashing against the shore.

When the hurricane arrived, the people of Galveston had no possibility of escape. Within hours the rising seas completely covered the island so that the only potential shelter was in taller, well-built structures. But even these failed to withstand the pounding of waves and debris. Cline later gave the following account of his ordeal:

> By 8 P.M. a number of houses had drifted up and lodged to the east and southeast of my residence, and these with the force of the waves acted as a battering ram against which it was impossible for any building to stand for any length of time, and at 8:30 P.M., my residence went down with about fifty persons who had sought it for safety, and all but eighteen were hurled into eternity. Among those lost was my wife, who never rose above the water after the wreck of the building.

Cline and his brother were luckier and grabbed onto floating debris that helped them stay afloat. After three hours, the floodwaters subsided and the Cline brothers were on solid land, among the survivors.

The horror did not end with the passage of the hurricane. There were still 6000 bodies that had to be dealt with. Some were taken out to sea on barges, but many washed back to shore. Ultimately, most of the bodies were cremated where they were found.

With our current ability to track and forecast the movement of approaching hurricanes, there is no reason there should be a repeat of this event in North America. Unfortunately, much greater death tolls continue to occur in poorer parts of the world. In 1972 Bangladesh (then part of Pakistan) was hit by a tropical cyclone that killed between 300,000 and 500,000 persons. More recently, another 70,000 were killed in the same region in 1991 by floodwaters brought on by a tropical cyclone.

▶ **Figure 1**
The Galveston hurricane of 1900 was the deadliest natural disaster in U.S. history.

Of course, Category 4 and 5 hurricanes are far more deadly and devastating than lower category hurricanes. The effects of Hurricane Camille (Category 5) in 1969 have already been described. Since then, there have only been two Category 5 hurricanes in the western Atlantic: Gilbert in 1988 and Andrew in 1992. Hurricane Gilbert brought sustained winds of 250 km/hr (155 mph) and gusts up to 350 km/hr (220 mph). It produced the lowest pressure ever recorded in the Western Hemisphere (850 mb), but more importantly, it killed 200 people (60 from a flash flood in Monterrey, Mexico, that swept five buses into a river). Though the storm appeared at one time to be headed straight for Brownsville, Texas, it suddenly changed course and did little damage within the United States.

On August 24, 1992, Hurricane Andrew devastated parts of south Florida. It approached southern Florida almost directly from the east and cut westward across the state, with its eye passing about 40 km south of Miami Beach. It completely leveled the town of Homestead, killed 24 people, and left 180,000 homeless in Florida before crossing into the Gulf of Mexico. It continued westward for two days and then took a sharp turn northward toward Louisiana. There it killed another four persons and caused hundreds of millions of dollars of damage. Until August 2002—just a few days before its tenth anniversary—Andrew was believed to have been a Category 4 hurricane. But recent improvements in the ability to infer ground level wind speeds from aircraft-based observations have revealed that Andrew was indeed a Category 5 event. Previously it was believed that surface winds in hurricane eye walls were about 75 percent as strong as those observed by surveillance aircraft at the 10,000-ft level. We now know that surface winds are normally about 90 percent as strong as those at 10,000 ft, which brings Andrew's estimated maximum surface wind speeds to 265 km/hr (165 mph). The great Galveston hurricane of 1900 that killed 6000 persons (see *Box 12–4, Special Interest: The Galveston Hurricane of 1900*) is believed to have been a Category 4 hurricane.

Without question, hurricanes are among the most exciting of natural phenomena, a fitting subject for concluding this section on weather disturbances. The remaining sections of this book will examine human activities and meteorology, and climate and climate change. In our next chapter we turn our attention to weather forecasting.

Summary

Hurricanes (and their counterparts such as typhoons and tropical cyclones) are extremely powerful storms that originate in tropical regions and migrate into the middle latitudes. They bring enormous destruction and loss of life to many coastal regions of the world. The hurricane that hit Galveston Island, Texas, in 1900 was the greatest single natural disaster to hit North America, with a death toll of 6000. But this figure pales in comparison to the hundreds of thousands of fatalities associated with individual tropical cyclones in southern Asia.

Most hurricanes begin their life cycles as uneventful tropical disturbances, small clusters of thunderstorms. When they intensify and organize into a rotating band of cloud cover and thunderstorm activity, they are called *tropical depressions*. Further intensification results in their being classified as tropical storms, or hurricanes if their sustained wind speeds exceed 120 km/hr. Because strong tropical storms can form only over oceans having high surface temperatures, tropical depressions most often become tropical storms and eventually hurricanes over the western portions of the ocean basins.

Hurricanes are smaller than mid-latitude cyclones but very much larger than tornadoes. They can last for a week or more and travel thousands of kilometers before dissipating. The heaviest thunderstorm activity occurs within bands of thick cloud cover that spiral toward the center of the system in a pinwheel pattern. The intensity of the storm increases toward its center until reaching the eye wall, the concentric zone of maximum activity that surrounds the eye. The eye of a hurricane is strikingly different from the rest of the hurricane because it is marked by generally clear skies, light winds, and higher air temperatures. Often it is hard to discern the true structure of a hurricane from above because the anticyclonically rotating outflow in the upper troposphere creates a blanket of cirrostratus clouds overlying the thicker cumulus.

Hurricanes can produce damage in several ways. Copious amounts of rain can bring intense floods, and strong winds can bring down structures. But the most serious threat posed by a hurricane is the storm surge, the elevated rise in sea level due to low atmospheric pressure and the piling up of water by strong winds. When the storm surge coincides with a high tide, the floodwaters (coupled with heavy surf) can penetrate considerable distances inland.

The National Hurricane Center of the National Weather Service uses a sophisticated network of satellites, research aircraft, and computer hardware and software to issue advisories on the likelihood of hurricane landfall. The erratic nature of hurricanes makes their prediction particularly difficult, but recent modernization at the National Weather Service has substantially increased forecast accuracy. In the next chapter we examine the process of forecasting everyday weather events.

Key Terms

hurricane p. 358
typhoon p. 358
cyclone p. 358
trade wind inversion p. 358
marine layer p. 358
eye p. 361
eye wall p. 361
tropical disturbance p. 362
easterly waves p. 362
tropical depression p. 363
tropical storm p. 363
storm surge p. 368
hurricane watch p. 374
hurricane warning p. 374
Saffir-Simpson scale p. 374

Review Questions

1. Describe the geographic distribution of hurricanes, typhoons, and cyclones. What environmental conditions at these locations favor the development of such storms?
2. Which region has the greatest incidence of major tropical storms?
3. What is the trade wind inversion, and what impact does it have on the formation of hurricanes?
4. Describe the size, sea level air pressure, and wind speed of a typical hurricane.
5. When are hurricanes most likely to form?
6. Describe the cloud and precipitation patterns associated with hurricanes, including those associated with the eye and eye wall.
7. Describe the various ways in which hurricanes differ from midlatitude cyclones.
8. What are tropical disturbances, and how are they influenced by easterly waves?
9. Describe the characteristics that distinguish tropical disturbances, tropical depressions, tropical storms, and hurricanes from each other.
10. What ocean surface characteristics are required for the intensification of storms into hurricanes and the maintenance of hurricanes?
11. Is there a "typical" path that hurricanes take after having formed? Explain.
12. What feature associated with hurricanes causes the greatest destruction to coastal regions? Is this also true of inland regions?
13. Why is the right-hand side of a hurricane (relative to its direction of movement) the most dangerous?
14. Where are tornadoes most likely to be imbedded in a hurricane?
15. What are hurricane watches and warnings? Are they exact corollaries to tornado watches and warnings?
16. Why are forecasters concerned with issuing hurricane advisories for areas that do not eventually get hit by a hurricane?
17. What is the highest hurricane category on the Saffir-Simpson scale? How frequently do hurricanes of that magnitude occur?

Critical Thinking

1. Why is it that hurricanes do not cross the equator?
2. If two hurricanes pass just to the west of Cuba over a 2-week period, what reasons might one have for expecting the second one to be weaker than the first?
3. How might previous drought conditions affect the intensity of a former hurricane as it passes over the southern United States?
4. El Niño conditions are believed to suppress hurricane development in the Atlantic. How might the phenomenon affect hurricane formation and movement in the Pacific?
5. It has been postulated that an increase in global temperatures could lead to an increase in the number and intensity of tropical storms and hurricanes. Global temperatures have been particularly high during the 1990s and early 2000s, and there has been an increase in Atlantic hurricane activity since 1995. Does this association prove the connection between temperature and hurricane activity? Explain why or why not.
6. If global warming continues, thermal expansion of the oceans and the melting of glaciers will lead to higher sea level. How would this affect the threat of storm surges relative to wind damage and flooding?
7. Experts believe that New York City is the third most dangerous city in the United States with regard to hurricanes, despite the fact that there has been no major hurricane-inflicted damage on the area (other than some wind damage and coastal erosion from Hurricane Gloria, in 1985). What factors are likely responsible for this vulnerability? After answering this question, refer to **http://www2.sunysuffolk.edu/mandias/38hurricane/** for an informative discussion on this issue.

Problems and Exercises

1. Compare the area of a hurricane that measures 600 km in diameter to a midlatitude cyclone having a diameter of three times greater (1800 km).
2. During the tropical storm season, use the Web sites described below to make note of the position of current systems and the probabilities of landfall at various coastal locations. Describe how successful the predictions proved to be.
3. Refer to the forecast for the upcoming tropical storm season at **http://tropical.atmos.colostate.edu/**. What existing conditions have led the forecast team to make its prediction? Also, use the Web site to determine how successful last year's forecast was.
4. Refer to **http://www.ncdc.noaa.gov/oa/climate/severeweather/hurricanes.html** and read the Special Reports on the past year's hurricane activity. Were there any tropical storms that were particularly noteworthy?

Quantitative Problems

A deeper understanding of hurricanes can be obtained by working out some numerical problems. These are available from this book's Web site, **http://www.prenhall.com/aguado**. Log on to the site and go to the Chapter 12 section for some thought-provoking problems.

Useful Web Sites

http://www.nhc.noaa.gov/index_special.html

Official web page for the National Hurricane Center Tropical Prediction Center.

http://www.wunderground.com/tropical/

Site opens with map showing current tropical activity and provides numerous links for satellite images, advisories, outlooks, and discussions.

http://www.nrlmry.navy.mil/sat-bin/tc_home

Comprehensive information provided by the Monterey Naval Research Laboratory.

http://cimss.ssec.wisc.edu/tropic/tropic.html

Archived and real time imagery, and a large amount of text information from the University of Wisconsin–Madison Tropical Cyclone Research Team.

http://hurricanes.noaa.gov/

Useful Web publication from the National Oceanographic and Atmospheric Administration, with news, features, satellite imagery, and current reports.

http://www.solar.ifa.hawaii.edu/Tropical/tropical.html

Offers current hurricane information and several data archives.

http://www.usatoday.com/weather/hurricane/whur0.htm

Some basic information from *USA Today*.

http://www.ncdc.noaa.gov/oa/climate/severeweather/hurricanes.html

Climatological information and data on past hurricanes from the National Climate Data Center.

http://tropical.atmos.colostate.edu/

Official Web site of the Tropical Meteorology Project at Colorado State University. Includes much information, including the seasonal tropical storm forecast and a description of the factors used to predict the upcoming season.

http://www2.sunysuffolk.edu/mandias/38hurricane/

Presents a comprehensive and interesting review of the hurricane of 1938 and what it tells us about the vulnerability of the New York area to future events.

Media Enrichment

Interactive Exercise

Tropical Cyclones

The destructive power of a tropical cyclone instills fear in people in its path and intrigues scientists interested in its dynamics. This module will guide you through the formation, structure, energy, and movement of tropical cyclones through the use of sketches, satellite imagery, and animations.

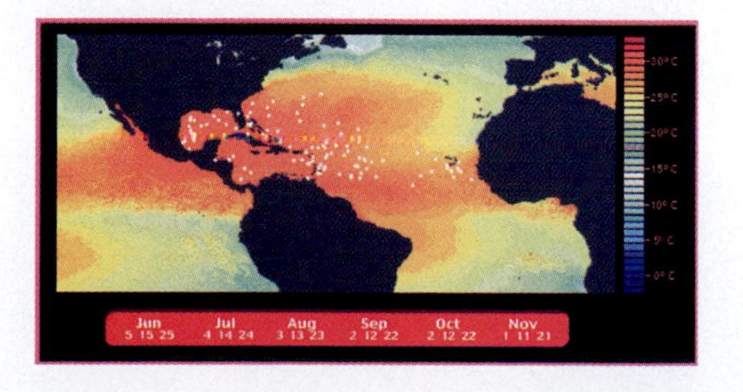

Weather in Motion

A Fly-Through of Hurricane Mitch

This movie opens with a view of this devastating hurricane obtained by satellite sensors. As the movie zooms in on the hurricane, the overlying cirrus clouds are stripped away so that the interior structure of the storm is evident. This movie was compiled using microwave, visible, and infrared imagery obtained by the Tropical Rainfall Measuring Mission (TRMM).

Weather in Motion

Hurricane Slice

This brief movie shows the pinwheel structure of a hurricane after clearing away the overlying cirrus.

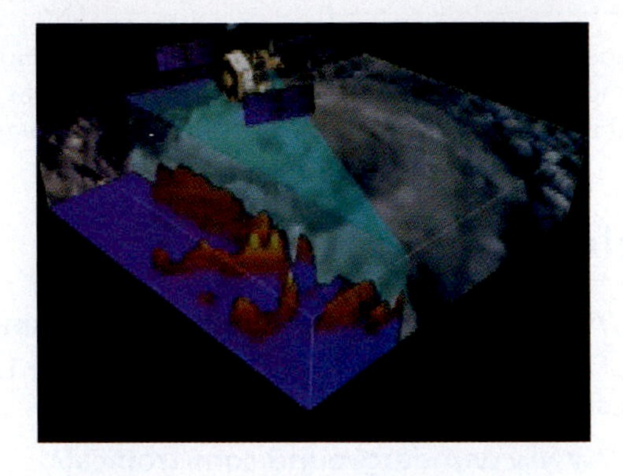

Weather in Motion

Hurricane Eye Wall

A view of a hurricane eye wall as seen from a hurricane chase plane. Note the abrupt transition from clear air within the hurricane eye and the massive wall of cloud encircling the eye.

Weather in Motion

Hurricane Dennis

A computer-enhanced movie showing the movement of Hurricane Dennis along the Atlantic Coast.

Weather in Motion
Storm Surge

The combination of low pressure and strong onshore winds can bring elevated sea levels and strong waves ashore. This scene shows what a moderate storm surge can look like.

Weather in Motion
Hurricane Damage

A view of severe inland damage in the wake of a major hurricane, as seen from a helicopter.

Weather in Motion
Interview with Chase Plane Pilot

Have you ever wondered what it must be like to fly into a hurricane? Take a look at this film clip and hear firsthand from an expert.

PART FIVE

Human Activities

Greenwich Village, New York, during the 1996 blizzard.

chapter 13

A wildfire near Lake Piru, California. The intense heat induced convection that formed the cumulus cloud above the plume of smoke.

Weather Forecasting and Analysis

Forecasters for the National Weather Service (NSW) spend a large portion of each workday fielding telephone inquiries about upcoming weather. Sometimes people want to know if their round of golf is likely to be rained out or if their outdoor plants might be vulnerable to overnight frost damage. But the most important aspects of a forecaster's job do not involve matters of simple convenience but questions that deal with life-and-death matters.

Such was the case for Mark Moede, meteorologist with the NWS Forecast Office in San Diego. In late August 1998 much of southern California was threatened by brush fires triggered by unusually heavy thunderstorm activity. Firefighters battling the blazes received constant updates on weather conditions that could either suppress or enhance the spread of the fire. On September 2 the situation reached its most critical stage. Pete Curran of the Orange County Fire Department maintained close contact with Moede for constant updates on a line of approaching thunderstorms. At issue was whether the thunderstorms would continue to strengthen and whether they would pass through the areas where the firefighters were working. If the storms passed that way, then strong winds, deadly lightning, and blinding rains would place Curran's crews in jeopardy. With the aid of Doppler radar and information from automated weather stations, Moede made the right call. He advised Curran to evacuate his crews from the eventual path of the storm, where winds in excess of 95 km/hr (60 mph) created an uncontrollable firestorm. Afterward, firefighters reported that the fire line they had evacuated had been completely burned over. If the call to evacuate had not been made, in all likelihood a number of firefighters would have been killed.

This chapter discusses the methods by which forecasters perform their job. We first look at some important issues regarding the general concept of weather forecasting and then discuss ways in which necessary data are obtained and processed. We then study the various types of weather maps and how they are used in weather analysis.

CHAPTER OUTLINE

Why Is Weather Forecasting Imperfect?

We've all had careful plans upset by a bad weather forecast and are understandably quick to find fault when actual conditions depart from the forecast. Implicit in such criticism is the premise that forecasts ought to be accurate and that there's no excuse for a miss. So why are forecasts often so far from correct? After all, with powerful computers, satellites, weather radar, and global communication networks, it seems as if making a good forecast ought to be easy. But however much the public might think so, this is definitely not the case—in fact, accurate weather forecasting is extraordinarily difficult.

To see why, imagine you want to forecast tomorrow's temperatures and think about just a few of the factors that you must consider. First, remember that the temperature structure of the atmosphere depends in part on absorption and emission of radiation (shortwave and longwave), which itself depends on the vertical and horizontal distribution of atmospheric gases, clouds, and so on. So, to compute the temperature at a point, you need to begin with detailed information about the composition of the atmosphere in three dimensions.

Of course, with water constantly shifting between the liquid, solid, and vapor phases, atmospheric composition is hardly constant, so you will need to forecast those changes over time. Remember also that as water changes phase, latent heat is added or removed from the atmosphere; thus, you will have to keep track of that as well as radiation transfer. But the phase changes are influenced by small-scale updrafts and downdrafts, so you will need to somehow forecast vertical motions as part of your overall effort. Furthermore, horizontal motion can't be neglected—you will need to allow for warm or cold air advection (Chapter 10).

With regard to wind, near the surface you face the problems of flow around complex terrain and somehow quantifying frictional effects between the atmosphere and surface. Above the friction layer things are less complicated, but not by much. The basic problem is one of continual interaction: the motions of the atmosphere change the motions that subsequently develop. In other words, after a short time the winds change the winds. So, even though you're only interested in temperature, you can't pretend the winds are unchanging, but instead you are forced into the business of forecasting atmospheric motion. Unfortunately, this is very difficult, because the atmosphere is dynamically unstable. By this we mean that small disturbances often grow into large features that eventually dominate the field of motion. Thus, whereas you might only care about motion on large scales for the purposes of heat advection, you must consider small-scale motions in order to know how the large-scale pattern will evolve.

Obviously, weather forecasting involves a set of interlocking problems, each difficult to solve in isolation, let alone in combination. In light of all these difficulties, it's remarkable that forecasts show any accuracy at all. Rather than wondering why they fail, we're more likely to wonder how they are able to succeed!

Weather forecasting by the U.S. government began in the 1870s, when Congress established a National Weather Service under the authority of the Army Signal Corps. In 1890 the organization was renamed the National Weather Bureau and transferred to the Department of Agriculture. There it remained until it was shifted to the Department of Commerce in 1942. The **National Oceanic and Atmospheric Administration (NOAA)** was established in 1970 to include a number of environmental agencies, including the **National Weather Service (NWS)**,* which reverted to its original name. The **Meteorological Service of Canada (MSC)**, located in Downsview, Ontario, assumes all forecasting duties for that country and provides local and regional information to its 14 **regional weather centres**.

Forecasting Methods

There is no single "correct" way to forecast the weather; depending on the length of forecast, the type of information desired, and how much is known about the current state of the atmosphere, any one of a number of methods might be used. One can even attempt a forecast in the absence of any data about current weather, provided that long-term information is available. For instance, a forecast of hot, muggy conditions with a chance of afternoon thundershowers in Orlando, Florida, in mid-August has a reasonably good chance of proving correct. Such prognoses based on long-term averages are known as **climatological forecasts**. Obviously, the reliability of a climatological forecast depends on year-to-year variability in weather conditions for the forecast day. Thus, a forecast based on climatology might have a reasonable chance of being accurate in Orlando during the summer, but only the truly daring would try this for Chicago in April, when almost any kind of weather can occur.

Another type of forecast, called a **persistence forecast**, relies completely on current conditions with no reference to climatology. Actually, a special case of persistence forecasting is used by all of us in everyday life. When we see clear skies and leave the umbrella behind, we're betting that the prevailing conditions will continue and are making a short-term forecast on that basis. This simple procedure might work for a little while but will eventually fail to catch changes in weather. A more sophisticated version might use a decrease in pressure over the past few hours

*The National Weather Service (NWS) is part of the National Oceanic and Atmospheric Administration (NOAA). Employees of the latter organization sometimes claim that the letters NOAA stand for National Organization for the Advancement of Acronyms.

to indicate the approach of a low-pressure system and its associated increase in cloud cover. In other words, one could assume persistence in a trend to make a guess regarding changes in weather. Here too, however, one extrapolates present behavior forward in time, again with certain knowledge that the forecast will fail when that behavior ceases. Of course, it is often precisely those breaks from past behavior we want to forecast; thus a method that can't provide such information is not terribly useful.

Until the 1950s, weather analysis and forecasting depended entirely on the experience of meteorologists and their interpretation of current conditions and the recent past. He or she (usually he in those days) would use past behavior for guidance in determining what a current weather system was most likely to do. In other words, the meteorologist would base a forecast largely on the comparison of the current situation to similar conditions encountered previously (but not necessarily in the immediate past). This approach led to development and use of numerous "rules of thumb," which attempted to capture repeatable patterns and relations between various weather elements. For example, winter precipitation over the eastern United States and Canada tends to be snow north of the 5640-m contour for the 500 mb level and rain south of that curve. In this so-called **analog approach**, one tries to recognize similarities between current conditions and similar well-studied patterns from before. There are many variations of the analog approach, some subjective (depending on the forecaster's expertise), others objective (depending on statistical relations). But all assume that what happened sometime in the past provides a clue about the future.

In the last few decades **numerical weather forecasting** has come to occupy a dominant position. The term is somewhat misleading, because the methods just described all produce numerical forecasts. What's different is that this method is based on computer programs that attempt to mimic the actual behavior of the atmosphere. That is, numerical weather models explicitly compute the evolution of wind, pressure, temperature, and other elements over time. By examining the output for a given point in time, one obtains a depiction of the three-dimensional state of the atmosphere for that moment. (This is in contrast to forecasting the surface values for just a few weather elements, as might be done with other methods.) The numerical models typically used in weather forecasting are very large and can only be run on the most powerful computers, so-called *supercomputers*. As reflects their importance in modern forecasting, these models are described in more detail in the section called "Forecast Procedures and Products" and in still more detail in the Chapter 13 appendix. For now, we simply want to draw a contrast between other methods and this physically based ("numerical") approach.

Types of Forecasts

The product, or result, of a forecast method can take a variety of forms, which we are calling the *type* of forecast. We are all familiar with quantitative forecasts, in which the "amount" of the forecast variable is specified. For example, a forecast that says "An inch of rain is expected" is a **quantitative forecast**. Similarly, forecasts of the expected high or low temperature are quantitative, because a value for the forecast variable is provided. In contrast, **qualitative forecasts** provide only a categorical value for the predicted variable. Examples of this are "rain/no rain," "hurricane/no hurricane," "above/below normal," or "cloudy/partly cloudy/mostly clear." In these examples, the predicted variable is assigned to a particular class, or category; hence it is a qualitative forecast.

In the preceding examples, the forecasts were provided without qualification. **Probability forecasts** are an alternative, in which the chance of some event is stated. Thus, for example, a categorical hurricane forecast might be stated as a probability, rather than as a certainty. Probability forecasts can take a variety of forms, the most common of which is undoubtedly the probability-of-precipitation forecast (PoP forecast). When the broadcaster says "The rain chance today is 70 percent," or

"There is a 60 percent chance of afternoon showers," he or she is reporting a PoP forecast. Note that these forecasts don't specify an amount of rain. Rather, the meaning of a PoP forecast is that a randomly chosen point in the forecast area has the stated probability of receiving measurable precipitation. For example, a 75 percent PoP forecast means that the odds of precipitation are 3:1, or equivalently, you have only one chance in four (25 percent) of staying dry throughout the forecast period.

Assessing Forecasts

Regardless of which method is used, or what form it takes, we obviously need some way of deciding how good a forecast is likely to be. Measures are needed, for example, in order to compare one forecast method to another or in trying to decide how much consideration should be given to a forecast when making plans. Most importantly, assessment measures are needed by those responsible for developing and administering forecasting programs. As attempts are continually made to improve data gathering and forecast procedures (at an ever-increasing cost), methods are needed for judging the value of changes, justifying future expenditures, and determining the return on investment. Although as consumers we don't hear much about forecast assessment, it is a routine and integral part of professional forecasting.

Over the years many evaluation measures and practices have been developed, each with particular advantages and disadvantages. To sort them out, we must first think about the purpose of the assessment. Do we want information about **forecast quality** or **forecast value?** The former refers to the agreement between forecasts and observations, whereas the latter refers to the utility of a forecast. These sound similar but are quite different. For example, a high-quality 100-percent-accurate forecast of rain might have zero value for scheduling crop irrigation, where knowing the amount of rainfall is essential. Because there's no simple relation between quality and value, different measures are needed for each. A second issue is the type of forecast: quantitative or qualitative, probability or unqualified, etc. Clearly, the appropriate assessment measure will vary with the type of forecast variable. Finally, there is the question of whether we want an absolute measure of performance or whether we are primarily interested in a relative, comparative measure.

Forecast *value* necessarily depends on the application of a forecast to a particular problem or decision. Most measures of value are based on loss/payoff tables, which attempt to capture the risks and rewards associated with various forecasts and responses to those forecasts. For example, knowing the cost of a ruined concrete job, the money earned when things go right, and the probability of a correct forecast, you could assign monetary value to the forecast. The details involve probability concepts that are beyond an introductory text; therefore, we won't discuss forecast value any further, except to note that a single forecast can have widely varying value, depending on how it is used.

With regard to the *quality* of a forecast, an obvious question to ask concerns **forecast accuracy**. That is, on average, how close is the forecast value to the true value? There are many ways to answer this simple question, each leading to a different accuracy measure. At the broadest level, we might want information about **forecast bias**, which concerns systematic over- or underprediction. A biased forecast method is one whose average forecast is above or below the true average. An unbiased method, by contrast, shows no tendency for over- or underprediction. Of course, that is not to say the method is perfect—it simply means the average overprediction is as large as the average underprediction, yielding an average error of zero. For example, if you wanted to predict the number of spots showing after the roll of a die, you could forecast 3.5 dots on every roll. Over many rolls (of a fair die), the average would be 3.5, agreeing with your predictions—the predictions are therefore unbiased. Of course, it's impossible for this method to yield even a single correct prediction. It's clear that although bias is a useful measure, we also need accuracy measures that don't allow positive errors to cancel negative errors. The simplest is the *mean absolute error (MAE)*, which ignores the sign (positive or negative) of the errors. That is, over- and underpredictions are treated the same, and we simply find the average error without regard to sign.

For laypeople using a forecast, accuracy is probably the most important quality issue. But professionals who develop forecasting methods are more likely to report **forecast skill**. Skill can be measured in various ways, but the concept is defined as the *improvement* a method provides over what can be obtained using climatology, persistence, or some other "no-skill" standard. For example, no measurable rainfall has been recorded for July in Jerusalem for the last 100 years. Any method that forecasts "no rain" for July is certainly going to be accurate most of the time. Likewise, if you predict that a hurricane will be present somewhere in the Atlantic Ocean next September 10, you have a 90 percent probability of success (using past behavior as a guide). But these forecasts have no skill, because they do not improve on "chance" (climatology in this case). Only if predictions were correct more than 90 percent of the time would we say that the forecast method possesses any skill. In the case of air temperature, we might compare the MAE of the forecast method to the MAE obtained using the climatological mean temperature. If the method is no better than climate, the skill score would be zero. The idea is that a climatology forecast requires no special knowledge of atmospheric behavior and thus does not contain any skill.

Data Acquisition and Dissemination

The starting point for almost all weather forecasting is information about the current state of the atmosphere. To know the future, we begin with information about the present. Thus, the first process in operational weather forecasting is the acquisition of the necessary data. For reasons that will become clear, this requires an international effort, even when making forecasts for "small" areas, such as individual countries.

The **World Meteorological Organization (WMO)**, under the auspices of the United Nations, oversees the collection of weather data across the globe from its 179 member nations. The WMO collects data from about 10,000 land observation stations, 7000 ship stations, 300 moored and drifting buoys with automatic weather sensors, and several weather satellites. It also obtains upper-air data from about 1000 weather balloon sites on a twice-daily basis and on a continuous basis from instruments aboard wide-bodied commercial aircraft. The data from all countries are sent to the three **World Meteorological Centers** at Washington, DC; Moscow, Russia; and Melbourne, Australia, which in turn disseminate the data to all members of the WMO.

The member nations of the WMO maintain their own meteorological agencies that obtain and process the data and issue regional and national forecasts. In the United States, the **National Centers for Environmental Prediction (NCEP)** of the Weather Service performs these tasks, while in Canada they are handled by the **Canadian Meteorological Centre** of the **Atmospheric Environment Service (AES)**.

Not surprisingly, the United States has a relatively dense network of surface observation stations. Of the approximately 1000 locations where surface conditions are recorded, about 120 are National Weather Service Offices; the rest are Federal Aviation Administration (FAA) airport sites. The Canadian AES operates about 270 surface stations. Together, these sites record temperature, humidity, pressure, cloud conditions (including type and height, and the percentage of sky obscured by cloud), wind direction and speed, visibility, the presence of significant weather, such as fog or rain, and accumulated precipitation readings at ground level. As part of an ongoing modernization program (see *Box 13–1, Special Interest: Modernization of the National Weather Service*), the FAA and NWS have installed a network of more than 800 automated sensors, called **Automated Surface Observing System**, or **ASOS**, for

13–1 Special Interest

Modernization of the National Weather Service

Over the last few decades, there has been a tremendous improvement in all aspects of weather forecasting. The gains in weather analysis have resulted partly from an improved surface data network and partly from better knowledge of meteorological principles. But the most striking improvements have come about in the technological front. The computers of the 1950s and 1960s, which had nowhere near the power of today's versions, ran only primitive models. Even more dramatically, until the launch of the TIROS I weather satellite in April 1960, there were no observations at all taken from space, and thus there were huge gaps in spatial coverage over both land and water.

By the 1970s the use of high-speed computers and sophisticated satellites became routine in daily operational forecasting. Geostationary satellites (those that orbit the equator at one revolution per day and thereby remain over a fixed location on the surface) and their polar orbiting counterparts (which orbit from pole to pole while the planet rotates beneath it) not only provide visual and infrared images of the clouds but also determine temperature and humidity characteristics at various depths of the atmosphere.

The combination of increased data and ever-improving data-processing capabilities enabled forecasters to rely heavily on model output to guide their predictions. But by the 1980s, it was decided that a huge overhaul of the Weather Service would be necessary to provide the highest possible level of forecasting and, in particular, severe weather warning. Thus began the current NWS modernization program.

One of the first steps in the modernization process was the improvement of computer hardware and the refinement of numerical models. Advances in computer hardware have been staggering in recent years and promise to improve even further. In May 2002 the NWS contracted to develop a new supercomputer that should be able to process 100 trillion calculations per second by the year 2009.

GOES-1 and GOES-2, the first in the current series of geostationary orbiting satellites, have been replaced by the currently operating GOES-10 and GOES-12, which provide excellent coverage over the western and eastern portions of North America, respectively. The newest generation of GOES satellites offers superior resolution and obtains better temperature and moisture information for various levels of the atmosphere.

Forecasters now also benefit from a modernized radar system. The NEXRAD generation of Doppler radar systems (described in Chapter 11) have been installed at 163 locations across the United States. In addition to the installed units already described, NOAA is now also experimenting with a different type of Doppler system. These **wind profilers** obtain measurements of horizontal winds for up to 72 different levels, between 0.5 and 10 km (0.3 to 6 mi) above ground level. They work by simultaneously emitting three radar beams. One of the beams travels directly upward from the surface. The other two beams are directed 16 degrees away from the vertical, one oriented toward the north and the other toward the east. Changes in the wavelength of the backscattered radar beams occur in response to the movement of various backscattering agents (such as dust, bugs, air molecules) and indicate the speed and direction of the wind. NOAA currently operates 29 profilers throughout the central United States. If the system proves valuable to forecasters, the network will expand to about 120 profilers.

Modernization has involved not only the application of new technology but also the streamlining of operations. Formerly there were 52 Weather Service Forecast Offices (WSFOs) and 204 smaller Weather Service Offices (WSOs). Today, the daily operations take place at 119 **Weather Forecast Offices (WFOs)**, providing a significant savings to the public.

measuring and recording these variables (Figure 13–1). The AES operates more than 100 data observation platforms across Canada similar to ASOS, called **Automated Weather Observation Systems (AWOS)**.

In addition to the surface observations made at these sites, the NWS launches hydrogen-filled balloons* carrying weather instrument packages called **radiosondes** (Figure 13–2). Twice a day—at 0000 and 1200 UTC (Universal Coordinated Time)†—about 750 radiosondes are launched worldwide, about 80 within the United States and Canada. Radiosondes continually observe and transmit to ground recording stations the pressure, air temperature, and wet bulb temperature (from which dew point temperatures are determined). Some radiosondes are tracked by radar as they ascend through the atmosphere, which enables a determination of the wind speed and direction of the middle and upper atmosphere. Radiosondes tracked by radar are called **rawinsondes**.

---

*Rather than using helium in the balloons, the Weather Service uses hydrogen, which it extracts locally, directly from the surrounding air.

†These times correspond to 1900 (7:00 P.M.) and 0700 (7 A.M.) Eastern Standard Time.

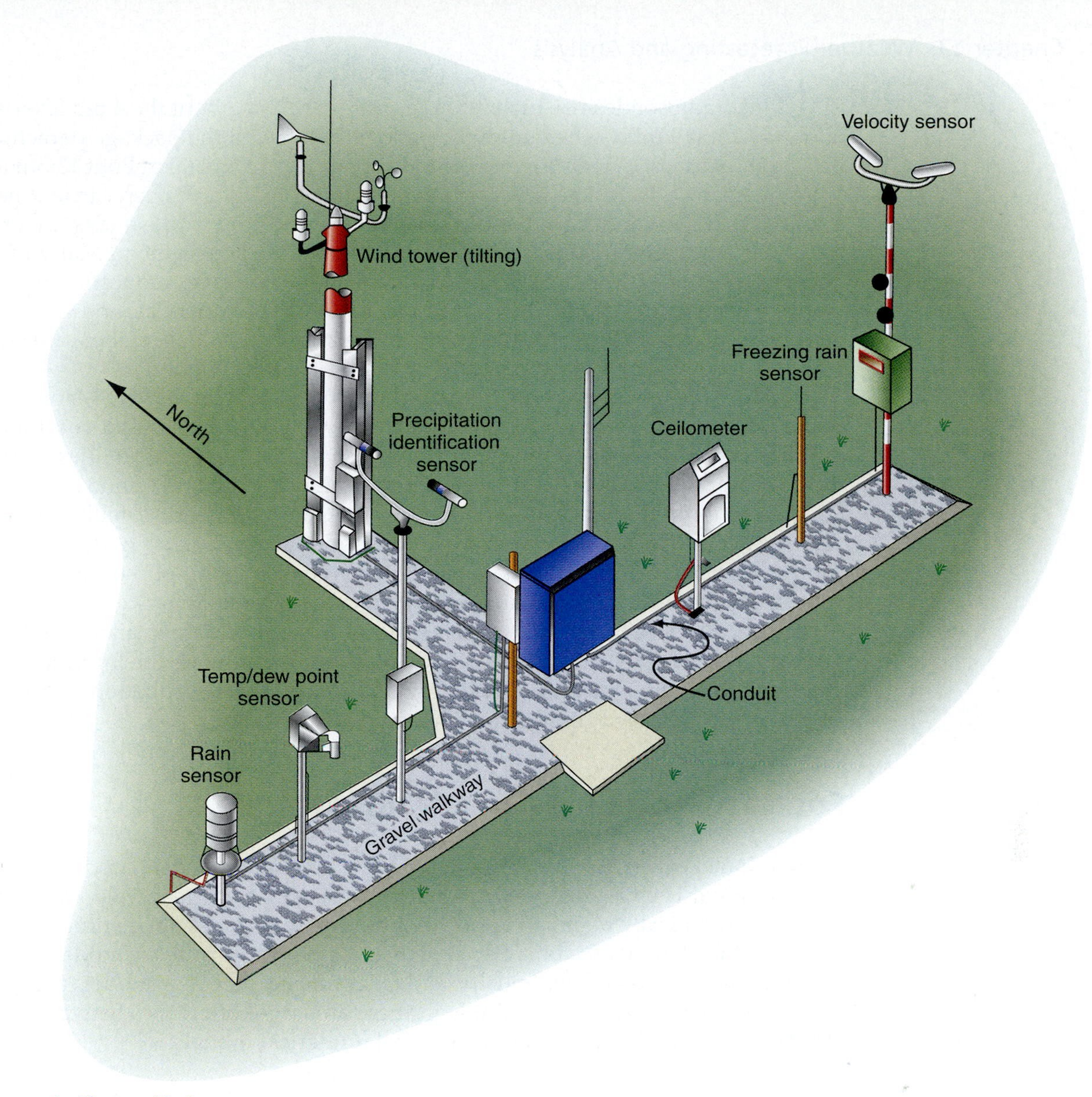

▲ **Figure 13–1**
A typical ASOS unit.

▲ **Figure 13–2**
A radiosonde launch outside the Topeka, Kansas, Weather Service Office.

Most radiosondes rise high into the stratosphere to about the 4 mb level—or about 30 km (19 mi)—whereupon the balloon bursts and the package parachutes back to the surface. Usually a complete radiosonde ascent will take about 1 hour and 50 minutes, by which time the instrument package may have been carried many kilometers downwind from the launch site. Interestingly, radiosondes are often found by non-meteorologists who happen to stumble across them. Many of the people who find them follow the instructions on the packages requesting that the package be dropped in a mailbox for return to the Weather Service, and the recovered radiosondes are then refurbished for subsequent reuse. Most of them are never recovered, however, landing in oceans or remote areas on land.

Upper-level information is obtained from other sources as well. Many wide-bodied commercial aircraft are equipped with weather sensors that continually monitor the atmosphere throughout their flights, and weather satellites supplement the upper-level database by determining temperatures and humidities at several depths throughout the atmosphere. Together, the aircraft and satellites provide data from locations far away from any radiosonde sites and play a particularly important role in gathering information over the oceans. In addition, surface data are collected by sensors on buoys and relayed to land via satellites.

Forecasters not only have a tremendous amount of data available to them, but they also have the ability to easily display and manipulate the information to suit the immediate forecasting need. The AWIPS system (described in Chapter 1) serves as the primary agent for data display and analysis.

Forecast Procedures and Products

As mentioned earlier, numerical models are the preeminent tool of modern weather forecasting. The various weather agencies around the world develop their own models and typically maintain a suite of models rather than a single program. For example, the National Weather Service currently has three primary models in use. The models are updated continually, and from time to time new models are introduced and older ones retired. Although there are large differences between models, the general procedure is the same for all numerical models.

Analysis Phase

First is the **analysis phase**, in which observations are used to supply values corresponding to the starting ("current") state of the atmosphere for all the variables carried in the model. Remember, the models are three-dimensional, meaning that values are needed throughout the depth of atmosphere, not just at the surface. Moreover, some of the models have a global domain covering the entire planet, and therefore they require starting values everywhere, over land and ocean.

Unfortunately, the network of weather stations and radiosonde launches is highly irregular and doesn't come close to providing even coverage. Part of what analysis accomplishes is converting those irregular observations into "uniform" initial values. Though only a preparatory step, this is a difficult task. There are millions of data values from a variety of sources (satellite, ships, and so on), representing various moments in time. None of the measurements is completely free of error, and many are subject to large error. It is necessary to remove as much error as possible while at the same time producing fields that are self-consistent. For example, when wind velocities are assigned, the resulting wind field must satisfy the conservation of mass.

Observed values also need to be consistent with the particular model being used. All the models are approximations to the real atmosphere, and it's important that the initial field not contain features that can't be represented by the model. Otherwise, early in the forecast period the model will adjust to the mismatch between the initial conditions and what it perceives to be plausible. This adjustment will be superimposed on whatever changes arise from processes within the scope of the model and will possibly contaminate the forecast. The various models therefore rely on different analysis methods. For all of them, though, it's a complicated procedure, involving the integration of the latest available observations, past values (such as snow cover), and even output from other models.

Prediction Phase

Fundamentally, the job of a numerical model is to solve the basic equations describing atmospheric behavior: the equation of motion, continuity equation, energy equation, etc. Collectively, these are known as the *governing equations*. Beginning with values delivered by the analysis phase, the model uses the governing equations to obtain new values a few minutes into the future. The process is then repeated, using the output from the first step as input for the next set of calculations. This procedure is performed over and over as many times as necessary to reach the end of the forecast period (24 hours, 48 hours, or whatever). This is called the **prediction phase** of the model run. Again, we emphasize that huge computational resources are needed for this. The governing equations cannot be solved directly but must be broken down into simple operations that computers can perform, such as multiplication and addition. This results in many billions of calculations for each time step, despite the fact that there are just a handful of fundamental atmospheric variables (temperature, pressure, wind velocity vector, density, and moisture).

Post-Processing Phase

The conditions forecast by the model at regular intervals (for example, every 12 hours) are represented in grid form for mapping and other display purposes. For example, in this **post-processing phase**, a series of maps might be produced for each of the 12-, 24-, 36-, and 48-hour periods, depicting the forecast distributions of

1. Sea level pressure and 1000 to 500 mb thicknesses
2. 850 mb heights and temperature
3. 700 mb heights and vertical velocities (such as the speed at which air rises or sinks)
4. 500 mb heights and absolute vorticity values
5. Precipitation amounts

These products are used in various ways, some of which are described later in this chapter. Speaking generally, forecasters study maps for each period and interpret the conditions that would likely be associated with such patterns. The maps are compared with one another and with corresponding maps from other models for the same time periods. Of course, the model forecasts differ from one another—the forecaster uses model output as guidance, weighting the results differentially according to what is known about each model's strengths and weaknesses, and supplementing model guidance with other analyses and observations.

Often the actual (final) forecast will not match any of the models exactly. For example, Figures 13–3a, b, and c show precipitation forecasts from three models for June 2, 1992. Although all three forecasts call for heavy rain in the Texas–Oklahoma area, the location and amount vary considerably from model to model. (Will central Oklahoma get 5 inches, or half an inch?) As can be seen in Figure 13–3d, the forecaster produced a manual forecast significantly different than any of the models, with the area of maximum rain displaced south and east of the models. Sadly, nature was perverse on that day, confounding both machine and human—heaviest rain was found in extreme southeastern Texas (Figure 13–3e). Regardless of that forecast's lack of success, the point we want to make is that model output is coupled with other information in producing official forecasts. Numerical models are certainly superior to purely subjective methods, but even better is a combination of model output and other information (including subjective judgment).

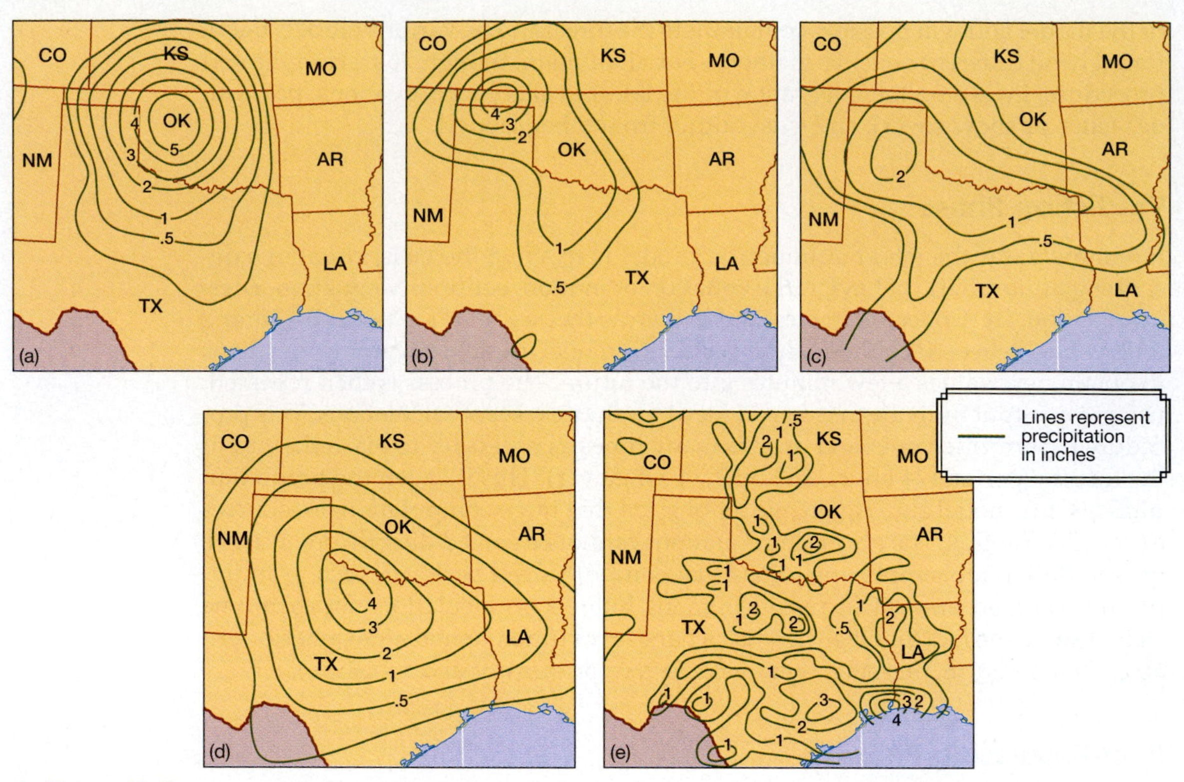

▲ **Figure 13–3**
Twenty-four-hour precipitation forecasts from three numerical models (a–c), the final (manual) forecast (d), and observed rainfall (e) for June 2, 1992.

It is interesting that experience and "rules of thumb" have not been rendered obsolete by computer forecasts, but instead are applied to those forecasts in hopes of improving on them. It is also interesting that as the numerical models evolve and change for the better, old rules must be constantly reevaluated in light of new model behavior. In other words, model improvements continually render old wisdom obsolete. But forecasters are partly to blame for their plight, because suggestions for change often come from quirks they identify and bring to the attention of modelers.

In addition to gridded fields of model variables, forecasts for a number of secondary variables are produced. Examples include maximum and minimum temperature, dew point, wind conditions, and the probability of precipitation. These are constructed using statistical relationships between model output and observed surface conditions from the past. These output products are called *model output statistics* (*MOS*) and are designed to capture the effect of topography and other factors that influence local weather conditions. Numerical models have only a limited ability to represent processes occurring near the surface, and they provide a rather generalized picture of the atmosphere. Thus, a statistical approach has considerable appeal. MOS is most effective for places where statistical relations were derived but is somewhat less effective at intermediate locations having a different topographic setting.

How good are today's forecasts? There's no single answer to this: It depends very heavily on the variable in question, the forecast lead time, the model used, and the place and season. For example, there's no doubt that temperature, wind, and pressure distributions are forecast far better than precipitation. An example of precipitation skill is shown in Figure 13–4, which gives skill scores from several models for various amounts of precipitation for spring and summer of 1995.* Skill scores

*In this figure, skill is measured by the equitable threat score, described in the appendix to this chapter. Here we simply note that this score, like that shown in Figure 13–4, measures the area of correct forecasts. A method that perfectly forecasts rain and rain-free areas would obtain a skill score of unity, representing 100 percent skill.

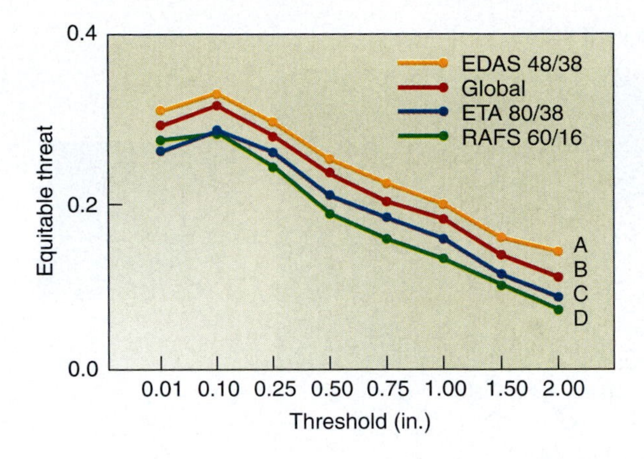

◀ **Figure 13–4**
Twenty-four-hour forecast skill (equitable threat score) for various NCEP models and precipitation amount. Forecasts are for warm season only (March through August) of 1995.

decline significantly with increasing precipitation amount, indicating the difficulty of forecasting heavy precipitation, which tends to be highly localized. Also evident are significant differences between models, although the skill rankings of the models are the same for all precipitation amounts. For example, model *A* always has the highest skill, whereas model *D* shows lowest skill at all precipitation levels. (This is not generally the case—using other measures of skill, the models would change position relative to one another.)

There can also be significant seasonal variations in skill, as indicated in Figure 13–5, which shows 1-inch forecasts for a decade (1984–1993). Clearly, winter precipitation is predicted with much more skill than summer rain. This is partly due to seasonal changes in the precipitation processes. As we go from large, synoptic-scale systems in winter to smaller-scale convection in the summer, it becomes more difficult to forecast the exact location of precipitation events. Second, winter precipitation tends to be much less intense, with storms lasting many hours. Finally, in the winter it is more likely that there will be just a few large centers of precipitation associated with fronts or mid-latitude cyclones. These might receive more attention from forecasters than numerous "popup" thunderstorms and other small disturbances that dot the summer landscape.

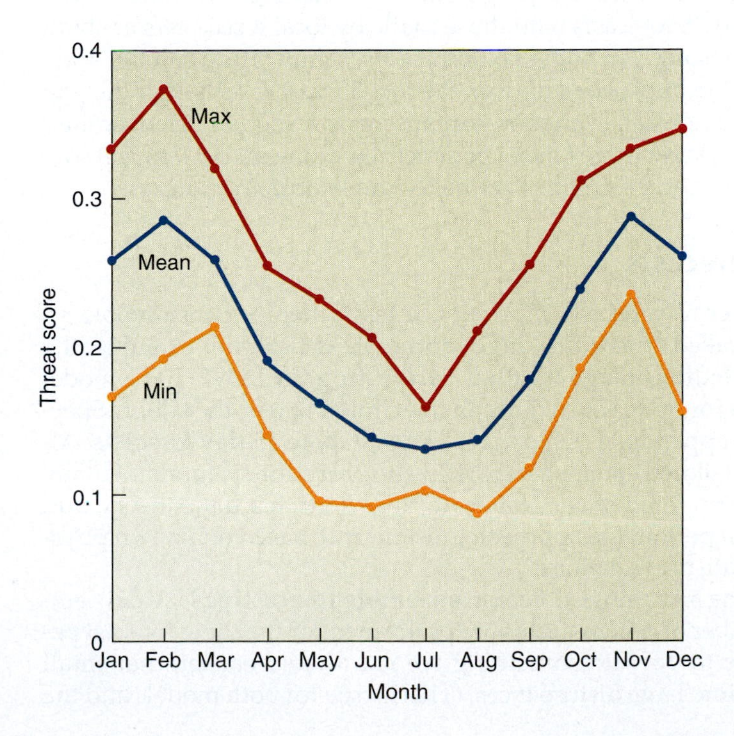

◀ **Figure 13–5**
Twenty-four-hour, 1-in. (2.5 cm) NCEP precipitation forecast skill averaged over 1983–1993. Skill is for the manual forecast, not any particular model.

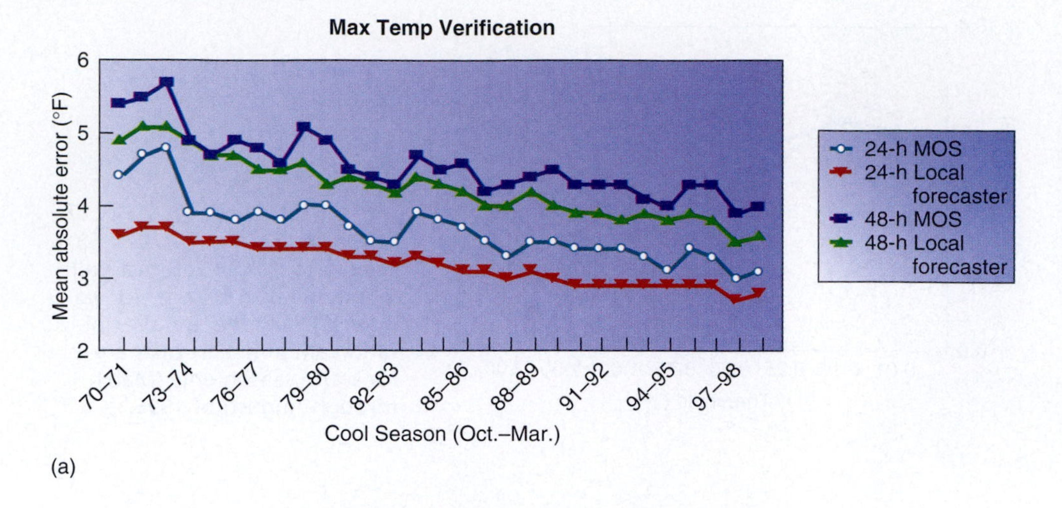

(a)

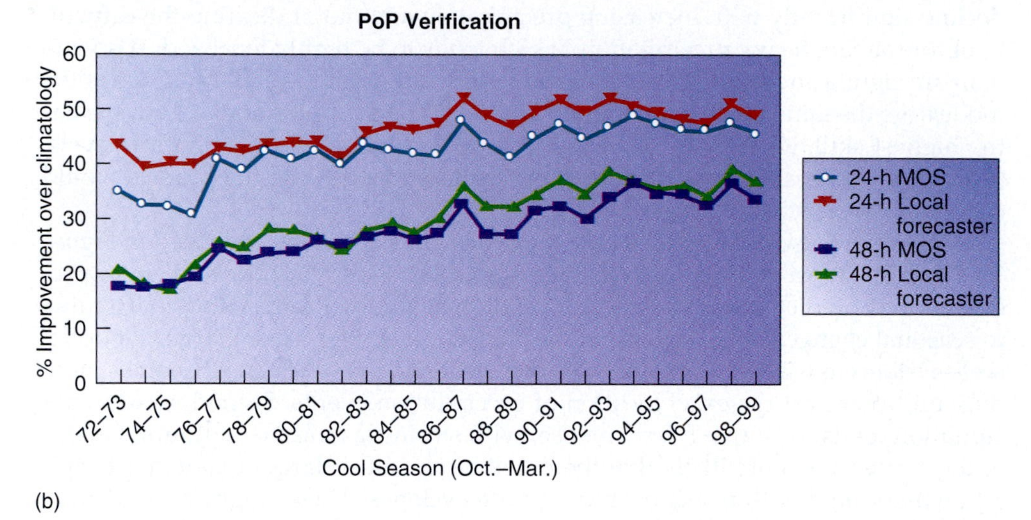

(b)

▶ **Figure 13–6** Cool-season (October–March) maximum temperature (a) and precipitation forecast (b) scores. The graphs compare objective (MOS) output with local forecasters for 24-hour and 48-hour forecasts.

A final point to be made is that despite incredible advances in computer technology, there is still a place for human judgment in forecasting. Consider Figure 13–6, which compares MOS forecasts with those made by local forecasters at about 95 stations in the United States. For both maximum daily temperature and PoP forecasts, we see considerable improvement over the last 30 years. Although the gap might have narrowed somewhat, even now humans enjoy an advantage over their silicon competitors. Local knowledge (and experience regarding model failings) continues to be useful and can add significant value to purely objective forecasts.

Medium-Range Forecasts

Going beyond "short-term" forecasts (72 hours or less), there is considerable attention directed at so-called **"medium-range" forecasts (MRF)**. For example, the **European Center for Medium-range Weather Forecasting (ECMWF)** has a model that generates forecasts for up to seven days. In the United States, the Global Spectral Model (Chapter 13 appendix) is used at NCEP to prepare 15-day forecasts. Although the models are tailored somewhat to MRF, the procedure is fundamentally the same as for short-term forecasting. That is, to the extent that data, theory, and computer resources will permit, the approach is numerical, based on known physical laws rather than statistical relations.

Rather than making just a single forecast, **ensemble forecasting** is widely employed, in which a number of different runs are performed for the same forecast period. The reasons have to do with something we mentioned earlier, that small disturbances can grow into large disturbances. (This is true for both models and the real atmosphere.) Thus, if two model runs are made with slightly different initial values, the results might be very different after a week or so. Discovered in 1963 by E. N. Lorenz, this behavior is now known to be typical of many natural and human systems and is usually called "chaotic behavior."

Chaos is a serious problem for weather forecasting because the initial conditions are never known exactly. Thus, for example, if an upper-level trough appears in the 15-day forecast, there is no way to know if it's "real" or if it arose because of errors in the initial data. Ensemble forecasting uses multiple runs starting with slightly different initial values. The different initial fields are created by introducing small changes (perturbations) in the best-guess field. There are several methods for assigning perturbations, but all attempt to mimic errors that might reasonably appear in the data. From this ensemble of initial conditions, an ensemble of forecasts is produced, each different from the other.

Figure 13–7 shows a ten-day ensemble for the Northern Hemisphere from NCEP. The map is a kind of 500 mb chart, except that only one contour is shown: the 5700-m contour. Plotted are the 17 NCEP ensemble members, plus the unperturbed (control) run. Notice that all ensemble members call for a trough in the western Pacific Ocean. The implication is that whatever errors might exist in the data, they don't affect the forecast there. Thus, we might have some confidence that a real trough will develop. But over central Asia, western Europe, and the north Atlantic, ensemble spread is much larger: Ensemble members show much less consistency with one another, suggesting we should be less confident about the forecast

Figure 13–7
Ten-day ensemble forecast from the NCEP MRF model. The 5700-m contour is plotted for all 17 ensemble members.

for these regions. (Of course, we would not base the final assessment on a single contour, and we would look at other variables in addition 500 mb height.)

Figure 13–7 illustrates the most important use of ensembles, which is to provide information about forecast uncertainty. With that, one has an estimate of reliability to go along with the forecast and won't pay much attention to forecasts deemed unreliable. Ensembles also can be used in other ways, including to generate the forecast itself. In particular, the mean of all ensembles can be treated as a forecast, even though it doesn't arise from a model run. Moreover, we might expect this forecast to be reasonably good, on the grounds that averaging will smooth away features found in just one or two members.

Yet another use of ensembles is to diagnose failings of the model. Suppose all the ensemble forecasts agree in some area, yet all depart from what is ultimately observed. The departure between the forecast and observation can't be explained by data error, because the ensemble suggests the forecast is insensitive to data error. This leaves model error as a likely culprit, meaning that one or more processes are poorly treated in the model and need improvement. The ensemble doesn't identify the problem, of course, but exposes the situation in which the problem influences the forecast. By studying the details of that situation, it might be possible to learn which aspect of the model needs work.

At the present time, there is little skill evident in MRF beyond a week or so, especially for precipitation. However, oftentimes they do produce accurate (and valuable) forecasts and can provide useful guidance about general tendencies. In addition, it has been shown that ensemble spread correlates with forecast error for as long as ten days, suggesting that estimates of forecast reliability also have value at medium range.

Long-Range Forecasts

Forecasts are also produced at still longer lead times, so-called **"long-range" forecasts**. In the United States, the **Climate Prediction Center (CPC)** of NCEP is charged with preparing forecasts for periods ranging from a week to the limits of technical feasibility. The methods used include climatology, statistics, numerical models, and subjective judgment. For example, because of its important role in the global climate system, sea surface temperature (SST) in the tropical Pacific is routinely forecast for up to a year in advance. The SST forecasts are based on a combination of three models:

1. An analog statistical model
2. A "canonical" statistical model based on correlations measured over space and time
3. A numerical model

In the numerical model, ocean and atmosphere are coupled so that the ocean responds to changes in the atmosphere, and vice versa. Output from the three models is combined statistically to yield the final forecast.

Another CPC product is the "seasonal outlook," a kind of forecast for an entire season, often misunderstood. In contrast to long-range forecasts that predict conditions for particular days, seasonal outlooks predict average conditions over a range of days. Figure 13–8 shows the format, in which seasons are classed as "above normal," "near normal," or "below normal." These classes are defined such that "above normal" includes the upper third of the distribution; that is, by definition "above normal" occurs about 33 percent of the time, when a variable is at the 66th percentile or higher. "Near normal" and "below normal" are defined similarly, as the middle and bottom thirds, respectively. Thus, without any special knowledge we could forecast "above normal" with 33 percent accuracy and likewise for "near normal" and "below normal." Suppose we believe "near normal" conditions are

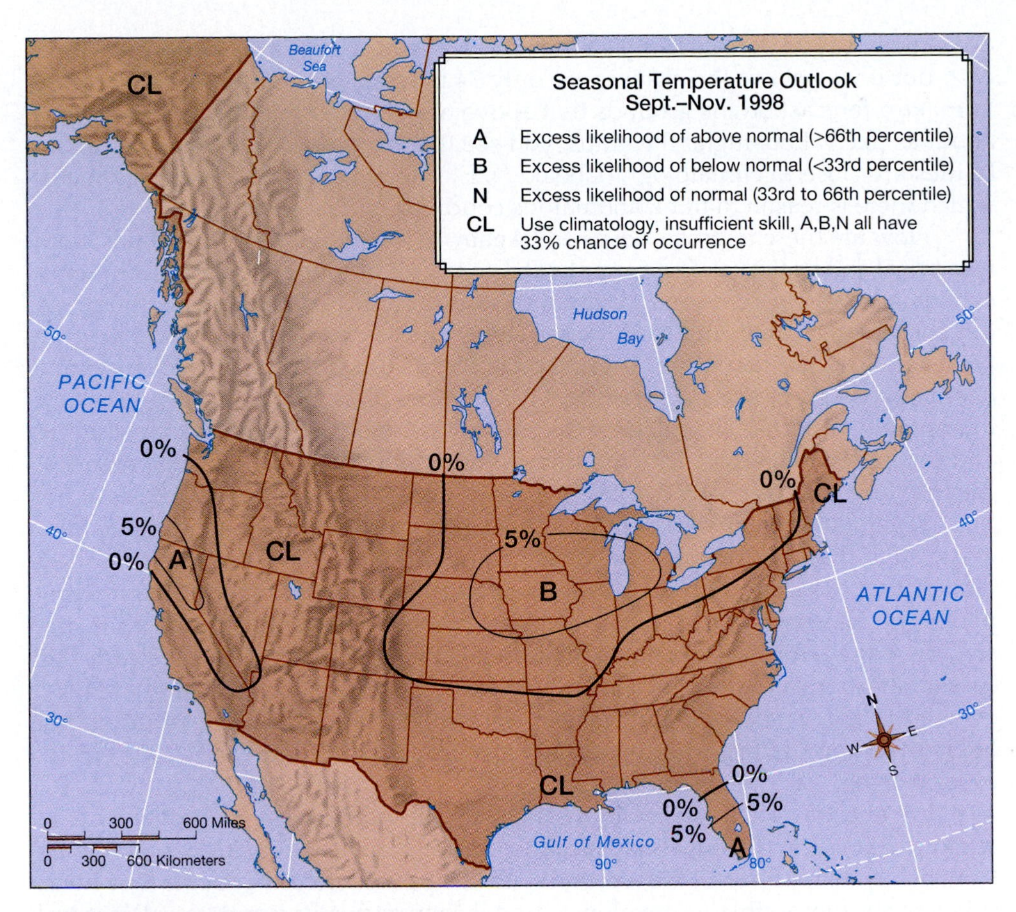

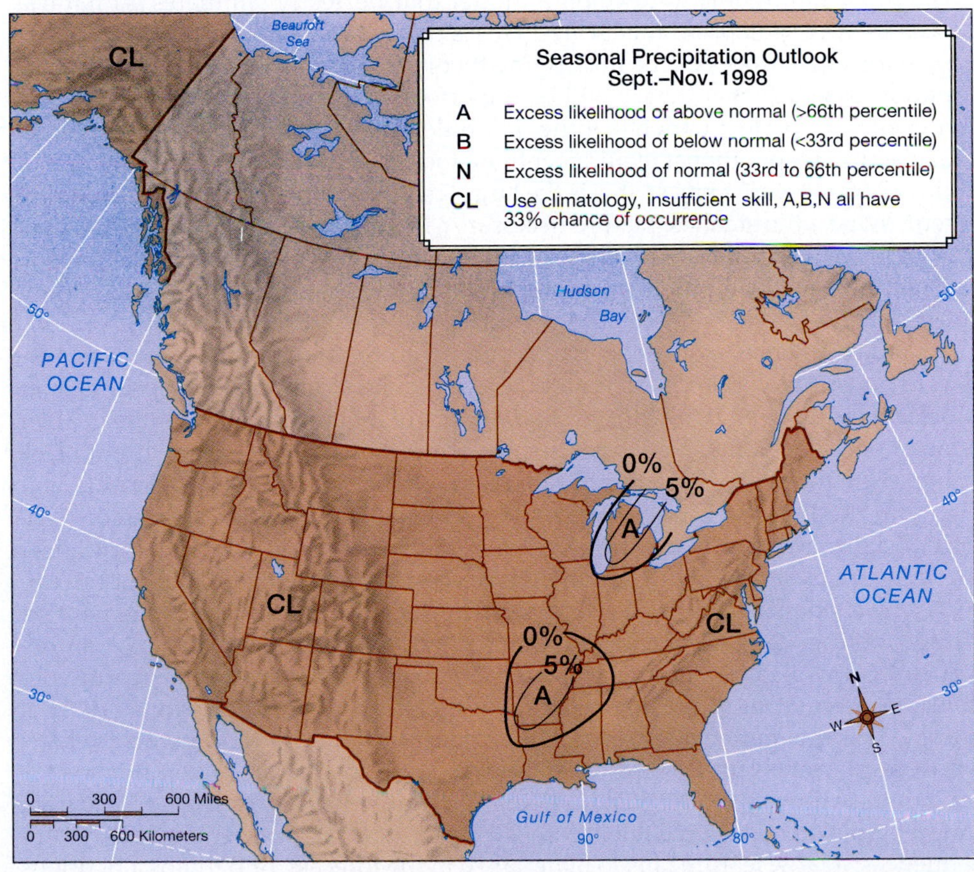

Figure 13–8 The seasonal outlook for September–November 1998, prepared 1 year earlier, September 15, 1997.

quite likely, say 45 percent likely. In that case we would issue a "near normal" forecast. But if we believe the chances are only 34 percent, we would probably refuse to make a forecast on the grounds that above or below normal conditions are just about as likely. Looking at the maps, you see that for most of the area, the probabilities are close to climatology (labeled "CL"). There are only a few regions for which there is reason to think anomalous conditions will occur.

How are these forecasts obtained? Again, by a combination of methods, both numerical and statistical. For example, SST forecasts from the coupled ocean-atmosphere model are used as boundary conditions for repeated runs of an atmospheric general circulation model. Another set of runs is made using SST based on persistence rather than physical principles. The result, once again, is an ensemble of atmospheric forecasts. The statistical techniques are mainly based on persistence in anomaly patterns. That is, departures from average are analyzed for stability and/or consistent patterns of evolution. Although the details are beyond an introductory text, the basic idea is not: The past is the key to the future.

Several things must be kept in mind when interpreting the long-range forecasts. First, they are forecasts for the entire period (season), not a particular day. One certainly does not expect above-average conditions throughout the forecast period, no matter what the "above average" probability might be. Also, at the present time, these forecasts do not exhibit much skill, and what skill exists varies by season, location, and the forecast variable (temperature vs. precipitation). Finally, even if the skill is relatively high (say 20 percent), the associated probability is not likely to be large. In fact, it is unusual to have probabilities much above 50 percent, even in areas of highest confidence.

We end this section by pointing out that although recent improvements in computer power and model sophistication are impressive, we are immeasurably far from being able to forecast the weather with 100 percent accuracy. In the first place, it is simply impossible to develop a model that perfectly simulates all natural processes at all spatial and time scales. This is partly a matter of insufficient knowledge—for some processes the underlying theory is not complete, so any computer program based on that theory will be imperfect (turbulence is a good example). But even with perfect understanding, it would not be possible to build a perfect model—the computational requirements are too great. Earth has a circumference of about 40,000 (4×10^4) km, so this is the largest size a complete model needs to represent. What's the smallest feature necessary? For the sake of discussion, let's say that in a perfect model, cloud drops 4 microns (4 μm) in diameter must appear, but nothing smaller. That gives a range of features covering 13 orders of magnitude in size.

Today's models span scales over roughly 3 orders of magnitude. To reach 4 orders of magnitude, about 5 km resolution, will require a computer a thousand times more powerful. Even that will leave us 9 orders of magnitude short of the goal, or a factor of one billion. Given this immense computational gulf, together with lack of complete theory, it's hard to imagine that model error will ever disappear.

But model error is only part of the problem; chaos is the other major difficulty. Earlier we briefly mentioned dynamic instability, in which disturbances grow in size over time. With that there is necessarily a propagation of energy upward from smaller scales to larger scales. To put this rather vague statement on firmer footing, consider a slowly moving river. As the water flows past an obstruction, a whirlpool or eddy is created. In this case the small-scale feature derives its energy from the large-scale flow; there is a downward propagation of energy. This is a routine occurrence in the atmosphere, but so is the reverse, where, for example, small updrafts grow into large cumulus towers. What this means for numerical weather forecasting is that one must have detailed and accurate initial values to make a reliable forecast. Even a small error might have dire consequences: a large feature fails to materialize in the forecast, or perhaps a fictitious feature unlike anything in the real atmosphere appears. This extreme sensitivity to initial conditions is characteristic of chaotic systems in general and the atmosphere in particular. Unless we are prepared to begin with perfect data, there is no hope for perfect prediction. Note that the problem of chaos is independent of any defects in the model, such as processes ignored or improperly treated. Even if we could build a perfect model, the presence of chaos rules out the possibility of a perfect forecast.

Weather Maps and Images

Although computers play a critical role in the analysis of weather, ultimately the meteorologist applies his or her knowledge to produce the forecast that gets issued to the general public. Probably no tool is as valuable to a forecaster as a weather map. And although newspapers and television news segments often just show their audience surface maps, weather forecasting requires the analysis of conditions in the middle and upper atmosphere as well. Not only do clouds exist well above the surface, but the middle and upper atmosphere are closely intertwined with the air near the surface. As a result, accurate weather analysis requires the use of a series of other maps, each representing different layers of the atmosphere.

Surface Maps

Surface maps of prevailing conditions (such as that shown in Figure 13–9) present a general depiction of sea level pressure distribution and the location of frontal boundaries. The pressure is shown by isobars drawn every 4 mb, with zones of locally highest and lowest pressure labeled *H* and *L*, respectively.

Large-Scale Features Even a nonprofessional can make a number of inferences from a brief inspection of a surface weather map. General wind speeds and directions obey the rules discussed in Chapter 4. That is, wind speed varies according

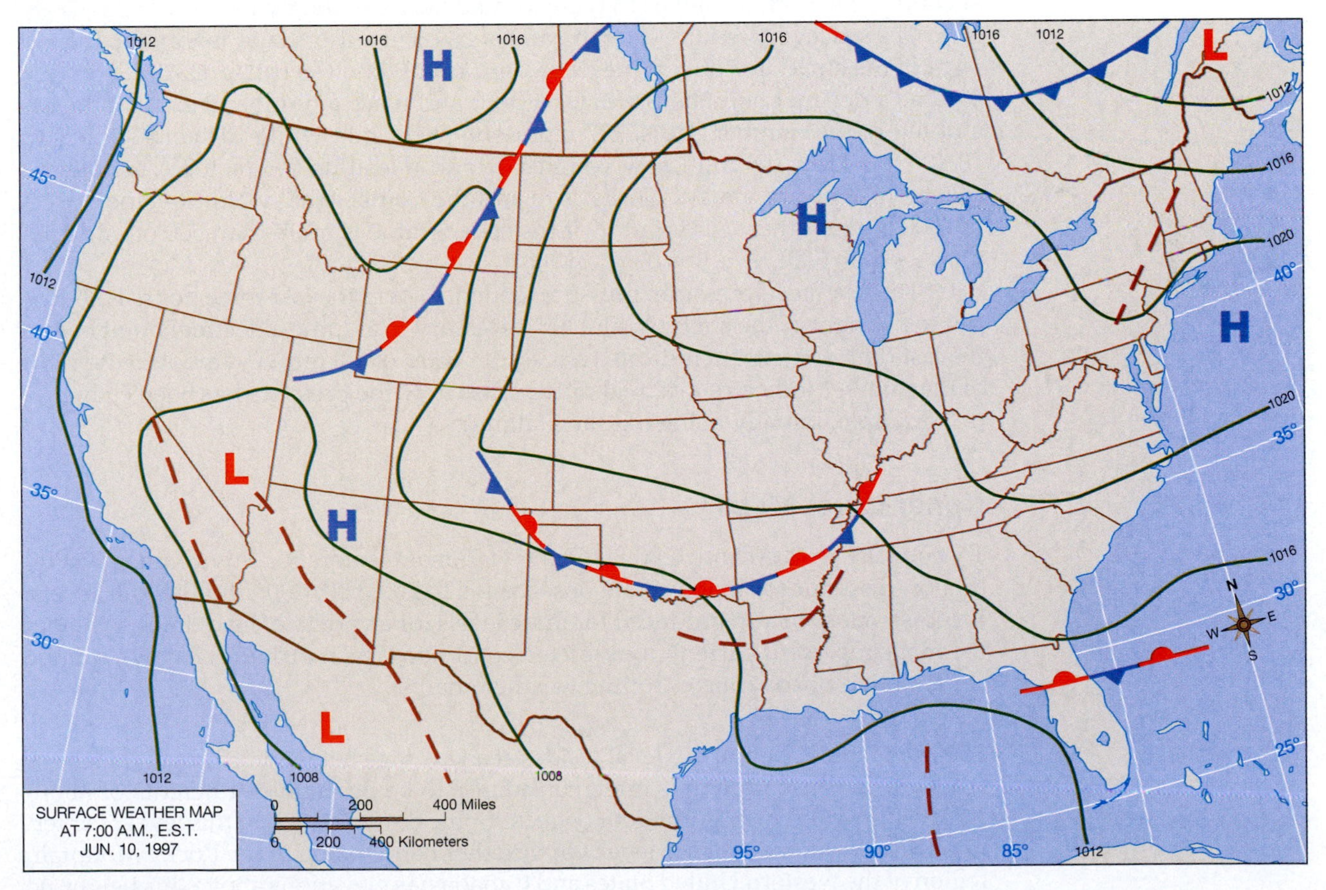

▲ **Figure 13–9**
A typical surface weather map.

to the spacing of the isobars, and Northern Hemisphere winds rotate clockwise out of high-pressure systems and counterclockwise into lows. High-pressure systems favor downward vertical motions that promote clear skies, whereas low pressure promotes updrafts that lead to adiabatic cooling and the formation of cloud cover. Surface maps become even more valuable when viewed in sequence. Because new maps are compiled every three hours, one can easily track the movement of individual weather systems as they move cross-country. By assuming that the systems will continue to behave as they have in the past few hours, their movement and intensification or dissipation over the next few hours can be inferred.

Station Models More detailed knowledge of the conditions at particular locations can be obtained from **station models**. Well over a dozen weather elements, including temperature, dew point, and pressure, are represented on each station model. A complete station model contains some information beyond the needs of most readers, so Figure 13–10 describes only the most important symbols.

The station model indicates cloud coverage by the amount of shading inside the central circle. A completely open circle indicates cloud-free conditions, a fully darkened one represents complete overcast; intermediate amounts of shading correspond to varying fractions of cloud cover. A line extending from the circle with tick marks or flags at the end is a wind "arrow," or "barb," showing the wind speed and direction. The free end of the arrow corresponds to the direction that the wind is blowing *from*. Thus, for instance, an arrow at the top of the circle indicates winds from the north, whereas one on the right side of the circle means that the wind is from the east. The wind speed is represented by the number of full or half tick-marks, and/or flags, as represented in Figure 13–10. An arrow having one complete tick-mark and one half tick-mark, for example, has winds between 15 and 20 mph (23 to 32 km/hr).

Temperature and dew point values (in degrees Fahrenheit) are listed to the top left and bottom left of the circle, respectively. Symbols representing common weather conditions (such as rain or fog) are located between the two.

The sea level pressure (in millibars) is given in a shorthand manner to the top right of the circle. To convert the three-digit number to the true pressure, you first assume a decimal point before the last digit and place a number 9 or 10 at the beginning of the number. Thus, 997 represents 999.7 mb, while 104 corresponds to 1010.4 mb. How do you know whether to add a leading 9 or a 10? A simple answer, which nearly always works, is to add the number that yields a value closest to 1000 mb (in other words, add 9 if the coded value is more than 500 or add 10 if it is less than 500).

The change in pressure (pressure tendency) over the last three hours is shown just to the right of the circle. Again, it is necessary to assume a decimal point before the last digit so that –10 indicates a pressure drop of 1.0 mb. A symbol to the right of the number indicates in a qualitative sense how the pressure has been changing (for example, initially rising but then falling).

Upper-Level Maps

Twice a day, at 0000 and 1200 UTC, the National Centers for Environmental Prediction disseminate maps of the observed 850, 700, 500, 300, and 200 mb levels. Forecast maps are also produced for those levels for a variety of lead times. Whether the maps represent current or predicted conditions, each provides its own unique combination of advantages for the weather analyst.

850 mb Maps The **850 mb level** resides at an average height of 1.5 km (1 mi) above sea level. Because friction is often considered to be negligible at heights of about 1.5 km above the surface, gradient or geostrophic flow can exist at this level over terrain with elevations near sea level. On the other hand, much of the Rocky Mountain region of the western United States and Canada has elevations above this height, so in those areas the 850 mb map actually represents near-surface conditions. At high

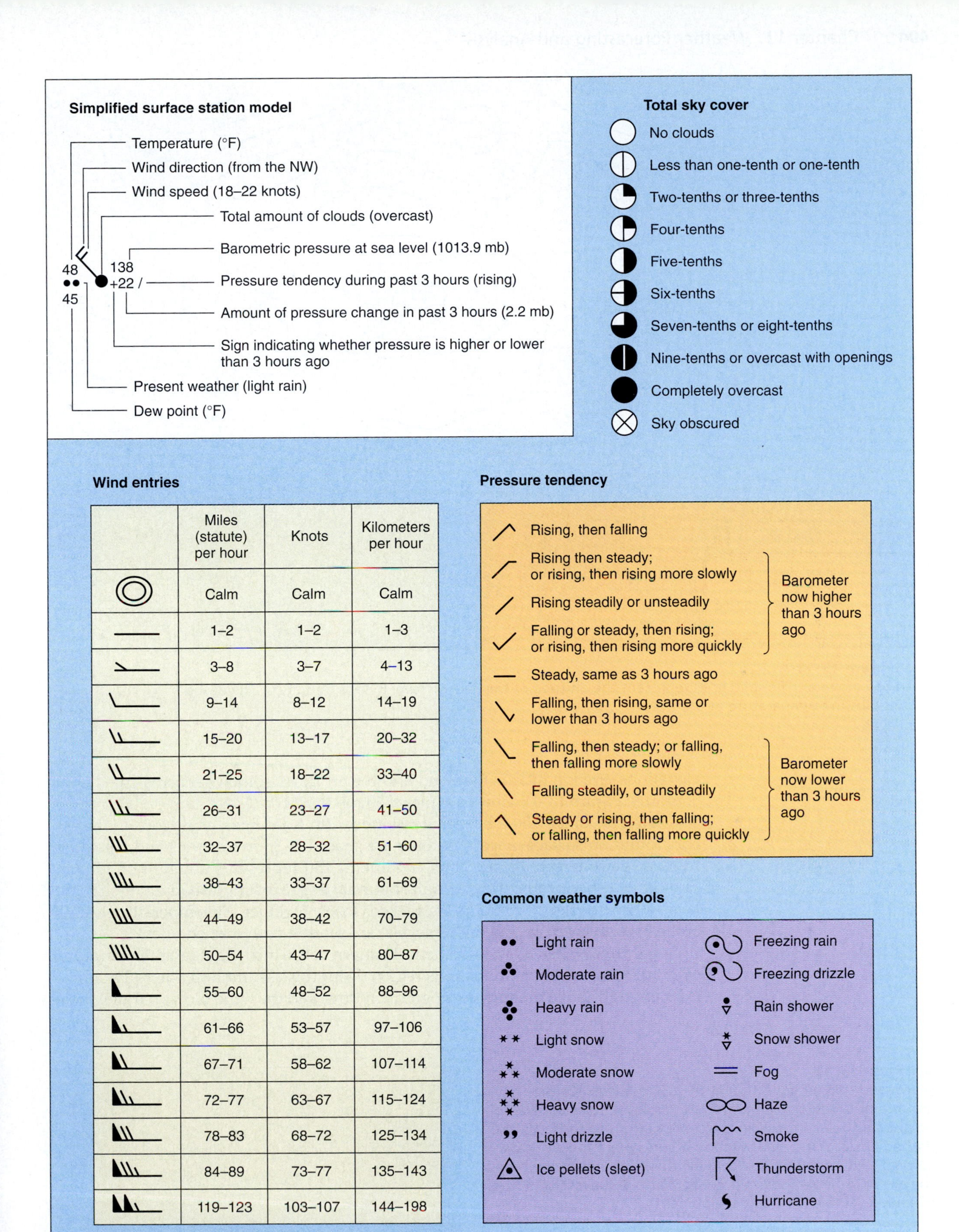

	Miles (statute) per hour	Knots	Kilometers per hour
	Calm	Calm	Calm
	1–2	1–2	1–3
	3–8	3–7	4–13
	9–14	8–12	14–19
	15–20	13–17	20–32
	21–25	18–22	33–40
	26–31	23–27	41–50
	32–37	28–32	51–60
	38–43	33–37	61–69
	44–49	38–42	70–79
	50–54	43–47	80–87
	55–60	48–52	88–96
	61–66	53–57	97–106
	67–71	58–62	107–114
	72–77	63–67	115–124
	78–83	68–72	125–134
	84–89	73–77	135–143
	119–123	103–107	144–198

▲ **Figure 13–10**
The arrangement of a surface station model and some important symbols.

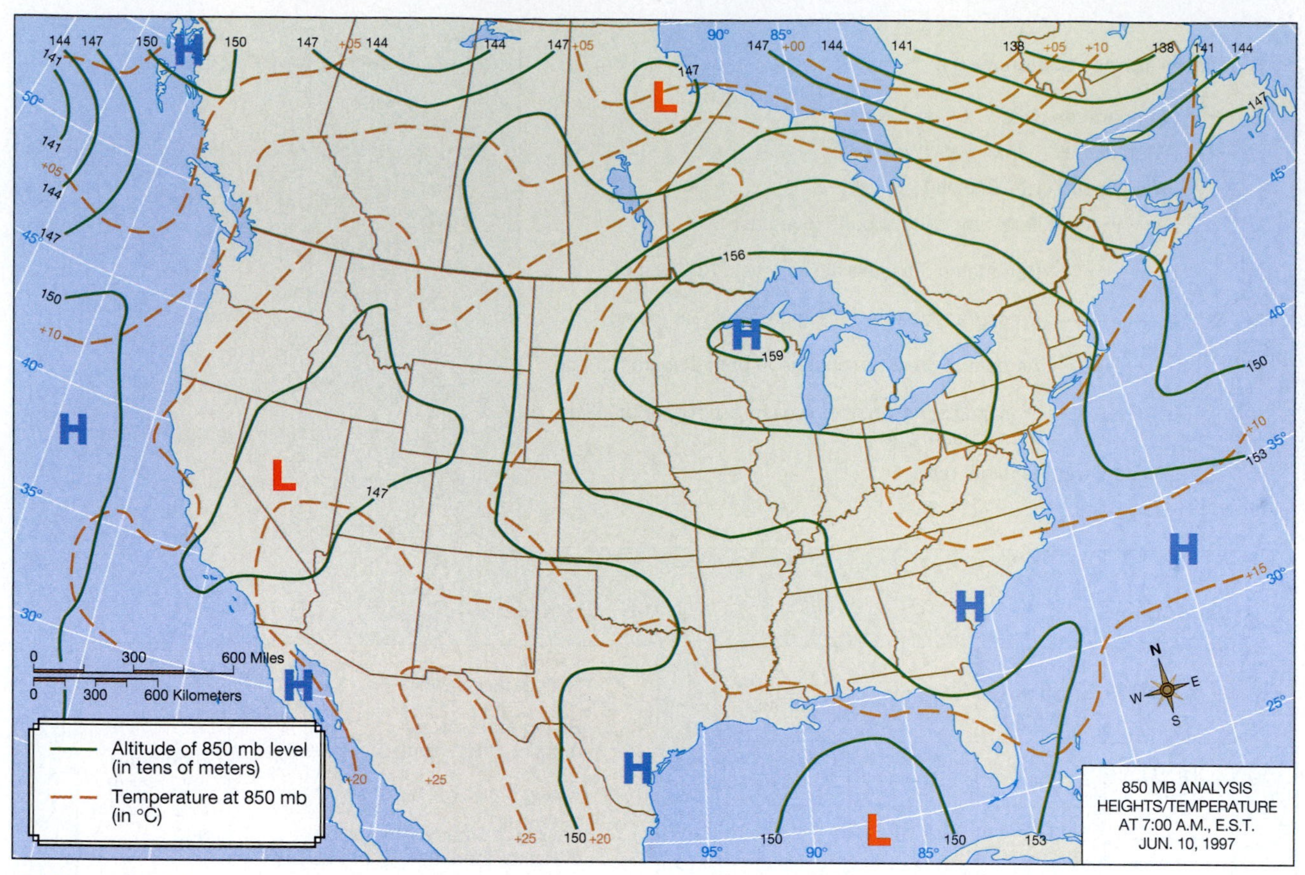

▲ **Figure 13–11**
An 850 mb map. The solid lines represent the altitude of the 850 mb level in tens of meters. The dashed lines plot the temperature at that level.

elevations, friction retards the wind and the air flows somewhat across the height contours. Figure 13–11 shows a typical 850 mb map.

Heights of the 850 mb level are plotted with solid lines, analogous to the isobars found on surface maps. Contours are spaced at 30-m intervals and labeled in units of decameters (10 m). Thus, a value of 150 represents a height of 1500 m. Though not specifically drawn, frontal boundaries are distinguishable at the 850 mb level, where the height contours are packed closely together. Air temperatures (in °C) at the 850 mb level (and all higher levels) are plotted with dashed lines.

Figure 13–12 shows a hypothetical pattern of 850 mb heights and temperatures over an assumed nonmountainous region. With this assumption, the 850 mb level is far enough above the surface for gradient flow, and the wind blows parallel to the

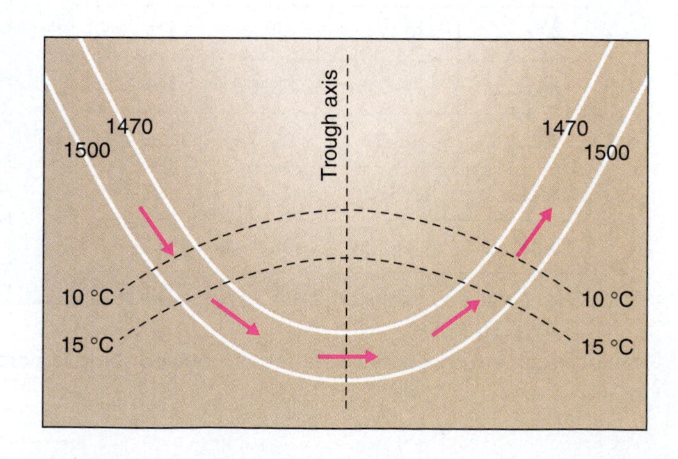

▶ **Figure 13–12**
Warm and cold air advection at the 850 mb level. Outside of mountainous regions, air at the 850 mb level flows parallel to height contours. If these lines cross temperature contours (the dashed lines), warm air advection occurs where the air flows toward colder regions, and cold air advection happens where the air flows toward warmer areas. Warm air advection at the 850 mb level favors cloud development.

height contours. As the air approaches the trough axis, it moves from a region below 10 °C to a region above 15 °C. Conversely, as the air flows away from the axis, it moves toward a colder region. Thus, the region upwind of the trough axis is marked by cold air advection, while warm air advection occurs downwind of the axis. These patterns are important because warm air advection at the 850 mb level indicates upward motion, which favors the formation of clouds and precipitation. Thus, in this example uplift is likely downwind of the trough, making the region favorable for cyclone development or intensification.

The distribution of temperatures at the 850 mb level provides forecasters with some useful rules of thumb. During the morning, for example, the 850 mb temperature often provides a good way to guess the daily maximum temperature over nonmountainous areas. At the 850 mb level, the air is far enough from the surface so that it does not undergo daily cycles of warming and cooling. Thus, during the summer, the maximum surface air temperature is usually about 15 °C (27 °F) greater than the 850 mb temperature, regardless of the time of day. During the winter, the difference will be about 9 °C (16 °F); during the fall and spring, about 12 °C (22 °F).

700 mb Maps Maps of the **700 mb level** (Figure 13–13) have many of the same applications that the 850 mb maps do. Like 850 mb maps, maps of observed conditions plot height contours (in decameters) and isotherms (°C), with solid and dashed lines, respectively. The 700 mb level occurs in the vicinity of 3 km (2 mi) above sea level and has a mean temperature of about –5 °C (23 °F).

Maps of the 700 mb level are best for observing the short waves that were shown in Chapter 10 to be so important in the formation and maintenance of midlatitude cyclones. They are also particularly valuable in predicting the movement of air mass thunderstorms, which usually move with about the same velocity as the 700 mb winds.

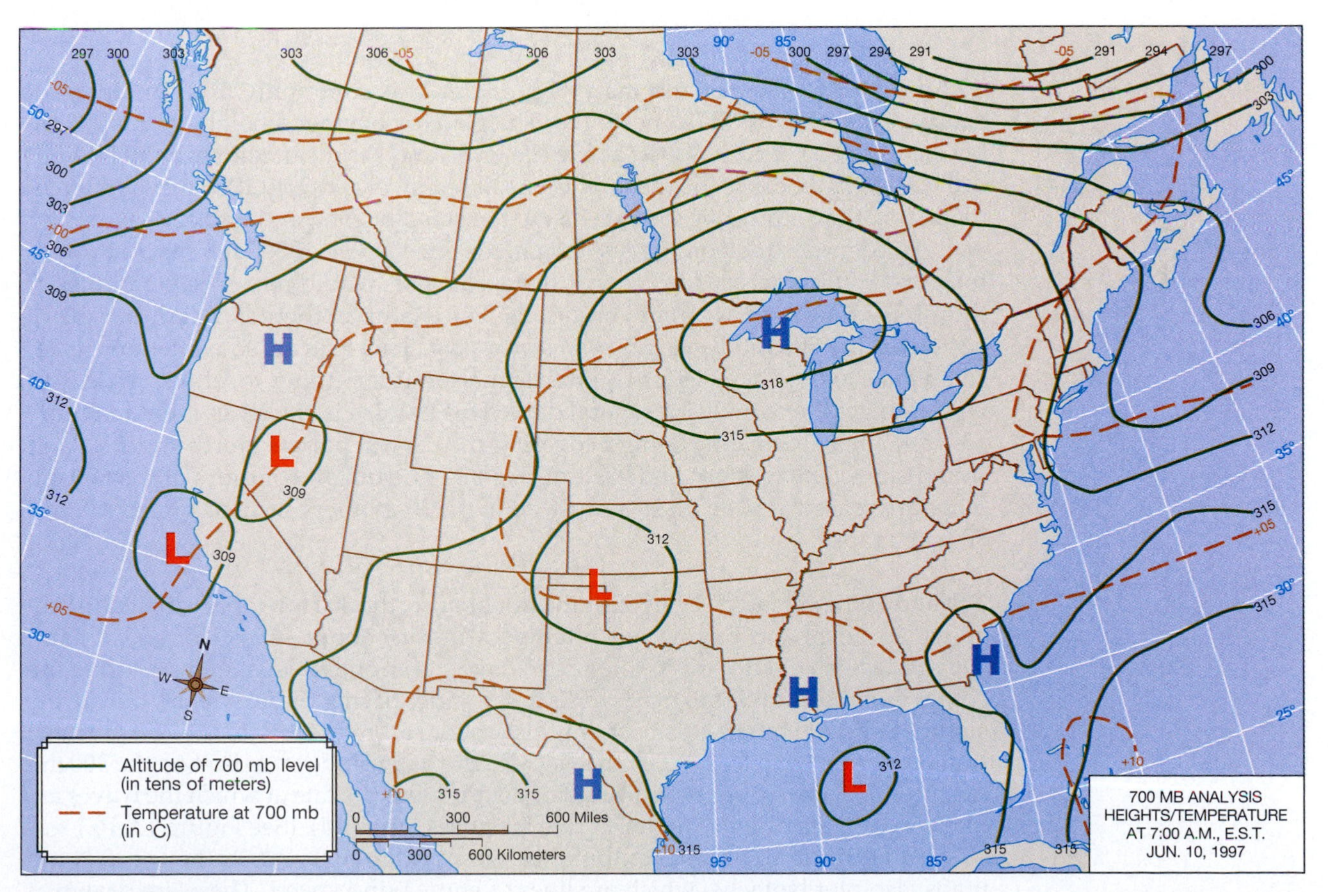

▲ Figure 13–13
A 700 mb map.

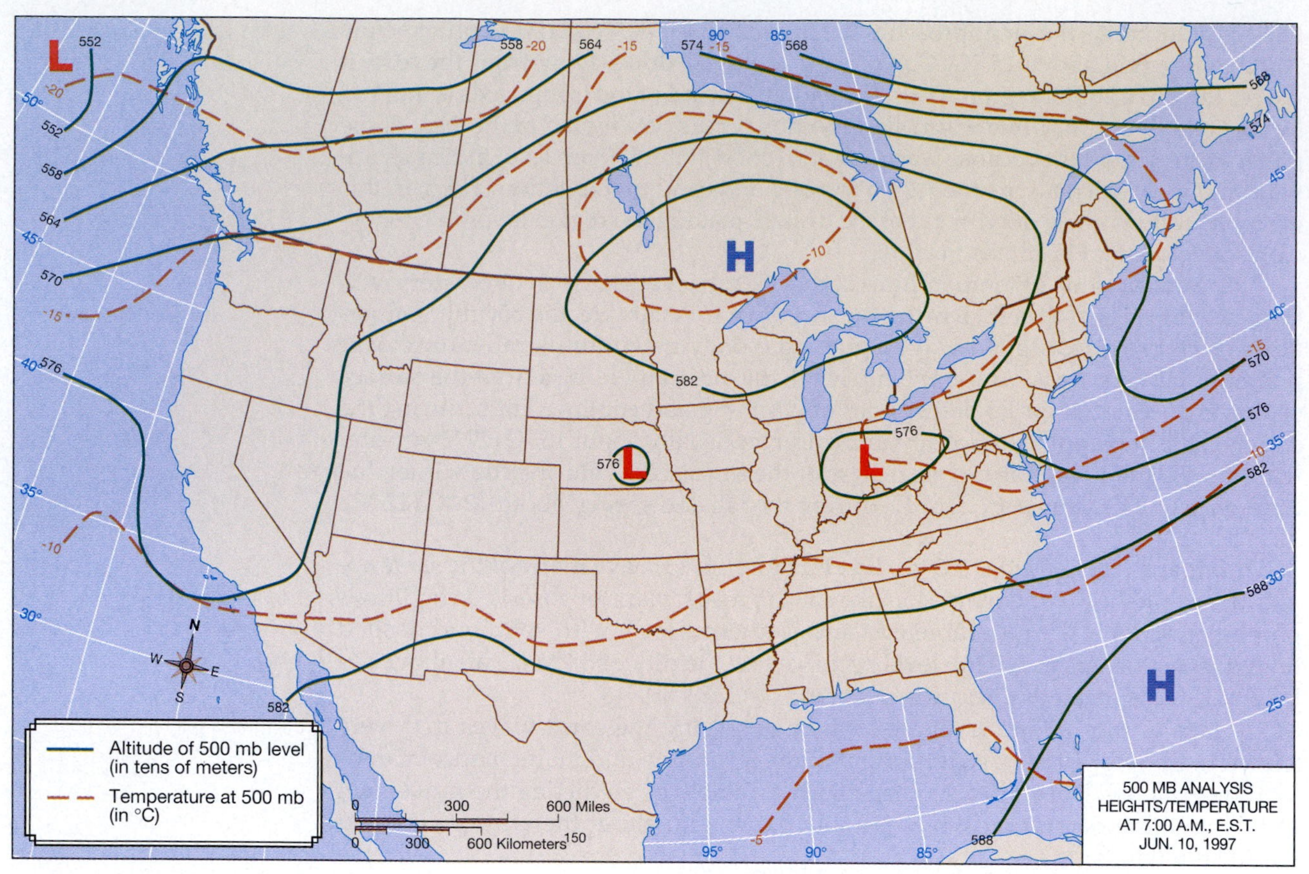

▲ **Figure 13–14**
A 500 mb map.

500 mb Maps The **500 mb map** (Figure 13–14) is commonly used to represent conditions in the middle atmosphere. The globally averaged height of the 500 mb level is about 5.6 km (18,000 ft) above sea level, and the mean temperature is about –20 °C (–4 °F). Because the mean pressure at sea level is nearly 1000 mb, about half the mass of the atmosphere exists below the 500 mb level, and half above.

For decades meteorologists have made special use of 500 mb maps. Prior to the advent of numerical forecasting, meteorologists would rely on several rules of thumb in predicting weather conditions. For example, there is a certain type of pattern at the 500 mb level called an *omega high.* This feature, so named for its resemblance to the Greek letter Ω, often signals that the upper-level pattern is likely to change only slowly for several days. This pattern is clearly evident in Figure 13–15, with a single omega high covering the central part of North America and troughs over the western and eastern regions. The height contours of the 500 mb maps are spaced at 60-m intervals instead of the 30-m intervals used for the 850 and 700 mb maps.

300 and 200 mb Maps Lying near the tropopause, the **300 mb** (approximately 9 km above sea level and having mean temperatures of about –45 °C, or –50 °F) and **200 mb levels** (12 km; –55 °C, or –65 °F) have the strongest jet streams. During the colder months, the 300 mb map works best for identifying the jet stream; during the summer, the 200 mb map is best. (In this section, all references to the 300 mb level pertain equally well to the 200 mb level.) Rossby waves show up best on the 300 mb maps, which makes the charts useful for determining the rate at which the waves are likely to migrate downwind (or in rare instances, upwind). (See Figure 13–16.)

In addition to height contours (at 120-m intervals) and isotherms, 300 mb maps also plot **isotachs**, which are lines of equal wind speed. These are drawn at intervals of 20 knots, beginning with the 10-knot isotach. Areas where the wind

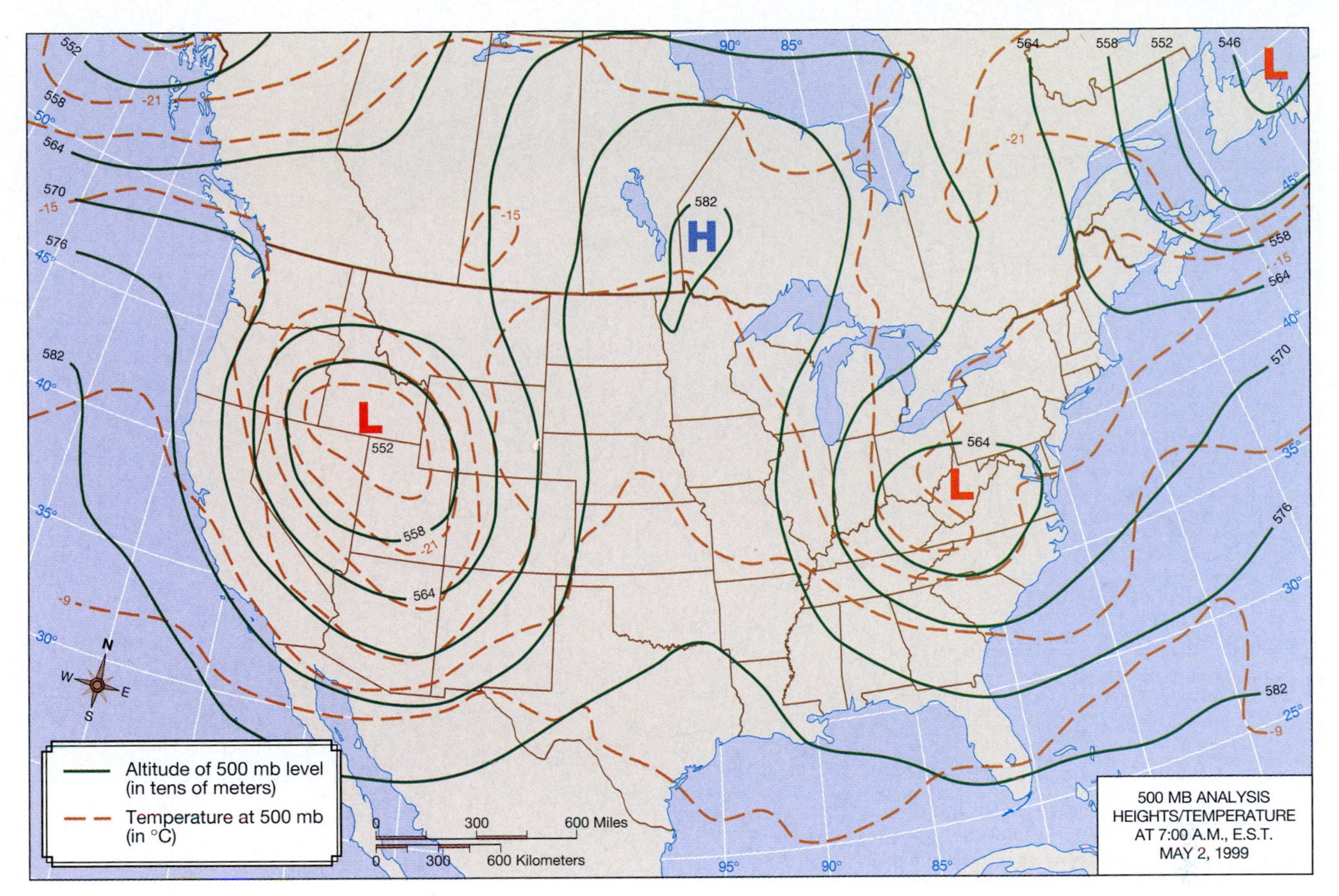

▲ **Figure 13–15**
An omega high.

speed is between 70 and 110 knots are indicated by shading. Superimposed areas having winds between 110 and 150 knots are unshaded, and zones having winds above 150 knots are again shaded. This makes it easy to note the regions of increasing and decreasing jet streams. This is important because air flowing into or out of local areas of high wind speeds (called *jets*, or *jet streaks*) generates local regions of upper level convergence and divergence, as shown in Figure 13–17.

Satellite Images

Anybody who watches the weather segment of any news broadcast is certainly familiar with radar and satellite imagery. **Visible images** (Figure 13–18a) view the atmosphere the way an astronaut in space would, simply by registering the intensity of reflected shortwave radiation. Obviously, these images are available only during the daytime.

Infrared images (Figure 13–18b) are based on measurements of longwave radiation *emitted* (not reflected) from below. If dense clouds are present, the source of the radiation will be the cloud top, otherwise the surface and lower atmosphere supply most of the upwelling radiation. Cumuliform clouds result from condensation associated with the adiabatic cooling of rising air. A deep cloud formed by air parcels rising great distances will therefore have a lower cloud-top temperature than will a mid-level cloud formed by just a kilometer or two of ascent. It follows that a satellite sensor will receive less radiation from a tall cloud than a low cloud and less from a low cloud than from a cloud-free region. Satellite images show the colder (higher) cloud top as whiter than warmer (low) clouds. Clear areas are very dark on these images.

Water vapor images (Figure 13–18c) provide a unique and often beautiful perspective on the atmosphere. Water vapor is a very effective radiator at wavelengths

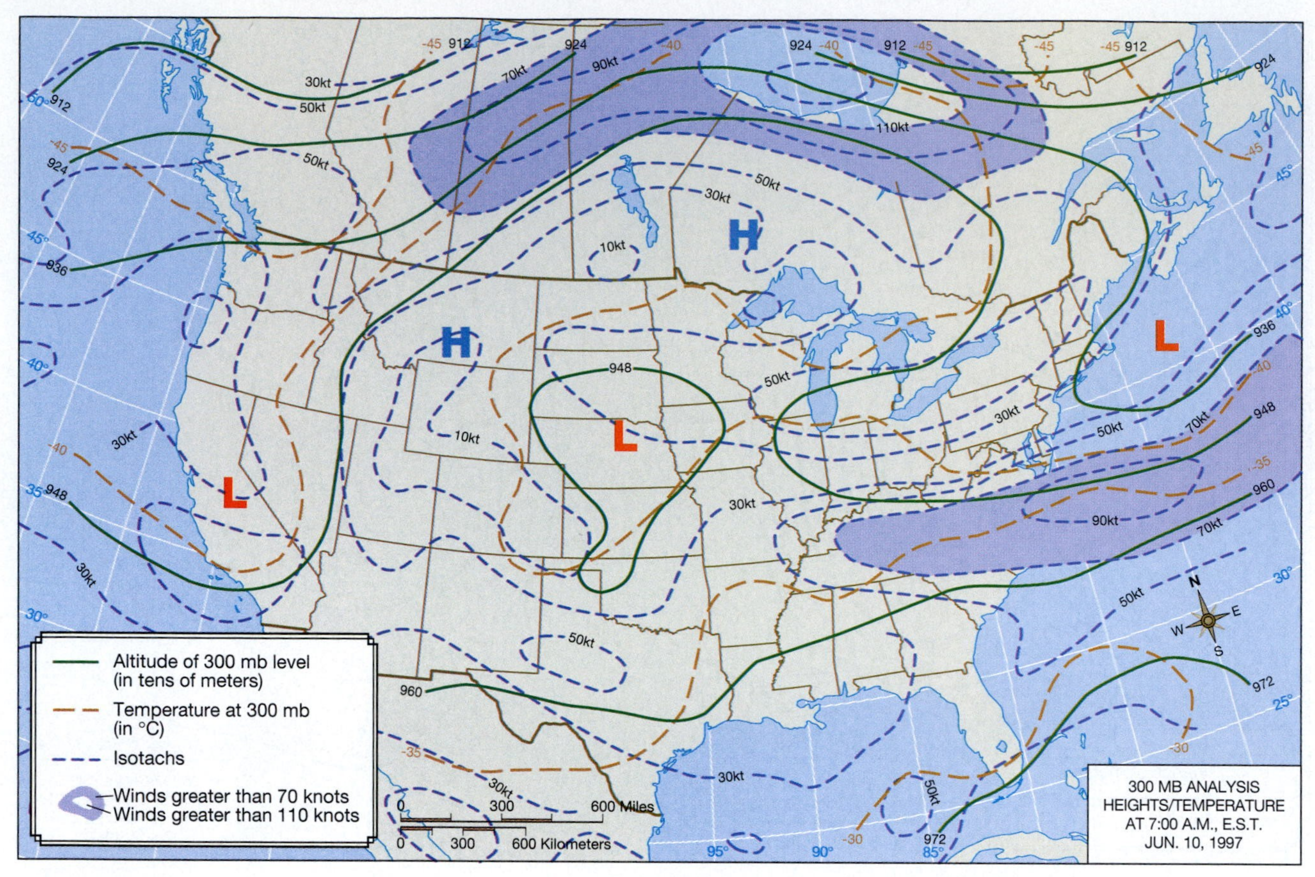

▲ **Figure 13–16**
A 300 mb map. The purple dashed lines are isotachs contoured at 20-knot intervals. Shaded areas have winds in excess of 70 knots. Open areas within the shaded regions have winds greater than 110 knots.

▶ **Figure 13–17**
Air flowing into and out of jet streaks (areas of locally fastest winds) creates patterns of upper-level convergence and divergence. Air entering the jet streak sets up convergence along its left flank (quadrant A), and divergence on its right (B). Exiting the jet streak, there is divergence to the left (quadrant C) and convergence to the right (D).

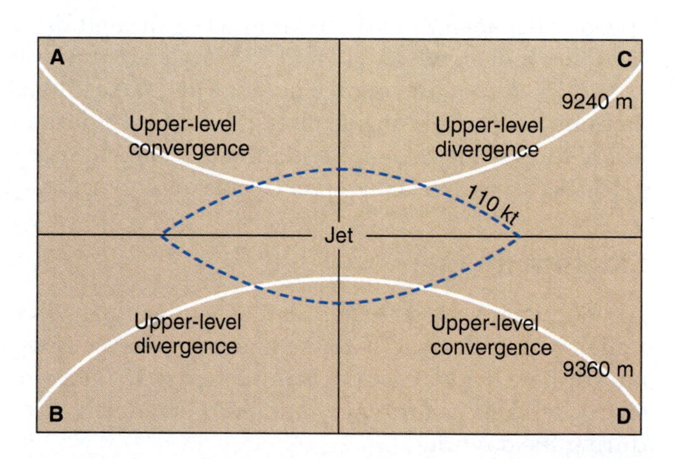

near 6.7 μm. Relative humidities above about 50% result in a high output of radiation in this part of the spectrum, and the sensors translate high values of this radiation into bright regions on the imagery. Water vapor images are particularly useful for tracking the flow of moisture across wide regions and helping to identify the location of frontal boundaries.

Radar Images

The utility of Doppler radar in tornado detection has already been described in Chapter 11, but the **radar images** generated by conventional systems have been extremely valuable for routine weather analysis for decades. Radar observes the internal cloud conditions by measuring the amount of radiation backscattered by

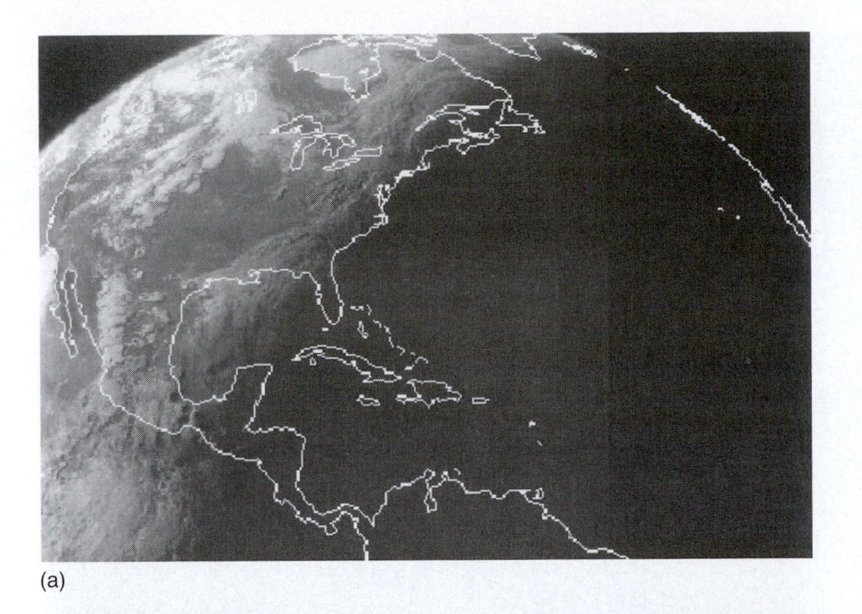

(a)

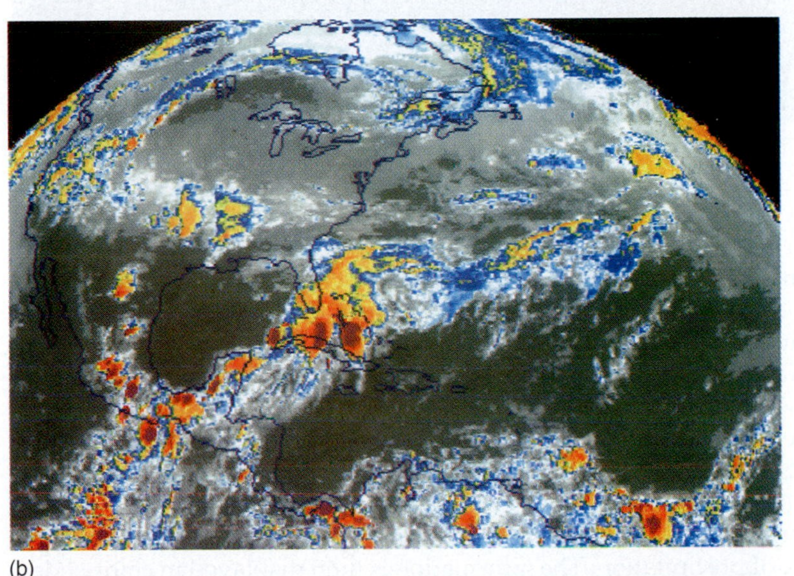

(b)

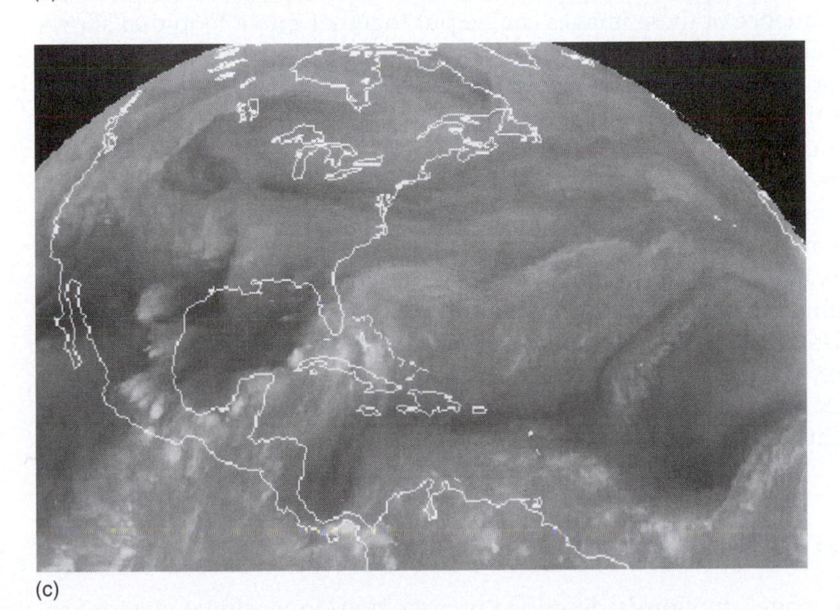

(c)

Figure 13–18
A visible (a), infrared (b), and water vapor (c) image obtained from the GOES 8 satellite.

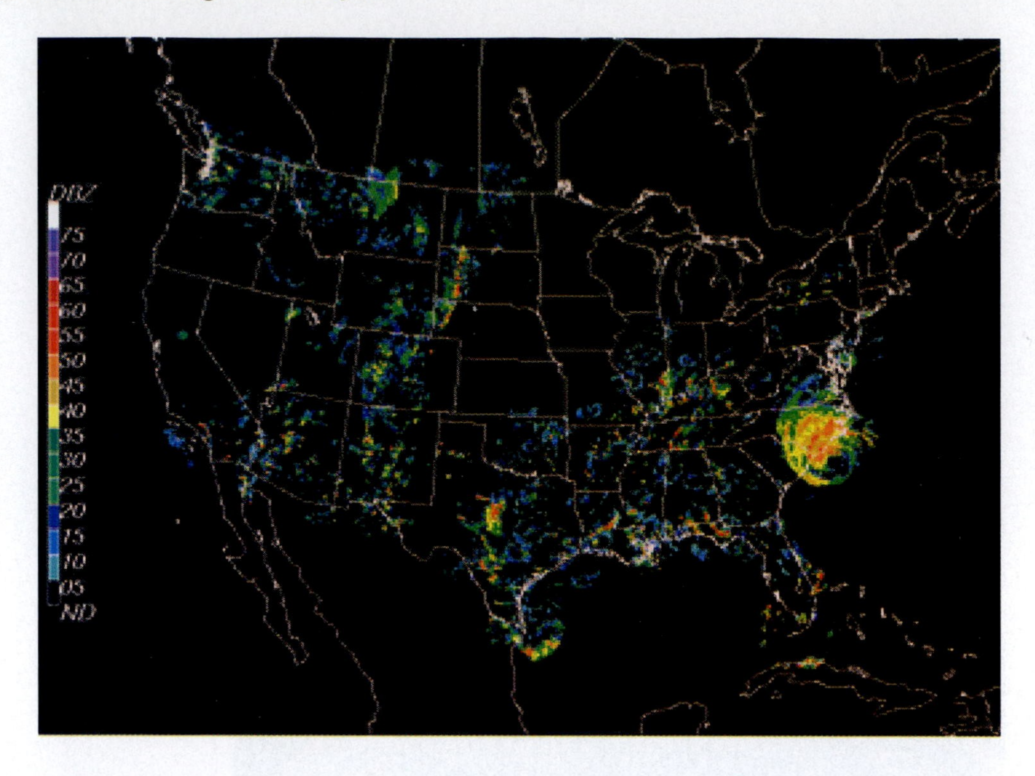

▶ **Figure 13–19**
A radar composite map.

precipitation (both liquid and solid). A transmitter sends out brief pulses of electromagnetic energy with wavelengths on the order of several centimeters. A receiver records the intensity of the echoed pulses (indicative of the number and size of droplets and crystals) and the time elapsed between pulse transmission and return (which indicates the distance to the scattering agent).

The radar continually emits these pulses as the transmitter/receiver rotates 360°, giving a two-dimensional representation of the cloud conditions surrounding the unit. After each rotation, the transmitter angle is increased slightly, and the radar scans a higher slice of the atmosphere. This procedure is repeated until a large volume of the surrounding atmosphere has been scanned, and the meteorologist can observe the distribution of the clouds, the heights of their tops and bases, and the relative intensity of precipitation. The information is then displayed in color-coded map form. A sequence of these images can be put together into a loop that shows the movement and changes in weather activity.

Each radar unit covers distances up to about 400 km (250 mi). The Weather Service assembles the returns from all the radar sites in the national network and compiles the information onto maps of the 48 contiguous United States and southern Canada (Figure 13–19).

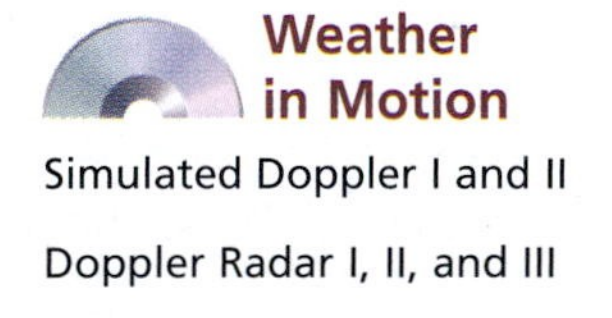

Weather in Motion

Simulated Doppler I and II

Doppler Radar I, II, and III

While it is relatively easy to determine the big picture of local activity from a radar image, a meteorologist has to beware of potential errors in its use. For instance, the curvature of Earth's surface causes a horizontally emitted beam of radiation to assume greater heights above the surface as it moves away far from the transmitter. Also, radar waves are refracted (bent) somewhat as they travel through the atmosphere. The extent to which the refraction occurs depends on the stability. Thus, the meteorologist not accounting for this effect can obtain false readings of cloud top height.

Thermodynamic Diagrams

The maps and images previously described provide two-dimensional views of atmospheric conditions, but they fail to provide detailed vertical information. Vertical profiles of temperature and dew point observed by radiosondes are plotted on

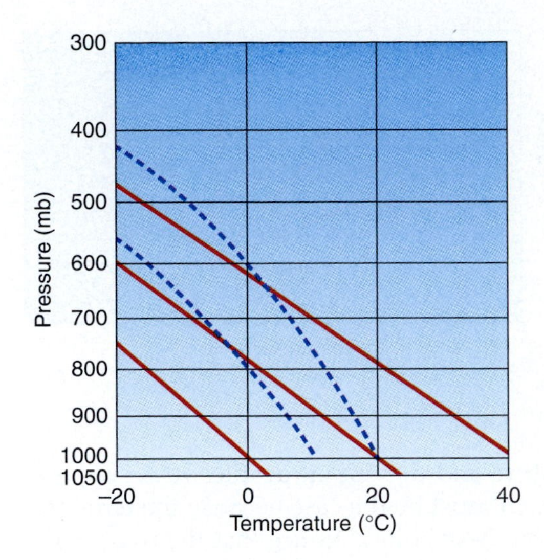

◀ Figure 13–20
A Stuve thermodynamic diagram. The vertical lines across the horizontal axis represent temperature (°C); the horizontal lines, pressure (note that the scale is not linear). The solid lines sloping to the left with height are dry adiabats. An unsaturated parcel of air moving up or down will have a decrease or increase in temperature determined by following the line. The slightly curved, dashed lines are wet adiabats, which show the change in temperature a saturated parcel would have as it rises or sinks.

thermodynamic diagrams (also called *pseudo-adiabatic charts*). The simplest thermodynamic diagram is the *Stuve* chart (Figure 13–20), on which the air temperature is scaled along the horizontal axis and the pressure on a nearly logarithmic vertical axis. The straight, solid lines that slant upward to the left are *dry adiabats.* These show the rate of temperature change for an unsaturated parcel of air that is lifted or lowered (in other words, they plot the dry adiabatic lapse rate). The dashed, slightly curved lines are *wet adiabats,* showing temperature changes experienced by a rising saturated parcel.

Soundings are plotted on the charts by marking the temperature and dew point data at numerous pressure levels and connecting the dots. Figure 13–21 shows an actual sounding for the Dallas–Ft. Worth airport on the morning of June 4, 1997.

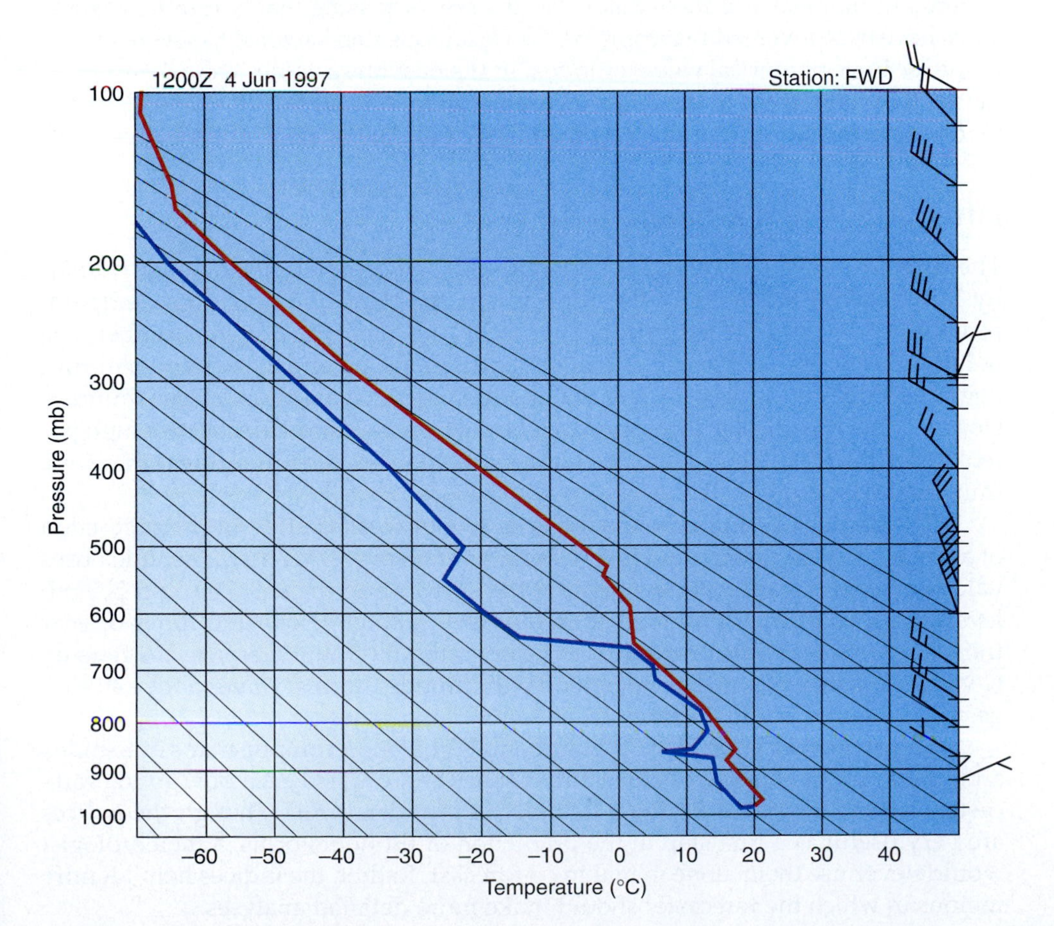

◀ Figure 13–21
An example of a sounding on a Stuve diagram. The heavy line on the right plots the temperature; the one on the left shows the dew point profile.

Figure 13–22
Stability, as indicated by the temperature profile on a Stuve diagram. Profile *A* is steeper than the dry adiabat, so the air is absolutely unstable. Profile *B* is less steep than the wet adiabat, so the air is absolutely stable. Profiles with slopes intermediate between the dry and wet adiabats (such as *C*) indicate conditionally unstable air.

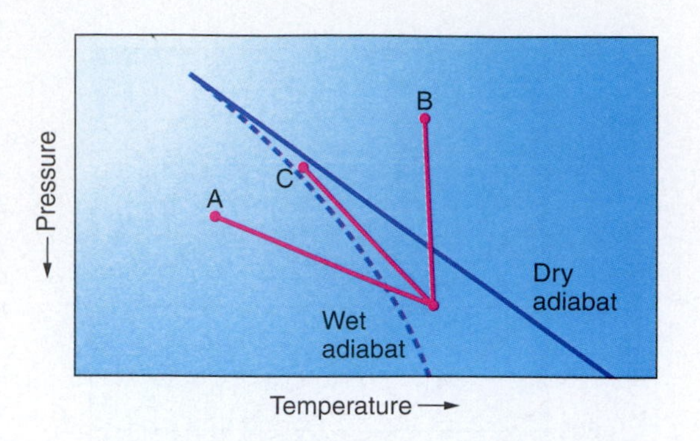

The heavy lines represent the temperature and dew point profiles (the temperature profile is the one on the right—which must be the case because the temperature is always equal to or greater than the dew point). Notice that the two profile lines nearly merge between about 800 and 660 mb, indicating that the layer is saturated and cloud covered. Throughout the rest of the atmosphere, the temperature is much higher than the dew point and the air is unsaturated.

The steepness of the temperature profiles relative to the dry and wet adiabats indicate the static stability for any portion of the atmosphere. Where the temperature decreases more rapidly with height than does the DALR (as in profile *A* in Figure 13–22), the air is absolutely unstable. Where it slants less steeply to the left than adjacent wet adiabats (profile *B*), the air is absolutely stable. Temperature lapse rates intermediate between the two adiabatic lapse rates (profile *C*) indicate conditionally unstable conditions, as we discussed in Chapter 5.

Displaying data on wind speed and direction at numerous heights (and thus the presence or absence of wind shear), thermodynamic diagrams provide a wealth of information to meteorologists, especially in the prediction of severe weather. Given the complexity of these diagrams, it's not surprising that a number of measures have been invented that combine the various elements related to severe weather into a single numerical value, or index. In the past these values were determined manually with the aid of the thermodynamic diagrams; today they are calculated by computers. The lifted index and the K-index are two such summary measures.

Lifted Index and K-Index

The **lifted index** was developed in the mid-1950s. It combines the average humidity in the lowest kilometer of the atmosphere, the predicted maximum temperature for the day, and the temperature at the 500 mb level into a single number. The magnitude and sign of the values together indicate the potential for thunderstorms; negative values indicate sufficient water vapor and instability to trigger thunderstorms. More specifically, lifted index values between –2 and –6 indicate a high potential for thunderstorms, whereas values less than –6 suggest a threat of severe thunderstorms.

The **K-index** is similar to the lifted index but works better for the prediction of air mass thunderstorms and heavy rain than for severe weather. The index uses values of temperature and dew point at the surface and the 850, 700, and 500 mb levels. Various rules of thumb for different geographic regions and times of year translate K-values to the probability of heavy rains and thunderstorms. In general, K-values less than 15 indicate no potential for thunderstorms; values above 40 suggest that they are highly likely.

The National Centers for Environmental Prediction compile maps of K-index and lifted index values across the United States and Canada each day, giving forecasters a quick reference for the possibility of thunderstorms. Though the indices are very useful as a first step in the prediction of thunderstorms, a meteorologist would never use them alone in making a forecast. Rather, the indices help identify regions in which the forecaster should make more detailed analyses.

Summary

Weather analysis and forecasting depends on the cooperative efforts of a large number of nations belonging to the World Meteorological Organization. The weather services of these nations obtain surface data from land and sea-based stations, the launching of radiosondes, and the remote sensing of atmospheric conditions by satellites. For decades, weather forecasting involved subjective analysis of the available data. But although today's meteorologists still rely heavily on their own interpretation of current weather patterns, they also make wide use of output from numerical models. There is a wide variety of these models, and they are undergoing constant evolution, but all attempt to forecast the atmosphere by applying known physical laws. Short-term forecasts (up to two days) have improved considerably in recent years in response to improved data acquisition and processing, as well as model improvements. Medium-range forecasts offer some degree of skill out to about a week. On the other hand, long-term forecasts for 30- and 90-day periods have not displayed a high degree of accuracy or skill.

While information is available in many different forms, the weather map continues to be one of the most important tools for forecasting. The NWS updates current surface maps for the contiguous 48 states and southern Canada every three hours, and forecast maps are published at 12-hour increments by the various models. Maps of observed conditions plot the distribution of pressure with isobars and the location of fronts with heavy lines. Station models provide detailed information on temperature, dew point, pressure, wind, and other conditions at large number of stations.

Though important, surface weather maps provide only part of the story and must be augmented by maps of atmospheric conditions at higher levels. In the United States, maps are routinely produced for the standard levels: the 850, 700, 500, 300, and 200 mb pressure levels. Each of these has its own set of advantages. For example, at the 850 mb level, the analysis of air flow relative to the temperature distribution helps identify areas of low-level uplift. The 700 mb map has many of the same uses that the 850 mb map does. In addition, it is particularly useful in predicting the movement of air mass thunderstorms. Of all the upper-atmosphere maps, 500 mb charts are probably the most widely used. Maps of the 300 and 200 mb levels are best for defining the number and position of Rossby waves and for analyzing jet stream patterns.

While weather maps provide two-dimensional views of the atmosphere at a number of levels, they do not provide much information on the vertical structure of the atmosphere at particular locations. This is where thermodynamic charts make an important contribution. Profiles of temperature and dew point identify moist and dry layers within the atmosphere along with the location of clouds. The steepness of temperature profiles relative to dry and wet adiabats also indicate the stability of the atmosphere. Over the years, meteorologists have formulated several numerical indices that condense the information from these diagrams into single numerical values. Values above or below certain critical numbers indicate the need for further analysis of the possibility of heavy precipitation or severe weather.

The material in this chapter has relied heavily on the principles discussed in the previous 12 chapters and has applied these principles to an activity of great importance to society, forecasting. The next chapter looks at another human aspect of meteorology—the effects of human activity on weather.

Key Terms

National Oceanic and Atmospheric Administration p. 386
National Weather Service (NWS) p. 386
Meteorological Service of Canada (MSC) p. 386
Regional weather centre p. 386
climatological forecasts p. 386
persistence forecasts p. 386
analog approach p. 387
numerical weather forecasting p. 387
quantitative forecasts p. 387
qualitative forecasts p. 387
probability forecasts p. 387
forecast quality p. 388
forecast value p. 388
forecast accuracy p. 388
forecast bias p. 388
forecast skill p. 389
World Meteorological Organization p. 389
World Meteorological Center p. 389
National Centers for Environmental Prediction (NCEP) p. 389
Canadian Meteorological Centre p. 389
Atmospheric Environment Service (AES) p. 389
Automated Surface Observing System (ASOS) p. 389
Automated Weather Observation Systems (AWOS) p. 390
radiosondes p. 390
rawinsondes p. 390
wind profilers p. 390
Weather Forecast Offices p. 390
analysis phase p. 392
prediction phase p. 393
post-processing phase p. 393
medium-range forecasts p. 396
European Center for Medium-Range Weather Forecasting (ECMWF) p. 396
ensemble forecasting p. 396
chaos p. 397
long-range forecasts p. 398
Climate Prediction Center p. 398
surface maps p. 401
station models p. 402
850 mb map p. 402
700 mb map p. 405
500 mb map p. 406
300 mb map p. 406
200 mb map p. 406
isotachs p. 406
visible images p. 407
infrared images p. 407
water vapor images p. 407
radar images p. 408
thermodynamic diagrams p. 411
lifted index p. 412
K-index p. 412

Review Questions

1. Briefly describe some of the variables that complicate weather forecasting.
2. Describe the basic characteristics of climatological forecasts, persistence forecasts, the analog approach, and numerical forecasting.
3. What are the distinguishing characteristics of quantitative, qualitative, and probability forecasts?
4. Explain the manner in which weather data are obtained and disseminated to agencies across the globe.
5. What are radiosondes and rawinsondes? What other sources of upper-atmosphere information are available to forecasters?
6. Describe the analysis, prediction, and post-processing phases in numerical forecasting.
7. What are model output statistics?
8. What are the primary characteristics of short-range, medium-range, and long-range forecasts? What types of information are needed in the preparation of the individual forecasts?
9. What is ensemble forecasting?
10. Describe the station model used for surface weather maps. How is the information presented on the station model? What measures must be used to convert the numerical data on the station model to real values?
11. Describe the characteristics of the 850 mb, 700 mb, 500 mb, 300 mb, and 200 mb maps that make each of them useful to forecasting.
12. What is the significance of cold air advection? Which weather map is most useful for locating it?
13. Which upper-level weather map would you use to locate short waves?
14. Why are omega highs significant? Which map is best for identifying them?
15. What is an isotach?
16. Which maps are most useful for locating the polar jet stream?
17. Describe the three types of satellite images discussed in this chapter. What characteristics make them useful?
18. Describe how radar works and how its information is presented.
19. Explain what a thermodynamic diagram does and how it is constructed.
20. Describe the lifted index and the K-index. How are they valuable to forecasters?

Critical Thinking

1. Why is it that the climatological, persistence, and analog approaches will never be entirely eliminated from the process of weather prediction?
2. If a forecast calls for a 70 percent chance of rain and no precipitation occurs, was the forecast actually wrong? What if this happens two days in a row? Ten days?
3. What would have to happen to the data acquisition network for the analysis phase of forecasting to be bypassed?
4. Are further improvements in weather forecasting likely to occur for large-scale phenomena or smaller-scale events? Explain your answer.

Problems and Exercises

1. Go to the Web site **http://www.wrh.noaa.gov/wrhq/nwspage.html (http://weatheroffice.ec.gc.ca/canada_e.html** in Canada), and click on the National Weather Service office nearest to you. Make note of the 24-hour forecast and follow up the next day to see if the forecast was correct. Do this for extended forecasts as well. In general, do you find the forecasts to be accurate?
2. During times of unusual or inclement weather, visit the Web sites listed below to obtain weather map, satellite, radar, and thermodynamic diagram information. Is the weather you are experiencing consistent with what you would have expected, based on the information obtained?
3. On a daily basis, make use of the available information from the Web sites listed below and make your own forecast (before reading the official forecast for your area). Then compare your forecast to that of the local weather office. Are your forecasts generally consistent with those of the professional meteorologists?
4. Read the forecast discussion on the web page of your local weather service office each day. These discussions explain the meteorologists' reasons for making their particular forecasts. (Warning: These can be a bit technical.)

Quantitative Problems

Weather forecasting is now strongly based on numerical techniques. This chapter is followed by an appendix that discusses some important aspects of numerical forecasting. The book's Web site offers several quantitative problems to help you better understand some of the nuances of numerical techniques.

Useful Web Sites

http://ww2010.atmos.uiuc.edu/(Gh)/guides/mtr/fcst/home.rxml

An excellent primer on forecasting from the University of Illinois.

http://www.crh.noaa.gov/lmk/soo/

Excellent information on forecasting tools available to meteorologists, courtesy of the Louisville, Kentucky, office of NWS. Has pages describing AWIPS, Doppler radar, and numerous training modules.

http://meted.ucar.edu/nwp/course/index.htm

Tutorials on the use of numerical weather prediction.

http://www.theweatherprediction.com/habyhints/index.html

An extensive array of forecasting tips and tidbits.

http://www.nrlmry.navy.mil/~kuciausk/esis/frames/image_interp_basic_frame.html

Excellent introduction to the use of satellite imagery in forecasting. Use the panel on the left side of the screen for navigation.

http://www.wrh.noaa.gov/wrhq/nwspage.html

Provides links to local NWS forecast offices and various support centers.

http://weatheroffice.ec.gc.ca/canada_e.html

An excellent first stop for forecast information in Canada.

http://weather.uwyo.edu/

Contains just about everything you need to make your own forecasts.

Media Enrichment

Weather in Motion

Simulated Doppler I and II

The first movie shows a horizontal slice of the thunderstorm simulation presented in Chapter 11. North–south winds are colored according to wind speed, with deep blue representing fast southward-moving wind, and deep red used for fast northward-moving wind. (Green shades indicate that air is not moving north or south.) As you view the movie, note the relationship between the streamlines and the colors.

While viewing the second movie, imagine a Doppler radar situated far to the south of the storm shown in the first movie. Southward-moving air would approach the radar and be rendered in blue. Similarly, northward-moving air would be shown in shades of red. This movie depicts what such a radar would observe. (It's the same sequence, viewed from above.) Areas where deep red adjoins deep blue correspond to areas of strong shear and would be interpreted as places for possible tornado formation.

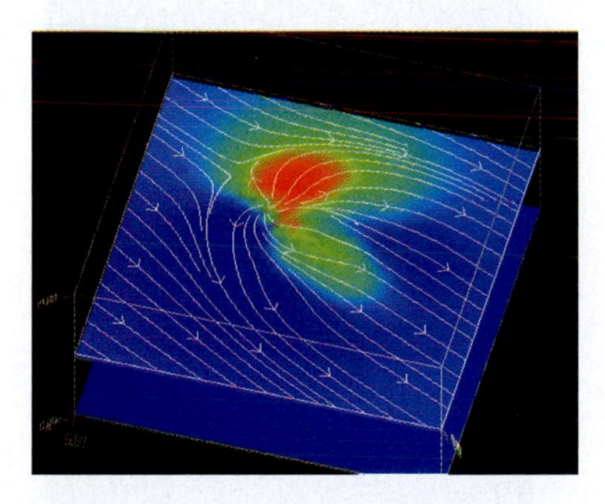

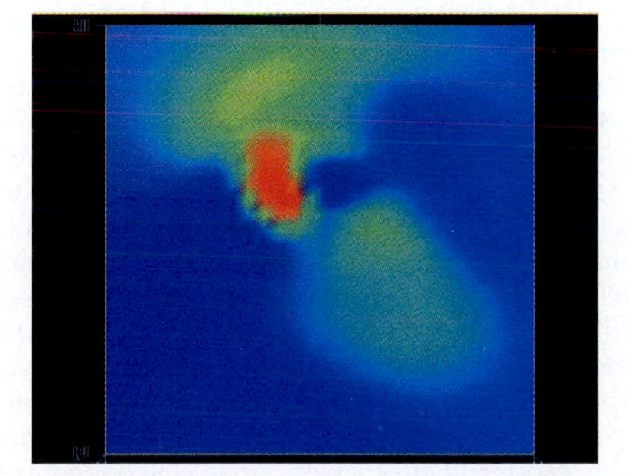

Weather in Motion

Doppler Radar I, II, and III

In these three movies a professional meteorologist gives a lesson on the use of Doppler radar. In the first movie, the radar displays four separate slices of a supercell storm, each at different heights. The top left panel displays a cross section of the lower portion of the supercell; moving clockwise, the other three panels show a progressively higher cut. The most significant feature of the radar display is the presence of a hook echo. Hook echoes frequently signal the presence or imminent formation of tornadoes. This was indeed the case with this storm, which spawned an F4 tornado.

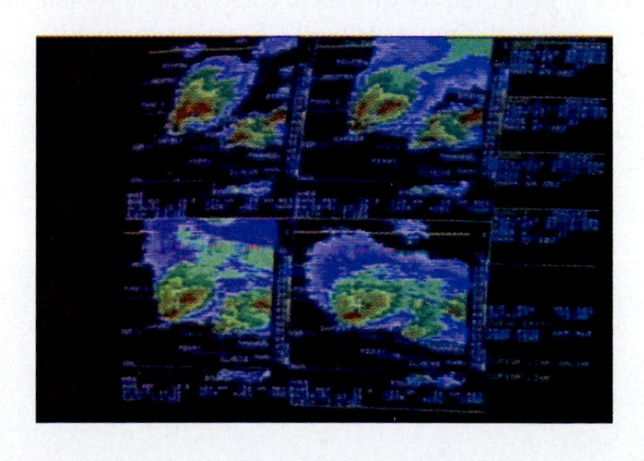

The second movie shows a different use of Doppler radar—one in which the direction and speed of movement of the air within the same storm are displayed. A rotational signature is clearly evident in this radar return display.

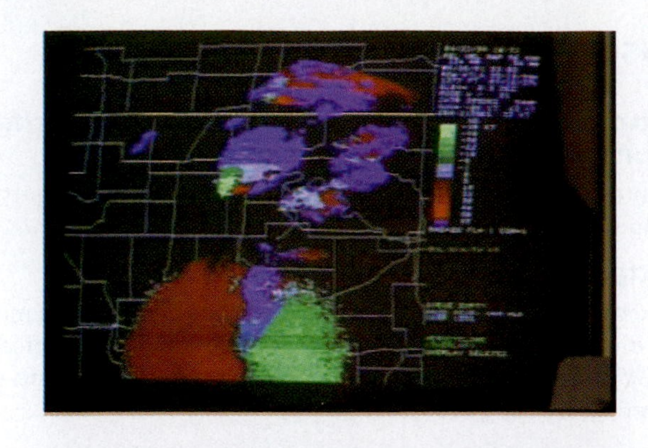

Forecasters undergo considerable training in the use and interpretation of radar imagery. This knowledge helps forecasters distinguish between significant features on the radar display and other less meteorologically important objects. The third movie shows a surprising feature on a radar display from Dodge City, Kansas.

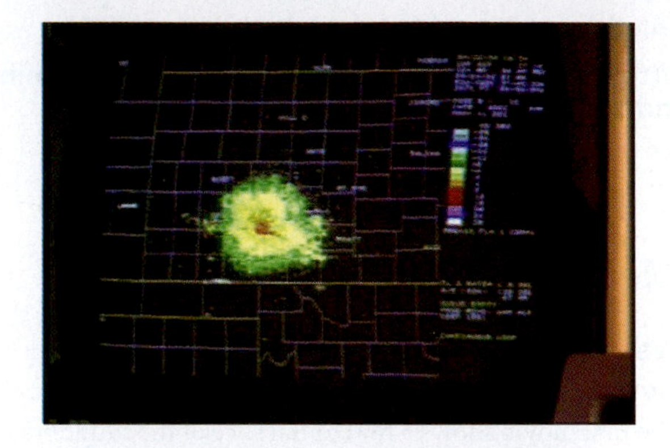

Weather in Motion

Ensemble Forecasts I and II

These two movies show global forecasts of the 500 mb height surfaces for a 2-week period, April 15, 2000, to April 30, 2000. The National Center for Atmospheric Prediction (NCEP) produces a set of 17 global forecasts each day. These forecasts are run with slightly different initial conditions.

The first movie shows all 17 individual forecasts, contoured at two heights: 5490 m (light blue) and 5730 m (red). Climatological values are shown in green, and the medium-range forecast (MRF) control run (no perturbations in the initial conditions) is shown in white. "Spaghetti" plots like this one may help forecasters to determine model skill. Similarities between ensemble members help forecasters assess the confidence they can place in the model forecast. In addition, differences between the ensemble and observations may help to expose model errors. Notice that as the animation runs, the various forecasts become increasingly different from one another, especially after 1 week or so (180 hours). Because the atmosphere displays chaotic behavior, the small differences in the initial conditions of each of the ensemble members grow with time, causing the forecasts to diverge.

The second animation shows the mean of all 17 ensemble members. The ensemble mean usually has greater skill than the individual model runs because averaging filters out the inconsistencies between ensemble members.

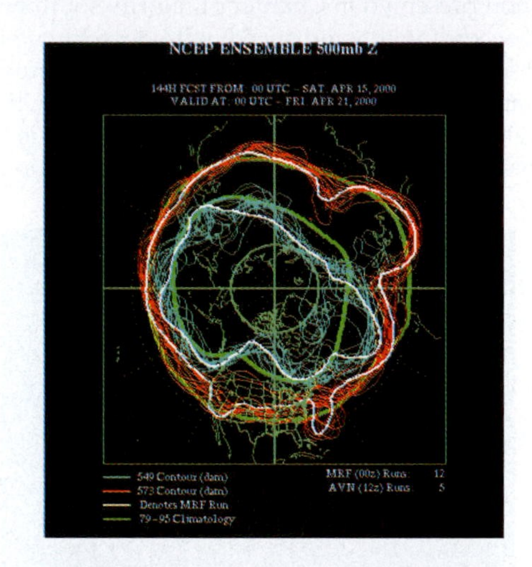

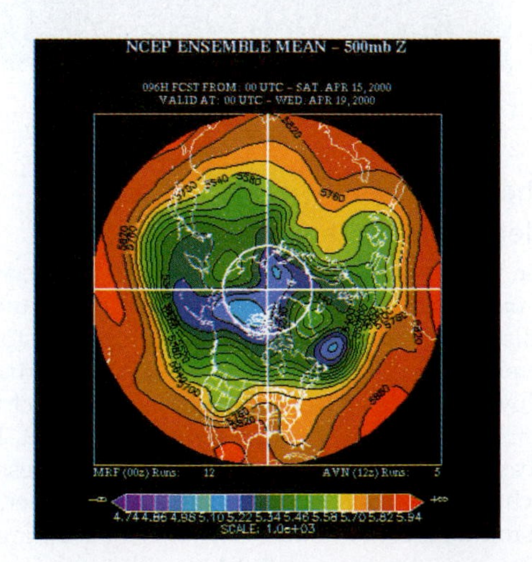

Chapter 13 Appendix

Numerical Forecast Models

Over the last four decades, weather forecasters have relied on several generations of computers and a variety of different models for guidance. As computers have increased in speed and capacity, the models have become increasingly complex, always straining the limits of computer power in an effort to achieve greater realism and accuracy. But even with today's computers, numerous compromises and approximations are necessary. In fact, it might be said that the spectrum of models arises not so much from differences in purpose or theory, but rather from differing approaches to the basic problems of abstraction and simplification. The result is tremendous variety in both the details and gross features of numerical models. In this appendix we first discuss the major features of numerical models, using the primary NCEP operational models as examples. We then describe some methods for assessing forecast quality.

Model Characteristics

At the present time, there are three numerical forecast models (ignoring variants): the **Global Spectral Model**, the **Nested Grid Model (NGM)**, and the **eta model**. The "eta" name comes from the vertical coordinate used. This model does not reference altitudes by height above the surface, but instead uses pressure, scaled in a particular way relative to surface pressure. Naming for the other two models will become clear below.

Scale Considerations

Thinking first about gross features, perhaps the most fundamental difference between models concerns the model **domain,** the region of the globe to be represented. Of course, if one wants to forecast for the entire globe, no decision is necessary—the computational domain must be the entire globe. But what if the goal is a European, or perhaps a North American, forecast? Where should the edges be? The boundaries of the domain require special treatment, usually in the form of strong assumptions about mass and energy transfer. To minimize their effect, one wants as large a domain as possible, with boundaries far outside the forecast region. Just how far outside depends in part on the forecast lead time. As lead time increases, locations farther removed from the forecast region begin to affect the forecast. Here, then, is a classic trade-off. A larger domain is preferable but requires more computer resources, which means some other aspect of the model must be compromised. Of the three NCEP models mentioned, only the Global Spectral Model has a global domain; the others have domains centered on the United States and cover only part of the Northern Hemisphere.

Another issue is spatial resolution. The fundamental equations governing atmospheric behavior are continuous in space, meaning that they describe the evolution of the atmosphere everywhere. If the governing equations were solved directly, they would yield a solution for an infinite number of points (we could find forecast values at every location within the domain). But the equations are far too complex to solve directly by analytical means; no such solution exists. Instead, the equations are written and solved in approximate form, with the result that forecast values are available only at widely spaced locations.

The approximation is accomplished in a number of different ways, but the result is always the same: there is some minimum size below which explicit representation is impossible. Current models are limited to a few tens of kilometers and larger; so, for example, nothing so small as an individual thunderstorm cloud can possibly appear. If such "sub-scale" phenomena are to be considered, they can only be expressed in terms of features that *are* resolved, an error-prone process called "parameterization." A cloud parameterization would typically use simple relations that compute ice and liquid water fractions on the basis of relative humidity and air temperature, without modeling the details of cloud growth. Obviously, one wants high resolution so that small-scale processes and phenomena can be modeled and appear in the forecast, but this can come only at the cost of more computation. (Roughly speaking, doubling the resolution leads to eight times the computation.) Increases in resolution can be subsidized by reducing the domain, but that creates its own problems, as previously described.

The resolution issue applies to the vertical coordinates as well as the horizontal coordinates. Modelers face difficult questions about how many vertical levels to include, as well as how they should be arranged (at what altitudes). As a point of comparison, the horizontal resolution of the Global Spectral Model is about 1° in latitude and longitude (about 60 mi). It has 28 levels in the vertical, ranging from the surface to the 2.7 mb level. These are far from equally spaced—there are 8 levels below 800 mb, with increasing spacing at lower pressures (higher altitudes). Operational versions of the eta model have been run with increasing resolution since its inception, ranging from 80 km initially to 29 km at the time of this writing. There are now 50 levels in the vertical. The NGM has 16 layers and two grids. The smaller inner grid, centered over the United States and Canada, has a resolution of about 80 km. It lies completely within a larger, coarser outer grid that extends the domain to much of the Northern Hemisphere. The advantage of this scheme is that most of the computing is confined to the region of interest—computer resources are not wasted on detailed features far outside the forecast area. But at the same time, those areas are not ignored, as they would be with a smaller domain.

Horizontal Representation

Yet another major difference among models is the horizontal representation. Many models adopt a **grid representation**, in which the domain is subdivided into a lattice of grid points. (The grids at various levels need not coincide, nor need they be the same for all variables.) The equations are solved only at the grid points. As we've said, the finer the grid, the higher is the model's resolution. Implicit in this is the idea that the grid captures horizontal variation in the atmosphere and that intermediate values can be inferred knowing values at nearby grid nodes.

The alternative is called a **spectral representation**. (Can you guess which representation is used in the Global Spectral Model?) Variables are represented as a series of "waves" in space, each having a characteristic wavelength. For example, there are waves in the model corresponding to Rossby waves, repeating just a few times around a parallel of latitude. Superimposed on these are other waves representing smaller-scale and larger-scale variations. To obtain the value for a particular variable at a point, its various wave functions are summed over all the wavelengths (the Global Spectral Model uses 126).

Spectral models have some advantages, especially for a global domain (the North and South Poles are easily handled in a spectral representation). There are other quite technical advantages as well, which we won't discuss. Instead we'll make a few general points about spectral modes. First, horizontal resolution is determined by the smallest of the "harmonics" present (smallest wavelength), so there is no escaping the problem of how to represent sub-scale processes. Second, not all of the variables can be represented in spectral terms. Only quantities subject to advection, such as heat and moisture, are treated this way. Other variables, such as radiation, must be computed on a point-by-point basis. Thus, the Global Spectral Model uses a grid for these "physical" quantities, and there is constant transformation of information from spectral to grid representation. (Fortunately, there are fast computational methods for this.) Finally, the spectral representation applies only to the horizontal—spectral models are layered in the vertical.

Physical Processes

Numerical models also differ greatly in their "physics," which basically includes all processes other than those related to motion ("dynamics"). A model's physics package includes purely atmospheric processes (such as condensation), atmosphere–surface interactions (such as friction between the atmosphere and ground), and purely surface processes (such as soil moisture or depth of snow). Consider, for example, just some of the physical processes included in the Global Spectral Model:

- **Radiation:** Shortwave absorption and scattering in three wavelength bands (including the effects of ozone, water vapor, and carbon dioxide), with multiple reflections between clouds and ground. Effects of clouds, water vapor, CO_2, and O_3 on longwave absorption and emission are explicitly modeled, including overlap in absorption bands. Radiative properties of clouds depend on cloud thickness, temperature, and moisture content.
- **Convection, Clouds, and Precipitation:** Stratiform (nonconvective) clouds, as might be found with fronts and tropical disturbances, are determined from relative humidity using a statistical relationship. Convective clouds are either precipitating (deep convection) or nonprecipitating (shallow convection). Temperature and moisture profiles are used to determine which occurs. Downdrafts and evaporation of precipitation are simulated, as are entrainment and detrainment of updraft and downdrafts. Precipitation arises from both deep convection and from large-scale condensation (when the air at a point becomes saturated, regardless of convection). Precipitation is evaporated into unsaturated air below the cloud; only that which survives the descent is deposited on the surface.
- **Surface Properties and Processes:** Sea surface temperature and sea-ice distributions are fixed for the model run. Sea-ice temperature is computed from heat exchange between the atmosphere and ocean below the ice. Surface albedo depends on zenith angle, snow cover, and vegetation type. Snow cover is determined by accumulation from falling snow, snow melt, and sublimation. Land surface evaporation consists of evaporation from the ground and plant canopy, as well as transpiration by the canopy. Precipitation not intercepted by the canopy is partitioned into soil moisture recharge and runoff. (Melting snow also contributes to soil moisture.)

Notice that many of the preceding are sub-scale processes; thus, parameterization is heavily used in the physics package. Different models not only include different processes but also employ different parameterizations for the same processes. Sometimes this is by necessity rather than preference, as parameterizations appropriate at one scale are not necessarily useful at another scale.

As we mentioned, there is a continual evolution of models (and no doubt many of the specifics in this appendix will have changed by the time you read this). For example, when it was introduced in 1973, the Global Spectral Model was the state of the art at NCEP. Today, the eta model is generally considered to occupy that position and is itself evolving. Is this effort at improving models justified—has forecasting skill improved? Definitely, as we will see later in this appendix.

The future of numerical prediction calls for ever-more realistic models. For example, NCEP plans a 5-km/150-level eta-like model in the near future. This is a daunting task: such a model will require 1000 times the computing power and 100 times the memory of today's model. Even a 5-km model will not permit explicit calculation of convection and other extremely important processes. Estimates suggest that this will require about 10,000 times the power of today's supercomputers. Whether or not this is possible depends on both the public's willingness to pay and the development of computer technologies not yet on the drawing board.

Measures of Forecast Accuracy and Skill

For some variables, such as temperature or precipitation amount, there are accuracy measures that practically suggest themselves, such as bias and mean absolute error (MAE). Bias is most easily defined as the difference between the average forecast value and the average observed value. It reveals any tendency for the method to give forecasts values above or below the true value (for example, too hot on average, too much rain on average). A mathematically equivalent definition of bias is simply the average error. For each forecast, we find the departure from observed and compute the average of those errors.

Figure 1 illustrates how bias can be used as a means to assess the accuracy of models. The figure plots the bias for 1-in. (2.5 cm) precipitation events predicted by NCEP for the last few decades. Considering the Day 1 forecasts, the figure shows that in the early 1960s and again in the mid-1970s errors were large, as forecasters overpredicted rain areas. Otherwise, from 1980 onward, there is remarkably little improvement in bias over the entire period. Bias for Day 2 forecasts are slightly below unity from about 1980 onward, indicating a slight tendency for underprediction.

Does the absence of bias mean a forecast method is perfect? Certainly not—it is always possible for overprediction (positive errors) to nearly balance underprediction (negative errors). As a matter of fact, most statistical forecast methods are totally unbiased, yet give far from perfect forecasts. To avoid this problem, one can use MAE, defined as the average of the absolute errors. Because absolute values are always positive, there is no cancellation of errors. MAE therefore provides information about how far an individual forecast is likely to be from true value, without regard to sign (positive or negative). If the MAE is 2 °C, we expect an individual forecast to be 2 °C away from the true value.

Note that in MAE, we averaged the absolute values to remove the sign of each error. We could accomplish the same thing in another way: by squaring each error before taking the average. This is done during the calculation of the **root-mean-square error (RMSE)**, which has some statistical advantages over MAE and is therefore more common. Obviously, as we sum only positive values, there is no cancellation of error. The root-mean-square error is just the square root of average squared error. Like MAE, it has the same units as the forecast variable (°C, millimeters of snow, and so on) and provides an estimate of the error expected for a single forecast.

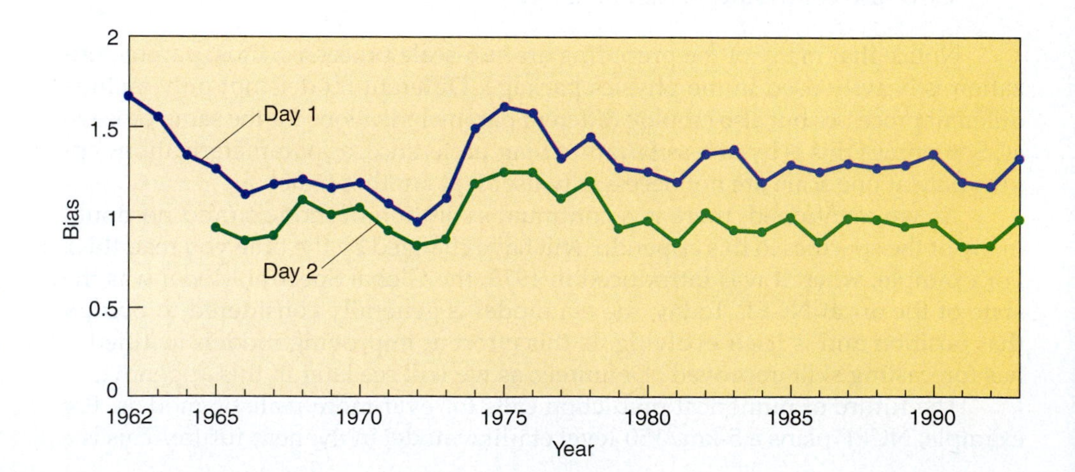

▶ **Figure 1**
Changes in accuracy (bias) for 1-in. (2.5 cm) precipitation forecasts for Day 1 and Day 2. Day 1 forecasts are for the period 12 to 36 hours ahead, whereas Day 2 represents the period 36 to 60 hours ahead. A bias value of unity occurs when the size of the forecast and observed precipitation areas are equal.

Table 1 • Rain Forecasting Accuracy

	Rain Observed	No Rain Observed	Total
Rain Forecast	10 km^2 (Correct)	30 km^2 (Error)	40 km^2
No Rain Forecast	20 km^2 (Error)	40 km^2 (Correct)	60 km^2
Total	30 km^2	70 km^2	100 km^2 Total Area

For a qualitative variable, the most common accuracy measure is the proportion of correct forecasts, known as the **hit rate**. In the case of a binary variable (just two classes), we would ask, of all the forecasts made, whether "yes" or "no," what percentage turned out correct? For example, suppose rain/no rain forecasts have been made for a 100 km^2 area, with the results shown in Table 1.

The same data are shown in map form in Figure 2a. Looking at either the table or the map, we see rain was correctly forecast for 10 percent of the region, and no rain was correctly forecast for 40 percent of the area, giving a hit rate of 0.5 (50 percent).

Somewhat similar is the **probability of detection (PoD)**, which is the proportion of occurrences that were correctly forecast. The difference is that nonevents are excluded in the PoD. Thus, for example, in Figure 2a, the PoD is 10/30 = 0.33 (one-third of the area receiving rain was correctly forecast). Obviously, because one can maximize the PoD by always forecasting "yes," the PoD is often accompanied by the **false alarm rate**, the proportion of "yes" forecasts that were wrong. In Figure 2a, 75 percent of the rain forecasts are false alarms (30/40).

Bias can also be computed for a binary variable; in this case, bias is simply the ratio of forecast to observed occurrences. Here, however, a perfect method has a bias value of 1. Values larger than unity imply a tendency to overpredict, whereas values less than 1 suggest underprediction. For our example (Figure 2a), the rain forecast area is too large by one-third: the bias is 1.33 = (40/30). As before, a perfect bias value does not indicate a perfect method. For example, if the rain forecast area were sliced down to 20 km^2, the size of forecast and observed areas would agree perfectly, even though the forecast would be wrong for a large part of the region.

Another widely used measure for binary variables is the **threat score**. This measure is similar to PoD in that it is concerned only with occurrences. However, it tries to penalize false alarms as well as missed occurrences (PoD responds only to the latter). Looking at Figure 2a, you can see the 10 km^2 area of correctly forecast occurrences is flanked by large areas of false alarm as well as missed occurrences. Given the size of those areas, the 33 percent PoD seems a rather inflated measure of success. The threat score accounts for this, and is given by

$$TS = \frac{\text{Correct}}{\text{Forecast} + \text{Observed} - \text{Correct}}$$

where the variables on the right-hand side are the area (or number of places) correct, area forecast, and area observed. The term *threat* is used because the denominator is the area "threatened" by an occurrence—either forecast, observed, or both. For Figure 2a, the threat score is 0.17 = 10/(40 + 30 − 10), half as large as PoD for the same map. Note that a perfect forecast will have a threat score of unity, because the forecast and observed areas will coincide perfectly.

Figure 3 illustrates the improvement in NCEP forecast model performance by plotting annual threat scores for 1 in. or more of precipitation. Unlike bias, threat scores over time have shown considerable improvement, reflecting better performance for the numerical models for both 1- and 2-day forecasts. Some of the variability is weather-related (wet years have higher scores), but significant upturns can be tied to new models coming online. Particularly striking is a narrowing of the gap between Day 1 and Day 2 forecasts, implying that Day 2 forecasts have improved more. In

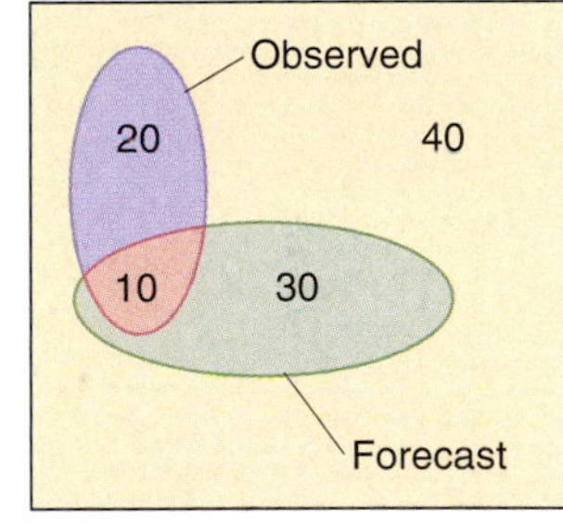

(a)

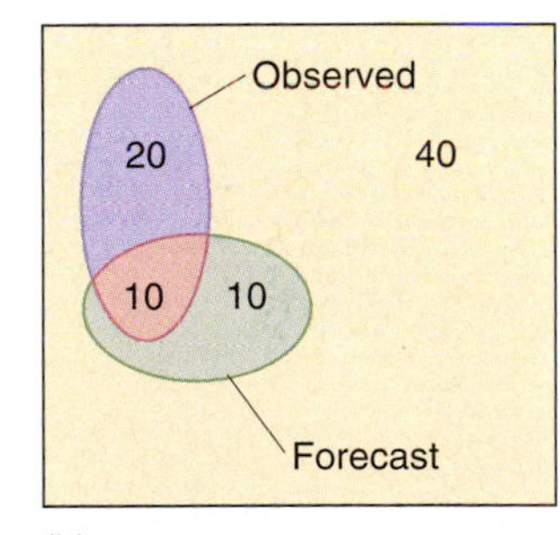

(b)

▲ **Figure 2**
Hypothetical distributions of observed and forecasted precipitation. Numbers in single ovals refer to percentages of times that precipitation was observed but not forecast, or vice versa. Numbers contained in both ovals indicate the percentage of times precipitation occurred in an area forecasted to have precipitation. Numbers outside the ovals are for areas in which precipitation was not observed or forecasted.

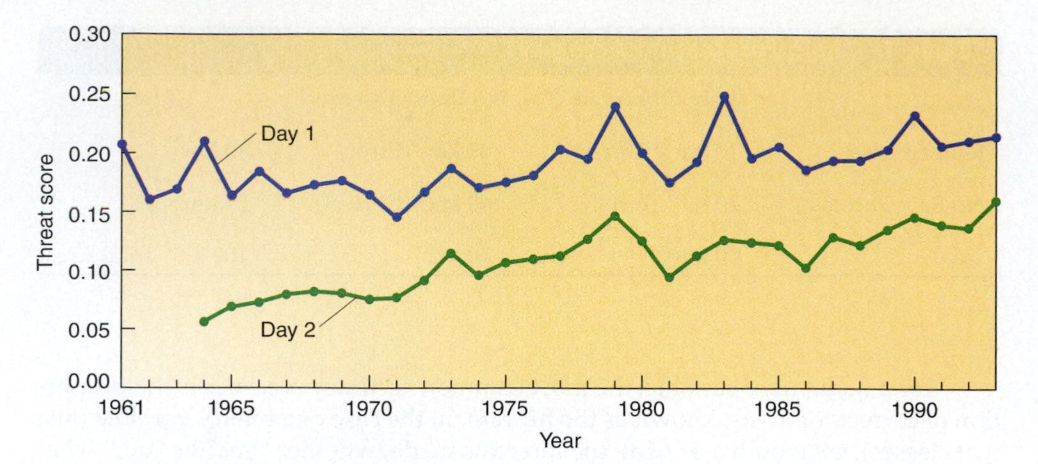

Figure 3
Change in skill (threat score) for 1-in. (2.5 cm) precipitation forecasts for Day 1 and Day 2. A perfect forecast has a skill score of unity.

fact, by the early 1990s the 2-day forecasts were about as good as the 1-day forecasts of the early 1960s. In that sense, predictive ability doubled over the period. Looking at changes in the two-day score alone, there is a threefold improvement.

It's a little more complicated to measure accuracy for probability forecasts. As we mentioned in the text, the most common probability forecast is a probability-of-precipitation, or PoP forecast. The PoP refers to the chance that a random place in the forecast area will receive 0.01 in. or more of precipitation in the forecast period. Equivalently, considering just a single location, a 60 percent chance of rain means that out of 10 forecast days, 6 will be rainy. A forecast might say, for instance, "the rain chance is 70 percent." How can we assess the accuracy of this statement? The forecast itself means that there is a 70 percent probability of rain for a randomly chosen location in the forecast area. Obviously, there is no way to assign accuracy on the basis of just a single precipitation measurement. (Failing to observe rain on a single day does not mean the forecast was wrong.) But if we were to count the number of times rain is recorded on many of these "70 percent" days, we could see if the actual frequency is above or below the forecast frequency, and thereby assess the method's accuracy. To fully verify the forecast method, we would compare all the forecast classes, not just the 70 percent class. In effect, we compare the forecast probability distribution with the observed probability distribution.

As we mentioned, forecast skill refers to the improvement provided by a forecast over and above some reference accuracy. The reference, or "no-skill," system is often taken to be persistence, or alternatively, climatology. But these aren't the only choices for the reference system; we could compare one numerical model with another, or with purely random values. For example, the threat score is often adjusted by the number of hits expected by random assignment of forecasts. In Figure 2a, rain was observed over 30 percent of the area. If we were to randomly forecast rain for various places, we would expect considerable success simply because so much of the area is wet. (If you forecast a single point by throwing a dart at the diagram, you have a 30 percent chance of being correct—assuming you can hit the diagram.) As the rain forecast area grows with no change in observed rain, the threat score increases. To account for this, we first find the expected area of correct forecasts, assuming purely random forecasting. This is given by

$$EC = \text{Forecast}\left(\frac{\text{Observed}}{\text{Total}}\right)$$

The term in parentheses is the proportion of the total area where rain fell—multiplying this by the forecast area gives the area expected to be correctly forecast. With this, the so-called **equitable threat score** is

$$ETS = \frac{\text{Correct} - EC}{\text{Forecast} + \text{Observed} - \text{Correct} - EC}$$

For Figure 2a, $EC = 40\ (30/100)$, or 12 km^2. That is, knowing nothing about the atmosphere, we'd expect rain to be correctly forecast for 12 km^2, given that we are forecasting rain to occur over 40 km^2. For this example, ETS is $(10 - 12)/(40 + 30 - 10 - 12)$ or -0.04, a negative number! Does this make sense? Well, as our forecast method generated fewer successes than would occur by chance, it certainly seems reasonable to treat this as "negative" skill.

Figure 2b shows a similar situation, but with a slightly smaller rain forecast area. The PoD is exactly the same as before (0.33), and the threat score is only slightly larger (0.25 vs. 0.17). However, where in Figure 2a we saw negative skill compared to chance, here the equitable threat score is 0.12. Although there are no more successful rain forecasts than before, this situation has fewer false rain forecasts, and therefore shows positive skill.

We see that there is no single measure of forecast quality. Rather, various measures provide different kinds of information, thus each has its own uses. The other side of this is that whether or not a forecast model shows improvement over time, or whether it is better than some other competing model, depends in part on the way we measure accuracy and skill. Careful assessment always requires that a number of measures be compared.

chapter 14

Massive fires over Indonesia in the summer and fall of 1997 caused extreme air pollution that spread into Malaysia and Thailand.

Human Effects: Air Pollution and Heat Islands

With 18 million people and 3.5 million cars, most of which are not equipped with modern emission-control devices, it is little wonder that Mexico City is one of the smoggiest places on Earth. Surrounded by mountains that confine the polluted air and subject to frequent temperature inversions that inhibit vertical dispersion of pollutants, Mexico City has all the right ingredients for a serious smog problem. But the spring of 1998 brought prolonged periods of unhealthful air quality that were extreme even for this city. The smog crisis was brought on primarily by a rash of forest fires in southern Mexico. To make matters worse, Popocatepetl volcano, 50 km (30 mi) southeast of the city, spewed tons of smoke and ash into the region. Wind transported the pollution south to Honduras and north all the way to Florida and Texas, where people were advised to stay indoors to mitigate health hazards.

Normally, Mexico City's worst smog occurs in January and February, when stagnant air traps automobile and other urban pollutants near the surface. Smog levels decline somewhat by springtime, although the slash-and-burn methods of field clearance used by farmers makes the air quality worse than it would be otherwise. But in 1998 a major drought, believed by many to have resulted from the strong El Niño, created particularly dry conditions that caused the fires to cover three times their normal area. By May the polluted air forced automobile drivers to use their headlights in the middle of the day in the state capital of Chiapas, and in Mexico City extreme ozone concentrations brought a dramatic surge in the number of people seeking medical care for respiratory problems. In response, the government exercised emergency powers that restricted automobile traffic and closed down many factories. However, that was not enough to make conditions tolerable for Pedro Chavez, who said, "You notice it in your eyes, in the tiredness you feel. Our children are getting sick more often. . . . If we could, we would leave, but this is where our business is."

The effects of human activities are not restricted to air quality degradation. We change the atmosphere in more subtle ways as well. For example, the construction of cities influences the way energy and water are exchanged near the surface. Every time a subdivision is laid out, natural soil and vegetation are replaced by concrete or asphalt. This greatly reduces the amount of water that can evaporate into the air and thereby increases the sensible heat flux (Chapter 3) to the atmosphere. We also build structures with vertical walls that receive sunlight at a more direct angle than the original absorbing surface. These processes work to increase the temperature of urban areas relative to their rural counterparts, creating the heat islands that we describe later in the chapter.

CHAPTER OUTLINE

Atmospheric Pollutants

Nowhere is the air entirely pristine. Small suspended solids and liquids (called *particulates*) enter the atmosphere from natural and human sources. Likewise, many gases that are considered pollutants also arise naturally from processes such as lightning-induced forest fires and volcanic eruptions. Nonetheless, natural dilution and removal of these gases and particulates makes them relatively unimportant to the air quality experienced by most people. More important are the effects of human activity, especially in and around urban and industrial centers where anthropogenic emissions are concentrated into much smaller areas. In this chapter all references to **air pollution** will concern the introduction of undesirable gases and

Weather Image

Atmospheric Pollution as Seen from Space Shuttle

Pollution over Amazon Basin

Kuwait Oil Field Fires

particulates by humans. The varying sources of particulates and other pollutants in the United States and their relative concentrations are shown in Figure 14–1.

In the most general sense, pollutants can be divided into two categories. Some, called **primary pollutants**, are emitted directly into the atmosphere. Others, called **secondary pollutants**, do not go directly into the atmosphere but result from one or more chemical transformations. Thus, a chemical emitted into the atmosphere may be innocuous in its original state but becomes a noxious gas or particulate after combining with other emissions or naturally occurring compounds. Several primary and secondary pollutants figure most prominently in the degradation of air quality.

Particulates

Industrial Pollution in Siberia

Particulates (also called *aerosols*) are solid and liquid materials in the air that are of natural or anthropogenic (human-made) origin. Though always small, particulates come in a wide range of sizes ranging from about 0.1 to 100 μm. Some of the particulates are primary pollutants put directly into the atmosphere, while others are secondary pollutants formed by the transformation of preexisting gases or from the growth of smaller particulates into larger ones by coagulation.

Sources of Particulates Particulates introduced directly into the air can originate from natural fires, volcanic eruptions, the ejection of salt crystals by breaking ocean waves—and as any sufferer of hay fever can tell you, by the entrainment of pollen by wind. Human activities, especially those involving combustion, produce primary and secondary particulates.

Some secondary particulates form by the coagulation of gases. This process is most rapid when the humidity is high, which creates an interesting situation. Recall from Chapter 5 that water droplets in nature always form on condensation nuclei—with large, hygroscopic aerosols being particularly effective at attracting water and lowering the relative humidity needed for droplet formation. Thus, the introduction of particulates, especially large ones, promotes the formation of fog or cloud droplets. At the same time, high humidities favor the conversion of certain gases into secondary particulates, which in turn promote the condensation of water vapor into liquid droplets. As a result, humid areas with a high concentration of industrial activities can become foggy at relative humidities considerably below 100%. This symbiotic relationship helped make the London type of smog ubiquitous in eastern, industrial North American cities in previous years.

Removal of Particulates Though particulates are always present in the air, no individual particulate stays in the atmosphere forever. As we have seen in Chapter 7, terminal velocities increase with the size of falling objects. Thus, particulates, which are always small, can remain suspended in the atmosphere for considerable lengths of time. Larger ones remain in the air for perhaps just a few hours, while smaller particulates can exist for weeks.

Several different processes remove particulates from the air. *Gravitational settling,* the process wherein they fall from the air (even if very slowly), effectively removes larger particulates. The smaller ones are less susceptible to this process because even very small eddies can keep them in suspension. Precipitation, on the other hand, removes both large and small particulates in two ways. First, the particulates that served as condensation nuclei in clouds are removed when the droplets that they are part of fall as rain or snow. Other particulates are removed by *scavenging,* the process in which falling droplets and crystals collide with particulates in their path. Upon collision, the precipitation incorporates the particulate and carries it to the surface. The scavenging of particulates largely explains why the air is so much cleaner and visibility is enhanced after a rainshower.

Effects of Particulates Particulates reduce visibility by increasing scattering of visible radiation, but their effect on visibility is of less importance than their impacts on health. Perhaps this isn't surprising, given that we are bathed in these tiny objects

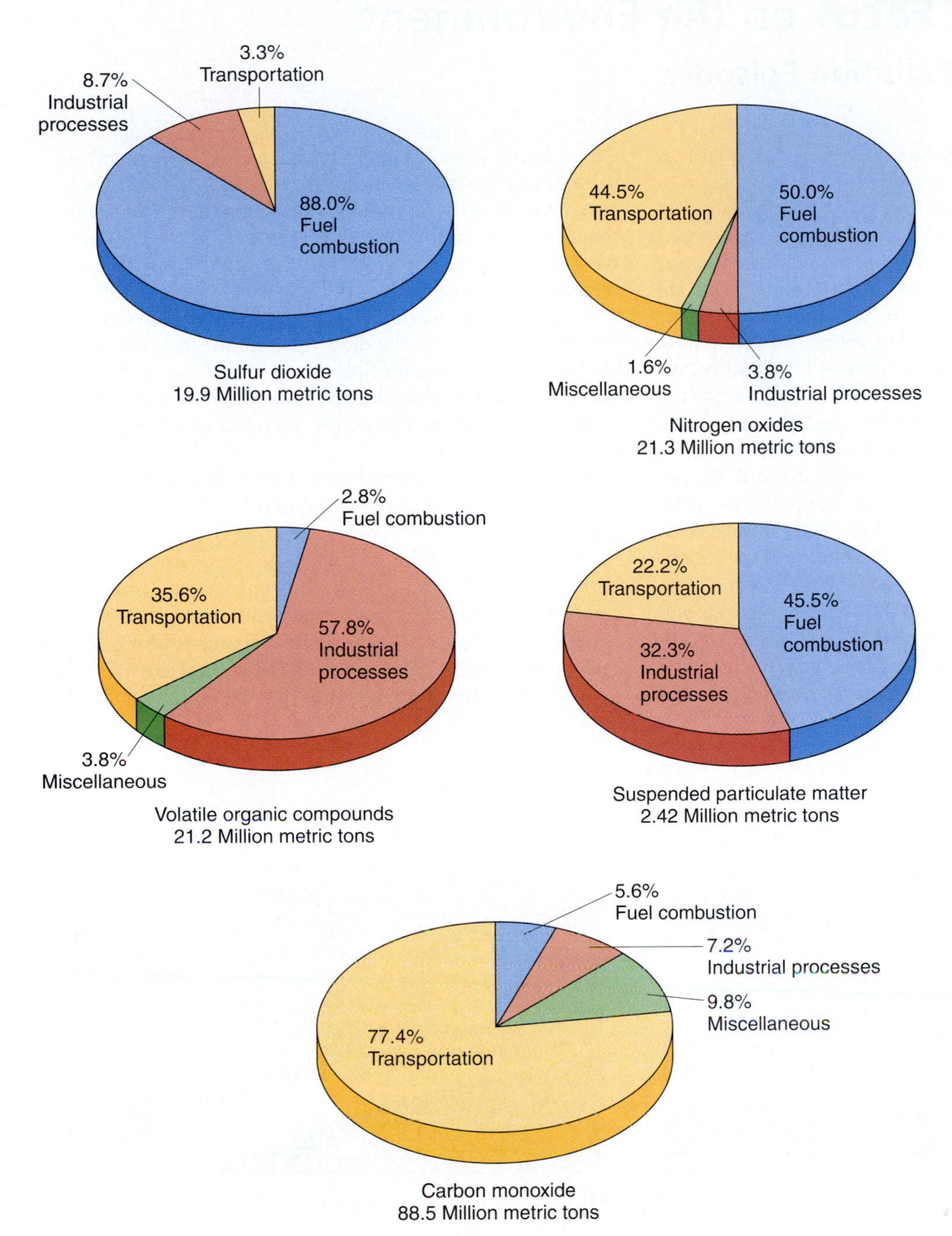

Figure 14–1 The sources of various pollutants in the United States.

every minute of the day. By 1987 it had become clear that a certain class of particulates—those smaller than 10 μm in diameter (called $\mathbf{PM_{10}}$)—most readily enter the lungs and bring about the most serious tissue damage. Although the lungs have cilia that can remove these small particulates, the particulate removal occurs very slowly—even on the order of several months.

A large body of research analyzing the effects of particulates has shown that a more specific class of particulates—those smaller than 2.5 μm (called $\mathbf{PM_{2.5}}$)—also present serious health problems. For this reason, the Environmental Protection Agency (EPA) in July 1997 revised its regulations regarding particulates so that in the future they will be based on these so-called *fine particles.* But the recent focus on $PM_{2.5}$ should not be taken to imply that larger particulates are not dangerous. For example, preliminary results of a study released in late 1997 have shown a strong correlation between hospital admissions in the Los Angeles basin and the levels of large particulates in the air. The increase in hospital admissions is nearly equally divided between patients with acute respiratory illness and those with cardiovascular disease.

14–1 Focus on the Environment

Severe Pollution Episodes

Though many of us live in places where poor air quality is a disturbing fact of life, much progress in solving the problem has been made in the developed world in recent decades, with the result that the most disastrous types of smog events are a thing of the past. Consider, for example, what is probably the most famous air pollution episode in history—the one that hit London, England, between December 5 and 9, 1952. In this 5-day period, a combination of stagnant, damp air and the burning of low-quality coal produced a lethal mixture of smoke and fog. An estimated 3500 to 4000 people—mostly children, elderly, and the already infirm—died as a direct result of the episode.

The most famous air pollution disaster in North America occurred in Donora, Pennsylvania, 50 km (30 mi) from Pittsburgh. Between October 26 and 31, 1948, sulfur, carbon monoxide, and heavy metal dusts emitted from the American Steel & Wire's Zinc Works mixed with a dense radiation fog to create what has been called the "Hiroshima of air pollution." Four days of intense smog took on even greater proportions by Saturday, the 30th. Fans at a high school football game were unable to see the events happening on the field. Others left the game early as word came that family members at home had died or were hospitalized from respiratory problems brought on by the smog. Those who tried to evacuate the town were unable to leave because near-zero visibility completely stalled traffic. By Sunday morning firefighters were bringing oxygen to people who were unable to breathe, but relief was only temporary as the departing firefighters felt their way over to the next victim requiring assistance. On Sunday morning, officials finally closed down the zinc works, and later that day the smog was finally washed away by rain—but only after 20 people had died and 7000 people had been hospitalized.

Except for its magnitude, this infamous event was not unique. Many industrial cities have endured severe air pollution as a result of local industrial manufacturing, smelting, petroleum refining, or other activities. However, it is widely believed that the Donora event was the principal catalyst in the enactment of antipollution legislation in the United States. Since 1948, economic changes, along with greater attention to environmental issues, have greatly improved air quality in many cities (Figure 1).

(a)

(b)

▲ **Figure 1**
Like some other former industrial centers, Pittsburgh's air quality has undergone a huge improvement due to the closure of foundries and factories. These photos show Pittsburgh in 1906 (a), and the same scene in 1986 (b).

Carbon Oxides

Carbon oxides (also called *oxides of carbon*) include **carbon monoxide** (CO) and **carbon dioxide** (CO_2). The latter has been discussed in Chapter 1 as one of the important variable gases that make up the atmosphere, and in Chapter 16 we expand on its possible role in climate change. Though important for its role in the transfer of energy within the atmosphere, high levels of CO_2 have no short-term deleterious effect on people or the environment. Thus, the gas is not strictly considered a pollutant. The same cannot be said for carbon monoxide, however.

Carbon monoxide is a colorless, odorless gas. In the natural environment it is released as a primary pollutant by volcanic eruptions, forest fires, bacterial action, and other processes. Though natural processes emit far more CO into the environment than do human activities, soil microorganisms consume it effectively and background values are very low. In cities, however, inputs can greatly exceed the rate of removal and unsafe concentrations can accrue. In the United States, the most important source of CO is the automobile (see Figure 14–1), which releases the gas as a byproduct of incomplete combustion. In well-maintained vehicles, carbon monoxide emissions are low, but poorly operating engines can cause CO concentrations to accumulate to unsafe levels. This is particularly true in confined areas, such as garages and tunnels. In the home, an improperly vented or malfunctioning furnace can release lethal doses of CO very quickly. Carbon monoxide is also released in home fires, where it probably is responsible for a high percentage of fire-related fatalities. Cigarette smoke also releases carbon monoxide as a byproduct sufficient to greatly increase its concentration in the bloodstream.

Carbon monoxide is extremely toxic. Even low levels can cause a person to immediately experience slowed reflexes, drowsiness, and a reduction or loss of consciousness. Exposure for 3 hours at 400 parts per million (ppm) is life threatening, and at 1600 ppm death comes within an hour. Over the long haul, it can contribute to heart disease. Table 14–1 highlights some effects of varying CO levels.

Unlike other pollutants that act mainly on the pulmonary system, carbon monoxide's toxicity arises from its effects on the bloodstream. Hemoglobin (the agent that gives red blood cells their characteristic color) absorbs oxygen in the lungs and circulates it throughout the body. Under ideal conditions the hemoglobin releases oxygen to cells and then returns to the lungs, whereupon the process is repeated again and again. Carbon monoxide in the bloodstream completely disrupts this process. As it turns out, hemoglobin has a two-hundredfold greater affinity for carbon monoxide than O_2. In other words, with carbon monoxide and oxygen both available in the lungs, the blood will far more readily absorb the carbon monoxide. Thus, the inhalation of carbon monoxide reduces the cardiovascular system's ability to circulate oxygen to the rest of the body.

Sulfur Compounds

Sulfur compounds in the atmosphere can occur in gaseous or aerosol form. The majority—roughly two-thirds—of all the sulfur compounds emitted into the atmosphere originate from natural processes. Steam vents, such as those at Yellowstone

Table 14–1 • Threshold Levels of Carbon Monoxide

Carbon Monoxide Concentration (ppm)	Comment
50	Maximum allowable OSHA dose for 8-hour exposure
200	Headache, fatigue, dizziness, nausea in 2–3 hours
400	Headache in 1–2 hours, life-threatening after 3 hours
800	Dizziness, nausea, and convulsions within 45 minutes; death in 2–3 hours
1600	Headache, dizziness, nausea in 20 minutes; death in 1 hour
3200	Headache, dizziness, nausea in 5–10 minutes; death in 25–30 minutes
6400	Headache, dizziness, nausea in 1–2 minutes; death in 10–15 minutes
12,800	Death within 1–3 minutes

National Park in Wyoming or Lassen National Park in California, provide interesting examples of the emission of sulfur compounds. The most important of these is the bacterial release of hydrogen sulfide (H_2S), a particularly noxious gas that smells like rotten eggs. Volcanic eruptions and sea spray also play an important role in releasing sulfur compounds. Fortunately, sulfur gases are readily dispersed in the atmosphere, so background concentrations are extremely low (on the order of one half a part per billion) and their environmental and health impacts are minimal.

Of the anthropogenic sulfur compounds released to the atmosphere, the most important are **sulfur dioxide** (SO_2) and **sulfur trioxide** (SO_3). These oxides of sulfur fall under the collective designation of SO_x. Sulfur dioxide is a primary pollutant released mainly by the burning of sulfur-containing fossil fuels, particularly coal and oil used for heating and electric power. Other industrial activities, such as petroleum refining and ore smelting, also contribute SO_2 (see Figure 14–1). Unlike natural processes, human activities tend to be concentrated over relatively small areas, allowing SO_x to attain high values over urban and industrial areas.

Sulfur dioxide is a colorless but highly corrosive gas that irritates human respiratory systems. High concentrations are associated with a number of lung problems, and even low concentrations can cause asthmatic subjects to experience severe bronchial constriction during exercise. Though widely blamed for causing respiratory problems, scientists are not sure what role high SO_2 concentrations directly play in their onset. It may be that the occurrence of respiratory illness during high SO_2 episodes is not due directly to the presence of the gas, but rather to the particulates that often accompany high sulfur dioxide concentrations.

Sulfur trioxide can be put directly into the air as a primary pollutant, but it more commonly builds up as a secondary pollutant following reactions involving SO_2. Sulfur trioxide is not by itself a major component of air pollution. However, it readily combines with water droplets to form *sulfuric acid*, H_2SO_4. If this process occurs near the surface, it forms **acid fog;** if it occurs in clouds, subsequent precipitation of the acid compound produces **acid rain.*** Not surprisingly, acid fog and rain are both capable of causing extreme environmental harm and through time can wear down human structures. Acid fog can be particularly dangerous to people because it is so easily inhaled. Buildings and monuments made of limestone are especially vulnerable to weathering from acid rain and fog (Figure 14–2).

Acid precipitation reaching the surface eventually feeds into the hydrologic system. Though some water falls directly onto lakes and rivers, the majority gets into

*The term *acid deposition* refers to the acidification of the surface environment by either contact with fog or the accumulation of acidic precipitation.

▶ **Figure 14–2**
Acid deposition can gradually wear down the surfaces of monuments and buildings.

them indirectly as soil or groundwater. Despite the indirect input into the surface waters, however, the water retains its acidity as it flows beneath the surface and eventually enters lakes and rivers. Through time, the surface water system becomes so acidic that it is inhospitable to life. At its worst, acidification can render lakes and rivers completely devoid of fish and aquatic birds. Unfortunately, this problem is neither hypothetical nor abstract. In the eastern United States, nearly 1200 lakes and 4700 streams have become acidified—some to the point where they no longer support any fish. In the Canadian province of Ontario, 1200 lakes are now essentially lifeless. Staggering as these figures may be, they pale in comparison to the 6500 lakes similarly impacted in Norway and Sweden.

As shown in Figure 14–3, acid precipitation is a much greater problem in the eastern United States and Canada than in the west, primarily because of the greater use of coal and heating oil. A huge proportion of the sulfur dioxide contributing to the acid rain originates from a relatively small number of sources. It is estimated that the 50 largest sulfur emitters in the region (all of which are power generating plants) account for half of the acid deposition.

Interestingly, one of the measures undertaken to improve air quality near sulfur-emitting industries and power utilities may have exacerbated the acid deposition problem farther downwind. To help in the dispersion of sulfur oxides from industrial plants, many industries and utilities have built large smokestacks to release the pollutants well above ground level (Figure 14–4). The idea behind the stacks is that, by releasing the smoke far above the surface, sulfur compounds will be transported considerable distances downwind before settling near the ground. While the stacks have successfully reduced sulfur concentrations near their source, they have had the unintended consequence of allowing the sulfur compounds to be transported hundreds of kilometers downwind, where they react to form acid deposition. Thus, the acid problem over much of the eastern United States and Canada is due to transported, rather than locally generated, pollutants. This has led to many years of litigation between states in the Midwest and Northeast and between the United States and Canada.

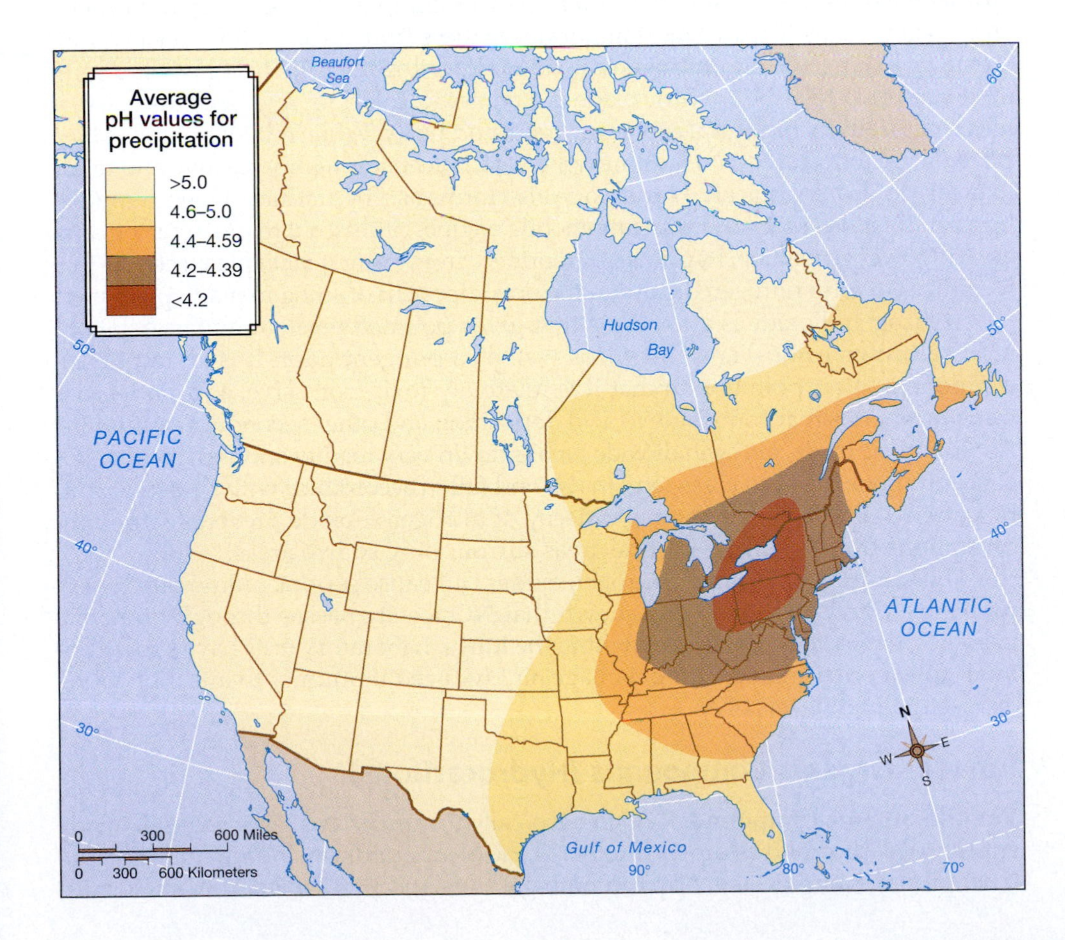

Figure 14–3 Acid precipitation is a widespread problem across eastern North America. The different colors indicate the average pH values for precipitation. Low pH values represent greater acidity. For reference, the pH for normal rainwater is 5.6. Tomatoes have a pH near 4.2.

▶ **Figure 14–4**
Smokestacks on manufacturing and power plants are designed to keep emissions away from the ground near the source. Unfortunately, pollutants are carried downwind for hundreds of kilometers, where they can exacerbate acid deposition.

While most acid deposition in eastern North America is associated with sulfur compounds, this is not always the case in other areas. Some acid deposition, especially in the western United States and Canada, is related to compounds made of nitrogen and oxygen.

Nitrogen Oxides (NO_x)

Nitrogen oxides (also called *oxides of nitrogen*) are compounds consisting of nitrogen and oxygen atoms. The two most important of these from an air pollution viewpoint are **nitric oxide** (NO) and **nitrogen dioxide** (NO_2). Together the two gases are commonly referred to as NO_x. Nitric oxide is a nontoxic, colorless, and odorless gas that forms naturally by biological processes in soil and water. While millions of tons of the material are introduced into the atmosphere each year, it is highly reactive and tends to break down very quickly. Nitric oxide also forms as a byproduct of high-temperature combustion associated with automobile engines, industrial manufacturing, and electric power generation. Its primary importance from an air quality perspective is that it oxidizes to form nitrogen dioxide, a major component of smog in many places.

Nitrogen dioxide is a toxic gas that gives polluted air its familiar yellow to reddish brown color (Figure 14–5), as well as a pungent odor. It is an important component in air pollution in that it is relatively toxic, corrosive, and undergoes transformations that contribute to acid deposition and other secondary pollutants. As with nitric oxide, nitrogen dioxide breaks down very readily and, as a result, NO_2 concentrations in urban areas tend to rise and fall in accordance with vehicular traffic patterns. Furthermore, the rapid decay of nitrogen dioxide prevents large concentrations from occurring in rural areas surrounding source areas.

Like sulfur compounds, nitrogen oxides can cause serious pulmonary health problems. Clinical studies have shown that NO_2 easily passes through bronchial passages and irritates tissue deep within the lungs. Laboratory tests have shown animals to experience severe lung damage and reduced immunity to infection when exposed to high levels of NO_2.

Volatile Organic Compounds (Hydrocarbons)

Volatile organic compounds (VOC), also called *hydrocarbons*, are materials made entirely of carbon and hydrogen atoms. These compounds, including methane, butane, propane, and octane, occur in both gaseous and particulate forms. Globally,

◀ **Figure 14–5**
Nitrogen dioxide gives polluted air a yellowish to reddish brown color, as in this photo of Hong Kong.

the vast majority of hydrocarbons arrive in the atmosphere via natural processes, including plant and animal emissions and decomposition. In the United States, industrial activities account for the greatest proportion of anthropogenic hydrocarbons, with automobiles also contributing a major share. The emissions associated with automobiles arise primarily from incomplete fuel combustion and the evaporation of gasoline (often while filling gas tanks).

Even in cities that have high VOC concentrations, there is little evidence that these chemicals have any direct adverse health impacts. Nonetheless, they are extremely important because in the presence of sunlight they recombine with nitrogen oxides and oxygen to produce photochemical smog.

Photochemical Smog

Weather Image
Ozone Concentration and Air Temperature in Los Angeles

If you have ever visited Los Angeles in the summer, you have probably heard the term **photochemical smog** and know what it feels like. Burning eyes, sore lungs, and a subtle but unpleasant odor accompany an atmosphere with poor visibility. Photochemical smog consists of secondary pollutants that include ozone (O_3), NO_2, peroxyacyl nitrate (PAN), formaldehyde, and other gases that occur in very minute quantities. As the name implies, this type of smog forms when sunlight triggers numerous reactions and transformations of gases and aerosols. Unlike the **London-type smog** found in many places where smoke combines with damp air (the word *smog*, in fact, originally derived from the terms *smoke* and *fog*), this **Los Angeles-type smog** usually involves dry air.

Ozone has been designated by the Environmental Protection Agency as the most important agent of photochemical smog. It can cause serious physical and environmental harm at surprisingly low concentrations, so low that the EPA established a concentration of only 0.12 ppm averaged over a 1-hour period as the maximum allowable without exceeding federal standards.

Exposure to ozone causes inflammation of air passages that can reduce lung capacity by as much as 20 percent. The EPA estimates that perhaps 20 percent of all respiratory-related hospital visits in the northeastern United States during the summer result from exposure to ozone. Although acute symptoms usually subside fairly quickly after ozone concentrations decline, research has shown that long-term exposure to ozone can cause permanent damage to lung tissue and impairment of the body's ability to resist bronchitis, pneumonia, and other diseases.

14–2 Focus on the Environment

The Counteroffensive on Air Pollution

Regulations designed to improve air quality have made a substantial impact on the lives of people in the United States and Canada. Although federal regulations regarding air pollution in the United States did not come into being until the 1950s, certain cities and states have had laws on the books regulating smoke emissions since the late nineteenth century. In some cases, such as in Pittsburgh, fairly stringent controls were in effect by the 1940s.

The first major U.S. initiative to clean up the nation's air was the 1963 Clean Air Act, which, among other things, expanded the role of the federal government in interstate air pollution control and authorized increased research and technical development initiatives. Major extensions of government involvement were subsequently added to the Clean Air Act by the adoption of numerous amendments in 1970, 1977, and 1990.

The original Clean Air Act and its amendments through 1977 established air pollution standards and created government agencies to ensure that those standards were being met. Maximum concentrations were established for PM_{10}, sulfur dioxide, carbon monoxide, nitrogen oxides, ozone, and lead (formerly an ingredient in gasoline but now outlawed by federal regulations). Individual states were required to establish agencies to ensure enforcement of the standards, and regions whose air quality fell short of the standards were designated nonattainment areas and required to take appropriate actions.

The act and its amendments also required automobile manufacturers to install emission reduction devices, such as the catalytic converter, that have lowered individual vehicle emissions by about 95 percent since the 1960s. The law also ordered the phaseout of open burning of refuse, the installation of filters on industrial smokestacks, and other emission-reducing measures. Overall, the act and its amendments have resulted in substantial reductions in pollutant levels in urban areas, despite a large increase in motor vehicle miles driven each year. Figure 1 shows the reduction in pollution levels.

Though the original act and the earliest amendments were monumental in their scope and effect, dozens of metropolitan areas still failed to meet the standards by 1990. In response, the U.S. Congress passed the most sweeping set of changes to the Act—the amendments of 1990. Consisting of 11 sets of provisions, the 1990 amendments established specific dates by which cities must meet ozone, PM_{10}, and CO standards, mandated corrective measures for areas of noncompliance, determined new emission standards for vehicles and stationary sources, and created new provisions to reduce acid deposition. Figure 2 shows the counties that have failed to meet ozone standards as of October 2002.

Recent research has shown that the standards adopted for various pollutants were not always based on the appropriate targets. For example, scientists now know that

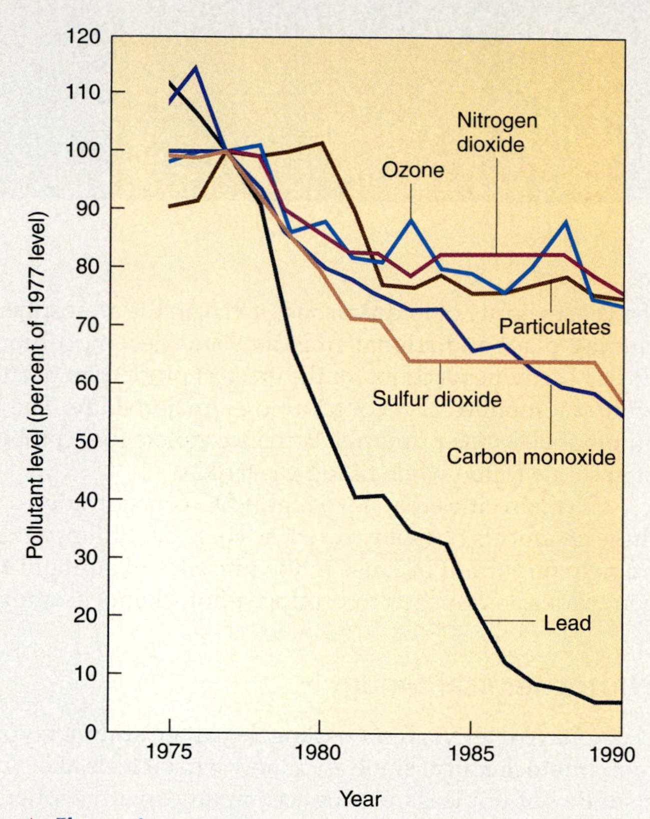

▲ **Figure 1**
Trends in pollutant levels in the United States relative to 1977.

Of course, ozone causes even greater problems for people with asthma and other preexisting pulmonary problems. For those people, ozone constricts lung passages to the point where breathing becomes nearly impossible, and the gas may contribute considerably to the 5000 fatalities in the United States each year from acute asthma attacks.

Not only are people directly harmed by ozone, but high levels also result in serious environmental degradation. Damage to agricultural crops by photochemical smog (mainly ozone) was identified in southern California grape fields in the late 1950s. Since that time, plant damage from oxidants (and to a lesser extent, PAN) has been widespread across North America, with conifers being particularly vulnerable.

1-hour exposures to ozone do not have as great an effect on human health as do longer-term exposures at lower levels. Thus, in July 1997 the EPA adopted the phasing-in of a new standard in which compliance will be based on 8-hour concentrations above 0.08 ppm, as opposed to the current 1-hour standard of 0.12 ppm. The EPA believes that the enactment of the new standards will annually prevent 15,000 premature deaths, 350,000 cases of aggravated asthma, and 1 million cases of decreased lung capacity in children. But the new standards did not get implemented without opposition. In May 1999 the District of Columbia Circuit Court of Appeals ruled against the EPA, which in turn appealed the decision to the U.S. Supreme Court. In February 2001 the Supreme Court reversed the lower court decision and determined that the EPA did not exceed its authority in issuing the new guidelines. However, it also declared that certain details of the regulations would have to be revieweds by the lower court. In March 2002, the issue was brought to rest when the appellate court ruled in favor of the EPA and declared the regulations binding.

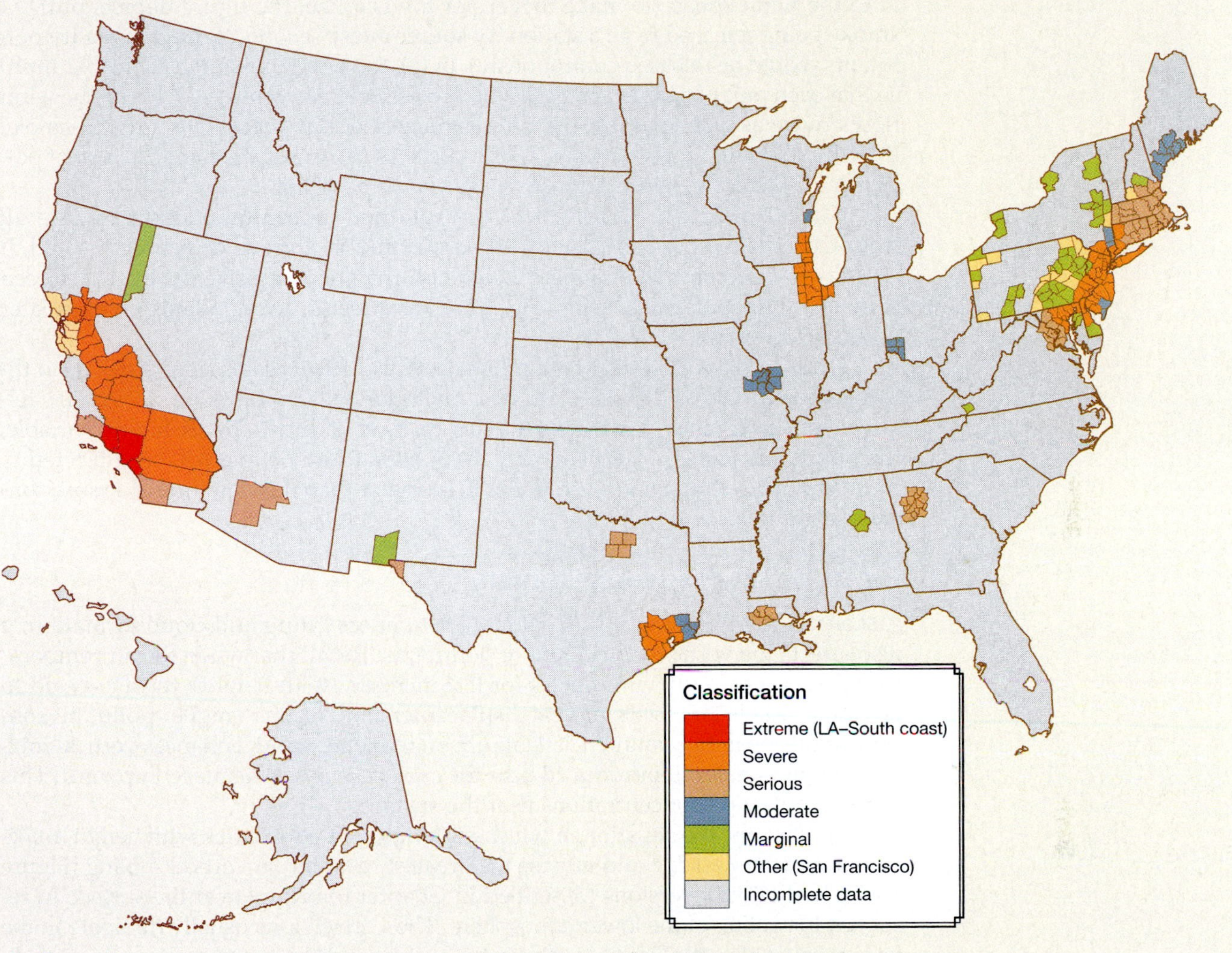

▲ **Figure 2**
Map of counties that have failed to meet ozone standards as of October 2002.

Atmospheric Controls on Air Pollution

As we have seen, a large portion of the chemicals that we consider pollutants occur naturally in the environment. These emissions do not create high concentrations, however, because their release into the atmosphere is over such a large area that they are immediately diluted. Urban emissions, on the other hand, are concentrated over much smaller areas and can thereby lead to significant pollution episodes.

Atmospheric conditions play a major role in determining pollution concentrations in several ways. Atmospheric stability and wind conditions control the vertical and horizontal dispersion of pollutants, and cloud conditions can influence the rate of photochemical reactions taking place. Furthermore, unusually cold or warm conditions encourage the increased use of heaters or air conditioners, which can increase emissions.

Effect of Winds on Horizontal Transport

Strong winds aid in the dispersal of pollutants two ways. First, they rapidly transport emissions from their source and spread them over a wide horizontal extent. Figure 14–6 illustrates how the concentration of pollution is inversely proportional to the wind speed (to make this easier to visualize, the figure depicts puffs of smoke being released from a stationary source every second, though in reality pollutants would be released continuously). In (a), the wind blows at 5 m/sec (11 mph) so that each puff of smoke travels 5 m before the next one is released. In (b), the wind flows twice as fast as in (a), and the distance between successive puffs of smoke likewise doubles. Thus, the greater wind speed in (b) causes the same amount of pollution to be diluted within twice as large a volume of air.

Greater wind speeds also lower the pollution concentration indirectly. Recall from Chapter 3 that air does not flow uniformly in a given direction. Instead, it contains small, swirling motions, called *eddies,* that mix the air vertically. This forced convection increases with wind speed and, as a result, strong winds favor greater vertical dispersion.

Short-term variations in wind direction also affect dispersion. If the wind direction varies only slightly through time, pollution will be concentrated within a relatively narrow area downwind of the source. If wind directions are highly variable, the pollutants will spread out over a wider area. More people will be subjected to the pollutants, but the concentration will be lower than it would under a more constant wind regime.

Effect of Atmospheric Stability

Just as the stability of the air (Chapter 6) influences lifting and cloud formation, it also affects the vertical movement of pollutants. Recall that when the air temperature decreases slowly with height (or if it increases with height), the air is said to be stable. Stable air resists vertical displacement and leads to higher pollutant concentrations near the ground. Unstable air, on the other hand, enhances vertical mixing, and any material introduced near the surface is easily displaced upward. This reduces pollution concentrations near the surface.

Inversions, the situation in which air temperature increases with height, make the air extremely stable and impose the greatest restraint on vertical mixing (Figure 14–7). Radiation inversions (described in Chapter 6) originate at the surface in response to cooling of the lower atmosphere. These inversions usually dissipate in the late morning after the Sun has warmed the surface and lower atmosphere. As a result, they tend to have the greatest impact on pollution concentrations in the early morning. These inversions are most important in areas subject to a London-type smog.

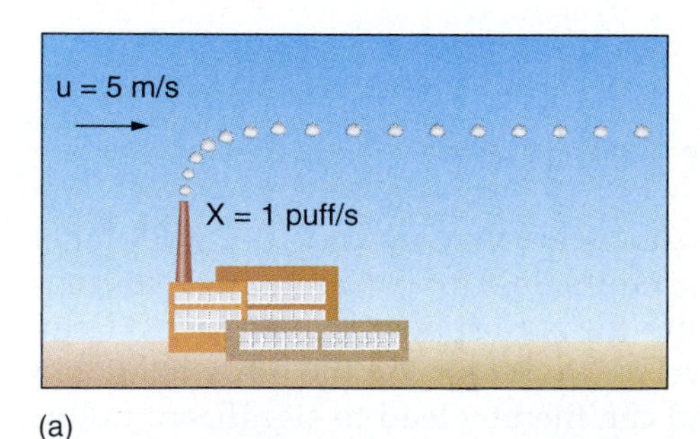

(a)

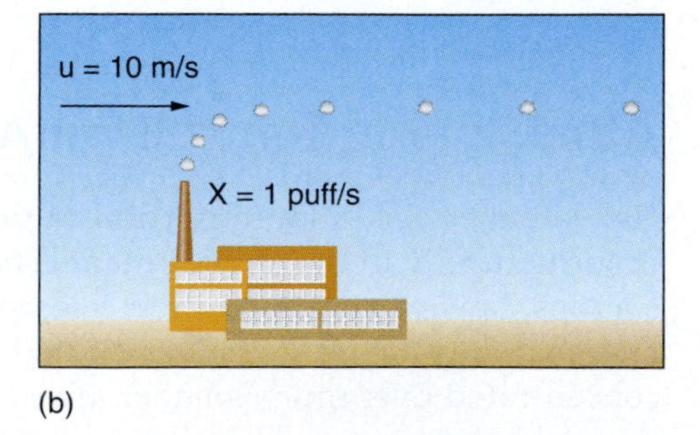

(b)

▲ **Figure 14–6**
The effect of wind speed on pollutant dispersal. In (a), the 5 m/sec wind moves individual puffs of smoke downwind slowly, so each successive puff is only 5 m behind the previous one. In (b), the 10 m/sec wind causes twice as great a distance between puffs, and thus only one-half the concentration of smoke.

◀ **Figure 14–7**
The base of an inversion is clearly evident above the smog layer in Los Angeles.

Subsidence inversions are often important where photochemical smog is the major problem. The base of a subsidence inversion marks the maximum height to which the air below can be easily mixed. An inversion with a base at 1000 m (0.6 mi) above the surface will result in a *mixing depth* of 1000 m, doubling the concentration of pollutants that would accompany a mixing depth of 2000 m (1.2 mi).

Just as the semi-permanent Hawaiian high-pressure system accounts for the mostly dry summers of southern California, subsidence from the same system also plays an important role in the region's poor air quality. As air rotates clockwise out of the eastern margin of the high, air in the middle troposphere descends and creates an inversion. During the summer, the base of the inversion level over Los Angeles typically occurs at about 700 m (2300 ft) above sea level, but the base of the inversion can also occur at lower levels and lead to particularly bad smog events.

Urban Heat Islands

Not all human impacts on the atmosphere are as dramatic as the pollution of the atmosphere. The well-known **urban heat island** is an excellent case in point. For centuries it has been known that urbanized areas often have higher temperatures than do adjacent countrysides. These differences can be quite dramatic, with temperatures in major metropolitan areas sometimes exceeding those of their hinterlands by as much as 12 °C (22 °F). Although the exact nature of the urban heat island varies from one city to another, in general the urban–rural temperature differences are greatest during the late evening and night and during the winter months.

Urban heat islands occur because of modifications to the energy balance (Chapter 3) that result when natural surfaces are paved and built upon, and when human activities release heat into the local environment. Though it is not possible to generalize the relative importance of these processes for every city, all of them probably play some role in causing the phenomenon to occur.

Atlanta Urban Heat Island I, II, and III

Several variables influence the magnitude of the heat island effect. Some are related to the local setting, others to the activities undertaken within it. But the most important are the size of the city itself and the density of its population, with large, densely populated cities having the largest heat island effect. Figure 14–8 illustrates the relationship between the population of North American cities and the maximum urban and rural temperature differences (note that the horizontal axis is plotted on a logarithmic scale).

The intensity of an urban heat island varies spatially across a city, with highest temperatures normally found within the city core. Figure 14–9 illustrates this effect by plotting temperatures over Vancouver, British Columbia, on a July evening. The downtown area is located on the southeastern part of the peninsula that juts into Burrard Inlet. Immediately to the northwest of downtown (on the northwestern part of

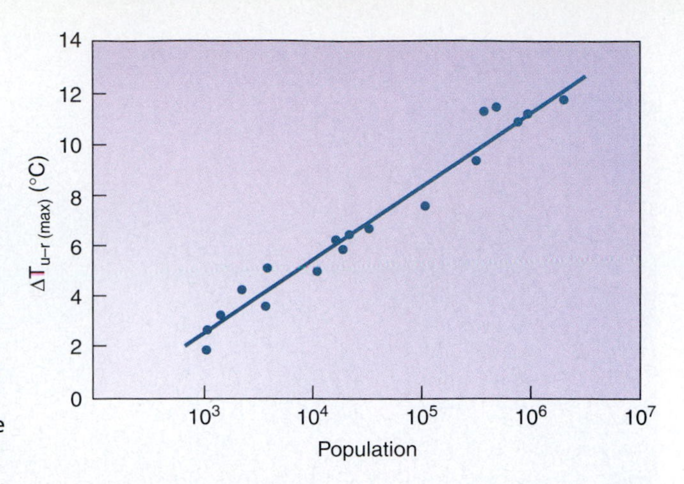

▶ **Figure 14–8**
Urban heat islands vary with city population. The vertical axis plots temperature differences between urban centers and surrounding areas in °C against city population. Note that the horizontal axis (population) is on a logarithmic scale.

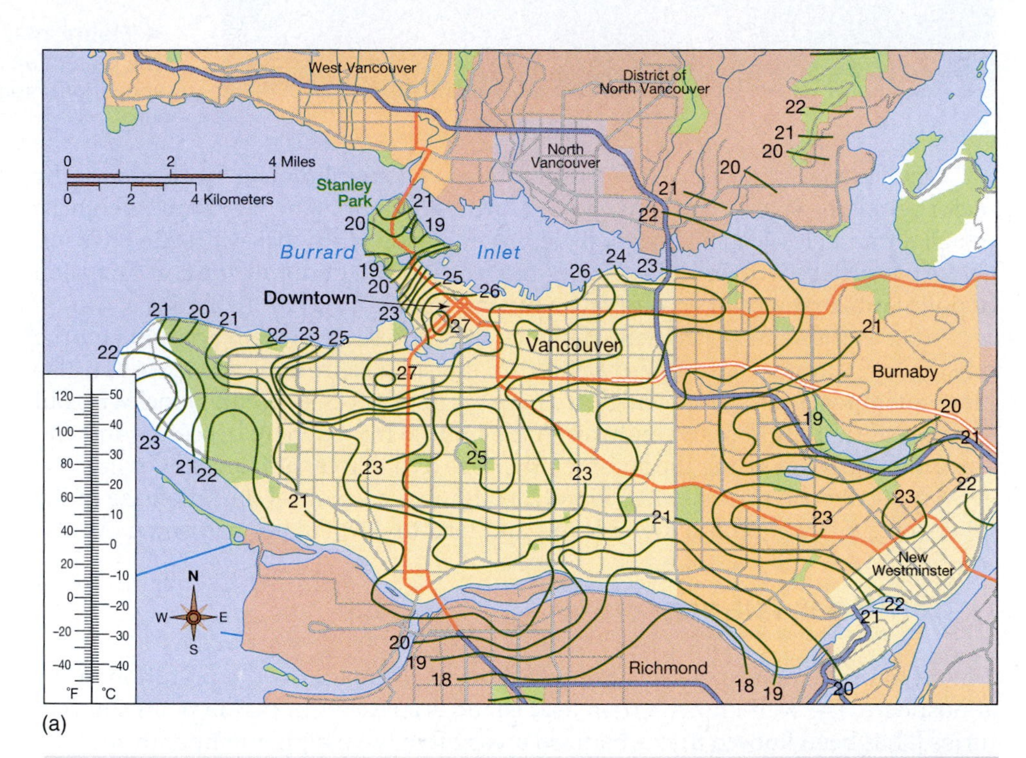

(a)

(b)

▶ **Figure 14–9**
(a) Temperatures in Vancouver, British Columbia, at 9 P.M. on July 4, 1972. Note the sizable temperature gradient between downtown and Stanley Park.
(b) Photo showing the Vancouver downtown area with Stanley Park in the background.

the peninsula) lies Stanley Park, a wooded area with few buildings. As expected, temperatures are greatest over the downtown region and decrease substantially over less-populated areas. Temperatures on the peninsula decrease dramatically between downtown and the middle of the park, with a difference of about 9 °C (16 °F) occurring over a distance of only about 1.5 km (1 mi). Wind speeds also play an important role. During windy conditions, cooler air from the surrounding countryside displaces the warmer urban air and thus reduces the magnitude of the urban heat island.

Radiation Effects

Urban particulates can also affect the intensity of the urban heat island through their effect on the radiation balance. Increased particulates associated with urban activity can absorb and scatter incoming solar radiation and also increase the amount of absorption and reradiation of longwave energy in the atmosphere. Although it is hard to generalize for all cities, it is believed that the particulates tend to decrease the amount of incoming solar radiation at the urban surface, but the increase in net longwave radiation probably offsets the reduction in absorbed solar energy. Thus, the direct effect of particulates on urban temperatures is probably negligible.

Particulates also affect the radiation balance indirectly. Recall that water droplets in the atmosphere form onto condensation nuclei. The increase in particulates due to human activity can increase cloud cover, as in the case of London, England, which has been shown to receive 270 fewer hours of bright sunlight annually than the surrounding area. Particulates have long been known also to increase precipitation downwind of urban regions. Interestingly, studies have also shown that precipitation can decrease downwind of major industrial centers, perhaps because cloud water is spread over many condensation nuclei, which lessens the chance of growth to precipitation size.

More important than the effect of increasing particulate concentrations is the impact that buildings have on the radiation balance. Consider the impacts that the construction of buildings with vertical walls might have on the receipt of solar radiation. When the sun is low in the sky—near sunrise and sunset, and during much of the day at high latitudes in the winter—direct sunlight that would otherwise reach the horizontal surface hits the vertical walls of buildings. This causes the angle of incidence to become closer to perpendicular and increases surface heating, which leads to a higher temperature.

The presence of buildings also affects the rate of heating by changing the surface albedo. Darker buildings, of course, absorb more sunlight than lighter ones, and, in general, urban surfaces (asphalt streets, roofing materials) have lower albedos than the natural surfaces they replace. The presence of buildings also affects the amount of absorption by causing multiple reflections to occur, as shown in Figure 14–10. As sunlight penetrates the urban landscape and hits the side of a building, some of the

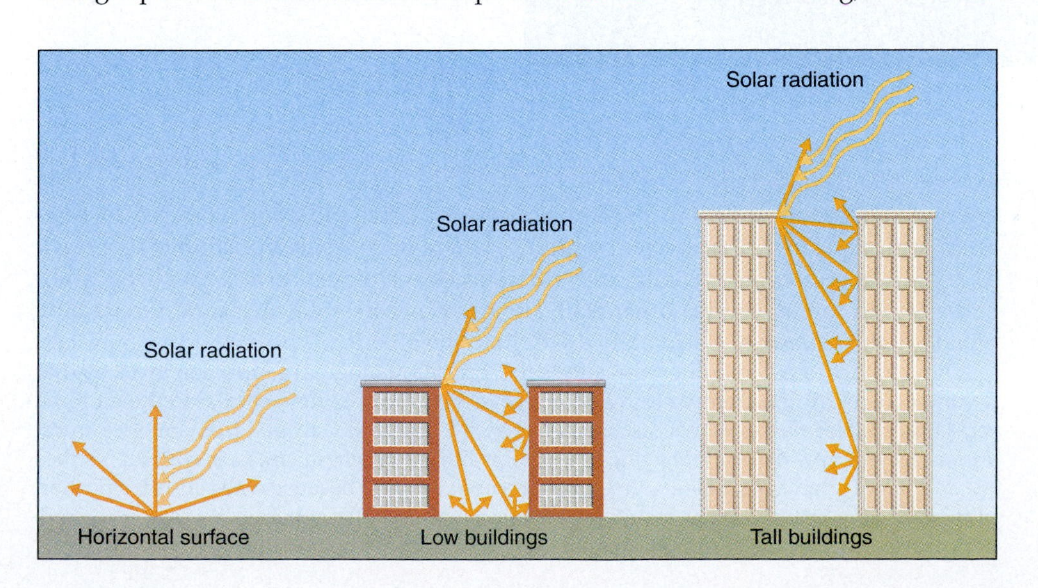

◀ **Figure 14–10**
Effect of buildings on solar radiation receipt. As incoming radiation contacts a building, some is scattered off in all directions and some is absorbed. The scattered radiation may in turn hit an adjacent building, where further absorption can take place. This lowers the urban albedo.

14–3 Focus on the Environment

Smog in Southern California

Los Angeles has long had a reputation for extremely bad air quality—and for good reason. Of all the cities in the United States, Los Angeles is the only one classified by the Environmental Protection Agency as an "extreme area" of noncompliance of ozone standards. A number of factors work together to make the air quality bad enough to earn this dubious distinction. As shown in Figure 1, Los Angeles occupies part of a basin bounded by mountains to the north and east that block the free movement of pollutants by the sea breeze, while the presence of a subsidence inversion during the warmer months restricts vertical dispersion. Add to that the typically cloud-free conditions during the midday period that trigger photochemical reactions. And finally there is the city's well-known love affair with the automobile, which contributes much of the estimated 2 million kilograms (2200 tons) of hydrocarbons and 1 million kilograms (1200 tons) of NO_x released each day into the four-county South Coast Air Basin.

Fortunately, a number of new regulations have improved the situation. For example, beginning in 1984 all automobiles were required to undergo biannual smog checks. More recently, regulations have been enacted requiring gas pump nozzles to have rubber sleeves to capture fumes that would otherwise escape into the air as people fill their tanks. Also, a cleaner-burning type of gasoline that releases fewer hydrocarbons has been phased in at all area gas stations. To illustrate how much progress has been made, consider the fact that during the 5-year period from 1976 to 1980, there was an average of 112 stage 1 smog alerts (ozone $\geq$ 0.20 ppm) each year in the South Coast Air Basin. During the period from 1995 to 1999, the average dropped to 7, with no smog alerts at all in 1999! In fact, in 1999 Los Angeles did not top the nation in the number of days in which ozone concentrations exceeded federal standards—that honor went to Houston, Texas.

During the summer, daily concentrations of photochemical smog vary on a regular basis in the course of a day. Prior to the morning rush hour, residual primary and secondary pollutants from the previous day leave a background level of air pollution. As traffic increases during the

◀ Figure 1
The topography of the Los Angeles basin.

energy is absorbed and some is scattered back as diffuse radiation. Some of the scattered radiation strikes an adjacent building, where once again a portion is absorbed. This process goes on repeatedly, with each successive reflection at least partially absorbed upon contact with another wall. This increases the total absorption so that the albedo of the urban area is actually lower than the albedo of the individual surfaces.

The presence of tall buildings also affects the transfer of longwave radiation in a way that favors higher nighttime temperatures. Essentially, the process is very similar to the multiple reflection of solar radiation just discussed. Longwave radiation emitted from an open, rural surface travels upward without being impeded by buildings. Urban areas, in contrast, reduce the amount of longwave radiation that freely escapes to space because walls absorb a portion of the outgoing radiation. The resultant reduction in longwave radiation loss slows the rate of nocturnal cooling and promotes higher daily minimum temperatures.

morning, emissions increase substantially. Early morning winds are usually weak, which leads to little movement of pollutants. At the same time, the low Sun angle and common presence of early morning fog and low clouds inhibit photochemical activity. The situation normally changes by late morning. A sea breeze usually develops along the coast and moves pollutants inland, while clearing skies and increasing Sun angles increase photochemical conversions.

As the sea breeze develops, a boundary called a **sea breeze front** separates the relatively clean marine air from the more polluted, drier air ahead. As the sea breeze front moves inland, it pushes the emissions eastward or northeastward. This often creates a strong gradient in ozone concentrations near the sea breeze front, with relatively clean air behind it and increasing concentrations to the east or northeast (Figure 2). By late afternoon, the cities in the eastern portion of the basin get the full onslaught of the advected pollutants, while local commuters add their own contribution to the photochemical smog. As a result, pollution levels can become extraordinarily high in the areas downwind of Los Angeles.

The area to the south of Los Angeles—including Orange and San Diego Counties—also has a significant air pollution problem. Usually the air pollution in San Diego, some 150 km (100 mi) to the south, is of local origin. During severe episodes, however, most of the pollution originates over Los Angeles and Orange Counties and gets carried into San Diego by the wind. These episodes often occur as Santa Ana winds die out. During the Santa Ana, the easterly winds force the basin's pollutants offshore. As the Santa Ana begins to weaken, a thermally induced low-pressure system over the eastern desert stretches into the San Diego area. This creates a northwesterly flow that transports pollution originating over Los Angeles and Orange County. Not surprisingly, the trend toward decreasing pollution in Los Angeles has also occurred in San Diego.

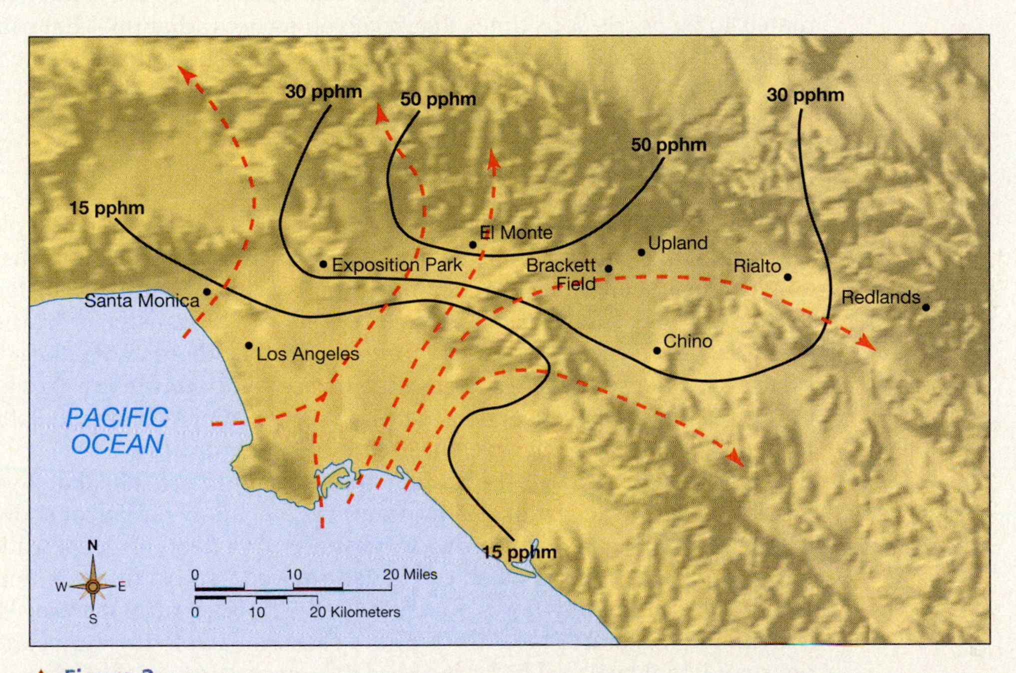

▲ **Figure 2**
The distribution of ozone between 4 and 5 P.M. on July 25, 1973. Highest concentrations (shown by solid lines) occur in the northeast, ahead of the sea breeze front. Dashed lines show the wind direction.

Changes in Heat Storage

As explained in Chapter 3, radiation is not the only mechanism that transfers energy from one place to another; conduction and convection are also important heat transfer mechanisms. In the middle of a sunny day, absorption of solar radiation warms the surface of the ground. Conduction within a very thin layer of the atmosphere and convection transfer much of this energy to the air. At the same time, a gradient develops in which soil temperature decreases with depth, and heat is conducted downward.

During the late afternoon, the surface begins to cool when energy losses by radiation and convection exceed the absorption of shortwave and longwave radiation. The soil temperature profile eventually reverses, with temperature increasing with depth. Thus, during the evening hours, heat that has been stored within the soil is transferred to the surface.

The same processes just described occur in the walls and roofs of urban buildings. As the surface is warmed during the day, a temperature gradient develops that conducts heat toward the building interior. When cooling occurs in late afternoon, the stored heat is released to the surface. What is different from the rural setting is that materials used in building construction have a much greater capacity to store heat than most natural surfaces. As a result, more stored heat is available for transfer to the lower atmosphere during the evening and nighttime than for natural surfaces, and nocturnal temperatures are increased.

The release of heat from buildings just described is supplemented by the anthropogenic heat produced for comfort (e.g., space heating) or as a byproduct of other activities (e.g., waste heat from a hot car engine). Anthropogenic heating is greatest during the winter, which partially explains why heat islands are most pronounced during the low sun season. Anthropogenic heat can be a surprisingly large component of the urban energy balance. In Vancouver, British Columbia (49° N), for example, the amount of anthropogenic heat released in the winter has been estimated to be nearly four times that received as net radiation. That estimate, based on 1970 data, probably understates the importance of anthropogenic heat currently released because the city has grown significantly since then.

Sensible and Latent Heat Transfer

In Chapter 3 it was shown that most of the global surface has a surplus of net allwave radiation on an annual basis. That surplus is transferred to the atmosphere as sensible and latent heat. When moisture is available near the surface, the transfer of energy as latent heat can exceed sensible heat transfer, indicating that most of the surplus is consumed by evaporation. On the other hand, if the surface is completely dry, surplus energy raises the surface temperature far above the air temperature, and sensible heat dominates. All other things being equal, the higher the ratio of sensible to latent heat, the greater the temperature.

Urbanization affects the routing of precipitation in a way that favors increased sensible heat transfer. Unlike natural surfaces that allow rainfall or snow melt to permeate the soil and be retained below ground, city streets and sidewalks are almost impervious to water. So when precipitation occurs, most of the water runs off the surface and ultimately flows out through the flood-control system. The reduction in available water increases the input of sensible heat to the atmosphere at the expense of latent heat and helps increase the temperature of the urban environment.

Urban Heat Islands and the Detection of Climate Change By now everyone has heard much discussion about the possibility of impending climatic change resulting from the anthropogenic emission of greenhouse gases. Many atmospheric scientists believe that this change has already begun to take place. To support this notion, they point to an overall warming of 0.3 to 0.6 °C (0.5 to 1 °F) occurring since the late nineteenth century for weather stations having records going back more than a century. However, we cannot use these recorded temperature changes at face value, because many of the data come from urban weather stations, and most cities having long-running weather stations have undergone considerable growth over the last century. Thus, we must contend with the problem of enhanced urban heat islands influencing the data, which means that records from large urban areas are not representative of the surrounding region. Atmospheric scientists are well aware of this source of bias in temperature records and routinely account for its effect, either by discarding contaminated data or by adjusting values downward for affected stations.

Summary

As the human population has grown in size and become more industrialized, societies have increased their impact on the atmospheric environment. The most dramatic effects result from the release of numerous gases and particulates. Although many of the emissions that we consider pollutants result from natural processes, rapid dispersion of these materials prevents them from causing negative impacts. In industrial and urban areas, on the other hand, these emissions are concentrated into smaller areas and often lead to serious problems.

Some atmospheric pollutants are released directly into the atmosphere (primary pollutants), while others result from transformations of other gases (secondary pollutants). Particulates are solid and liquid aerosols that can be produced as either primary or secondary pollutants. Recent research has shown that the health effects of the smallest particulates are most critical because they are easily lodged in lung tissue. Pollutant gases include carbon oxides, sulfur compounds, nitrogen oxides, volatile organic compounds (also called hydrocarbons), and photochemically formed gases (the most notable of which is ozone). Each gas presents its own set of health problems, ranging from reduced immunity to permanent lung damage to cardiovascular problems.

Up until the middle part of the twentieth century, efforts to control these emissions were instituted on a local scale. Beginning in 1955, the U.S. government began to enact laws designed to improve air quality across the nation and reduce the number of major health problems that result from air pollution. The original Clean Air Act and its amendments have had a dramatic impact on air quality. This legislation has required the formation of local agencies to monitor pollution levels and ensure compliance with federal standards. It has also required the automobile industry, power utilities, ore smelting plants, and manufacturing industries to reduce the amount of their emissions. While reductions in many pollutants have been dramatic, numerous cities in the United States still have not met clean air goals.

The amount of air pollution does not depend entirely on the activities of people; atmospheric conditions also affect the dispersal of pollutants. If winds are strong and constantly shift direction, pollutants are distributed over a larger area and concentrations decrease. Statically unstable air also favors the dilution of gases and particulates by enhancing vertical mixing. On the other hand, stable conditions and, in particular, the presence of an inversion, can greatly restrict vertical motions and concentrate pollutants near the ground.

Human impacts on the atmosphere are not restricted to pollution. The urban heat island is a well-known phenomenon in which changes in the surface (such as the replacement of vegetated surfaces with concrete and asphalt), the existence of buildings with vertical walls, and the release of heat as a byproduct of human activity combine to increase temperatures. These increases are most notable during the evening and nighttime hours and in the winter season.

We have now looked at the natural processes that make up daily weather, and the ways humans analyze, predict, and alter the resultant patterns. The last two chapters of this book concern the longer-term state of the atmosphere—the climate. Chapter 15 looks at general patterns across the globe, and Chapter 16 examines past and possible future changes in climate.

Key Terms

air pollution p. 425
primary pollutants p. 426
secondary pollutants p. 426
particulates p. 426
PM_{10} p. 427
$PM_{2.5}$ p. 427
carbon oxides p. 428
carbon monoxide p. 428
carbon dioxide p. 428
sulfur dioxide p. 430
sulfur trioxide p. 430
acid fog p. 430
acid rain p. 430
nitrogen oxides p. 432
nitric oxide p. 432
nitrogen dioxide p. 432
volatile organic compounds (hydrocarbons) p. 432
photochemical smog p. 433
London-type smog p. 433
Los Angeles-type smog p. 433
urban heat island p. 437
sea breeze front p. 441

Review Questions

1. Explain the distinction between primary and secondary pollutants.
2. What are particulates and how are they introduced into the atmosphere?
3. What are the two processes most responsible for removing particulates from the atmosphere?
4. What are PM_{10} and $PM_{2.5}$? Does one pose a greater health risk than the other?
5. List the most important gases that contribute to air pollution.
6. What are the primary sources of carbon monoxide in the atmosphere? If these sources are nonanthropogenic, why is it that CO is considered a pollutant?
7. In what way does carbon monoxide harm the human body?
8. What are the pimary sources of suflur dioxide and sulfur trioxide in the atmosphere?
9. Would a person be more likely to notice the presence of high CO or high SO_2 contents in the ambient air?
10. Which primary pollutant is most likely to promote the formation of acid fog or acid rain?
11. Why is it that nitric oxide is much less common in the atmosphere than nitrogen dioxide?
12. Describe the general composition of volatile organic compounds.

13. How do London-type and Los Angeles-type smog differ from each other?
14. Which pollutant gases cause a noticeable coloration of the atmosphere? Which have a characteristic odor?
15. Describe the various atmospheric controls that affect the concentration of air pollutants.
16. How does the construction of buildings in cities alter the radiation exchange near the surface and contribute to the urban heat island effect?
17. Describe the way in which heat storage in cities differs from that of surrounding rural areas.
18. What effect does urbanization have on the exchange of sensible and latent heat?
19. Does the urban heat island manifest itself equally during the day and night? Are there seasonal differences in the magnitude of the heat island effect?

Critical Thinking

1. There has been much improvement in air quality for North America as a whole. Further improvements will only arise from measures that may be costly, both directly through the application of technology and by reductions in certain economic activities. Do you personally believe that further improvements can be brought about at a cost that people are willing to bear?
2. Visit the Web page at **http://www.gsfc.nasa.gov/gsfc/earth/terra/co.htm**, and form your own conclusions about whether air pollution is now primarily a local or a global process.
3. It is believed that global warming over the last few decades might have been even greater, were it not for the effect of certain types of air pollution. Explain how this could come about.

Problems and Exercises

1. Examine the map in Box 14–2, Focus on the Environment: The Counteroffensive on Air Pollution. How is the air pollution situation in your area? Is the information on the map consistent with your perception of the local air quality? What factors do you think lead to the type of air quality that your area has?
2. Visit the Web page at **http://www.epa.gov/air/partners.html**, for a listing of EPA, state, and local agencies that provide data on air quality. Refer to one of the agencies in your area for a daily report on the air quality in your area. How do the changes in air quality fluctuate with different weather patterns where you live?
3. If you live in or near an urban area, make note of the daily maximum and minimum temperatures in your region. Do you detect a significant urban heat island? How does the magnitude of this heat island compare for maximum and minimum temperatures? Are there seasonal differences?

Quantitative Problems

The companion Web site to this textbook, **http://www.prenhall.com/aguado**, has some interesting problems related to the concentration of pollutants and precipitation pH values. We suggest you solve those problems to further your understanding of air pollution.

Useful Web Sites

http://www.epa.gov/air/EPA

Environmental Protection Agency Web page with wide range of information on various types of air pollution.

http://www.cdc.gov/nceh/airpollution/default.htm

National Center for Environmental Health, an agency of the Centers for Disease Control. Includes numerous links to reports on the effect of air pollutants on human health.

http://www.tc.gc.ca/programs/Environment/AirPollution/menu.htm

Information on air pollution in Canada.

http://www.ces.ncsu.edu/depts/pp/notes/Ozone/ozone.html

Describes the effects of ozone on plants.

http://www.epa.gov/oar/oaq_caa.html
Presents comprehensive information on the Clean Air Act.

http://www.nrdc.org/air/pollution/default.asp
A Web page of the Natural Resources Defense Council, a non-profit public interest group that advocates environmental causes.

http://www.gsfc.nasa.gov/gsfc/earth/terra/co.htm
Reports on how NASA has been monitoring global air pollution. Includes some very interesting movies.

Media Enrichment

Weather Image

Atmospheric Pollution as Seen from Space Shuttle

This cross section of Earth's atmosphere at sunset and earth limb (24.5° S, 43.5° E) displays an unusual layering believed to be caused by temperature inversions that concentrate aerosols into narrow, horizontal layers. The white layer, possibly containing volcanic particles, represents the top of the stratosphere. Smoke created by biomass burning is found in the troposphere (the purple layer).

Weather Image

Pollution over Amazon Basin

This NASA space shuttle image, taken September 18, 1991, is centered over the Bolivian Altiplano (20.0° S, 65.0° W), looking northeast into the lower elevations of Bolivia and Brazil. Thick haze is present over the southern Amazon Basin due to seasonal biomass burning. This haze is trapped in the lower atmosphere by a temperature inversion.

Weather Image

Kuwait Oil Field Fires

This Landsat-5 Thematic Mapper image shows oil wells burning in Kuwait at the end of the Persian Gulf War. Fires appear as red dots, and smoke plumes stretch along the Persian Gulf coast from Kuwait to Saudi Arabia. The smoke plumes were relatively narrow—25 to 60 km—and extended to a height of 3 to 4 km in the atmosphere. Although it was believed that the smoke from these fires would have some effect on global climate, effects were limited to the Gulf region.

Weather Image

Industrial Pollution in Siberia

This photograph, taken April 2, 1992, shows industrial pollution (soot-blackened snow) around the Siberian city of Troitsk (54.0° N, 61.0° E), an industrial city located east of the Ural Mountains. Troisk is considered one of the most polluted cities in the region. Due to the large amount of soot in the atmosphere, respiratory diseases among citizens of Troisk are chronic.

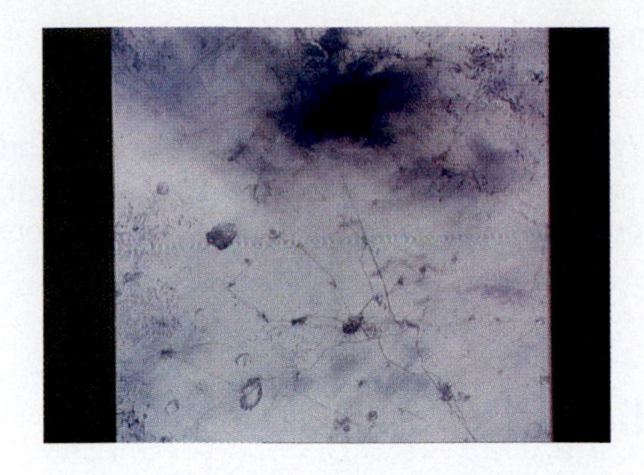

Weather Image

Ozone Concentration and Air Temperature

This graph shows the relationship between photochemical smog (measured as ozone in parts per billion) and temperature. Photochemical smog is created when sunlight triggers numerous reactions and transformations of gases and aerosols. These reactions are more intense and occur more often at high temperatures. As this graph for 1985 shows, when the daily maximum temperature was below 70 °F in Los Angeles, ozone amounts seldom exceeded the national standard of 120 ppb. When temperatures increase, however, ozone concentrations tend to increase as well. Note that ozone alerts were common at temperatures above 95 °F.

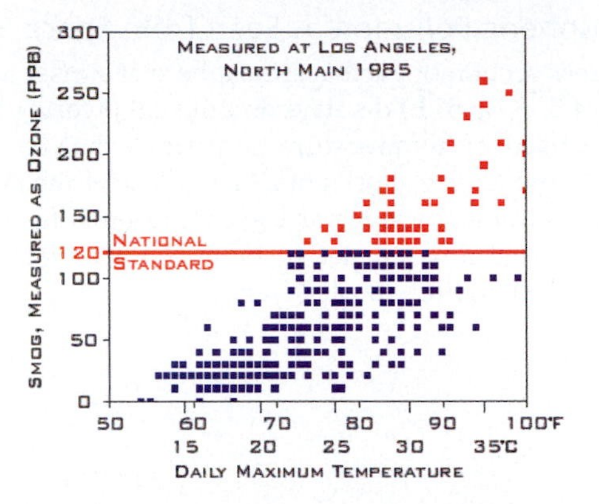

Weather Image

Atlanta Urban Heat Island I, II, and III

These satellite images were taken of Atlanta in 1997 as part of a study on the effects of Atlanta's urban expansion on the area weather and air quality. Researchers found that urban Atlanta experienced temperatures that were warmer by 4.5 to 5.5 °C (8 to 10 °F) than outlying rural areas, making a good example of a heat island.

The first scene, obtained on May 11, 1997, is a black-and-white image showing daytime thermal infrared data (oriented with north at the top). Surfaces that are "hot" or "warm" appear in varying shades of white to light gray, whereas cooler surfaces are darker. Most of the buildings and other urban surfaces are hotter than the surrounding suburban and rural areas during the daytime.

The second picture, a false-color image, was obtained using a multispectral sensor flown on an aircraft. Trees and other vegetation appear red. Buildings, streets, and other urban land covers appear white, blue-green, or black.

The third picture is a GOES 8 satellite image showing a regional perspective of the urban heat island. The thermal channels of the GOES 8/10 geostationary satellites measure outgoing longwave radiation. In the absence of clouds, the satellite data can be used to monitor hourly changes in thermal characteristics of the surface. This image was taken at 10:30 P.M. EDT on May 6, 1997, and shows surface temperatures in degrees Fahrenheit. Atlanta is warmer than the surrounding area by several degrees.

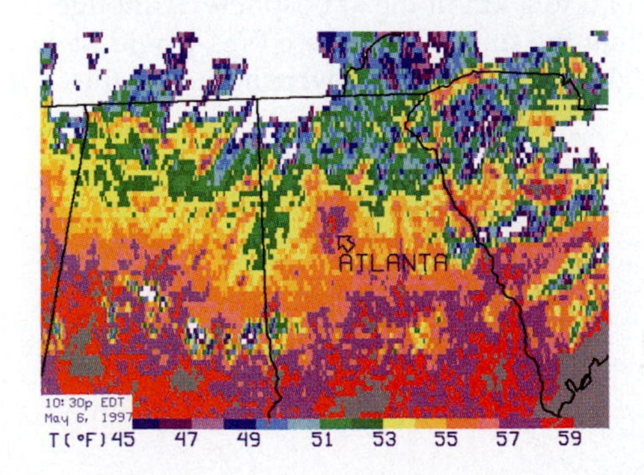

PART SIX

Current, Past, and Future Climates

Stonehenge in Salisbury Plain, England. About 4000 years old, the arrangement of the stones probably served to track the seasonal changes in Earth–Sun geometry.

chapter 15

Rain forest in Venezuela.

Earth's Climates

Although we don't always know in advance what type of weather will occur at any particular place on a particular day, we do have some idea of what type of weather is considered normal for a given location. For example, Florida typically has hot, humid summers and mild winters—although cold spells do occur from time to time. But there are times when truly anomalous weather events occur in some particular location, events that defy our expectations. These may take the form of unusual heat or cold, but they may also appear as severe weather events uncommon to a particular region. Such an event occurred in Salt Lake City, Utah, on August 11, 1999, when an F2 tornado roared through the downtown area.

Tornadoes are not unknown in Utah; they occur on average about twice a year, but they almost always are classified as "weak" tornadoes, F0 or F1. The Salt Lake City tornado was an F2 that covered a 3-km (2-mi) path. The tornado brought winds in excess of 180 km/hr (110 mph) to the downtown area, toppling trees, knocking over tractor trailers, and damaging more than 120 homes. The Delta Center (the home of the Utah Jazz basketball team) and the Salt Palace Convention Center were among the more notable structures damaged. David Gross, who was inside the Convention Hall, reported that, "The roof opened and it [the wind] ripped off a door. It was over in 15 or 20 seconds, but it seemed like a lot longer than that."

One man was killed by flying debris in a large tent set up for an outdoor retailers convention. The victim was believed to be the first fatality from a tornado in Utah's recorded history. One hundred other people were injured, and 40 of them required hospitalization.

Of course, the climate we associate with an area involves more than just severe weather; it concerns the entire range of weather conditions in the area. Climate deals with long-range conditions, as opposed to the passage of daily weather occurrences. In this chapter we examine the issues involved in classifying climate, and we describe the broad climatic regions across the globe.

CHAPTER OUTLINE

Defining Climate

We define **climate** formally as the statistical properties of the atmosphere. This is consistent with the notion that the details of an individual event or moment in time are not of interest—instead, climate is concerned with the long-term behavior, or expected (typical) conditions. Thus, for example, the mean (average) temperature is a climatological value, and by comparing mean temperatures one gains information about differences in climate. But climate is more than average values. For example, two places might have similar long-term average rainfall. But if one place typically has both very high and very low values, while the other tends to receive nearly the same total year in and year out, we would surely say they have different climates. This year-to-year variability is another statistical property and is thus a measure of climate.

In addition to variability, we might be interested in the degree to which above-average or below-average values tend to come in runs or be clustered in successive years. For example, do periods of drought tend to be followed by wet periods? Or are extended episodes of extreme heat and cold a common occurrence? These are questions about another statistical property, namely the correlation between values in successive years. Again, this is part of climate. The point we want to make is that climate consists of all statistical properties. For our purposes, a consideration of average values will be sufficient, but a complete description of climate would include much more.

Although the delineation of distinct climates may seem like a very straightforward endeavor, establishing the criteria by which climates are delineated requires considerable subjectivity. Consider what you would do if called on to devise a scheme by which Earth's surface would be covered by distinct climatic zones, each having properties that set them apart from the others. The job would require you to set distinct boundaries that separate one climate zone from another. Yet in nature such clear boundaries are rare. Thus, there is a considerable difference in temperature and precipitation along the east coast of North America from Florida to the Maritime Provinces of Canada, and you certainly would not want to put St. John's, Newfoundland, in the same climatic zone as Tallahassee, Florida. But exactly where would you draw the lines that separate the various climates? And how would you decide how many climates? Too many would make the system too complex; too few would fail to capture the patterns you would want to identify.

Climatologists over the years have made numerous attempts to establish useful climatic classification schemes. Some are based on the obvious properties of temperature and precipitation. Others use the frequency with which air mass types occupy various regions, differences in energy budget components, or seasonal characteristics of the water balance at the surface. Each has its own advantages, depending on the purpose of the classification. Agriculturalists, for example, would probably be most interested in using a classification that yields information on water availability relative to plant needs, reflecting gains and losses of water in the soil column (precipitation, evapotranspiration,* runoff, and losses to groundwater).

The Koeppen System

For many people, a climatic classification scheme based on temperature and precipitation is useful because it yields information on the two meteorological variables of greatest general interest. The most widely used systems based on these variables have followed on the work of Vladimir Koeppen, a German citizen of Russian ancestry. The **Koeppen system** was developed over the period from 1918 to 1936 in a process of almost continual revision and refinement. Koeppen looked at the world distribution of natural vegetation types, located the boundaries that separated them, and determined what combinations of monthly mean temperature and precipitation were associated with those boundaries. Thus, although its climatic types are determined by temperature and precipitation, Koeppen's system is inherently tied to natural vegetation. Contrary to what one might assume, it does not begin with the idea of "natural" temperature/precipitation regimes. The boundaries are associated with plant associations, which may or may not coincide with what seem to be obvious or striking gradients in temperature and precipitation.

This chapter will use a version of the Koeppen system as modified by Trewartha. Various versions of the Koeppen system apply different names for each of the climates and may have slightly varying criteria for distinguishing them. Thus, our maps and descriptions are necessarily somewhat different than some other portrayals of the Koeppen scheme. Moreover, this scheme is totally descriptive and does not attempt to explain the causes of the various climates.

Koeppen used a multi-tiered classification system having primary climates delineated by capital letters ranging from A through E. These five broad categories tend to arrange themselves across Earth's surface in response to the latitude, degree of continentality, and location relative to major topographic features. In addition to these climates, the version we are using includes an additional one for mountain environments, designated by H. The main climate groups (designated by first letter only) can be briefly described as follows:

- A—**Tropical.** Climates in which the average temperature for all months is greater than 18 °C (64 °F). Almost entirely confined to the region between the equator and the Tropics of Cancer and Capricorn.

*The combination of water directly evaporated at the ground surface and that absorbed by plants and evaporated through their leaves.

- B—**Dry.** Potential evaporation exceeds precipitation.
- C—**Mild Mid-latitude.** The coldest month of the year has an average temperature higher than –3 °C (27 °F) but below 18 °C (64 °F). Summers can be hot.
- D—**Severe Mid-latitude.** Winters have at least occasional snow cover, with the coldest month having a mean temperature below –3 °C (27 °F). Summers are typically mild.
- E—**Polar.** All months have mean temperatures below 10 °C (50 °F).

Climatic zones A, C, D, and E are based on temperature characteristics. The A climates (tropical) tend to straddle the equatorial regions; C, D, and E climates usually occur sequentially further from the Tropics and toward the polar regions. The sole primary climate zone that considers precipitation is the B climate, designating deserts and semi-deserts.

The A through E climates are subdivided into smaller zones represented by a second letter, which are further subdivided. We will use subdivisions only through the third letter; thus, each individual climate is represented by a three-letter combination describing its temperature and precipitation characteristics. The descriptions for each of the three-letter climates are presented in Table 15–1 and their

Table 15–1 • Climate Types According to Koeppen

Type	Subtype	Letter Code	Characteristics
A—Tropical	Tropical wet	Af	No dry season
	Tropical monsoonal	Am	Short dry season
	Tropical wet and dry	Aw	Winter dry season
B—Dry	Subtropical desert	BWh	Low-latitude dry
	Subtropical steppe	BSh	Low-latitude semi-dry
	Mid-latitude desert	BWk	Mid-latitude dry
	Mid-latitude steppe	BSk	Mid-latitude semi-dry
C—Mild	Mediterranean	Csa	Dry, hot summer
	Mid-latitude	Csb	Dry, warm summer
	Humid subtropical	Cfa	Hot summer, no dry season
		Cwa	Hot summer, brief winter dry season
	Marine west coast	Cfb	Mild throughout year, no dry season, warm summer
		Cfc	Mild throughout year, no dry season, cool summer
D—Severe Mid-latitude	Humid continental	Dfa	Severe winter, no dry season, hot summer
		Dfb	Severe winter, no dry season, warm summer
		Dwa	Severe winter, winter dry season, hot summer
		Dwb	Severe winter, winter dry season, warm summer
	Subarctic	Dfc	Severe winter, no dry season, cool summer
		Dfd	Extremely severe winter, no dry season, cool summer
		Dwc	Severe winter, winter dry season, cool summer
		Dwd	Extremely severe winter, winter dry season, cool summer
E—Polar	Tundra	ET	No true summer
	Polar ice cap	EF	Perennial ice
H—Highland	Highland	H	Highland

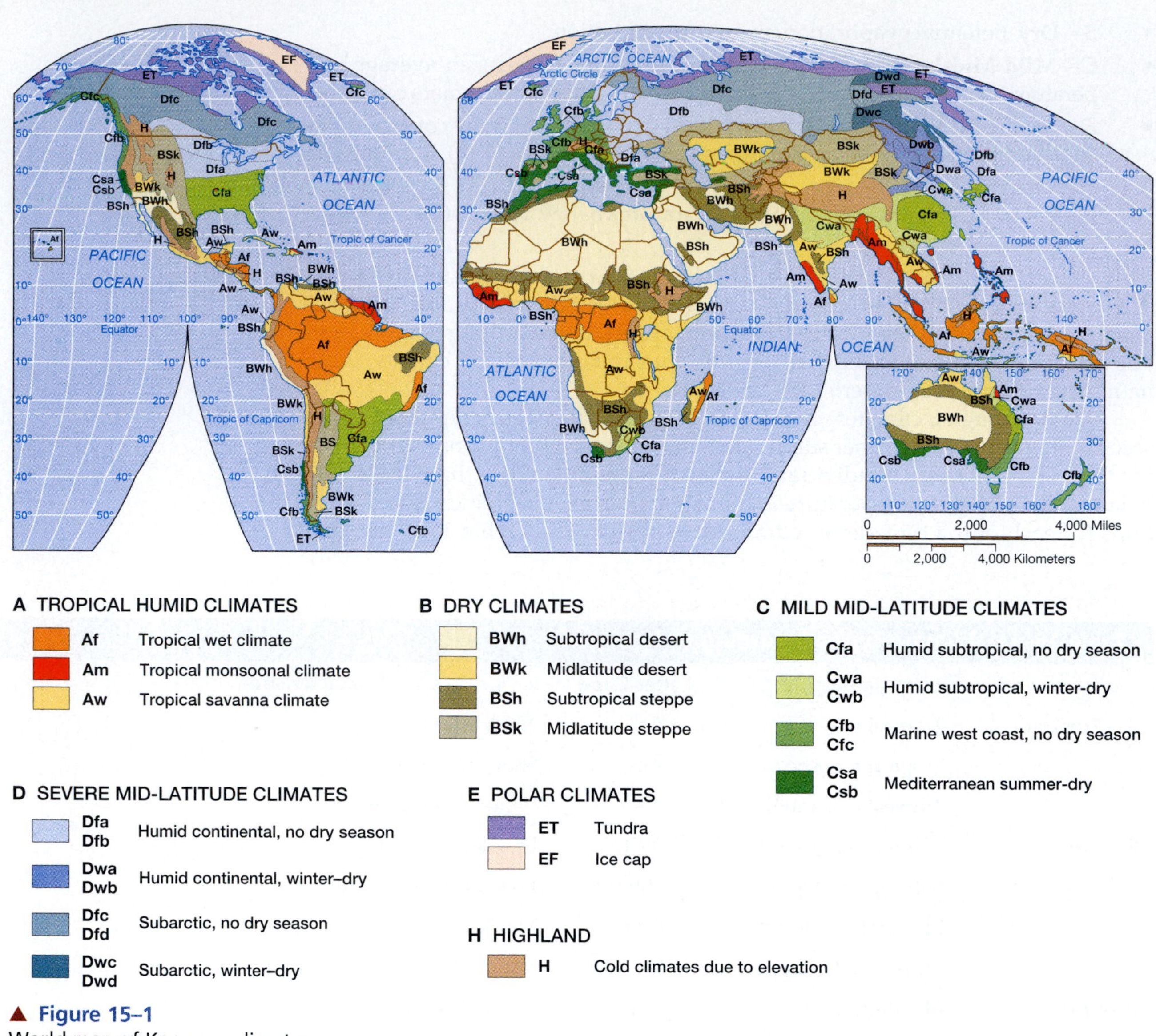

▲ **Figure 15–1**
World map of Koeppen climates.

distributions are depicted in Figure 15–1. While this system has been taught to countless numbers of students over the years, there is little insight to be gained by memorizing the exact temperature and precipitation values that define the climate boundaries. Thus, the critical values are omitted from the table.

For the A climates, the second letters *f*, *m*, and *w* indicate if and when a dry season occurs. **Af** climates have no dry season at all. **Am** climates are the monsoonal climates in which a short dry season is normally experienced while the rest of the year is rainy. **Aw** climates have a distinct dry season, usually coinciding with the seasonal presence of the subtropical high pressure of the Hadley circulation. The dry climates are divided into two classes: the true deserts (BW) and the semi-deserts (BS). The second letter of the C and D climates signifies the timing of the dry season. The letter *f* indicates no dry season at all (as with the A climates), while *s* and *w* represent dry summers and winters, respectively. The second letter of the E climates (capitalized) distinguishes polar tundra (ET) regions from areas covered by glaciers (EF). The third letter of each climate represents the temperature regime. As you can see, the meaning of the third symbol varies among the major groups.

Tropical Climates

The name of this climatic group could not be more straightforward or accurate. Tropical climates exist almost entirely between the Tropics of Cancer and Capricorn. The tropical group consists of three climates, each of which is warm year-round, with only minor—and in some cases, minimal—variation in temperature throughout the year. The three climates are distinguished by their different degrees of precipitation seasonality. The **tropical wet climate** has significant rainfall every month of the year, the **tropical wet and dry climate** has a pronounced dry season, and the **monsoonal climate** undergoes relative dryness for 1 to 3 months but receives sufficient moisture that vegetation need not be adapted to seasonal drought. All three tropical climates are dominated by the seasonal movement of the Hadley cells, described in Chapter 8.

Tropical Wet (Af)

The three largest tropical wet climates are found in the Amazon Basin of South America, over western equatorial Africa, and on the islands of the East Indies (see Figure 15–1). As the map shows, the majority of these locations exist within about 10° on either side of the equator, though some are found as far poleward as 20°. Tropical wet climates have no dry period because their position near the equator puts them under the constant influence of the intertropical convergence zone. For this reason, precipitation is almost always convectional, with strong solar heating of the surface triggering brief but heavy thundershowers in the mid- to late afternoon. In the more poleward areas in which Af climates occur, rain often results from orographic uplift of the predominant trade winds. The tropical wet climate of the Atlantic coast of Central America serves as an excellent example of this phenomenon.

Figure 15–2 presents two **climographs** (a climograph depicts monthly mean temperatures and precipitation, with line and bar graphs plotted simultaneously) for Singapore and for Belém, Brazil, typical tropical wet climate stations. Notice that the rainfall is distributed nearly uniformly throughout the year for most Af locations (although Belém has greater seasonality than Singapore), and that all months average at least 22 cm (5 in.) of precipitation. Even more striking is the uniformity of average monthly temperatures, which vary in these examples by only about 2 °C (4 °F).

While temperatures are often high throughout the year, these climates are not among the hottest on Earth. The ever-present moisture availability at the surface

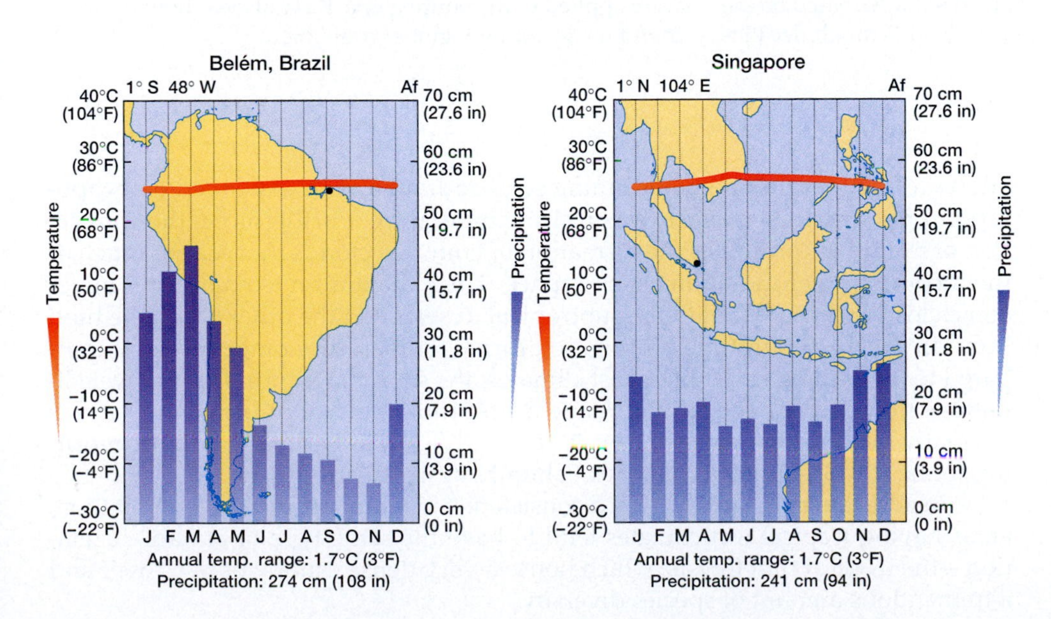

◀ **Figure 15–2**
Climographs for Belém, Brazil, and Singapore, representative of tropical wet climates. The bars plot the monthly mean precipitation (scaled on the right vertical axis). The lines represent mean monthly temperature (scaled on the left).

15–1 Special Interest

Different Climate Classification Schemes for Different Purposes

The Koeppen system is undoubtedly the most widely used for distinguishing and mapping the climates of the world, in large part because it is based on easily obtainable data and attempts to have boundaries that coincide with observable vegetation borders. But there really is nothing magical about it—it is simply one of many schemes that have been developed over the years. Like all the others, it is a human construct, created with certain goals in mind. It follows, therefore, that the climates that emerge from Koeppen's classification rules are also human constructs, rather than objective entities waiting to be discovered by Koeppen and his followers. In other words, although the mechanics of climatic classification involve the development and use of objective rules, in a larger sense climate classification is a decidedly subjective process. What constitutes a "good" classification depends very much on judgment and purpose. By no means, then, should one take Koeppen's scheme as "correct" or even "best." In fact, even its admirers readily admit to several shortcomings.

One of the most important shortcomings of the system is that it is based on vegetation boundaries that have been associated with monthly values of mean temperature and precipitation. This is problematic because these two variables alone do not directly determine the geographic limits of natural vegetation. A superior system would be based on the factors that play a more direct role in determining the geographic limits of vegetation, the most prominent of which are precipitation and potential evapotranspiration.

In tandem, the opposing effects of evapotranspiration and precipitation determine the *water balance*. Wherever evapotranspiration exceeds precipitation, the amount of moisture in the soil is reduced, as its loss is not offset by soil moisture inputs. When precipitation exceeds evapotranspiration, the soil moisture is replenished until it reaches the maximum amount of water that can be retained by the soil against the force of gravity (field capacity). Typical plots of monthly water budgets reflecting these inputs and outputs are shown in Figure 1.

Thornthwaite's classification system, based on the principle of the water balance, evolved through decades of work that culminated in its final form in 1955. The system uses four criteria for delimiting climate regions. The first criterion is a *moisture index* that compares the amount of average precipitation each month to the potential evapotranspiration. The latter value derives from a formula using mean temperatures and the monthly values of average period of daylight (a function of latitude) for each station to establish a monthly moisture index. These monthly values are then summed to produce an annual moisture index, the value of which distinguishes arid, semiarid, subhumid, humid, and perhumid climates. These divisions are based on arbitrary percentage changes in the moisture index (20 percent changes for the humid climates, 33 percent changes for the dry climates), not from associations with plants or other nonclimatic phenomena.

The second criterion is the *thermal efficiency* of a location, or the total amount of potential evapotranspiration. The remaining two criteria are based on the seasonality of precipitation and potential evapotranspiration. When combined, the four criteria create a more physically based climate classification scheme than that of Koeppen.

So why hasn't the Thornthwaite system supplanted Koeppen's as the most popular? In part because of its greater complexity. Compared to Koeppen's system, the water-balance computations needed by Thornthwaite's method are laborious, and the resulting regions are therefore quite removed from the underlying climatic data. Thus, the pattern of climates that emerge is harder to interpret in terms of large-scale processes. Also, although the basic concept of potential evapotranspiration is widely accepted, the method developed by Thornthwaite does not follow from physical principles, but instead relies on data collected mainly in the eastern United States. The data were used to construct empirical (observation-based) equations for potential evapotranspiration. Unfortunately, Thornthwaite did not publish details regarding how the equations were developed, and there are questions as to how well they work in other areas. Thus, if Thornthwaite's method were applied to the entire globe, it is not clear how meaningful the resulting regions would be.

allows a large portion of the incoming solar radiation to be expended on evaporation rather than increasing the surface temperature. Furthermore, the convection of humid air promotes the formation of cumulus clouds that scatter much of the incoming solar radiation back to space. Thus, maximum temperatures never come close to those found in the subtropical deserts. On the other hand, the high humidity retards nighttime cooling, so diurnal temperature ranges are low compared to drier climates. Unlike most climates, the diurnal range in tropical wet climates often exceeds the annual range. Minimum and maximum temperatures normally range from about the low 20s Celsius (low 70s Fahrenheit) in the morning to the low 30s Celsius (high 80s Fahrenheit) in the afternoon.

In addition to the unsurpassed consistency in temperature and precipitation, areas having tropical wet climates tend to have the same type of natural vegetation—the tropical rain forests, which house a very dense canopy of tree cover and a tremendous amount of species diversity.

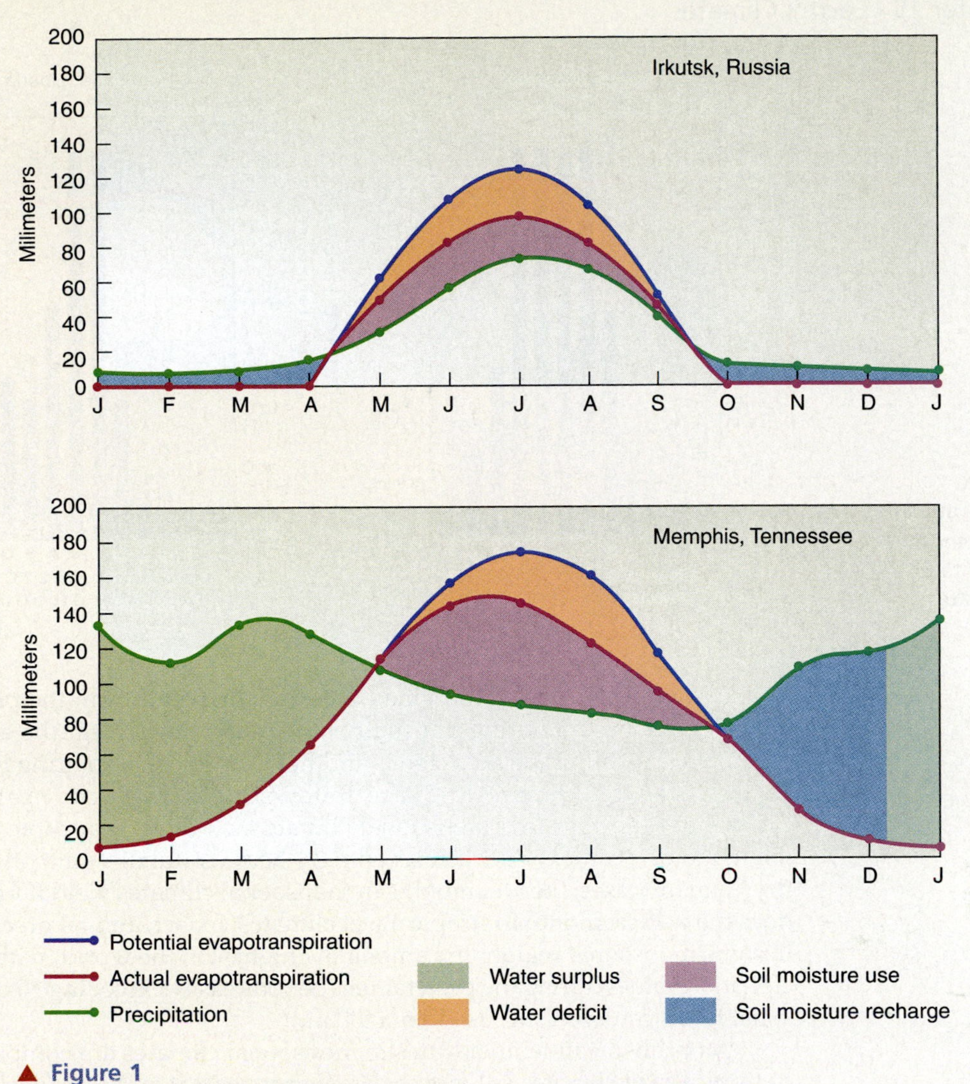

▲ **Figure 1**
Water balance diagrams for Irkutsk, Russia, and Memphis, Tennessee.

Of course, the Koeppen and Thornthwaite systems are not the only systems that have been devised. Most of the others are designed for more specific applications than either of these two. Some, for example, have identified regions of human comfort, whereas others consider the dominance of different types of air masses. More sophisticated ones have used energy budget considerations. Regardless of the premises on which they are based, each has its own set of advantages and disadvantages.

Monsoonal (Am)

Monsoon

The monsoonal climate can be thought of as a transition between the tropical wet to the tropical wet and dry climates. Monsoonal climates usually occur along tropical, coastal areas subjected to predominant onshore winds that supply warm, moist air to the region throughout most of the year. Such areas are found along northeastern South America, southwest India, near the eastern Bay of Bengal, and in the Philippines.* These areas do not extend nearly as far inland as the tropical wet climates, because their existence depends largely on the effect of speed convergence that occurs as offshore air reaches the coast. Rainfall in these climates is also enhanced by orographic uplift. Thus, localized convergence from surface heating is much less a factor in causing precipitation than it is in the tropical wet climates. During the low

*Note that the designation *monsoonal climate* is not synonymous with locations subject to the reversal of the winds discussed in Chapter 8.

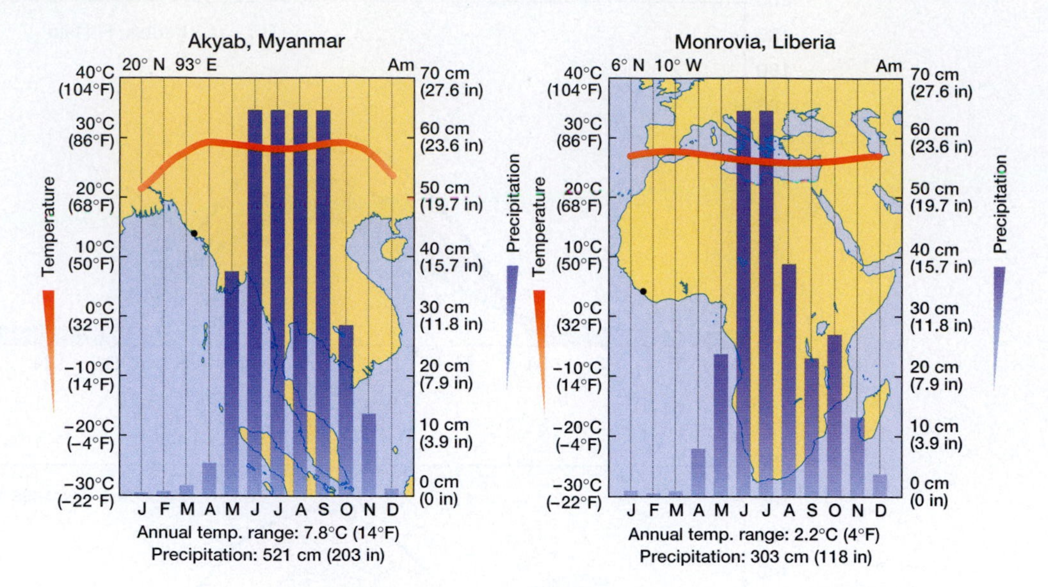

▶ **Figure 15–3**
Climographs for Akyab, Myanmar, and Monrovia, Liberia, representative of monsoonal climates.

sun season, some precipitation may occasionally result from the passage of midlatitude cyclones migrating unusually far equatorward. Near the end of summer and into early fall, tropical cyclones and hurricanes can also bring heavy deluges.

As shown in Figure 15–3, precipitation does not occur nearly as steadily throughout the year in a monsoonal climate as it does in a tropical wet climate. Some months can experience exceedingly heavy rainfall while others are nearly dry. In many cases, the wet months in monsoonal climates yield far more rain than does the wettest month for tropical wet climates. In fact, annual precipitation totals in some monsoonal regions are among the highest in the world, with monthly precipitation values during the peak rainfall periods easily exceeding 80 cm (33 in.). Seasonal totals can even surpass 10 m (400 in.)!

Monthly mean temperatures in monsoonal climates are comparable to those of tropical wet climates; all months are warm and there is little variation between months. Whatever variation in temperature there is appears to be associated with the timing of precipitation. Typically, the warmest months occur just prior to the onset of the main precipitation season. At this time there is relatively little cloud cover, which allows a greater amount of solar radiation to reach the surface.

Despite the presence of a brief dry season, monsoonal climates usually support dense forests. In these environments, the soil maintains sufficient moisture to maintain the lush vegetation even in the absence of heavy rain for part of the year. In other words, the annual total precipitation is large enough that plants do not experience pronounced moisture stress during the dry season; therefore vegetation generally does not require adaptation to drought. Thus, the Am climate is said to have a "non-compensated" dry season, unlike the tropical wet and dry climate. While not as luxuriant or as abundant in species diversity as the tropical wet environments, these locales contain much more living matter than those of the drier, tropical wet and dry climates.

Tropical Wet and Dry (Aw)

Tropical wet and dry climates often occur along the poleward margins of the Tropics and border dry climates on one side and tropical wet climates on the other. They are most extensive in South and Central America and southern Africa. Because they are farther from the equator, they undergo much greater seasonality in precipitation and temperature than do the tropical wet and the monsoonal climates.

As with the other two climates of the humid Tropics, tropical wet and dry climates owe their existence largely to the Hadley cell. During the high sun season, the intertropical convergence zone favors the formation of afternoon thundershowers. As the position of the overhead sun shifts to the opposite hemisphere,

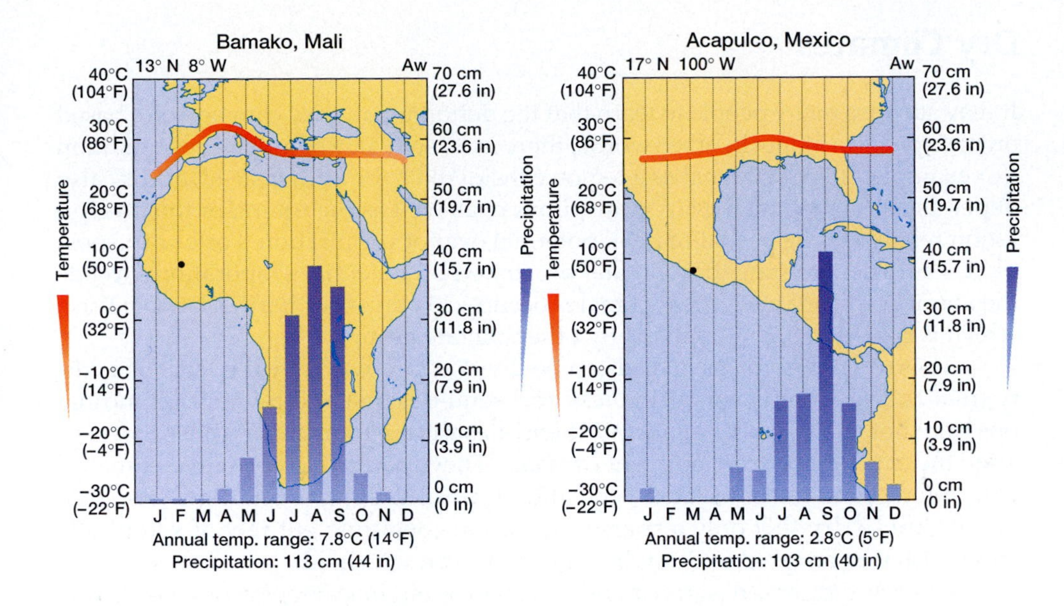

Figure 15–4 Climographs for Bamako, Mali, and Acapulco, Mexico, representative of tropical wet and dry climates.

however, the subtropical high arrives to bring descending air and the resultant lack of precipitation. These periods of dryness are more pronounced and longer lasting than those of the monsoonal climate because their distance farther from the equator puts them closer to the mean position of the subtropical high.

Localized convection by solar heating within the ITCZ is not the only process that brings precipitation to tropical wet and dry climates. Tropical depressions can bring widescale precipitation. Along coastal areas, occasional tropical storms and hurricanes can increase average accumulations. Figure 15–4 illustrates the seasonality of temperature and precipitation for typical tropical wet and dry climates. In Acapulco, for example, each of the months between May and October receive an average of at least 12 cm (5 in.) of rain. September is by far the wettest month, with an average of about 36 cm (15 in.) of rain. This is largely due to the occasional passage of tropical storms and hurricanes that can dump huge amounts of precipitation. Although most years go by without the passage of these storms, their occasional occurrence increases the monthly average. During the fall, the monthly precipitation decreases, until the dry months of February–April. On an annual basis, these climates receive less precipitation than do either the tropical wet or the monsoonal climates.

Unlike the other two topical humid climates, the tropical wet and dry undergo considerable year-to-year variability. Thus, drought episodes can reduce even further the amount of precipitation received in the dry season—often with fatal consequences, as we have witnessed in the Sahel of Africa (see Chapter 8). Likewise, unusually wet rainy seasons can lead to severe flooding and erosion.

Within the year, monthly mean temperatures exhibit more variability than is found in the other tropical climates, but the variability is nonetheless low compared to most others. The annual temperature range is usually between about 3 and 10 °C (5 and 18 °F). Diurnal variations are likewise greater in these parts of the tropical humid environment than in the wetter regions. This is especially true during the dry season, when the absence of clouds facilitates greater daytime heating and nighttime cooling, and daily lows and highs might range between 15 and 30 °C (59 and 86 °F). During the rainy season, the combination of high humidity and cloud cover reduces the diurnal ranges to values similar to those of the tropical wet regions (about 10 °C, or 18 °F).

Tropical wet and dry climates are associated with a natural vegetation type unique among the tropical regions—the **savanna**. This vegetation consists mainly of grasses interspersed with widely separated trees or clumps of trees. While it is tempting to attribute to the lack of forest to the presence of the dry season, many ecologists doubt the causality of this relationship. Instead they believe that the vegetation complex results from numerous factors, including recurrent fire, waterlogged soils, and the development of hard layers within the soil.

Dry Climates

It may surprise many people to learn that the definition of a dry climate is not based on precipitation alone. In other words, there is no set value of annual precipitation (for example, 10 cm) that makes a region considered to be arid. Instead, aridity also depends on the potential evapotranspiration, as well as the timing of the precipitation relative to the period of peak potential evapotranspiration. For our purposes, it is sufficient to state that climates are dry when potential evaporation exceeds the annual precipitation. These climates occupy 30 percent of Earth's land surface, which is considerably more than any other climate group.

The dry regions of the world can be divided two ways: by the level of aridity (that is, true deserts versus the less arid semi-deserts) and by their latitudinal position (hot versus cooler dry areas). **Semi-deserts** are transitional zones that separate the true deserts from adjacent climates. They are also called **steppe** climates, with reference to the associated vegetation type consisting of short grasses. True deserts are so dry that only a sparse vegetation consisting entirely of xerophytic species (that is, adapted to drought conditions) can take hold.

Deserts and semi-deserts can be classified as either *subtropical* or *mid-latitude.* Subtropical dry regions extend across wide expanses of land between the latitudes of about 10° to 30° in either hemisphere and result from large-scale sinking air motions during most of the year. Notable examples include the desert of southern California–Arizona–Baja California, the Sahara Desert of North Africa and the Arabian Peninsula, and most of the Australian interior.

Mid-latitude deserts and semi-deserts usually occur to the east of major topographic barriers that create a strong rainshadow or over interior continental regions well removed from moisture sources. They are found over large portions of the western United States and interior Asia.

The two-tiered system of categorization yields four types of dry climates: **subtropical desert, subtropical steppe, mid-latitude desert**, and **mid-latitude steppe**.

Subtropical Deserts (BWh)

As shown in Figure 15–1, the most extensive areas of desert exist in the subtropical regions, particularly within the western portions of the continents. The most important factor in the formation of the subtropical deserts is the subsidence associated with the Hadley circulation. As we saw in Chapter 8, the subtropical highs of the Hadley circulation do not appear as continuous bands at the surface, but rather as semi-permanent cells. Although the Hawaiian and Bermuda–Azores high-pressure systems at the surface contract and weaken during the winter, subsidence nonetheless dominates within the middle troposphere. This causes stable conditions to exist in the middle troposphere, which restricts uplift and inhibits precipitation, creating the major subtropical deserts.

Some deserts occur in subtropical regions as narrow strips along the west coast of continents, adjacent to cold ocean currents. The Atacama Desert along the west coast of Chile has the lowest average precipitation on Earth and provides an excellent example of this phenomenon. As air flows out of the subtropical high pressure system in the eastern part of the South Pacific, it flows over the cold Peru current, and the lower portion of the atmosphere is cooled. This cooling can bring the air temperature down to the dew point so that the air becomes damp and foggy. It also lowers the environmental lapse rate and causes the air to be extremely stable. Though the air advected off the coast is damp, stable conditions suppress uplift so that several years can go by without any precipitation at all in some areas. But even under these difficult conditions, humans have a way of adapting. In some parts of South America and Baja California, villagers in remote areas use a simple but ingenious system for capturing drinking water from the fog: Nylon mesh screens, each about one square meter in area, are extended between supporting posts so that passing fog condenses onto their surfaces. As the fog accumulates on

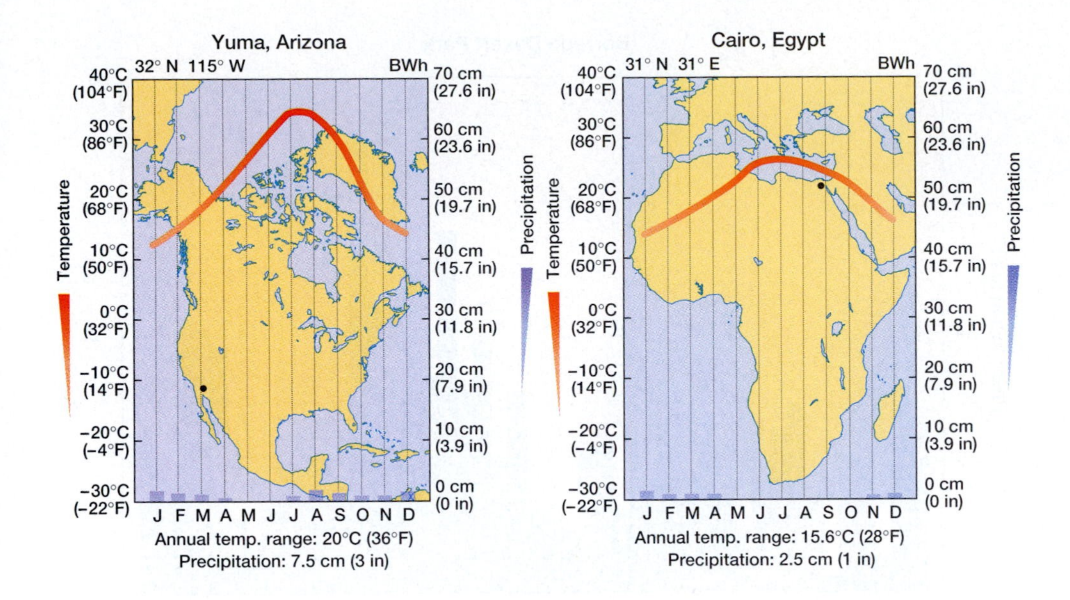

Figure 15–5 Climographs for Yuma, Arizona, and Cairo, Egypt, representative of subtropical deserts.

the mesh screens, the surplus drips down into collecting systems and provides enough drinking water to maintain small communities.

Over many subtropical deserts, the precipitation that does occur often comes in the form of localized showers from summertime convectional activity. This is not the case for all subtropical deserts, however, as illustrated in Figure 15–5. Yuma, Arizona, and Cairo, Egypt, are both located at about the same latitude, in the midst of subtropical deserts. While both receive scant precipitation over the course of the year, Cairo's precipitation occurs mainly in the winter while Yuma's is about equally divided between the late summer and winter months. As is the case over much of southern Arizona, August usually marks the peak of what is locally known as the *Arizona monsoon*. While having little similarity to the wet season of the Asian monsoon, the Arizona version does undergo a shift in the airflow that brings damp air into the area in late summer. This influx of moisture is subject to strong surface heating that can lift the air sufficiently to trigger isolated thunderstorms. While these thunderstorms can be intense and cause flash flooding, they are neither frequent nor strong enough to make the area anything other than the true desert that it is. The winter precipitation at Yuma results from the passage of mid-latitude cyclones. While these systems can bring precipitation across a wide swath of the country, often in the form of heavy rainshowers or snowstorms, over the desert the moisture supply is usually too low to allow much precipitation to fall. This is because Pacific moisture is blocked by the western mountains, and Gulf of Mexico air does not usually flow this far westward.

Daytime summer temperatures in subtropical deserts can be extremely high. In fact, the hottest locations in the world are all found in these regions. During the summer, the combination of low humidities, high sun, and clear skies allows high inputs of solar radiation to be absorbed at the surface. Furthermore, the lack of soil moisture (except after recent rainshowers) causes the ground temperatures to become extremely high as little heat is expended in the evaporation of water. Thus, it is not uncommon for daytime temperatures to reach as high as 45 °C (113 °F) or higher. After the Sun sets, the clear skies and low humidities that led to rapid heating also allow the air to cool considerably. As a result, diurnal temperature ranges can be very large.

The same applies to the annual temperature ranges. At Baghdad, Iraq, for example, the mean monthly temperature in August is 35 °C (95 °F), which is 25 °C (45 °F) higher than the January mean of 10 °C (50 °F). This type of temperature range is greater than that found in any other climate of the Tropics or subtropics. As we will see, however, mid-latitude deserts commonly have even greater temperature ranges.

There is a widely heard axiom in regional climatology that areas of low annual precipitation also have the greatest amount of interannual variability. This

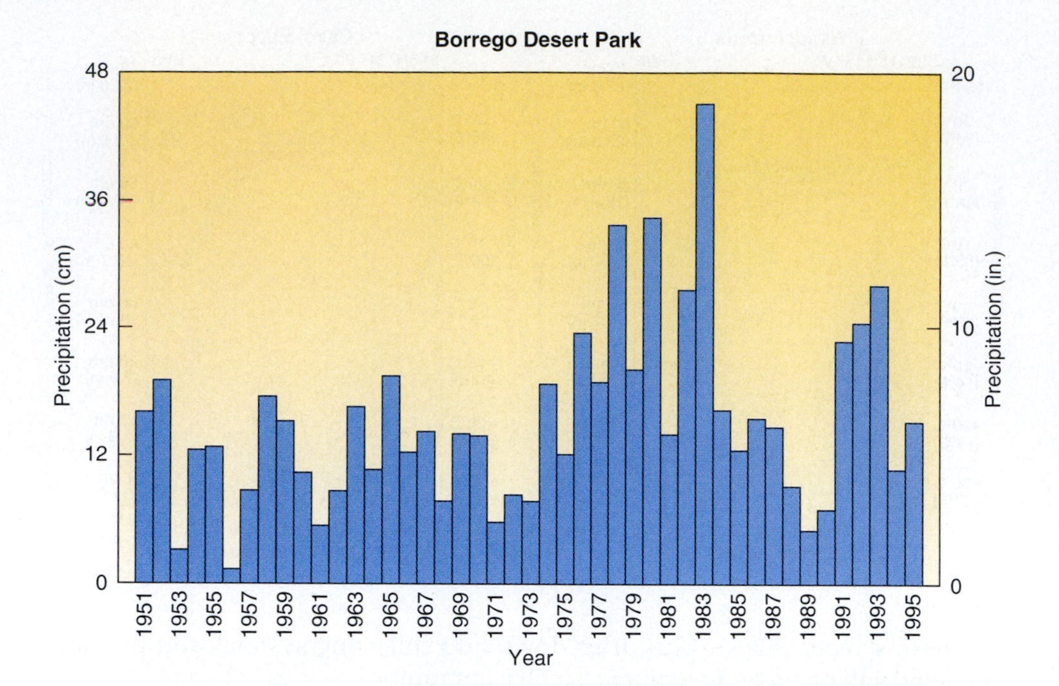

▶ **Figure 15–6**
Annual precipitation at Borrego Springs, California, 1951–1995. Dry climates such as this one normally have a wide year-to-year variability in precipitation.

certainly applies to subtropical deserts, where many years can go by with only minimal rainfall, only to be followed by a season having numerous rainshowers, delivering several years' worth of precipitation. Thus, the annual average is a poor indicator of how much rain is likely to fall in given year. Figure 15–6 plots annual rainfall at Borrego Springs, California, over a 45-year period of record as an example.

Subtropical Steppe (BSh)

All the conditions that distinguish a subtropical desert also apply to subtropical steppe climates—though to a lesser degree. Like their more extreme desert counterparts, subtropical steppes are marked by aridity, high year-to-year variations in precipitation, extreme summer temperatures, and large annual and daily temperature ranges. It is therefore not surprising that subtropical steppe climates commonly border the subtropical deserts. And though they are transitional between deserts and non-arid regions in terms of their climatological characteristics, they are not necessarily narrow buffer zones. Much of the southwestern United States and northern Mexico, for example, is occupied by subtropical steppe.

Cloncurry, Australia, and Monterrey, Mexico (Figure 15–7), illustrate the greater precipitation totals, somewhat cooler conditions, and lower annual temperature ranges associated with these climates. These two examples also reveal a distinct seasonality in the precipitation regime typical of subtropical steppes located on the equatorward side of deserts. In these regions, precipitation occurs more often during the summer months than during the winter, as a result of localized convection and tropical disturbances. In contrast, steppe regions on the poleward side of subtropical deserts experience most precipitation in the winter in response to the passage of mid-latitude cyclones.

Mid-latitude Deserts (BWk)

Mid-latitude deserts result from extreme continentality. Such regions occur deep within continental interiors or downwind of orographic barriers that cut off the supply of moisture from the ocean. The greatest expanse of mid-latitude desert occurs in Asia—which is not surprising considering the immense size of that continent. The two major areas of mid-latitude desert in Asia are found just east of the Caspian Sea and north of the Himalayas. Both are far to the east of the Atlantic Ocean, so much of the moisture associated with eastward-moving cyclones is depleted before

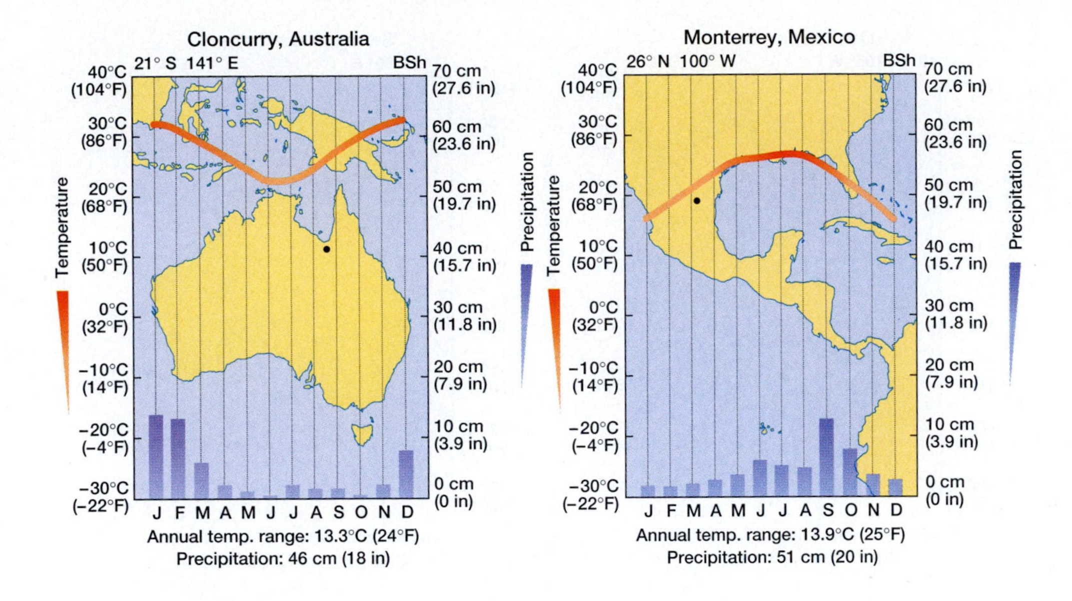

◀ **Figure 15–7** Climographs for Cloncurry, Australia, and Monterrey, Mexico, representative of subtropical steppe climates.

reaching either area. Mountains to the south block the northward flow of moisture out of the Indian Ocean.

The second greatest expanse of mid-latitude desert occurs in the western United States. The mid-latitude desert extends southward and merges with the subtropical desert of the Southwest. Although the mid-latitude desert lies poleward of the subtropical desert, it should not be assumed that it is neither as dry nor as hot in the summer as the desert to the south. In fact, Death Valley, California, one of the hottest and driest places in the world, lies within the mid-latitude desert.

Mid-latitude desert in the Southern Hemisphere is confined to a narrow strip in South America, east of the Andes. A quick look at Figure 15–1 reveals that the mid-latitudes of the Southern Hemisphere are almost completely covered by ocean. Thus, it is rare to see any type of land climate in this region, let alone a mid-latitude desert.

Figure 15–8 shows two typical climographs for mid-latitude deserts. Mid-latitude deserts have a greater range of temperatures, both on a daily and annual basis, than do their subtropical counterparts. While both types of deserts become extremely hot during summer days, mid-latitude deserts have more rapid nighttime and winter cooling. With more precipitation and lower potential evapotranspiration, they are generally more humid than subtropical deserts. Thus, although vegetation is adapted to dry conditions, ground cover is likely to be continuous, without large patches of bare ground common in hotter (subtropical) desert regions.

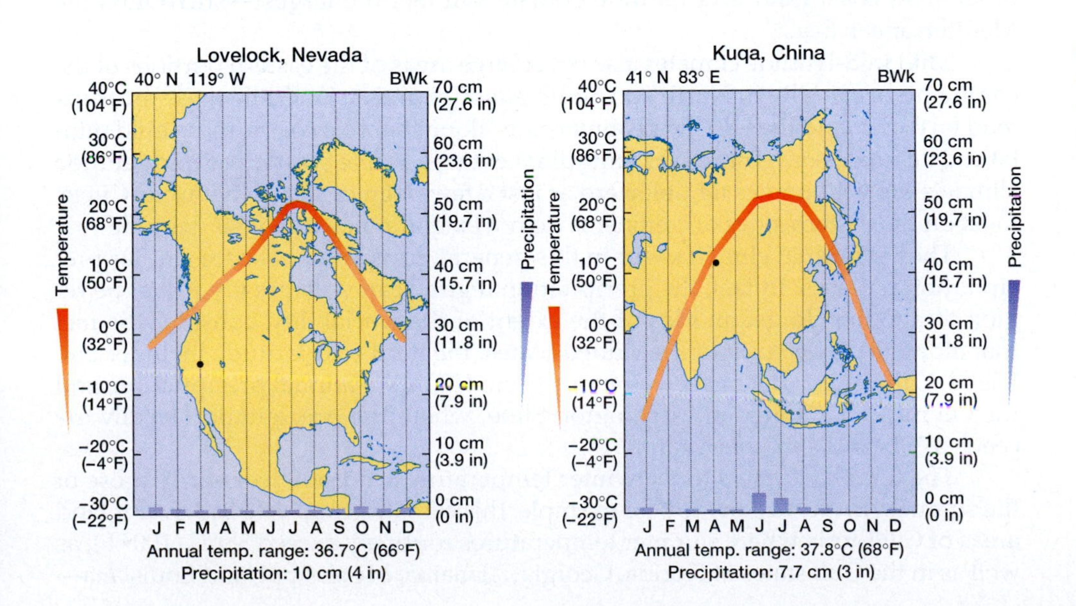

◀ **Figure 15–8** Climographs for Lovelock, Nevada, and Kuqa, China, representative of mid-latitude desert climates.

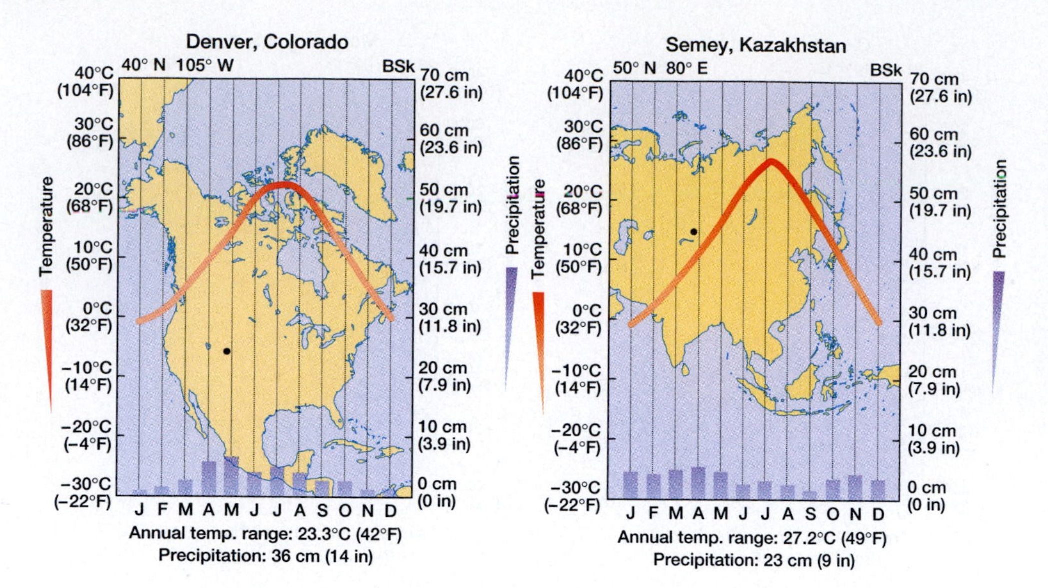

▶ Figure 15–9 Climographs for Denver, Colorado, and Semey, Kazakhstan, representative of mid-latitude steppe climates.

Mid-latitude Steppe (BSk)

The mid-latitude steppe accounts for most of the arid regions of western North America. It flanks the northern portion of the mid-latitude desert and merges with the subtropical steppe to the south. A large swath of steppe also extends from the Great Plains, east of the Rocky Mountains, all the way from northeast Mexico into western Canada. The mid-latitude steppe region of Asia is in many places a fairly narrow strip of land surrounding the true deserts.

Mid-latitude steppes have the same temperature characteristics as the mid-latitude deserts. The primary difference between the two is the greater amount of precipitation in the steppes, which commonly totals about 50 cm (20 in.) annually (Figure 15–9).

Mild Mid-latitude Climates

The mild mid-latitude climates are located in parts of the latitude range between 30° and 60° in either hemisphere. They occur as long, narrow strips of land along the west coasts of North and South America and southern Australia. In North and South America, the eastern border of the climate is delimited by a mountain. Another west coast mild mid-latitude climate—in fact the largest—surrounds the Mediterranean Sea.

Mild mid-latitude climates also cover large areas of the eastern portions of the continents, especially in North and South America and Asia. While these areas extend farther inland than do their counterparts along the west coasts, these mild mid-latitude climates have narrower latitudinal extents. While west coast mid-latitude climates can be found as far poleward as just a few degrees shy of the Arctic Circle, most on the eastern sides of continents do not extend poleward of 40° latitude.

The individual climates within this general group do not share similar precipitation patterns. In fact, the precipitation regime can vary greatly from one region to another. To get an idea of the extent of this variability, consider the fact that along the west coast of the United States, the mild mid-latitude includes San Diego, California, which receives about 25 cm (10 in.) of annual precipitation, and the Olympic rain forest in Washington State, where the precipitation locally exceeds 375 cm (150 in.) per year.

The term *mild* refers to the winter temperatures and not necessarily those of the summer. In North America, for example, this climate group is found over inland areas of California where summer temperatures routinely exceed 38 °C (100 °F), as well as in the Gulf states of Florida, Georgia, Alabama, Mississippi, and Louisiana—

places hardly noted for their mild summers. Thus, in effect, *mild* refers to little or no snowcover.

This climate group is subdivided into three climates, two of which exist along the west coasts of continents and the other on the eastern sides. **Mediterranean** climates can be found along the west coasts between about 25° and 40° latitude. Within about the same range of latitude on the eastern side of continents are the **humid subtropical** climates. The **marine west coast** climates lie adjacent to and poleward of the mediterranean climates.

Mediterranean (Csa, Csb)

Mediterranean climates are the only ones that have a distinct summer dry season and a concentration of precipitation in the winter. Figure 15–10 clearly depicts this pattern for two typical sites, Athens, Greece, and Los Angeles, California. The summer aridity in mediterranean climates is attributable to the presence of semi-permanent subtropical high-pressure systems offshore. For North America, the Hawaiian high-pressure system to the west "blocks" the eastward migration of summertime mid-latitude cyclones and deflects them to the north. This deflection of the storms, coupled with subsidence along the eastern portion of the subtropical high, deprives coastal southern California of the uplift mechanisms necessary for precipitation. During the winter months, the Hawaiian high weakens, shrinks in size, and migrates toward the equator. This opens the way for the passage of Pacific cyclones.

Annual precipitation increases with latitude and with elevation along windward slopes in mediterranean climates. The latitudinal gradient results from the more frequent passage of mid-latitude cyclones at the higher latitudes. At the same time, the mountain slopes induce orographic uplift of the predominantly westerly airflow.

While the climographs in Figure 15–10 reveal the mean precipitation patterns for the two sites, they do not reveal the extreme variability inherent with winter rainfall. Indeed, some winters can be quite dry, while others bring an onslaught of sequential storms that produce major flooding and hillside erosion. This climatic feature has been particularly prominent since the 1970s in southern California, which has experienced several record drought years along with extremely wet winters.

Winter temperatures in mediterranean climates are usually mild, especially right along the coast. Inland temperatures occasionally drop below freezing and sometimes threaten fruit-growing areas with widespread crop damage. On the other hand, precipitation results mainly from the passage of mid-latitude cyclones, which transport relatively warm, maritime polar air. Thus, precipitation in the lower elevations falls almost exclusively as rain and not snow.

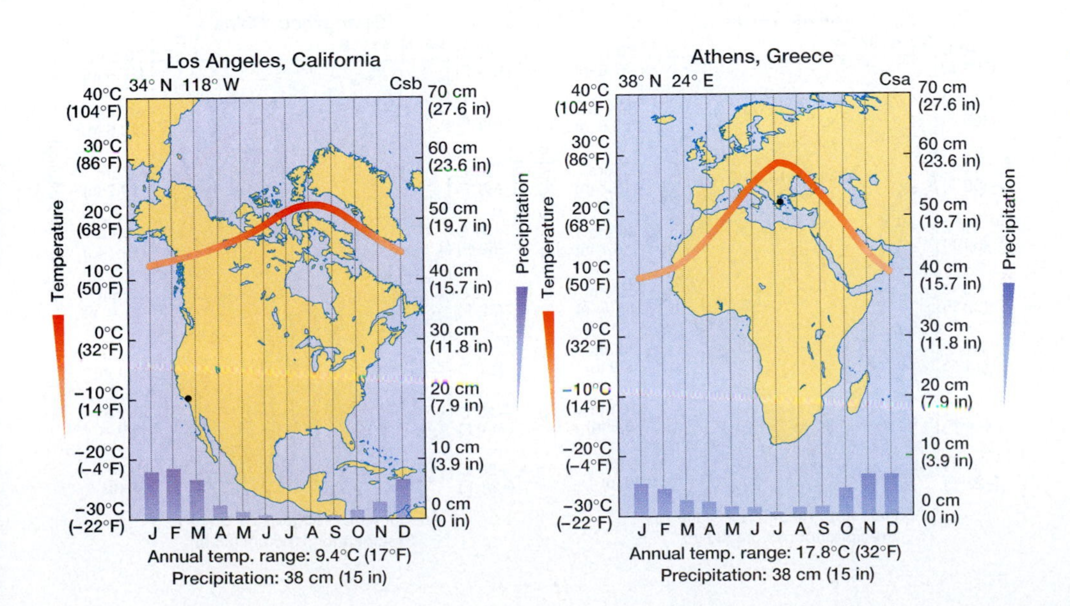

Figure 15–10 Climographs for Los Angeles, California, and Athens, Greece, representative of mediterranean climates.

Summer temperatures range from mild to hot, with daily highs typically decreasing toward the coast and higher latitudes. Central California provides a good example of how variable the summer temperatures can be in mediterranean climates. San Francisco is noted for particularly cool summers, while about 120 km (70 mi) to the northeast, Sacramento experiences hot, dry conditions.

Humid Subtropical (Cfa, Cwa)

Humid subtropical climates occur within the lower middle latitudes of eastern North America, South America, and Asia. Typical temperature and precipitation patterns are shown in Figure 15–11.

Though they exist in the middle latitudes, these climates have a distinct tropical feel during their long summers. Located to the west of large semi-permanent anticyclones and warm ocean currents, the prevailing winds circulate hot, humid air into these climatic zones. Summer daytime temperatures are usually in the lower 30s Celsius (high 80s to low 90s Fahrenheit), and dew points in the mid-20s Celsius (mid-70s Fahrenheit) help retard nighttime cooling. Thus, hot muggy conditions remain throughout the day and night, especially along the equatorward boundaries of the climates. Fortunately, afternoon convectional thundershowers are common in these areas and bring temporary relief from the extreme heat.

Winter temperatures are typically lower than those of mediterranean climates farther to the west because of their greater continentality, and subfreezing temperatures are not uncommon. The occurrence of frost and snow decreases toward the lower latitudes, but even south Florida is not completely immune.

Humid subtropical areas receive abundant precipitation, ranging from about 75 to 250 cm (30 to 100 in.) per year. Over most areas the maximum precipitation is concentrated in the summer, but this generalization does not always hold. Over most of the southeastern United States, for example, summer is the wettest season. But the area extending from east Texas into Tennessee and Kentucky has a winter precipitation maximum. Regardless of which season receives most precipitation, summer is always the season of moisture deficit, because of greater potential evapotranspiration.

Precipitation during the summer is largely convectional in nature and tends to be scattered and brief. Winter precipitation, on the other hand, is usually triggered by the passage of mid-latitude cyclones. Over the coastal areas, tropical storms and hurricanes are capable of bringing extreme rainfall from time to time during the late summer and early fall.

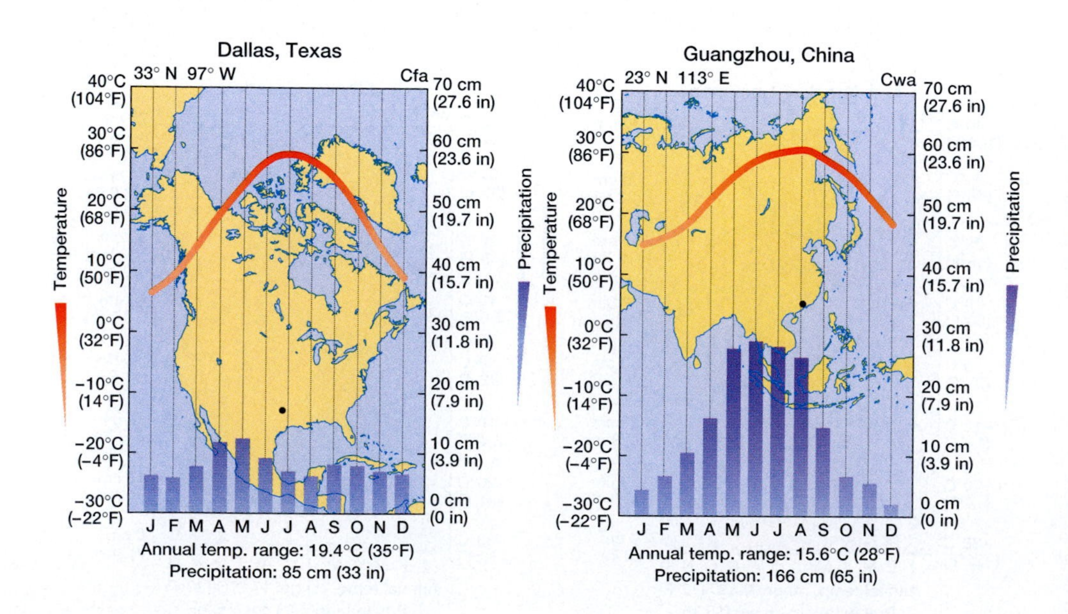

▶ **Figure 15–11** Climographs for Dallas, Texas, and Guangzhou, China, representative of humid subtropical climates.

Marine West Coast (Cfb, Cfc)

Marine west coast climates (Figure 15–12) normally occur poleward of mediterranean climates and, as the name would suggest, along the west coasts of climates. But inspection of Figure 15–1 shows that the latter is not always the case, as large areas having the climate are found along southeastern Australia and southeast Africa. While not truly located on west coasts, those areas are located east of fairly narrow strips of land that don't greatly modify the maritime nature of the predominantly westerly flow. This exposure to air passing over cold ocean currents and their location in the path of eastward migrating mid-latitude cyclones gives these climates their characteristic features.

Both summers and winters are typically mild. Extremely high summer temperatures are certainly not unknown in these climates, but they are the exception rather than the norm. From northern California through Alaska, for example, low and mid-level cloud decks frequently keep daytime temperatures down and often bring drizzle or light rain to coastal regions. Along the lower elevations, the percentage of days experiencing precipitation can be high, but the total amount of rainfall is usually light. On the other hand, the Coast Ranges parallel the Pacific coastline and create a major barrier to the westerly flow. This creates a strong orographic effect that causes some areas to have extremely heavy precipitation amounts that rival or exceed those of the Tropics. The Olympic Peninsula of Washington State is an excellent example of this phenomenon, where extremely heavy rainfalls are sufficient to support a lush, mid-latitude rain forest.

Just as maritime conditions moderate temperatures during the summer, they also allow for mild winter conditions at surprisingly high latitudes. Even as far north as Sitka, Alaska (57° N), the coldest month of the year has a mean temperature above the freezing point of water. At many lower latitude locations, low-elevation snowfall is a rarity; when it does occur, it usually melts away within a short time. In Europe, where the marine west coast climate extends farther inland than it does in North America, snow is more frequent and remains on the ground for a longer period of time.

Given the fact that both summer and winter temperatures are mild, it follows that these climates have low annual temperature ranges. This contrasts sharply with the situation for climates in the eastern portion of continents at the same range of latitudes. These climates have moderately high summer temperatures, but when combined with the extreme cold of winter, they yield extremely high temperature ranges of the severe mid-latitude group of climates.

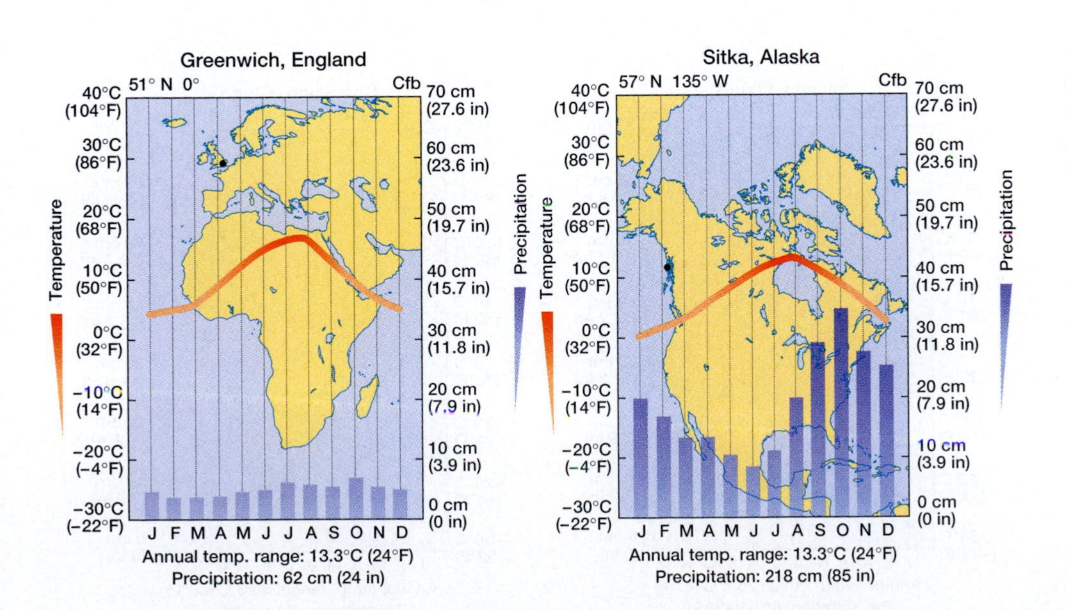

Figure 15–12 Climographs for Greenwich, England, and Sitka, Alaska, representative of marine west coast climates.

Severe Mid-latitude Climates

The severe mid-latitude climate group includes two climates, humid continental and subarctic, both of which are marked by very cold winters. These climates require large continental areas within the high-middle latitudes—between about 40° and 70°. Thus, they are restricted to Europe, Asia, and North America and are not found at all in the Southern Hemisphere. As expected from climates resulting from strong continentality, they exhibit large annual temperature ranges.

Both of the severe mid-latitude climates receive precipitation throughout the year and have no true dry season. In many locations, there is greater precipitation in the summer than winter. Precipitation during the summer can result from local convection or by the passage of mid-latitude cyclones; winter precipitation (mostly as snow) results almost entirely from cyclonic activity.

Humid Continental (Dfa, Dfb, Dwa, Dwb)

A huge segment of the population of the United States, Canada, eastern Europe, and Asia live in **humid continental** climates (Figure 15–13), including the inhabitants of New York, Chicago, Toronto, Montreal, Moscow, Warsaw, and Stockholm. This is the more temperate of the two severe mid-latitude climates, normally found between about 40° N and 55° N in the eastern parts of continents. Summers are warm and often hot. New York City, for example, has an average high temperature in August of 29 °C (84 °F). But winter is another matter altogether, with a mean February maximum temperature of 4 °C (40 °F). It should be noted that New York City is located near the southern boundary of the climate and on the coast, so temperatures are milder than those of most other locations with a humid continental climate.

Mean annual precipitation in these climates usually ranges between 50 and 100 cm (20 to 40 in.). Over the United States and southern Canada, there is a conspicuous decrease in precipitation with increasing latitude and distance from the Atlantic shoreline, reflecting a reduced moisture content in the atmosphere.

Subarctic (Dfc, Dfd, Dwc, Dwd)

Worm Climates

Subarctic climates occupy the northernmost extent of the severe mid-latitude regions, with more than half of the area of Alaska and Canada having this type of climate. In North America, the coniferous forest that dominates the region is referred to as the **boreal forest**; in Asia, it goes by the name **taiga**. Summer temperatures are somewhat lower than those of the adjacent humid continental regions,

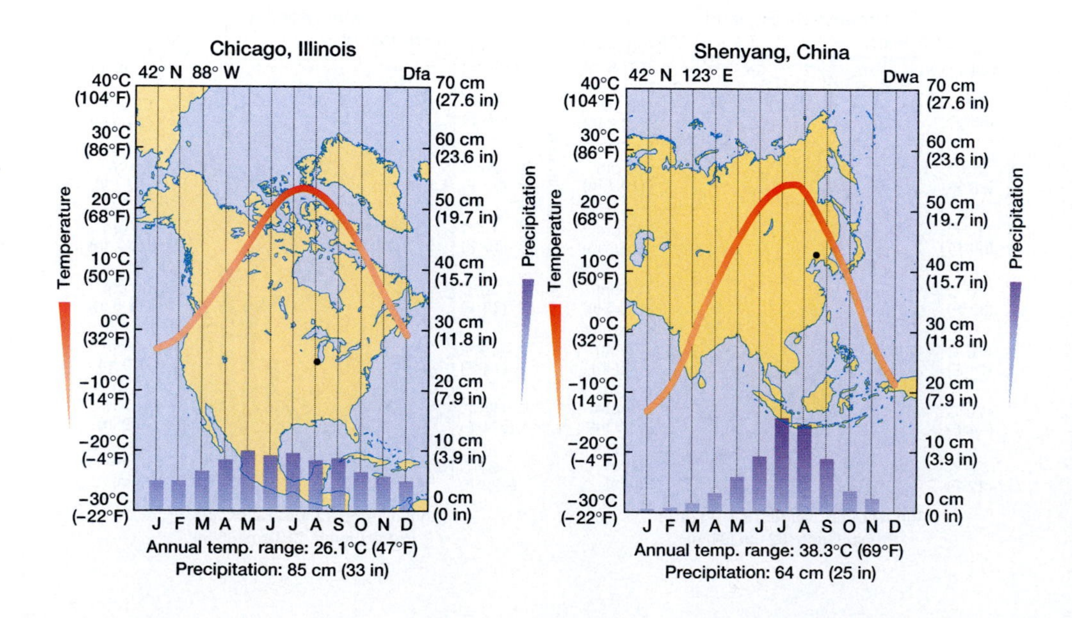

Figure 15–13 Climographs for Chicago, Illinois, and Shenyang, China, representative of humid continental climates.

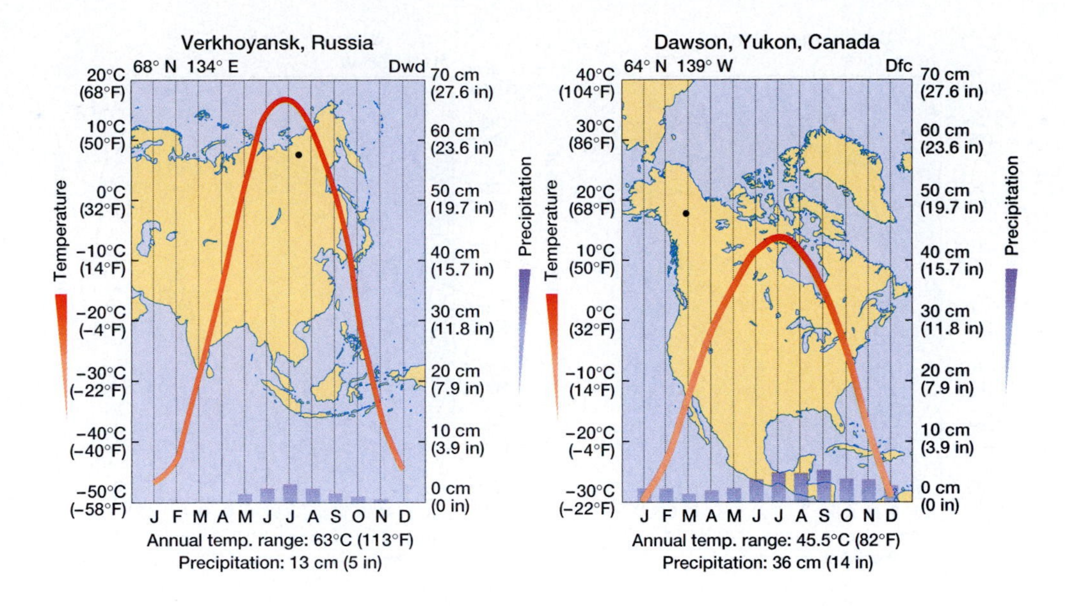

Figure 15–14 Climographs for Verkhoyansk, Russia, and Dawson, Yukon, Canada, representative of subarctic climates.

but the major difference occurs in winter, when mean monthly temperatures can be extremely low. At many locations, the monthly mean temperature can remain below the freezing level for up to seven months. Thus, winter is long and separated from the brief summer only by a short-lived autumn and spring. Figure 15–14 highlights the very large annual temperature ranges that result from the mild summers and severe winters.

Typically, precipitation is greater in the summer than winter, mostly because of the more poleward displacement of mid-latitude cyclone tracks in summer. Nonetheless, annual precipitation is usually low, ranging from about 12 to 50 cm (5 to 20 in.).

Polar Climates

Polar climates exist in the highest latitudes—typically areas poleward of about 70°. Such regions occur in the Northern Hemisphere, across northern Canada, Alaska, Asia, and coastal Greenland. In the Southern Hemisphere they are almost entirely confined to the continent of Antarctica. As we mentioned earlier, the climate boundaries of the Koeppen system were set to coincide with boundaries of differing natural vegetation types. In this case, polar climates begin at the high latitude boundaries of the subarctic climates, where vast expanses of coniferous forest give way to a generally treeless landscape. The polar climate group consists of two distinct types. The most equatorward and milder of the two is the **tundra**. At the most poleward regions of the globe lie the true ice cap climates.

Tundra (ET)

Tundra climates are named for the associated vegetation type that consists primarily of low-growing mosses, lichens, and flowering plants, with few woody shrubs and trees. The boundary dividing the tundra from the subarctic climate occurs where the mean monthly temperature of the warmest month does not exceed 10 °C (50 °F). In the Northern Hemisphere, this occurs in the general vicinity of 60° N. This climate is almost entirely missing from the Southern Hemisphere, where there is only minimal land coverage at this latitude.

Tundra climates have severe winters in which the Sun rises only briefly each day and never gets very high above the horizon. Under these conditions, radiational cooling at the surface leads to low temperatures and strong stability—both factors that inhibit precipitation. Thus, as shown in Figure 15–15, low winter precipitation

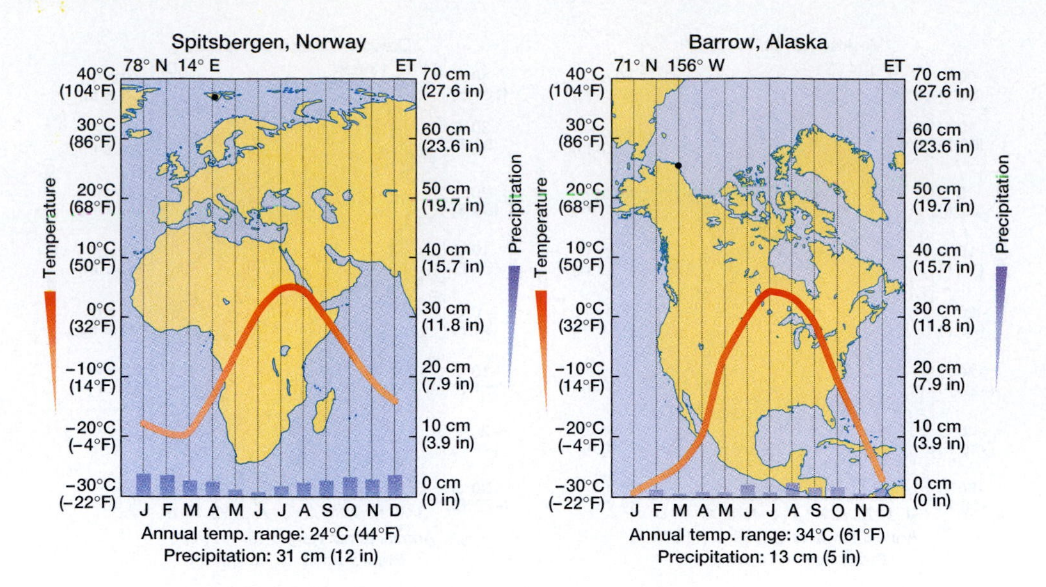

▶ **Figure 15–15** Climographs for Spitsbergen, Norway, and Barrow, Alaska, representative of tundra climates.

is the rule. Surprisingly, perhaps, the winters in the tundra regions are often less severe than those of the adjacent and lower-latitude subarctic climates. This is because the tundra regions of North America, Greenland, and Asia are located nearer to major water bodies and thus have a lesser degree of continentality. (Even when ice-covered, an ocean has a somewhat moderating effect on winter temperatures.)

During the summer, tundra regions have very long periods of daylight, but again the Sun never gets very high above the horizon. As a result, temperatures are normally mild in the mid-afternoon, not very much greater than those of the predawn period. Thus, despite the fact that annual temperature ranges are fairly high, daily temperature ranges are low.

One very conspicuous feature of tundra regions is the existence of **permafrost,** a perennially frozen layer below the surface. During the winter, conditions are so cold that the entire soil is completely frozen, in some places to a depth of several hundred meters. When summer arrives, there is enough warming right at the surface to melt the uppermost soil, and a relatively shallow layer of thawed soil overlies the completely impermeable frozen layer several tens of centimeters below. As a result, any rain or snowmelt that occurs at the surface is unable to permeate deeply into the soil, and the upper part of the surface becomes saturated. This combination of water-logged soil near the surface and a solid ice layer below precludes the establishment of any deeply rooted plants, and only low-growing vegetation can take hold.

Ice Cap (EF)

As the name rightfully suggests, **polar ice cap** areas exist where ice covers the ground throughout the entire year. They are confined to the Greenland interior and most of Antarctica. They exist where the mean temperature of the warmest month does not go above the melting temperature for ice, 0 °C (32 °F). Ice accumulations over these regions can exceed several kilometers, so in addition to being found at high latitudes they also occupy high elevations—a situation perfect for maintaining extraordinarily low winter temperatures (Figure 15–16).

The cold air overlying polar ice caps (also called *continental glaciers*) becomes extremely dense and frequently flows down the margins of the ice under the force of gravity. When funneled through narrow canyons, these katabatic winds (discussed in Chapter 8) can become extremely strong.

Most areas of ice cap receive little precipitation because of the intense cold. On the other hand, all precipitation falls as snow and has the potential to accumulate

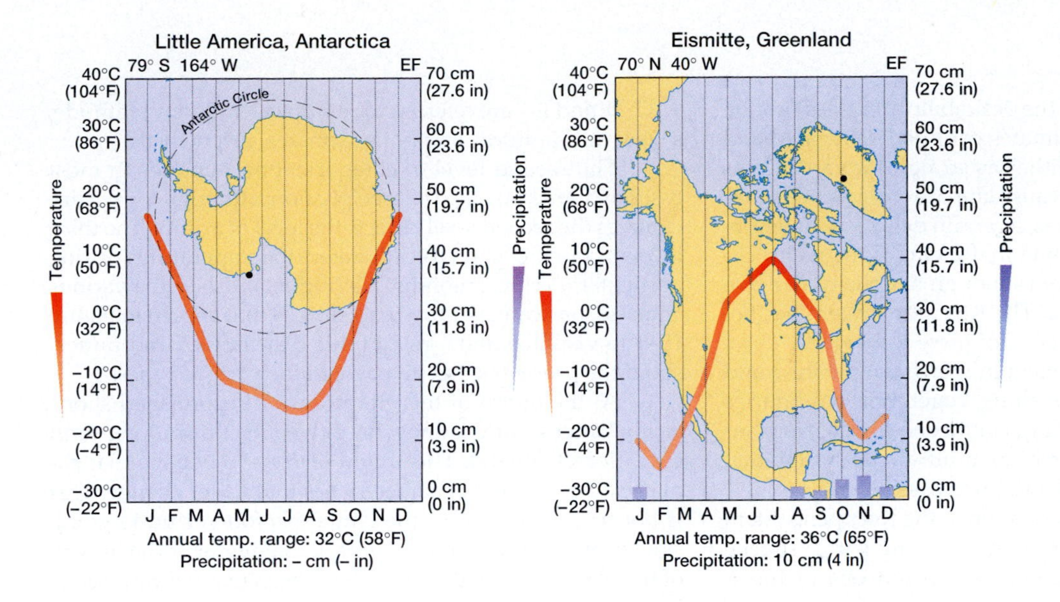

◀ **Figure 15–16** Climographs for Little America, Antarctica, and Eismitte, Greenland, representative of ice cap climates.

onto existing ice. Except along the margins, there is no melt and only a small amount of sublimation to remove the ice. Yet some ice does get removed. Beneath the surface, the individual ice crystals merge into an almost solid block of ice. When subjected to the constant pressure of the overlying mass, the ice gradually deforms and expands outward toward the margins of the ice sheet. At the continental margins, the ice often breaks off into large icebergs that float in the Arctic and Antarctic waters.

Mammoth Antarctic Icebergs

Highland Climates (H)

Highland climates are unique among those in the Koeppen system because their distribution is not governed by geographic location, but rather by the topography. These climates are those found in large mountain or plateau areas. As we saw in Chapter 1, the temperature tends to decrease with height in the troposphere. Thus, in high mountains there can be large changes in mean temperature over short distances, simply as a function of elevation. The temperature situation becomes more complicated when one considers that changes in slope angle and aspect over short distances influence the intensity of solar radiation received.

Precipitation type and intensity also vary spatially across highlands. Mountain slopes can enhance precipitation on their windward sides and simultaneously create a rainshadow downwind. Elevation difference also affects the ratio of precipitation falling as snow versus rain, with higher elevations favoring a greater amount of snow. Mount Kilimanjaro in Tanzania provides a striking example of this phenomenon. Located very near the equator at 3° S, Kilimanjaro possesses an extraordinarily wide range of climates from its base near sea level to its peak at 5895 m (19,340 ft). Near the surface, the mountain has a tropical wet and dry climate similar to the surrounding area. But the climate changes with height and the mountain eventually becomes covered with perennial ice over its higher reaches. This is an example of *vertical zonation,* the layering of climatic types with elevation in mountainous environments. Thus, although we use a single designation H, it must be understood that this category contains an extremely rich collection of climates. Obviously, the various climates within the H group are highly localized, with a spatial arrangement not related to large-scale atmospheric processes. As such, the group can be thought of as kind of a climatic "grab bag," consisting of climates that would not necessarily be expected by considering latitude and continental location alone.

Summary

People have long recognized the desirability of a classification system by which distinct climates could be delineated and mapped. This pursuit is neither as straightforward nor as easy as one might expect, because climate consists of a number of different elements. Thus, a certain degree of ingenuity is necessary to create a scheme that includes the major climatic variables and still retains enough simplicity to make the climates comprehensible. The Koeppen system has become the most widely used of all of these techniques.

Koeppen devised a system in which temperature and precipitation characteristics along vegetation boundaries were used as criteria for distinguishing one climate from another. The scheme is a hierarchical, multi-tiered system that uses combinations of capital and lowercase letters.

The first level of categorization in the Koeppen system uses a capital letter from A through E, along with a special category, H, for highland climates. All but one of the A through E climates are based on temperature characteristics. The one exception, the B climates, represent areas of aridity and are called *dry climates*. The temperature-based climates—A, C, D, and E—are referred to as tropical, mild mid-latitude, severe mid-latitude, and polar climates, respectively.

The second level of categorization depends in most cases on the seasonality of precipitation, although for the dry climates the second level reflects the level of aridity. The third level of classification typically refers to temperatures during particularly critical months. Having established the various climates, one can refer to a map such as that shown in Figure 15–1 to visualize their geographic distribution and understand the processes that give rise to the various types.

At the outset of this chapter we distinguished short-term changes in the atmosphere (weather) from long-term conditions (climate). This might lead us to believe that climates are permanent characteristics that never change. This is not true, however, as the climate is forever undergoing slow changes through time. In fact, looking over the history of the planet, we see that there have been many time periods in which the climates were far different from what we know today. The next chapter of this book addresses those temporal changes in climate.

Key Terms

climate p. 449
Koeppen system p. 450
tropical wet p. 453
tropical wet and dry p. 453
monsoonal p. 453
climograph p. 453
Thornthwaite's classification system p. 454
savanna p. 457
semi-desert p. 458
steppe p. 458
subtropical desert p. 458
subtropical steppe p. 458
mid-latitude desert p. 458
mid-latitude steppe p. 458
mediterranean p. 463
humid subtropical p. 463
marine west coast p. 463
humid continental p. 466
subarctic p. 466
boreal forest p. 466
taiga p. 466
tundra p. 467
permafrost p. 468
polar ice cap p. 468
highland p. 469

Review Questions

1. Describe what is meant by *climate*.
2. Describe the general criteria by which the Koeppen system delineates climates.
3. The first order grouping of climates in the Koeppen system is based mainly on temperature. Which climate type departs from that rule?
4. Describe the geographical distribution of tropical climates. What features distinguish this particular group?
5. Briefly describe the fundamental differences between Af, Am, and Aw climates.
6. Of the three types of tropical climates, which occupies the smallest portion of Earth's land surface?
7. Despite their low latitudes, tropical climates are not among the hottest on Earth? Why not?
8. Where are the various dry climates located, and what geographical characteristics cause them to occur where they do?
9. Describe the four types of dry climates and explain how they differ from each other.
10. What factor other than annual precipitation is involved in a climate being defined as dry?
11. Describe the various types of mild mid-latitude climates and their distribution. Why is it that two of them locate mostly along the west coast of continents, while the other tends to be on the eastern side?
12. Are the mild mid-latitude climates really mild? Explain.
13. What are the two types of severe mid-latitude climates, and how do they differ?
14. Why are severe mid-latitude climates missing from the Southern Hemisphere?
15. Describe the three types of polar climates and their distributions.

Critical Thinking

1. Although the Koeppen system is intended to create distinct climate classes, its boundaries are based on the boundaries between vegetation types. Is this really a problem? Can you think of any alternative methods to delineate climates?
2. Do you anticipate that global warming, if it continues as expected, will substantially alter the location of the boundaries of the various Koeppen climates?
3. Western Kansas is located near the junction of B, C, and D climates. Do you suppose the a person driving around this region would notice substantial climatic differences as she crossed from one climate zone to another? What does this tell us about the applicability of large-scale classification schemes to smaller scale analysis?

Problems and Exercises

1. Check Figure 15–1 and determine the type of climate you live in. Then log on to **http://ggweather.com/normals/index.htm** and find the climate information for the location nearest to where you live. Create a climograph for that location and compare it to the example given in the book for the type of climate you live in. How closely do they match?
2. Go to the Web page for the weather service office nearest to where you live. Compare the monthly temperature and precipitation values observed there over the last year to the average for that location. Were they markedly different? How much variance do you expect to encounter between monthly observed values and climatological averages?

Useful Web Sites

http://www.fao.org/WAICENT/FAOINFO/SUSTDEV/Eldirect/climate/Elsp0002.htm

Offers a large number of climate maps, compiled by the United Nations Food and Agriculture Organization.

http://ggweather.com/normals/index.htm

A great source of information for monthly average temperature and precipitation data.

http://www.intellicast.com/Almanac/

Descriptive climate information for each month.

http://iri.columbia.edu/climate/cid/latest/

A Web based newsletter providing information on seasonal climate and its impacts on society.

http://www.ncdc.noaa.gov/oa/climate/research/monitoring.html

Online access to many reports from the National Climate Data Center.

http://www.ncdc.noaa.gov/oa/climate/severeweather/extremes.html

Provides data on extreme weather events.

http://www.cpc.ncep.noaa.gov/

Home page of the NOAA Climate Prediction Center.

Media Enrichment

Interactive Exercise

Global Climate Controls

As described in this chapter, global climate patterns result from the interplay of numerous factors (for example, latitude, elevation, and ocean currents). The text treated each climate type in turn, attempting to explain general precipitation and temperature characteristics in terms of these factors. This exercise takes a different perspective in that it focuses on the controls themselves. Example climographs are used to isolate causative factors from each other, thereby allowing you to examine the role of each factor one-by-one. See how latitude, land vs. water, geographic position, mountains, ocean currents, and prevailing wind influence climates around the world.

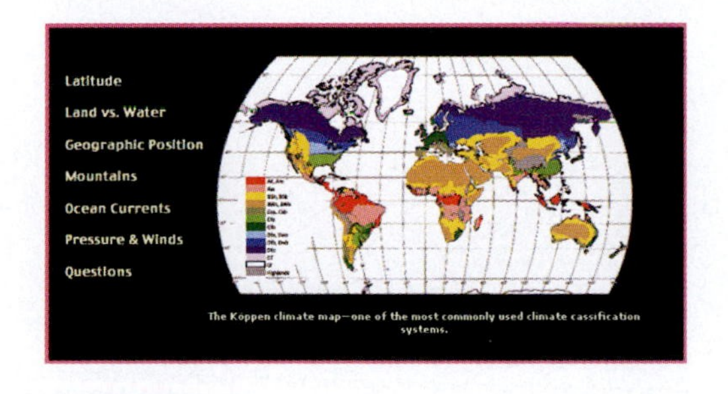

Weather in Motion

Monsoon

Seasonal changes in the mean wind (indicated by the length and direction of arrows) and precipitation (color coded) conditions over the monsoon region are illustrated for the period April through December. Initially, the Indian subcontinent is dry, with heavy precipitation confined to Indonesia and southeastern China. Following the northward march of solar declination, low-level winds begin ferrying moisture to India and southeast Asia, resulting in widespread areas of heavy rainfall for June though August. The wind field weakens and reverses in October, bringing a return to dry conditions.

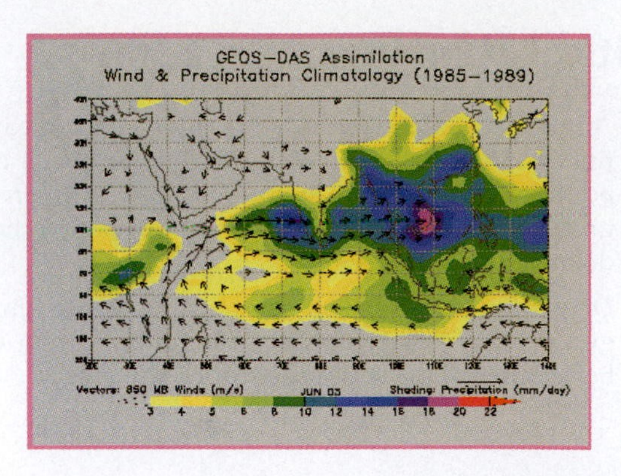

Weather in Motion

Worm Climates

This movie shows the seasonal progression of soil temperature at four locations having D climates. The first two, Green Bay, Wisconsin, and Fargo, North Dakota, do not have permafrost. For both, we see a surface layer that freezes and thaws seasonally, underlain by unfrozen ground. (Predictably, the depth of frost is greater in the colder climate.) But Churchill, Manitoba, and Fairbanks, Alaska, are both permafrost climates, with an active layer above permanently frozen ground. Notice that the active layer is thinner at the colder climate, where summer warmth is short-lived and does not penetrate deeply. Notice also that, in each case, temperature changes are greatest near the surface and the amplitude of the seasonal oscillation decreases with depth. Look closely and you'll see soil temperatures follow behind those at the surface, with the lag increasing downward. Thus while it might be winter at the surface, some depth is experiencing its maximum temperature for the year. With sufficient climatological training a worm could move up and down, enjoying perpetual summer!

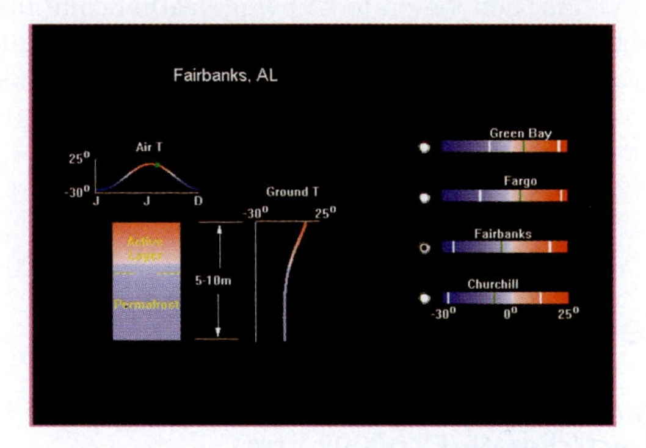

Weather Image

Mammoth Antarctic Icebergs

Two huge icebergs broke away from the Ross Ice Shelf of Antarctica in March 2000. The larger of the two (named B-15) is about the size of Connecticut (300 km by 37 km) and contains about six times as much water (3.4 trillion gallons) as the system of reservoirs serving New York City. Though enormous, B-15 is not the largest ever cited. That honor goes to a 1927 iceberg—330 km by 100 km. Icebergs of this size are estimated to appear every 50 to 100 years. Like their smaller cousins, they result from continental ice that has been pushed far over the ocean before breaking off. Similar conditions do not appear in the Arctic, so icebergs like this are exclusively a Southern Hemisphere phenomenon.

chapter 16

An ancient bristlecone pine, White Mountains, California.

Climate Changes: Past and Future

Imagine you live on the southern margin of Greenland in the thirteenth century. Among the Viking pioneers, your family's livelihood revolves around fishing, hunting, and domestic animals—chiefly sheep, which feast on grasses flourishing in the long summer days. But gradually, over decades, climatic conditions deteriorate from what has been the normal pattern. Winter storms become more frequent and severe, with heavy snows far more common. The growing season shrinks as the number of frost-free days decreases. Increasing sea ice limits fishing and forces long-used sailing routes to be abandoned in favor of more southerly paths. Eventually winter dominates the entire year and food shortages are the rule. Faced with these changes, your family retreats to warmer Scandinavian climates, leaving behind generations of deceased relatives, buried in what will soon become permafrost.

Thinking about this, one cannot help but wonder if there are implications for the present. Although modern societies certainly have greater adaptive capability than the Viking settlers, we are hardly immune to the effects of climate change. (After all, a cooperative climate is still necessary for almost all food production.) It is therefore not surprising that the topic of climatic change has received tremendous scientific and media attention in recent years. In large part this attention arises out of concern for possible human impacts on climate, but also because of interest in natural climatic change, as might be associated with El Niños and volcanic eruptions. In this chapter we will review past climates, the factors that might be responsible for climate change, the methods used to infer past climates, and the use of general circulation models to study potential human effects.

CHAPTER OUTLINE

Defining Climate Change

We saw in Chapter 15 that climate is defined as the statistical properties of atmospheric variables, including temperature, precipitation, and wind. Thus, climate change can be defined as a change in any statistical property of the atmosphere, such as a change in mean temperature. We also saw in Chapter 15 that climate is more than just the average, or mean, value. Year-to-year variations, seasonal variations, and the tendency for above- or below-normal years to occur in sequence are also a major component of climate. Thus, changes in climate may occur even though the mean values of precipitation, temperature, and wind remain the same over time. For example, even with no change in annual average precipitation, changes in the timing of drought and heavy rainfall years could have profound consequences for people and would certainly be considered climatic change.

Consider how changes in climate might occur, thinking first in very general terms. In some ways Earth's climate is like a system that responds to the configuration of external factors, often called *boundary conditions*. In the case of global climate, boundary conditions include the intensity of sunlight, the arrangement of continents and oceans, the composition of the atmosphere, etc. When one or more of these external factors changes, the system adjusts accordingly. Thus, for example, if the Sun's output were to increase, we would expect an increase in global average temperature. Some years might be cooler than before, but on average we would expect a warmer climate, which is of course a change in a statistical property. External conditions might vary too quickly for any sort of climatic equilibrium to be obtained, but we can nevertheless think of the external conditions as driving climatic change. Looking at things this way, climate change can be defined as the response of the Earth–atmosphere system to changes in boundary conditions.

Before adopting this very appealing (and widely used) view of climate change, we need to ask two questions. First, we wonder if a given set of boundary conditions uniquely determines Earth climate. Putting it differently, we could ask "Is more than one climate possible for a single set of boundary values?" Interestingly, both theory and observations suggest the answer might be "yes." This behavior, which is known to exist in systems far less complicated than Earth's climate, is termed **intransitivity**. If Earth's climate is intransitive, that obviously complicates the problem of knowing if boundary conditions have forced an observed change and also makes it difficult to ascribe changes to particular causes.

A second question concerns the way we detect climatic change. Imagine a perfect instrument, one that could measure the state of the atmosphere at any moment in time throughout Earth's history. Measurements from such a device could be used to compute statistical properties for various periods, which in turn would provide evidence for climatic change. We might, for example, average temperature values over successive 100-year periods to get an idea of century-to-century change. We wonder, "Is it possible for statistics computed like this to change over time, even with no change in boundary conditions?" The answer, again, is "yes." Using elementary statistical theory, it's easy to show that statistical measures computed like this are subject to variation independent of those arising from changes in boundary conditions. Thus, for example, even though the true mean is unchanging, the individual 100-year means rise and fall from one century to the next. (Other statistics behave similarly.) In physical terms, the problem arises because short, unpredictable variations associated with weather events spawn variations on longer time scales.* Short-term departures from average become large-term departures and contaminate whatever changes are caused by external conditions. So, if we observe the 100-year mean temperature changing, it's not clear that we are seeing climatic change—it might be an artifact of working with finite-length samples. Obviously, there is no way around this problem; using longer averaging periods reduces the sampling error, but increases the likelihood of blurring truly distinct climates (those arising from changing external conditions).

In light of these difficulties, we'll need to broaden our view of climate change to include all change, whether driven by boundary values or not. In other words, in this book we don't distinguish between change generated by internal processes and change generated by external processes. Regardless of the "instrument" we use (thermometer, ice accumulation in glaciers, tree-ring width), if the statistical values change over time, we will call that climatic change. Admittedly, this is not ideal (change now depends in part on the sample length), but it does have great practical appeal and avoids the complications just described.

The Time Scales of Climate Change

Meteorologists are frequently asked a very straightforward question: Is the climate warming? Although the question is simply phrased, no simple response can be given, in part because the answer depends on the time scale involved. Consider the hypothetical trend in temperature for some location, shown in Figure 16–1. If one is concerned with the temperature trend over the full 150 years, there is an obvious warming trend. On the other hand, somebody who is 20 years old could report that temperatures have fallen throughout her entire lifetime.

In real life, the situation is actually more complex than this example, because changes in climate occur at once on many different time scales, not just two. Some changes take place on time scales of hundreds of millions of years, others over hundreds or tens of thousands of years, some corresponding to centuries, and some even over the course of mere decades. There is no hard-and-fast relationship between the size of a climate change and its time scale; however, as a general rule, oscillations

*We discussed how this happens in Chapter 13 under the topic of chaos.

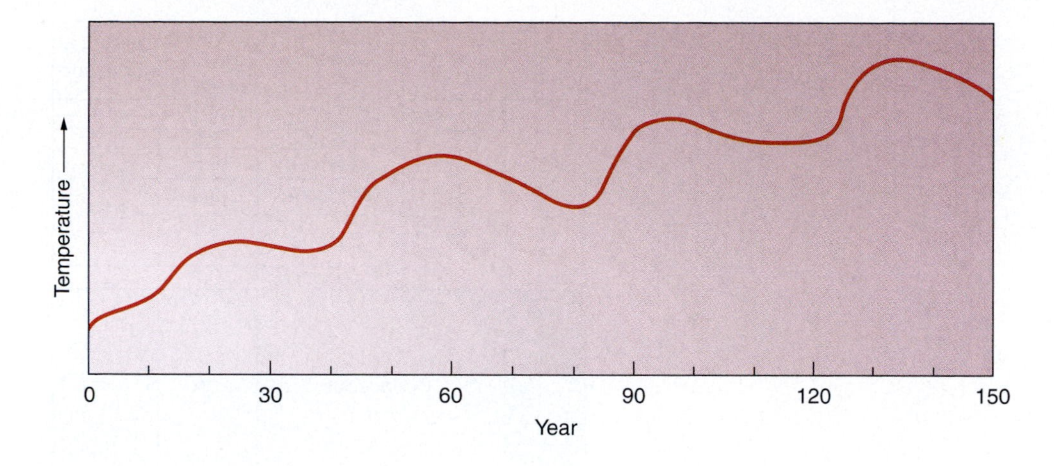

Figure 16–1
A hypothetical plot of temperature through time.

that take place over longer time periods typically have greater magnitudes than those occurring on shorter time scales.

The warming question is also confounded by the fact that climate change varies spatially across the globe. Thus, for example, a temperature trend in the Northern Hemisphere might be absent, or even of opposite direction, in the Southern Hemisphere. Again, there is no fixed relation, but tropical latitudes typically change less than high latitudes. Significant longitudinal variations can also occur, so that large differences emerge *within* a latitude band. This might happen, for example, if the mean position of Rossby waves shifts east or west. In contrast to changes over time, there is a widespread tendency for changes on small spatial scales to be larger than those measured across big areas. Thus, we see that answers to questions about future climate change must be referenced to both a spatial and a temporal scale. Similarly, in looking at past climates, very different pictures emerge depending on the area considered and time scale adopted.

Past Climates

Earth scientists have devised a widely used scheme to divide the planet's natural history into distinct time frames. The geologic column shown in Figure 16–2 uses a hierarchical system dividing time into *eras, periods,* and *epochs.* These time segments are not based on climatic characteristics but rather on geologic and fossil evidence indicative of past environmental conditions and events. Thus, while climatic episodes are sometimes associated with particular eras, periods, or epochs, the time segments should not be considered to have uniform climatic conditions. In addition, significant climate events sometimes cross boundaries in the column, so the divisions do not mark starting or ending times. For our purposes, the terminology is best used as a kind of calendar, to identify points in time, rather than to identify particular climates or events. With this in mind, we can discuss some significant episodes of Earth's climatic history.

Warm Intervals and Ice Ages

Like Viking settlers, we tend to think of our own time as "normal," but this is really not the case at all. Looking at a broad span of Earth's history, we would have to describe the present climate as highly unusual, because most of the time our planet has been considerably warmer than it is today. Unlike today, when the Arctic Sea is mostly frozen all year and huge ice sheets cover the bulk of Antarctica and Greenland, for most of its life Earth has been largely free of permanent (year-round) ice. A more accurate depiction is one of a warm planet, punctuated by perhaps seven relatively brief **ice ages.** Though varying, the warm times persist for hundreds of millions of years to billions of years, whereas the ice ages last on the order of tens of millions of years to perhaps a hundred million years. All of human existence,

Era	Period	Epoch	Duration in millions of years	Millions of years ago
CENOZOIC	Quaternary	Holocene	0.01	0.01
		Pleistocene	1.8	1.8
	Tertiary	Pliocene	3.5	5.3
		Miocene	18.5	23.8
		Oligocene	9.9	33.7
		Eocene	21.1	54.8
		Paleocene	10.2	65
MESOZOIC	Cretaceous		79	144
	Jurassic		62	206
	Triassic		44	250
PALEOZOIC	Permian		36	286
	Carboniferous	Pennsylvanian	39	325
		Mississippian	35	360
	Devonian		50	410
	Silurian		30	440
	Ordovician		65	505
	Cambrian		39	544
PRECAMBRIAN				

▶ **Figure 16–2**
The geologic column.

including the historical period, has been spent in the most recent of these great ice ages. Thus, if someone asks if an ice age is coming, the answer is "No, it's already here."

The earliest known ice age dates to about 2.3 billion years ago, followed by three between 900 and 600 million years ago (MYA). Other ice ages are dated at about 440 MYA and 300 MYA. Our own, the latest, covers the last 15 million or so years.

There is no evidence of ice ages during the first 2 billion of Earth's 4.5 billion years. Of the last 2.5 billion years, only 10 to 20 percent have been ice ages, according to one estimate. A competing, more liberal view allots 50 percent of this time to ice ages. But in either case, it's clear that the majority of Earth's history has been marked by conditions warmer than those of today. Especially prominent in this regard was the warmth of the mid-Cretaceous period, extending from about 120 to 90 MYA. In those times, dinosaurs roamed beyond the Arctic Circle. Coral reefs, which thrive only in warm waters, grew up to 15° latitude poleward of their present locations, as did a number of continental plant communities. With very little water held in land ice, sea level was perhaps 150 to 200 m higher, flooding about 20 percent of continental areas. On a seasonal basis climates were probably more equitable, especially at middle and higher latitudes. Global average temperature is thought to have been anywhere from 5 °C to 15 °C warmer than at present, and the equator-to-pole temperature gradient perhaps 15 °C smaller. Though by no means identical, other warm times undoubtedly had many similar traits.

All ice ages stand in stark contrast to "warm ages" such as the mid-Cretaceous. At one extreme, nearly the entire planet might have been ice-covered during the ice age of about 700 MYA, in a condition aptly called "snowball Earth." New evidence for a global snowball comes from indicators of biological productivity, which are thought to show a decline to almost zero with growing ice volume, followed by a slow rebound during warmer climates that followed. Short of nearly global ice cover, it's hard to see why productivity would be so low. Other interpretations are possible, however, and the iceball hypothesis remains controversial insofar as the amount of ice is concerned. Regardless of how the controversy is resolved, it is clear that all ice ages have abundant year-round ice. For example, there is no dispute that during the most recent ice age, ice sheets advanced to within 40° of the equator, covering the ground to a depth of several kilometers. With so much water locked in land ice, sea level is much lower during ice ages. Differences between ice ages and warm ages vary greatly by latitude, with high latitudes showing greater changes than the Tropics. For example, during an ice age, polar sea surface temperatures (SSTs) might be 10 °C colder, but tropical SSTs only 1 °C to 5 °C colder. Beyond these generalities, little can be said with certainty about the six earliest ice ages. On the other hand, we have substantial information regarding conditions during the most recent ice age.

The Current Ice Age

One of Earth's recent epochs, the **Pleistocene,** is often referred to as *The Ice Age.* However, as we have already seen, Earth has undergone at least several ice ages, and we are in one today. Moreover, the term *Pleistocene Ice Age* is misleading, because the roots of the last glaciation go back considerably farther than the start of the Pleistocene. In fact, the most recent ice age had its origins some 55 MYA, when the global climate began to cool following a warm episode (Figure 16–3a). The first major ice accumulation occurred in Antarctica about 34 MYA. But the cooling was very gradual, so that even as recently as 20 MYA the climate was still warm enough that areas of forest could be found on Antarctica. By about 14 MYA, east Antarctica was glaciated; by 10 MYA, the Antarctic sheet had reached its present size. Finally, by about 5 MYA a continental ice sheet covered nearly all of Greenland; thus, one would have to say the last ice age was firmly established.

Within our ice age, climate has been anything but uniform, with numerous oscillations clearly evident (Figure 16–3b). Beginning about 2.5 MYA, these oscillations began increasing in amplitude, and about 800,000 years ago (Figure 16–3c) the amplitude increased dramatically, becoming about twice as large as in the preceding million years. These oscillations in temperature and ice cover are called **glacial/interglacial cycles.** As can be seen in the diagram, they are quite irregular. For most cycles ice volume increases slowly and then terminates rapidly in a warming event.

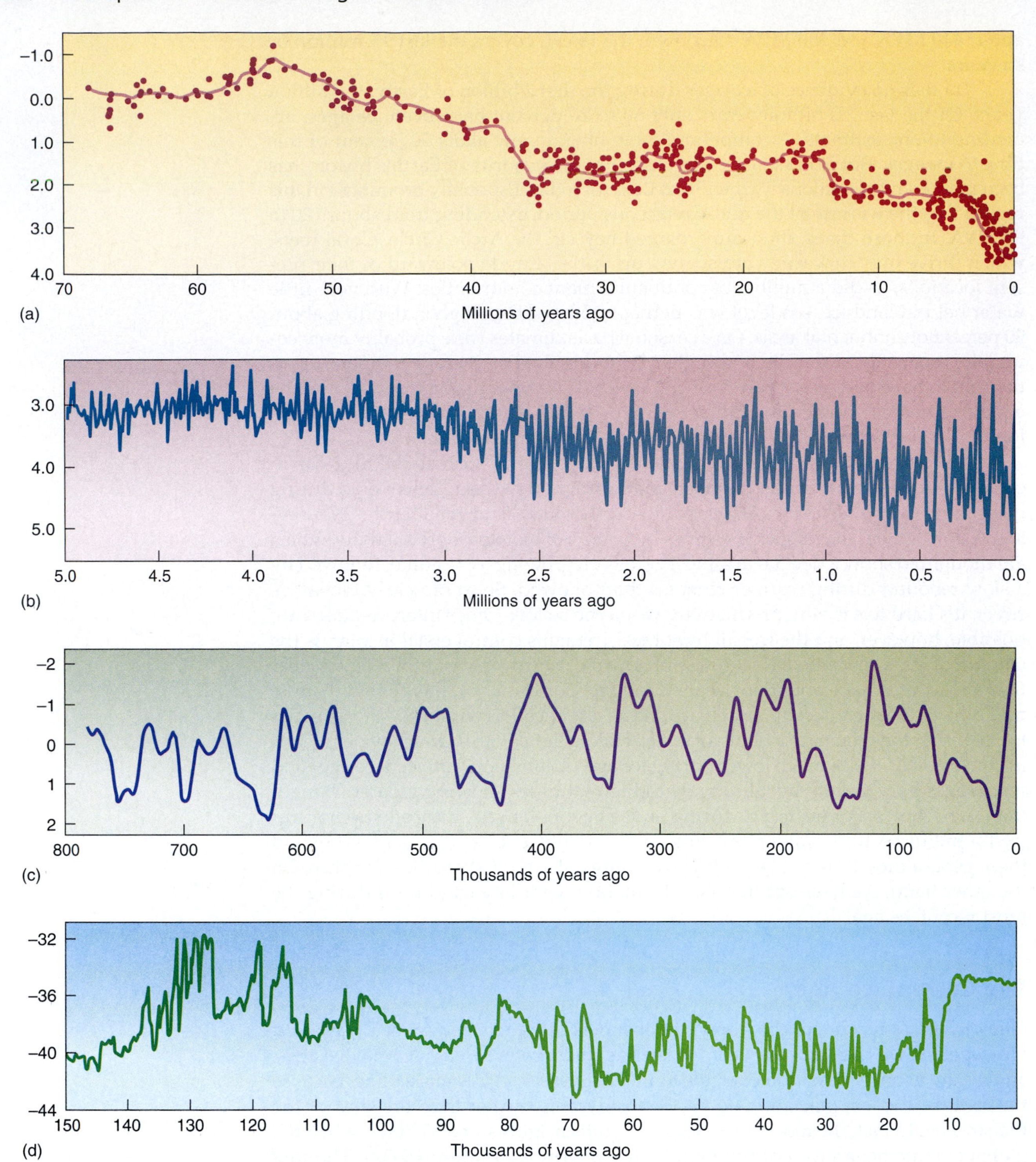

▲ **Figure 16–3**
Indicators of climate for various parts of the last 70 million years. Curve (a) is a smoothed composite of global average ocean temperature and ice volume. Curves (b) and (c) indicate mainly global changes in ice volume. Curve (d) reflects climatic conditions over the North Atlantic, but is broadly consistent with global changes, as can be seen by comparison with later portions of (c).

In addition, neither ice growth nor decay is uniform; "quivering change" is a better descriptor, with short-term oscillations superimposed on the longer cycles. The last three-quarters of a million years have been dominated by cycles lasting about 100,000 years, with shorter-term quivering present on a subdued scale. Throughout the entire **Quaternary,** the current geologic period, there have been about 30 cycles altogether, with associated global temperature changes of perhaps 5 °C.

Ice volume changes have been largest in the Northern Hemisphere, with the size of ice sheets growing and shrinking by a factor of 3 or so with each glacial cycle. Although the Antarctic sheet has changed far less, Antarctic air temperature changes are believed to match those of north polar latitudes (about 10 °C cooler during a glacial). Evidence from mountain snow lines suggests that other Southern Hemisphere locations change temperature as much as their northern counterparts between glacial and interglacial times, and there is other evidence implying that the timing of major warming and cooling events have been generally in step over the last 150,000 years. When the two hemispheres are out of phase, it appears that the Southern Hemisphere leads the Northern Hemisphere by just a little more than 1000 years. But even accepting these differences in timing, we must conclude that the Southern Hemisphere, as a whole, has participated in glacial/interglacial cycles just as the rest of the planet has.

As is clear from Figure 16–3, the planet is now in a warm interglacial, rivaled only a few times in the last 2 million years. Interestingly, one of those times was the last interglacial, which reached its peak about 125,000 years ago and might have set the record for Pleistocene warmth. Global sea level was about 6 m higher than now, and there is evidence that mid-latitude continental areas were 1 to 3 °C warmer. In contrast, sea surface temperatures were not too different from what they are now. Between these two warm periods sits the most recent glaciation, an event that reached its maximum about 20,000 years ago.

The Last Glacial Maximum

Following the last interglacial, ice volume increased, but not uniformly. There were two main pulses of glaciation, one about 115,000 years ago, and another about 75,000 years ago. It seems that most ice was added to polar caps during the first pulse and to ice caps in North America and Eurasia during the later pulse. In the depths of the last glaciation, around 20,000 years ago, many aspects of the Earth–atmosphere system were different. Most dramatically, of course, land ice covered much more area, as seen in Figure 16–4.

Retreat of Continental Ice Sheets

In North America, it is certain that ice reached about as far south as present-day St. Louis, but only to the latitude of New York and Seattle on the East and West Coasts. Paradoxically, there is considerable doubt about the polar boundary of the northern sheet, with some arguing that much of the Arctic rim was unglaciated. Regardless, tremendous quantities of water were transferred from ocean to land, building sheets 3500 to 4000 m thick. Given enough time, this would be enough to depress the continental crust by more than 800 m. When the ice melted, the land surface gradually expanded upward toward its original level. (Even now, continents have yet to fully rebound from glacial depression.) The Laurentide ice sheet in North America was in some ways equivalent to a huge mountain range, running from the Rocky Mountains to the Atlantic. Sea level was about 120 m lower than it is now, so that a land bridge existed between Siberia and Alaska. (At equilibrium, movement of water from the ocean would cause oceanic floors to rise by about 35 m.) There were also significant changes in sea ice, especially in the Antarctic Ocean, where winter sea ice covered about twice the area it now does. Of course, the sea ice changes had little effect on sea level, because water did not move between the land and ocean reservoirs.

On land, temperature changes varied greatly by proximity to the ice sheets and to the ocean. For example, in western North America, maritime air masses kept temperatures to within 4 °C to 5 °C of modern values. This contrasts sharply with the area that is now Tennessee and South Carolina, where temperatures were 15 °C to 20 °C colder than they are now. These two examples might represent the extremes for mid-latitude changes—a number of other mid-latitude locations were in the range of 5 °C to 8 °C cooler. Temperature changes in the Tropics were smaller, perhaps 4 to 5 °C. Snowline was about 1000 m lower, which translates into a temperature decrease of 5 °C to 6 °C for elevations above 2000 m.

Most places were not only colder, but they seem also to have been drier. This is especially true for the high latitudes, where precipitation amounts were about

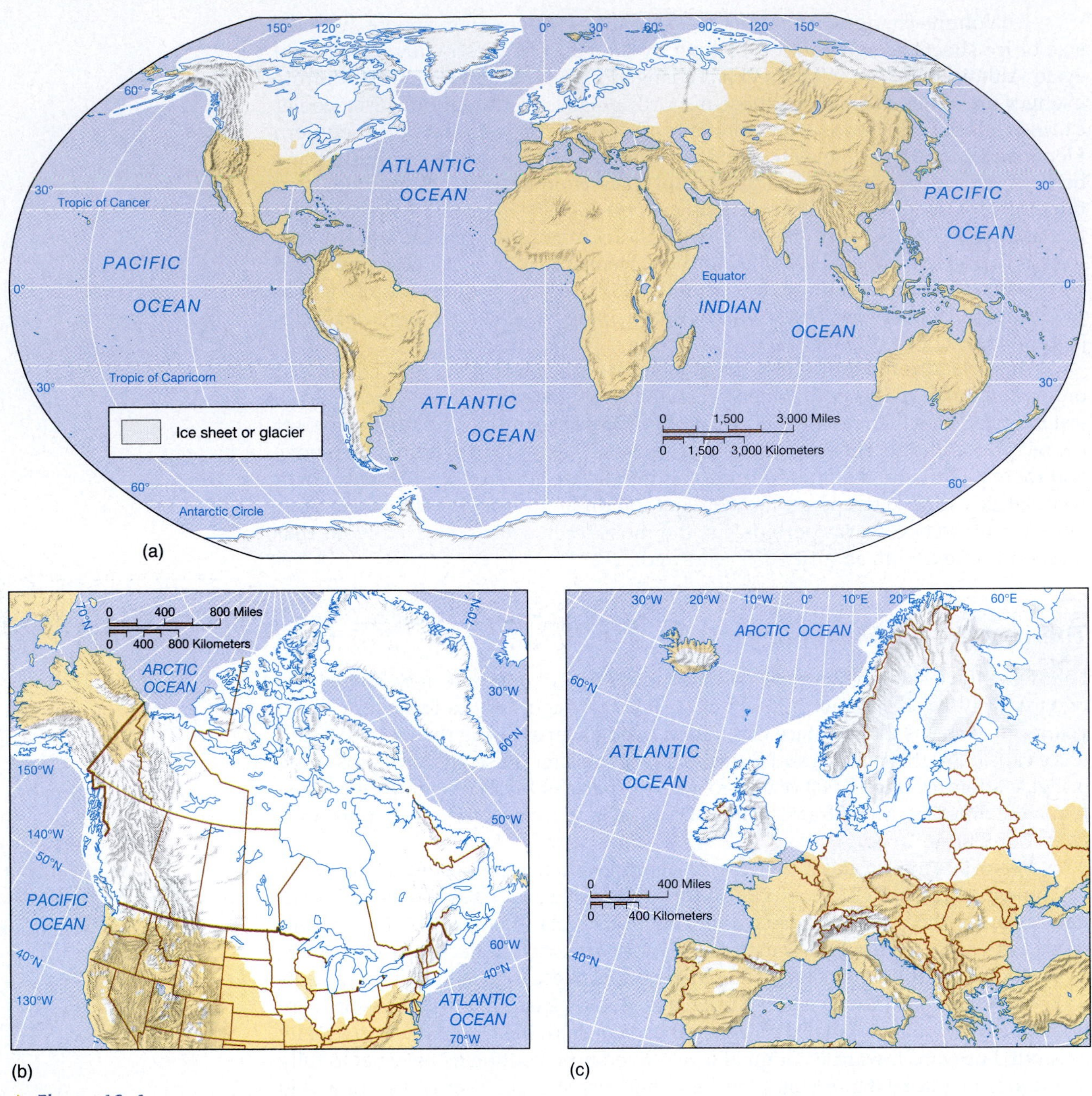

▲ **Figure 16–4**
Map showing the maximum extent of ice at time of last glaciation.

50 percent below today's. Some desert areas of both South America and Africa were larger, and lake levels were lower in tropical Africa and Central America. A cold desert covered much of the section of western Europe that wasn't under ice. Although increased dryness was the norm, a number of wetter areas could be found on all continents, as reflected in lake level changes and other indicators. Precipitation changes were undoubtedly determined to a large degree by circulation changes. Where prevailing winds shifted such that they blew from high latitudes (as for eastern North America), conditions were drier. What little evidence exists suggests that wind speeds were greater, as might be expected with a stronger equator-to-pole temperature gradient.

It must be emphasized that this glacial period was hardly uniform. In fact, abrupt climate changes were very common throughout this period, with polar temperatures changing by 5 °C to 8 °C over the course of just decades to centuries (Figure 16–3d). Because changes such as these are not confined to the glacial period, they are covered in more detail in a later section.

The Holocene

Like other glacial-to-interglacial transitions, warming following the last glacial maximum was fast, at least compared to times of cooling. Warming began about 15,000 years ago (prior to the start of the **Holocene**), only to be interrupted about 2000 years later when colder conditions returned. The cold event was most extreme (almost glacial) in the North Atlantic area, but it can be found in records as far away as Antarctica. Called the **Younger Dryas,** it lasted for about 1200 years. Following the Younger Dryas, starting about 11,800 years ago, came another period of abrupt warming, with temperatures in Greenland increasing by about 1 °C/decade, bringing climate into the interglacial we enjoy today. This was interrupted by yet another abrupt and short cooling 8200 years ago. Aside from that, the early Holocene was warm, perhaps even more so than today by 1 °C in global average temperature. Although the North American Great Plains area was drier, much of monsoonal Africa and Asia was wetter. (Remains of large grazing animals are found in areas of modern desert, for example.) Calculations suggest that summer monsoon circulations were stronger, contributing moisture to land areas not reached by those winds today. Sea level was a few meters higher than now, consistent with a warmer planet.

Following the early Holocene warmth, global temperatures have declined somewhat, but the declining temperatures have been modulated by a number of events of various durations and expressions, depending on the time and place. For example, there is evidence that the period A.D. 900–1200 was warm in the North Atlantic. Called the **Medieval Warm Period,** it coincides with the Viking settlement of Greenland. Mountain glaciers in Europe advanced before and after, but not during, this time, and there is evidence for glacial retreat elsewhere as well (including the Canadian Rockies). But there is also much physical and historical information indicating that this was not a global event of any significance.

Less ambiguous is another celebrated event, the so-called **Little Ice Age.** Spanning from 1450 to 1850, this was a cold period for western Europe. During these years, alpine glaciers advanced as temperatures fell by about 0.5 °C to 1 °C. Historical records indicate that this seemingly small decrease in mean temperature had a considerable effect on living conditions throughout Europe. Shortened growing seasons led to reductions in agricultural productivity, especially in northern Europe. In contrast to the Medieval Warm Period, the Little Ice Age is expressed in mountain records from around the world. Although in no sense is it a true "ice age," it does represent the largest temperature change during historical times and is considered a global event.

Of course, there have been other climate changes on smaller scales involving both precipitation and temperature. Consider, for example, Figure 16–5, which shows moisture conditions for coastal Virginia/North Carolina from A.D. 1200 to the present. Evident are numerous excursions from average conditions, some persisting for decades. It is particularly interesting to note that the driest three years of the entire 800-year period coincide with the disappearance of the Lost Colony of Roanoke sometime after August 1587. Similarly, the drought of 1606–1612 brought the driest 7-year period of the record, by happenstance coinciding with establishment of the Jamestown colony in 1607. American readers will recall that the Jamestown colony suffered tremendous mortality (60 percent died in the first year) and was nearly abandoned, in large part because of malnutrition. Standard explanations for these catastrophes mention poor planning, mismanagement, bad relations with native societies, and similar social factors. In light of the extreme droughts, one must guess that environmental factors were at least as important.

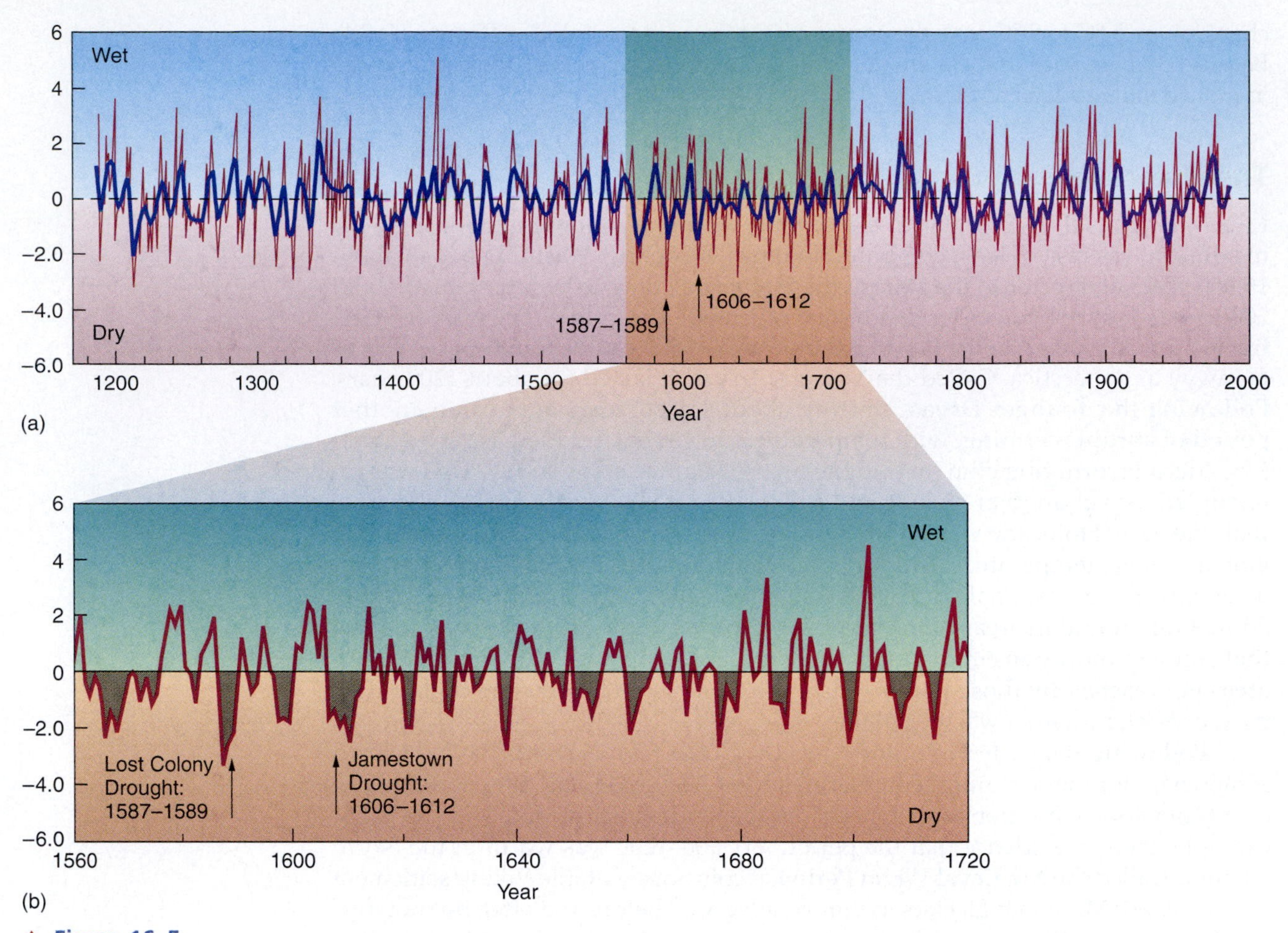

▲ Figure 16–5
A longtime series representing July moisture conditions for the Tidewater region of Virginia and North Carolina.

The Last Century

During the middle of the last century, a growing network of meteorological stations was established around the world. Though there are many problems in the data set, such as the movement of meteorological instruments at a given site, the data give firsthand accounts of temperature and precipitation patterns. As described in *Box 16–1, Focus on the Environment: Historic Warmth in the 1990s and 2000s*, there have been two episodes of general warming during the century separated by a two-decade period without a discernable trend.

There is a very important question regarding the rising temperatures of the last few decades. As we have seen, changes in global climate occur on wide variety of time scales, each with its own range of variation. We must now ask if this recent rise in temperature is just part of the natural variability in climate, or if it marks the onset of human-induced warming from the emission of "greenhouse gases" into the atmosphere. At present, it is impossible to conclude one way or the other with any certainty. But most atmospheric scientists believe that this warming may be largely the result of increases in greenhouse gases, and others are now fully convinced.

Millennial-Scale Oscillations

The preceding sections mostly took a sequential view, tracking climate through time. Another way of looking at climate is to focus on individual time scales of change, without regard to the state of the system (warm vs. cold, etc.). When this is done, persistent oscillations emerge for every time period examined so far, going back 500,000 years. The oscillations are called *millennial-scale* because they appear at intervals of roughly 6000, 2600, 1800, and 1450 years. During the most recent glacial period, the oscillations are quite similar to one another; each begins with a rapid increase in temperature, taking just a few decades to centuries to develop. Air temperatures over Greenland increase by 5 °C to 8 °C, and North Atlantic sea surface temperatures (SSTs) increase by about 3 °C. Temperatures remain high for 1000 to 2000 years, after which they return quickly to former low values. Thus, rather than rounded cycles, these oscillations are more like square waves, with climate jumping between two states.

Oscillations of similar length occurred during other glacials, as well as during interglacials and transitions between glacials and interglacials. Oddly, they are largest during interglacial-to-glacial transitions, with SST changes in the range of 4 °C to 4.5 °C. During the interglacials, the oscillations are just as persistent, but much smaller, with North Atlantic SST warming and cooling by about 0.5 °C to 1 °C. They appear throughout the Holocene and are expressed most recently as the Little Ice Age cooling.

Millennial-scale oscillations are significant because they suggest that the Earth–atmosphere system has a tendency to flip back and forth between warm and cold states, independent of whatever the long-term climate is doing. Thus, rather than a regular progressive change that might follow changes in some boundary condition, it's as if the system suddenly reorganizes itself and moves quickly to another mode. No one knows why this happens, although a widely held view is that it arises from processes internal to the system, most likely involving feedbacks between atmosphere and ocean. The alternative would call for something external to turn on and off at more-or-less regular intervals, with effects that are amplified during transitional and glacial times. Regardless of the cause, the size and speed of change is a matter of some concern. If the planet really does have this bipolar behavior, a switch from the present warm state to the other could have profound effects on many social systems.

Before concluding this section, we must stress again that climatic changes are not restricted to mean values; there can also be changes in the frequency of rare events. Thus, for example, computer models suggest that many areas may experience an increase in the incidence of heavy precipitation and drought in association with an increase in temperature. A recent study has shown that just such an increase has occurred over North America during the last century. The percentage of annual precipitation accounted for by heavy rainfall events has risen substantially over most of the United States in conjunction with the observed increase in temperature. Extreme events now yield about 12 percent of annual precipitation, whereas in 1910 they accounted for about 9 percent.

Factors Involved in Climatic Change

As we have seen, Earth's climate has undergone significant changes of varying magnitudes and time scales over the course of its existence. The big question, of course, is *why?* Several possible causes are easy to identify. These include variations in the intensity of radiation emitted by the Sun, changes in Earth's orbit, land surface changes, and differences in the gaseous and aerosol composition of the atmosphere. Each operates on a different time scale, as we will now see.

While examining the list of factors effecting climate change, it is important to understand that many of them do not operate independently. This means, first of all, that they operate simultaneously. Thus, for example, while one agent might be leading to warming, another might counteract or enhance that warming. Second, it means that agents of change might interact with one another, so that the effects are not merely additive. For example, the effect of tropospheric aerosols produced by humans might differ depending on whether or not they are overlaid by a plume of stratospheric volcanic aerosols. The issues of simultaneous and interacting factors will become especially evident in the discussion on changes in Earth's orbital characteristics.

16–1 Focus on the Environment

Historic Warmth in the 1990s and Early 2000s

There is no question that global temperatures have increased during the twentieth century, as shown in Figure 1. The values on the vertical axes indicate the difference in temperature for a given year relative to the "climatic average" for 1880–2001. Every year from 1880 through the mid-1920s experienced below-normal temperatures, but temperatures began to increase at about 1900. Following the late 1930s, most years had above-normal temperatures, with marked warming occurring since the late 1970s. Over the entire twentieth century global surface temperatures increased at a rate of 0.6 °C (1.1 ° F) per century, but during the last quarter of the century the rate of increase was much greater, at 2.0 °C (3.6 ° F) per century.

The last decade of the twentieth century and the first 3 years of the twenty-first century were remarkably warm. Every one of the 13 years between 1990 and 2002 ranked among the 18 warmest for as long as records have been kept, and 9 of the 10 warmest years on record all occurred during that period. The years 1998, 2002, and 2001 were the three warmest years on record. The overall warmth of the period would have been even greater were it not for the eruption of Mt. Pinatubo in the Philippines in 1991. Stratospheric aerosols from the eruption reduced the amount of solar radiation reaching the surface of the Northern Hemisphere during 1992 and 1993, making them the 2 coolest years of the decade (though still warmer than the long-term average). And although we don't have reliable records for the years prior to 1880, a recent analysis using techniques described in this chapter indicates that 5 of the years of the 1990s were probably the warmest for North America over the last six centuries! Interestingly, though land and ocean temperatures both increased significantly in the 1990s, they were not perfectly in sync with each other.

Human health threats due to rising temperatures may be more serious than temperature data alone would suggest. A recently published study looked at trends in both temperature and apparent temperature (based on the combination of temperature and humidity, as described in Chapter 3) in the United States between 1949 and 1995. The study found that the incidence of extreme apparent temperatures has increased more dramatically than the incidence of extreme temperatures alone. To make matters worse, the increase in nighttime apparent temperature has been greater than the increase in daytime apparent temperatures, meaning that the high apparent temperatures, instead of dropping at sunset, persisted overnight. This is important because experts believe that the most dangerous health threat occurs not from very high daytime temperatures, but from the persistence of high apparent temperatures for several days without intervening cool nights. The results of the study are even more noteworthy when one takes into account that data from the record warm years of 1997 through 1999 were not included.

The big question with regard to warming is whether it is largely the result of human activities or natural variation. While the answer is not known for certain, each new warm year lends further support to the general circulation models' predictions of a warmer atmosphere in response to increasing concentrations of greenhouse gases. In 2001 the Intergovernmental Panel on Climate Change (IPCC) issued its third assessment report, which summarized the current knowledge about global warming. This prestigious panel of scientists, under the auspices of the World Meteorological Organization (WMO) and the United Nations Environment Programme, stated in its report that "the warming over the past 100 years is very unlikely to be due to internal variability alone. . . ." It further stated that "most of the observed warming over the last 50 years is likely to have been due to the increase

Variations in Solar Output

As we've implied earlier, Earth's climate is quite sensitive to the Sun's output. Considering that the amount of energy emitted by the Sun is not truly constant, this mechanism of climate change has considerable theoretical appeal. For example, some changes in the solar output, on the order of 0.1 to 0.2 percent, appear to be related to the occurrence of sunspots. As mentioned in Chapter 2, sunspots are relatively cold regions of the photosphere, about the size of Earth's diameter. The abundance of sunspots rises and falls on several time scales, including the very striking 10.7-year cycle. (As is customary, we will refer to this as the 11-year cycle.)

Satellite measurements show that when measured over the course of a few weeks, solar radiation decreases as sunspots increase. This is consistent with sunspots being cold—when more of the Sun is covered by cold regions, less radiation is emitted. But at longer time scales, such as those of a complete 11-year cycle, increasing sunspots are correlated with more radiation. Clearly, at these longer time scales, there must be solar changes that compensate for the increased area of sunspots. (The most likely explanation is an increase in surrounding bright areas.) The link between sunspot activity and solar output has led to considerable speculation that such phenomena could account for some of the climatic changes that

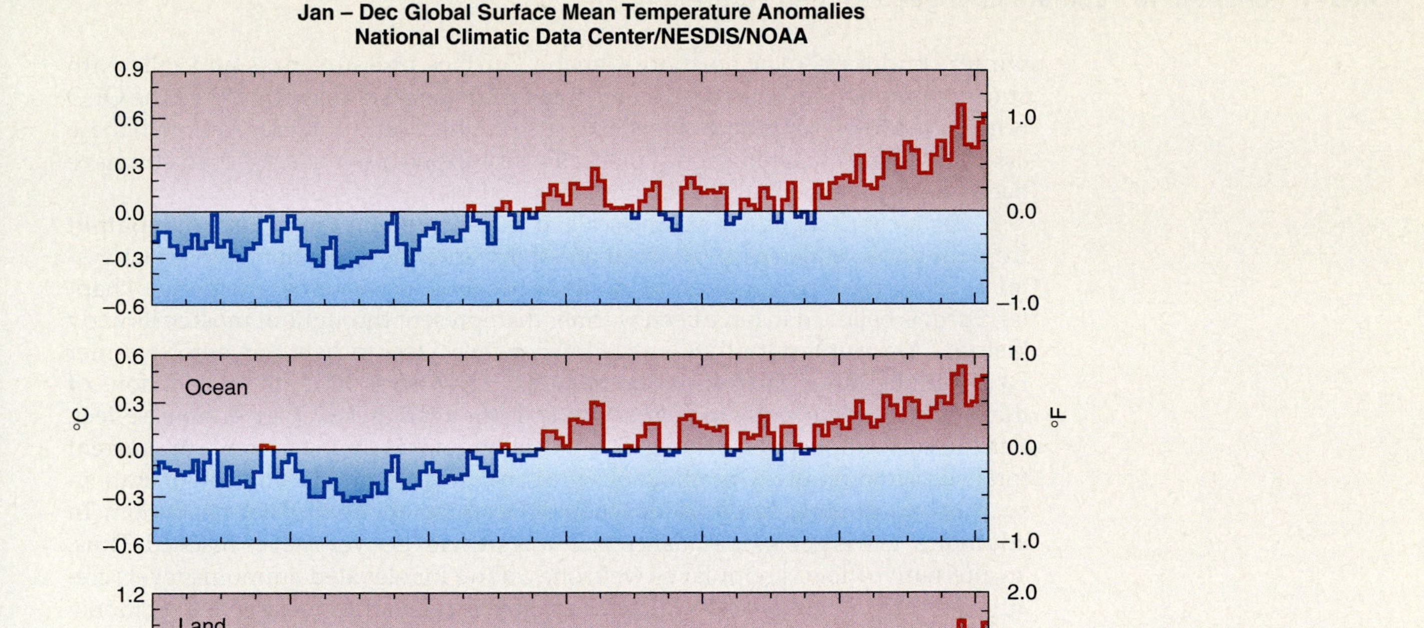

◀ **Figure 1**
Average global air temperatures from 1880 to 2002 for the entire Earth surface (a), over the oceans (b), and over land (c).

in greenhouse gas concentrations." In June 2002, the U.S. Environmental Protection Agency (EPA) further concluded that warming over the last several decades is "likely due mostly to human activities. . . ." The report added that even if greenhouse emissions were to be reduced in upcoming years, the effects of their increase in the atmosphere over the last few decades would not be soon reversed.

have occurred on Earth. For example, droughts in the Great Plains of the United States have shown some tendency to recur at an interval that roughly corresponds to a double sunspot cycle, and earlier research has noted similar periodicities in Nile River flows. In addition, air temperatures over eastern North America rise and fall by about 0.2 °C in apparent synchroneity with the 11-year cycle. Some supporting evidence for the connection between climate and solar activity is also given by the fact that the **Maunder Minimum,** the period of minimal sunspot activity between about 1645 to 1715, coincided with one of the coldest periods of the Little Ice Age (see *Box 2–3, Physical Principles: The Sun* on page 50). However, there have been other episodes in which variations in sunspot activity did not coincide with changes in climate, and alternative explanations have been offered for the apparent 22-year climate changes.

The equivocal evidence for a Sun–climate connection became stronger in the late 1980s, when several scientists observed that the relationship between tropospheric conditions and sunspot activity was much stronger when the direction of stratospheric winds over the Tropics was taken into account. These winds tend to reverse their direction in approximately two-year cycles in a pattern known as the **quasi-biennial oscillation (QBO).** When the QBO is in its west-to-east mode, for example, there appears to be a relationship between the number of sunspots and winter conditions over northern Canada. Surface pressure rises and falls with sunspot number, and the mean storm track shifts north and south. When the QBO is in its east-to-west phase, however, no such connection is evident. Although these associations are intriguing and statistically very strong, no causal mechanisms have been proven to explain these relationships.

On an entirely different time scale, it is believed that the rate of solar output from the Sun has increased by about one-third since the formation of the solar system. This brings up an interesting paradox because, as we saw earlier in this chapter, Earth is believed to have been warmer than present throughout most of its early history. This counterintuitive association between a less radiant Sun and a warmer Earth is called the "early faint Sun paradox." Two very different explanations of the paradox are currently in vogue. One calls for a massive CO_2 greenhouse effect, with an early atmosphere having CO_2 partial pressures up to ten times the current total surface pressure. Organic molecules are much more difficult to generate in such an atmosphere; thus, this idea makes the appearance of life hard to fathom. In addition, there is geologic evidence that says such levels were never reached. Thus, an alternative view is popular as well, one calling for elevated ammonia levels creating an early greenhouse. Disputed for many years on the grounds that ammonia would be broken up by ultraviolet radiation, this was considered an implausible explanation. However, in 1997 it was shown that high-altitude shielding by other gases may have allowed ammonia to accumulate at lower levels. Regardless of which (if either) hypothesis is correct, it's interesting that the Sun–climate connection runs from the extreme of wondering whether there is any effect at all (sunspots) to wondering why the effect was formerly so weak (early faint Sun).

Changes in Earth's Orbit

In Chapter 2 we saw that the seasons occurred primarily because of the tilt of Earth's axis relative to the Sun. If we imagine a plane on which Earth makes its revolution around the Sun, we can see that the axis of rotation is oriented 23.5° from the perpendicular to the plane (that is, it has an obliquity of 23.5°). The orientation of the axis is constant through the course of the year, so no matter where Earth is relative to the Sun, its axis points toward the North Star, Polaris. For the six months following the March equinox, the Northern Hemisphere is inclined toward the Sun, and during the rest of the year the Southern Hemisphere has a greater exposure to the Sun. This is the primary factor in causing the seasons. Obviously, if the axis of rotation were greater than 23.5°, this effect would be stronger and lead to a greater seasonality. Likewise, the effect would disappear if the axis was exactly perpendicular to the plane of the orbit.

We also saw that Earth's orbit is elliptical, rather than circular, so that on about January 4 Earth is about 3 percent closer to the Sun than on July 4 (see Figure 2–9). This change in Earth–Sun distance causes the planet as a whole to receive about 7 percent more solar radiation at the top of the atmosphere in early January (perihelion) than during early July (aphelion). It should be readily apparent that a greater eccentricity would result in greater differences in incoming radiation available at the top of the atmosphere during the course of a year.

As far as the Northern Hemisphere is concerned, the Earth–Sun distance is largest during the summer and smallest during the winter, causing winters to be somewhat warmer and summers to be cooler than they otherwise would be. Thus, not only is the amount of eccentricity important with regard to seasonality, but so is the timing of the minimum and maximum Earth–Sun distances with respect to the equinoxes and solstices (as will soon be explained).

In sum, there are three astronomical factors that influence the timing and intensity of the seasons: eccentricity in the orbit, the tilt of Earth's axis off the perpendicular to the plane of the orbit, and the timing of aphelion and perihelion relative to the timing of the equinoxes. As it happens, these three factors all change slowly over time on a variety of time scales.

Eccentricity The **eccentricity** of Earth's orbit changes cyclically on several time scales, with a cycle of about 100,000 years being especially prominent. Though the Earth–Sun distance at aphelion is currently about 3 percent greater than at perihelion, the relative distance has varied between about 1 and 11 percent over the last 600,000 years. Over about the last 15,000 years, there has been a steady decrease in eccentricity, which will continue for about another 35,000 years.

Obliquity The tilt of the Earth's axis, **obliquity,** also varies cyclically, but with a dominant period of about 41,000 years, during which it varies between 22.1° and 24.5° off the perpendicular. While the range of the axis tilt may seem small, it is capable of producing substantial differences in summer and winter insolation. In particular, high-latitude regions can undergo changes in available solar radiation at the top of the atmosphere about 15 percent due to variations in obliquity. The most recent peak in obliquity occurred roughly 10,000 years ago. Thus, we are about midway in the half cycle from maximum to minimum obliquity.

Precession Though the summer solstice for the Northern Hemisphere currently occurs near the time of aphelion, this changes through time because the axis wobbles on a 27,000-year cycle. In other words, the axis of rotation gyrates so that in about 13,500 years it will point to a different star, Vega, instead of Polaris (Figure 16–6). This change in orientation of the Earth axis, called **precession,** directly alters the timing and intensity of the seasons. Combined with changes in the orientation of the elliptical orbit, the result is a 23,000-year cycle in radiation. If orientation of the axis toward Vega were to exist today along with the current timing of aphelion and perihelion, the winter solstice for the Northern Hemisphere would nearly coincide with aphelion. The resultant increase in seasonality would cause warmer summers and cooler winters in the Northern Hemisphere. At the same time, the Southern Hemisphere would experience less seasonality because its summer solstice would occur near aphelion.

It is important to note that the importance of precession on influencing radiation receipts depends on the magnitude of the eccentricity of the orbit. Low values of eccentricity (nearly circular orbits) mitigate the importance of precession; greater eccentricity amplifies it. Given the current trend toward decreasing eccentricities, we can expect that this effect will be relatively small over the next 50,000 or so years.

These three cycles are collectively called the **Milankovitch cycles,** in honor of the early-twentieth-century astronomer who expounded on their potential influence on Earth's climate. Most scientists believe that the Milankovitch cycles have played an important role in the expansion and retreat of glaciers during the Quaternary because of the way they work together to influence seasonality. Large glaciers, such as those currently occupying most of Greenland and Antarctica, are most likely to expand when seasonality is low. With less seasonality, warmer winter temperatures foster a greater amount of snowfall over much of the ice sheets due to a greater availability of water vapor. Cooler summers also promote glaciation because the rate of melt along the margins of the ice sheets is slowed.

Orbital Variations

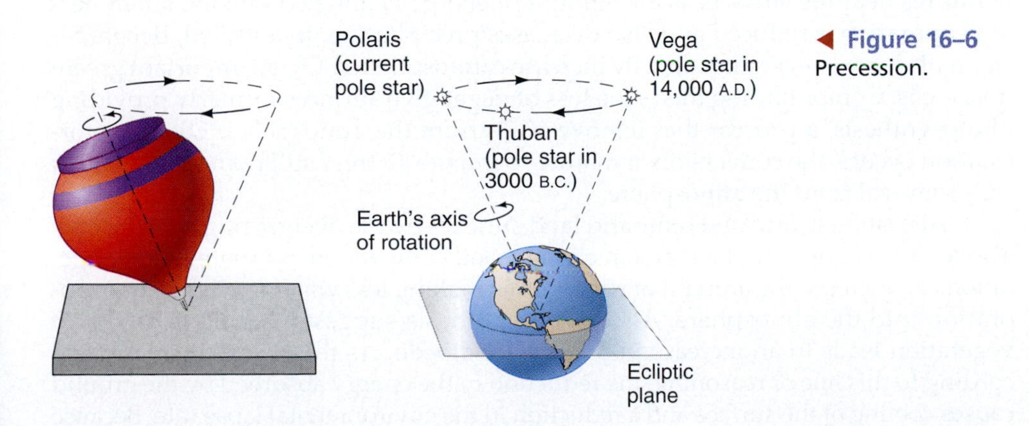

◀ **Figure 16–6** Precession.

Considerable observational evidence for the role of Milankovitch cycles is seen in the climate record of the Pleistocene. Using radiation in the northern middle latitudes as a surrogate for orbital forcing, there is good agreement between the timing of ice advances and retreats. In addition, when the variability in the climate record is broken down according to various time scales, the three main Milankovitch periods emerge as containing most of the climate variation. For these and other reasons, changing orbital parameters are widely accepted as driving glacial/interglacial cycles. To the extent that episodes of glacial advance and retreat are affected by Milankovitch cycles, the scenario is for us to return to another episode of glacial advance sometime during the next few millennia. But the Milankovitch theory is not without its problems, chief of which is one that concerns the eccentricity cycle. The 100,000-year cycle has the smallest effect on radiation reaching the planet but emerges as the strongest in the climate record. If this is the true cause, processes yet to be fully explained must amplify the relatively small radiation cycle. A related question concerns the observed amplitude increase in the 100,000-year climate cycle. If eccentricity cycles are an explanation, why did they suddenly become more important about 800,000 years ago? Finally, there is the question of precession, which is out of phase between the Northern and Southern Hemispheres. Precession effects that would cause cooler conditions in one hemisphere should produce the opposite response in the other hemisphere. Yet, as we have seen, climate changes have been similar in both hemispheres, not opposed to one another.

Changes in Land Configuration and Surface Characteristics

Many climatologists believe that climatic changes occurring over the longest time spans were at least partly in response to changes in the size and location of Earth's continents. Support for an older view, that continental drift was the primary forcing variable, has eroded in the face of quantitative estimates of change obtained from computer simulations. The breakup of **Pangaea** (the early supercontinent) and the slow movement of the resultant continents undoubtedly caused major climatic changes, even if not as large as observed in the geologic record. Such would have to be the case, of course, because all the factors that affect temperature and other climate variables (such as latitude and continentality) were themselves greatly affected by the movement of the continents. Though dramatic, the climatic changes resulting from continental displacement would be extremely slow. Like changes in position, episodes of continental mountain building have almost certainly produced significant climate change. For example, computer modeling suggests that the presence of large mountain regions (the Rockies, Himalayas, Andes) would amplify Rossby waves during the winter season and promote enhanced monsoon circulations in the summer, both of which are consistent with observational data.

At shorter time scales, modification of Earth's surface, especially by human activity, can greatly influence the disposition of solar radiation. One such activity is deforestation, in which large tracts of land are cleared of trees. The loss of vegetation reduces evapotranspiration from the surface. This in turn leads to higher temperatures near the surface, as the amount of energy channeled into the latent heat of evaporation is reduced and also decreases precipitation. In addition, decomposition of cleared vegetation directly increases atmospheric CO_2, an important greenhouse gas. Compounding this is the loss of vegetative surfaces formerly providing photosynthesis, a process that removes CO_2 from the atmosphere. Even if reforestation occurs, the conversion of mature to immature trees still reduces the rate of CO_2 removal from the atmosphere.

Alteration of arid and semi-arid land surfaces by the overgrazing of cattle may also lead to changes in the regional climate. Soil compaction associated with vegetation can increase the amount of runoff, thus making less water available for evaporation into the atmosphere. Also, one hypothesis suggests that the removal of vegetation leads to an increase in the surface albedo. As the albedo increases, according to this line of reasoning, the reduction in the energy absorbed by the ground causes cooling of the surface and a reduction in the environmental lapse rate. Because a lowered lapse rate would make the air more stable and less susceptible to convectional precipitation, overgrazing could enhance the vulnerability of these regions to drought. On the other hand, the reduction in vegetation due to overgrazing could have the opposite effect, in which a reduction in evaporation leads to a general warming of the surface and an increase in the environmental lapse rate. Of course, the instability associated with the increased lapse rate will have little effect if there is not enough water available to yield precipitation. Yet another effect arises from the change in surface roughness, which alters the momentum transfer between atmosphere and ground.

Changes in Atmospheric Turbidity

Atmospheric **turbidity** refers to the amount of suspended solid and liquid material (aerosols) contained in the air. Some aerosols are released into the atmosphere through natural processes, such as large-scale volcanic eruptions; others are released by human activity, such as smokestack emissions. Aerosols can enter the atmosphere directly as in the examples just named. They can also enter the atmosphere indirectly as a result of chemical processes in which certain gases—usually sulfates, but also nitrates and hydrocarbons—react in the presence of sunlight to form solid and liquid aerosols. Some aerosols are no more than tiny clumps of molecules, whereas others are millimeters across. This is a huge range in size, roughly comparable to the difference between tennis balls and planets. Regardless of their source, composition, or size, however, prolonged variations in aerosol contents can have important ramifications for climate.

Existing both in the stratosphere and the troposphere, aerosols directly affect the transmission and absorption of both solar and infrared radiation. By absorbing incoming sunlight they can cause a heating of the atmosphere around the aerosols, but they can also increase the amount of backscattering and thereby reduce the amount of radiation reaching the surface. The relative effects of aerosol absorption vs. backscatter are hard to assess and depend on numerous factors, including the albedo of the aerosols and the underlying surface.

To a lesser extent, aerosols can increase the absorption of outgoing longwave radiation that would otherwise escape to space and thereby increase the longwave radiation radiated from the atmosphere to the surface. In this way, they can have the direct effect of increasing nighttime temperatures.

Aerosols can also affect climate indirectly, through their ability to serve as cloud condensation nuclei. A cloud with a greater number of condensation nuclei might contain as much liquid water as one with few nuclei, but it will have more droplets of mostly smaller sizes. Clouds containing a large number of small droplets are less likely to yield precipitation and, therefore, are likely to persist for longer periods. In addition, the total droplet surface area of such clouds is greater, and this greater surface area increases the reflectivity of the cloud (assuming the same amount of liquid water). Thus, tropospheric aerosols might give rise to more extensive, longer-lived, brighter clouds.

Support for the indirect effect of aerosols comes from satellite images of ship trails, which are tracks of low-level clouds produced by ocean-going vessels (Figure 16–7). In the relatively clean ocean atmosphere, aerosols enhance condensation, creating long cloud trails that stand out against the surrounding clear sky. With their high albedo, these clouds reduce the amount of solar radiation reaching the surface. Additional support for indirect aerosol effects comes from a recent study of pollution tracks on land, which appear on satellite images as plumes of brighter clouds downwind of cities. By analyzing data from a number of sensors, it was discovered that aerosols substantially reduce precipitation in downwind areas by suppressing both ice formation and droplet coalescence. (Recall that these are important precipitation growth mechanisms.)

Moving beyond these generalities, let us now look in more detail at the climatological effects of tropospheric and stratospheric aerosols as agents of climatic change.

▶ **Figure 16–7**
Ship tracks appear as white streaks embedded in a low-level cloud deck of speckled light gray clouds. The image shows an area just offshore of the northwestern United States (colored green), with small amounts of cloud-free ocean (blue).

Tropospheric Aerosols Natural sources of tropospheric aerosols include the spraying of salt particles by ocean waves and bubbles, soot and gases from fires, the erosion of soil by wind, the dispersal of spores and pollen, the emission of sulfides by marine plankton, and other biospheric processes. (In some cases, the original emission is gaseous but is converted to liquid or solid.) Of course, volcanic eruptions can suddenly discharge a huge amount of material into the troposphere, but it settles and precipitates out soon after the eruption and thus exerts no long-term effects on climate (unless many volcanoes erupt sequentially).

Although we have no direct long-term records of tropospheric aerosols, human activities have unquestionably increased their concentrations. This is especially true over industrialized land areas. Unlike some gases, however, which have very long residence times (Chapter 1), individual aerosols do not have long life spans in the troposphere. As a result, aerosols produced over urban or industrial centers settle out before they can be widely dispersed across the globe, thereby leading to concentrations that vary widely both temporally and spatially. As a result, their climatic effects tend to be isolated to areas near the pollution sources.

Until recently it was not known whether the direct effects of tropospheric aerosol increases would lead to an overall warming or cooling near the surface. Numerical models now tell us that increased aerosol contents have the net effect of reducing surface temperatures globally. In fact, the cooling due to anthropogenic aerosols appears to have been about the same order of magnitude as the warming that would theoretically accompany observed increases in greenhouse gases over recent decades. Thus, if the addition of greenhouse gases into the atmosphere has helped promote warming during the last century, its effect may have been masked by the cooling effects of aerosols. But we cannot expect the aerosol effect to offset the greenhouse gas effect indefinitely. If there were to be a stabilization in the release of anthropogenic emissions, atmospheric CO_2 contents could take up to 200 years to reach a new equilibrium level, while aerosol levels would respond almost immediately. Thus, the greenhouse effect would continue to increase while the aerosol effect would stabilize and cease to offset the opposing effect. Despite the possibly beneficial effects of aerosols on climate change, it should be remembered that aerosols exert their own negative effects on humans and the environment (Chapter 14).

Stratospheric Aerosols Unlike tropospheric aerosols, those in the stratosphere result primarily from natural processes; human activities are not as important. The stratosphere maintains a background level of aerosols introduced by the upward diffusion of sulfur gases from the troposphere, which then undergo gas-to-particle conversion. Though background levels remain fairly constant through time, significant increases can occur in the months following volcanic eruptions.

Stratospheric aerosols can remain in the stratosphere longer than their tropospheric counterparts for two reasons: First, they tend to be smaller and therefore have lower terminal velocities. More importantly, the most effective agent of aerosol removal, precipitation, is not important in the stratosphere.

As previously stated, aerosols affect the energy balance by absorbing and backscattering solar radiation and by absorbing and radiating longwave radiation. The reduction in solar radiation reaching the surface would favor a lowering of tropospheric air temperatures, while the absorption of outgoing longwave radiation and the increase in downward emission would promote surface warming. So which actually dominates in the case of stratospheric aerosols? It turns out that the relative importance of shortwave and longwave radiation changes depending on the size of the aerosols. If the stratospheric aerosols are small, the reduction in solar radiation reaching the surface exceeds the gain in longwave radiation. In fact, the aerosols are small and promote lower temperatures for up to several years after volcanic eruptions.

Several recent volcanic eruptions have provided scientists with useful observational data on the effects of stratospheric aerosols. Mount St. Helens in Washington underwent a major eruption on May 18, 1980. Despite the massive ejection of solid material (much of one side of the mountain was blown away), it caused little if any major impact on hemispheric weather because it released relatively small amounts of sulfuric gases, which can ultimately transform to aerosols. Though the daytime sunlight was blocked out downwind of the eruption for a few days, no effects were noted much beyond that time frame.

The April 4, 1982, eruption of El Chichón in Mexico was far more violent than that of Mount St. Helens and, more significantly, it released a particularly large amount of sulfuric gases. The result was an increase in atmospheric albedo, which decreased global temperatures by about 0.2 °C (0.4 °F) for several months. The Mt. Pinatubo (Philippines) eruption of June 12, 1991, is believed to have released about twice as much sulfuric gas as did El Chichón, along with the predictable effect of an even greater reduction in global temperatures. The eruption increased the atmospheric albedo both directly (by enhanced aerosol backscatter) and indirectly (by an increase in the reflectivity of clouds). The cooling effect was greatest by about August 1992, when the global mean tropospheric temperature decreased by about 0.73 °C (1.6 °F) from that of June 1991 (despite the fact that August is climatologically warmer than June in the Northern Hemisphere). Figure 16–8 shows the change in aerosol content over the globe prior to (a) and during four periods (b), (c), and (d) following the eruption. Aerosol contents increased 100-fold during the peak concentration.

Changes in Radiation-Absorbing Gases

The issue of potential climatic warming due to increases in carbon dioxide and other greenhouse gases has been the subject of intense scientific scrutiny and political controversy. At its essence, the subject is really quite simple. The burning of fossil fuels and the clearing of forested areas has led to a steady increase in the carbon dioxide content of the atmosphere (see Figure 1–5).

Carbon dioxide is an effective absorber of longwave radiation. Thus, increases in its content through time has the potential to exert a warming effect on Earth's climate: As CO_2 builds up, there can be an increase in the absorption of outgoing radiation and the establishment of a higher equilibrium temperature in the lower troposphere. On the other hand, the stratosphere loses energy mainly by emission of longwave radiation. Because increasing CO_2 leads to greater emission, the stratosphere (and upper troposphere) are expected to cool as CO_2 levels grow. Given the

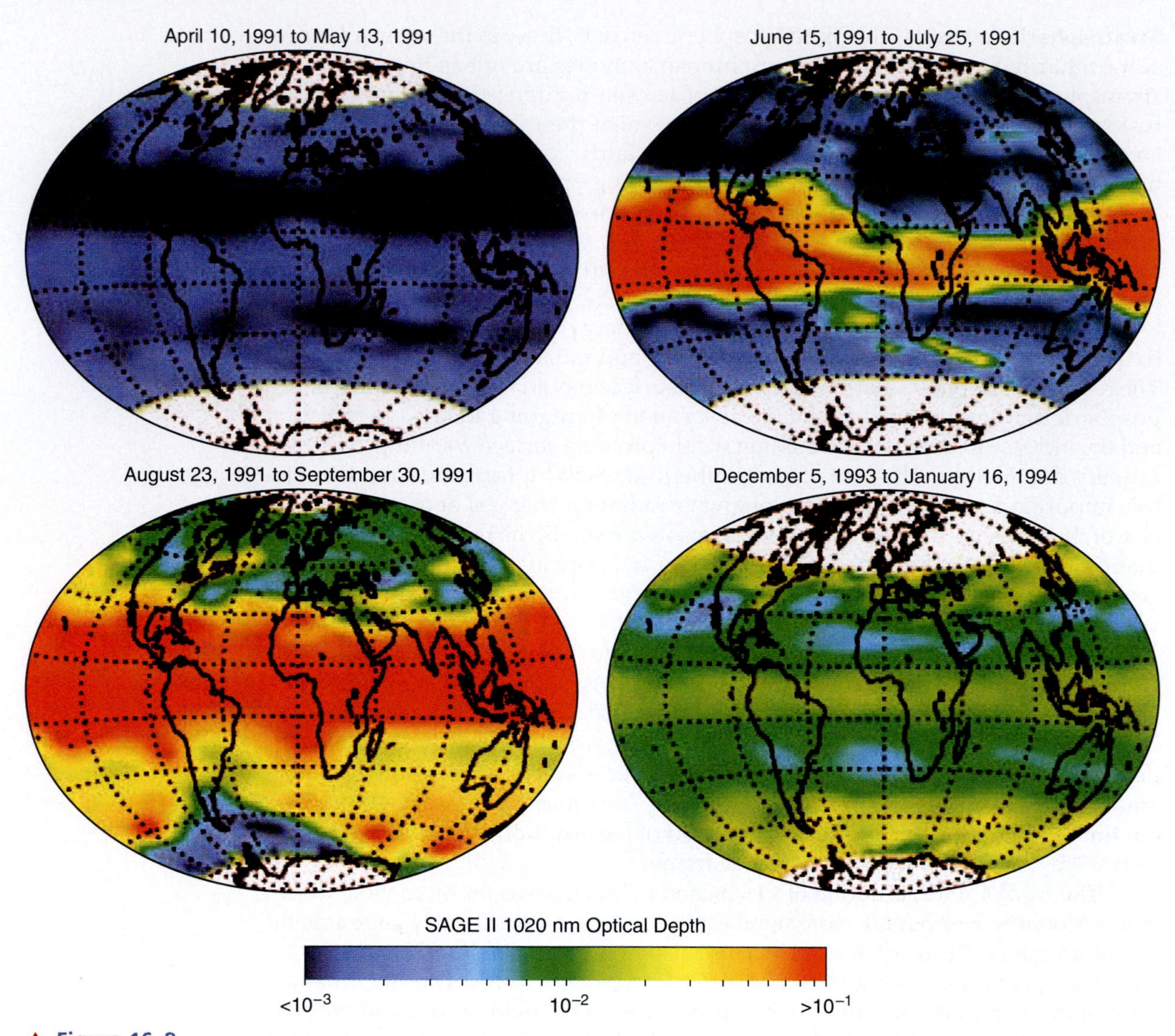

▲ **Figure 16–8**
The average aerosol content over the globe prior to and after the 1991 Mt. Pinatubo eruption. Redder areas indicate greater aerosol contents.

complexity of the atmosphere, it should not be surprising that there are a host of processes and conditions that are not known well enough for us to establish with certainty the exact outcomes of the buildup. What follows is an overview of some of the important components of the greenhouse gas question.

Anthropogenic Contributions of CO_2 Although much of the media discussion about the current increase in greenhouse gases centers on carbon dioxide, CO_2 is only one of several anthropogenic greenhouse gases that absorb outgoing longwave radiation. Methane (CH_4), nitrous oxide (N_2O), and chlorofluorocarbons (CFCs) are also effective absorbers whose contents are currently increasing in the atmosphere. In fact, only little more than half of the increase in net radiation accounted for by these gases is due to carbon dioxide. The other gases combine to produce nearly as great an effect as carbon dioxide.

As we will see, the potential for these gases to promote a global warming is only partly due to their absorption of longwave radiation. They also can promote warming through the indirect effect of increasing water vapor contents in the atmosphere. That is, if their absorption of radiation increases the atmospheric temperature, net evaporation should increase and thereby raise the water vapor content of the air. This would lead to even greater absorption, because water vapor is even more effective at absorbing longwave radiation than the anthropogenic trace gases.

Though water vapor is an important contributor to any potential increase in greenhouse warming, it will be excluded from the rest of this discussion because its increased level in the atmosphere is an indirect response to human activities.

Since the middle of the nineteenth century, there has been an exponential increase in the input of carbon dioxide to the atmosphere by fossil fuel consumption. Not surprisingly, most emissions have come from the developed nations, with 90 percent originating in the Northern Hemisphere. Between 1980 and 1989, fossil fuel combustion released about 5.4 billion tons (gigatons, Gt) of carbon (plus or minus 0.5 Gt) into the atmosphere each year. This is supplemented by another 2 Gt per year resulting from deforestation. Deforestation adds CO_2 to the atmosphere through direct burning of the logged trees, decomposition of biomass, and other processes. Moreover, the removal of trees reduces the ability to remove subsequent inputs of carbon dioxide by photosynthesis.

Exchange of CO_2 between the Atmosphere and Ocean At the current rate, humans are adding enough carbon dioxide to increase the atmospheric CO_2 content by 3.5 parts per million (ppm) per year over the current value of about 370 ppm. But the actual rate of increase is only about half as large. So where does the lost carbon dioxide go? About half enters the ocean, with the rest unaccounted for.

There is a constant exchange of carbon dioxide between the ocean and atmosphere as gas molecules cross back and forth across the ocean surface. The rate of exchange depends on several variables, including wind speed, water temperature, and the difference between the CO_2 content of the atmosphere and surface layer of the ocean. Overall, the atmospheric CO_2 pressure is higher than the ocean values, so there is a net transfer into the ocean, making the oceans a net sink for the greenhouse gas. Obviously, carbon cannot accumulate indefinitely in the upper ocean, or else it would cease to receive atmospheric carbon. Photosynthesis by phytoplankton and other marine plants removes carbon, whereupon it enters the marine food chain. Eventual settling of plant and animal remains transfers carbon downward, acting like a kind of biological pump. One of the important issues in predicting future atmospheric CO_2 levels is whether the net rate of carbon dioxide removal from the atmosphere into the ocean will change in the future.

If the removal rate were to decrease, there would be an accelerated increase in atmospheric carbon dioxide levels. One estimate is that increasing ocean temperatures will lead to a future reduction in the ocean's ability to take in atmospheric carbon dioxide, causing a 5 percent increase in the rate of atmospheric CO_2 growth. But this is clearly a very complicated question, depending as it does on both purely physical processes and on the vigor of the biological pump. One can imagine that climate changes might perturb marine ecosystems in ways that overwhelm the effects of changing solubility.

As mentioned earlier, carbon dioxide is not the only greenhouse gas released to the atmosphere by human activities. Unlike carbon dioxide, several of these come from sources quite different from fossil fuel burning. The important point to keep in mind here is that collectively they are about as important as CO_2 with regard to potential climatic change (despite lesser mention in press reports). Yet, even so, changes in the concentration of these gases are just part of the puzzle of climatic change. And our ability to predict future changes is further complicated by the interconnections of different components of the atmospheric systems. These interconnections are known as **feedbacks**.

Feedback Mechanisms

Consider a simple system that consists of only two variables, and assume that changes in either one affect the state of the other. If there is a change in the first variable, it will produce a change in the second, which will in turn affect the first variable.

Some feedbacks tend to inhibit further changes in the system, while others support further changes in the system. Those that inhibit further change, called

negative feedbacks, are self-regulating. In other words, when the second variable responds to initial change in the first variable, its response will suppress further change in the first. Feedbacks that amplify change in the initial variable are called **positive feedbacks**. When a positive feedback situation exists, the response of the second variable causes the initial change to grow (you can also think of this as a "snowball effect").

Of course, the Earth system involves many feedbacks, some positive, some negative. The situation is further confounded by the fact that the various feedbacks operate at vastly different rates: some are instantaneous, whereas others operate over thousands of years. In some cases relationships seem relatively straightforward, but for others it is difficult to determine if a single connection between two variables is positive or negative. With these points in mind, we will examine some of the important feedback mechanisms related to climatic change.

Ice-Albedo Feedback

Much of Earth's surface is occupied by large continental ice sheets and floating sea ice (Figures 16–9 and 16–10). If the atmosphere were to cool along the margins of these ice masses, there could be an expansion of the ice-covered area. Similarly, any warming would likely lead to melting and a retreat of the ice margin. Because ice has a higher albedo than most other natural surfaces, an expansion of the ice would lead to a reduction in the amount of insolation absorbed by the surface, which in turn would lead to further cooling (a positive feedback). Likewise, a retreat of the ice associated with a warming climate would increase the amount of surface no longer covered by the reflective ice and, therefore, lead to enhanced warming. Thus, there is a positive feedback mechanism that favors continued warming or cooling once such a temperature trend is initiated. It is largely for this reason that middle and high latitudes show larger changes than low latitudes. Where there is little prospect for ice (the Tropics), this feedback does not operate. Of course, ice advances and retreats do not continue unchecked, because other feedback mechanisms prevent the ice from totally covering the globe or from completely disappearing.

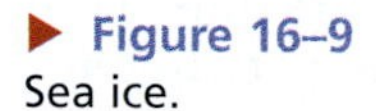

▶ **Figure 16–9**
Sea ice.

◀ **Figure 16–10**
The distribution of Northern Hemisphere sea ice, as obtained by satellite based radar imagery.

Evaporation of Water Vapor

Another positive feedback mechanism involved in climatic change concerns the water vapor content of the atmosphere. As we pointed out in Chapter 5, more water can exist as a vapor in warm air than in cold air. If the climate were to warm, there would likely be an increase in evaporation from the surface and an increase in the atmospheric water vapor content. Because water vapor is an extremely effective absorber of radiation emitted by the surface, higher water vapor contents should increase the atmospheric absorption of thermal energy and thus amplify the initial warming. In fact, computer models tell us that water vapor feedback contributes about half of the warming associated with increasing carbon dioxide, methane, nitrous oxide, and chlorofluorocarbons.

Ocean–Atmosphere Interactions

One obvious relationship between the oceans and the atmosphere relates to the effect that warming temperatures has on sea levels. Increases in global temperatures can raise the mean sea level, primarily by the expansion of warmer waters. Increases in atmospheric temperature can also contribute to rising sea levels by causing glaciers to melt and release water back to the oceans. A recent study estimates that increasing temperatures associated with a projected doubling of carbon dioxide within the next half century could cause the average sea level to increase by 19 cm (8 in.), but the rise in sea level would not be uniform throughout the oceans, and the rise could be as much as 35 cm (14 in.) off the coast of Europe.

On longer time scales, 500 to 1000 years, melting of the West Antarctic ice sheet becomes a possibility, which would raise sea levels by 5 m or more. Unlike any other ice sheet, the West Antarctic sheet is a concern because its base is below sea level. It is therefore susceptible to melting by ocean waters eroding its periphery. Until recently, this idea was not widely accepted, in large part because there was no evidence that such a thing had ever happened before. In mid-1998, however, information gleaned from sediments beneath the ice sheet strongly suggested that part or all of the sheet collapsed sometime in the last 2 million years, most likely in the interglacial period 400,000 years ago. Sea level changes, whether large or small, feed back on climate patterns through a number of mechanisms, including effects on surface water currents and the movement of deep water below the surface, the proportion of land and ocean, and marine and terrestrial productivity, to name a few.

A very important feedback, only recently recognized, concerns global-scale ocean circulations. As we discussed in Chapter 8, there is a large flow of water northward in Atlantic surface waters that delivers tremendous heat to the overlying atmosphere. After cooling and sinking in the North Atlantic, water flows southward at great depth to the rim of Antarctica, where it joins deep water forming at the edge of the ice cap. Other branches of deep water flow northward into the Indian and Pacific Oceans, eventually rising and returning to the southern ocean at the surface. This huge, beltlike circulation system is partly driven by temperature differences, but also by salinity variations; hence, it is called **thermohaline circulation**.

There is now convincing evidence to suggest that this pattern is not steady, but instead makes rapid jumps from one mode of operation to another. Rates and locations of sinking water change abruptly, which in turn affect other aspects of the pattern. The new configuration lasts for some time before changing abruptly again. Numerous reorganizations of such ocean circulation have been found in a marine record running from about 60,000 years ago until about 10,000 years ago. Because they appear at nearly the same times as large swings in Greenland air temperature, they are associated with the millennial-scale climate oscillations described earlier. In this way, changes in ocean circulation are implicated in abrupt climate jumps, at least during the last glacial period. By implication, the same forces are responsible for millennial-scale changes in other periods, though that has yet to be demonstrated.

Feedback from atmosphere to ocean arises as changing climate alters patterns of runoff, precipitation, and evaporation, which in turn influence ocean temperature and salinity. If these changes are large enough, ocean circulation shifts to a new mode, which in turn influences global climate. If true, this feedback mechanism provokes a concern that relatively modest global warming might cause another rapid reorganization perhaps 100 years from now, when world population will demand about three times today's food supply. If the resulting climate change turns out to be as large as the one 8200 years ago (which occurred when the climate was even warmer than now), the results could be disastrous.

It should be noted here that the huge mass and high specific heat of the oceans, along with their vertical and horizontal mixing, create a strong thermal inertia. This means that warming of the ocean in response to increasing atmospheric temperatures would be very slow. Thus, oceanic temperature changes would be retarded in response to global warming, and there could be a time lag of several decades before any increase in atmospheric temperature would be observed in the oceans. Once the warming is realized, it would modify atmospheric pressure and wind distributions.

Atmosphere–Biota Interactions

Changes in climate are linked with land–vegetation patterns. The influence of climate on vegetation is easy to comprehend—banana trees do not grow in Greenland and there is no tundra vegetation in the Amazon Basin. Less obvious, perhaps, is the fact that vegetation likewise influences the climate in a number of ways. By transpiring moisture to the air, the presence of plants affects the moisture content of the atmosphere and the likelihood of precipitation. Plant assemblages also affect the surface albedo and the transfer of heat at the surface.

One of the most important feedback mechanisms involved in the possible warming of the atmosphere involves the influence of CO_2 concentration on photosynthetic rates. It has been known for some time that many plant species undergo accelerated growth and enhanced photosynthesis in an environment rich in carbon dioxide, and some species also do better under warmer conditions. This could create a negative feedback (with potentially positive results as far as people are concerned) in which increasing CO_2 contents and temperatures allow plants to suppress further increases in the greenhouse gas.

Scientists believe that this "fertilization" process may already be appearing, especially in the latitudes between about 45° and 70° N. Satellite observations indicate that the region's supply of green vegetation is increasing and that its growing season is beginning about a week earlier and ending a week later than it did prior to the 1980s. These effects would be consistent with those expected from increased warmth and CO_2 levels observed in the area. At the same time, the seasonal oscillations in Northern Hemisphere carbon dioxide contents described in Chapter 1 have undergone changes in the intensity and timing of their cycles. The spring decrease in carbon dioxide associated with the leafing of deciduous plants is also occurring about a week earlier, and the difference between springtime maxima and late-summer CO_2 minima is increasing. Changes in the seasonal cycle of CO_2 are most conspicuous near the Arctic, where the magnitude of the oscillations has increased by 40 percent over the past few decades.

On the other hand, a lack of nutrients can become a limiting factor in plants' ability to response to an enriched CO_2 atmosphere. This has recently been demonstrated in the Alaskan Arctic, where tundra vegetation was artificially subjected to twice the carbon dioxide level of the normal atmosphere. At first, the vegetation responded to the fertilization with increased growth rates. But by the third year, elevated CO_2 ceased to have any effect.

In a separate experiment, it appeared that high CO_2 contents might even *harm* tropical rainforest plants by causing a loss of nutrients in the soil. And another recent study has shown that trees in tropical rain forests have been growing, maturing, and dying more rapidly than in the past. This leads to a more open environment that favors replacement of trees with vines requiring greater amounts of sunlight. Ironically, the vines are less effective at photosynthesis than the trees they replace, and they don't even consume enough CO_2 to offset that released by the decaying trees.

Increases in atmospheric CO_2 could lead to other unwelcome results. It is possible, for example, that certain weeds would benefit more from higher carbon dioxide concentrations than would agricultural plants. Insects and other pests might also enjoy a warmer environment associated with increased CO_2 becoming more troublesome to agriculturalists than they are today.

General Circulation Models

General circulation models are mathematical representations of the atmosphere run on supercomputers. The models couple known laws of physics with prescribed initial and boundary conditions of the atmosphere to compute the evolution of the atmosphere through time. By averaging the output over a number of years, an estimate of climatic response can be obtained. These models help us understand the types of equilibrium and nonequilibrium conditions the atmosphere could assume in response to designated changes (such as a doubling of the carbon dioxide content).

At present, there are several such models in operation at various research institutions and universities worldwide. Although the models are all based on the same laws of physics, they each have different ways of dealing with processes that cannot be explicitly represented based on physical laws. For example, they all must make certain assumptions about what sets of conditions would lead to the formation of clouds and the height at which those clouds would exist. Variations in some of these underlying assumptions result in different regional outputs by the various models. Another important difference concerns the oceans. Some use fixed ocean temperatures, some allow SSTs to change but have no ocean currents, some have only fixed ocean currents, and some allow the oceans to respond fully to changes in the atmosphere. The appendix to Chapter 13 deals with many of the issues involved in weather forecast models that are similar to GCMs.

In 2001 the Intergovernmental Panel on Climate Change (IPCC) used the results of recent GCM runs to predict the range of possible outcomes from elevated atmospheric CO_2 concentrations. Their models indicated a "worst-case" outcome resulting in an increase in global average temperature of as much as a 5.8 °C (10.4 °F)

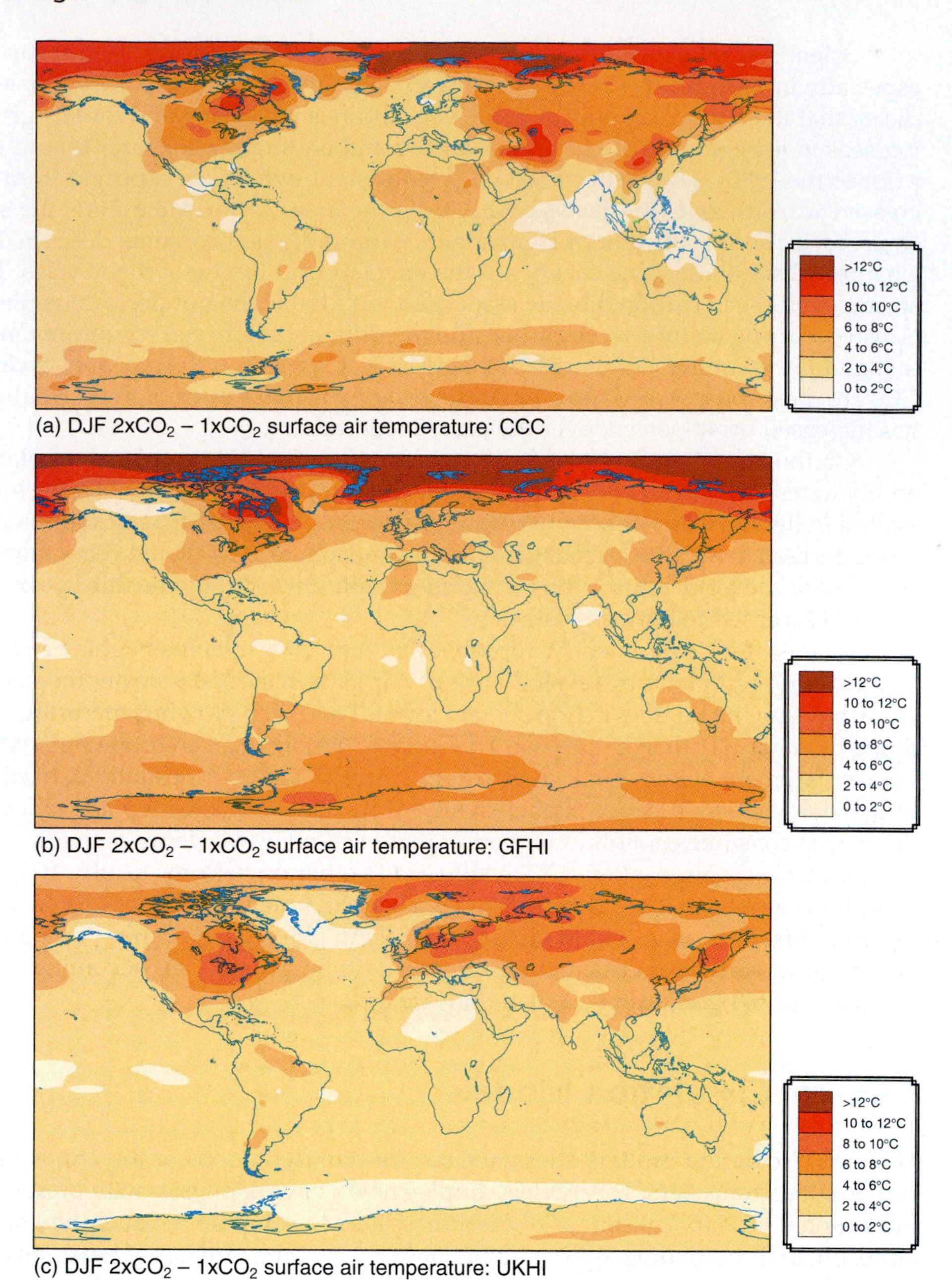

▶ **Figure 16–11**
Temperature changes projected by three GCMs in response to a doubling of atmospheric carbon dioxide.

in the year 2100, relative to the 1990 value. In its previous report issued in 1995, the panel predicted a worst-case increase of 3.5 °C (6.3 °F). The difference in the two forecasts arose because the new estimate incorporated the reduction in sulfuric particulates in the atmosphere due to environmental regulations. These particulates, though responsible for acid and other environmental hazards, have suppressed global warming by reducing the amount of insolation reaching the surface. The reduction of these aerosols may reduce the degree to which they suppress further warming.

Figure 16–11 illustrates the climatological changes in temperature for December through February projected by three GCMs in response to a presumed doubling of carbon dioxide. The models all predict substantial warming, particularly in the high latitudes. Although the various models are in reasonably close agreement with regard to worldwide changes in temperature, there is considerable variation in their regional patterns. This is clearly evident over North America, where the location of greatest temperature increase is highly variable from model to model. Another clear difference can be seen in western Asia, where the GFHI model indicates little increase in temperature, in distinct contrast to the output of the other two models.

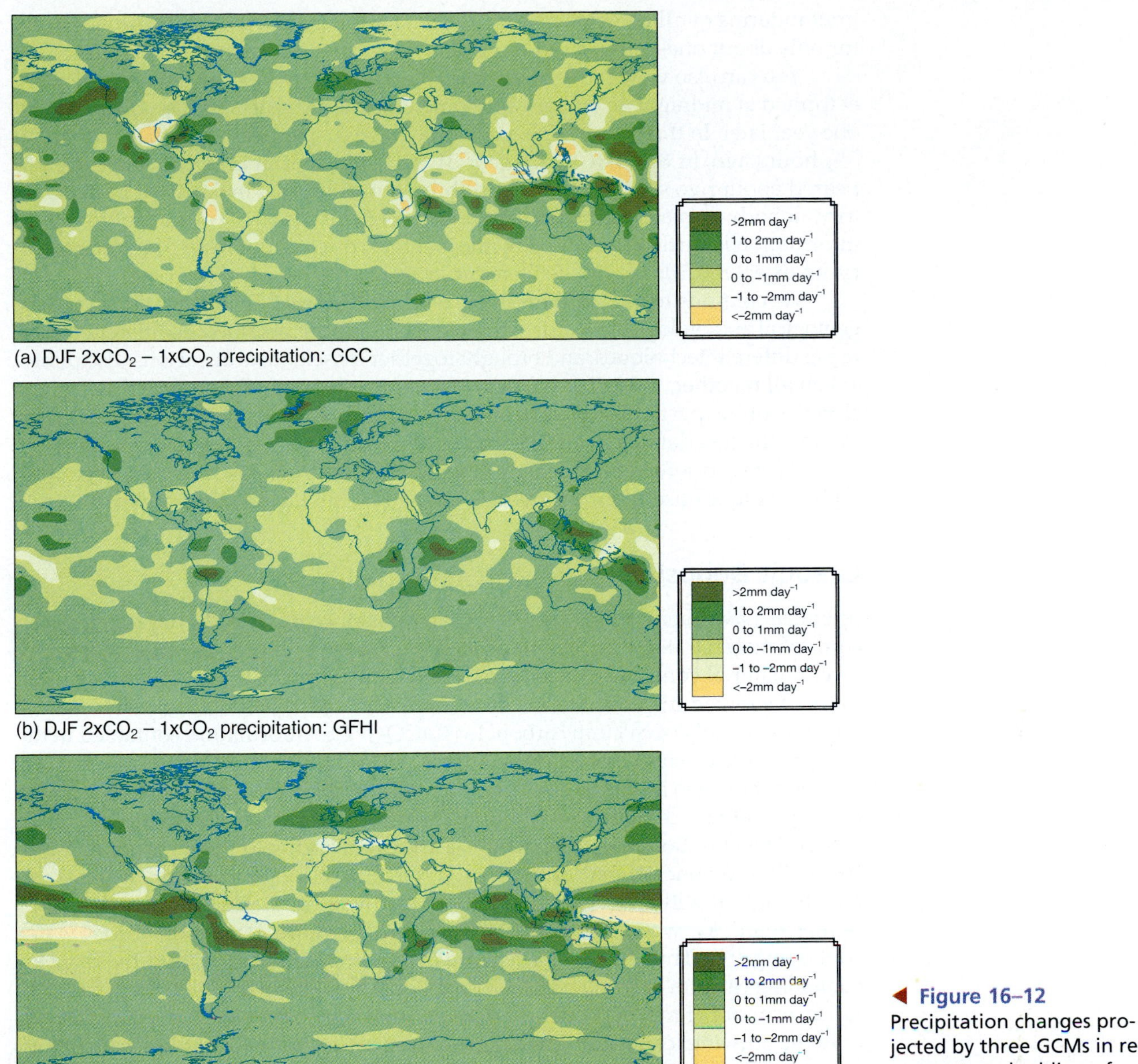

(c) DJF $2xCO_2 - 1xCO_2$ precipitation: UKHI

◄ Figure 16–12 Precipitation changes projected by three GCMs in response to a doubling of atmospheric carbon dioxide.

There is even less consensus among the models for the global distribution of precipitation in response to a doubling of CO_2 (Figure 16–12). All three models show large areas of enhanced precipitation over the Tropics, but in the mid-latitudes there is considerable variability. Again, using North America as an example, there is good agreement between the GFHI and UKHI predictions, but the two differ substantially from the CCC model output.

Methods for Determining Past Climates

Though there are huge gaps in our knowledge of past climates, scientists have learned much about the climatic history of Earth. It is really quite remarkable that we are able to describe past climates as well as we can. Consider the fact that although Earth formed about 4.5 billion years ago, people are believed to have inhabited the planet for only about 2 million years. While 2 million years is an extremely long period of time relative to our own experiences, it is actually very brief in terms of all geologic history. In fact, people have walked the face of Earth for only about one-half of a thousandth of the planet's existence.

You can also view the relative time spans another way: Imagine that the planet formed at midnight on January 1 of some vast cosmic year, and it is now midnight one year later. In this scenario people would have first appeared on the scene about $3\frac{1}{2}$ hours ago, at 8:27 P.M., December 31. The first thermometer would have appeared about two seconds ago. Thus, throughout the vast majority of Earth's history, nobody was around to observe and record the climate. The lack of firsthand information is further worsened by the fact that, throughout most of human history, no direct records of climate were preserved.

We continue to gain insights into past climates based on information left in the geological and biological records. The evidence has been obtained in different places, using different techniques, and brought together by scientists in various disciplines. Taken all together, the information gives us some idea of what **paleoclimates**, the climates of the past, must have been like. Not surprisingly, the further back we go in time, the less detailed is the information provided by the various indicators.

We now briefly describe some methods for studying past climates, beginning with techniques useful for uncovering change on the longest time scales.

Oceanic Deposits

Scientists have been drilling into the ocean floor since the 1970s. They extract deep cores of material that has been deposited over very long time periods, with more recent material constantly burying older material previously laid down. Included in the deposited material are the bones and shells of plankton and other animal life, made largely of calcium carbonate ($CaCO_3$). The information contained in the oxygen in the calcium carbonate is most important for determining past climates.

Most oxygen atoms have an atomic weight of 16 (^{16}O), but a small percentage of oxygen atoms contain two additional neutrons per atom, making their atomic weight 18 (^{18}O). Both *isotopes* exist in ocean water, and both get incorporated into the shells and bones of marine animals. If the ratio of ^{18}O to ^{16}O in the water is relatively high, it will also be high in the sea life living in that water. Because ^{16}O is lighter than ^{18}O, water containing ^{16}O evaporates more readily than does water containing ^{18}O. Thus, when glaciers are expanding, the oceans (and the oxygen-containing calcium carbonate in shells and bones) will have relatively high $^{18}O/^{16}O$ ratios, as more of the ^{16}O water is removed from the ocean and deposited as snow onto the growing ice sheets. When the organisms die, they sink to the ocean bottom where their calcium carbonate is deposited. The ocean bottom thereby maintains a record of climate through the varying ratios of $^{18}O/^{16}O$ in its layers. Scientists extract cores of the ocean-bottom material, note the isotope ratios, and infer past changes in global ice volume. Information obtained from sea cores was instrumental in overturning a long-held, but mistaken, idea that there were four glaciation episodes during the Pleistocene.

Other core indicators of past climates comes in the form of material removed from land surfaces by glaciers. Icebergs shed by continental glaciers carry rocks, pebbles, and other debris equatorward, and this material is deposited on the ocean floor when the iceberg melts. Called **ice-rafted debris**, this material was a key in discovering millennial-scale oscillations.

Ice Cores

Scientists have also determined $^{18}O/^{16}O$ ratios for deep **ice cores** obtained from the Greenland and Antarctic ice sheets and from alpine glaciers at lower latitudes. When snow is deposited onto existing glaciers, the oxygen in the H_2O may be either ^{18}O or ^{16}O. Snow that falls under relatively warm conditions contains a higher ratio of the heavier isotope. On the ice caps, scientists from Europe and the United States have been drilling through the ice and extracting cores nearly 3 km (1.8 mi) deep to infer past temperature patterns.

16–2 Focus on the Environment

Plant Migrations and Global Change

As global temperatures have warmed and cooled, plant communities have responded by migrating poleward during periods of glacial retreat and equatorward as glaciers advanced. For example, large tracts of species such as spruce that once inhabited much of what is now the northern and northeastern United States have shifted northward into Canadian regions formerly covered by glacial ice.

The rate at which plant communities can migrate in response to changing climates is important to their survival. If plant communities are too slow to adapt to changing conditions, the vegetation type faces possible extinction. Because of the possibility of future, rapid increases in global temperatures, scientists are concerned about the potential for major changes in Earth's plant communities. But recently obtained evidence suggests that plants migrate more rapidly in response to climate changes than previously believed. Thus, the threat of extinctions may be less menacing than once feared.

On the other hand, it is clear that some species in a plant community might be better able to expand their boundaries than others. Such changes in the rate of migration for individual species would lead to a change in the plant community, in which the overall composition of the plants in the new environment is different from what it previously was. Moreover, the expansion of suburbs and agriculture into previously undisturbed areas has caused some plant communities to become fragmented. In other words, instead of having extensive areas of a particular plant community, those vegetation associations now occur in isolated patches. The fragmentation of these communities hinders their ability to migrate in response to climate change and could make some future migrations impossible.

An interesting example of rapid vegetation change due to increasing temperatures has been observed in the grasslands of northeastern Colorado. A trend toward increasing nighttime temperatures has caused the average date of the last killing frost to occur earlier in the spring. This has put the normally dominant grass species—the blue grama, which historically has accounted for 90 percent of the ground cover—at a competitive disadvantage compared to various weeds. The weeds that are taking over the grasslands landscape are far more susceptible to drought. This is highly significant to ranchers who rely on the grass cover to supply most of their livestock's food needs.

Oxygen isotope studies have shown that at the end of the Pleistocene, the temperature over the Greenland ice sheet warmed about 9 °C (16 °F) over several decades. It was not known, however, whether the Arctic warming preceded or followed climatic changes in the Tropics. A 1999 study based on nitrogen and argon isotope ratios ($^{15}N/^{14}N$ and $^{40}Ar/^{36}Ar$)* in the Greenland ice sheets strongly suggests that Arctic warming probably preceded tropical warming or increased precipitation (or both) by two to eight decades. If true, this finding is consistent with the notion that changes in thermohaline circulation can be an important mechanism for triggering rapid climatic shifts.

In addition to the temperature data obtained from isotope ratios, ice cores provide information on the past chemistry of the atmosphere and on the incidence of past volcanic eruptions. As new snow falls onto a glacier, bubbles of the ambient air become permanently trapped in the ice. The concentration of carbon dioxide and other trace gases in these bubbles yields a long-term record of their varying levels in the air. Among the more interesting results from chemical analyses at the cores is the strong correlation between past temperatures and carbon dioxide concentrations. Past periods of high temperature coincide with high CO_2 concentrations, whereas glacial periods coincide with reduced CO_2 concentrations. For some scientists, this result supports the notion that there may be future climate warming as a result of the emission of greenhouse gases. Others see this only as evidence that climate change influences land and marine ecosystems.

Ice sheets also provide valuable information about major volcanic eruptions. When such eruptions occur, some of the dust ejected into the atmosphere settles on the tops of glaciers. Researchers can determine when heavy volcanic activity occurred by noting the depth of the dust layers within the cores. The chemical composition of aerosols deposited in glacial ice also provides information on past events, with high acidities implying increased volcanic activity. This information may prove to be valuable in determining the importance of volcanic activity as a causal factor for climatic change.

*A description of how these ratios are used is beyond the scope of this discussion.

Remnant Landforms

All of Earth's landforms are the end result of processes that build up and wear down features at the surface. The largest features, such as mountains and valleys, are produced by *tectonic* forces, those that produce a deformation of Earth's crusts. Once formed, all such features become subject to erosional processes that remove material at the surface and transport it to other locations. When the forces transporting the material are no longer capable of moving the material, *deposition* occurs. There are several mechanisms for eroding and depositing material, including the movement of water, the slow-moving ice sheets expanding across the surface, wave action along coastlines, wind, and floating icebergs carrying land debris. Each of these mechanisms leaves certain telltale characteristics, and trained field scientists can use the evidence to infer climatic conditions at the time of erosion or deposition.

Features Associated with Ice and Water Obviously, glaciers can exist only under cold climatic conditions. Fortunately for the earth scientists, erosional and depositional features caused by the movement of glaciers often leave very distinct signatures on the landforms that last long after the glaciers have fully retreated. Some of the features associated with glaciation are related to erosion. For example, as alpine glaciers expand down preexisting valleys, they widen out the lower reaches of the valleys. This can transform typical V-shaped valleys (looking up or down the valleys), associated with those cut by running water, to U-shaped valleys. Glaciers also often leave scratch marks on solid rock walls and valley floors, or polish exposed rocks to a very smooth finish.

Unlike running water, flowing ice sheets are capable of moving very large-sized sediment as effectively as small particles of sand, silt, and clay. When the ice sheet melts away, its deposits (called *till*) can contain a very wide assortment of sediment sizes, from microscopically small clays to very large boulders. Thus, poorly sorted deposits often exist along the past margin of an ice sheet. Where glaciers terminate in an ocean, land materials are rafted away and eventually deposited on the seafloor as icebergs melt.

Streams and rivers also provide useful evidence for reconstructing past climates. Depositional features can be particularly useful in this regard, because the size of material that can be transported by running water depends on the speed of flow. Rapidly moving streams can carry large rocks. If flow of water in a stream begins to slow down, the maximum size of sediment it can transport decreases, and the largest material it contains will be deposited. This allows us to infer that layers made up of very large sediment must have been deposited by large streamflows. Thus, by examining the layering, or *stratigraphy,* of stream banks or road cuts, one can get some idea of the sequencing of high and low precipitation episodes (Figure 16–13).

Waves along coastlines also leave distinct features that can be left behind after sea level rises or lowers. Thus, by examining the elevation of terraces or submerged coastal platforms, we can infer how much glacial ice has accumulated or melted.

Coral Reefs **Coral reefs** are hard ridges extending from the ocean floor to just below the water surface, along the shallow margins of warm, tropical oceans. They form by the growth of small marine organisms (coral) having hard shells, composed largely of calcium. Colonies of coral live together atop the hard material left behind by past coral. When these coral die, their shells remain at the top of the reef and provide the foundation upon which subsequent coral grow.

Coral reefs have been useful in several ways for the inference of past conditions. Because they exist along shallow waters, relic coral reefs can provide information on the location of past sea levels. Moreover, because the chemical composition of growing coral is affected by the water temperature, analysis of the changing chemistry of coral reefs with depth provides useful information on past conditions. They have been particularly useful in providing information on past El Niño events.

◀ **Figure 16–13**
Layers of sediment composed of large material can indicate the occurrence of heavy precipitation in the past.

Past Vegetation

Climate is by no means the only important factor affecting the distribution of vegetation, but it does exert a strong influence on the distribution of vegetation communities. When a vegetation community occupies a region, some of its pollen and spores can be deposited and preserved indefinitely in lake beds or bogs. This preserved pollen can be extracted and identified by *palynologists,* who might then subject it to a technique called **radiocarbon dating**. Radiocarbon dating provides a good estimate of the age of material younger than about 50,000 years and allows a determination of the distribution of vegetation species that existed at various times during the past. This information can then be displayed in pollen diagrams (Figure 16–14) that depict the sequence of past vegetation assemblages. Because many types of vegetation stands are identified with particular climate types, these diagrams provide useful information for deciphering the climatic history of an area.

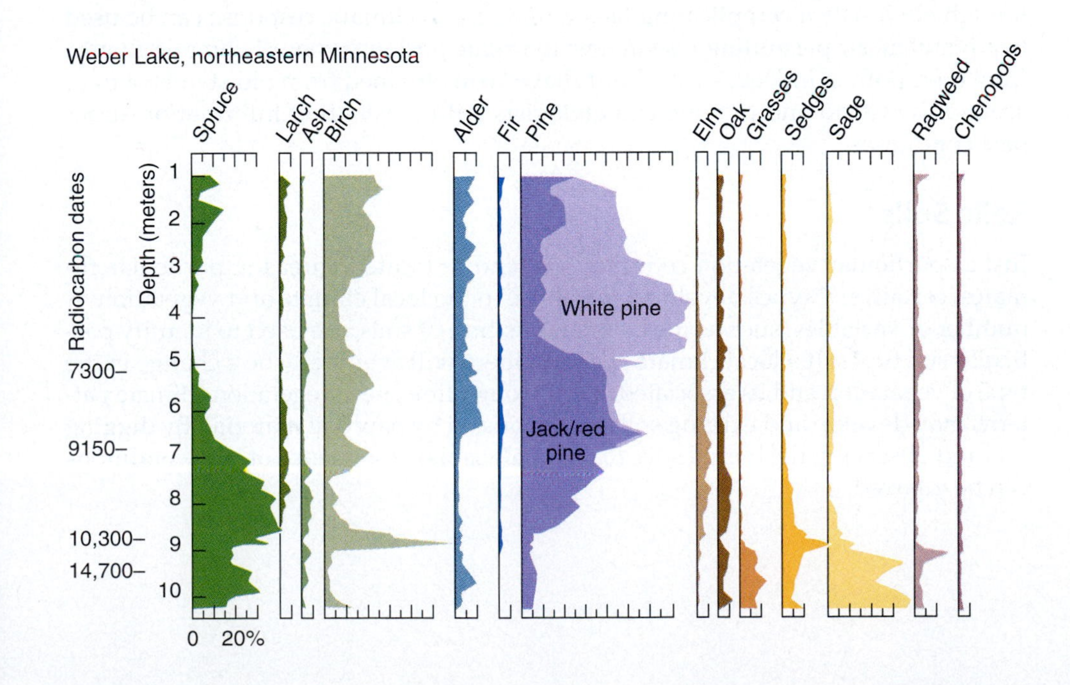

◀ **Figure 16–14**
Pollen diagrams provide information on past vegetation at a site, which is useful for determining past climates.

▶ **Figure 16–15**
Tree rings.

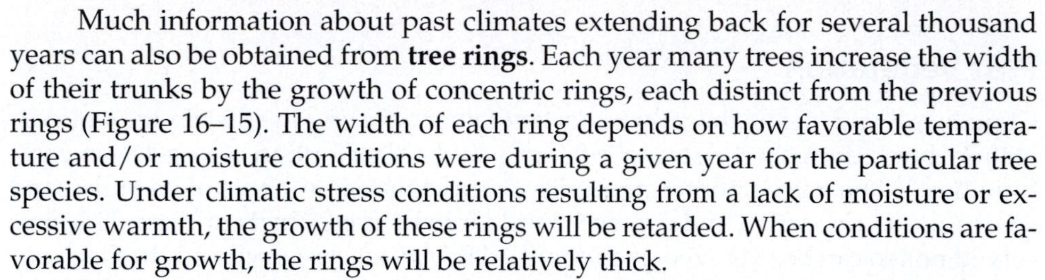

Much information about past climates extending back for several thousand years can also be obtained from **tree rings**. Each year many trees increase the width of their trunks by the growth of concentric rings, each distinct from the previous rings (Figure 16–15). The width of each ring depends on how favorable temperature and/or moisture conditions were during a given year for the particular tree species. Under climatic stress conditions resulting from a lack of moisture or excessive warmth, the growth of these rings will be retarded. When conditions are favorable for growth, the rings will be relatively thick.

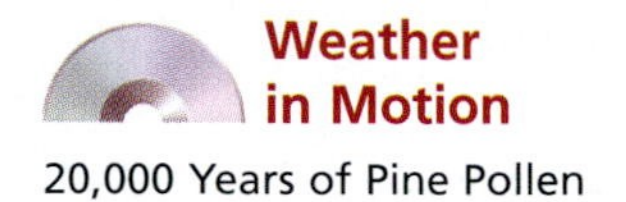

20,000 Years of Pine Pollen

Some tree species are sensitive to temperature variations, whereas other species are affected primarily by changing moisture conditions. In either case, the extraction of cores from the trunks of very old trees (the oldest ones date back more than a millennium) yields a continuous record of annual tree growth, which correlates with precipitation and/or temperature. The correlation depends on the species and on the environmental setting. At high elevations, for example, low temperatures might slow annual growth, while for the same species at lower elevation, warm temperatures might create slower growth through increased moisture stress. Although obviously a complicating factor, differential climatic response can be used to advantage by permitting researchers to isolate various types of climatic change from one another. Indeed, such records have been obtained from old stands of trees around the world and provide climatologists with a wealth of information about past conditions.

Relic Soils

Just as particular vegetation communities tend to locate in areas of particular climates, certain soil types develop in response to the local climate and vegetation. A number of variables, such as the color and texture of soils, are used to identify particular soil types. If a local climate changes, there will eventually be a change in the type of vegetation and its associated soil. Through time, new vegetation–climate patterns may develop and existing soils can be buried by new soil material. By digging pits and observing the buried soils (called **paleosols**), a sequence of past conditions can be inferred.

Summary

The climate of Earth has undergone numerous changes that have occurred over differing time scales, and there is every reason to expect that the climate will continue to change long into the future. Climatic changes have occurred with differing magnitudes over varying time frames. In general, the longer period oscillations have had more extreme changes than those over shorter time intervals.

At the longest time scales, Earth has been mostly warm, with little permanent ice. But at least seven times the planet has experienced an ice age, the most recent of which continues to the present. Within an ice age, glaciers grow and shrink in what are called glacial/interglacial cycles. Over the last 5 million years, these cycles have varied considerably, both in amplitude and length. Present times correspond to an interglacial, which followed the last glacial maximum of 20,000 years ago. Within this period (the Holocene), the most notable changes were a large abrupt cooling 8200 years ago and another much smaller one from 1400 to 1850 (the Little Ice Age). Over the historical period, temperature records suggest global warming by a few tenths of a degree Celsius. Throughout both glacial and interglacial times, there have been rapid changes in climate, as Earth jumps back and forth between two more-or-less stable states (millennial-scale oscillations).

We know of several processes that can cause the climate to change. For example, changes in solar radiation receipts over hundreds of thousands of years are caused by variations in Earth's orbit around the Sun—the Milankovitch cycles. Radiation receipts also vary in response to changes in radiation emitted by the Sun. Changes in the land surface of the planet may have been responsible for shifts in climate over both very long and relatively short time scales. Atmospheric turbidity increases from volcanic eruptions are believed by some scientists to have affected climate, although this viewpoint is highly controversial. The absorption of solar and thermal radiation by aerosols and radiation-absorbing gases may be important determinants in recent and future climatic conditions. All of these factors that influence climatic conditions do not act separately, but as part of complex feedback mechanisms connecting many parts of the Earth system. The effects of these feedbacks can be examined by the use of General Circulation Models (GCMs).

Most of what we know about past climates has been inferred through indirect evidence of past conditions. Such evidence includes remnant landforms, botanical information, the presence of old soils, oceanic deposits, and cores obtained from the Greenland and Antarctic ice sheets. Hypotheses regarding future climatic conditions and the impacts of increasing carbon dioxide and other gases can be analyzed with the aid of general circulation models.

Key Terms

intransitivity p. 476
ice ages p. 477
Pleistocene p. 479
glacial/interglacial cycles p. 479
Quaternary p. 480
Holocene p. 483
Younger Dryas p. 483
Medieval Warm Period p. 483
Little Ice Age p. 483
millenial-scale oscillations p. 485
Maunder Minimum p. 487
quasi-biennial oscillation p. 487
eccentricity p. 489
obliquity p. 489
precession p. 489
Milankovitch cycles p. 489
Pangaea p. 490
turbidity p. 491
feedbacks p. 495
negative feedbacks p. 496
positive feedbacks p. 496
thermohaline circulation p. 498
general circulation models p. 499
paleoclimates p. 502
ice-rafted debris p. 502
ice cores p. 502
coral reefs p. 504
radiocarbon dating p. 505
tree rings p. 506
paleosols p. 506

Review Questions

1. Explain how it is possible that the global climate can be both cooling and warming at the same time.
2. How does the present climate compare to past climates over the course of geologic history? Is it correct to say that the ice age is over?
3. Describe the frequency at which glacial/interglacial cycles have occurred during the Pleistocene.
4. What time frames constitute the Quaternary period and the Pleistocene and Holocene epochs?
5. What magnitude of mean temperature differences coincided with the various glacial/interglacial episodes of the Quaternary?
6. Which regions of North America experienced the greatest temperature differences from current values during the last glacial episode of the Pleistocene?
7. List the factors that can lead to climatic change. At what time scales do each of these occur?
8. Describe the factors that can lead to variations in the amount of solar radiation available at the top of Earth's atmosphere.
9. What evidence is there that variations in sunspot activity do or do not lead to climate changes on Earth?
10. How do changes in eccentricity and obliquity and precession interact to influence Earth's climate? What time scales apply to each?
11. What types of occurrences on Earth can affect atmospheric turbidity? How do turbidity differences affect global temperatures?
12. Describe positive and negative feedbacks and provide examples of each.

13. Explain how cores taken from ocean deposits and ice sheets can be used to infer past climate conditions.
14. Describe two types of remnant landforms that can provide information on past climates in a region.
15. What is radiocarbon dating, and how does it help in determining past climates?
16. How are tree rings used as indicators of past climates?
17. Describe how pollen samples obtained from old soils provide information on past climates.

Critical Thinking

1. The opening to this chapter discussed the concept of intransitivity. What would be the implications if Earth's climate were indeed found to be intransitive? Do you think that the question of whether the climate is intransitive can ever be answered?
2. The vast majority of atmospheric scientists now believe that at least some of the warming of recent decades is due to the activities of humans, but some are not convinced that the effect of this warming will be problematic for society. How would you argue for and against this viewpoint?
3. In addition to changes average temperature and precipitation, it is believed that the frequency of unusual events (such as droughts and floods) might be more common as a result of global warming. What regions of North America are more vulnerable to economic losses from these events than from increased temperatures?
4. As recently as 20,000 years ago a continental glacier extended as far south as the central United States. Describe what impacts the ice might have had on the average position and magnitude of the polar jet stream.

Problems and Exercises

1. View the Web site at **http://www.pewclimate.org/** and check to see if it includes any new press releases. If so, what are the major findings contained in them? Are these findings actually new or do they expand upon on known information?
2. Open the Weather in Motion module, "Retreat of Continental Ice Sheets," on this book's CD-ROM. Move the cursor on the map so that you can observe the retreat of the Northern Hemisphere glaciers over the last 21,000 years. Did the ice retreat uniformly across North America? Where in the Northern Hemisphere did the ice first begin to retreat?
3. Check your favorite newspaper or news magazine to see if there are any articles describing new findings about climate change or political issues related to the topic. How would you rank the importance of climate change relative to other major political issues?

Quantitative Problems

The atmosphere is an extremely complex system with numerous feedbacks. Atmospheric scientists use sophisticated models to better understand the interrelationships between its different components. The Quantitative Problems section for this chapter on the book's Web site at **http://www.prenhall.com/aguado** allows you to use a very simplified model to see how some of atmospheric variables may respond to changes in other variables.

Useful Web Sites

http://www.ipcc.ch/

Contains a vast amount of interest from the latest and previous reports of the Intergovernmental Panel on Climate Change. An extremely valuable resource for anybody interested in any aspect of climate change.

http://www.pewclimate.org/

Web page for the Pew Center for Global Climate Change, an advocacy group to promote action on issues related to the topic.

http://www.usgcrp.gov/

Comprehensive site of the U.S. Global Change Research Program. Includes numerous links, including summary reports of major research initiatives.

http://www.gcrio.org/ipcc/qa/cover.shtml

A useful primer on climate change from the U.S Global Change Research Information Office.

http://www.exploratorium.edu/climate/index.html

An interesting and well-designed site with pages dedicated to various aspects of global climate change.

http://www.unep.ch/iuc/submenu/infokit/factcont.htm

Published by the United Nations Environment Programme, this page links to 30 climate information pages discussing various aspects of climate change and its impacts.

http://www.climatechange.gc.ca/english/index.shtml

The Climate Change Web site produced by the Canadian government.

http://www.iisd.org/climate/climate_canada/

Web-based newsletter providing information on climate change as it affects Canada.

Media Enrichment

Weather in Motion

Retreat of Continental Ice Sheets

This movie illustrates the transition from the last glacial maximum to the present, showing changes in sea level and contintental ice over the last 21,000 years. Drag the mouse left/right to rotate Earth and up/down to move forward and backward in time. Note that at the outset nearly all of Canada and considerable portions of the northeastern United States were under ice. Asia and North America were joined by a land bridge, thanks to lower sea level. By 11,000 years ago, the western portion of North America is ice-free, and we see total ice extent not too different from today by 7000 before present (BP). By then, rising sea level accompanying deglaciation had opened the Bering Strait.

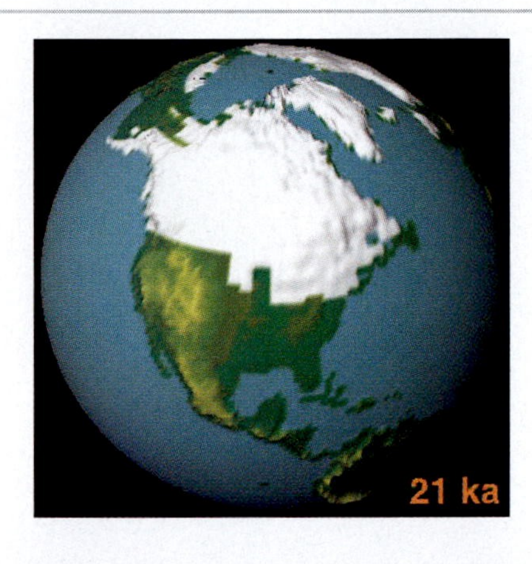

Weather in Motion

Orbital Variations

Solar radiation received by the planet is affected by three primary changes in Earth orbit: eccentricity, obliquity, and precession. In the interest of clarity, these movies show those variations greatly exaggerated. Notice how eccentricity influences the total radiation received on various days of the year. For example, the June solstice gets more or less radiation depending on the flatness of the orbit. By contrast, obliquity (the amount of tilt) influences the spatial distribution of radiation on a given day and the seasonal distribution at any latitude. Precession (a change in direction of tilt) determines the position of the equinoxes and solstices along the orbit. In "Precession II," you see the December solstice moving along the orbit, making one complete circuit during the precessional cycle of 20,000 years. The three orbital parameters vary simultaneously on separate time scales, creating very complicated patterns of change (click "All Parameters" to see this).

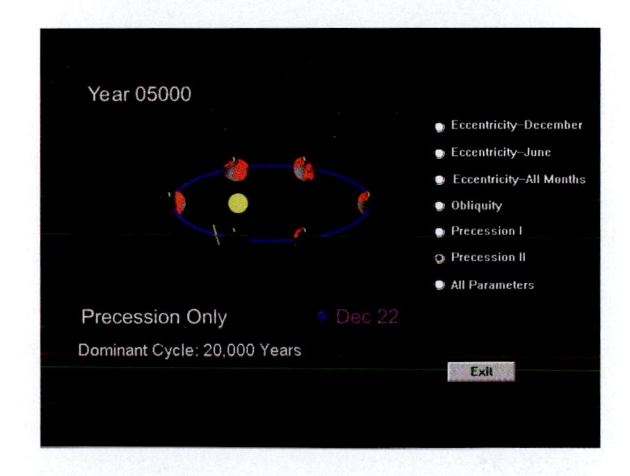

Weather in Motion

20,000 Years of Pine Pollen

The relative abundance of pollen from various plant species provides a useful indicator of past climatic conditions. This movie, in effect, shows the change in the distribution of pine forest over eastern North America during the most recent deglaciation. During the glacial maximum, pine forests were mainly found over the southeastern United States. With the onset of warming, they slowly came to occupy more northerly climes, achieving greatest abundance about 7000 years BP in a continous swath from the eastern seaboard to the Great Plains. Since then, pines have given way to other species, though they are still found in much of the region.

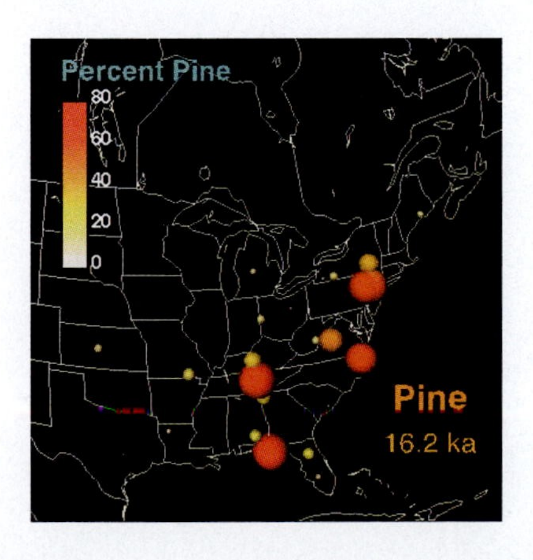

PART SEVEN

Special Topics and Appendices

A rainbow, one of many optical phenomena resulting from the systematic bending of light rays.

chapter 17

A mirage in southern New Mexico.

Atmospheric Optics

Imagine yourself driving along a straight, two-lane highway on a hot, sunny, summer afternoon. The car ahead of you is moving just a little too slowly, so you take a look into the oncoming traffic lane and you decide that it is safe to pass. But as you cross over into the lane of oncoming traffic, another car appears, seemingly out of nowhere. Though you are surprised by the sudden appearance of the other vehicle, you have enough time to get back into your own lane—perhaps a bit unnerved but otherwise no worse off. But you can't help but wonder why you didn't see that car before you started to pass. Were your eyes playing tricks on you? Or perhaps the visibility was not as great as you thought it was. The answer could very well be that the atmosphere altered the path of the visible radiation reflected off the oncoming car so that the rays were deflected away from your eyes, and that it was not until you got sufficiently close to the vehicle that the light was able to meet with your eyes. Sometimes the atmosphere can indeed make objects appear to be in a different position from where they really are or even alter their appearance entirely. This final chapter describes the processes by which the atmosphere affects the path of visible radiation passing through it and the resultant images we see. We refer to these topics collectively as **atmospheric optics.**

CHAPTER OUTLINE

Clear Air Effects

In Chapter 3 we saw that the atmosphere scatters and absorbs incoming solar radiation. Both processes have an important effect on the energy obtained by the surface and the atmosphere. In addition to scattering and absorbing, the atmosphere also *refracts* solar radiation, where **refraction** is defined as the bending of rays as they pass through the atmosphere.

Refraction occurs whenever radiation travels through a medium whose density varies or whenever it passes from one medium to another having a different density. Refraction in air occurs because radiation speed varies with density—the denser the atmosphere, the slower the radiation. To visualize how differential speeds cause the radiation to bend, imagine two people in a canoe, with one of them paddling on the right side of the canoe and the other on the left. If the person on the right paddles more vigorously, the canoe steers to the left. The same thing happens to a wave of electromagnetic energy passing through the atmosphere. Under most circumstances the density of the atmosphere decreases with height above the surface. This causes the radiation to refract slightly, forming an arc with the concave side oriented downward (Figure 17–1). The rate at which density changes with height

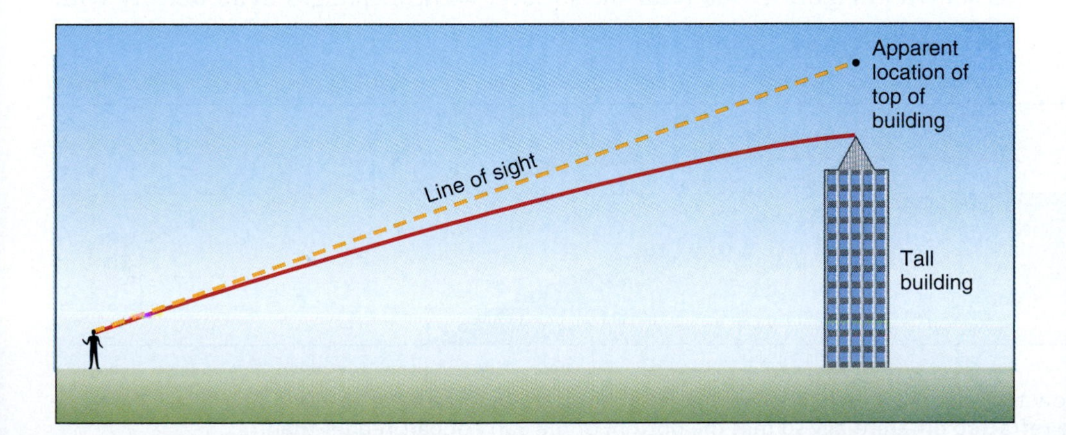

Figure 17–1 Refraction due to air density differences causes light to be refracted. In this case, the light from the top of the tall building is bent downward, so its path is concave downward. The light reaches the viewer's eye at an angle slightly greater than what it would without refraction, making the top of the building appear higher up than it really is.

varies considerably from place to place and from day to day because of differences in the temperature profile above the surface (the relationship between temperature and density was discussed in Chapter 4). Thus, the amount and direction of refraction vary with atmospheric conditions. Let's look now at several notable results of refraction.

Refraction and the Setting or Rising Sun

Refraction of incoming solar radiation is greatest when the Sun is low over the horizon, because the low solar angle causes the rays to pass through a greater amount of atmosphere (as described in Chapter 3). At sunset, refraction is sufficient to cause direct rays to be visible even after the Sun has dropped below the horizon (this is also true just prior to sunrise). In Figure 17–2 the Sun is positioned below the horizon. Without an atmosphere this would bring nightfall, but refraction causes the Sun to appear to be just above the horizon. When the Sun is positioned slightly farther below the horizon, its direct rays cannot be seen at the surface, but diffuse radiation can illuminate the sky to create **twilight** conditions. A further lowering of the Sun puts it far enough below the horizon to bring total nighttime. The period of twilight varies; it is longest during the high Sun season and increases with latitude.

In addition to slightly shifting the apparent position of the Sun near the horizon, refraction can also affect its apparent shape and color. You may have noticed that the Sun near dawn or sunset seems to have horizontal bands of different colors, with redder coloration near the bottom. This occurs because longer wavelength colors (such as reds and oranges) undergo less refraction than do shorter wavelength colors (such as blue and green). The shorter wavelengths, undergoing a greater amount of refraction, concentrate near the top of the apparent sun, while the longer wavelengths locate near the bottom. Under some atmospheric conditions the Sun appears momentarily to be capped by a bright green spot, known as the **green flash** (Figure 17–3).

Mirages

We are all familiar with movie and cartoon images of a parched, emaciated man crawling through a hot desert who perceives an oasis on the horizon, only to be disappointed by the reality of it being a mere **mirage.** Such mirages are caused not by excessive optimism, but rather by the refraction of visible light when the temperature decreases rapidly with increasing height. And contrary to what some people think, the false oasis is not the only type of mirage; that term applies to any apparent upward or downward displacement of an object due to refraction.

To understand how mirages form, let's first consider a midday situation in which the air temperature near the surface decreases rapidly with increasing height, as shown in Figure 17–4a. Near the surface, vertical changes in air density with

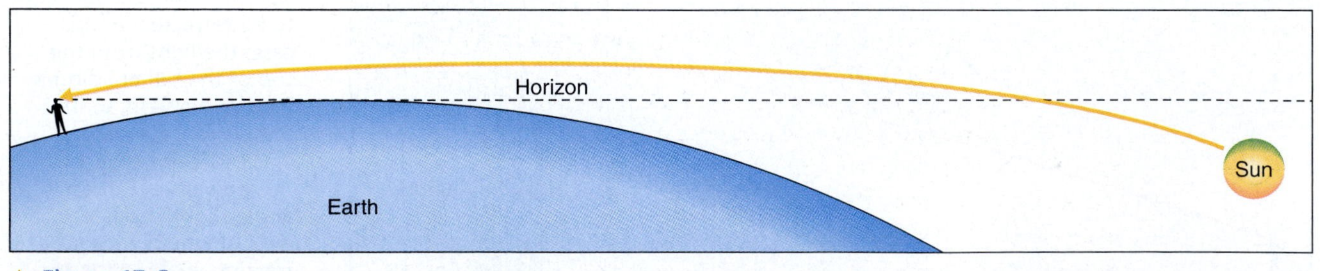

▲ **Figure 17–2**
When the Sun is slightly below the horizon, it can still be seen at Earth's surface because of refraction. The various wavelengths are refracted differentially so that the bottom of the Sun appears redder than the top.

◀ **Figure 17–3**
A green flash.

height are controlled primarily by vertical temperature gradients. With a very high temperature near the surface, density is low. The steep temperature gradient in this example results in increasing density with height, so the path of rays moving through the atmosphere are curved upward, as indicated by the yellow lines on Figure 17–4b.

In Figure 17–4b the viewer at the left (let's call her Lauren) perceives distant objects to be slightly lower than they actually are. This is because the light rays reflected off distant objects approach Lauren's eyes at an angle slightly below horizontal. Thus, a person standing at position *A* (we'll call him William) appears slightly shorter than he really is. Despite the minor distortion in the perceived height of William, Lauren has no trouble seeing him in his entirety.

Now look at how one type of mirage develops as William moves over to position *B*. The lower portion of his body appears to have disappeared because the light reflected off his legs is bent all the way to the ground, where it is absorbed before it can reach Lauren. As William walks farther away, more and more of his body disappears from the bottom upward until he completely vanishes from sight at position *C*.

A different type of mirage appears from still more intense heating near the surface, such as that over an asphalt road on a hot afternoon. The heated air in the shallow layer just above the surface has an extremely steep temperature profile, while the air immediately above the shallow layer is somewhat cooler and has a less steep

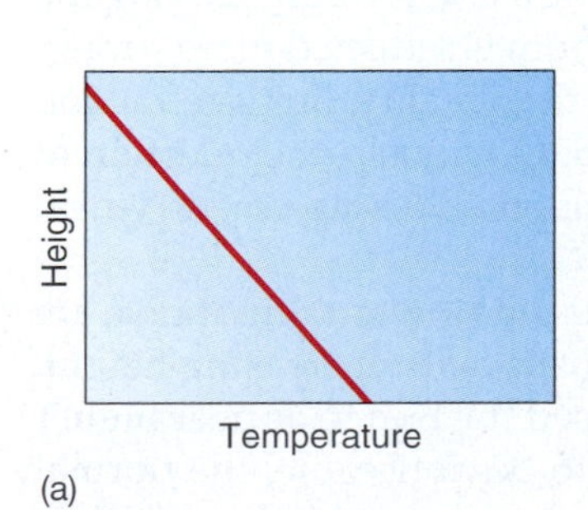

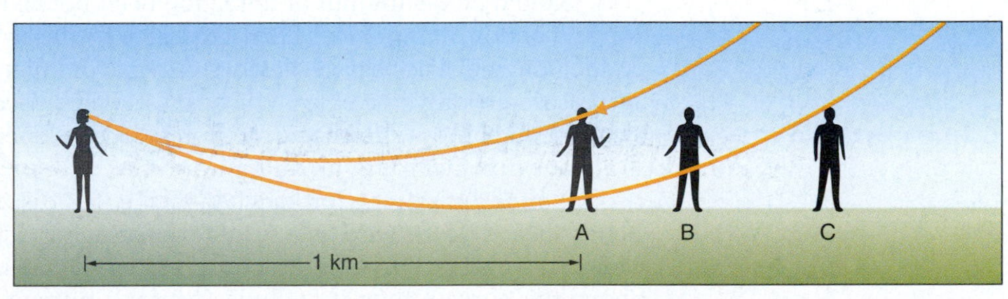

▲ **Figure 17–4**
A steady, steep drop in temperature with height (a) can cause a refraction pattern in which rays are bent concave upward (b). Though there is some distortion of the perceived image, a person standing at position *A* can be viewed in his or her entirety by the person on the left. As the person on the right moves to position *B*, the visible light reflected off the lower legs does not reach the viewer on the left because it is absorbed at the ground. At position *C*, the person's image disappears entirely to the viewer on the left. Note the extreme vertical exaggeration of the diagram.

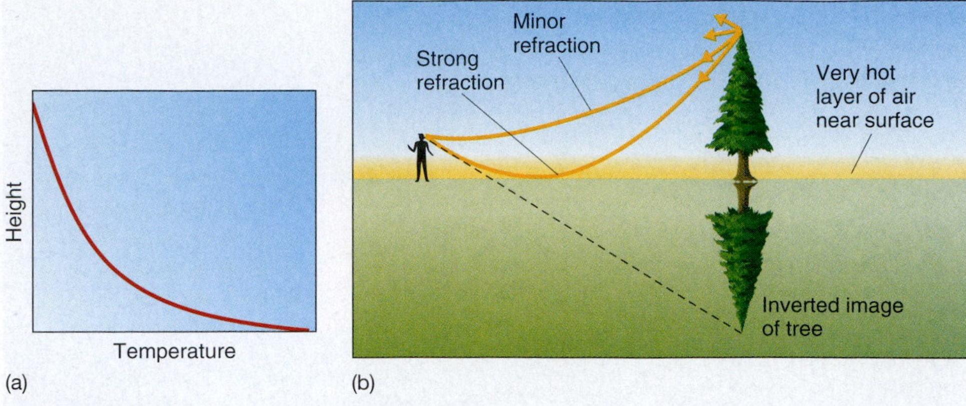

▲ **Figure 17–5**
If there is a very steep decrease in temperature immediately near the surface along with a lesser temperature gradient just above (a), a two-image inferior mirage can occur (b). Diffuse, visible radiation off the top of the tree goes in many directions. Some of the light passes directly to the viewer with minimal refraction (the upper arc from the treetop to the viewer). Some of the light approaches the surface, where intense refraction (the lower arc) gives a second image of the tree. This creates an inverted image beneath the normal image.

vertical temperature profile (Figure 17–5a). The steeper temperature gradient of the lower layer causes it to refract air more strongly than does the air above it. When this happens, a two-image **inferior mirage** can be seen, in which the viewer perceives not only a true image of an object, but also an inverted image directly below. This can be seen in Figure 17–5b, where light is reflected off the top of the tree in all directions. Some of the light is directed toward the viewer's eyes after undergoing only a small amount of refraction. This produces the regular image of the tree. But some of the reflected light is also directed toward the ground, where the steep temperature gradient causes very strong refraction. This light is refracted upward and reaches the viewer's eye from below, making the top of the tree appear below the ground and upside-down as if the viewer were looking at the tree's reflection on the calm surface of a pond. Light reflected off all parts of the tree undergo this type of refraction to produce a mirror image in the "pond."

So what actually happens when one sees a mirage that resembles a puddle of water? In that case a viewer perceives an inferior mirage of the sky, which looks very similar to a water surface. This chapter's opening photo of the car on the hot desert road shows an inverted image of the car apparently immersed in a puddle of water.

Note that the amount of refraction need not be the same for light leaving the top and bottom of the object. If light reflected by the bottom is refracted more strongly, the image will appear vertically stretched, or taller than reality. But if refraction increases vertically, the object will appear compressed. Obviously, the question of whether an object is stretched or compressed is independent of displacement, which is always downward for an inferior mirage.

A **superior mirage** forms when images are displaced upward. Light rays are bent concave downward as a result of decreasing density with increasing height. This is the normal situation described earlier that caused the Sun to appear higher above the horizon than it really is. But for a mirage to be noticeable, the normal density gradient must be enhanced by a temperature profile in which warm (less dense) air lies above cold air. As with inferior mirages, there may be stretching or compression of the image, depending on how refraction varies with altitude. Here, however, there is compression when refraction decreases with altitude, and stretching when refraction increases vertically. In extreme situations small objects (boats, houses, etc.) can be lifted and stretched, giving the appearance of floating cities or mountains, as in Figure 17–6.

◀ **Figure 17–6**
A superior mirage.

Cloud and Precipitation Optics

Refraction is not limited to clear air, nor is it the only process that can create interesting optical effects. In the remainder of this chapter we will see how refraction and two other processes produce some very familiar (and not so familiar) optical effects.

Rainbows

One of the most striking features to appear in the atmosphere is the familiar **rainbow** (Figure 17–7). Rainbows are sweeping arcs of light that exhibit changes in color from the inner part of the ring to the outer part. Rainbows only appear when rain is falling some distance away, and with a clear sky above and behind the viewer that

◀ **Figure 17–7**
Rainbows. Note that the brighter primary rainbow is surrounded by a dimmer secondary rainbow.

allows sunlight to reach the surface unobstructed. You might have observed that rainbows are always located in exactly the opposite direction of the Sun, In other words, if the Sun is to your back in the southwest, your shadow will point toward the center of the rainbow to the northeast.

The brightest and most common rainbows are **primary rainbows**. These rainbows are always the same size, so that at that horizon the angular distance from one end to the other extends about 85 degrees of angle (for visualization purposes, think of this as an angle that extends almost all the way from due north to due east). In a primary rainbow, the shortest wavelengths of visible light (violet and blue) appear at the innermost portion of the ring, and the longer wavelengths (orange and red) frame the outermost portion. A primary rainbow is often surrounded by a less distinct **secondary rainbow** that covers about 100 degrees of arc at the horizon and has the reverse color scheme of the primary rainbow (that is, the reds appear on the inner portion of the ring and the blues on the outer portion). Of course, if the precipitation shaft is not large enough or is too far in the distance, only a partial rainbow will appear.

The big question, of course, is how do these form? The answer lies in the way in which sunlight is refracted (bent) and reflected as it enters and penetrates a raindrop. When light passes through a medium of varying density, it is subject to refraction. The same phenomenon occurs as light penetrates a boundary separating substances of dissimilar density, such as air surrounding a raindrop. Let's first look at how this creates a primary rainbow. As sunlight enters a raindrop, it undergoes some refraction, with longer wavelengths being refracted less than shorter wavelengths. The refracted light penetrates the raindrop, with the majority exiting at the opposite side from which it entered. However, a small portion of the light hitting the back of the raindrop is reflected back from the interior of the surface, penetrates the droplet once again, and is refracted a second time as it exits the front side of the droplet at a position somewhat lower than where it entered the droplet. This process is shown for two hypothetical raindrops in Figure 17–8a. Because each wavelength is refracted differentially, only one particular wavelength of light exiting a raindrop is directed toward a particular viewer at a particular location. Thus the upper raindrop directs red light toward the viewer, while the lower raindrop directs violet light to that person. Because the lower raindrop appears at a lower angle above the horizon, its violet light forms the lower (inner) side of the ring, and the red light from the upper raindrop appears at the upper (outer) portion. The red light is refracted 42.3° and the violet light is refracted 40.6°. As a result, the ring is only 1.7° wide.

Secondary rainbows are formed in much the same manner as are primary rainbows, except that two reflections occur at the back of the raindrop, as shown in Figure 17–9. This results in a reverse color scheme than that of the primary rainbow,

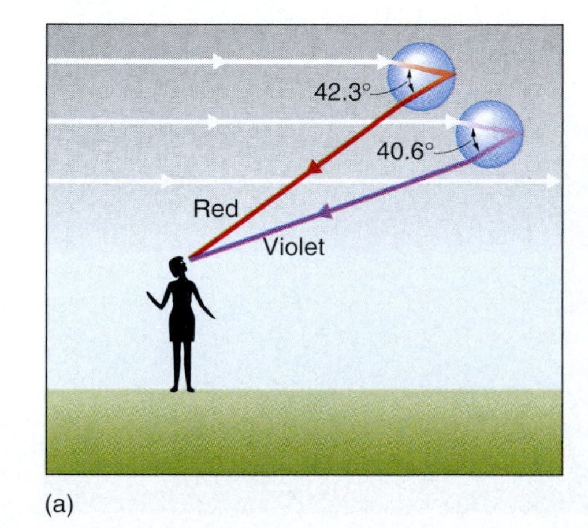

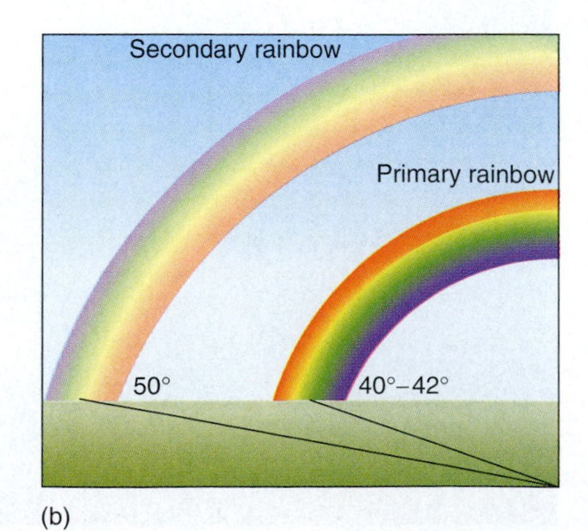

Figure 17–8 Sunlight from behind the viewer undergoes reflection and refraction (a) to produce a primary rainbow. The amount of total refraction is different for each wavelength, causing the familiar color separation of a rainbow. A viewer at ground level observes two concentric arcs creating a primary and a secondary rainbow (b).

◀ **Figure 17–9**
A secondary rainbow requires two reflections within raindrops.

with longer wavelengths of light situated on the inner portion of the band and with the shorter wavelengths on the outer portion. Sunlight exits the same side of the raindrop as it enters but is directed downward at a 50° angle, thus putting the top of the rainbow at 50° above the horizon (Figure 17–8b).

Halos, Sundogs, and Sun Pillars

Cirrostratus clouds produce circular bands of light that surround the Sun or Moon, called **halos** (shown in Figure 17–7), with radii of 22° or 46° (46° halos are less common and not as bright as 22° halos). Unlike rainbows, whose appearance requires that the Sun be directly behind the viewer, halos occur when ice crystals are between the viewer and the Sun or Moon. Figure 17–10a illustrates the refraction within ice crystals that produce a 22° halo. Sunlight (or moonlight) passes through the sides of column-shaped and platelike ice crystals, in which each of the six edges form a 60° angle. The ice crystal acts as a prism that refracts the sunlight 22°. Crystals that are 22° away from the sight path of the Sun or Moon and have the necessary orientation will refract light toward the viewer. Because ice crystals are so numerous and randomly aligned within the cloud, a sufficient number will direct the light toward the observer to make the halo bright enough to be visible from the ground. Figure 17–10b shows how column-shaped

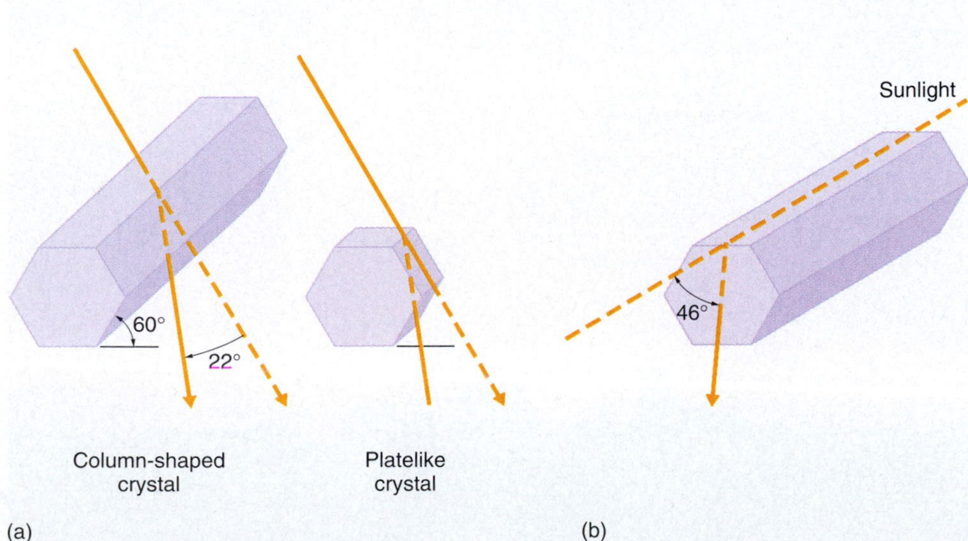

◀ **Figure 17–10**
Column-shaped and platelike crystals refract light to produce a 22° halo (a). Refraction where ice crystals have 90° angles produces a 46° halo (b).

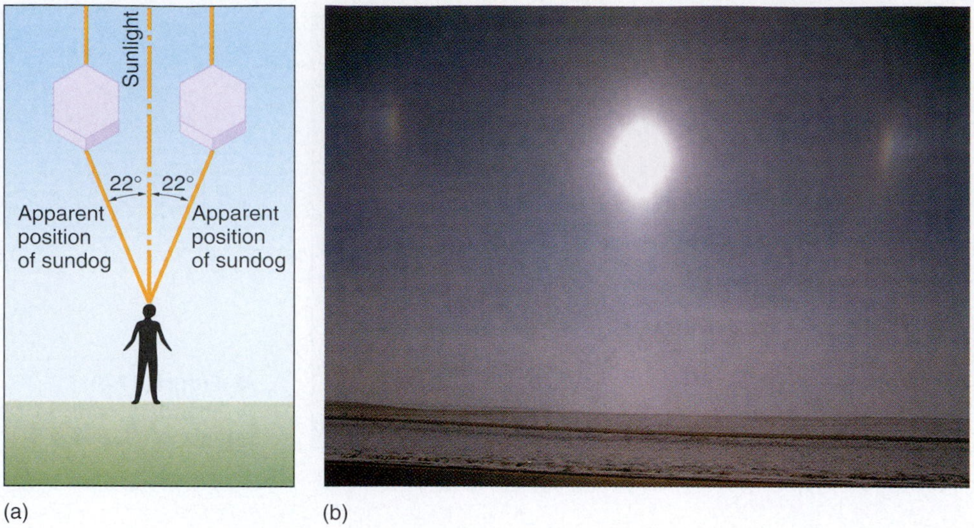

Figure 17–11 Nearly horizontally oriented ice crystals refract light from a setting or rising Sun to produce sundogs (a), so named because of the way they accompany the Sun (b).

ice crystals refract light at 46° when the crystal is oriented lengthwise toward the incoming light. The 46° halo can only result from column-shaped crystals, not by platelike crystals.

Platelike ice crystals larger than about 30 μm across tend to align themselves horizontally. If the Sun is slightly above the horizon and behind these crystals, bright spots appear 22° to the right and left of the Sun (Figure 17–11). These **sundogs** (or *parhelia*) often appear as whitish spots in the sky, but sometimes they exhibit color differentiation, with redder colors located on the side of the sundog nearest the sun and the blues and violets located on the outer side. Platelike crystals between a low Sun and an observer can also *reflect* (as opposed to refracting) sunlight off their tops and bottoms to produce **sun pillars** (Figure 17–12). The many ice crystals are aligned almost, but not exactly, horizontally, with each reflecting a portion of the incoming light differently to produce the apparent columns stretching upward and downward from the Sun.

Figure 17–12 A sun pillar.

Coronas and Glories

Coronas and glories are optical phenomena resulting from the bending of light as it passes around water droplets (**diffraction**). The **corona** (Figure 17–13) is a circular illumination of the sky immediately surrounding the Moon—or in rarer instances, the Sun. Clouds having uniform droplet sizes cause highly circular coronas that concentrate shorter wavelength (bluish) colors on their innermost portions and longer (redder) wavelengths on their outer margins. When the cloud contains a wide assortment of droplet sizes, the illumination appears white and irregularly shaped. The size of the corona is also related to droplet size, with larger droplets producing smaller coronas.

If you are ever in an aircraft flying above a cloud deck, look for the plane's shadow on the clouds. You may see a series of rings called a **glory** (Figure 17–14). Glories occur when sunlight entering the edge of a water droplet (Figure 17–15) is first refracted, then reflected off the inside of the back of the droplet, and refracted again as it exits the droplet. In this regard, the process that produces a glory is very similar to that which creates a primary rainbow. However, a glory requires that the combination of these processes redirect the incoming light a full 180°, so that the viewer sees the returned light from the top of the cloud with the Sun behind her. In order to accomplish the additional bending, diffraction must occur along the edge of the droplet as the light makes its return back toward the Sun.

▲ Figure 17–13
A corona.

◀ Figure 17–14
A glory.

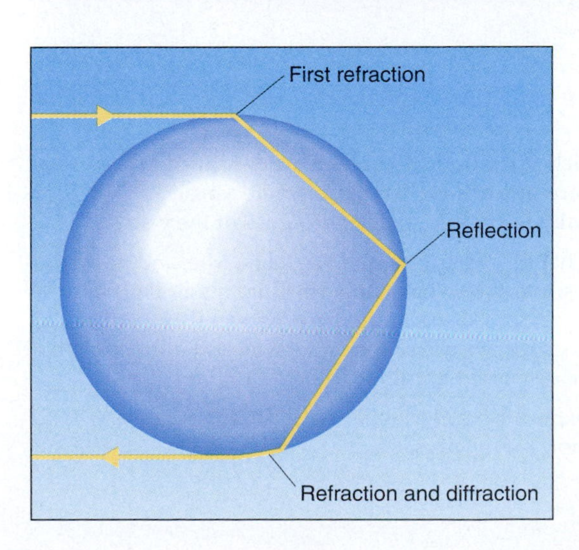

◀ Figure 17–15
Glories require diffraction along the edge of a cloud droplet as the sunlight exits the droplet. The bending from refraction returns the sunlight almost 180° from the direction at which it entered the droplet.

Summary

As rays of light travel through the atmosphere, they undergo a certain amount of bending, or refraction, as a result of air density differences away from the surface. Refraction causes some interesting sunrise and sunset effects. First, the bending of solar radiation allows the Sun to be "visible" even though it is truly below the horizon. Also, the fact that different wavelengths of sunlight are refracted differentially causes a setting or rising sun to display a sequence of horizontal bands with different colors. At its most extreme, refraction can cause only the green light from the Sun that is below the horizon to reach a viewer, resulting in a green flash.

Cloud droplets and ice crystals, along with precipitation, can produce their own unique optics. Rainbows form by the combination of refraction and reflection in raindrops that cause multicolored bands to be seen when the viewer stands between the Sun and the drops. Refraction within ice crystals causes halos and sundogs, while reflection on ice-crystal surfaces can produce sun pillars that seem to emanate from the setting or rising Sun. Diffraction by water droplets creates a corona around the Sun or Moon, while a combination of refraction, reflection, and diffraction produce glories.

Key Terms

atmospheric optics p. 513
refraction p. 513
twilight p. 514
mirage p. 514
green flash p. 514
inferior mirage p. 516
superior mirage p. 516
rainbow p. 517
primary rainbow p. 518
secondary rainbow p. 518
halo p. 519
sundog p. 520
sun pillar p. 520
diffraction p. 521
corona p. 521
glory p. 521

Review Questions

1. What is refraction and why is it related to variations in atmospheric density?
2. Describe the way refraction alters the apparent position of the setting or rising Sun.
3. Do longer or shorter wavelengths of light undergo greater refraction when passing through the atmosphere? How does the differential refraction cause an apparent banding of the Sun near the horizon?
4. Which type of vertical temperature gradients promote the appearance of superior and inferior mirages?
5. How do some mirages create the appearance of standing water on hot days?
6. Explain why the Sun must be behind you when you see a rainbow.
7. Describe the difference in the way primary and secondary rainbows form.
8. How does the color pattern of a secondary rainbow differ from that of a primary rainbow?
9. In addition to refraction, what process must occur within raindrops to produce a rainbow?
10. Why is it that some halos at 22° angles and others at 60°?
11. How are sundogs formed? Describe the color patterns associated with them.
12. Describe the formation of sun pillars. Does refraction play a role in their formation?
13. Explain how coronas are formed around the Sun or Moon. What factor or factors determine their size?
14. What are glories, and how are they formed? Are they the result of refraction alone or is another process also involved?

Critical Thinking

1. Consider the way the apparent position of the Sun sweeps across the sky over the course of the day (see Chapter 2). How will the period of twilight vary between summer and winter where you live? Will twilight conditions generally last longer in the Tropics or in the high latitudes? After answering this question, go to Problems and Exercises question #1, and check to see if your answer was correct.
2. Can falling ice crystals produce rainbows? Explain why or why not.
3. Can altostratus clouds produce halos? Explain why or why not.
4. Which of the optical phenomena described in this chapter are most likely to occur where you live? Are they equally likely to appear at all times of the year?
5. In Chapter 3 we discussed Rayleigh, Mie, and nonselective scattering. What similarities and dissimilarities exist between those scattering processes and the optical effects caused by refraction, reflection, and diffraction that were discussed in this chapter?
6. Explain why superior mirages do not occur over land on hot, sunny days.

Problems and Exercises

1. Refer to the Web site **http://aa.usno.navy.mil/faq/docs/RST_defs.html#top**, and look up the definitions of civil, nautical, and astronomical twilight. Then use the available tables to determine the length of day where you live for March 21, June 21, September 21, and December 21. Does the length of day show significant differences using each of the three definitions of twilight? How do these differences vary through the year?
2. On hot, sunny days, look for the presence of mirages. Are they equally apparent in all directions? If not, why do you think that might be the case? Also, check to see how long they remain visible? Do they still persist at sunset?

Useful Web Sites

http://aa.usno.navy.mil/faq/docs/RST_defs.html#top

Provides tables of sunrise and sunset for any location, incorporating the effects of twilight. Also defines three types of twilight.

http://www.weather-photography.com/gallery.php?cat=optics

Offers photos depicting many different types of optical phenomena.

http://ww2010.atmos.uiuc.edu/(Gh)/guides/mtr/opt/home.rxml

An interesting site from the University of Illinois with much information on various aspects of atmospheric optics.

http://www.sundog.clara.co.uk/atoptics/phenom.htm

Extensive information on the effects of ice crystals, water droplets, and other topics.

http://virtual.finland.fi/finfo/english/mirage.html#refr

A Finnish site with explanations and photos of different types of mirages.

http://www.polarimage.fi/

Contains numerous photographic examples of the phenomena discussed in this chapter, as well as other interesting images.

Appendix A

Units of Measurement and Conversions

SI Units	
I. Basic	
Length	meter (m)
Mass	kilogram (kg)
Time	second (s)
Electrical Current	ampere (A)
Temperature	kelvin (K)

II. Derived

Force	newton ($N = kg\ m/s^2$)
Pressure	pascal ($Pa = N/m^2$)
Energy	joule (J = N m)
Power	watt (W = J/s)
Electrical Potential Difference	volt (V = J/C)
Electrical Charge	coulomb (C)

III. Some Useful Conversions

Length

1 centimeter = 0.39 inches

1 meter = 3.281 feet
= 39.37 inches

1 kilometer = 0.62 miles

1 inch = 2.54 centimeters

1 foot = 30.48 centimeters
= 0.305 meters

1 mile = 1.61 kilometers

Mass/Weight

1 gram = 0.035 ounces

1 kilogram = 2.2 pounds

1 ounce = 28.35 grams

1 pound = 0.454 kilograms

Speed

1 meter/second = 2.24 miles/hour
= 3.60 km/hour

1 mile/hour = 0.45 meters/second
= 1.61 km/hour

Temperature

Celsius Temperature = (°F−32)/1.8
= K−273.15

Fahrenheit Temperature = 1.8°C + 32

Kelvin Temperature = °C + 273.15

Energy

1 joule = 0.239 calories

1 calorie = 4.186 joules

Appendix B

The Standard Atmosphere

Altitude (km)	Temperature (°C)	Pressure (mb)	p/p_0*	Density (kg/m³)	ρ/ρ_0*
30.00	–46.60	11.97	0.01	0.02	0.02
25.00	–51.60	25.49	0.03	0.04	0.03
20.00	–56.50	55.29	0.05	0.09	0.07
19.00	–56.50	64.67	0.06	0.10	0.08
18.00	–56.50	75.65	0.07	0.12	0.09
17.00	–56.50	88.49	0.09	0.14	0.12
16.00	–56.60	103.52	0.10	0.17	0.14
15.00	–56.50	121.11	0.12	0.20	0.16
14.00	–56.50	141.70	0.14	0.23	0.19
13.00	–56.50	165.79	0.16	0.27	0.22
12.00	–56.50	193.99	0.19	0.31	0.25
11.00	–56.40	226.99	0.22	0.37	0.30
10.00	–49.90	264.99	0.26	0.41	0.34
9.50	–46.70	285.84	0.28	0.44	0.36
9.00	–43.40	308.00	0.30	0.47	0.38
8.50	–40.20	331.54	0.33	0.50	0.40
8.00	–36.90	356.51	0.35	0.53	0.43
7.50	–33.70	382.99	0.38	0.56	0.45
7.00	–30.50	411.05	0.41	0.59	0.48
6.50	–27.20	440.75	0.43	0.62	0.50
6.00	–23.90	472.17	0.47	0.66	0.54
5.50	–20.70	505.39	0.50	0.70	0.57
5.00	–17.50	540.48	0.53	0.74	0.60
4.50	–14.20	577.52	0.57	0.78	0.63
4.00	–11.00	616.60	0.61	0.82	0.67
3.50	–7.70	657.80	0.65	0.86	0.70
3.00	–4.50	701.21	0.69	0.91	0.74
2.50	–1.20	746.91	0.74	0.96	0.78
2.00	2.00	795.01	0.78	1.01	0.82
1.50	5.30	845.59	0.83	1.06	0.86
1.00	8.50	898.76	0.89	1.11	0.91
0.50	11.80	954.61	0.94	1.17	0.95
0.00	15.00	1013.25	1.00	1.23	1.00

*p/p_0 = ratio of air pressure to sea level value; ρ/ρ_0 = ratio of air density to sea level value.

Appendix C

Weather Map Symbols

Explanation of Codes

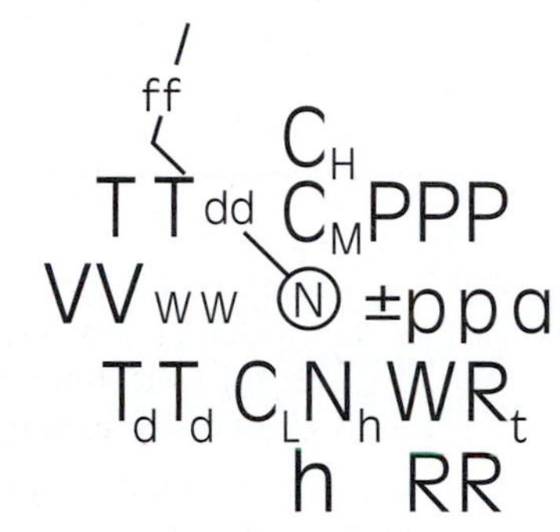

Symbol station model

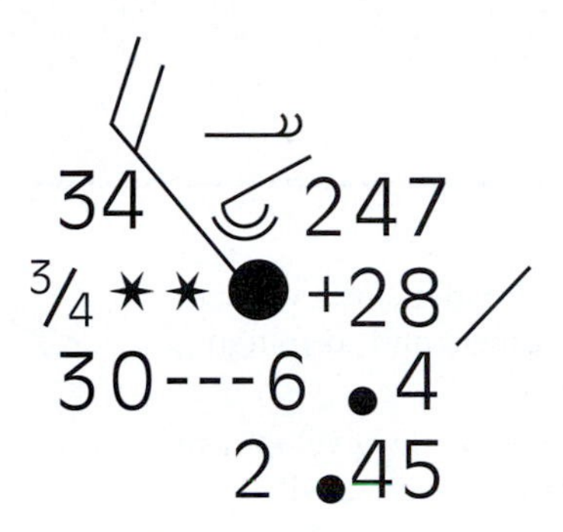

Sample report

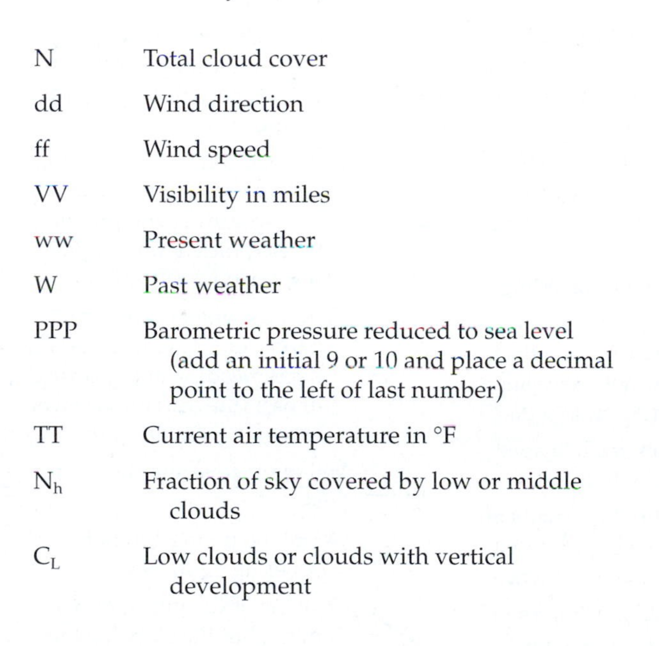

N	Total cloud cover
dd	Wind direction
ff	Wind speed
VV	Visibility in miles
ww	Present weather
W	Past weather
PPP	Barometric pressure reduced to sea level (add an initial 9 or 10 and place a decimal point to the left of last number)
TT	Current air temperature in °F
N_h	Fraction of sky covered by low or middle clouds
C_L	Low clouds or clouds with vertical development
h	Height in feet of the base of the lowest clouds
C_M	Middle clouds
C_H	High clouds
T_dT_d	Dewpoint temperature in °F
a	Pressure tendency
pp	Pressure change in mb in preceding 3 hr (+28 = +2.8)
RR	Amount of precipitation in last 6 hr
R_t	Time precipitation began or ended (0 = none; 1 = <1 hr ago; 2 = 1–2 hr ago; 3 = 2–3 hr ago; 4 = 3–4 hr ago; 5 = 4–5 hr ago; 6 = 5–6 hr ago; 7 = 6–12 hr ago; 8 = >12 hr ago; 9 = unknown)

Air Pressure Tendency

- Rising, then falling; same as or higher than 3 hr ago
- Rising, then steady; or rising, then rising more slowly
- Rising steadily, or unsteadily
- Falling or steady, then rising; or rising, then rising more rapidly
- Steady; same as 3 hr ago
- Falling, then rising; same as or lower than 3 hr ago
- Falling, then steady; or falling, then falling more slowly
- Falling steadily, or unsteadily
- Steady or rising, then falling; or falling, then falling more rapidly

Cloud Abbreviations

St	stratus
Fra	fractus
Sc	stratocumulus
Ns	nimbostratus
As	altostratus
Ac	altocumulus
Ci	cirrus
Cs	cirrostratus
Cc	cirrocumulus
Cu	cumulus
Cb	cumulonimbus

Cloud Types

- Cu of fair weather, little vertical development and seemingly flattened
- Cu of considerable development, generally towering, with or without other Cu or Sc, bases all at same level
- Cb with tops lacking clear-cut outlines, but distinctly not cirriform or anvil-shaped; with or without Cu, Sc, or St
- Sc formed by spreading out of Cu; Cu often present also
- Sc not formed by spreading out of Cu
- St or StFra, but no StFra of bad weather
- StFra and/or CuFra of bad weather (scud)
- Cu and Sc (not formed by spreading out of Cu) with bases at different levels
- Cb having a clearly fibrous (cirriform) top, often anvil-shaped, with or without Cu, Sc, St, or scud
- Thin As (most of cloud layer semitransparent)
- Thick As, greater part sufficiently dense to hide sun (or moon), or Ns
- Thin Ac, mostly semitransparent; cloud elements not changing much and at a single level
- Thin Ac in patches; cloud elements continually changing and/or occurring at more than one level
- Thin Ac in bands or in a layer gradually spreading over sky and usually thickening as a whole
- Ac formed by the spreading out of Cu or Cb
- Double-layered Ac, or a thick layer of Ac, not increasing; or Ac with As and/or Ns
- Ac in the form of Cu-shaped tufts or Ac with turrets
- Ac of a chaotic sky, usually at different levels; patches of dense Ci usually present
- Filaments of Ci, or "mares' tails," scattered and not increasing
- Dense Ci in patches or twisted sheaves, usually not increasing, sometimes like remains of Cb; or towers or tufts
- Dense Ci, often anvil-shaped, derived from or associated with Cb
- Ci, often hook-shaped, gradually spreading over the sky and usually thickening as a whole
- Ci and Cs, often in converging bands, or Cs alone; generally overspreading and growing denser; the continuous layer not reaching 45° altitude
- Ci and Cs, often in converging bands, or Cs alone; generally overspreading and growing denser; the continuous layer exceeding 45° altitude
- Veil of Cs covering the entire sky
- Cs not increasing and not covering entire sky
- Cc alone or Cc with some Ci or Cs, but the Cc being the main cirriform cloud

Height of Base of Lowest Cloud

Code	*Feet*	*Meters*
0	0–149	0–49
1	150–299	50–99
2	300–599	100–199
3	600–999	200–299
4	1000–1999	300–599
5	2000–3499	600–999
6	3500–4999	1000–1499
7	5000–6499	1500–1999
8	6500–7999	2000–2499
9	8000 or above or no clouds	2500 or above or no clouds

Cloud Cover

No clouds

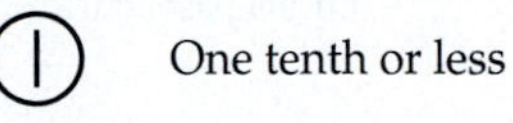

One tenth or less

Two tenths or three tenths

Four tenths

Five tenths

Six tenths

Seven tenths or eight tenths

Nine tenths or overcast with openings

Completely overcast (ten tenths)

Sky obscured

Wind Speed

	Miles per hour	*Kilometers per hour*
	Calm	Calm
	1–2	1–3
	3–8	4–13
	9–14	14–19
	15–20	20–32
	21–25	33–40
	26–31	41–50
	32–37	51–60
	38–43	61–69
	44–49	70–79
	50–54	80–87
	55–60	88–96
	61–66	97–106
	67–71	107–114
	72–77	115–124
	78–83	125–134
	84–89	135–143
	119–123	192–198

Fronts

Fronts are shown on surface weather maps by the symbols below. (Arrows—not shown on maps—indicate direction of motion of front.)

Cold front (surface)

Warm front (surface)

Occluded front (surface)

Stationary front (surface)

Warm front (aloft)

Cold front (aloft)

Weather Conditions

Cloud development NOT observed or NOT observable during past hour	Clouds generally dissolving or becoming less developed during past hour	State of sky on the whole unchanged during past hour	Clouds generally forming or developing during past hour	Visibility reduced by smoke
Light fog (mist)	Patches of shallow fog at station, NOT deeper than 6 feet on land	More or less continuous shallow fog at station, NOT deeper than 6 feet on land	Lightning visible, no thunder heard	Precipitation within sight, but NOT reaching the ground
Drizzle (NOT freezing) or snow grains (NOT falling as showers) during past hour, but NOT at time of observation	Rain (NOT freezing and NOT falling as showers) during past hour, but NOT at time of observation	Snow (NOT falling as showers) during past hour, but NOT at time of observation	Rain and snow or ice pellets (NOT falling as showers) during past hour, but NOT at time of observation	Freezing drizzle or freezing rain (NOT falling as showers) during past hour, but NOT at time of observation
Slight or moderate dust storm or sandstorm, has decreased during past hour	Slight or moderate dust storm or sandstorm, no appreciable change during past hour	Slight or moderate dust storm or sandstorm has begun or increased during past hour	Severe dust storm or sandstorm, has decreased during past hour	Severe dust storm or sandstorm, no appreciable change during past hour
Fog or ice fog at distance at time of observation, but NOT at station during past hour	Fog or ice fog in patches	Fog or ice fog, sky discernible, has become thinner during past hour	Fog or ice fog, sky NOT discernible, has become thinner during past hour	Fog or ice fog, sky discernible, no appreciable change during past hour
Intermittent drizzle (NOT freezing), slight at time of observation	Continuous drizzle (NOT freezing), slight at time of observation	Intermittent drizzle (NOT freezing), moderate at time of observation	Continuous drizzle (NOT freezing), moderate at time of observation	Intermittent drizzle (NOT freezing), heavy at time of observation
Intermittent rain (NOT freezing), slight at time of observation	Continuous rain (NOT freezing), slight at time of observation	Intermittent rain (NOT freezing), moderate at time of observation	Continuous rain (NOT freezing), moderate at time of observation	Intermittent rain (NOT freezing), heavy at time of observation
Intermittent fall of snowflakes, slight at time of observation	Continuous fall of snowflakes, slight at time of observation	Intermittent fall of snowflakes, moderate at time of observation	Continuous fall of snowflakes, moderate at time of observation	Intermittent fall of snowflakes, heavy at time of observation
Slight rain shower(s)	Moderate or heavy rain shower(s)	Violent rain shower(s)	Slight shower(s) of rain and snow mixed	Moderate or heavy shower(s) of rain and snow mixed
Moderate or heavy shower(s) of hail, with or without rain, or rain and snow mixed, not associated with thunder	Slight rain at time of observation; thunderstorm during past hour, but NOT at time of observation	Moderate or heavy rain at time of observation; thunderstorm during past hour, but NOT at time of observation	Slight snow, or rain and snow mixed, or hail at time of observation; thunderstorm during past hour, but NOT at time of observation	Moderate or heavy snow, or rain and snow mixed, or hail at time of observation; thunderstorm during past hour, but NOT at time of observation

Weather Conditions

Haze	Widespread dust in suspension in the air, NOT raised by wind, at time of observation	Dust or sand raised by wind at time of observation	Well-developed dust whirl(s) within past hour	Dust storm or sandstorm within sight of or at station during past hour
Precipitation within sight, reaching the ground but distant from station	Precipitation within sight, reaching the ground, near to but NOT at station	Thunderstorm, but no precipitation at the station	Squall(s) within sight during past hour or at time of observation	Funnel cloud(s) within sight of station at time of observation
Showers of rain during past hour, but NOT at time of observation	Showers of snow, or of rain and snow, during past hour, but NOT at time of observation	Showers of hail, or of hail and rain, during past hour, but NOT at time of observation	Fog during past hour, but NOT at time of observation	Thunderstorm (with or without precipitation) during past hour, but NOT at time of observation
Severe dust storm or sandstorm has begun or increased during past hour	Slight or moderate drifting snow, generally low (less than 6 ft)	Heavy drifting snow, generally low	Slight or moderate blowing snow, generally high (more than 6 ft)	Heavy blowing snow, generally high
Fog or ice fog, sky NOT discernible, no appreciable change during past hour	Fog or ice fog, sky discernible, has begun or become thicker during past hour	Fog or ice fog, sky NOT discernible, has begun or become thicker during past hour	Fog depositing rime, sky discernible	Fog depositing rime, sky NOT discernible
Continuous drizzle (NOT freezing), heavy at time of observation	Slight freezing drizzle	Moderate or heavy freezing drizzle	Drizzle and rain, slight	Drizzle and rain, moderate or heavy
Continuous rain (NOT freezing), heavy at time of observation	Slight freezing rain	Moderate or heavy freezing rain	Rain or drizzle and snow, slight	Rain or drizzle and snow, moderate or heavy
Continuous fall of snowflakes, heavy at time of observation	Ice prisms (with or without fog)	Snow grains (with or without fog)	Isolated starlike snow crystals (with or without fog)	Ice pellets or snow pellets
Slight snow shower(s)	Moderate or heavy snow shower(s)	Slight shower(s) of snow pellets, or ice pellets with or without rain, or rain and snow mixed	Moderate or heavy shower(s) of snow pellets, or ice pellets, or ice pellets with or without rain or rain and snow mixed	Slight shower(s) of hail, with or without rain or rain and snow mixed, not associated with thunder
Slight or moderate thunderstorm without hail, but with rain, and/or snow at time of observation	Slight or moderate thunderstorm, with hail at time of observation	Heavy thunderstorm, without hail, but with rain and/or snow at time of observation	Thunderstorm combined with dust storm or sandstorm at time of observation	Heavy thunderstorm with hail at time of observation

Glossary

A

Absolute Humidity Mass of water vapor per unit volume of air, usually expressed in grams per cubic meter (g/m^3).
Absolute Vorticity The sum of vorticity relative to the surface and vorticity arising from Earth's rotation.
Absorption A process in which radiation is captured by a molecule. Unlike reflection, absorption represents an energy transfer to the absorbing molecule.
Acceleration A change in velocity; a change in speed or direction, or both.
Acceleration of Gravity Acting alone, gravity would accelerate all objects by the same rate, about 9.8 m/sec/sec. *See* gravity.
Adiabatic Term for processes in which no heat is added or removed. For example, a rising air parcel cools adiabatically as it expands.
Advanced Weather Interactive Processing System (AWIPS) A system for the display and manipulation of weather information at Weather Service Offices.
Advection Horizontal transport of some atmospheric property (heat, moisture, etc.).
Aerosols Small, suspended particles in the atmosphere.
Aggregation The process in which ice crystals join together to form snowflakes. If these snowflakes melt, they fall as rain.
Air Mass A large body of air having little horizontal variation in temperature and moisture.
Air Mass Thunderstorms Relatively small, short-lived thunderstorms that do not produce very strong winds, large hail, or tornadoes.
Albedo The fraction of solar radiation arriving at a surface that is reflected.
Aleutian Low A semi-permanent cell found in the North Pacific in Winter.
Altocumulus A mid-level layered cloud with some rolls or patches of vertical development.
Altostratus A mid-level layered cloud.
Aneroid Barometer A device used to measure air pressure. The barometer's elastic chamber expands and contracts in response to the surrounding pressure.
Antarctic Circle A line of latitude 66.5° S, the southern limit of which experiences 24 hours of daylight or darkness.
Aphelion Earth's position when it is farthest from the Sun (July 4).
Arctic Circle A line of latitude 66.5° N, the northern limit of which experiences 24 hours of daylight or darkness.
Arctic Oscillation An oscillation in the temperature distribution of Arctic sea surface temperatures in the North Atlantic. Sometimes called the North Atlantic Oscillation.
Atmosphere The gases, droplets, and particles surrounding Earth's surface.
Atmospheric Window The range of wavelengths (about 8 to 12 μm) that are not readily absorbed by the gases of the atmosphere.
Aurora Borealis or Aurora Australis An illumination of the sky found in the high northern (borealis) or southern (australis) latitudes, which is produced as charged particles arriving from the Sun react with the upper atmosphere.
AWIPS Acronym for Advanced Weather Interactive Processing System, the computer system used by NWS forecasters for the display of weather maps, satellite and radar imagery, and other types of data.

B

Banner Clouds A cloud formed near the top of a topographic barrier by orographic uplift.
Barometer An instrument for measuring air pressure.
Beam Spreading The process whereby a beam of radiation is distributed over a larger horizontal area as the angle of incidence departs from vertical. Reduces the intensity of radiation absorption by the surface.
Bergeron Process The primary mechanism for precipitation formation outside the Tropics; this process involves the coexistence of ice crystals and supercooled water droplets.
Bermuda Azores High Low A semi-permanent cell found in the Atlantic in summer.
Blackbody An object or substance that is perfectly efficient at absorbing and radiating radiation. Blackbodies do not exist in nature, but represent an ideal.

C

Calorie Amount of heat required to raise the temperature of 1 gram of water 1 °C (about 4.2 J).
Carbon Dioxide An important variable gas in the atmosphere, made up of one atom of carbon bound to two atoms of oxygen. An important greenhouse gas.
Celsius Scale The temperature scale that designates 0° as the freezing point and 100° as the sea level boiling point of water.
Charge Separation The separation of positive and negative ions into different parts of a cloud. A necessary precursor for lightning.

Chromosphere The layer of the Sun immediately surrounding the photosphere.

Cirrocumulus A high cloud composed of ice that is generally layered but with some rolls or pockets of vertical development.

Cirrostratus A high, layered cloud consisting of ice crystals.

Cirrus A high cloud made up entirely of ice crystals.

Climate The statistical properties of the atmosphere, including measures of average conditions, variability, etc.

Cloud An area of the atmosphere containing sufficient concentration of water droplets and/or ice crystals to be visible.

Cloud Seeding An attempt to stimulate precipitation by introducing certain materials into existing clouds.

Cold Cloud A cloud whose temperature is below 0 °C from top to bottom.

Cold Front Transition zone between cold and warm air masses; forms when a cold air mass advances on a warm air mass.

Collector Drop A relatively large falling raindrop that collides and coalesces with smaller, slower-moving droplets beneath it.

Condensation Change from vapor to liquid phase. Condensation releases the energy required for evaporation. *See* latent heat of vaporization.

Condensation Nuclei Small, airborne particles that enhance condensation. Without condensation nuclei, condensation would occur only at very high relative humidity (at about 200% or more), while condensation nuclei allow condensation to occur at or slightly below 100% relative humidity.

Conduction Heat transfer from molecule to molecule, without significant movement of the molecules.

Convection Heat transfer by fluid flow (movement of a gas or liquid).

Convergence Horizontal motions of air, the result of which is a net inflow of air (with more air imported than exported). Causes rising or sinking motions.

Conveyor Belt Model The modern description of air flow through mid-latitude cyclones.

Cool Cloud A cloud whose lower reaches have temperatures above 0 °C and whose upper portions are below 0 °C.

Coriolis Force An imaginary deflective force arising from Earth's rotation that is necessary to account for motions measured relative to the surface.

Core The interior of the Sun, where nuclear fusion produces energy that is ultimately radiated to Earth.

Corona A circular illumination of the sky immediately surrounding the Moon, or in some instances the Sun, caused by diffraction.

Cumulonimbus A cumulus cloud with very deep vertical development extending into the lower stratosphere, distinguished by an anvil at its top consisting of ice crystals.

Cumulus Any cloud having substantial vertical development.

Cutoff Low An upper-level area of low pressure that takes on a circular flow distinct from the general flow around it.

Cyclogenesis The beginning of cyclone formation.

Cyclone A region of low pressure relative to the surrounding area.

D

Dalton's Law This law of physical science states that the pressure of a combination of gases is equal to the sum of the partial pressures of each of the gases.

Dart Leader A zone of ionized air that serves as a conduit for a lightning stroke subsequent to the initial one in a lightning flash.

Density The mass of a substance per unit volume, expressed as kilograms per cubic meter (kg/m^3) in the International System of Units (SI).

Deposition Change from the vapor phase to the solid phase (frost is an example). Deposition releases the energy of vaporization and fusion. *See* latent heat.

Dew Point The temperature to which the air must be cooled to become saturated.

Dew Point Temperature (also called *dew point*) The temperature at which saturation will occur, given sufficient cooling.

Diabatic Processes that involve the addition or removal of heat. For example, air in contact with a cold surface loses heat diabatically by conduction.

Diffluence A type of horizontal divergence that occurs when streamlines spread apart in the downstream direction.

Diffuse Solar Radiation Sunlight that is scattered downward to the surface. *See* scattering.

Direct Solar Radiation Sunlight that passes through the atmosphere without absorption or scattering.

Divergence Horizontal motions of air, the result of which is a net outflow of air (with more air exported than imported). Causes rising or sinking motions.

Doppler Radar A type of radar that can measure horizontal motions as well as the internal characteristics of clouds.

Dry Adiabatic Lapse Rate (DALR) Temperature decrease experienced by a rising unsaturated parcel (about 1 °C/100 m). Sinking parcels warm at the same rate. The DALR is a constant.

Dry Line A boundary between humid air and denser dry air. A favored location for thunderstorm development.

Dynamic Low A low-pressure system created by divergence in the middle or upper troposphere.

E

East Greenland Drift An ocean current that flows southward in the North Atlantic.

Ecliptic Plane The imaginary surface swept by Earth's orbit around the Sun.

El Niño A recurrent event in the tropical eastern Pacific in which sea-surface temperatures are significantly above normal. The inverse event (cold sea-surface temperatures) is called a *La Niña*.

Electromagnetic Radiation Energy emitted by virtue of an object's temperature. Radiation is unique in that it does not require a transfer medium and can travel through a vacuum. The energy transfer is accomplished by oscillations in an electric field and a magnetic field.

Emissivity The property of a substance or object that expresses, as a fraction or percentage, how efficient it is at emitting radiation.

Entrainment The incorporation of surrounding, unsaturated air into a cloud.

Environmental Lapse Rate (ELR) The rate of vertical temperature decrease in the air column. The value is highly variable, depending on local conditions. For the troposphere, the global average is about 0.65 °C/100 m.

ENSO An acronym for the El Niño Southern Oscillation phenomenon. Involves the interaction of Tropical Pacific Sea Surface Temperatures and atmospheric pressures.

Equation of State The equation relating air pressure to temperature and density.

Equinoxes The two days of the year on which Earth's axis is not tilted toward or away from the Sun. On the equinoxes every latitude receives 12 hours of sunlight, and the Sun is overhead at the equator. The equinoxes occur March 21–22 and September 22–23.

Escape Velocity The rate of movement required of an air molecule to escape Earth's gravity.

European Center for Medium-Range Weather Forecasting Weather forecasting agency for the European Union.

Evaporation The change in phase of liquid water to water vapor.

Evapotranspiration The combined processes of evaporation and transpiration; the delivery of water to the atmosphere by vegetation and by direct evaporation from wet surfaces.

Eye The center of a hurricane, marked by generally clear skies and light winds.

Eye Wall The portion of a hurricane immediately adjacent to the eye; usually the region of highest windspeed and most intense precipitation.

F

Fahrenheit Scale A temperature scale that assigns values of 32° to the freezing point of water and 212° to the sea level boiling point of water.

Fetch Distance traveled by wind over a uniform surface, such as a water body.

First Law of Thermodynamics Most generally, the law that states that energy is a conserved property. In a meteorological context it states that heat added to a gas results in some combination of expansion of the gas and an increase in its internal energy.

Flares Intensely hot eruptions on the solar surface.

Foehn Wind A synoptic scale wind that flows downslope and warms by compression.

Fog Air that is adjacent to the surface and contains suspended water droplets, usually formed by diabatic cooling.

Force The product of mass and acceleration, as expressed in Newton's Law ($F = ma$).

Freezing Rain A form of precipitation in which rain droplets freeze as they fall below an inversion and pass into air having a temperature below 0 °C.

Friction Force that acts to slow wind but does not change its direction. Friction develops between the atmosphere and surface and between layers of air moving at different velocities.

Front A transition zone between two dissimilar air masses (that is, air masses with differing temperature, moisture, or density).

Frost A coating of ice crystals on a surface when the air adjacent to the surface becomes saturated at temperatures below 0 °C.

Frost Point The temperature at which saturation occurs, provided that that temperature is less than 0 °C.

Frozen Dew A coating of ice on a surface that occurs when a layer of dew freezes as temperatures drop below 0 °C.

Fujita Scale The scale for categorizing tornado intensity.

Funnel Cloud A column of rapidly rotating air similar to a tornado, except that the column has not extended to the ground.

G

Gamma Rays Electromagnetic radiation at wavelengths far shorter than those of visible light (from about 0.0000001 μm to 0.000001 μm). Gamma rays, which make up only a tiny proportion of the Sun's energy, are absorbed hundreds of kilometers above the surface.

General Circulation A term that refers to planetary-scale winds and pressure, features that appear in the time-averaged state.

Geopotential Height Loosely defined, the altitude at which atmospheric pressure takes on a particular value, as in "500 mb height." Because pressure reflects the mass of overlying atmosphere, geopotential heights reflect the potential energy atmosphere above that height.

Geostrophic Flow An idealized condition in which the upper-level air flows at constant speed and direction, parallel to straight isobars. There is no acceleration in geostrophic flow, and frictional forces are negligible.

Glory A series of rings formed around the shadow of an aircraft on the top of cloud, formed by refraction, reflection, and diffraction.

Gradient A change in some quantity (temperature, moisture, pressure) over space. For example, the temperature gradient is the rate of change of temperature per unit distance, and might be expressed in degrees Celsius per kilometer (°C/km).

Gradient Wind Wind flowing parallel to curved isobars. Frictional forces are negligible. With gradient flow there is a constant adjustment between the pressure gradient force and Coriolis force, causing the wind to change speed and direction as it flows along the isobars.

Graupel Ice crystals that have grown by riming to produce a spongy, somewhat translucent particle.

Gravity The force that attracts objects to Earth's surface. Although the acceleration of gravity is constant, the force of gravity varies from object to object. The force of gravity per unit volume of air is directly proportional to density.

Graybodies Bodies or substances that are not 100 percent efficient at absorbing or radiating energy. In reality, all bodies are graybodies.

Green Flash The brief appearance of green light near the top of the Sun sometimes observed at sunrise or sunset.

Greenwich Mean Time Also called *universal time* (*UT*). An international reference for time-keeping used for weather observations, satellite imaging, etc. It corresponds to local time at 0° longitude, a meridian passing through Greenwich, England.

H

Hadley Cell A somewhat idealized, large-scale wind and pressure pattern found in tropical latitudes of both hemispheres. Air rises above the equator, flows poleward to about 25° latitude, subsides, and flows back to the equator at low levels.

Hail Precipitation in the form of ice crystals, almost always associated with thunderstorms. Hail falls rapidly to the surface and thus does not melt during its descent.

Hawaiian High A semi-permanent cell found in the Pacific, most notably during the summer months.

Halo A circular band of light surrounding the Sun or Moon, caused by ice crystal refraction.

Heat The kinetic energy of atoms or molecules comprising a substance.

Heat Index A measure of apparent temperature used for warm conditions, incorporating temperature and humidity.

Heterogeneous Nucleation The condensation of liquid droplets or the deposition of ice crystals onto condensation or ice nuclei.

Heterosphere The high atmosphere (above 80 km or so), where gases are not well mixed, but rather are stratified according to molecular weight. Vertical motions are too weak to overcome gravitational settling, so heavier gases are found beneath lighter gases.

High Another term for *anticyclone*.

Homogeneous Nucleation The condensation of water droplets or deposition of ice crystals without condensation or ice nuclei.

Homosphere The lowest 80 km of the atmosphere, in which the relative abundance of the permanent gases is constant.

Horse Latitudes Areas associated with the oceanic subtropical highs, generally characterized by clear skies and light winds.

Humidity An expression of the amount of water vapor in the air.

Hurricane An intense tropical cyclone (warm-core low), with sustained winds of at least 120 km/hr.

Hydrostatic Equilibrium When the vertical pressure gradient force is balanced by the force of gravity. Because the forces balance, there is no acceleration upward or downward.

Hygrometer An instrument that measures humidity.

Hygroscopic Nuclei Airborne particles having an affinity for water, serving as condensation nuclei.

Hygrothermograph An instrument that records humidity and temperature.

I

Ice Nuclei Particles onto which ice crystals can form when the air becomes saturated. In the absence of freezing nuclei, water droplets freeze only at very low temperatures (near –40 °C). Ice nuclei allow ice to form at relatively "high" temperatures (around –10 °C).

Icelandic Low A semi-permanent cell found in the North Atlantic in Winter.

Ideal Gas Law (also known is the equation of state) Important law describing the relationship between pressure, temperature, and density.

Infrared Radiation Electromagnetic radiation at wavelengths longer than visible radiation, from about 0.7 μm to 1000 μm.

Insolation Incident, or incoming, solar radiation.

Interglacial A warmer segment of time between episodes of glaciation.

Inversion *See* temperature inversion.

Ionosphere Region in the upper atmosphere from about 80 to 500 km where charged particles (ions) are relatively abundant.

Ions Electrically charged atom or group of atoms.

Isobar A line on a weather map connecting points of equal pressure. Moving along an isobar, there is no change in pressure. The pressure gradient force acts perpendicular to isobars.

Isotherm A line on a weather map connecting points of equal temperature. Moving along an isotherm, there is no change in temperature. Temperature gradients are perpendicular to isotherms.

J

Joule Basic unit of energy in the International System of Units (SI). A joule (J) is the energy needed to accelerate 1 kg at a rate of 1 m/sec/sec across a distance of 1 m. A joule is equivalent to about 0.25 calories. *See* calorie.

K

Katabatic Winds Air flow down a slope under the influence of gravity.

Kelvin Scale An absolute temperature scale, where a value of 0 K implies an absence of thermal energy. The Kelvin scales assigns 100 units between the melting and sea level boiling points of water.

Kinetic Energy Energy of motion.

Kilopascal A unit of pressure equal to 1000 pascals or 0.1 millibars.

L

La Niña The opposite pattern to an El Niño, in which below-normal sea surface temperatures exist in the tropical eastern Pacific.

Land Breeze A wind that blows from the land toward the water along the coastal zone during the night and early morning.

Latent Heat (1) Energy present in water vapor, used in converting water from liquid to gas. Latent heat is released upon condensation. (2) Energy associated with the change of phase of a substance. *See* latent heat of fusion *and* latent heat of vaporization.

Latent Heat Flux Heat transfer that occurs whenever water vapor moves from one place to another. Energy used to evaporate water travels with the water vapor and is released upon condensation.

Latent Heat of Fusion Also called *latent heat of melting*. Energy released when a substance freezes, consumed when a substance melts. For water, the latent heat of fusion is about 334,000 J/kg.

Latent Heat of Vaporization Also called *latent heat of condensation*. Energy consumed when a substance evaporates, released when a substance condenses. For water, the latent heat of vaporization is about 2,500,000 J/kg.

Leader A column of ionized air that approaches the surface and precedes cloud-ground lightning.

Lenticular Cloud A lens-shaped cloud that usually forms downwind of topographic barriers.

Level of Free Convection The level to which conditionally unstable air must be lifted so that it can continue to rise due to its own buoyancy.

Lifting Condensation Level Altitude to which an air parcel would need to be lifted for condensation to occur.

Limb-Darkening The phenomenon in which the edge of the Sun appears darker than its center.

Little Ice Age A period in Earth's history from about 1400 to 1850 characterized by low temperatures.

Long-Range Forecast A weather prediction extending beyond seven days.

Long Waves *See* Rossby waves.

Longwave Radiation Another term for *infrared radiation*.

Low Another term for *cyclone*.

M

Mammatus A feature on parts of some cumulonimbus clouds marked by round, downward-extending protrusions.

Mature Stage The stage in an air mass thunderstorm marked by heavy storm activity, with strong updrafts, lightning, and heavy precipitation.

Maunder Minimum A period of Earth's history between about 1645 and 1715 characterized by minimal sunspot activity.

Mean Free Path The average distance traveled by molecules before colliding with adjacent molecules; increases with altitude.

Mechanical Turbulence (also called *forced convection*) Mixing of the air caused by horizontal movements (wind).

Medium-Range Forecast Weather forecasts for predictions three to seven days in advance.

Mercury Barometer The standard instrument for the measurement of atmospheric pressure.

Meridional Wind Wind flowing north–south parallel to a line of longitude. Actual winds are seldom completely meridional, but usually have both a meridional and zonal component.

Mesocyclone A rotating region within a cumulonimbus cloud where tornadoes often form.

Mesoscale A scale of meteorological phenomena typically having horizontal extents of several tens of kilometers.

Mesoscale Convective Complex (MCC) A type of mesoscale convective system having an oval or nearly circular shape.

Mesoscale Convective System (MCS) A general clustering of thunderstorms.

Mesosphere Region of the atmosphere from about 50 km to 80 km, characterized by decreasing temperature with increasing altitude.

Meteorological Service of Canada The official meteorological agency for Canada.

Meteorology The science that studies the atmosphere.

Microburst A small but severe downburst, whose wind shear is capable of causing air crashes.

Microscale The smallest scale of meteorological phenomena, such as that which might surround a leaf.

Microwave Radiation Electromagnetic radiation with wavelengths between about 0.1 and 300 μm. Weather radars use microwave radiation for imaging.

Mid-latitude Cyclone A low pressure system characterized by the presence of frontal boundaries.

Mie Scattering Scattering of visible radiation caused by particulates.

Milankovitch Cycles Variations in Earth's orbital characteristics having periodicities of tens of thousands of years.

Millibar A unit of atmospheric pressure, abbreviated as mb. Sea level pressure is about 1013 mb.

Mirage The apparent displacement of an object's true position due to refraction.

Mixed Layer That part of the lower atmosphere in which vertical motion (convection) is strong enough for even dispersal of pollutants.

Mixing Ratio A measure of atmospheric moisture: the mass of water vapor per unit mass of dry air, usually expressed in grams per kilogram (g/kg).

Moist Adiabatic Lapse Rate Another term for the *Saturated Adiabatic Lapse Rate* (*SALR*).

Monsoon A regional circulation pattern in which there is a seasonal reversal of wind and pressure, generally characterized by onshore flow during the summer and offshore flow during the winter.

Monsoon Depressions Areas of low pressure superimposed in the southeasterly air flow out of the Bay of Bengal.

Mountain Breeze A breeze that flows down a hill at night.

N

Nacreous Clouds Multicolored, pearlescent clouds found in the stratosphere. Also called *mother-of-pearl clouds*, these consist of ice crystals or supercooled water.

National Weather Service The official meteorological agency for the United States.

NCDC Stands for National Climate Data Center.

NCEP The initials of the National Centers for Environmental Prediction.

Nested Grid Model A particular numerical weather prediction model.

Neutral Stability A condition in which a lifted parcel of air does not return to its original position nor continue to rise. Neutral stability occurs when the environmental lapse rate is equal to the appropriate adiabatic rate, so that temperatures inside the parcel match those of the surroundings.

Newton's Second Law An expression of the conservation of momentum that is stated as: net force equals mass times acceleration ($F = ma$).

NEXRAD Stands for Next Generation Weather Radar, a network of Doppler radar units established by the U.S. National Weather Service.

NHC Stands for National Hurricane Center.

Nimbostratus A low, layered cloud that yields light precipitation.

NOAA Stands for National Oceanic and Atmospheric Administration.

Noctilucent Cloud A type of cloud that exists in the mesosphere, visible just after sunset (or before sunrise), when the surface and the lower atmosphere are in Earth's shadow.

Nonselective Scattering Scattering of radiation in which all wavelengths are scattered about equally. This type of scattering causes clouds to appear white.

North Atlantic Oscillation *See* Arctic Oscillation.

Northeaster or Nor'easter A winter weather condition of the Atlantic Coast of the United States and Canada associated with the passage of mid-latitude cyclones. The strong northeasterly winds are usually coupled with blizzard conditions.

NSSFC Stands for National Severe Storms Forecast Center.

Nuclear Fusion The thermonuclear process in which extreme heat and pressure cause atoms to combine, forming a different (heavier) element. A small part of the original mass is converted to tremendous quantities of energy and released to the environment.

Numerical Weather Prediction Weather prediction based on equations representing physical processes (as opposed to statistical relations).

NWS The initials of the National Weather Service.

O

Obliquity The degree of tilt of Earth's axis relative to the ecliptic plane, currently about 23.5°.

Occluded Front A front found in the late stages of a mid-latitude cyclone.

Ocean Current The horizontal movement of surface waters caused by prevailing winds.

Omega High A pressure pattern depicted on upper-level weather maps by a pattern resembling the Greek letter Ω.

Orographic Lifting Rising motions caused by airflow over a mountain range or other topographic barrier.

Outgassing The emission of gases that accompanies volcanic eruptions.

Overrunning Warm air sliding over a dense cold air mass; the characteristic flow associated with a warm front.

Oxides of Carbon A general class of air pollutants consisting of oxygen and carbon.

Ozone Molecules consisting of three oxygen atoms, most abundant in the middle and upper stratosphere.

Ozone Hole Ozone depletions found at high latitudes (especially over Antarctica) in the spring of each year.

Ozone Layer The portion of the stratosphere where ozone is relatively abundant, reaching a few parts per million.

P

Pacific Decadal Oscillation An alternating pattern of sea surface temperature in the Pacific that reverses itself over periods of several decades.

Particulates *See* aerosols.

Pascal The standard unit of pressure in most scientific applications, equal to 1 N/m^2.

Perihelion Earth's closest approach to the Sun (January 4).

Permanent Gases Those gases whose relative abundance is constant within the homosphere.

Persistence Forecast A weather forecast made by assuming some existing trend continues into the future.

Photochemical Smog Secondary air pollutants formed by chemical reactions in the presence of sunlight.

Photodissociation Splitting of molecules into atoms or submolecules by radiation. For example, in the thermo sphere, ultraviolet radiation dissociates molecular oxygen (O_2) into atomic oxygen (O).

Photosphere That part of the Sun that emits most of the energy reaching Earth. It is the "visible" part of the Sun, a layer representing about 0.05 percent of the solar radius.

Photosynthesis The growth process of green plants, whereby water and carbon dioxide are converted to carbohydrate, releasing oxygen.

$PM_{2.5}$ Designation given to particulates smaller than 2.5 μm in diameter. Major attention has recently been given to this class of particulates as possibly the most damaging to human health.

PM_{10} Designation given to particulates with diameters smaller than 10 μm, which are believed to have major health consequences for humans.

Polar Easterlies Low-level winds originating in the polar highs, a feature of the general circulation of the atmosphere, often very weak or absent.

Polar Front Transition zone between cold polar air and warmer air of the mid-latitudes.

Polar Front Theory The theory postulated in the early part of the twentieth century describing the formation, development, and dissipation of mid-latitude cyclones. Many of the features of the theory are still considered valid.

Polar Highs Low-level anticyclones of the Arctic and Antarctic. A feature of the general circulation of the atmosphere, often absent or weakly developed.

Polar Jet Stream A jet stream found in the upper troposphere above the polar front, a result of the strong temperature contrast across the front.

Polaris The North Star

Potential Energy Energy possessed by virtue of an object's position above some reference level. Potential energy is available for conversion to kinetic energy.

Potential Instability The condition in which a layer of air can become statically unstable if lifted sufficiently.

Power The rate at which work is done or energy expended. The standard unit is the watt, equal to 1 J/sec.

Precession The wobble of Earth's axis that has a periodicity of about 27,000 years.

Precipitable Water Vapor A measure of the total water vapor content of the atmosphere. The depth of water that would result if all the water in the column were to condense. Global average precipitable water vapor is about 2.5 cm.

Precipitation Liquid water or ice that falls to Earth's surface. Rain is considered precipitation, but dew is not.

Precipitation Fog A type of fog that develops when falling raindrops evaporate enough water vapor into the air to saturate it.
Pressure Force exerted per unit area. In most sciences the standard unit of measurement is the pascal (Pa), equal to 1 N/m^2. In daily meteorological applications, however, the millibar (mb) is frequently used in the United States and the kilopascal in Canada.
Pressure Gradient Force A force that arises from spatial variation in pressure. Acting alone, the pressure gradient force would cause air to blow from an area of high pressure toward an area of low pressure. The vertical pressure gradient force is always present but is nearly balanced by gravity most of the time. Much weaker horizontal pressure gradients are the ultimate cause of wind.
Primary Pollutants Substances that pollute the atmosphere upon release. *See* secondary pollutants.
Psychrometer An instrument for measuring atmospheric moisture.
Pyranometer An instrument for measuring solar radiation.

R

Radar A device that uses microwave radiation for imaging the atmosphere.
Radiation Another term for *electromagnetic radiation.*
Radiation Fog A low-level cloud formed diabatically when the atmosphere loses heat by radiation upward.
Radiosonde An instrument package carried by balloon, used to measure vertical profiles of temperature, moisture, and pressure. Measurements are radioed to the ground from the instrument cluster.
Rain Precipitation arriving at the surface in the form of liquid drops, usually between 0.5 and 5 mm. Outside of the Tropics, rain usually begins in the ice stage and melts before reaching the surface. Rain that freezes on contact with the surface, forming a layer of ice, is called *freezing rain.*
Rainbow A wide, sweeping band of light caused by refraction of sunlight by rain drops.
Rain Shadow An area on the lee (downwind) side of a mountain barrier having relatively low precipitation.
Rawinsonde A radiosonde tracked by radar to provide wind information.
Rayleigh Scattering The scattering of radiation by agents substantially smaller than the radiation's wavelength. In the case of the atmosphere, this applies to the scattering of visible radiation by air molecules.
Reflection A process in which radiation arriving at a surface bounces back, without being absorbed or transmitted. Reflection does not heat the reflector, because there is no net energy transfer to the surface.
Refraction The bending of light within a medium or as it passes from one medium to another. Refraction results from density differences within/between the transfer media.
Relative Humidity The measure of the amount of water vapor in the air as a fraction of saturation, often expressed as a percentage. Because the saturation point is temperature-dependent, relative humidity depends on both the moisture content and the temperature of the air.
Return Stroke Synonymous with *lightning stroke.*
Ridge An elongated axis of high pressure.
Riming The growth of a falling ice particle as it collides with nearby water droplets that freeze onto the particle.
Rossby Waves Also known as *long waves.* Waves in the mid-latitude westerlies having wavelengths on the order of thousands of kilometers. Often a series of Rossby waves circle the planet, forming a pattern of ridges and troughs.

S

Saffir-Simpson Scale A scheme for classifying the intensity of hurricanes.
Santa Ana Wind A local name for a foehn wind in California.
Saturated Adiabatic Lapse Rate (SALR) Rate of temperature change for a rising saturated parcel of air. The value ranges from about 4 to 10 °C/km, depending mainly on temperature.
Saturation The maximum amount of water that can exist in the atmosphere as a vapor. More precisely, saturation occurs when a flat surface of pure water is in equilibrium with the overlying atmosphere. The evaporation rate equals the condensation rate, so the vapor content of the air is unchanging. The saturation point increases with increasing temperature.
Saturation Mixing Ratio The mixing ratio of the atmosphere when it is saturated.
Saturation Specific Humidity The specific humidity of the atmosphere when it is saturated.
Saturation Vapor Pressure The vapor pressure of the atmosphere when it is saturated.
Scalar A quantity or property that possesses magnitude but has no direction. Examples are temperature, pressure, and density.
Scattering The dispersion or redirection of radiation by gases, dust, water drops, ice, and other particulates. Scattering does not heat the atmosphere, because there is no energy transfer to the scattering agent.
Sea Breeze A flow of air from the water toward land along a coastal region.
Sea Level Pressure The pressure that would presumably exist at a point if it were at sea level. This involves a conversion of observed surface air pressure.
Secondary Pollutants Pollutants that form in the atmosphere from reactions involving anthropogenic or natural substances.
Semi-desert A type of dry climate that annually receives enough precipitation to distinguish it from a true desert.
Semipermanent Cell Large area of high or low pressure present throughout the year, usually with size and location changing seasonally.
Sensible Heat Heat transfer produced by the movement of warm air.
Severe Thunderstorm A thunderstorm that produces either very strong winds, large hail, or tornadoes.
Severe Thunderstorm Warning An advisory issued by a local office of the National Weather Service indicating that severe thunderstorms are occurring or imminent.

Severe Weather Warning Advisory issued when severe weather is observed. In the case of a hurricane warning, the potential for landfall exists within a 24-hour period.

Severe Weather Watch Advisory issued when atmospheric conditions are favorable for severe weather. In the case of a hurricane watch, the potential for landfall exists outside a 24-hour period.

Short Wave A small wave in the mid-latitude westerlies. Often superimposed on Rossby waves, they move more quickly, and thus travel through the large-scale pattern.

Shortwave Radiation Electromagnetic energy having wavelengths shorter than about 4 μm.

Siberian High A semi-permanent cell found in North Asia during Winter.

Sleet Precipitation in the form of ice pellets, resulting when raindrops freeze before reaching the surface.

Snow Frozen, crystalline precipitation that forms and remains in the ice stage throughout its descent.

Solar Altitude The angle between the horizon and the Sun. When the Sun is overhead, the solar altitude is 90°. *See* zenith angle.

Solar Constant The amount of radiation reaching the top of the atmosphere when Earth is at its average distance from the Sun. This is not a pure constant, but rises and falls with changes in solar emission. Its value is about 1376 W/m^2.

Solar Declination The latitude of overhead Sun; the place where one would go to find the Sun directly overhead at noon.

Solar Wind A continuous stream of particles (mostly protons and electrons) emitted by the Sun, traveling about one-third to one-half the speed of light.

Solstices The two times each year that mark the northern and southern limits of the latitude of overhead sun. On the June Solstice (approximately June 21) Northern Hemisphere latitudes have their longest day of the year. On the December Solstice (approximately December 22), Southern Hemisphere latitudes have their longest day of the year.

Source Region A large area of land or ocean of more-or-less uniform characteristics, above which an air mass can form.

Southern Oscillation The reversal of surface pressure patterns over the tropical Pacific associated with El Niño events.

Specific Heat The amount of energy required to raise the temperature of a given mass of a substance by a given amount.

Specific Humidity A measure of atmospheric moisture. The mass of water vapor per unit mass of air, usually expressed in grams per kilogram (g/kg).

Speed A scalar property representing the rate of motion.

Speed Convergence The compaction of air due to decreasing wind speed in the downwind direction.

Speed Divergence The spreading of air due to increasing wind speed in the downwind direction.

Squall Line A linear band of thunderstorms, often found several hundred kilometers ahead of a cold front.

Stable Air Air that, when displaced vertically, returns to its initial position. Stable air resists uplift.

Standard Atmosphere The mean structure of the atmosphere with regard to temperature and pressure.

Static Stability The condition of the atmosphere that inhibits or favors vertical displacement of air parcels.

Station Model A plotting on weather maps for individual locations depicting current temperature, dew point, pressure, and other meteorological information.

Stationary Front A transition zone between dissimilar air masses (a front) showing little or no tendency to move.

Steam Fog Fog that forms when cold air moves over a warmer water surface.

Stefan-Boltzmann Law A law for blackbody emission that states that the total energy emitted over all wavelengths is proportional to the fourth power of absolute temperature.

Steppe Another term for *semi-desert*.

Stepped Leader A narrow zone of ionized air that serves as a conduit for an initial lightning stroke.

Storm Surge A potentially damaging influx of coastal waters brought about by high winds and low pressures associated with hurricanes.

Stratocumulus A low, layered cloud having superimposed rows or cells of vertical development.

Stratopause Upper limit of the stratosphere; the transition between the stratosphere and mesosphere.

Stratosphere A layer of the atmosphere between about 16 and 50 km, characterized by generally increasing temperature with increasing altitude.

Stratus A cloud with a layered structure.

Streamlines Lines that depict the path of wind. Air parcels are envisioned to flow along streamlines. *See* confluence *and* diffluence.

Structure The layering of the atmosphere combined with the reduction in density with altitude.

Stuve Diagram A particular type of thermodynamic diagram used for plotting temperature and moisture profiles.

Sublimation Change from a solid into a vapor without passing through the liquid phase. Also used to describe the reverse (vapor to solid).

Subpolar Low A belt of low pressure in the three-cell model, between the polar easterlies and mid-latitude westerlies.

Subtropical High A semi-permanent cell that occupies large areas of the mid-latitude oceans, especially in the warm season.

Subtropical Jet A jet stream common in the upper troposphere on the poleward side of the Hadley cells, produced by the conservation of angular momentum.

Suction Vortex A zone of intense rotation within a large tornado that often causes the most devastation.

Sulfur Oxides A general class of pollutants consisting of sulfur and oxygen.

Sun Pillars Bands of light stretching vertically from the Sun caused by reflection off almost horizontally aligned ice crystals.

Sundogs Paired bright spots found 22° to the right or left of the Sun caused by ice crystal refraction.

Sunspots Magnetic storms of the Sun, appearing as dark (Earth-sized) spots on the photosphere.

Supercell Thunderstorm A very large thunderstorm formed from an extremely powerful updraft.
Supercooled Water Water existing in the liquid phase with a temperature less than 0 °C.
Supersaturation A relative humidity greater than 100%, when the atmosphere is more than saturated with water vapor. Requires a very clean atmosphere, where condensation nuclei are lacking.
Synoptic Scale The scale of meteorological phenomena having areas on the order of hundreds or thousands of square kilometers.

T

Teleconnection Relationship between weather or climate patterns at two widely separated locations.
Temperature An index of the average kinetic energy of the molecules comprising a substance.
Temperature Gradient Temperature change per unit distance. A strong temperature gradient implies that temperature changes rapidly over a short distance.
Temperature Inversion Condition in which temperature increases with increasing altitude.
Terminal Velocity The final speed obtained by an object falling through the atmosphere, when friction with the surrounding air balances the force of gravity.
Thermal Low Low-pressure cell produced by heating of the surface.
Thermistor An object whose electrical resistance changes with temperature, thus allowing temperature to be determined by measuring changes in electrical current.
Thermodynamic Diagram A diagram showing the relationship between pressure, temperature, density, and water vapor content, such that characteristics of air parcels can be determined as they *ascend* and *descend*.
Thermoelectric Effect A theory of lightning formation in which separation of charge is produced by positive ions migrating from warmer particles to colder ice crystals.
Thermohaline Circulation A movement of surface waters in the oceans due to variations in temperature and salt content.
Thermometer Instrument used to measure temperature.
Thermosphere Outermost reaches of the atmosphere, beginning at about 80 km, characterized by increasing temperature with increasing altitude and by extremely low density.
Thornthwaite's Classification System The most widely used system for classifying general climate zones.
Threat Score A measure of precipitation forecast skill that considers the area correctly forecast relative to that under threat of precipitation.
Three-cell Model A generalized description of global-scale circulation that calls for three large cells in each hemisphere. The cells rotate on a vertical plane with axes parallel to latitude lines, thereby moving heat and moisture in a north–south direction.
Thunder Sound produced when lightning discharges heat the surrounding air, causing pressure waves to emanate outward.
Tibetan Low A semi-permanent cell found in southern Asia in Summer.
Tipping-bucket Gage A type of automated rain gage.
Tornado Warning An advisory issued by a local office of the National Weather Service indicating that a tornado is occurring or imminent.
Tornado A rotating column of air with extreme horizontal winds.
Trade Wind Inversion A temperature inversion (layer of air having increasing temperature with altitude) commonly found at subtropical and tropical latitudes.
Trade Winds Prevailing lower troposphere winds of the Tropics, associated with Hadley circulation. Strongest in the respective winter season, the trades blow from the northeast in the Northern Hemisphere and from the southeast in the Southern Hemisphere.
Transpiration Transfer of water to the atmosphere by vegetation, mostly by water evaporating and escaping plant tissues through leaf pores.
Tropic of Cancer A line of latitude at 23.5° N, the northern limit of solar declination.
Tropic of Capricorn A line of latitude at 23.5° S, the southern limit of solar declination.
Tropical Depression A closed zone of low pressure with wind speeds less than 60 km/hr.
Tropical Disturbance A disorganized group of thunderstorms with weak pressure gradients and little or no rotation.
Tropical Storm A storm that originates in tropical regions and has wind speeds between 60 and 120 km/hr.
Tropopause Upper limit of the tropopause; the transition between the troposphere and stratosphere.
Troposphere The lowest temperature layer of the atmosphere, from the surface to about 16 km, characterized by generally decreasing temperatures with increasing altitude.
Trough An elongated axis of low pressure.
Turbidity Loosely speaking, the "dustiness" of the atmosphere, including the effect of all particulates that reduce visibility.

U

Ultraviolet Radiation Electromagnetic radiation at wavelengths too short to be visible, from about 0.001 to 0.4 μm.
Unstable Air Air that experiences a buoyant force following a vertical displacement, causing it to rise. Uplift is promoted by instability.
Upwelling Movement of ocean or lake water from lower levels toward the surface.
Urban Heat Island Increased local temperatures that result from urbanization.

V

Valley Breeze A low-level movement of air in an upslope direction, developing during the daylight hours as result of solar heating.
Vapor Pressure A measure of atmospheric moisture, the partial pressure exerted by water vapor.

Variable Gases Gases present in amounts that vary greatly in abundance, either vertically, horizontally, or seasonally. Water vapor is the most important variable gas.

Vault An apparently empty area on a radar display, where moist air enters a supercell thunderstorm. Water droplets are abundant but are too small to provide a strong radar echo.

Vector A quantity or property possessing both magnitude and direction. Examples are wind velocity, the Coriolis force, and the pressure gradient force.

Veering Wind Wind that changes direction in a clockwise sense.

Velocity A vector property that includes speed and motion of an object or substance.

Vertical Pressure Gradient Force The upward-directed force that arises because pressure always decreases with increasing altitude.

Visible Radiation Electromagnetic radiation between about 0.4 and 0.7 μm, which is detectable by the human eye.

Volatile Organic Compounds Carbon-hydrogen molecules, both anthropogenic and naturally produced, which can be gaseous or particulate. A precursor to photochemical smog.

Vorticity The turning of an object (such as an air parcel), usually with respect to the vertical direction. This is important to meteorology because of its association with areas of divergence and convergence.

W

Walker Circulation An east–west circulation pattern of the Tropics, characterized by several cells of rising and sinking air connected by horizontal motions along more-or-less parallel lines of latitude.

Wall Cloud Thick cloud beneath a rotating thunderstorm, a place where severe weather often develops.

Warm Advection Heat carried by airflow across isotherms from warm to cold.

Warm Core High High-pressure cells with higher temperature than surrounding air. Also called *warm core anticyclones*, they originate from atmospheric motion, not from differences in heating.

Warm Core Low Low-pressure cells that are warmer than surrounding air, produced by heating. Also called *warm core cyclones*.

Warm Front Transition zone between two air masses of different temperature, produced when warm air advances on and overruns cold air.

Waterspout A rather weak whirlwind (narrow rotating column of air), forming over a water surface. Rising and condensing air makes the waterspout visible.

Water Vapor Water in its gaseous phase, not to be mistaken for small water droplets. Colorless and odorless, it seldom amounts to more than a few percent of the total atmospheric mass.

Watt The SI unit of power, abbreviated as W, with dimensions of energy per unit time; 1 W = 1 J/sec.

Wave Cyclone Low-pressure storm common in mid-latitudes, typically formed in a favored position along a Rossby wave.

Wavelength Distance between successive peaks of a wave, or successive troughs, or between any two corresponding points along a wavetrain.

Weather Forecast Office A United States Weather Service facility that issues local and area forecasts.

Weighing-bucket Gage A type of automated gage for the measurement of precipitation.

West Greenland Drift A branch of a current flowing southward in the North Atlantic.

Westerlies Winds belts found in the middle latitudes of both hemispheres that have a strong west-to-east component.

Wet Bulb Depression The difference between air temperature and wet bulb temperature. Large values (large differences between dry and wet bulb temperatures) are indicative of low humidity.

Wet Bulb Temperature The minimum temperature achieved as water evaporates from a wick surrounding a thermometer's bulb.

Wien's Law Law for blackbody emission that states that the wavelength of maximum emission is inversely proportional to absolute temperature.

Wind The horizontal movement of air.

Wind Chill Temperature Index An index of apparent temperature used for cold conditions that incorporates air temperature and wind speed.

Wind Profiler A type of Doppler radar unit that provides information about vertical changes in wind speed and direction.

Wind Shear Large change in wind direction or speed over a short distance.

Wind Vane An instrument for measuring or indicating wind direction, often a pivoting arrow that continually points into the wind.

WMO Stands for the World Meteorological Organization, the agency of the United Nations charged with collecting and disseminating meteorological data.

X

X-Ray Radiation Electromagnetic radiation at wavelengths between about 0.000001 μm and about 0.001 μm.

Z

Zenith Angle The angle between the Sun and the vertical direction. When the Sun is overhead, the zenith angle is zero. *See* solar altitude.

Zonal Wind (Flow) Wind flowing east–west, parallel to a line of latitude. Actual winds are seldom completely zonal but usually have both a meridional and zonal component.

Zone Forecast A weather forecast issued at regular intervals for a particular region.

Credits

Chapter 1

Photographs
Part Opener 1, Edward Aguado
Chapter Opener 1, AFP/CORBIS
1–1, NASA/Johnson Space Center
1–3a,b, Space Science and Engineering , University of Wisconsin-Madison
1–5, George Holton/Photo Researchers, Inc.
1–7, Photo by Tom L. McKnight
1–10, Joe Towers/Corbis/Stock Market
1–12, Pekka Parviainen/Science Photo Library/Photo Researchers, Inc.
1–16, NASA Headquarters
1–19, NASA Headquarters
1–21a-c, Edward Aguado
1–B2–1, NASA Goddard Laboratory for Atmospheres
1–B3–1, National Geophysical Data Center
1–B4–3, Richard Carson/Reuters/Getty Images, Inc.

Illustrations
1–4, 1–6, From National Oceanographic and Atmospheric Agency
1–15, 1–17, 1–18, 1–20, Modified from Chaisson and McMillan, *Astronomy Today*, Second Edition, ©1997, Prentice Hall, Inc., Upper Saddle River, NJ

Chapter 2

Photographs
Chapter Opener 2, AP/Wide World Photos
2–4a, Julian Baum & Nigel Henbest/Science Photo Library/Photo Researchers, Inc.
2–4b, National Optical Astronomy Observatories
2–8, NASA/Weatherstock
2–B3–2a, National Optical Astronomy Observatories
2–B3–4, Photo by George East, courtesy Sky & Telescope

Illustrations
2–1, Modified from Nebel and Wright, *Environmental Science*, Sixth Edition, ©1998, Prentice Hall, Inc., Upper Saddle River, NJ
2–6, Modified from Chaisson and McMillan, *Astronomy Today*, Second Edition, ©1997, Prentice Hall, Inc., Upper Saddle River, NJ
Box 2–2, Figure 1; Box 2–3, Figures 1 & 2b, From Chaisson and McMillan, *Astronomy Today*, Second Edition, © 1997, Prentice Hall, Inc., Upper Saddle River, NJ
Box 2–3, Figure 3, Data from Solar-Terrestrial Physics Division, National Geophysical Data Center, Boulder, Colorado, and John A. Eddy, "The Maunder Minimum," *Science*, Vol. 192, pp. 1189–1202.

Chapter 3

Photographs
Chapter Opener 3, Mike R. Tuten
3–1, NASA/Goddard Space Flight Center/Laboratory for Atmospheres
3–4, NASA/Johnson Space Center
3–6, Larry Cameron/Photo Researchers, Inc.
3–11b, Robert J. Erwin/Photo Researchers, Inc.
3–20, Edward Aguado
3–22, NOVALYNX Corporation, Grass Valley, CA
3–23, Qualimetrics, Inc.
3–24, Edward Aguado

Illustrations
3–9b, J. N. Howard, *Proc. I.R.E.*, v. 47, 1959, and R. M. Goody and G. D. Robinson, *Quart. J. Roy. Meteorol. Soc.*, v. 77, 1951, as modified by R. G. Fleagle and J. A. Businger, *An Introduction to Atmospheric Physics*, Second Edition, © 1980, Academic Press, New York, NY
3–16, 3–18, Modified from Christopherson, *Geosystems, An Introduction to Physical Geography*, Third Edition, ©1997, Prentice Hall, Inc., Upper Saddle River, NJ

Chapter 4

Photographs
Chapter Opener 4, Jon Burbank/The Image Works
4–4, Charles D. Winters/Photo Researchers, Inc.
4–5a, Edward Aguado
4–5c, Edward Aguado
4–22, Edward Aguado

Illustrations
4–5b, Modified from Moran and Morgan, *Meteorology, The Atmosphere and the Science of Weather*, Fifth Edition, © 1997, Prentice Hall, Upper Saddle River, NJ

Chapter 5

Photographs
Part Opener 2, Melvin Grubb/Grubb Photo Service, Inc.
Chapter Opener 5, © John Biever/SI/NFL Photos
5–8, Michael Brisson
5–11a, NOVALYNX Corporation, Grass Valley, CA
5–11b, NOVALYNX Corporation, Grass Valley, CA
5–15a, Rod Planck/Photo Researchers, Inc.
5–15b, F. Stuart Westmorland/Photo Researchers, Inc.
5–16, ©Arjen & Jerrine Verkaik/Skyart
5–17, National Environmental Satellite, Data, and Information Center/NOAA
5–18, Gary Randall/Getty Images, Inc.

Illustrations
5-6, National Climatic Data Center

Chapter 6

Photographs
Chapter Opener 6, Howard Bluestein/Photo Researchers, Inc.
6–1, Copyright Howard B. Bluestein
6–3a, Edward Aguado
6–3b, Edward Aguado
6–3c, Edward Aguado
6–4a, David Weintraub/Photo Researchers, Inc.
6–4b, George Ranalli/Photo Researchers, Inc.
6–16, Joyce Photographics/Photo Researchers, Inc.
6–17a, Edward Aguado
6–18, Maurice Nimmo/Frank Lane Picture Agency/CORBIS
6–19, John Foster/Science Source/Photo Researchers, Inc.
6–20, Arjen and Jerrine Verkaik/Skyart Productions
6–21, A & J Verkaik/Skyart Productions
6–22, Pekka Parviainen/Science Photo Library/Photo Researchers, Inc.
6–25, John Sohlden/Visuals Unlimited
6–26, David R. Frazier/Photo Researchers, Inc.
6–28, Day Williams/Photo Researchers, Inc.
6–29, Howard B. Bluestein/Photo Researchers, Inc.
6–30, Kent & Donna Dannen/Photo Researchers, Inc.
6–33, Pekka Parviainen
6–34, Pekka Parviainen/Science Photo Library/Photo Researchers, Inc.
6–B2–1a, Chuck O'Rear/Woodfin Camp & Associates
6–B2–1b, Chuck O'Rear/Woodfin Camp & Associates
6–23, ©Warren Faidley/Weatherstock
6–24, Claudia Parks/Corbis/Stock Market
6–31, Richard J. Green/Photo Researchers, Inc.
6–32, ©Warren Faidley/Weatherstock®

Illustrations
6-15, From Christopherson, *Geosystems*, 4th edition, ©2000, Prentice Hall, Upper Saddle River, NJ
6–17b, Adapted from Gedezelman, *The Science and Wonders of the Atmosphere*, © 1980, Wiley, Inc., New York, NY

Chapter 7

Photographs
Chapter Opener 7, Howard B. Bluestein/Photo Researchers, Inc.
7–1, Courtesy of Spaceimaging.com
7–6, Doug Millar/Science Source/Photo Researchers, Inc.
7–8a, Clyde H. Smith/Peter Arnold, Inc.
7–8b, Scott Camazine/Photo Researchers, Inc.
7–8c, Richard C. Walters/Visuals Unlimited
7–10, Joe Traver/Getty Images, Inc - Liaison
7–12, Howard Bluestein/Photo Researchers, Inc.
7–16, The Image Works
7–18a, Edward Aguado
7–18b, Qualimetrics, Inc.
7–20, National Oceanic and Atmospheric Administration

Illustrations
7–2, Adapted from McDonald, *The Physics of Cloud Formation, Advances in Geophysics*, v. 5, Academic Press, New York, NY, as modified by Rogers, *A Short Course in Cloud Physics*, © 1979, Pergamon Press, New York, NY
7–13, Adapted from Browning and Foote, *Q.J.R. Meteorol. Soc.*, London, as modified by Eagleman, *Meteorology: The Atmosphere in Action*, Second Edition, © 1985, Wadworth, Inc., Belmont, CA
7–14, Adapted from NOAA
7–19, Courtesy of David Legates

Chapter 8

Photographs
Part Opener 3, Carl Purcell/Science Source/Photo Researchers, Inc.
Chapter Opener 8, C. S. Gray/The Image Works
8–1, ©Dick Keen/Visuals Unlimited
8–3, Image produced by M. Jentoft-Nilsen, F. Hasler, D. Chesters (NASA/Goddard) and T. Nielsen (Univ. of Hawaii)./NASA Headquarters
8–9, NCDC at NOAA
8–16, O. Brown, R. Evans, and M. Carle, University of Miami Rosenstiel School of Marine and Atmospheric Science, Miami, Florida
8–17a,b, Tokai University Research and Information Center
8–21a, Edward Aguado
8–21b, Downey/Getty Images, Inc - Liaison
8–23, NASA/Johnson Space Center
8–26, NOAA/Science Photo Library/Photo Researchers, Inc.

Illustrations
8–4, Modified from Christopherson, *Geosystems, An Introduction to Physical Geography*, Third Edition, © 1997, Prentice Hall, Inc., Upper Saddle River, NJ
8–6, Adapted from Miller et al., *Elements of Meteorology*, © 1983, Merrill Publishing Co., Columbus, OH
8–14, Modified from Trujillo and Thurman, *Essentials of Oceanography*, Sixth Edition, © 1999, Prentice Hall, Inc., Upper Saddle River, NJ
8–25, Modified from Peixoto and Oort, *Physics of Climate*, New York: American Physics Inst., © 1992, p. 419
8–27, Courtesy of NASA/JPL/California Institute of Technology

Chapter 9

Photographs
Chapter Opener 9, Photo by Tom L. McKnight
9–7a, NOAA Central Library Photo Collection
9–7b, Weather Services International, a Landmark Company
9–B1–1b, d, f, National Oceanic and Atmospheric Administration/Seattle
9–B1–2b, National Oceanic and Atmospheric Administration/Seattle
9–B1–3b, d, f, National Oceanic and Atmospheric Administration/Seattle

Illustrations
9–1, Modified from Moran and Morgan, *Meteorology, The Atmosphere and the Science of Weather*, Fifth Edition, © 1997, Prentice Hall, Inc., Upper Saddle River, NJ
9–8, Modified from Christopherson, *Geosystems, An Introduction to Physical Geography*, Third Edition, © 1997, Prentice Hall, Inc., Upper Saddle River, NJ

Chapter 10

Photographs
Part Opener 4, James Wiesmueller
Chapter Opener 10, Airphoto - Jim Wark
10–11c, 10–12c, 10–13c, 10–14c, 10–18b, National Oceanic and Atmospheric Administration/Seattle

Illustrations
10–8, Adapted from Petterson, *Weather Analysis and Forecasting*, Vol. 1, Second Edition, © 1956, McGraw-Hill, New York, NY

Chapter 11

Photographs
Chapter Opener 11, Kennan Ward Photography
11–1, NOAA Central Library Photo Collection
11–3, Ed Degginger/Color-Pic, Inc.
11–5, Daniel L. Osborne, University of Alaska/Detlev Van Ravenswaay/Photo Researchers, Inc.
11–6, Victor Pasko
11–8, Larry Miller/Photo Researchers, Inc.
11–9, National Oceanic and Atmospheric Administration/Seattle
11–12, Weather Services International, a Landmark Company
11–14, Howard B. Bluestein, Professor of Meteorology
11–16b, National Oceanic and Atmospheric Administration/Seattle
11–19a, Sheila Beougher/Getty Images, Inc - Liaison
11–19b, A & J Verkaik/Skyart Productions
11–21, Arjen Verkaik/Skyart Productions
11–31, Photo by Jeff Mitchell REUTERS/Landov LLC
11–B4–1, National Oceanic and Atmospheric Administration/Seattle

Illustrations
11–2, 11–13, 11–16a, Adapted from Wallace and Hobbs, *Atmospheric Science, An Introductory Survey*, © 1977, Academic Press, New York, NY
11–4, Adapted from Zajac and Weaver, Cooperative Institute for Research in the Atmosphere, Colorado State University.
11–7, Adapted from Gedzelman, *The Science and Wonders of the Atmosphere*, © 1980, Wiley, Inc., New York, NY
11–10, Image from Roger Edwards/NOAA Storm Prediction Center
11–15, Adapted from Djuric, *Weather Analysis*, © 1994, Prentice Hall, Inc., Upper Saddle River, NJ
11–18, 11–26, 11–27, 11–28, Source: NOAA
11–22, From Bluestein, *Tornado Alley*, 1999, Oxford University Press
11–23, 11–24, Adapted from Brandes, in Church et al., eds., *The Tornado: Its Structure, Dynamics, Prediction, and Hazards*, © 1993, American Geophysical Union, Washington, DC
11–25, Modified from McKnight, *Physical Geography, A Landscape Appreciation*, © 1996, Prentice Hall, Inc., Upper Saddle River, NJ
11–29, Adapted from Fujita, *Journal of Atmospheric Sciences*, © 1981, American Meteorological Society, Boston, MA, as modified by Battan, *Fundamentals of Meteorology*, © 1984, Prentice Hall, Inc., Upper Saddle River, NJ
11–30, National Weather Service.

Chapter 12

Photographs
Chapter Opener 12, Pablo Bartholomew/Getty Images, Inc - Liaison
12–1, NASA Headquarters
12–10a,b, National Geographic Image Collection
12–B4–1, National Oceanic and Atmospheric Administration/Seattle

Illustrations
12–2, Modified from McKnight, *Physical Geography, A Landscape Appreciation*, Fifth Edition, © 1996, Prentice Hall, Inc., Upper Saddle River, NJ
12–3, 12–6, Modified from Moran and Morgan, *Meteorology: The Atmosphere and the Science of Weather*, Fifth Edition, © 1997, Prentice Hall, Inc., Upper Saddle River, NJ
12–4, Adapted from Miller, *Science*, v. 157, 1967, American Association for the Advancement of Science, as modified by Eagleman, *Meteorology: The Atmosphere in Action*, Second Edition, © 1985, Wadsworth, Inc., Belmont, CA
12–7, Source: NOAA
12–9, Adapted from Novlan and Gray, *Monthly Weather Review*, © 1974, American Meteorological Society, Boston, MA, as modified by unpublished National Weather Service Training Center pamphlet

Chapter 13

Photographs
Part Opener 5, David Pollack/Corbis/Stock Market
Chapter Opener 13, AP/Wide World Photos
13–2L,M,R, Edward Aguado
13–18a, b, c, National Oceanic and Atmospheric Administration/Seattle
13–19, Weather Services International, a Landmark Company

Illustrations
13-1, Adapted from DeFelice, *Introduction to Meteorological Instrumentation and Measurement*, ©1998, Prentice Hall, Inc., Upper Saddle River, NJ
13–3, Adapted from Olson et al., *Weather and Forecasting*, v. 10, 1995

13–4, 13–5, Adapted from Mesinger, *Bulletin of the American Meteorological Society,* v. 77, 1996
13–7, 13–8, Source: NOAA
13–19, Weather Services International (Billerica, MA)
Appx 1, 3, Adapted from Olson et al., *Weather and Forecasting,* v. 10, 1995

Chapter 14

Photographs
Chapter Opener 14, Paula Bronstein/Getty Images, Inc - Liaison
14–2, Adam Hart-Davis/Science Photo Library/Photo Researchers, Inc.
14–4, Martin Bond/Science Photo Library/Photo Researchers, Inc.
14–5, Tony Craddock/Science Photo Library/Photo Researchers, Inc.
14–7, Bruce Fier/Getty Images, Inc - Liaison
14–9b, Edward Aguado
14–B1–1a, Carnegie Library of Pittsburgh
14–B1–1b, Photo by Arthur G. Smith
14–B3–1a, The MapFactory/Photo Researchers, Inc.

Illustrations
14–1, From Nebel and Wright, *Environmental Science,* Sixth Edition, © 1998, Prentice Hall, Inc., Upper Saddle River, NJ
14–3, Modified from Moran and Morgan, *Meteorology: The Atmosphere and the Science of Weather,* Fifth Edition, © 1997, Prentice Hall, Inc., Upper Saddle River, NJ
14–8, 14–9a, Adapted from Oke, *Boundary Layer Climates,* © 1978, Methuen & Co., Ltd., New York, NY
14–10, Adapted from Hidore and Oliver, *Climatology, An Atmospheric Science,* © 1993, Macmillan Publishing Co., New York
Box 14–2, Figure 1, From Nebel and Wright, *Environmental Science,* Fifth Edition, © 1993, Prentice Hall, Inc., Upper Saddle River, NJ
Box 14-2, Figure 2, From Environmental Protection Agency.
Box 14–3, Figure 2, Adapted from Blumenthal et al., *Atmospheric Environment,* vol. 12, © 1978.

Chapter 15

Photographs
Part Opener 6, Robert Llewellyn
Chapter Opener 15, Karl Weidmann/Photo Researchers, Inc.

Illustrations
15–1, 15–2, 15–3, 15–4, 15–5, 15–7, 15–8, 15–9, 15–10, 15–11, 15–12, 15–13, 15–14, 15–15, 15–16, Modified from McKnight, *Physical Geography, A Landscape Appreciation,* Fifth Edition, © 1996, Prentice Hall, Inc., Upper Saddle River, NJ
15–6, Source: NOAA
Box 15–1, Figure 1, Adapted from Critchfield, *General Climatology,* Third Edition, © 1974, Prentice Hall, Inc., Upper Saddle River, NJ

Chapter 16

Photographs
Chapter Opener 16, Dennis Flaherty/Photo Researchers, Inc.
16–7, Phil Durkee/The Naval Postgraduate School
16–9, Simon Fraser/Science Photo Library/Photo Researchers, Inc.
16–10, NASA/Johnson Space Center
16–13, Edward Aguado
16–15, Stephen J. Krasemann/Photo Researchers, Inc.

Illustrations
16–2, From Hamblin, *Earth's Dynamic Systems, Eighth Edition,* © 1998, Prentice Hall, Inc., Upper Saddle River, NJ
16–3, P. J. Bartlein, 1997. "Past Environmental Changes: Characteristic Features of Quaternary Climate Variations." In B. Huntley, W. Cramer, A. V. Morgan, H. C. Prentice, and J. R. M. Allen (eds.), Past and Future Rapid Environmental Changes: The Spatial and Evolutionary Responses of Terrestrial Biota. *NATO ASI Series,* v. 147, pp. 11–29. Berlin: Springer-Verlag
16–4, Modified from McKnight, *Physical Geology, A Landscape Appreciation,* Fifth Edition, © 1996, Prentice Hall, Inc., Upper Saddle River, NJ
16–5, D. W. Stahle, M. K. Cleaveland, D. B. Blanton, M. D. Therrell, and D. A. Gay, 1998. "The Lost Colony and Jamestown Droughts." *Science,* v. 280, 24 April 1998, 564–567
16–6, Modified from Chaisson and McMillan, *Astronomy Today,* Second Edition, © 1997, Prentice Hall, Inc., Upper Saddle River, NJ
16–8, NASA
16–10, From Jet Propulsion Laboratory.
Box 16–1, Figure 1, Source: NOAA
16-11 & 12, Adapted from *Climate Change, The IPCC Scientific Assessment,* Houghton, Jenkins, and Ephraums, eds., © 1990, Cambridge University Press, Cambridge, England

Chapter 17

Photographs
Part Opener 7, © Charles O'Rear/CORBIS
Chapter Opener 17, Kent Wood/Photo Researchers, Inc.
17–3, Pekka Parviainen
17–6, Pekka Parviainen/Science Photo Library/Photo Researchers, Inc.
17–7, Jim Steinberg/Photo Researchers, Inc.
17–11b, Arjen & Jerrine Verkaik/Skyart Productions
17–12, Warren Faidley/Weatherstock
17–13, Pekka Parviainen/Science Photo Library/Photo Researchers, Inc.
17–14, Warren Faidley/Weatherstock

Index